SPECTROSCOPY WITH POLARIZED LIGHT

Solute Alignment by Photoselection, in Liquid Crystals, Polymers, and Membranes

Distribution: VCH Verlagsgesellschaft mbH, P.O. Box 1260/1280, D-6940 Weinheim,
Federal Republic of Germany

USA and Canada: VCH Publishers, Inc., 303 N.W. 12th Avenue, Deerfield Beach, FL 33441–1705, USA

SPECTROSCOPY WITH POLARIZED LIGHT

Solute Alignment by Photoselection, in Liquid Crystals, Polymers, and Membranes

Josef Michl

Erik W. Thulstrup

VCH

Josef Michl
Department of Chemistry
University of Utah
Salt Lake City, Utah 84112

Erik W. Thulstrup
Department of Chemistry
Royal Danish School of Educational Studies
DK-2400 Copenhagen NV, Denmark

Library of Congress Cataloging-in-Publication Data

Michl, Josef, 1939-
Optical spectroscopy of partially aligned solutes.

Includes bibliographies and index.
1. Optical rotary dispersion. 2. Spectrum analysis.
I. Thulstrup, Erik Waaben, 1941– . II. Title.
QD473.M48 1986 543′.0858 86-19122
ISBN 0-89573-346-3

Printed in the United States of America.

ISBN 0-89573-346-3 VCH Publishers
ISBN 3-527-26516-3 VCH Verlagsgesellschaft

Dedication

To our respective families, who suffered valiantly, if not always quietly, as the book was being born.

Preface

Although the fact that optical properties of most molecules exhibit directional dependence has been well known among spectroscopists for many decades, it has been only in more recent years that chemists and biologists in general have become interested in its possible exploitation. The greatly increased activity in the last dozen years has concentrated primarily on investigations of linear dichroism of partially aligned molecular assemblies produced in stretched polymers, nematic liquid crystals, and lipid bilayers, or by the application of strong electric fields.

The purpose of this book is to provide an introduction for those interested in the application of the orientational dependence of optical properties of molecules to the solution of problems in chemistry, biology, or polymer science. While electric dichroism has already been the subject of several books, to our knowledge this is the first book devoted to the subject in its general form. In view of the extensive prior coverage, we do not treat electric dichroism in much detail.

Because of our background, we have written the present text from the point of view of someone using uniaxial solvents as a tool to orient small molecules, rather than the point of view of someone using small molecules as probes to investigate the structure of uniaxial solvents. However, we are aware of the need for an increased communication and collaboration between the two groups of scientists involved in these endeavors; it is hoped that the book will promote such a development.

The book is written at a level suitable for graduate students and researchers who have been exposed to elementary quantum mechanics, optics, and spectroscopy. The readers who do not specialize in spectroscopy and have a limited background in mathematics should skip some of the chapters on first reading, as indicated below.

In Chapter 1, we provide a condensed review of the fundamentals of optical spectroscopy, one of whose purposes is to establish the notation to be used in subsequent chapters, but which can also be used as a refresher course. Sections 1.1 and 1.2 should be skipped by those with a good background in physics and quantum theory. In Section 1.3.1 we introduce the concept of "transition moment", whose experimental determination is dealt with in most of the rest of the book. The fundamental formulas are stated with appeal to physical intuition but without derivations. In Section 1.3.2 we provide a quantitative treatment of the most important case, one-photon absorption and emission, and in Section 1.3.3 we provide a quantitative treatment of chiroptical spectroscopy. These two sections can be skipped by those readers who are primarily interested in the applications rather than the theory of optical spectroscopy.

Before going on to the discussion of measurements whose evaluation yields the molecular transition moments, we attempt to make these fundamental quantities somewhat more tangible by (i) considering the ways in which they can be obtained in an *a priori* fashion from molecular structures and wavefunctions and (ii) illustrating their relation to observable spectral properties on the example of pyrene. This is done in Chapter 2; Sections 2.1 and 2.2 could be skipped by those with limited interest in the theoretical aspects of spectroscopy.

The next two chapters lay the groundwork for the interpretation of spectral measurements on partially aligned samples in terms of the transition moments. Chapter 3 describes the experimental techniques most commonly used for the preparation of samples containing partially aligned solutes and for the spectroscopic measurements. A quantitative description of alignment is given in Chapter 4; some of it has not been published in this form before, particularly the treatment of photoselection in Section 4.6.

With this background, the quantitative evaluation of various kinds of spectroscopic measurements on partially aligned samples is outlined in the next three chapters, and it is expected that the reader will choose the sections of particular interest to him or her.

Chapter 5 is devoted to processes of rank two, the foremost among which is linear dichroism. The evaluation of the experimental results requires certain assumptions and many different sets of approximations have been proposed in the literature. At the moment, newcomers to the field are faced with a confusing multitude of possible evaluation methods, often without clearly stated limitations. We put the various proposed procedures on a common footing and outline their areas of applicability.

In Chapters 6 and 7, we discuss ways in which other types of directional dependence of optical properties can be exploited. Chapter 6 deals with processes of higher rank which can be effectively reduced to rank two; these are the natural and magnetic optical activity. Chapter 7 treats processes of rank four: photoluminescence, photodichroism, two-photon absorption, ordinary and resonant Raman scattering.

Finally, Chapter 8 provides examples of applications selected from electronic and vibrational spectroscopy. This is the only chapter in which we provide detailed references to the original literature. In the other chapters, annotated bibliography is provided at the end for those interested in further detail.

In the appendices, we summarize some of the mathematical apparatus used in the derivation of the results which were used in the text. Appendix I provides information for the most common molecular point symmetry groups: the character tables, the form taken by the electric dipole, magnetic dipole, and electric quadrupole operators, and the direct products of irreducible representations. Appendix II contains an introduction to tensors and to their use in orientational averaging, while Appendix III briefly describes the use of matrix calculus to describe the action of various optical elements on polarized light passing through them.

We would like to express our gratitude to J. H. Eggers, who originally introduced us to the subject of linear dichroism, and to Jan Linderberg and Frank E. Harris, who have patiently guided us through the intricacies of tensor calculus. Financial support for our research was provided by the National Institute of Health, the Danish Natural Science Research Council, and the NATO Office of Scientific Research.

Frans Langkilde, Juliusz G. Radziszewski, and John W. Downing have performed experiments and made calculations needed in the production of several of the drawings and made useful comments on the text. We are very indebted to them and to the many additional friends who have agreed to critically read portions of the book and provided valuable suggestions: Jörg Fleischhauer, Georg Hohlneicher, Hans Peter Jensen, Martin Klessinger, Martin McClain, Bengt Nordén, Yngve Öhrn, Claudio Puebla, and Bruno Samorì.

Finally we are grateful to Philip Willden, Rosemary Laufer, Jean Eden, Becky Cunningham, and Judith Raiguel, without whose skill in deciphering the hieroglyphics in our notes and converting them into a superbly typed text the draft would never have become a book, and to Alexis Kelner, who transformed our rough sketches into pieces of art.

Salt Lake City, July 1986—J. Michl and E. W. Thulstrup

Contents

1

Introduction and Theoretical Background

About a thousand years ago, Saint Olaf, King of Norway, visited a peasant by the name of Rødulf. According to a Viking saga as recorded in the "Flatey book" (Figure 1.1), the visit was celebrated by a large feast. After the meal it became increasingly difficult for the guests to conceal their own excellence. One of Rødulf's sons, Sigurd, claimed that he could "distinguish the movements of heavenly bodies, the sun and the moon" even when they could not be seen. When the King was ready for bed, the sky was clear. He asked Sigurd about the next day's weather, and Sigurd answered: "Snow!" "It does not look like that to me," said the King and went to bed. The next day the weather was as Sigurd had forecast, and not much sky could be seen. The King called Sigurd before him and challenged the young man to demonstrate his skill and tell how high the sun was in the sky. Sigurd unhesitatingly indicated an exact position. The King held up his sunstone and saw how it shone and indicated the same position as Sigurd had said.

How did Sigurd know, and how was the King able to check the correctness of his answer? Neither could see the sun, and what is a sunstone anyway? The Flatey book doesn't say much about it because obviously every Viking knew then. In 1966, a Danish archeologist, Torkild Ramskou, who wondered a great deal about sunstones, wrote an article in a popular Danish magazine on archeology describing their apparently magic powers, and noted that a scientific explanation ought to be possible. Independently, two employees of Scandinavian Airlines System wrote to Ramskou that an instrument, the Twilight Compass, which was used for flights in polar regions where a magnetic compass is useless, seemed to work in very much the same way as sunstones. In a second article, Ramskou explained the working of the Viking "solsten," or sunstone, as follows:

Figure 1.1 A page from the Flatey book telling the story of King Olaf and Sigurd. Reproduced from M. Thorhallsson and J. Thordarson, Flatey, Iceland, 1380–1390.

Daylight arriving from any point in the sky is partially linearly polarized in a direction perpendicular to the plane defined by the point at which the sun is located, the point which is being observed, and the eye of the observer. The Twilight Compass consists primarily of a polarizing filter which makes it possible to analyze the polarization of the light from the zenith when the sun is between 7° below and 30° above the horizon. The Vikings' Twilight Compass, the sunstone, may have been a small crystal of a dichroic mineral such as cordierite $[Mg_2Al_3(AlSi_5O_{18})]$, found in Norway. The intensity of light transmitted through a dichroic crystal depends on the angle between the electric vector of the light, or its polarization direction, and the axis of the crystal. The eyes of many insects and possibly some humans (like Sigurd) are also dichroic and can thus also serve in this way as navigational aids.

Actually, most of the light around us is partially polarized. Scattering of sun rays is one source of this polarization; reflection is another. Most reflecting surfaces in nature are horizontal and produce light with prevalent polarization of its electric vector in the horizontal direction. Glasses containing a layer which transmits only vertically polarized light will remove most of the reflected light and are well known as polarizing sunglasses.

The modern history of polarized light begins with Rasmus Bartholin, a Danish scientist who sent an assistant to Iceland to collect minerals with interesting optical properties. One of the numerous kinds of stones brought back by the assistant earned a place in history. It was the doubly refracting Icelandic spar, whose properties were described by Bartholin in a publication which appeared in 1669 (Figure 1.2) and which was important for both Newton

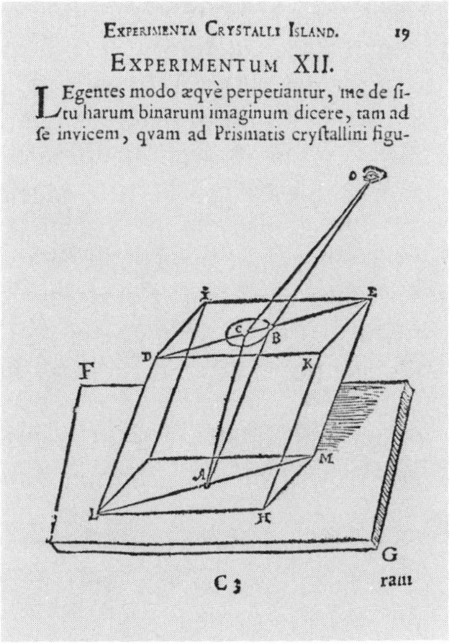

EXPERIMENTA CRYSTALLI ISLAND. 19

EXPERIMENTUM XII.

LEgentes modo æqvè perpetiantur, me de ſitu harum binarum imaginum dicere, tam ad ſe invicem, qvam ad Priſmatis cryſtallini figu-

C 3 ram

Figure 1.2 Double refraction in Icelandic spar. Reproduced from R. Bartholin, *Experiments with the Double Refracting Iceland Crystal*, Copenhagen, 1669.

and Huygens in their investigations of the nature of light. This crystal separates natural light into two mutually perpendicularly polarized components and sends them in different directions, leading to the doubling of an image. It is used to this day for the construction of linear polarizers. However, it was only the introduction of the efficient, inexpensive, large, and sturdy polarizing H sheets by Land in 1938 that led to the current large-scale availability of linearly polarized light. Finally, it should be mentioned that lasers are becoming increasingly important as new sources of polarized light.

Linearly polarized light is of interest to us presently because it represents a powerful tool for the investigation of oriented assemblies of molecules, such as crystals, and partially oriented assemblies of molecules, such as those found in lipid bilayers. In the process of being absorbed by a molecule, the electric vector of plane-polarized incident light interacts with the so-called transition moments of the molecule responsible for the particular transition in question. The electric dipole transition moment is the most important among these. The probability of electric dipole absorption is given by the square of the scalar product of the transition moment vector with the electric vector of the light and is therefore a function of the square of the cosine of the angle between them. This leads to an orientational dependence of the absorption of linearly polarized light, and similarly of its emission. It allows one to use spectroscopy with linearly polarized light to investigate the nature of molecular transitions as well as the nature of partial or complete orientation of molecular assemblies.

The direction of the transition moment vector in the molecular framework is frequently referred to as the absolute polarization direction of the transition, or simply its absolute polarization. A relative polarization of two transitions refers to the angle formed by their respective transition moment vectors.

A sample is said to exhibit linear dichroism if it absorbs light to different degrees depending on its linear polarization. In general, all samples which are partially or fully oriented will exhibit linear dichroism in one or another region of the spectrum. Linear dichroism in absorption and in emission is only one of several orientation-dependent optical properties, albeit perhaps the most important one.

Many naturally occurring solid samples actually are partially oriented, and it is frequently difficult to find one which is perfectly isotropic, just as it is difficult to produce light which is not partially polarized. On the other hand, gases and common liquids are isotropic, and ordinary fluid solutions do not exhibit linear dichroism.

A very high degree of molecular orientation is found in crystals, and much of the work in polarized spectroscopy has concentrated on either single crystals or mixed crystals. In the latter, the orientation of the host molecules is used to induce an orientation of guest molecules.

The present text deals with samples which typically exhibit only a much lower degree of orientation, such as oriented polymers, liquid crystals, and lipid bilayers. Specifically, we shall consider the spectra of solute molecules contained in such samples and partially aligned by interaction with the anisotropic environment, as well as spectra of solutes partially aligned by other means. The same methods can be applied to the study of the partially aligned media themselves or, for that matter, to the highly aligned molecules contained in crystals, but we have decided to omit the latter two cases in order to keep the size of the text within reasonable bounds.

The investigation of partially aligned solutes with linearly polarized light goes back at least to 1888, when Ambronn described the linear dichroism of dye-stained cell membranes (Figure 1.3). After World War I, dyed stretched polymer sheets became commercially available as linear polarizers due to the work of Land (Section 8.10.1). Since World War II, the number of investigations of this type of samples by polarized optical spectroscopy has increased rapidly, hundreds of papers on the subject have appeared, and several quantitative evaluation methods have been proposed. Most of them concentrate on linear dichroism in the UV-visible region, which is the easiest to measure. The great activity in the field reflects the increased interest in the assignment of molecular transitions and state symmetries, as well as the importance attributed to the detailed understanding of the nature of partial molecular orientation, primarily in naturally occurring samples.

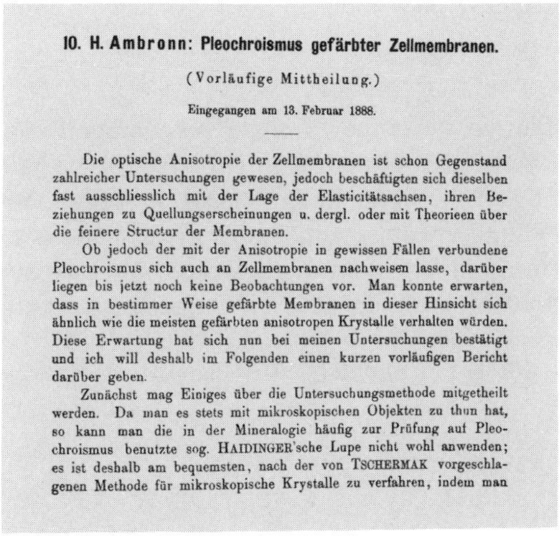

10. H. Ambronn: Pleochroismus gefärbter Zellmembranen.

(Vorläufige Mittheilung.)

Eingegangen am 13. Februar 1888.

Die optische Anisotropie der Zellmembranen ist schon Gegenstand zahlreicher Untersuchungen gewesen, jedoch beschäftigten sich dieselben fast ausschliesslich mit der Lage der Elasticitätsachsen, ihren Beziehungen zu Quellungserscheinungen u. dergl. oder mit Theorieen über die feinere Structur der Membranen.

Ob jedoch der mit der Anisotropie in gewissen Fällen verbundene Pleochroismus sich auch an Zellmembranen nachweisen lasse, darüber liegen bis jetzt noch keine Beobachtungen vor. Man konnte erwarten, dass in bestimmer Weise gefärbte Membranen in dieser Hinsicht sich ähnlich wie die meisten gefärbten anisotropen Krystalle verhalten würden. Diese Erwartung hat sich nun bei meinen Untersuchungen bestätigt und ich will deshalb im Folgenden einen kurzen vorläufigen Bericht darüber geben.

Zunächst mag Einiges über die Untersuchungsmethode mitgetheilt werden. Da man es stets mit mikroskopischen Objekten zu thun hat, so kann man die in der Mineralogie häufig zur Prüfung auf Pleochroismus benutzte sog. HAIDINGER'sche Lupe nicht wohl anwenden; es ist deshalb am bequemsten, nach der von TSCHERMAK vorgeschlagenen Methode für mikroskopische Krystalle zu verfahren, indem man

Figure 1.3 The first article describing the linear dichroism of a dyed membrane. Reproduced from H. Ambronn, *Ber. Deutsch. Botan. Ges.* **6**, 85 (1888).

Other types of optical measurements on partially oriented samples have been much less popular so far. Nevertheless, we shall discuss them in some detail and hope that the availability of a general and relatively simple theoretical description will spur further experimental activity. For this reason, this opening chapter presents a brief elementary survey of the principles underlying those optical spectroscopic methods which presently appear promising for application to partially aligned solutes. This also allows us to introduce the nomenclature which we shall need in the subsequent chapters, and it may further be useful to those readers who desire to read a refresher course on the fundamentals of optical spectroscopy before plunging into the study of the effects of molecular alignment. Parts of it are probably too condensed for readers who have had little prior exposure to optical spectroscopy, and additional introductory texts are suggested in Section 1.4.

It is important to note that much additional information on the static and dynamic behavior of solutes in partially oriented solutions has been obtained from other types of spectroscopy, in particular from magnetic resonance (NMR, ESR). A description of these methods and results lies outside the scope of the present book.

1.1 Polarized Light

We have seen how polarized light occurs in nature, and we shall now describe it in simple mathematical terms.

1.1.1 Linearly Polarized Light

In vacuum, the electric vector of a linearly polarized electromagnetic wave at any point in space is given by

$$\mathbf{E}(t) = \mathbf{E}_0 \sin(2\pi\nu t + \theta) \tag{1.1}$$

where $\mathbf{E}_0$ is a constant vector in the YZ plane perpendicular to X, the direction of propagation of the light. The direction of $\mathbf{E}$ will be referred to as the polarization direction of the light. In (1.1), $(2\pi\nu t + \theta)$ is the phase at time t, θ is the phase at $t = 0$, and ν is the frequency in cycles per second (cps, Hz). Angular frequency ω in radians per second is larger by the factor 2π, $\omega = 2\pi\nu$.

As a function of position along the X axis, the electric vector is given by

$$\mathbf{E}(t,X,0,0) = \mathbf{E}_0 \sin[2\pi(\nu t - X/\lambda) + \theta] = \mathbf{E}_0 \sin[2\pi\nu(t - X/c) + \theta)] \tag{1.2}$$

where $\lambda = c/\nu$ is the wavelength, with $c = 3 \times 10^{10}$ cm/s standing for the speed of light in vacuum, and where the phase is θ for $(t,X) = (0,0)$.

The light is characterized by the wave vector $\mathbf{K} = (2\pi/\lambda)\boldsymbol{\varepsilon}_X$, where $\boldsymbol{\varepsilon}_X$ is a unit vector in the propagation direction X. It should be noted that the definition of $\mathbf{K}$ used by certain other authors does not contain the factor 2π. With the definition used here, $\mathbf{K}\hbar$ is the magnitude of the linear momentum of the photon ($\hbar = h/2\pi$, where h is Planck's constant, $h = 6.6256 \times 10^{-27}$ erg s). The relations between the length of the wave vector $|\mathbf{K}|$, the wavelength λ, the frequency ν, the angular frequency ω, and the wavenumber $\tilde{\nu}$ are

$$\begin{aligned} \nu &= \tilde{\nu}c = c/\lambda = c|\mathbf{K}|/2\pi \\ \omega &= 2\pi\tilde{\nu}c = 2\pi c/\lambda = c|\mathbf{K}| \end{aligned} \tag{1.3}$$

Visible light is composed of radiation with wavelengths between 700 nm (red light) and 400 nm (violet light) corresponding to frequencies between 4×10^{14} Hz and 7×10^{14} Hz and wavenumbers between 14 300 cm^{-1} and

25 000 cm^{-1}. Infrared (IR) light and ultraviolet (UV) light are found at longer and shorter wavelengths, respectively.

In addition to the electric field, an electromagnetic wave also carries a magnetic field. The latter is given by

$$\mathbf{H}(t,X,0,0) = \mathbf{H}_0 \sin[2\pi(\nu t - X/\lambda) + \theta] \quad (1.4)$$

where $\mathbf{H}_0$ is a constant vector in the YZ plane perpendicular to $\mathbf{E}_0$.

Figure 1.4 shows the electric and magnetic vectors of an electromagnetic wave as a function of time and distance.

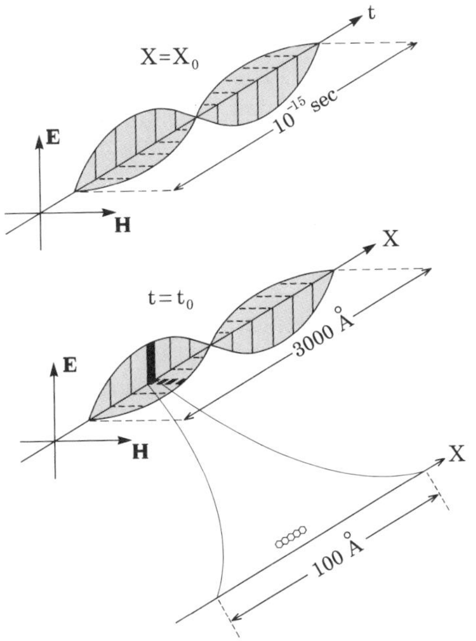

Figure 1.4 The variation of the electric (**E**) and magnetic (**H**) fields of an electromagnetic wave of UV light in time (top) and space (bottom). A comparison with the size of a typical molecule is shown.

Both the electric field **E**(t,**R**) and the magnetic field **H**(t,**R**) of a general light wave in a space point $\mathbf{R} \equiv (X,Y,Z)$ at time t can be derived from a single quantity known as the vector potential **A**(t,**R**). This can be expressed in several forms. In the so-called Coulomb gauge (and in Gaussian units),

$$\mathbf{E}(t,\mathbf{R}) = -(1/c)\,\partial\mathbf{A}(t,\mathbf{R})/\partial t \quad (1.5)$$

$$\mathbf{H}(t,\mathbf{R}) = \nabla \times \mathbf{A}(t,\mathbf{R}) \quad (1.6)$$

The vector potential is normally defined as a complex quantity, but we shall need to work with only the real part. A sign ambiguity exists in the definition; the convention we shall follow is the one normally used in classical optics. The opposite choice, common in quantum mechanics, would lead to a change of sign or complex conjugation in certain intermediate expressions in Section 1.3, but with the same ultimate results. The most obvious difference is the appearance of the argument $\omega t - \mathbf{K} \cdot \mathbf{R}$, where the opposite convention would have $\mathbf{K} \cdot \mathbf{R} - \omega t$; a related difference will be the appearance of $\boldsymbol{\varepsilon}_U^*$, where the opposite convention would have $\boldsymbol{\varepsilon}_U$.

For a plane wave of light characterized by the wave vector **K** directed along X and linearly polarized along U, we have

$$\mathbf{A}(t,\mathbf{R}) = \boldsymbol{\varepsilon}_U A_0 \cos(2\pi\nu t - \mathbf{K} \cdot \mathbf{R}) = \boldsymbol{\varepsilon}_U A_0 \cos 2\pi(\nu t - X/\lambda) \tag{1.7}$$

where $\boldsymbol{\varepsilon}_U$ is a unit vector along direction U and the amplitude A_0 is proportional to both $|\mathbf{E}_0|$ and $|\mathbf{H}_0|$ (in Gaussian units, these amplitudes are equal). The direction U can be any direction perpendicular to **K** and thus perpendicular to X. For example, U = Y and U = Z would be two choices frequently used in actual experiments, but $U = (Y - Z)/\sqrt{2}$ (which means $\boldsymbol{\varepsilon}_U = (\boldsymbol{\varepsilon}_Y - \boldsymbol{\varepsilon}_Z)/\sqrt{2}$) is equally acceptable (we shall occasionally use this shorthand notation for directions in the following).

Substitution of (1.7) into (1.5) and (1.6) yields (1.2) and (1.4), respectively.

1.1.2 Circularly and Elliptically Polarized Light

In-Phase Superposition of Two Linearly Polarized Waves. Consider two light waves, 1 and 2, with identical amplitudes $|\mathbf{E}_0^{(1)}| = |\mathbf{E}_0^{(2)}|$, frequencies $\nu_1 = \nu_2$, and directions of propagation X, but with linear polarizations that are mutually orthogonal, $\mathbf{E}_0^{(1)} \cdot \mathbf{E}_0^{(2)} = 0$ (Figure 1.5). If the phases of the two waves are identical, $\theta_1 = \theta_2 = \theta$, their superposition will produce a new linearly polarized wave:

$$\mathbf{E}(t,X,0,0) = (\mathbf{E}_0^{(1)} + \mathbf{E}_0^{(2)}) \sin[2\pi(\nu t - X/\lambda) + \theta] \tag{1.8}$$

with an amplitude which is $\sqrt{2}$ times larger than that of either of the original waves, and with a direction of polarization which forms an angle of 45° with the polarization directions of either of the two waves: If waves 1 and 2 are polarized along Y and Z, respectively, the resulting wave will be polarized along $(Y + Z)/\sqrt{2}$.

Superposition of Linearly Polarized Waves Differing in Phase. If wave 2 is not in phase with wave 1 but is delayed by a quarter wave, as can be achieved, e.g., by insertion of a quarter-wave plate into its path, we have $\theta_2 = \theta_1 - \pi/2$. Now, the superposition of the two waves results in a circularly polarized wave:

$$\mathbf{E}(t,X,0,0) = \mathbf{E}_0^{(1)} \sin[2\pi(\nu t - X/\lambda) + \theta_1] + \mathbf{E}_0^{(2)} \cos[2\pi(\nu t - X/\lambda) + \theta_1] \tag{1.9}$$

which has a constant amplitude $|\mathbf{E}_0^{(1)}| = |\mathbf{E}_0^{(2)}|$, but with a direction of the electric vector which at a given point X_0 rotates in time with frequency ν (Figure 1.5).

Circular polarization is said to be right-handed if the direction of rotation is clockwise when viewed against the direction of propagation and left-handed if the sense of the rotation is opposite. When the position of the endpoint of the electric vector is viewed at a given time t as a function of distance along X, it forms a left-handed helix if the light polarization is left-handed and a right-handed helix if it is right-handed. A photon of left-handed circularly polarized

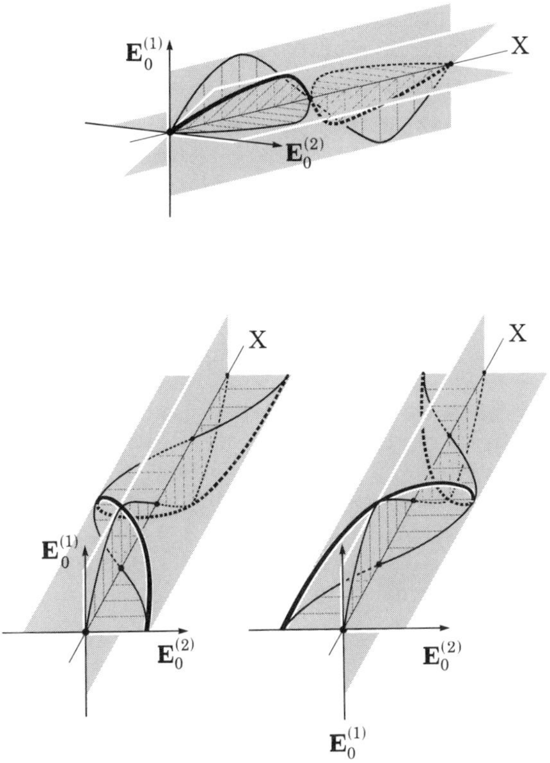

Figure 1.5 Superposition of two electromagnetic waves of mutually perpendicular linear polarization (only the electric field vector **E** is shown). Top: identical phase. Bottom: phase difference of $\pi/2$. The resultant electromagnetic waves are shown in thick lines.

light propagating in the positive direction of the X axis is associated with angular momentum $\hbar$ along the $+X$ direction, and a photon of right-handed light propagating in the same direction, with angular momentum $-\hbar$.

The general result of a superposition of linearly polarized waves with phases which do not differ by an integral multiple of $\pi/2$ is elliptical polarization. The ellipticity of the light is the angle whose tangent is equal to the ratio of the minor and major axes of the ellipse traced at a fixed X_0 by the endpoint of the electric vector of the light.

In order to extend our notation to circularly and elliptically polarized light, we shall permit the unit vector $\boldsymbol{\varepsilon}_U$, which characterizes the polarization properties of the light wave, to be complex. Thus, while $\boldsymbol{\varepsilon}_Y$ and $\boldsymbol{\varepsilon}_Z$ were the polarization vectors of light propagating along X and linearly polarized along Y and Z, respectively, $(\varepsilon_Y + i\varepsilon_Z)/\sqrt{2}$ will stand for the polarization vector of right-hand circularly polarized light, and $(\boldsymbol{\varepsilon}_Y - i\boldsymbol{\varepsilon}_Z)/\sqrt{2}$ for that of left-handed circularly polarized light propagating in the positive sense of X. Here and elsewhere, $i = \sqrt{-1}$ is the imaginary unit.

For reasons that will become clear later, we shall have opportunity to deal only with circularly polarized light propagating along Z. Then, for propagation in the positive sense of Z, the polarization vector is $(\boldsymbol{\varepsilon}_X + i\boldsymbol{\varepsilon}_Y)/\sqrt{2} = \boldsymbol{\varepsilon}_+$ for right-handed circular polarization and $(\boldsymbol{\varepsilon}_X - i\boldsymbol{\varepsilon}_Y)/\sqrt{2} = \boldsymbol{\varepsilon}_-$ for left-handed circular polarization. In the sign convention opposite to the one adopted here, our $\boldsymbol{\varepsilon}_+$ would stand for left-handed circular polarization and our $\boldsymbol{\varepsilon}_-$ for right-handed circular polarization. This sometimes leads to confusion; we shall employ our convention consistently throughout.

The form of the real part of the vector potential for a plane wave of monochromatic light in the case of general elliptical polarization will be given in (1.31) (Section 1.3.2).

1.2 Molecular States and Symmetry

A state of a quantum-mechanical system such as a molecule is described by its wavefunction. The wavefunction is a function of the space and spin coordinates of all particles in the system and of time. For our purposes, it is not necessary to consider the translation and rotation of the molecule as a whole, and we shall consider only internal degrees of freedom.

1.2.1 Stationary States

In a general case, the state of a system and the values of the observable properties change with time in a complicated manner. However, there is a set of states, the so-called stationary states, whose wavefunctions change in time in a simple manner and whose observable properties do not change in time at all. The wavefunctions of stationary states can be written in the form

$$\Psi(\mathbf{q},\mathbf{Q},t) = \Psi(\mathbf{q},\mathbf{Q})e^{-2\pi itE/h} = \Psi(\mathbf{q},\mathbf{Q})e^{-iEt/\hbar} \qquad (1.10)$$

where the vector $\mathbf{q}$ represents the coordinates of all the electrons, the vector $\mathbf{Q}$ represents the coordinates of all the nuclei, and t is time. In the absence of outside magnetic field, the time-independent wavefunctions $\Psi(\mathbf{q},\mathbf{Q})$ can be and usually are chosen real, and we shall do so unless specifically stated otherwise. These wavefunctions are eigenfunctions of the molecular Hamiltonian operator $\hat{H}(\mathbf{q},\mathbf{Q})$, and their energies E are its eigenvalues. In the rest of Section 1.2 we shall deal only with the time-independent part of the stationary wavefunction, but it is important to remember the existence of the time-dependent part, known as the phase. For instance, the phase plays a role in transitions between different stationary states, as will be briefly described in Section 1.3.

The Born-Oppenheimer Approximation. For the description of processes discussed in the following, the dependence of the molecular wavefunction on the electronic and vibrational coordinates is of primary interest. Furthermore, in most cases the Born-Oppenheimer approximation provides a satisfactory basis for the description. In this approximation, a stationary state wavefunction describing the molecular degrees of freedom of interest to us is written as

$$\Psi_{j,v}(\mathbf{q},\mathbf{Q}) = \psi_j(\mathbf{q},\mathbf{Q})\chi_{j,v}(\mathbf{Q}) \qquad (1.11)$$

where j characterizes the electronic state and v characterizes the vibrational sublevel of that state (Figure 1.6).

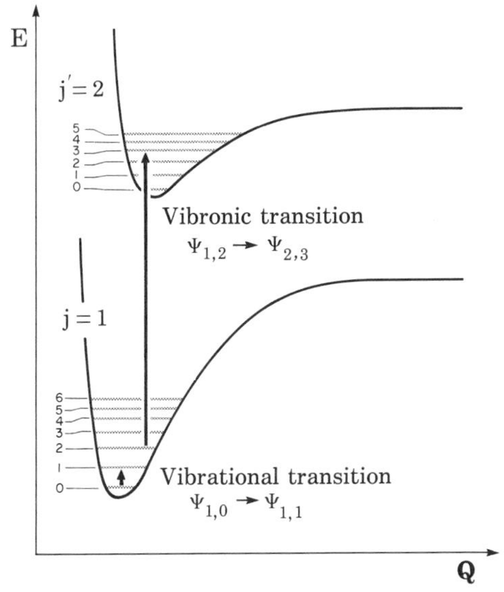

Figure 1.6 Potential energy curves and vibrational levels for two electronic states of a molecule.

The electronic wavefunction $\psi_j(\mathbf{q},\mathbf{Q})$ is an eigenfunction of the electronic Hamiltonian $\hat{H}_{el}(\mathbf{q},\mathbf{Q})$ defined for a particular geometry $\mathbf{Q}$ as an operator containing the potential energy of electrostatic electron-nuclear attraction, electron-electron and nucleus-nucleus repulsion as well as the kinetic energy of the electrons. Thus, $\psi_j(\mathbf{q},\mathbf{Q})$ is a function of the electronic coordinates $\mathbf{q}$ and contains $\mathbf{Q}$ as a parameter. For each $\mathbf{Q}$, there are an infinite number of stationary electronic wavefunctions $\psi_j(\mathbf{q},\mathbf{Q})$ which differ in their label j and energy $E_j(\mathbf{Q})$. Only some of them represent bound states; those above the ionization limit represent ionized states. There is a different electronic wavefunction for a given (j-th) state for each value of the parameter $\mathbf{Q}$.

The vibrational wavefunctions $\chi_{j,v}(\mathbf{Q})$ are eigenfunctions of the vibrational Hamiltonian $\hat{H}_{vib}(j,\mathbf{Q})$, which is defined for a particular electronic state j as an operator containing the potential energy $E_j(\mathbf{Q})$ dictating the nuclear motions and the kinetic energy of the nuclei. For every electronic state j, there is a different potential energy, therefore a different $\hat{H}_{vib}(j,\mathbf{Q})$, and therefore different vibrational eigenfunctions $\chi_{j,v}(\mathbf{Q})$. For each j, there are an infinite number of vibrational eigenfunctions $\chi_{j,v}(\mathbf{Q})$ which differ in their label v and energy $E_{j,v}$. Only some of them represent bound states; those above the dissociation limit represent dissociative states.

The form chosen for the description of the positions of the nuclei through the internal coordinates $\mathbf{Q}$ are usually the "normal coordinates," which describe the displacement of the nuclei from their equilibrium positions in the electronic state j along directions defined by the normal vibrational modes. They will be described in somewhat more detail in Section 2.1.

The individual vibrational wavefunctions $\chi_{j,v}(\mathbf{Q})$ belonging to the same electronic state j are mutually orthogonal,

$$\langle \chi_{j,v}(\mathbf{Q}) | \chi_{j,v'}(\mathbf{Q}) \rangle_{\mathbf{Q}} = \delta_{vv'} \tag{1.12}$$

where $\delta_{vv'}$ is the Kronecker delta symbol: $\delta_{vv'} = 1$ if $v = v'$ and $\delta_{vv'} = 0$ if $v \neq v'$. The subscript **Q** shows that the integration indicated by the symbol $\langle | \rangle$ is to be performed over the vibrational coordinates.

For a given nuclear configuration **Q**, the electronic wavefunctions of different electronic states are also mutually orthogonal:

$$\langle \psi_j(\mathbf{q},\mathbf{Q}) | \psi_{j'}(\mathbf{q},\mathbf{Q}) \rangle_{\mathbf{q}} = \delta_{jj'} \tag{1.13}$$

Here, the subscript **q** indicates integration over the coordinates of the electrons.

1.2.2 Molecular Symmetry and Transition Moments

The equilibrium geometries of many molecules possess certain elements of symmetry, and this imposes strict separate restrictions on the vibrational and electronic wavefunctions in the Born-Oppenheimer approximation (1.11). (Note that the molecules are not permitted to rotate, so the question of rotational-vibrational coupling does not come up.)

For instance, if the equilibrium geometry of a molecule possesses a plane of symmetry, that is, if a reflection in this plane converts the nuclear geometry of the molecule into itself, then such reflection cannot have any effect on observable properties. The square of the absolute value of the wavefunction, which provides information about the probability of finding the particles contained in the molecule at various positions in space, is one such observable. The net effect of reflection on a nondegenerate wavefunction can therefore be only no change at all or multiplication by a complex unity. Since a repetition of the reflection operation recreates the starting situation, the two possibilities are multiplication by $+1$ or by -1. In the former case, the wavefunction is called symmetric, and in the latter antisymmetric with respect to the symmetry plane. Similar statements can be made about other types of symmetry operations such as rotation and inversion through a center of symmetry.

Transition Moments. Many observable properties can be calculated from wavefunctions by the evaluation of a "transition moment" integral of the type

$$\langle \Psi_{j,v}(\mathbf{q},\mathbf{Q}) | \hat{A} | \Psi_{j',v'}(\mathbf{q},\mathbf{Q}) \rangle_{\mathbf{q},\mathbf{Q}} = A(jv,j'v') \tag{1.14}$$

where $\hat{A}$ is an operator corresponding to the desired observable property. Usually, the operator $\hat{A}$ will not be a scalar but rather a vector or a tensor. If it is a vector operator, the integral $\mathbf{A}(jv,j'v')$ in (1.14) will be a vector. It will have three components, which we shall express in a molecule-fixed system of axes x,y,z. They are obtained by performing the integration in (1.14) separately for each of the three components of the operator $\hat{\mathbf{A}}$. Similarly, for a tensor operator $\hat{\mathsf{A}}$ the integral $\mathsf{A}(jv,j'v')$ will be a tensor (see Appendix II).

We shall assume that $\Psi_{j,v}$ and $\Psi_{j',v'}$ have the Born-Oppenheimer form (1.11), with j and j′ characterizing electronic states and v and v′ their vibrational sublevels. If a property of a stationary state j,v is desired, we set $j' = j$ and $v' = v$ in (1.14). Properties involving more than one stationary state, such as probabilities of transitions between stationary states, involve either $j \neq j'$ or $v \neq v'$ or both. Transitions in which $j \neq j'$ are called electronic transitions, and those in which $j = j'$ but $v \neq v'$ are called vibrational transitions (Figure 1.6). Transitions in which both $j \neq j'$ holds and v and v′ are specified are often referred to as vibronic (vibrational-electronic).

Transition moments between real wavefunctions will always be real quantities if the operator $\hat{A}$ is real (e.g., equal to the electric dipole moment operator $\hat{\mathbf{M}}$) and pure imaginary quantities if the operator $\hat{A}$ is imaginary (e.g., equal to the magnetic dipole moment operator $\hat{\mathcal{M}}$).

The symmetry properties of the two wavefunctions and the operator which enter into the integral A(jv,j′v′) combine to make these integrals vanish for many choices of j, v, j′, and v′. Only when the integrand is symmetric ("even") with respect to all symmetry operations, that is, when it is totally symmetric, can the integral have a nonzero value. This usually happens only for some of the components of vector operators $\hat{\mathbf{A}}$ and tensor operators $\hat{A}$. For instance, when the integrand is antisymmetric ("odd") with respect to reflection in a plane, the integration can be divided into two parts, one on each side of the plane (a "nodal plane"). The results of integration on each side are equal but opposite in sign and therefore cancel when the total integral is evaluated (Figure 1.7).

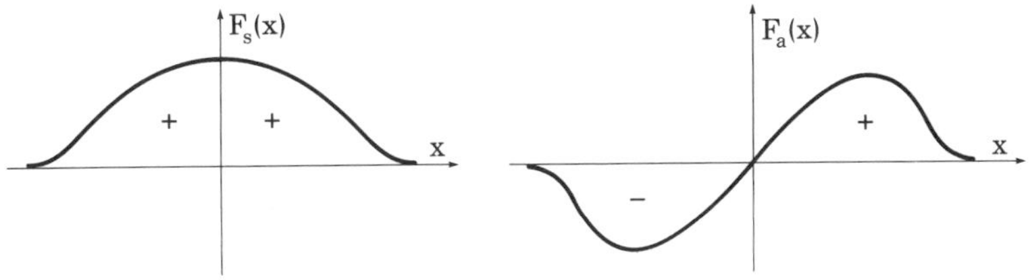

Figure 1.7 Examples of functions symmetric (left) and antisymmetric (right) with respect to mirroring in a plane perpendicular to x. Note that $\int_{-\infty}^{+\infty} F_a(x)\,dx = 0$.

Electric Dipole Transition Moments. An illustration is provided by the electric dipole moment operator $\hat{\mathbf{M}}$, which is a vector operator with coordinates $\hat{M}_x$, $\hat{M}_y$, $\hat{M}_z$ in the molecular frame. If the three coordinates of the transition moment integral **M**(jv,j′v′) are calculated between states $\Psi_{j,v}$ and $\Psi_{j',v'}$, which have the same symmetry property with respect to reflection in a symmetry plane (xy) of the molecule, i.e., which are both symmetric or both antisymmetric, only the components of $\hat{\mathbf{M}}$ which are symmetric with respect to the plane, $\hat{M}_x$ and $\hat{M}_y$, lead to a totally symmetric integrand. Therefore, only M_x(jv,j′v′) and M_y(jv,j′v′) can be different from zero and contribute to the transition moment vector. In other words, the transition moment is directed along some direction which lies in the symmetry plane. In the electric dipole mechanism of light absorption, only the square of the projection of the electric vector of the light into this direction will count. When probed with a beam of linearly polarized light, this orientational dependence of the transition probability will be revealed, and the transition is said to be polarized along this particular in-plane direction.

Similarly, if the two wavefunctions $\Psi_{j,v}$ and $\Psi_{j',v'}$ have different symmetry properties with respect to mirroring in the xy plane, only the component $\hat{M}_z$ leads to a totally symmetric integrand, so that only $M_z(jv,j'v')$ can be different from zero, and the transition will be polarized along z, perpendicular to the symmetry plane.

If a molecule possesses two symmetry planes perpendicular to each other, say, xz and yz, transitions between states with the same symmetry property with respect to mirroring in the two planes will be polarized along the direction z, which is symmetric with respect to both planes (Figure 1.8). If the two states have the same properties with respect to yz but not with respect to xz, the transition will be polarized along y, which is symmetric with respect to yz and antisymmetric with respect to xz. If the two states have the same properties with respect to xz but different properties with respect to yz, the transition will be x-polarized. Finally, if the two states differ in their symmetry properties with respect to both planes, all three components of the transition moment will vanish. Such an electric dipole transition is said to be symmetry forbidden.

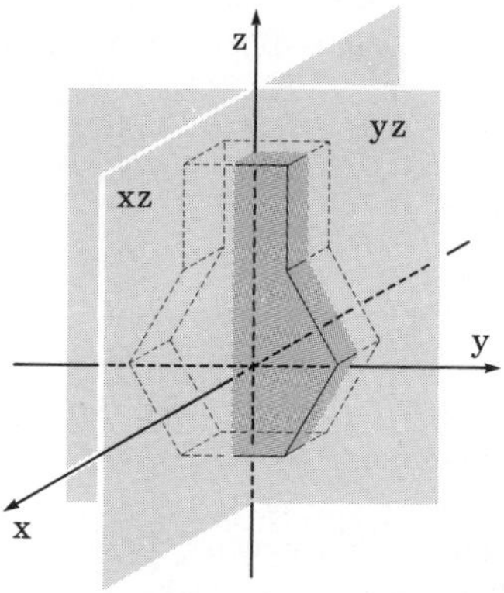

Figure 1.8 A molecule with symmetry planes xz and yz. Transition moments can lie only along the x, y, or z directions.

Planar Molecules. A very important example is provided by the electronic transitions of planar molecules such as aromatic hydrocarbons. These molecules have the molecular plane as a symmetry element, and the electronic wavefunctions are usually constructed from one-electron functions, molecular orbitals, which are either symmetric (σ orbitals) or antisymmetric (π orbitals) with respect to the molecular plane yz in Figures 1.8 and 1.9. The electronic state is symmetric with respect to the plane if it has an even number of electrons in π orbitals; otherwise it is antisymmetric. Thus, allowed electric dipole transitions between states with the same number of electrons in π orbitals will be polarized in the

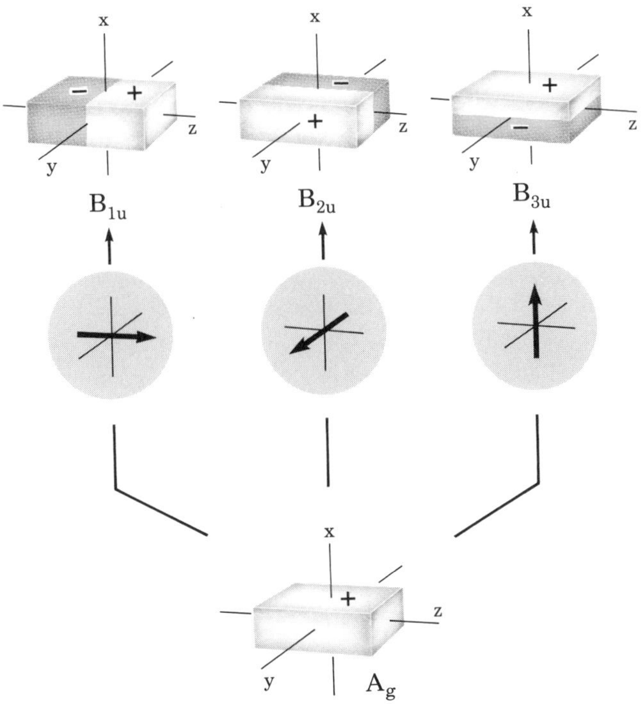

Figure 1.9 A molecule with symmetry planes xy, xz, and yz. The totally symmetric ground state (bottom) can absorb a photon of linear polarization shown (center) to produce states with one nodal plane (top). The states are assigned the labels of the irreducible representations of the D_{2h} group which characterize them.

molecular plane, while those which change the number of π electrons by one will be polarized perpendicular to the plane. Examples of transitions of the first type are $\sigma\sigma^*$ amd $\pi\pi^*$ transitions, while the second type may be $\sigma\pi^*$, $\pi\sigma^*$, or $n\pi^*$ transitions. When the molecule has an additional symmetry plane, perpendicular to the molecular plane, the possible electric dipole transition moment directions are limited to the three molecular symmetry axes, as shown in Figure 1.9. For instance, the polarization of an electric dipole $\pi\pi^*$ transition in such a molecule is restricted to one of only two different directions, y or z.

Molecules with Rotational Symmetry. Another important symmetry operation is rotation around an axis C_n by an angle $2\pi/n$. In molecules possessing one twofold axis of symmetry ($n = 2$), electric dipole transitions will be polarized either along the axis or in a direction perpendicular to it. If a threefold or higher order axis is present ($n \geqslant 3$), electric dipole transitions into some excited states may be forbidden, others may be polarized along the C_n axis, and still others may be evenly polarized in the plane perpendicular to the C_n axis. With even (or isotropic) polarization in a plane, it is the square of the projection of the electric vector of the incident light into this plane which counts and not the square of the projection into a particular direction in the plane. These two concepts must be carefully distinguished although they are sometimes both simply referred to as in-plane polarization.

In order to obtain even polarization in a plane for a transition from a totally symmetric ground state, the excited state must be doubly degenerate. The two components of the doubly degenerate state can be chosen so that the transitions from a totally symmetric ground state into them are linearly polarized in two mutually perpendicular directions in a plane perpendicular to the C_n axis ($\Psi_{j',v'}$ and $\Psi_{j'',v''}$), but other choices are also possible. For instance, the choice $(\Psi_{j',v'} \pm i\Psi_{j'',v''})/\sqrt{2}$ describes the two components of the degenerate excited state in such a way that transitions into them have mutually opposite circular polarizations (left-handed, right-handed) and would be suitable, say, for the description of magnetic circular dichroism. Regardless of the particular choice made, the absorptions into the two components of the excited state mutually complement their directional properties in such a way that all in-plane directions for the electric vector of the incident light become equivalent.

If several higher order axes are present, triple degeneracy is possible. An allowed transition from a totally symmetric ground state into a triply degenerate state is unpolarized ("isotropic," "isotropically polarized"). In such a case, contributions to the absorption due to the three components of the excited state complement each other in such a way that the direction of the electric dipole of the incident light no longer matters.

1.2.3 Group Theory

A formalization of the simple symmetry considerations outlined above is provided by group theory (Appendix I). The set of all symmetry operations of a molecule forms a point symmetry group. The elements are usually associated with symbols such as C_n^m (rotation by $2\pi m/n$ around an n-fold rotation axis), σ_h (mirroring in a plane perpendicular to a rotation axis), σ_v (mirroring in a plane parallel to a rotation axis), i (inversion through a center of symmetry), E (identity operation), S_n (rotation by $2\pi/n$ around an axis followed by mirroring in a plane perpendicular to the axis). Some examples of groups and their symmetry elements are C_s (one symmetry plane), C_n (one n-fold axis), D_2 (three mutually perpendicular C_2 axes), C_{2v} (two mutually perpendicular planes and a C_2 axis parallel to both, see Figure 1.8), D_{2h} (three mutually perpendicular planes, center of symmetry, and three C_2 axes, see Figure 1.9). Examples of molecules belonging to these point groups are shown in Figure 1.10: 1-fluoronaphthalene (C_s), one of the conformers of isobutane (C_3), gas-phase (partially twisted) biphenyl (D_2), 1,4-difluoronaphthalene (C_{2v}), and naphthalene (D_{2h}).

Appendix I lists the symmetry operations in these and several other point groups. Each operation labels a column in what is known as the character table of the group. Wavefunctions of molecules belonging to a particular symmetry group are classified according to their responses to the symmetry operations (see Figure 1.9). Each possible set of responses is described by a so-called irreducible representation which corresponds to one row of the character table. Each irreducible representation is labeled by a symbol such as A,B ("one-

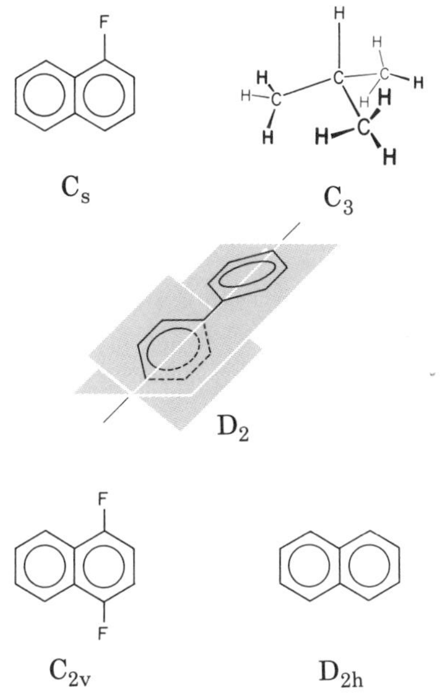

Figure 1.10 Examples of molecules and their point symmetry groups.

dimensional"), E ("two-dimensional"), or T ("three-dimensional") and frequently with subscripts for further classification, such as u and g for odd (ungerade) and even (gerade), respectively, relative to the operation of inversion i. It is customary to use a capital letter when referring to the irreducible representation which describes the symmetry of a many-electron or total (vibronic) wavefunction (e.g., B_{1g}) and to use lowercase letters when describing the symmetry properties of a one-electron wavefunction (orbital) or a vibrational wavefunction (e.g., b_{1g}).

The entries in the table, called characters, provide information on the result of the operation listed at the head of the column on a wavefunction belonging to the irreducible representation listed at the head of the row. For one-dimensional representations (A,B), +1 means that the operation converts the wavefunction into itself and −1 means that the operation converts the wavefunction into the negative of itself. The situation is somewhat more complicated for doubly degenerate wavefunctions which transform according to the two-dimensional E representations. Here, a symmetry operation can also convert one member of the degenerate pair of wavefunctions into the other or into a mixture of both members.

As we have already noted, in order for the integral A(jv,j′v′) defined in (1.14) not to vanish, its integrand has to transform according to the totally symmetric representations. Here, a symmetry operation can also convert one member of everywhere on its line in the character table of the molecular point symmetry group. The symmetry of the integrand is obtained as a direct product of three irreducible representations, those of $\Psi_{j,v}$, $\Psi_{j',v'}$, and the operator $\hat{A}$. The symmetry properties of some of the most important operators are listed on the right-hand side of the character tables.

It is easy to find the irreducible representation Γ given as a direct product of two irreducible representations, Γ_1 and Γ_2, if at least one of them is one-dimensional. To obtain the character of the direct product representation $\Gamma = \Gamma_1 \times \Gamma_2$ for any given symmetry operation, the characters of the representations Γ_1 and Γ_2 for this operation are multiplied. An example is provided in Figure 1.9. If the symmetry of $\Psi_{j,v}$ is A_g as shown at the bottom of the figure, and the symmetry of $\Psi_{j',v'}$ is B_{1u}, B_{2u}, or B_{3u} as shown on top, the symmetry of the component of a vector operator $\hat{\mathbf{A}}$ which makes a nonvanishing contribution to $A(jv,j'v')$ has to be that of unit vectors in the direction z, y, or x, respectively.

Finding the direct products of irreducible representations of higher dimensionality is more complicated. This and additional information on the fundamentals of group theory is available in Appendix I, which also contains three collections of tables for the most frequently encountered point symmetry groups. These are (i) character tables, (ii) tables of patterns of electric dipole and magnetic dipole transition moment vectors as well as the two-photon absorption and Raman polarizability transition tensors, and (iii) tables of direct products of irreducible representations.

1.3 A Review of Optical Spectroscopy

It has been mentioned above that the excitation of molecules by light occurs via the interaction of molecular transition moments with the electric and, to a much smaller degree, magnetic fields of the light. How does this interaction, which leads to absorption and emission of light, occur? A detailed answer is provided by quantum electrodynamics, but for cases in which the number of photons in the electromagnetic field interacting with the molecule is very large, the simpler semiclassical description is quite adequate. Since this is the case encountered in most of the experiments discussed presently, we shall content ourselves with the semiclassical description, in which the states of the molecules are quantized but the electromagnetic field is viewed in the classical sense of Maxwell.

We shall first provide a simplified and qualitative view of the basics, including one-photon absorption and emission, two-photon absorption, and Raman scattering (Section 1.3.1). Subsequently, we shall describe the most important case, one-photon absorption and emission, in a more quantitative fashion (Section 1.3.2). Finally, we shall address briefly the issue of optical activity (Section 1.3.3).

1.3.1 A Qualitative Description of Light–Molecule Interactions

Let us imagine an isolated molecule initially in a state described by a stationary wavefunction $\Psi_0(\mathbf{q},\mathbf{Q},t)$ of the type (1.10) and energy E_0, exposed to electromagnetic radiation of frequency ν propagating in the direction characterized by the unit vector ε_K. Let the molecule possess another stationary state characterized by another wavefunction of type (1.10), $\Psi_f(\mathbf{q},\mathbf{Q},t)$, and by energy E_f.

The electromagnetic perturbation due to the incoming wave causes a mixing of the ground state of the molecule Ψ_0 with its excited states. The resulting nonstationary state of the molecule, described by a linear combination of Ψ_0, Ψ_f, and other excited states, $c_0\Psi_0 + c_f\Psi_f + \cdots$, will in general have nonvanishing electric and magnetic moments, even if the pure stationary states Ψ_0, Ψ_f, etc., do not. The most important among these are the electric dipole and quadrupole moments and the magnetic dipole moment. These moments will oscillate in time at angular frequencies given by the beat frequencies between the time-dependent phase factors $e^{-iE_0t/\hbar}$, $e^{-iE_ft/\hbar}$, etc., associated with the wavefunctions Ψ_0, Ψ_f, etc. [cf. equation (1.10)].

The beat frequencies depend on the differences between molecular state energies. In the case of Ψ_f, the angular beat frequency will be $(E_f - E_0)/\hbar$ radians s^{-1} (Figure 1.11). This corresponds to $(E_f - E_0)/h$ Hz.

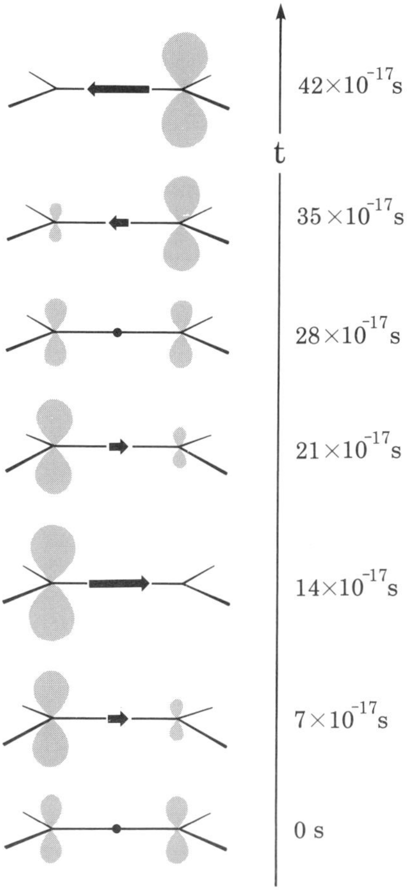

Figure 1.11 Time dependence of the charge density of the 1:1 superposition state of Ψ_0 and Ψ_f in the $\pi\pi^*$ transition of ethylene (schematic). Only the contribution from the optical electron is shown. The arrow in the center of the bond indicates the instantaneous dipole moment.

One-Photon Absorption and Emission. If we select the frequency ν_f (in Hz) of the electromagnetic radiation to be such that $2\pi\hbar\nu_f = h\nu_f = E_f - E_0$, Ψ_f will mix with Ψ_0 far more than any other excited wavefunction will. The corresponding wavenumber is $\tilde{\nu}_f = \nu_f/c$. Then the oscillating electric and magnetic moments will be in resonance with the oscillating electric and magnetic vectors of the light. The field will exchange energy with the molecule by either acting on its oscillating moments or being acted on by those moments. Light is absorbed if the molecule accepts energy from the electromagnetic field, and stimulated emission occurs if it provides energy to the field. The latter is the basis of laser action but will be of limited interest to us here.

In order to understand spontaneous emission, one must go beyond the semiclassical description and introduce the perturbation of the molecular stationary state by fluctuations of the electromagnetic field of the vacuum. Even in this case, the same set of molecular transition moments is responsible for the emission.

How much energy is actually transferred to the molecule from the electromagnetic field will depend on its intensity. In very intense fields saturation can occur, that is, the molecular wavefunction will contain equal parts of Ψ_0 and Ψ_f. Further absorption will then be canceled exactly by further stimulated emission, so that no additional net exchange of energy will occur between the molecule and the field. It is easy to achieve this situation in NMR or EPR spectroscopy, but very difficult in optical spectroscopy, which will be of interest to us in this book, so we need not be concerned with it.

This simple description of light absorption and emission accounts for two well-known facts. First, a transition of energy $E_f - E_0$ will require photons of frequency $\nu_f = (E_f - E_0)/h$. Second, the strength of the molecule–light interaction will depend on the strength of the electric and magnetic fields of the light wave, i.e., its intensity, on the amplitude of the electric and magnetic moments induced in the molecule, and on the mutual orientation of the field vectors and the molecular moments.

The maximum instantaneous values of the oscillating molecular moments are characterized by the quantities introduced in (1.14), the transition moments. The ones we need to consider are the electric dipole transition moment $\mathbf{M}(0f)$, the electric quadrupole transition moment $\mathsf{Q}(0f)$, and the magnetic dipole transition moment $\mathcal{M}(0f)$. Their contributions to absorption and emission probabilities are proportional to the square of the projection of $\mathbf{M}(0f)$ into the direction of the electric vector of the light, of $\mathcal{M}(0f)$ into the direction of the magnetic vector of the light, and of the "double projection" of the tensor $\mathsf{Q}(0f)$ into the direction of light propagation and into the direction of its electric vector. Thus, for pure electric dipole, pure magnetic dipole, and pure electric quadrupole interactions, the transition probabilities for light linearly polarized along U and propagating along K are proportional to the first power of the light intensity and can be written as

$$\begin{aligned} W_{el.dip.} &= \text{const.} \times \tilde{\nu}|\boldsymbol{\varepsilon}_U^* \cdot \mathbf{M}(0f)|^2 \\ &= \text{const.} \times \tilde{\nu}\boldsymbol{\varepsilon}_U \cdot \mathbf{M}(f0)\mathbf{M}(0f) \cdot \boldsymbol{\varepsilon}_U^* \\ &= \text{const.} \times \tilde{\nu}[\mathsf{M}(0f)]_{UU^*} \end{aligned} \qquad (1.15)$$

$$\begin{aligned} W_{\text{mag.dip.}} &= \text{const.} \times \tilde{\nu}|(\boldsymbol{\varepsilon}_K \times \boldsymbol{\varepsilon}_U^*) \cdot \mathcal{M}(0f)|^2 \\ &= \text{const.} \times \tilde{\nu}(\boldsymbol{\varepsilon}_K^* \times \boldsymbol{\varepsilon}_U) \cdot \mathcal{M}(f0)\mathcal{M}(0f) \cdot (\boldsymbol{\varepsilon}_K \times \boldsymbol{\varepsilon}_U^*) \\ &= \text{const.} \times \tilde{\nu}[\mathfrak{M}(0f)]_{K^* \times U, K \times U^*} \end{aligned} \qquad (1.16)$$

$$\begin{aligned} W_{\text{el.quad.}} &= \text{const.} \times \tilde{\nu}^3|\boldsymbol{\varepsilon}_K \mathsf{Q}(0f)\boldsymbol{\varepsilon}_U^*|^2 \\ &= \text{const.} \times \tilde{\nu}^3\boldsymbol{\varepsilon}_K^*\boldsymbol{\varepsilon}_U \mathsf{Q}(f0)\mathsf{Q}(0f)\boldsymbol{\varepsilon}_K\boldsymbol{\varepsilon}_U^* \\ &= \text{const.} \times \tilde{\nu}^3[{}^{(4)}\mathsf{Q}(0f)]_{K^*UKU^*} \end{aligned} \qquad (1.17)$$

In the case of linearly polarized light, the complex conjugation shown can be ignored. When elliptically or circularly polarized light is used, as in chiroptical spectroscopy, the vector $\boldsymbol{\varepsilon}_U$ is complex, and it is no longer possible to dismiss the complex conjugation. The description of natural optical activity requires a simultaneous consideration of the electric dipole, magnetic dipole, and electric quadrupole moments induced in the molecule by the electric field of the light wave, rather than a consideration of these moments one at a time as in (1.15)–(1.17). The description of magnetically induced optical activity requires a consideration of the perturbing effect of a static magnetic field on the energies and wave-functions of the ground and excited states in the expression for $W_{\text{el.dip.}}$ given in (1.15).

As shown on the right-hand sides of (1.15)–(1.17), squares of absolute values can be also viewed as elements of the electric dipole absorption tensor $\mathsf{M}(0f) = \mathbf{M}(f0)\mathbf{M}(0f)$ and the magnetic dipole absorption tensor $\mathfrak{M}(0f) = \mathcal{M}(f0)\mathcal{M}(0f)$, both of which are of rank 2, and of the electric quadrupole absorption tensor ${}^{(4)}\mathsf{Q}(0f) = \mathsf{Q}(f0)\mathsf{Q}(0f)$, which is of rank 4 (see Appendix II for a brief review of tensor calculus; some authors would use the term "order of a tensor" where we use "rank of a tensor"). Also, natural and magnetic activity are described in terms of tensors: the rotatory strength tensor $\mathsf{R}(0f)$ and the magnetic rotatory strength tensors $\mathsf{A}(0f)$, $\mathsf{B}(0f)$, and $\mathsf{C}(0f)$. Qualitative and quantitative aspects of the two kinds of chiroptical spectroscopy are discussed in more detail in Section 1.3.3.

For linearly polarized light, the subscripts of the tensor element to be taken in (1.15)–(1.17) are seen to be defined by the direction of the electric vector in the first case (U,U), by the direction of the magnetic vector in the second case (K × U, K × U), and by both the light propagation direction and the direction of the electric vector in the third case (K,U,K,U). We shall see in the following that this is a very characteristic form for spectroscopic intensities. The anisotropy of the optical properties is described by a tensorial quantity such as $\mathsf{M}(0f)$, and the magnitude actually measured is obtained by taking a projection of the tensor in a direction specified by the directional properties of the polarized light. The rank of the tensor shall be used in the following for classification purposes. Processes described by tensors of rank 2 are discussed in Chapter 5; processes described by tensors which are inherently of higher rank but can be reduced to rank 2 are discussed in Chapter 6; and processes described by tensors of rank 4 are treated in Chapter 7.

Emission of light is described by tensors entirely analogous to those used to describe absorption as long as the probability that a molecule becomes excited, and thus capable of emitting, is independent of molecular orientation. A practical realization of such a situation would be, for instance, chemiluminescent emission.

If the emission is stimulated by prior absorption of a photon (photoluminescence), the situation becomes more complicated. Now, the second-rank absorption process and the second-rank emission process must be coupled into a photoluminescence process of overall rank 4, as described in Section 7.2. A similar situation occurs if a molecule successively absorbs two photons (photoinduced dichroism).

Similar coupling of an even larger number of one-photon events leads to processes of even higher rank, but we do not address these in any detail.

Excited State Relaxation. Saturation is hard to achieve in optical spectroscopy because of the fast decay rate of the excited state. This is due both to spontaneous emission (particularly important for electronically excited states) and to nonradiative relaxation to other excited states and to the ground state by a variety of intramolecular mechanisms. The rate of spontaneous emission determines the natural or radiative lifetime.

Within the spirit of the present qualitative discussion, we introduce the overall exponential decay by making the energy of the excited state, E_f, a complex number. Then the beat frequency will be complex as well, i.e., it will correspond to a periodic vibration damped by an exponentially decaying amplitude factor. A perfectly sharp resonance with monochromatic light is then no longer possible; strong interaction (resonance) will occur for a range of frequencies ν which correspond to the real part of the energy difference $E_f - E_0$. The shape of the function which describes the rate of energy transfer between the molecule and the radiation field as a function of the frequency of the latter is then Lorentzian, and the line shape is said to be homogeneously broadened by the finite lifetime of the excited state.

In reality, the situation is complicated by processes such as collisions with neighboring molecules which destroy the coherence of the molecule-radiation interaction by affecting the phase factor in the molecular wavefunction. These processes thus increase the damping and broaden the spectral line further without actually shortening the radiative lifetime, since they leave the molecule excited. We shall briefly return to this topic in Section 1.3.2 in connection with the Franck-Condon principle. Collisions and other intermolecular interactions can actually also shorten the excited state lifetime, introducing further complications. In condensed media, rotational fine structure is normally not resolved because frequent collisions with the surrounding molecules perturb the rotational states of the solute. The resulting line shape contains information on the nature of the perturbed rotational motion.

Real samples consists of a very large number of molecules and absorption, emission and saturation, as well as line broadening, must then be described in a statistical fashion. If the distribution of molecular velocities in the gas phase or, of more immediate interest to us, the distribution of molecular environments in the condensed phase is statistical (Gaussian), the absorption line will acquire a Gaussian profile (inhomogeneous broadening). The effect of molecular velocities on the excitation energy is known as the Doppler effect, and the effect of the environment, as the site effect. Under the circumstances of interest to us here, (i) UV-visible absorption and emission lines are broadened much more strongly by the site effects of the orienting environment than by lifetime effects, and the latter can be neglected; and (ii), IR and Raman line shapes are dominated by the rotational motion of the solute to a considerable degree, but lifetime effects cannot always be safely ignored. A quantitative description of the Lorentzian and Gaussian line shapes will be given in Section 1.3.2.

Virtual Excitation. If the beat frequency with which the molecular electric and magnetic moments oscillate is not equal to the light frequency or very close to it, the molecule may still be strongly perturbed by the field, and large oscillating electric and magnetic moments may be induced in it. Now, however, no net exchange of energy between the field and the molecule by absorptive or emissive processes will be possible to the first order, since there is no resonance between the oscillatory behavior of these moments and the oscillation of the field of the light wave. This situation is known as virtual excitation.

In a better approximation, it can be seen that virtual excitation can serve as a launching pad for a variety of processes in which energy is exchanged between the field and the molecule after all, and we shall mention two of them.

First, since molecular polarizability is not strictly linear, a light wave of frequency ν will induce not only electric and magnetic moments oscillating at the same frequency ν, but also weaker ones oscillating at 2ν and even higher multiples of ν. Thus, if the molecular beat frequency is double (a multiple of) the light frequency, simultaneous absorption of two (several) photons can occur.

Second, if the molecular beat frequency is modulated by a vibrational motion, the molecule can emit light at a frequency of a side band, i.e., $\nu \pm \Delta\nu$, where $\Delta\nu$ is the frequency of the vibrational motion (Raman scattering).

Two-Photon Absorption. As indicated above, very intense radiation fields, such as those produced by lasers, can cause simultaneous absorption of two or even a larger number of photons ("two-photon absorption," etc.). In general, the two photons do not have to have the same energy. Very roughly speaking, one photon is needed to mix small amounts of all excited states Ψ_j into the ground state Ψ_0 to produce a virtual state, an energetic perturbed nonstationary state of the molecule. The mixing of an excited state Ψ_j into the virtual state will be favored if the transition moments for the excitation $0 \rightarrow j$ have a large magnitude and if their orientation relative to the electric vector of the light is favorable. The promotion from the virtual state to the final state Ψ_f, for which the energy of the other photon is required, will be similarly favored if the transition moments for $j \rightarrow f$ are large and favorably oriented. Contributions to the excitation amplitude from all excited states Ψ_j will have to be summed in a coherent fashion, with weighting factors which reflect the mismatch between the light frequency ν on the one hand and ν_j, the beat frequency of Ψ_j and Ψ_0, on the other hand. The role played by both photons is, of course, the same in reality since they are absorbed simultaneously, and even if they have different wavenumbers, $\tilde{\nu}_1$ and $\tilde{\nu}_2$, they have to enter into the expression for the probability of absorption in a symmetrical fashion.

A more quantitative analysis along the lines indicated in Section 1.3.2 for the simpler case of one-photon absorption shows that in the electric dipole approximation the probability of molecular transition from the electronic state Ψ_0 to a higher energy electronic state Ψ_f by simultaneous absorption of two photons of wavenumbers $\tilde{\nu}_1$ and $\tilde{\nu}_2$ with linear polarizations $\boldsymbol{\varepsilon}_U$ and $\boldsymbol{\varepsilon}_V$, respectively, to yield an excited state of energy $\tilde{\nu}_1 + \tilde{\nu}_2$ is proportional to the product of the intensities of the two light beams and to the square of the "double projection" of the two-photon absorption tensor into the directions U and V:

$$
\begin{aligned}
W &= \text{const.} \times \tilde{\nu}_1\tilde{\nu}_2|\boldsymbol{\varepsilon}_U^*\mathsf{T}(0f,\tilde{\nu}_1)\boldsymbol{\varepsilon}_V^*|^2 \\
&= \text{const.} \times \tilde{\nu}_1\tilde{\nu}_2\boldsymbol{\varepsilon}_U\boldsymbol{\varepsilon}_V\mathsf{T}^*(0f,\tilde{\nu}_1)\mathsf{T}(0f,\tilde{\nu}_1)\boldsymbol{\varepsilon}_U^*\boldsymbol{\varepsilon}_V^* \\
&= \text{const.} \times \tilde{\nu}_1\tilde{\nu}_2[^{(4)}\mathsf{O}(0f,\tilde{\nu}_1)]_{UVU^*V^*} \qquad (1.18)
\end{aligned}
$$

The second-rank two-photon transition moment tensor $\mathsf{T}(0f,\tilde{\nu}_1)$ plays a role analogous to that of the electric dipole transition moment vector $\mathbf{M}(0f)$ in ordinary one-photon absorption. The two-photon absorption tensor $^{(4)}\mathsf{O}(0f,\tilde{\nu}_1) = \mathsf{T}^*(0f,\tilde{\nu}_1)\mathsf{T}(0f,\tilde{\nu}_1)$ is of rank 4 and is analogous to the one-photon absorption tensor $\mathsf{M}(0f) = \mathbf{M}(f0)\mathbf{M}(0f)$ which was of rank 2 [cf. (1.15)].

The second-rank tensor $\mathsf{T}(0f,\tilde{\nu}_1)$ contains $\tilde{\nu}_1$ and $\tilde{\nu}_2$ in a symmetrical fashion. It can be written in the form anticipated from our qualitative argument, containing an infinite sum of contributions from all electronic states of the molecules [see (2.20)].

Raman Scattering. As indicated above, the interaction of a photon of incident light ($h\nu_1$) with a molecule does not have to result in absorption or transmission. Another possible outcome is scattering of the photon. If its energy is not changed in the process, this referred to as Rayleigh scattering. It is of limited interest in the present context, since we concentrate on dilute solutions and the effect will then normally be dominated by the solvent.

If the energy of the photon is changed in the scattering process to $h\nu_2$, the phenomenon is known as the Raman effect; if the molecule also exhibits significant absorption at $\tilde{\nu}_1$, it is referred to as resonant Raman effect. Because of the frequency shift, lines due to different components in a mixture are separated in the spectrum, and measurements on solutes in quite dilute solutions are readily possible. The energy difference $\Delta h\nu$ is taken up by the molecule (Stokes lines, $\tilde{\nu}_2 = \tilde{\nu}_1 - \Delta\tilde{\nu}$) or released by the molecule (anti-Stokes lines, $\tilde{\nu}_2 = \tilde{\nu}_1 + \Delta\tilde{\nu}$) and usually has the form of vibrational energy, although electronic Raman scattering is also known. Once again, we can think of the Raman scattering event as a process in which the molecule is perturbed by the interaction with the electromagnetic wave into a virtual excited state. Its amplitude can be expressed as a coherent sum over all electronic excited states j and is related to molecular polarizability. In the electric dipole approximation, the terms in the sum will contain the transition moments $\mathbf{M}(0j)$. If the virtual excited state changes as the molecule vibrates, an interaction will be set up between electronic and vibrational motion, and it is thus hardly surprising that the quantity which determines Raman scattering intensities, the Raman polarizability tensor $\alpha'(\tilde{\nu},f)$, is related to the change in the molecular polarizability tensor with respect to displacement along the vibrational mode f of frequency $\Delta\tilde{\nu}$ which is involved in the transition.

This can be stated more quantitatively. If certain simplifying conditions are satisfied ($\tilde{\nu}_1$ is much larger than $|\tilde{\nu}_1 - \tilde{\nu}_2|$ and much less than any electronic transition frequency, and the initial state is not degenerate), the nonresonant Raman intensity of the f-th vibration is proportional to the intensity of the incident light beam of wavenumber $\tilde{\nu}_1$ polarized along $\boldsymbol{\varepsilon}_U$. When measured for scattered light of wavenumber $\tilde{\nu}_2$ polarized along $\boldsymbol{\varepsilon}_V$, it can be written as

$$\begin{aligned} W &= \text{const.} \times \tilde{\nu}_2^4 |\boldsymbol{\varepsilon}_U^* \alpha'_j(\tilde{\nu}_1,f)\boldsymbol{\varepsilon}_V|^2 \\ &= \text{const.} \times \tilde{\nu}_2^4 \boldsymbol{\varepsilon}_U \boldsymbol{\varepsilon}_V^* \alpha'_j(\nu_1,f)\alpha'_j(\nu_1,f)\boldsymbol{\varepsilon}_U^* \boldsymbol{\varepsilon}_V \\ &= \text{const.} \times \tilde{\nu}_2^4 [{}^{(4)}O_j(\tilde{\nu}_1,f)]_{UV^*U^*V} \end{aligned} \quad (1.19)$$

where $\alpha'(\nu_1,f)$ is known as the Raman scattering tensor. The tensor ${}^{(4)}O_j(\tilde{\nu}_1,f) = \alpha_j'^*(\tilde{\nu}_1,f)\alpha'_j(\tilde{\nu}_1,f)$ is of rank 4.

As can be anticipated from the qualitative argument given above, the second-rank tensor $\alpha'_j(\tilde{\nu}_1,f)$ can be written in a form which contains an infinite sum over all electronic states of the molecule [see (2.24)].

1.3.2 A Quantitative Treatment of One-Photon Absorption and Emission

Among the light–molecule interactions described qualitatively in the preceding section, one-photon absorption and emission are the most important. We shall now discuss them with a little more rigor and indicate briefly the derivation of the fundamental results (1.15)–(1.17). The reader is referred to the literature listed in Section 1.4 for additional detail and for the derivation of expressions (1.18) and (1.19) based on a quantitative treatment of two-photon absorption and Raman scattering.

The Lambert-Beer Law. When a collimated monochromatic light beam of incident intensity I_0 and wavenumber $\tilde{\nu}$ passes through an absorbing homogeneous isotropic sample, it is attenuated. Let us assume that light absorption is the only attenuating mechanism, and let the molar concentration of the solute be c (mol/L) so that the number of molecules per cubic centimeter is $n' = cN/1000$, where N is Avogadro's number, $N = 6.0225 \times 10^{23}\ \text{mol}^{-1}$. After passage through a layer 1 cm thick, the light intensity $I(\tilde{\nu})$ is given by the Lambert-Beer law:

$$I(\tilde{\nu}) = I_0(\tilde{\nu})10^{-\varepsilon(\tilde{\nu})cl} = I_0(\tilde{\nu})e^{-\sigma(\tilde{\nu})n'l} = I_0(\tilde{\nu})e^{-\alpha(\tilde{\nu})l} \quad (1.20)$$

where $\varepsilon(\tilde{\nu})$ is the decadic molar extinction (absorption) coefficient, $\sigma(\tilde{\nu})$ is the absorption cross section of the solute in the particular solvent used, and $\alpha(\tilde{\nu})$ is the absorption coefficient of the sample. The negative of the exponent in (1.20), $\varepsilon(\tilde{\nu})cl$, is called the absorbance or optical density of the sample, and we shall denote it $E(\tilde{\nu})$. If several noninteracting solutes are present, they contribute to the absorbance in an additive fashion. One such contribution may come from the solvent if it absorbs at $\tilde{\nu}$. The fraction $I(\tilde{\nu})/I_0(\tilde{\nu})$ is known as the transmittance $T(\tilde{\nu})$.

The decadic molar extinction coefficient $\varepsilon(\tilde{\nu})$ is the quantity commonly used by chemists in the UV-visible region. It is usually referred to simply as "the extinction coefficient." Traditionally, its unit is $L\ mol^{-1}\ cm^{-1} = 1000\ cm^2\ mol^{-1}$ and is understood but not explicitly stated on spectra. An alternative quantity is the absorption cross section $\sigma(\tilde{\nu})$, usually expressed in units of square centimeters per molecule and related to $\varepsilon(\tilde{\nu})$ in its usual units by

$$\sigma(\tilde{\nu}) = 2303\varepsilon(\tilde{\nu})/N = 3.823 \times 10^{-21}\varepsilon(\tilde{\nu})$$

In the IR region, chemists usually use natural logarithms and define absorptivity as $\alpha(\tilde{\nu})/c$.

The absorption coefficient $\alpha(\tilde{\nu}) = 2.303\varepsilon(\tilde{\nu})c$ traditionally has the units cm^{-1}. It can be combined with the wavenumber-dependent index of refraction $n(\tilde{\nu})$ into the "complex index of refraction" $N(\tilde{\nu})$ which we define as

$$N(\tilde{\nu}) = n(\tilde{\nu}) - (i/4\pi\tilde{\nu})\alpha(\tilde{\nu}) \qquad (1.22)$$

where i is the imaginary unit. The minus sign in this definition is compatible with the signs convention adopted in Section 1.1.1; the opposite convention would require a plus sign in (1.22). The dependence of the ordinary index of refraction $n(\tilde{\nu})$ on light frequency is known as dispersion [strictly speaking, the term dispersion refers to $dn(\tilde{\nu})/d\tilde{\nu}$, but loosely the curve $n(\tilde{\nu})$ is referred to as the dispersion curve].

In a partially oriented sample, $N(\tilde{\nu})$ will depend on the propagation direction and the state of polarization of the light. Even in an isotropic sample, $N(\tilde{\nu})$ will depend on the state of polarization of the light unless it is linear (or unpolarized), provided that either the sample is optically active or an outside magnetic field with a component in the light propagation direction is present. We shall use the symbols $N_Z(\tilde{\nu})$ and $N_Y(\tilde{\nu})$ for the complex index of refraction for light polarized linearly along Z or Y, respectively. We shall use $N_L(\tilde{\nu})$ and $N_R(\tilde{\nu})$ for the complex index of refraction for left-handed and right-handed circularly polarized light, respectively. As described in more detail in Section 1.3.3, the difference $N_Z(\tilde{\nu}) - N_Y(\tilde{\nu})$ is a measure of linear birefringence (the real part) and linear dichroism (the imaginary part), and the difference $N_L(\tilde{\nu}) - N_R(\tilde{\nu})$ is a measure of circular birefringence (the real part) and circular dichroism (the imaginary part).

In samples for which $N(\tilde{\nu})$ depends on the polarization state of light, the Lambert-Beer law (1.20) does not hold in general but may still hold for light of particular polarizations.

In isotropic optically active samples, (1.20) holds only for circularly polarized light [either $N_L(\tilde{\nu})$ or $N_R(\tilde{\nu})$ applies]. Light of other polarizations behaves as if it were composed of a left-handed and a right-handed circularly polarized part, each following (1.20) with its own complex refractive index. Since the difference between $N_L(\tilde{\nu})$ and $N_R(\tilde{\nu})$ is very small in practice, it is usually adequate to assume that (1.20) holds for the total intensity as well.

In anisotropic samples the state of light polarization generally changes as the light propagates, making the situation very complicated (Appendix III). In uniaxial samples, the particular cases in which light propagates without changing its polarization acquire overwhelming practical importance and will be the only ones discussed in the following chapters.

(i) Light propagating along the unique axis Z behaves as if it were propagating in an isotropic solution. If the sample is optically inactive and no magnetic field along Z is present, light of any polarization propagates without changes in polarization and (1.20) applies. If the sample is optically active or if a magnetic field along Z is present, only circularly polarized light propagates without changes in polarization and obeys (1.20). Light of other polarizations still obeys (1.20) to a high degree of accuracy.

(ii) Light propagating in a direction perpendicular to Z does not change its polarization if it is linearly polarized along Z or perpendicular to Z, and obeys (1.20) with $N_Z(\tilde{\nu})$ or $N_Y(\tilde{\nu})$, respectively, provided that the sample is optically inactive and that no magnetic field with a component along the light propagation direction is present. The attenuation of incident light of polarization other than linear along Z or Y is quite complicated, since the birefringent sample changes its polarization into elliptical. The attenuation of unpolarized light occurs as if it were composed of a Z-polarized and a Y-polarized part, each following (1.20) with its own complex index of refraction, $N_Z(\tilde{\nu})$ and $N_Y(\tilde{\nu})$. Since these can differ considerably, it is not safe to assume that (1.20) holds for the total intensity. An example of a situation of this kind is worked out in Section 3.2.1.

Samples with small absorbance $E(\tilde{\nu}) = \varepsilon(\tilde{\nu})cl$ are said to be optically thin. For $E(\tilde{\nu})$ up to about 0.2, it is usually possible to expand the exponential in (1.20) and keep only the term linear in $\varepsilon(\tilde{\nu})$. Under these conditions, the fraction of light which is absorbed is proportional to $\varepsilon(\tilde{\nu})$,

$$(I_0 - I)/I_0 \cong 2.303\varepsilon(\tilde{\nu})cl$$

For practical application of the Lambert-Beer law, it is important to keep in mind the assumptions under which it is valid. Real systems may exhibit deviations from (1.20) if their absorption lines are narrow relative to the bandwidth of the "monochromatic" light used for the measurement, if they scatter light, if the walls of the container reflect light, if the light intensity is too high, leading to ground-state depletion or two-photon absorption, etc. Suitable corrections are then necessary.

Oscillator Strength. Transitions of interest in this book are of the electronic (vibronic) or vibrational types (Figure 1.6). In principle, both may have fine structure due to the presence of individual rotational, and in the former case also vibrational, components. We have already mentioned that rotational structure is normally not resolved in solutions, and we shall therefore ignore it. Even vibrational structure is often resolved only poorly, so that one does not observe lines but broad bands. The integrated intensity of a band is then the quantity of interest.

In vibrational spectroscopy, integrated intensity is usually defined by

$$A = \frac{1}{c}\int_{\text{band}} \alpha(\tilde{\nu})\,d\tilde{\nu} \tag{1.23}$$

and its currently favored unit is $\mathrm{km\,mol^{-1}}$. The integration is over the spectral region corresponding to the band in question.

In electronic spectroscopy, integrated intensity is normally expressed in terms of the dimensionless oscillator strength f of the transition, defined for a measurement on an isotropic sample as

$$f = (mc/\pi e^2 n_0)\int_{\text{band}} \sigma(\nu)\,d\nu = (2303mc^2/\pi n_0 Ne^2)\int_{\text{band}} \varepsilon(\tilde{\nu})\,d\tilde{\nu}$$

$$= \frac{4.319 \times 10^{-9}}{n_0}\int_{\text{band}} \varepsilon(\tilde{\nu})\,d\tilde{\nu} \tag{1.24}$$

where m is the mass of the electron, n_0 is the mean index of refraction of the solution in the region of the band, $\varepsilon(\tilde{\nu})$ is in its traditional units of $\mathrm{L\,mol^{-1}\,cm^{-1}}$, and $\tilde{\nu}$ is in $\mathrm{cm^{-1}}$. The inclusion of the index of refraction n_0 in this definition is a crude attempt to compensate for solvent effects (see Section 1.3.2). Some authors omit n_0 in the definition of f which then is solvent-dependent. The integration in (1.24) is over the spectral region corresponding to the band in question.

The reasons for this particular definition are historical and need not be discussed here. Very strong UV transitions have oscillator strengths of the order of 1, and the Kuhn-Thomas sum rule states that the sum of the oscillator strengths of all electronic transitions originating in the ground state is equal to the number of electrons in the molecule.

The use of (1.23) and (1.24) is straightforward if different absorption bands do not overlap. If they do, they must be separated before their oscillator strengths can be determined. Polarized spectroscopy on oriented samples frequently helps with this task.

In both vibrational and electronic spectroscopy the integration is sometimes not done over $d\tilde{\nu}$, as shown in (1.23) and (1.24), but rather over $d(\ln \tilde{\nu})$. The resulting integrated intensities are usually labeled Γ in vibrational spectroscopy and, with an appropriate numerical factor in front, D in electronic spectroscopy. We shall refer to the dipole strength D in Section 1.3.3. [equation (1.100)].

A complete description of an electronic or vibrational spectral band in an ordinary (isotropic) absorption spectrum is usually given by specifying its oscillator strength f or integrated intensity A, respectively, plus its band-shape function $g'(\tilde{\nu})$. The latter is a normalized function which describes the spectral shape of the absorption band,

$$\int_0^{\infty} g'(\tilde{\nu})\,d\tilde{\nu} = \int_{\text{band}} g'(\tilde{\nu})\,d\tilde{\nu} = 1 \tag{1.25}$$

The function $g'(\tilde{\nu})$ has the dimension of length, and since the standard unit for $\tilde{\nu}$ is $\mathrm{cm^{-1}}$, the unit for $g'(\tilde{\nu})$ is cm. We have chosen to work with wavenumbers

$\tilde{\nu}$ throughout, since they are commonly used in optical spectroscopy. Theoretical derivations are normally performed using the frequency ν and lead to line-shape functions $g'(\nu)$ which have the dimension of time and are normalized so that $\int_0^\infty g'(\nu)\,d\nu = 1$. Since the standard unit for ν is s^{-1}, the unit for $g'(\nu)$ is seconds. The two functions have the same shape and are related by multiplication with the speed of light, $cg'(\nu) = g'(\tilde{\nu})$. For this reason, c will appear in somewhat unexpected places in those equations which contain $g'(\tilde{\nu})$, starting with (1.38).

Relation of Absorption Intensity to Molecular Quantities. In order to obtain a theoretical expression for the integrated absorption strength of a transition in a set of equally oriented molecules in terms of the molecular wavefunctions of the initial and final states, the following steps are taken: (i) time-dependent perturbation theory is used to derive a general expression for the weight of the final state in the superposition state produced by the light wave; (ii) the perturbation Hamiltonian operator $\hat{H}'$ which describes the light–molecule interaction is derived, and its form is substituted into the result of (i); (iii) it is recognized that the light beam used cannot be strictly monochromatic and that an integration over light frequency needs to be performed; and (iv) the result is expressed in a convenient form. Once this is obtained, it is possible to derive the usual expressions involving electric and magnetic transition moments by expanding an exponential in a series.

(i) Time Evolution of the Molecular Wavefunction. The superposition state produced by the perturbing effect of the radiation can be written as a linear combination of the stationary states given in (1.10), which are eigenfunctions of the unperturbed Hamiltonian $\hat{H}$. It has to satisfy the time-dependent Schrödinger equation so that we can write

$$(\hat{H} + \hat{H}')\sum_n c_n(t)\Psi_n(\mathbf{q},\mathbf{Q})e^{-iE_nt/\hbar} = i\hbar\frac{\partial}{\partial t}\sum_n c_n(t)\Psi_n(\mathbf{q},\mathbf{Q})e^{-iE_nt/\hbar} \tag{1.26}$$

This simplifies to

$$\sum_n c_n(t)\hat{H}'\Psi_n(\mathbf{q},\mathbf{Q})e^{-iE_nt/\hbar} = i\hbar\sum_n \frac{\partial c_n(t)}{\partial t}\Psi_n(\mathbf{q},\mathbf{Q})e^{-iE_nt/\hbar} \tag{1.27}$$

Multiplication from the left by $\Psi_f^*(\mathbf{q},\mathbf{Q})$ and integration over all space and spin coordinates yields

$$\sum_n c_n(t)\langle\Psi_f(\mathbf{q},\mathbf{Q})|\hat{H}'|\Psi_n(\mathbf{q},\mathbf{Q})\rangle e^{-iE_nt/\hbar} = i\hbar\frac{\partial c_f(t)}{\partial t}e^{-iE_ft/\hbar} \tag{1.28}$$

Only one term survived on the right-hand side, since $\langle\Psi_f(\mathbf{q},\mathbf{Q})|\Psi_n(\mathbf{q},\mathbf{Q})\rangle = \delta_{fn}$.

If the molecule is in the state $\Psi_0(\mathbf{q},\mathbf{Q})$ at time $t = 0$, we have $c_0(0) = 1$ and $c_n(0) = 0$ if $n \neq 0$. Even after the light is turned on, the ground state will continue to dominate the wavefunction of the superposition state, since the intensity of the light is assumed to be small. Then, $c_0(t) \cong 1$ and $c_0(t) \gg c_n(t)$ for $n \neq 0$. Therefore, only the term with $n = 0$ will be kept in the summation, and we set $c_0(t) = 1$.

Integration from time $t = 0$ to time $t = \tau$ then yields

$$c_f(\tau) = \frac{-i}{\hbar}\int_0^{\tau} \langle \Psi_f(\mathbf{q},\mathbf{Q})|\hat{H}'|\Psi_0(\mathbf{q},\mathbf{Q})\rangle e^{i(E_f - E_0)t/\hbar}\, dt \qquad (1.29)$$

Before we can evaluate the integral, an explicit form needs to be substituted for $\hat{H}'$, since it depends on time.

(ii) The Electromagnetic Perturbation Operator. The perturbation Hamiltonian H' contains a contribution from the interaction of each charged particle present in the molecule with the electric and magnetic fields of the light wave. Using q_j for the charge of the j-th particle, $\mathbf{r}_j$ for its position, and m_j for its mass, one can write in a good approximation

$$\hat{H}' = \sum_j \frac{q_j}{m_j c} \mathbf{A}(t,\mathbf{r}_j) \cdot \hat{\mathbf{p}}_j \qquad (1.30)$$

where $\mathbf{A}(t,\mathbf{r}_j)$ is the real part of the vector potential at the position of the j-th particle and $\hat{\mathbf{p}}_j = -i\hbar\nabla_j$ is the linear momentum operator. The components of the gradient operator ∇_j are $\partial/\partial x_j$, $\partial/\partial y_j$, and $\partial/\partial z_j$. A more complete expression for $\hat{H}'$ would also contain terms of the form $(q_j^2/2m_jc^2)\mathbf{A}(t,\mathbf{r}_j) \cdot \mathbf{A}(t,\mathbf{r}_j)$, of importance for nonlinear optical effects, and terms explicitly dependent on spin. The latter require a relativistic treatment and will be introduced in (1.50) in an ad hoc fashion.

For a plane wave of strictly monochromatic light with wavevector $\mathbf{K}$ and polarization characterized by the unit vector $\boldsymbol{\varepsilon}_U$, the real part of the vector potential $\mathbf{A}(t,\mathbf{r}_j)$ can be written as

$$\mathbf{A}(t,\mathbf{r}_j) = \tfrac{1}{2}A_0[\boldsymbol{\varepsilon}_U e^{i(\omega t - \mathbf{K}\cdot\mathbf{r}_j)} + \boldsymbol{\varepsilon}_U^* e^{-i(\omega t - \mathbf{K}\cdot\mathbf{r}_j)}] \qquad (1.31)$$

For linearly polarized light, $\boldsymbol{\varepsilon}_U = \boldsymbol{\varepsilon}_U^*$ and the unit vector $\boldsymbol{\varepsilon}_U$ can be placed in front of the brackets to that (1.31) simplifies to (1.7). Note that the roles of $\boldsymbol{\varepsilon}_U$ and $\boldsymbol{\varepsilon}_U^*$ would be interchanged if we had chosen the opposite sign convention in Section 1.1.

We now substitute (1.31) into (1.30) and (1.30) into (1.29) and integrate over time from $t = 0$ to $t = \tau$:

$$c_f(\tau) = \frac{A_0}{2\hbar c}\left[\left\langle \Psi_f(\mathbf{q},\mathbf{Q})\left|\sum_j \frac{q_j}{m_j} e^{i\mathbf{K}\cdot\mathbf{r}_j}(\boldsymbol{\varepsilon}_U^* \cdot \hat{\mathbf{p}}_j)\right|\Psi_0(\mathbf{q},\mathbf{Q})\right\rangle \frac{1 - e^{2\pi i(\nu_f - \nu)\tau}}{2\pi(\nu_f - \nu)}\right.$$
$$\left. + \left\langle \Psi_f(\mathbf{q},\mathbf{Q})\left|\sum_j \frac{q_j}{m_j} e^{-i\mathbf{K}\cdot\mathbf{r}_j}(\boldsymbol{\varepsilon}_U \cdot \hat{\mathbf{p}}_j)\right|\Psi_0(\mathbf{q},\mathbf{Q})\right\rangle \frac{1 - e^{2\pi i(\nu_f + \nu)\tau}}{2\pi(\nu_f + \nu)}\right] \qquad (1.32)$$

where $\nu_f = |E_f - E_0|/h$ is the previously mentioned beat frequency between $\Psi_0(\mathbf{q},\mathbf{Q},t)$ and $\Psi_f(\mathbf{q},\mathbf{Q},t)$. The function $(1 - e^{ixt})/x$ is nearly zero everywhere except for small x. Therefore, $c_f(\tau)$ will differ significantly from zero only if the denominator in one of the fractions is very small. This happens in two cases which correspond to the fulfillment of the resonance condition between the light frequency ν and the beat frequency $+\nu_f$ or $-\nu_f$.

First, if $\nu = \nu_f$, that is, if $E_f - E_0 = h\nu$, we have $E_f > E_0$, so that energy is taken up by the molecule. Only the first term in the brackets in (1.32) need be kept in such a case.

Second, if $\nu = -\nu_f$, that is, if $E_0 - E_f = h\nu$, we have $E_0 > E_f$, so that energy is given up by the molecule. Only the second term in the brackets in (1.32) need be kept then. In the following, we shall consider light absorption. The results for stimulated emission which originate in the second term in the brackets are analogous.

The weight of the final state in the superposition state after time τ is thus given by

$$|c_f(\tau)|^2 = \frac{A_0^2}{4\hbar^2 c^2} \left| \left\langle \Psi_f(\mathbf{q},\mathbf{Q}) \left| \sum_j \frac{q_j}{m_j} e^{i\mathbf{K}\cdot\mathbf{r}_j}(\varepsilon_U^* \cdot \hat{\mathbf{p}}_j) \right| \Psi_0(\mathbf{q},\mathbf{Q}) \right\rangle \right|^2 \frac{\sin^2 \pi(\nu_f - \nu)\tau}{\pi^2(\nu_f - \nu)^2} \qquad (1.33)$$

The weight is the largest in the limit $\nu \to \nu_f$, when the fraction at the end of (1.33) reaches the value $\lim_{x\to 0} (\sin^2 x\tau/x^2) = \tau^2$.

(iii) Integration over Light Frequency. To obtain a result which corresponds to physical reality, it must be recognized that the light wave cannot be made sufficiently monochromatic and the measurement sufficiently short to probe the peak of the frequency function $[\sin^2 \pi(\nu_f - \nu)\tau]/\pi^2(\nu_f - \nu)^2$. If one could do so, $\nu = \nu_f$, and $|c_f(t)|^2$ would grow with the square of time.

In reality, even for highly "monochromatic" light, a wide frequency band centered at ν will be present and the radiation density $\rho(\nu)$ will have an essentially constant value $\rho(\nu_f)$ over the frequency range where $[\sin^2 \pi(\nu_f - \nu)\tau]/\pi^2(\nu_f - \nu)^2$ is significantly different from zero. The time dependence of the weight of the final state must then be obtained by integration over the frequency range. This yields τ, as is readily seen: While the amplitude of the main peak of the frequency at $\nu = \nu_f$ increases as τ^2, the width of the peak decreases as $1/\tau$. The weight of the final state thus increases linearly with time, i.e., the absorption probability is constant in time as long as the ground-state depletion is negligible.

(iv) Einstein Coefficients for Absorption and Stimulated Emission. The amplitude of the vector potential A_0 of a collimated beam of light of frequency ν is related to the intensity of the radiation by

$$A_0^2 = \frac{2I(\tilde{\nu})}{\pi c \tilde{\nu}^2} \qquad (1.34)$$

This leads to

$$|c_f(\tau)|^2 = \frac{I(\nu_f)}{2\pi\hbar^2 \tilde{\nu}^2 c^3} \left| \left\langle \Psi_f(\mathbf{q},\mathbf{Q}) \left| \sum_j \frac{q_j}{m_j} e^{i\mathbf{K}\cdot\mathbf{r}_j}(\varepsilon_U^* \cdot \hat{\mathbf{p}}_j) \right| \Psi_0(\mathbf{q},\mathbf{Q}) \right\rangle \right|^2 \tau \qquad (1.35)$$

The probability P_f that the absorption from the ground state to the f-th state will take place in unit time is given by $|c_f(\tau)|^2/\tau$, i.e., by the factor in front of τ in (1.35). This can be written as a product of the radiation density $\rho(\tilde{\nu}_f) = I(\tilde{\nu}_f)/c$ and the Einstein transition probability coefficient for induced absorption, $B_{0\to f}$.

$$P_f = |c_f(\tau)|^2/\tau = \rho(\tilde{\nu}_f)B_{0\to f} \qquad (1.36)$$

$$B_{0\to f} = \frac{1}{2\pi\hbar^2 \nu_f^2} \left| \left\langle \Psi_f(\mathbf{q},\mathbf{Q}) \left| \sum_j \frac{q_j}{m_j} e^{i\mathbf{K}\cdot\mathbf{r}_j}(\varepsilon_U^* \cdot \hat{\mathbf{p}}_j) \right| \Psi_0(\mathbf{q},\mathbf{Q}) \right\rangle \right|^2 \qquad (1.37)$$

Using now the empirical normalized line-shape function $g_f'(\tilde{\nu})$ introduced in (1.25), the probability P_f becomes

$$P_f = |c_f(\tau)|^2 g_f'(\nu)/\tau = |c_f(\tau)|^2 g_f'(\tilde{\nu})/c\tau \tag{1.38}$$

For a sample of noninteracting ground-state molecules containing n′ molecules per cubic centimeter, with negligible population of the excited state, the decrease of light intensity per centimeter will be

$$dI(\tilde{\nu})/dl = -\alpha_f(\tilde{\nu})I(\tilde{\nu}) = -hc\tilde{\nu}n'P_f(\tilde{\nu}) \tag{1.39}$$

Thus, the relation of the absorption coefficient $\alpha_f(\tilde{\nu})$ to the probability $P_f(\tilde{\nu})$ is

$$\alpha_f(\tilde{\nu}) = hc\tilde{\nu}n'P_f(\tilde{\nu})/I(\tilde{\nu}) \tag{1.40}$$

and substitution from (1.35) yields

$$\alpha_f(\tilde{\nu}) = \frac{n'}{\hbar c^2 \tilde{\nu}} \left| \left\langle \Psi_f(\mathbf{q},\mathbf{Q}) \middle| \sum_j \frac{q_j}{m_j} e^{i\mathbf{K}\cdot\mathbf{r}_j}(\boldsymbol{\varepsilon}_U^* \cdot \hat{\mathbf{p}}_j) \middle| \Psi_0(\mathbf{q},\mathbf{Q}) \right\rangle \right|^2 \frac{g'(\tilde{\nu})}{c} \tag{1.41}$$

The dependence of the absorption coefficient on the molecular orientation is contained in the polarization vector $\boldsymbol{\varepsilon}_U$. As U is varied, the matrix element on the right-hand side may acquire different values, since it contains $\boldsymbol{\varepsilon}_U^*$ (in the sign convention opposite to that adopted in Section 1.1 it contains $\boldsymbol{\varepsilon}_U$ instead). The integration is over the space and spin coordinates of all the charged particles in the molecule.

For emission from an initial state $\Psi_0(\mathbf{q},\mathbf{Q},t)$ to a final state $\Psi_f(\mathbf{q},\mathbf{Q},t)$ we obtain a similar result with $e^{i\mathbf{K}\cdot\mathbf{r}_j}$ replaced by its complex conjugate, $e^{-i\mathbf{K}\cdot\mathbf{r}_j}$, and with $\boldsymbol{\varepsilon}_U^*$ replaced by $\boldsymbol{\varepsilon}_U$. The probability coefficients for absorption and stimulated emission are therefore the same, $B_{0\to f} = B_{f\to 0}$.

The operator in the matrix element in (1.37) consists of a sum of one-particle contributions. The sum runs over both nuclei, with $q_j = Z_j|e|$ and $m_j = M_j$ dependent on the atomic number Z_j and mass M_j, and electrons, with $q_j = -|e|$ and $m_j = m$. In addition to the numerical factor q_j/m_j, each term in the sum consists of a product of two parts. One is an exponential whose absolute value is unity and which assigns a complex phase factor to each point in real space. The phase factor is a function of distance in the molecule along the direction of light propagation. It originates in the $\mathbf{K}\cdot\mathbf{R}$ term in (1.7) and (1.31) and reflects the variation of the electric and magnetic fields at any given instant over the space occupied by the molecule. The other term, $\boldsymbol{\varepsilon}_U^* \cdot \hat{\mathbf{p}}_j = -i\hbar\boldsymbol{\varepsilon}_U^* \cdot \nabla_j$, is an operator which takes the gradient of the wavefunction $\Psi_0(\mathbf{q},\mathbf{Q})$ in the direction U*: For linearly polarized light (real U), it measures how steeply the wavefunction changes in the direction of the electric vector of the light.

The assumption $c_0(t) = 1$ for $t < \tau$, used in the derivation, makes it impossible to account for finite linewidths. A derivation of line shapes from first principles is difficult, and we shall use an empirical approach based on normalized shape functions [cf. (1.25)]. For the moment, we recognize that the widths are finite.

Electric Dipole, Magnetic Dipole, and Electric Quadrupole Transition Moments. Significant contributions to the integral in (1.37) arise only from that part of space in which the molecular wavefunctions $\Psi_0(\mathbf{q},\mathbf{Q})$ and $\Psi_f(\mathbf{q},\mathbf{Q})$ are not negligible, i.e., where the molecule is located. Let us choose the origin of the coordinate system somewhere within the molecule. For molecules of usual size, say up to several dozen angstroms across, the length of $\mathbf{r}_j$ is then much less than λ anywhere within the region which contributes to the integral for light of interest presently, i.e., ultraviolet or longer wavelengths (Figure 1.4). Therefore, $\mathbf{K}\cdot\mathbf{r}_j \leqslant |\mathbf{K}|\,|\mathbf{r}_j| = (2\pi/\lambda)|\mathbf{r}_j| \ll 1$, and the expansion of the phase factor $e^{i\mathbf{K}\cdot\mathbf{r}_j}$ into an infinite series converges rapidly:

$$e^{i\mathbf{K}\cdot\mathbf{r}_j} = 1 + i\mathbf{K}\cdot\mathbf{r}_j - \tfrac{1}{2}(\mathbf{K}\cdot\mathbf{r}_j)^2 + \cdots \tag{1.42}$$

where each successive term is smaller by a factor given approximately by the ratio of the size of the molecule to the wavelength of the light, about 10^{-3}.

Electric Dipole Transition Moment. In the so-called electric dipole approximation, only the first term in the expansion (1.42) is kept: The phase factor $e^{i\mathbf{K}\cdot\mathbf{r}_j}$ is assumed to be equal to $+1$ everywhere. Physically, this corresponds to the assumption that the electric and magnetic fields of the light wave are constant throughout the molecule at any one instant (Figure 1.4). Then the integral in (1.41) becomes

$$\boldsymbol{\varepsilon}_U^* \cdot \left\langle \Psi_f(\mathbf{q},\mathbf{Q}) \middle| \sum_j \frac{q_j}{m_j}\,{}_j \middle| \Psi_0(\mathbf{q},\mathbf{Q}) \right\rangle \tag{1.43}$$

and the transition moment vector is an element of the velocity operator $\hat{\mathbf{V}}$,

$$|e|\hat{\mathbf{V}} = -\frac{|e|}{m}\sum_{l=1}^{n}\hat{\mathbf{p}}_l + |e|\sum_{k=1}^{N}\frac{Z_k}{M_k}\hat{\mathbf{P}}_k \tag{1.44}$$

where Z_k is the charge and M_k the mass of the k-th nucleus. The first sum in (1.44) runs over all electrons, and the second sum over all nuclei. The linear momentum operator $\hat{\mathbf{p}}_l = -i\hbar\boldsymbol{\nabla}_l$ applies to the l-th electron, $\hat{\mathbf{P}}_k = -i\hbar\boldsymbol{\nabla}_k$ to the k-th nucleus. This is the so-called dipole velocity formulation.

It is more common to use the dipole length formulation. This is equivalent in principle but need no longer be so if approximate wavefunctions are used. From the commutation properties of the operators $\hat{\mathbf{p}}$ and $\hat{\mathbf{r}}$, one obtains for exact wavefunctions

$$\begin{aligned}\boldsymbol{\varepsilon}_U^* \cdot \left\langle \Psi_f(\mathbf{q},\mathbf{Q}) \middle| \sum_j \frac{q_j}{m_j}\hat{\mathbf{p}}_j \middle| \Psi_0(\mathbf{q},\mathbf{Q}) \right\rangle &= 2\pi i\nu_f\boldsymbol{\varepsilon}_U^* \cdot \langle \Psi_f(\mathbf{q},\mathbf{Q})|\textstyle\sum q_j\hat{\mathbf{r}}_j|\Psi_0(\mathbf{q},\mathbf{Q})\rangle \\ &= 2\pi i\nu_f\boldsymbol{\varepsilon}_U^* \cdot \langle \Psi_f(\mathbf{q},\mathbf{Q})|\hat{\mathbf{M}}|\Psi_0(\mathbf{q},\mathbf{Q})\rangle \end{aligned} \tag{1.45}$$

In the region of absorption, we can set $\nu = \nu_f$. This brings us to the usual form of the transition moment vector, which is a matrix element of the dipole moment operator $\hat{\mathbf{M}}$ defined by

$$\hat{\mathbf{M}} = -|e|\sum_{l=1}^{n}\hat{\mathbf{r}}_l + |e|\sum_{k=1}^{N}Z_k\hat{\mathbf{R}}_k \tag{1.46}$$

where $\mathbf{r}_l$ is the position vector of the l-th electron and $\mathbf{R}_k$ is the position vector of the k-th nucleus. The operator $\hat{\mathbf{M}}$ is real, so that $\mathbf{M}(0f) = \mathbf{M}(f0)$ if the wavefunctions Ψ_0 and Ψ_f are real.

Using (1.41), the absorption coefficient $\alpha(\tilde{\nu})$ of the f-th transition can now be written in terms of the matrix elements of $\hat{\mathbf{M}}$:

$$\alpha(\tilde{\nu}) = \frac{4\pi^2\tilde{\nu}}{\hbar} n'|\boldsymbol{\varepsilon}_U^* \cdot \mathbf{M}(0f)|^2 \frac{g'(\tilde{\nu})}{c} \tag{1.47}$$

where $\mathbf{M}(0f)$ is the matrix element of $\hat{\mathbf{M}}$ between the initial and final states,

$$\mathbf{M}(0f) = \langle \Psi_f(\mathbf{q},\mathbf{Q})|\hat{\mathbf{M}}|\Psi_0(\mathbf{q},\mathbf{Q})\rangle \tag{1.48}$$

We shall be interested in measurements performed on partially oriented samples. Then, the relative orientation of the transition moment vector $\mathbf{M}(0f)$, associated firmly with the molecular framework, and of the polarization vector of the light $\boldsymbol{\varepsilon}_U$ is of essential importance for absorption probability. Because of the form of (1.43), we must set $U = Y$ or Z for absorption of light linearly polarized along Y or Z, respectively, so that the important quantity is $|M_Y(0f)|^2$ or $|M_Z(0f)|^2$. Unless these are equal for the particular molecular orientation chosen, linear dichroism will result. The quantity governing the absorption of right-handed circularly polarized light propagating in the positive sense of Z is $|M_X(0f) - iM_Y(0f)|^2/2$, and that for left-handed light is $|M_X(0f) + iM_Y(0f)|^2/2$. Circular dichroism will result if they are not equal, as in the presence of a magnetic field.

For an absorption measurement performed on an isotropic sample, averaging over all orientations produces an expression for the oscillator strength of the f-th transition:

$$f_f = \frac{4\pi m \nu_f}{3\hbar e^2} |\mathbf{M}(0f)|^2 = 4.702 \times 10^{-7}\tilde{\nu}_f|\mathbf{M}(0f)|^2 \tag{1.49}$$

with $\mathbf{M}(0f)$ in the units of Debye (10^{-18} esu cm).

Magnetic Dipole and Electric Quadrupole Transition Moments. At times, it is necessary to consider how retention of the second term in (1.42), $i\mathbf{K} \cdot \mathbf{r}_j$, will affect the absorption probability obtained from (1.41). In practice, this is most important if the electric dipole contribution vanishes by symmetry.

The second term contributes the following to the integral in (1.41):

$$\left\langle \Psi_f(\mathbf{q},\mathbf{Q}) \left| \sum_j \frac{q_j}{m_j} (i\mathbf{K} \cdot \hat{\mathbf{r}}_j)(\hat{\mathbf{p}}_j \cdot \boldsymbol{\varepsilon}_U^*) \right| \Psi_0(\mathbf{q},\mathbf{Q}) \right\rangle \tag{1.50}$$

Using $\mathbf{K} = (2\pi\nu/c)\boldsymbol{\varepsilon}_K$, we obtain

$$2\pi i\nu \left\langle \Psi_f(\mathbf{q},\mathbf{Q}) \left| \sum_j \frac{q_i}{m_j c} (\hat{\mathbf{r}}_j)_K(\hat{\mathbf{p}}_j)_{U^*} \right| \Psi_0(\mathbf{q},\mathbf{Q}) \right\rangle \tag{1.51}$$

where $(\mathbf{r}_j)_K$ is the coordinate of $\mathbf{r}_j$ in the direction of light propagation $\boldsymbol{\varepsilon}_K$ and $(\mathbf{p}_j)_{U^*}$ is the coordinate of $\mathbf{p}_j$ in the direction of the electric vector of the light

ε_U^*. Using

$$(\mathbf{r}_j)_K(\mathbf{p}_j)_{U^*} = \tfrac{1}{2}[(\mathbf{r}_j)_K(\mathbf{p}_j)_{U^*} - (\mathbf{p}_j)_K(\mathbf{r}_j)_{U^*}] + \tfrac{1}{2}[(\mathbf{r}_j)_K(\mathbf{p}_j)_{U^*} + (\mathbf{p}_j)_K(\mathbf{r}_j)_{U^*}]$$
$$= \tfrac{1}{2}(\mathbf{r}_j \times \mathbf{p}_j)_{K \times U^*} + \tfrac{1}{2}[(\mathbf{r}_j\mathbf{p}_j)_{KU^*} + (\mathbf{r}_j\mathbf{p}_j)_{U^*K}] \quad (1.52)$$

we decompose the contribution of the second term into two parts referred to as the magnetic dipole and electric quadrupole contributions. The former corresponds to the component of the angular momentum, defined as the vector product of $\mathbf{r}_j$ and $\mathbf{p}_j$, in the direction defined by the unit vector $\varepsilon_K \times \varepsilon_U^*$. The latter corresponds to the sum of the KU* and U*K components of the direct product of $\mathbf{r}_j$ and $\mathbf{p}_j$ (Appendix II).

For the magnetic dipole contribution, we obtain

$$2\pi i\nu(\varepsilon_K \times \varepsilon_U^*) \cdot \left\langle \Psi_f(\mathbf{q},\mathbf{Q}) \left| \sum_j \frac{q_j}{2m_j c} \hat{\mathbf{r}}_j \times \hat{\mathbf{p}}_j \right| \Psi_0(\mathbf{q},\mathbf{Q}) \right\rangle \quad (1.53)$$

The operator $\sum_j (q_j/2m_jc)(\hat{\mathbf{r}}_j \times \hat{\mathbf{p}}_j)$ represents the orbital part of the magnetic dipole moment operator $\hat{\mathcal{M}}$. A relativistic treatment would add a spin part, so that the complete expression for $\hat{\mathcal{M}}$ is

$$\hat{\mathcal{M}} = \frac{-|e|}{2mc} \sum_{l=1}^{n} (\hat{\mathbf{r}}_l \times \hat{\mathbf{p}}_l + 2\hat{\mathbf{s}}_l) + \sum_{k=1}^{N} \left[\frac{Z_k|e|}{2M_k c} (\hat{\mathbf{R}}_k \times \hat{\mathbf{P}}_k) + \gamma_k \hat{\boldsymbol{\sigma}}_k \right] \quad (1.54)$$

where $\hat{\mathbf{s}}_l$ is the spin angular momentum operator of the l-th electron, $\hat{\boldsymbol{\sigma}}_k$ is the spin angular momentum operator of the k-th nucleus, and γ_k is its gyromagnetic ratio. This operator is purely imaginary so that $\mathcal{M}(0f) = -\mathcal{M}(f0)$ if the wavefunctions Ψ_0 and Ψ_f are real.

The magnetic dipole contribution to the integral in (1.41) is thus given by

$$2\pi i\nu(\varepsilon_K \times \varepsilon_U^*) \cdot \langle \Psi_f(\mathbf{q},\mathbf{Q}) | \hat{\mathcal{M}} | \Psi_0(\mathbf{q},\mathbf{Q}) \rangle = 2\pi i\nu(\varepsilon_K \times \varepsilon_U^*) \cdot \mathcal{M}(0f) \quad (1.55)$$

and once again the dependence of this quantity on the polarization vector ε_U is obvious.

For the electric quadrupole contribution, we obtain

$$\pi i\nu\varepsilon_K \cdot \left\langle \Psi_f(\mathbf{q},\mathbf{Q}) \left| \sum_j \frac{q_j}{m_j c} (\hat{\mathbf{r}}_j\hat{\mathbf{p}}_j + \hat{\mathbf{p}}_j\hat{\mathbf{r}}_j) \right| \Psi_0(\mathbf{q},\mathbf{Q}) \right\rangle \cdot \varepsilon_U^* \quad (1.56)$$

Using the commutation properties of $\hat{\mathbf{r}}_j$ and $\hat{\mathbf{p}}_j$ and setting $\nu = \nu_f$ in the region of absorption, this expression is converted to

$$-(2\pi^2\nu^2/c)\varepsilon_K \cdot \left\langle \Psi_f(\mathbf{q},\mathbf{Q}) \left| \sum_j \hat{\mathbf{r}}_j\hat{\mathbf{r}}_j \right| \Psi_0(\mathbf{q},\mathbf{Q}) \right\rangle \cdot \varepsilon_U^* \quad (1.57)$$

It is common to convert the operator into the traceless form by subtracting $\mathsf{1}(\hat{\mathbf{r}}_j \cdot \hat{\mathbf{r}}_j)/3$ along the diagonal, where $\mathsf{1}$ is the unit tensor. The electric quadrupole moment operator is then defined by

$$\hat{\mathsf{Q}} = -|e| \sum_{l=1}^{n} [\hat{\mathbf{r}}_l\hat{\mathbf{r}}_l - \mathsf{1}(\hat{\mathbf{r}}_l \cdot \hat{\mathbf{r}}_l)/3] + |e| \sum_{k=1}^{N} Z_k[\hat{\mathbf{R}}_k\hat{\mathbf{R}}_k - \mathsf{1}(\hat{\mathbf{R}}_k \cdot \hat{\mathbf{R}}_k)/3] \quad (1.58)$$

where it is assumed that quadrupolar nuclei are absent. This operator is real so that Q(0f) = Q(f0) if the wavefunctions Ψ_0 and Ψ_f are real. Some authors add a multiplicative factor of $\frac{3}{2}$ in the definition (1.58).

The electric quadrupole contribution to the integral in (1.41) is thus given by

$$-(2\pi^2\nu^2/c)\boldsymbol{\varepsilon}_K \cdot \langle \Psi_f(\mathbf{q},\mathbf{Q})|\hat{Q}|\Psi_0(\mathbf{q},\mathbf{Q})\rangle \cdot \boldsymbol{\varepsilon}_U^* = -(2\pi^2\nu^2/c)\boldsymbol{\varepsilon}_K \mathrm{Q}(0f)\boldsymbol{\varepsilon}_U \qquad (1.59)$$

and also depends on the polarization vector $\boldsymbol{\varepsilon}_U$.

When electric dipole, magnetic dipole, and electric quadrupole contributions are all considered simultaneously, the expression (1.41) for the absorption coefficient in terms of molecular quantities and an empirical line shape is no longer reduced to (1.47) but rather is replaced by

$$\alpha(\tilde{\nu}) = \frac{n'}{\hbar c^2 \tilde{\nu}} |2\pi i\nu \boldsymbol{\varepsilon}_U^* \cdot \mathbf{M}(0f) + 2\pi i\nu(\boldsymbol{\varepsilon}_K \times \boldsymbol{\varepsilon}_U^*) \cdot \mathcal{M}(0f) - (2\pi^2\nu^2/c)\boldsymbol{\varepsilon}_K \mathrm{Q}(0f)\boldsymbol{\varepsilon}_U^*|^2 \frac{g'(\nu)}{c}$$

$$= \frac{4\pi^2\tilde{\nu}}{\hbar c} n'|\boldsymbol{\varepsilon}_U^* \cdot \mathbf{M}(0f) + (\boldsymbol{\varepsilon}_K \times \boldsymbol{\varepsilon}_U^*) \cdot \mathcal{M}(0f) + i\pi\tilde{\nu}\boldsymbol{\varepsilon}_K \mathrm{Q}(0f)\boldsymbol{\varepsilon}_U^*|^2 g'(\tilde{\nu}) \qquad (1.60)$$

For the oscillator strength f_f of the f-th transition, measured on an isotropic sample with unpolarized light, averaging over all orientations yields

$$f_f = \frac{4\pi m \nu_f}{3\hbar e^2}\left[|\mathbf{M}(0f)|^2 + |\mathcal{M}(0f)|^2 + \frac{3\pi^2\tilde{\nu}^2}{10}\sum_{u,v}|\mathrm{Q}_{uv}(0f)|^2\right] \qquad (1.61)$$

The quantity $|\mathbf{M}(0f)|^2$ usually dominates within the brackets in (1.61). It is sometimes called the dipole strength of the transition, D(0f) [cf. (1.100)].

If lines or bands due to several transitions f overlap in the spectrum, as is often the case in electronic spectra and sometimes also in vibrational spectra, their contributions must be added. We shall assume throughout that this happens in a noncoherent fashion, in which the squares of the transition moments are added (as opposed to taking squares of sums of transition moments). Then the absorption coefficient is given by

$$\alpha(\tilde{\nu}) = \frac{4\pi^2\tilde{\nu}}{\hbar c} n' \sum_f g_f'(\tilde{\nu})|\boldsymbol{\varepsilon}_U^* \cdot \mathbf{M}(0f) + (\boldsymbol{\varepsilon}_K \times \boldsymbol{\varepsilon}_U^*) \cdot \mathcal{M}(0f) + i\pi\tilde{\nu}\boldsymbol{\varepsilon}_K \mathrm{Q}(0f)\boldsymbol{\varepsilon}_U^*|^2 \qquad (1.62)$$

and the decadic molar extinction coefficient can be obtained readily since $\varepsilon(\tilde{\nu}) = \alpha(\tilde{\nu})/2.303c$ and $n' = cN \times 10^{-3}$, so that $\varepsilon(\tilde{\nu}) = \alpha(\tilde{\nu})N/2303n'$. This expression is suitable for interpretation of linear dichroism (real $\boldsymbol{\varepsilon}_U$) and circular dichroism (complex $\boldsymbol{\varepsilon}_U$) if solvent effects can be neglected.

Note that the expression inside the absolute value bars in (1.62) is complex even if $\boldsymbol{\varepsilon}_U$ is real (linearly polarized or unpolarized light). The first term, containing $\mathbf{M}$(0f), then represents the real part. The second term, containing the purely imaginary vector $\mathcal{M}$(0f), and the third term, containing Q(0f), together represent the imaginary part. The difference is due to the fact that the term with $\mathbf{M}$(0f) originates in the leading term in the expression in (1.42), which is real (unity), while the terms with $\mathcal{M}$(0f) and Q(0f) originate in the next term in (1.42), which is purely imaginary ($i\mathbf{K} \cdot \mathbf{r}$). This alternation continues as one proceeds to the successively higher terms in (1.42), but these have no practical importance today.

Since the square of the absolute value of a complex number $a + ib$ is $a^2 + b^2$ and contains no cross-terms of the type ab, equations (1.15)–(1.17) do not require much modification even when the transition $0 \to f$ has nonvanishing values of $\mathbf{M}(0f)$, $\mathcal{M}(0f)$, and $Q(0f)$ simultaneously, as can happen when the molecular symmetry is low. The contribution from $\mathbf{M}(0f)$ is still all given by (1.15), since the real–imaginary cross-terms of the type

$$\varepsilon_U^* \cdot \mathbf{M}(0f)\mathcal{M}(f0)(\varepsilon_K \times \varepsilon_U) \qquad \text{and} \qquad \varepsilon_U^* \cdot \mathbf{M}(0f)[\varepsilon_K Q(f0)\varepsilon_U]$$

disappear when the square of the absolute value is taken. The only required modification will be to add to the pure magnetic dipole [(1.16)] and pure electric quadrupole [(1.17)] contributions to the rate of absorption a contribution from a cross-term of the type $(\varepsilon_K \times \varepsilon_U^*) \cdot \mathcal{M}(0f)[\varepsilon_K Q(f0)\varepsilon_U]$. However, this term is normally ignored, since it averages to zero for most orientation distributions of interest, such as isotropic or uniaxial.

The situation is more complicated if the measurements are performed with elliptically or circularly polarized light. Then, ε_U and ε_U^* are complex so that $\varepsilon_U^* \cdot \mathbf{M}(0f)$ is no longer real but rather has nonvanishing real and imaginary parts. Similarly, $(\varepsilon_K \times \varepsilon_U^*) \cdot \mathcal{M}(0f)$ and $i\pi\tilde{v}\varepsilon_K Q(0f)\varepsilon_U^*$ are no longer purely imaginary but rather have nonvanishing real and imaginary parts as well. As a result, the cross-terms between $\mathbf{M}(0f)$ on the one hand and $\mathcal{M}(0f)$ and $Q(0f)$ on the other hand begin to contribute. As a matter of fact, a differential absorption measurement using left-handed and right-handed circularly polarized light will depend only on these cross-terms (cf. (1.96)]. The existence of such differential absorption is known as natural circular dichroism (CD). Along with the corresponding dispersion it is referred to as natural optical activity and will be addressed in Section 1.3.3. The presence of the first power of $\mathbf{M}(0f)$ in the expression for natural optical activity is important in that it shows that the effect is limited to molecules which are chiral (not superimposable on their mirror image; $\hat{\mathbf{M}}$ is odd while $\hat{\mathcal{M}}$ and $\hat{Q}$ are even with respect to the parity operation).

Solvent Effects. So far, the theory has been developed for isolated molecules in vacuum. In practice, we shall be concerned with dilute solutions in "inert" media. Under such conditions, the solute molecules can often be assumed not to interact with each other, but the presence of the "inert" solvent can no longer be neglected.

In its general form, the problem of solvent effects on the interaction of a molecule with a radiation field is very complicated. An approximate treatment is based on the concept of effective or internal fields. These are the electric and magnetic fields experienced by the molecule in the actual solvent environment. They are a sum of the macroscopic electromagnetic field as it would be in the absence of the solvent, and electric and magnetic fields due to the solvent molecules.

It is customary to equate the effective magnetic field with the magnetic field in vacuum, since the magnetic permeability of nonmagnetic materials such as the usual solvents is essentially identical to that of the vacuum. For the effective electric field one writes $\mathbf{E}_{\text{eff}}(\tilde{\nu}) = \alpha'(\tilde{\nu})\mathbf{E}$, where $\alpha'(\tilde{\nu})$ is the effective polarizability of the solvent. In calculating the intensity of the radiation, we use the macroscopic field propagating through the solvent with velocity $c/\mathrm{n}(\tilde{\nu})$, where $\mathrm{n}(\tilde{\nu})$ is the usual refractive index of the solvent. The light frequency is constant, but its wavelength and wavenumber depend on the medium in which it propagates. The wavenumber $\tilde{\nu}$ used in this text refers to the vacuum, and the values in air are nearly the same.

A procedure similar to that outlined above for a molecule in vacuum then yields

$$\varepsilon(\tilde{\nu}) = \frac{4\pi^2\tilde{\nu}N}{2303\hbar c}\sum_{\mathrm{f}} \mathrm{g}_{\mathrm{f}}'(\tilde{\nu}) \times \left| \frac{\alpha'(\tilde{\nu})}{\sqrt{\mathrm{n}(\tilde{\nu})}} \boldsymbol{\varepsilon}_{\mathrm{U}}^{*} \cdot \mathbf{M}(0\mathrm{f}) + \sqrt{\mathrm{n}(\tilde{\nu})}\left[(\boldsymbol{\varepsilon}_{\mathrm{K}} \times \boldsymbol{\varepsilon}_{\mathrm{U}}^{*}) \cdot \mathcal{M}(0\mathrm{f}) + i\pi\tilde{\nu}\boldsymbol{\varepsilon}_{\mathrm{K}}\mathrm{Q}(0\mathrm{f})\boldsymbol{\varepsilon}_{\mathrm{U}}^{*}\right]\right|^2 \qquad (1.63)$$

and this is our fundamental equation for the interpretation of linear and circular dichroism on oriented samples in which all molecules are oriented alike. It represents the starting point in Chapter 5 and much of Chapter 7.

The difficulty comes when a specific expression is to be used for $\alpha'(\tilde{\nu})$. The simplest procedure is to assume that the electric field $\mathbf{E}_{\text{eff}}(\tilde{\nu})$ acting on the molecule is the average effective field over the entire solvent in the presence of a field $\mathbf{E}$ in the medium. In this fashion, Lorentz derived the result

$$\alpha'(\tilde{\nu}) = \frac{\mathrm{n}^2(\tilde{\nu}) + 2}{3} \qquad (1.64)$$

for the regions in which the solvent is transparent. The field $\mathbf{E}_{\text{eff}}(\tilde{\nu}) = \mathbf{E}[\mathrm{n}^2(\tilde{\nu}) + 2]/3$ is referred to as the Lorentz effective field. For a solvent with a refractive index $\mathrm{n} = 1.5$, this yields a correction factor $[\alpha'(\tilde{\nu})]^2/\mathrm{n}$ equal to about 1.3. This predicts extinction coefficients 1.3 times larger in such a solution than in the gas phase. The inclusion of the mean refractive index over the region of the band, n_0, in the denominator of the expression (1.24), which relates oscillator strength to extinction coefficients measured in solution, is a crude attempt to compensate for the solvent effect. At least approximately, then, the use of (1.24) yields the oscillator strength as defined for an isolated molecule, related to molecular properties by (1.49).

Various attempts to improve on (1.64) have been published, but their discussion lies outside the scope of the present treatment. Fortunately, in most of the work on oriented molecules, only the ratios of intensities are of interest, and in these the solvent correction tends to cancel. The main reason the cancellation is only approximate is the difference in the refractive indices of uniaxial media for light whose electric vector is directed along the unique axis, $\mathrm{n}_{\parallel}(\tilde{\nu})$, and for light whose electric vector is directed perpendicular to it, $\mathrm{n}_{\perp}(\tilde{\nu})$. The straightforward use of the Lorentz effective field then yields different correction factors for measurements using different polarizations, and these do not cancel when a ratio is taken. Fortunately, many of the uniaxial orienting media have very small birefringence $n_{\parallel}(\tilde{\nu}) - n_{\perp}(\tilde{\nu})$ in the UV, visible, and IR regions, and the cancellation is still nearly exact (stretched polyethylene, certain liquid crystals).

An additional and potentially more serious complication in measurements in which ratios of polarized spectra are taken is the weak dependence of transition energies on the refractive index. For both reasons, it is highly desirable to work with materials with minimum birefringence.

Absorption and Dispersion Line Shapes. We have already mentioned in Section 1.3.1 that absorption intensity of a transition f is distributed over a range of wavenumbers in a way which we describe by the empirical line-shape function $g_f'(\tilde{\nu})$ [cf. (1.25)]. As already stated, this function is dictated by a variety of intramolecular and intermolecular factors, and we make no attempt to derive it from first principles.

Now, we shall describe quantitatively the two shapes which are expected in particularly simple physical situations and which are also frequently encountered in practice: the Lorentzian and Gaussian line shapes. At the same time, we note that the complex refractive index $N(\tilde{\nu})$ defined in (1.22) is an analytic complex function so that its real and imaginary parts are not independent. The relations between the negative of the imaginary part $\alpha(\tilde{\nu})/4\pi\tilde{\nu}$, which describes the absorption spectrum and has the line-shape function $g'(\tilde{\nu})$, and the real part, $n(\tilde{\nu})$, which describes the optical dispersion and has the line-shape function $g''(\tilde{\nu})$, are known as the Kramers-Kronig transforms. In their general form, the transforms between the real (f_1) and imaginary (f_2) parts of an analytic complex function $f(x) = f_1(x) + if_2(x)$ of a real variable x are

$$f_1(x_0) = \frac{-2}{\pi} ⨍_0^{\infty} \frac{x f_2(x)}{x^2 - x_0^2}\, dx \qquad f_2(x_0) = \frac{-2}{\pi} x_0 ⨍_0^{\infty} \frac{f_1(x)}{x^2 - x_0^2}\, dx \tag{1.65}$$

where ⨍ stands for the Cauchy principal value of the integral, $f_1(x)$ is assumed to be an even and $f_2(x)$ an odd function of x. In order to provide a concrete representation of the relation between the absorption and dispersion line shapes $g'(\tilde{\nu})$ and $g''(\tilde{\nu})$, we shall state them for both Lorentzian and Gaussian lines.

Lorentzian Line Shapes. The Lorentzian shape is expected in simple cases when the line width is limited by the radiative and/or nonradiative processes determining the lifetime ("longitudinal" relaxation) and by dephasing ("transverse" relaxation). This type of line broadening is the same for all molecules in an assembly and is called homogeneous broadening. The normalized absorption shape $g'(\tilde{\nu})$ for a line centered at $\tilde{\nu}_0$ is

$$L'(\tilde{v}) \cong (\tilde{\gamma}/\pi)/[(\tilde{v}_0 - \tilde{v})^2 + \tilde{\gamma}^2] \tag{1.66}$$

and the corresponding dispersion shape $g''(\tilde{\nu})$ is

$$L''(\tilde{\nu}) = (\tilde{\nu}_0/2\pi)(\tilde{\nu}_0^2 - \tilde{\nu}^2)/[(\tilde{\nu}_0^2 - \tilde{\nu}^2)^2/4 + \tilde{\gamma}^2\tilde{\nu}^2] \tag{1.67}$$

Here, $\tilde{\gamma}$ is the damping parameter in cm^{-1} and is equal to half the line width at half height. It is a sum of a part due to longitudinal relaxation processes and a part due to transverse relaxation processes. In the absence of the latter (in the limit of zero temperature), the lifetime of the excited state is equal to $1/2c\tilde{\gamma}$.

The complex function from which both $L''(\tilde{\nu})$ and $L'(\tilde{\nu})$ are obtained as the real part and the negative of the imaginary part, respectively, is

$$L(\tilde{\nu}) = L''(\tilde{\nu}) - iL'(\tilde{\nu}) = (\tilde{\nu}_0/\pi)/[(\tilde{\nu}_0^2 - \tilde{\nu}^2)/2 + i\tilde{\gamma}\tilde{\nu}] \tag{1.68}$$

This function arises in the theory of a damped oscillator. It originates in the sum

$$L(\tilde{\nu}) = \frac{1}{\pi}\left[\frac{1}{\tilde{\nu}_0 - \tilde{\nu} + i\tilde{\gamma}} + \frac{1}{\tilde{\nu}_0 + \tilde{\nu} - i\tilde{\gamma}}\right] \tag{1.69}$$

where the first fraction is the resonant and the second the antiresonant term analogous to those we have already encountered in (1.32). In (1.32), they occurred without the damping constant $\tilde{\gamma}$, since at that point we were assuming an infinite lifetime for the excited state. Its energy then was real rather than complex, as it is for a finite lifetime. In going to (1.33), we kept the resonant term, which is a measure of how close the light frequency is to the beat frequency between the ground and excited states, and dropped the antiresonant term.

When dealing with dispersion phenomena, we may be interested in regions far from resonance, and in general we cannot afford to drop either term. Near resonance, the antiresonant term is frequently neglected. The line-shape functions $L'(\tilde{\nu})$ and $L''(\tilde{\nu})$ then acquire the commonly used simple forms

$$L'(\tilde{\nu}) \cong (\tilde{\gamma}/\pi)/[(\tilde{\nu}_0 - \tilde{\nu})^2 + \tilde{\gamma}^2] \tag{1.66a}$$

$$L''(\tilde{\nu}) \cong (1/\pi)(\tilde{\nu}_0 - \tilde{\nu})/[(\tilde{\nu}_0 - \tilde{\nu})^2 + \tilde{\gamma}^2] \tag{1.67a}$$

Gaussian Line Shapes. The Gaussian line shape is expected in simple cases when the broadening is dominated by the Doppler effect or by statistical fluctuations in the environment. Different molecules in the assembly then absorb at different frequencies, and this type of line broadening is referred to as inhomogeneous broadening.

The normalized absorption shape $g'(\tilde{\nu})$ for a line centered at $\tilde{\nu}_0$ is

$$G'(\tilde{\nu}) = \frac{1}{\tilde{a}\sqrt{\pi}} e^{-(\tilde{\nu} - \tilde{\nu}_0)^2/\tilde{a}^2} \tag{1.70}$$

Strictly speaking, in order to make $G'(\tilde{\nu})$ an odd function of $\tilde{\nu}$, one must add the "antiresonant" term $-(1/\tilde{a}\sqrt{\pi})\exp[-(\tilde{\nu} + \tilde{\nu}_0)^2/\tilde{a}^2]$ to the right-hand side, but this is normally so small in the region of interest ($\tilde{\nu} > 0$) that it can be safely ignored in practice. Using the Kramers-Kronig transformation (1.65), the dispersion shape $g''(\tilde{\nu})$ is found to be

$$G''(\tilde{\nu}) = \frac{2}{\tilde{a}\pi}[I_- e^{-(\tilde{\nu} - \tilde{\nu}_0)^2/\tilde{a}^2} + I_+ e^{-(\tilde{\nu} + \tilde{\nu}_0)^2/\tilde{a}^2}] \tag{1.71}$$

where

$$I_\pm = \int_0^{\tilde{a}(\tilde{\nu}_0 \pm \tilde{\nu})} e^{x^2}\, dx \tag{1.72}$$

In (1.70) and (1.71), $\tilde{a}$ is the line-width parameter in cm^{-1}. The half-line width at half height is given by $\tilde{a}\sqrt{\ln 2}$.

Comparison of Absorption and Dispersion Line Shapes. The Lorentzian and Gaussian shapes for absorption and dispersion are shown in Figure 1.12. Both absorption lines have been chosen to have a height of 1.0 and full width at half height of 0.03 cm^{-1}. It is obvious that the dispersion line shapes $g''(\tilde{\nu})$ are far broader than the absorption line shapes $g'(\tilde{\nu})$. Thus, contributions from different transitions overlap far more extensively in the former case, making dispersion measurements more difficult to interpret in terms of molecular properties. This remains true even for lines whose shapes are more complicated than the simple Lorentzian and Gaussian considered here.

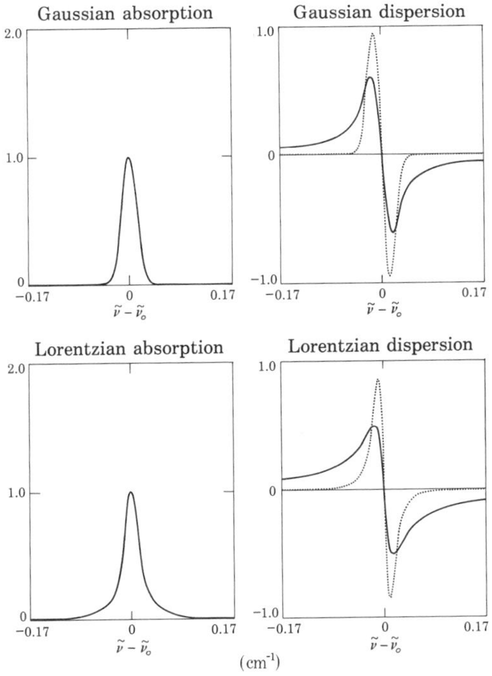

Figure 1.12 Absorption (left) and dispersion (right) line shapes of a Gaussian (top) and Lorentzian (bottom) type. Both absorption peaks have been chosen to have unit height and the same width at half height. The dotted curves on the right represent the first derivatives of the absorption line shapes (their vertical scale has been reduced by a factor of 50).

Figure 1.12 also shows the shapes of the first derivatives $dg'(\tilde{\nu})/d\tilde{\nu}$ of the absorption as dotted lines. These are somewhat similar to the dispersion line shapes $g''(\tilde{\nu})$ but much narrower. They are of importance in magnetic circular dichroism (Section 1.3.3).

The Complex Refractive Index in Terms of Molecular Properties. It is obvious from the preceding that expression (1.62) for the absorption coefficient can be generalized to account for the dispersion of the refractive index as well. The result for the complex refractive index $N(\tilde{\nu})$ is simple only for low-density gases for which solvent corrections of the kind introduced in (1.63) are negligible and the refractive index is close to unity. Although not of much practical use for our present purposes, the result is of interest for comparison with the expression for the complex rotation to be discussed in Section 1.3.3. For an isotropic dilute sample, the complex index of refraction, defined in (1.22), is given in the electric dipole approximation by

$$\begin{aligned} \mathrm{N}(\tilde{\nu}) &= \mathrm{n}(\tilde{\nu}) - (i/4\pi\tilde{\nu})\alpha(\tilde{\nu}) \\ &= 1 + (\mathrm{n}'\pi/\hbar c)\sum_{\mathrm{f}} (|\mathbf{M}(0\mathrm{f})|^2/3)[\mathrm{g}_{\mathrm{f}}''(\tilde{\nu}) - i\mathrm{g}_{\mathrm{f}}'(\tilde{\nu})] \end{aligned} \qquad (1.73)$$

This relation contains both the expression for the absorption coefficient $\alpha(\tilde{\nu})$ given in (1.62) [note that isotropic averaging of $\boldsymbol{\varepsilon}_{\mathrm{U}}^* \cdot \mathbf{M}(0\mathrm{f})$ yields $|\mathbf{M}(0\mathrm{f})|^2/3$] and an analogous expression for the real refractive index,

$$\mathrm{n}(\tilde{\nu}) = 1 + (\mathrm{n}'\pi/\hbar c)\sum_{\mathrm{f}} [|\mathbf{M}(0\mathrm{f})|^2/3]\mathrm{g}_{\mathrm{f}}''(\tilde{\nu}) \qquad (1.74)$$

Note that the summation of variously shaped contributions to absorption starts with a zero background, while a similar summation of contributions to the dispersion curve starts with a background of unity. This is not obvious in Figure 1.12, where only the shapes of $\mathrm{g}'(\tilde{\nu})$ and $\mathrm{g}''(\tilde{\nu})$ are displayed.

Spontaneous Emission. We have already stated that absorption and stimulated emission have the same probability coefficients, $\mathrm{B}_{0\to\mathrm{f}} = \mathrm{B}_{\mathrm{f}\to 0}$. We shall now see that spontaneous emission probability is also governed by analogous quantities, $|\boldsymbol{\varepsilon}_{\mathrm{U}} \cdot \mathbf{M}(0\mathrm{f})|^2$, $|(\boldsymbol{\varepsilon}_{\mathrm{U}} \times \boldsymbol{\varepsilon}_{\mathrm{K}}) \cdot \mathcal{M}(0\mathrm{f})|^2$, and $|\boldsymbol{\varepsilon}_{\mathrm{U}}\mathrm{Q}(0\mathrm{f})\boldsymbol{\varepsilon}_{\mathrm{K}}|^2$, where U and K now describe the properties of emitted photons. The only difference is the absence of complex conjugation on the light polarization vector $\boldsymbol{\varepsilon}_{\mathrm{U}}$. Thus, for linearly polarized light, there is no difference.

If the number of molecules with energy E_{f} is N_{f}, and the number of molecules with energy E_0 is N_0, then at equilibrium

$$\mathrm{N}_{\mathrm{f}}/\mathrm{N}_0 = \mathrm{e}^{-(h\nu_{\mathrm{f}}/\mathrm{kT})} \qquad (1.75)$$

where k is the Boltzmann constant and T is absolute temperature. At equilibrium, the total rate of absorption is $\mathrm{N}_0\mathrm{B}_{0\to\mathrm{f}}\rho(\nu_{\mathrm{f}})$, and this is equal to the total rate of emission $\mathrm{N}_{\mathrm{f}}[\mathrm{A}_{\mathrm{f}\to 0} + \mathrm{B}_{\mathrm{f}\to 0}\rho(\nu_{\mathrm{f}})]$, where $\mathrm{A}_{\mathrm{f}\to 0}$ is the Einstein transition probability coefficient for spontaneous emission [i.e., independent of $\rho(\nu_{\mathrm{f}})$]. Thus,

$$\rho(\nu_{\mathrm{f}}) = \mathrm{A}_{\mathrm{f}\to 0}\mathrm{e}^{-(h\nu_{\mathrm{f}}/\mathrm{kT})}/[\mathrm{B}_{0\to\mathrm{f}} - \mathrm{B}_{\mathrm{f}\to 0}\mathrm{e}^{-(h\nu_{\mathrm{f}}/\mathrm{kT})}] \qquad (1.76)$$

Using $\mathrm{B}_{0\to\mathrm{f}} = \mathrm{B}_{\mathrm{f}\to 0}$, and comparing with Planck's radiation distribution law,

$$\rho(\nu_{\mathrm{f}}) = (8\pi \mathrm{h}\nu_{\mathrm{f}}^3/c^3)/[\mathrm{e}^{(h\nu_{\mathrm{f}}/\mathrm{kT})} - 1] \qquad (1.77)$$

one obtains

$$\mathrm{A}_{\mathrm{f}\to 0} = (8\pi \mathrm{h}\nu_{\mathrm{f}}^3/c^3)\mathrm{B}_{\mathrm{f}\to 0} \qquad (1.78)$$

The rate of spontaneous emission is seen to be a rapidly increasing function of the transition frequency ν_{f}. It is very important in UV-visible spectroscopy, less so in IR spectroscopy, and plays a negligible role in the microwave and radio frequency regions. Its physical origin is not readily described in the semiclassical language adopted here. Roughly speaking, the perturbation which causes spontaneous emission is the random fluctuation of the electromagnetic field in vacuum around its zero average value. When radiation is quantized, such fluctuation is demanded by the uncertainty principle.

Relation (1.74) can be used to derive a series of increasingly less accurate expressions for the natural radiative lifetime τ_{rad} in terms of the intensity of the emitting transition:

$$\begin{aligned} 1/\tau_{rad} &= 2.88 \times 10^{-9} n_0^2 \langle \tilde{\nu}_e^{-3} \rangle^{-1} \int \frac{\varepsilon(\tilde{\nu})}{\tilde{\nu}}\, d\tilde{\nu} \\ &\cong 2.88 \times 10^{-9} n_0^2 \int \frac{(2\tilde{\nu}_0 - \tilde{\nu})^3}{\tilde{\nu}}\, \varepsilon(\tilde{\nu})\, d\tilde{\nu} \\ &\cong 2.88 \times 10^{-9} \tilde{\nu}_0^2 n_0^2 \int \varepsilon(\tilde{\nu})\, d\tilde{\nu} = n_0^3 \tilde{\nu}_0^2 f/1.5 \end{aligned} \qquad (1.79)$$

where n_0 is the average refractive index of the solution in the region of the emitting transition, $\langle \tilde{\nu}_e^{-3} \rangle$ is the mean value of $\tilde{\nu}^{-3}$ over the emission spectrum, and $\tilde{\nu}_0$ is the wavenumber of the 0-0 band in cm^{-1}, often identified as the wavenumber of the "mirror symmetry point" between absorption and emission. The extinction coefficient $\varepsilon(\tilde{\nu})$ is in its usual units of L mol^{-1} cm^{-1}.

The Franck-Condon Principle. In the usual Born-Oppenheimer approximation, molecular wavefunctions are written as products of electronic, vibrational, rotational, and translational parts (cf. Section 1.2). In general, electronic motion with a typical period of 10^{-15} s is much faster than vibrational motion with a typical period of 10^{-13} s. As a result of this, light of frequencies adequate for electronic excitation oscillates far too fast for the nuclei to follow its electric vector faithfully. The transition moment established in the molecule by interaction with the field and oscillating in resonance with the field is primarily due to a periodic displacement of the positions of electrons in the molecule with essentially no contribution from the displacement of nuclei. Therefore, during an electronic excitation the probability of reaching a vibronic level selected by the frequency used is highest if its nuclear wavefunction is the same as that of the initial vibronic level. The excitation probability decreases with the square of the overlap of the two (normalized) nuclear wavefunctions, which provides a measure of the degree to which they are alike, and drops to zero if they are orthogonal. This statement is known as the Franck-Condon principle (see below; for additional detail, see Section 2.1.1).

Sometimes it is stated that the principle is a result of the fact that the electronic excitation occurs so fast that nuclei cannot change their positions during the time of transition from the electronic ground state to the electronic excited state. This is somewhat misleading, since the interaction of the molecule with the field and therefore the excitation process continues as long as the light source is turned on. In the absence of other perturbations, interaction produces a superposition state, that is, one in which the wavefunction $\Psi_0(\mathbf{q},\mathbf{Q},t)$ and

$\Psi_f(\mathbf{q},\mathbf{Q},t)$ are mixed with coefficients c_0 and c_f, respectively. A subsequent measurement of the energy of the produced state can only give the answer E_0 (molecule did not absorb light) or E_f (molecule did absorb light) with probabilities given by c_0^2 and c_f^2, respectively. In real samples, which consist of a large number of colliding molecules, the buildup of the superposition state never continues uninterrupted for long even if the light source is coherent in time, since intermolecular collisions destroy the coherence of the excitation process. A description of the behavior of the sample is then usually provided by means of its density matrix and is most easily visualized in terms of the gyroscopic model. These subjects lie outside the scope of the present book, and the reader is referred to the references in Section 1.4.

The potential governing the nuclear motion is in general different in the initial electronic state, ψ_0, and the final electronic state, ψ_f. Since one normally starts with a vibrational eigenstate, say the lowest energy vibrational state of the ground electronic state of the molecule, the probability that different vibrational levels of the final state will be observed when a measurement of its energy is made is given by the squares of the overlap of the initial vibrational function $\chi_{j,v}$ and the vibrational eigenfunctions of the final state $\chi_{j',v'}$. These are referred to as the Franck-Condon factors (Figure 1.13). If neither the equilibrium geometry nor the vibrational frequencies change upon excitation, the only nonvanishing factors will be $|\langle\chi_{j,v}|\chi_{j',v}\rangle|^2 = 1$, and only the $0 \rightarrow 0$, $1 \rightarrow 1, \ldots, v \rightarrow v$ vibrational transitions will be observed in absorption or emission. This is rarely the case. More commonly, the equilibrium geometries and the frequencies differ in the electronic states j and j′, and several vibrational

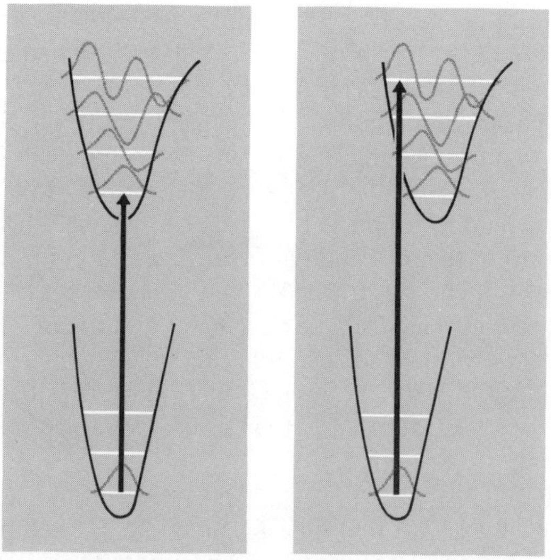

Figure 1.13 Franck-Condon allowed (left) and forbidden (right) electronic transitions. The vibrational wavefunctions are indicated schematically, and the most intense vibronic component in each case is shown.

components v → v′ will be observed for each v. If the difference is small, the v → v transition remains more intense than v → v + 1, v → v + 2, etc. ("Franck-Condon allowed transition"). If it is large, one of the v → v + n peaks in the spectrum is the strongest, and the peak v → v may even be very hard to observe ("Franck-Condon forbidden transition"). These results are discussed in a more quantitative fashion in Chapter 2.

Effect of Static Fields. Static electric and magnetic fields applied from the outside have two kinds of influence on the observed spectra.

The first of these is due to their *orienting effect.* Unless the molecules are clamped rigidly in a medium of very high viscosity, they will be partially aligned due to the interaction of their permanent electric dipole moments and/or anisotropic electric polarizability with an applied electric field. They will be similarly aligned, albeit usually to a much smaller extent, by the interaction of their permanent magnetic dipole moments and/or anisotropic magnetic susceptibility with an applied magnetic field. The consequences of molecular orientation for spectroscopic observables have already been mentioned. They are discussed briefly in Section 1.3.4 and then in detail in Chapters 5–7.

If the electric and magnetic properties of the solute molecules are sufficiently well known, and if there are no other sources of anisotropy, the orientation distribution resulting from the application of an electric (or magnetic) field can be predicted explicitly. We shall take up the subject in Section 4.6.2.

The second spectral influence of an applied outside field is due to its *intrinsic effect* and is caused by the modification of the molecular vibronic state energies and wavefunctions by the field. This effect remains even if the molecules are clamped rigidly and cannot reorient.

The usual ways in which the combined action of the two effects is observed are (i) the measurement of linear birefringence with the field transverse to the light propagation direction, known as the Kerr effect in the case of electric field and as the Cotton-Mouton effect in the case of magnetic field; (ii) the measurement of linear dichroism, with the field again transverse, known as electric linear dichroism and magnetic linear dichroism, respectively (the latter is rarely measured due to its weakness); (iii) the measurement of unpolarized absorption intensity in the presence and in the absence of the outside field, known as electrochromism (normally with transverse field) and magnetic differential absorption (with longitudinal field, parallel to the light propagation direction; this is predicted to be weak and so far remains unobserved); and (iv) the measurement of circular dichroism, with longitudinal magnetic field, known as magnetic circular dichroism.

The effects observed in an electric field tend to be dominated by contributions from the orienting effect. Measurements on rigid samples suppress this effect but are performed relatively rarely. The orienting effect of a magnetic field is far smaller, and indeed the best known intrinsic effect is magnetic circular dichroism, frequently measured on fluid samples. In preparation for the discussion of chiroptical spectroscopy in Section 1.3.3, we shall now concentrate on this effect. In order to provide a broader perspective, however, we shall elegant way of treating these effects theoretically utilizes time-reversal as well as space-inversion symmetries. Our treatment will be much more pedestrian.

Longitudinal Magnetic Field. Our discussion of the physical significance of the various diagonal terms and cross-terms which appear in the squared absolute value on the right-hand side of (1.62), both for linearly polarized light and for chiral light (elliptically or circularly polarized), only applied in the absence of outside magnetic field. In the presence of a longitudinal magnetic field, with a component B_K in the direction of light propagation K, two considerations are required. First, state energies change. This is particularly important for degenerate states since the Zeeman effect splits the degeneracy linearly. In difference spectra such as MCD, this introduces band shapes which have the appearance of a derivative. Second, transition probabilities change for all states. This produces ordinary band shapes and can be understood in a straightforward manner from (1.62). It is now no longer possible to assume that the initial wavefunction Ψ_0 and the final wavefunction Ψ_f are real and that the matrix elements $\mathbf{M}(0f)$ and $Q(0f)$ are real while $\hat{\mathcal{M}}(0f)$ is purely imaginary. This invalidates the arguments we used to show that the diagonal terms $|\varepsilon_U^* \cdot \mathbf{M}(0f)|^2$, $|(\varepsilon_K \times \varepsilon_U^*) \cdot \mathcal{M}(0f)|^2$, and $|i\pi\tilde{\nu}\varepsilon_K Q(0f)\varepsilon_U^*|^2$, as well as the term $2\,\mathrm{Re}\,\{\pi\tilde{\nu}(\varepsilon_K \times \varepsilon_U^*) \cdot (\mathcal{M}(0f)/i)[\varepsilon_K Q(f0)\varepsilon_U]\}$, where Re stands for "the real part of," are the only ones which contribute to the absorption coefficient for linearly polarized light. By the same token, it is also no longer true that the cross-terms involving $\varepsilon_U^* \cdot \mathbf{M}(0f)$ and either $(\varepsilon_K \times \varepsilon_U^*) \cdot \mathcal{M}(0f)$ or $i\pi\tilde{\nu}\varepsilon_K Q(0f)\varepsilon_U^*$ are the only ones which contribute to optical activity.

A proper way to proceed is to rederive (1.62) in the presence of a magnetic field, using a new vector potential which incorporates its presence. This is relatively complicated, and we choose a more approximate but simpler procedure. As will be shown in detail in Section 1.3.3, the way in which a magnetic field with a longitudinal component B_K causes the wavefunctions and thus the transition moments to become complex is readily understood in terms of perturbation theory, using only the Zeeman operator, $-\varepsilon_K \cdot \hat{\mathcal{M}} B_K$. The perturbed wavefunctions Ψ_0 and Ψ_f will contain admixtures of all other molecular wavefunctions Ψ_j. The coefficients with which these enter are given to the first order in the magnetic field by $B_K\langle\Psi_j|\varepsilon_K \cdot \hat{\mathcal{M}}|\Psi_0\rangle/\Delta E_{j0}$ and $B_K\langle\Psi_j|\varepsilon_K \cdot \hat{\mathcal{M}}|\Psi_f\rangle/\Delta E_{jf}$, respectively, where ΔE is the energy difference of the two states which are being mixed by the perturbation. Since $\hat{\mathcal{M}}$ is purely imaginary and Ψ_0, Ψ_j, and Ψ_f are the zero-field wavefunctions, i.e., have been chosen real, these coefficients are purely imaginary. Since they are of first power in B_K, they will change sign if the direction of the magnetic field is reversed (note that the direction is defined as proceeding from the north to the south pole). In second order in the magnetic field, the mixing coefficients have the form $B_K^2\langle\Psi_j|\varepsilon_K \cdot \hat{\mathcal{M}}|\Psi_k\rangle\langle\Psi_k|\varepsilon_K \cdot \hat{\mathcal{M}}|\Psi_0\rangle/\Delta E_{jk0}$. These will be real and independent of the sense of the magnetic field.

Insertion of the perturbation theory expressions for the magnetically perturbed wavefunctions in terms of the field-free unperturbed wavefunctions into the expressions for the matrix elements $\mathbf{M}(0f) = \langle f|\hat{\mathbf{M}}|0\rangle$, $\mathcal{M}(0f) = \langle f|\hat{\mathcal{M}}|0\rangle$, and $Q(0f) = \langle f|\hat{Q}|0\rangle$ in (1.62) will lead to a fairly complicated expression. A term such as $\varepsilon_U^* \cdot \mathbf{M}(0f)$ will now be replaced by a sum of terms which can be arranged by the power of B_K. The most important term will not contain B_K at all and represents the original behavior in the absence of the magnetic field.

Those terms which contain the first power of B_K will be imaginary if the unperturbed term was real, such as $\boldsymbol{\varepsilon}_U^* \cdot \mathbf{M}(0f)$ for linearly polarized light, and they will be real if the unperturbed term was imaginary, such as $(\boldsymbol{\varepsilon}_K \times \boldsymbol{\varepsilon}_U^*) \cdot \mathcal{M}(0f)$ and $i\pi\tilde{\nu}\varepsilon_K Q(0f)\varepsilon_U^*$ for linearly polarized light. They will change their sign upon reversal of the direction of the magnetic field (linear magnetic field effects).

Those terms which contain the second power of B_K will behave like the unperturbed terms. They will be real if the unperturbed term was real and imaginary if the unperturbed term was imaginary. They will be independent of the direction of the magnetic field (quadratic magnetic field effects).

Now, we can take a qualitative look at the optical phenomena to be expected when a magnetic field is imposed, starting with (1.62). Since even the strongest available magnetic fields are rather weak when expressed in atomic units, the effects to be expected will be weak.

Looking first at the *linear magnetic effects*, we find that the extra imaginary unit i which comes with the first power of B_K interchanges the role of the diagonal terms and the cross-terms when the product implied by the absolute value squared is multiplied out. The measurement of linear effects is not impeded by the orienting effect of the magnetic field, since the molecular orientation will be the same regardless of the sense of the field while the linear effects will change sign. However, strictly speaking, the measurement is not being performed on the same orientation distribution as one would have in the absence of the field unless the sample is rigid. Since the orienting effect of magnetic field is so weak, the difference is normally not even detectable and can be safely ignored.

Magnetic Differential Absorption (MDA). For light of linear polarization, we find that terms containing the coefficient $B_K\boldsymbol{\varepsilon}_K \cdot \hat{\mathcal{M}}(jf)/\Delta E_{jf}$ attached to "cross-products" such as $\boldsymbol{\varepsilon}_U^* \cdot \mathbf{M}(0j)[(\boldsymbol{\varepsilon}_K \times \boldsymbol{\varepsilon}_U) \cdot \mathcal{M}(f0)]$, which previously corresponded to optical activity, now contribute to the absorption coefficient $\alpha(\tilde{\nu})$. The presence of these terms is responsible for magnetic differential absorption. The effect is limited to chiral molecules, since $\hat{\mathbf{M}}$ is present in the first power. Closer analysis shows that in isotropic solutions MDA does not disappear and is independent of the polarization of the light used.

Magnetic Circular Dichroism (MCD). Similarly, for light of circular polarization, we find that terms containing the coefficient $B_K\boldsymbol{\varepsilon}_K \cdot \mathcal{M}(jf)/\Delta E_{jf}$ attached to "diagonal" products such as $\boldsymbol{\varepsilon}_U^* \cdot \mathbf{M}(0j)\mathbf{M}(f0) \cdot \boldsymbol{\varepsilon}_U$, which previously corresponded to absorption and canceled out in optical activity, now contribute to the latter. The presence of these terms is responsible for magnetic circular dichroism. The effect occurs for all molecules, since $\hat{\mathbf{M}}$ is present in the second power. Closer analysis shows that it does not disappear upon isotropic averaging.

While MDA has apparently not yet been observed, MCD has received considerable attention.

Intrinsic *quadratic magnetic effects* are weaker and have no practical importance at present. They will be difficult to separate from the orientation effects unless the measurement is performed on a rigid sample. Since the presence of B_K^2 at a particular term, such as $\boldsymbol{\varepsilon}_U^* \cdot \mathbf{M}(0j)[(\boldsymbol{\varepsilon}_K \times \boldsymbol{\varepsilon}_U) \cdot \mathcal{M}(k0)]$, implies that there is no change in its real or purely imaginary nature relative to what it was at zero magnetic field, it is clear that the diagonal terms which contributed to the absorption coefficient for linearly polarized light in (1.62) in zero field, such as $|\boldsymbol{\varepsilon}_U^* \cdot \mathbf{M}(0f)|^2$, will now have counterparts which again contribute to absorption. Those cross-terms which contributed to the differential absorption of circularly polarized light (CD) in (1.62) in zero field, such as $\boldsymbol{\varepsilon}_U^* \cdot \mathbf{M}(0f)[(\boldsymbol{\varepsilon}_K \times \boldsymbol{\varepsilon}_U) \cdot \mathcal{M}(0f)]$, will now have counterparts which again contribute to optical activity. The latter contain $\hat{\mathbf{M}}$ to the first power and will be present only for chiral molecules (the corresponding dispersion phenomenon is referred to as the quadratic Faraday effect).

Longitudinal Electric Field. The intrinsic effects of the presence of a longitudinal electric field on the absorption coefficient can also be understood by considering the effects on state energies (Stark effect) and transition probabilities using (1.62) and applying perturbation theory. Once again, a proper semiclassical treatment would require the incorporation of the outside electric field into the vector potential at the outset and repetition of the whole derivation. Instead, we shall only indicate the effects the electric perturbation will have by considering the Stark Hamiltonian, $-\hat{\mathbf{M}} \cdot \mathbf{E}$. Here, $\mathbf{E}$ is the outside electric field, which we take as directed along the light propagation direction K, so that the perturbation Hamiltonian is $-\boldsymbol{\varepsilon}_K \cdot \hat{\mathbf{M}} E_K$. The molecules are again assumed to be fixed in space so that the electric field does not affect their orientation.

There are two obvious differences relative to the case of the longitudinal magnetic field. First, $-\boldsymbol{\varepsilon}_K \cdot \hat{\mathbf{M}} E_K$ is a real operator, and it does not share the intriguing ability of the magnetic perturbation operator to add imaginary to real terms and vice versa, no matter in which power it occurs. Thus, the electric perturbation cannot convert an effect observed with linearly polarized light or unpolarized light (real $\boldsymbol{\varepsilon}_U$, ordinary absorption) into one observed only with circularly polarized light (complex $\boldsymbol{\varepsilon}_U$, circular differential effect). A "diagonal" term linear in the electric field such as one containing the product $[E_K \boldsymbol{\varepsilon}_K \cdot \mathbf{M}(jf)/\Delta E_{jf}][\boldsymbol{\varepsilon}_U^* \cdot \mathbf{M}(0j)][\boldsymbol{\varepsilon}_U \cdot \mathbf{M}(f0)]$ will represent a contribution to ordinary absorption just like its "parent," $|\boldsymbol{\varepsilon}_U^* \cdot \mathbf{M}(0f)|^2$, does. Similarly, a "cross-term" linear in the electric field such as one containing the product $[E_K \boldsymbol{\varepsilon}_K \cdot \mathbf{M}(jf)/\Delta E_{jf}][\boldsymbol{\varepsilon}_U^* \cdot \mathbf{M}(0j)][(\boldsymbol{\varepsilon}_K \times \boldsymbol{\varepsilon}_U) \cdot \mathcal{M}(f0)]$ will represent a contribution to circular differential absorption like its "parent," $\boldsymbol{\varepsilon}_U^* \cdot \mathbf{M}(0f)(\boldsymbol{\varepsilon}_K \times \boldsymbol{\varepsilon}_U) \cdot \mathcal{M}(f0)$, does. The same will hold even for effects which depend on the second and higher powers of the electric field.

The second difference relative to the case of the longitudinal magnetic field has to do with the fact that the operator $\hat{\mathbf{M}}$ is odd with respect to the parity operation, while $\hat{\mathcal{M}}$ is even. We have already pointed out that effects described by terms which contain an odd number of matrix elements of $\hat{\mathbf{M}}$ vanish except for chiral molecules and are of equal magnitude but opposite sign for a pair of enantiomers, while no such restriction applies to terms in which $\hat{\mathbf{M}}$ occurs an even number of times. Adding $\hat{\mathcal{M}}$ any number of times as a result of magnetic perturbation had no effect on this requirement for chirality in the sample.

Obviously, this will be different for the electric perturbation: Each time a power of $\hat{\mathbf{M}}$ is added, the requirement will change. Of the terms mentioned above, $[E_K\varepsilon_K \cdot \mathbf{M}(jf)/\Delta E_{jf}][\varepsilon_U^* \cdot \mathbf{M}(0j)][\varepsilon_U \cdot \mathbf{M}(f0)]$ will require a chiral sample and will give the opposite results for two enantiomers, while the "parent" $|\varepsilon_U^* \cdot \mathbf{M}(0f)|^2$ is responsible for ordinary electric dipole absorption, a nonchiral phenomenon. On the other hand, $[E_K\varepsilon_K \cdot \mathbf{M}(jf)/\Delta E_{jf}][(\varepsilon_U^* \cdot \mathbf{M}(0j)][(\varepsilon_K \times \varepsilon_U) \cdot \mathcal{M}(f0)]$ will not require a chiral sample and will give the same results for two enantiomers, while the "parent" $\varepsilon_U^* \cdot \mathbf{M}(0f)(\varepsilon_K \times \varepsilon_U) \cdot \mathcal{M}(f0)$ is responsible for natural optical activity, a chiral phenomenon.

As in the case of the longitudinal magnetic field, effects dependent on an odd power of the electric field will change sign when the field direction is inverted, and those dependent on an even power will not. Thus, the former will be easily separated from the orienting effects of the electric field, and the latter will not. Unfortunately, for most orientation distributions of interest, the linear effects average to zero.

Magnetic and Electric Fields—A Juxtaposition. A survey of the implications of (1.62) for the effect of magnetic and electric fields is given in Table 1.1. It is assumed that molecular orientation is unaffected by the perturbing field. Circular differential effects are labeled with a dagger, chiral effects with a double dagger. The strengths of the effects, and thus their detectability and importance, decrease with the order of the perturbation (number of operators in parentheses) and increase with the fraction of times $\hat{\mathbf{M}}$ (as opposed to $\hat{\mathcal{M}}$ or $\hat{Q}$) appears outside parentheses. It should be noted that some of the contributions vanish upon isotropic or uniaxial averaging in a way that depends on the field orientation (longitudinal or transverse). For instance, magnetic circular dichroism requires longitudinal magnetic field. These matters are best viewed in terms of time-reversal and space-inversion symmetries, and the interested reader is referred to Section 1.4.

The ranks of the tensors which describe the phenomena listed in Table 1.1 are given by the number of Cartesian unit vectors in the terms which describe them. As seen in (1.62), one such vector is associated with the electric dipole transition moment, $\varepsilon_U^* \cdot \mathbf{M}(0f)$, and two are associated with the magnetic dipole, $(\varepsilon_K \times \varepsilon_U^*) \cdot \mathcal{M}(0f)$ and the electric quadrupole, $\varepsilon_K Q(0f)\varepsilon_U^*$, transition moments. The vector ε_K is associated with the perturbation by the longitudinal magnetic field, $-\varepsilon_K \cdot \hat{\mathcal{M}} B_K$, or electric field, $-\varepsilon_K \cdot \hat{\mathbf{M}} E_K$. The vector product $\varepsilon_K \times \varepsilon_U^*$ represents an axial vector, and this permits an effective reduction of the rank of the tensors in which it occurs.

Thus, ordinary electric dipole absorption represents an effect of rank 2, ordinary magnetic dipole absorption can be reduced to rank 2, electric quadrupole absorption represents an effect of rank 4, natural optical activity represents an effect of rank 3, and magnetic optical activity an effect of rank 4, both of which can be reduced to rank 2, etc. We shall devote Chapter 5 to the detailed discussion of electric dipole absorption and emission processes which are truly of rank 2. In Chapter 6 we briefly consider magnetic dipole absorption and concentrate on more complicated effects which can be reduced to a second-rank tensor, namely natural and magnetic circular dichroism. The

Table 1.1
Optical Absorption (Dispersion) Effects Induced by Magnetic and Electric Fields[a]

Field Dependence	Effect of Field Reversal	"Diagonal" Terms in (1.62)	"Cross-Terms" in (1.62)
B^2	$\times 1$	$(\hat{\mathcal{M}}\hat{\mathcal{M}})\hat{\mathbf{M}}\hat{\mathbf{M}}$ $[(\hat{\mathcal{M}}\hat{\mathcal{M}})\hat{\mathcal{M}}\hat{\mathcal{M}}$ $(\hat{\mathcal{M}}\hat{\mathcal{M}})\hat{\mathcal{M}}\hat{Q}$ $(\hat{\mathcal{M}}\hat{\mathcal{M}})\hat{Q}\hat{Q}]$ MAGNETOCHROMISM (Cotton-Mouton effect)	$(\hat{\mathcal{M}}.\hat{\mathcal{M}})\hat{\mathbf{M}}.\hat{\mathcal{M}}^{\dagger\ddagger}$ $(\hat{\mathcal{M}}.\hat{\mathcal{M}})\hat{\mathbf{M}}\hat{Q}^{\dagger\ddagger}$ QUADRATIC MAGNETIC CIRCULAR DICHROISM (Quadratic Faraday effect)
B	$\times(-1)$	$(\hat{\mathcal{M}})\hat{\mathbf{M}}\hat{\mathbf{M}}^{\dagger}$ $[(\hat{\mathcal{M}})\hat{\mathcal{M}}\hat{\mathcal{M}}^{\dagger}$ $(\hat{\mathcal{M}})\hat{\mathcal{M}}\hat{Q}^{\dagger}$ $(\hat{\mathcal{M}})\hat{Q}\hat{Q}^{\dagger}]$ MAGNETIC CIRCULAR DICHROISM (Faraday effect)	$(\hat{\mathcal{M}})\hat{\mathbf{M}}.\hat{\mathcal{M}}^{\ddagger}$ $(\hat{\mathcal{M}})\hat{\mathbf{M}}\hat{Q}^{\ddagger}$ MAGNETIC DIFFERENTIAL ABSORPTION[b]
None	$\times 1$	$\hat{\mathbf{M}}\hat{\mathbf{M}}$ $[\hat{\mathcal{M}}\hat{\mathcal{M}}$ $\hat{\mathcal{M}}\hat{Q}$ $\hat{Q}\hat{Q}]$ el. dip. mag. dip. el. quadr. ORDINARY ABSORPTION (Ordinary dispersion)	$\hat{\mathbf{M}}.\hat{\mathcal{M}}^{\dagger\ddagger}$ $\hat{\mathbf{M}}\hat{Q}^{\dagger\ddagger}$ NATURAL CIRCULAR DICHROISM (Rotatory dispersion)
E	$\times(-1)$	$(\hat{\mathbf{M}})\hat{\mathbf{M}}\hat{\mathbf{M}}^{\ddagger}$ $[(\hat{\mathbf{M}})\hat{\mathcal{M}}\hat{\mathcal{M}}^{\ddagger}$ $(\hat{\mathbf{M}})\hat{\mathcal{M}}\hat{Q}^{\ddagger}$ $(\hat{\mathbf{M}})\hat{Q}\hat{Q}^{\ddagger}]$ ELECTRIC DIFFERENTIAL ABSORPTION (Pockels effect)	$(\hat{\mathbf{M}})\hat{\mathbf{M}}.\hat{\mathcal{M}}^{\dagger}$ $(\hat{\mathbf{M}})\hat{\mathbf{M}}\hat{Q}^{\dagger}$ ELECTRIC CIRCULAR DICHROISM
E^2	$\times 1$	$(\hat{\mathbf{M}}\hat{\mathbf{M}})\hat{\mathbf{M}}\hat{\mathbf{M}}$ $[(\hat{\mathbf{M}}\hat{\mathbf{M}})\hat{\mathcal{M}}\hat{\mathcal{M}}$ $(\hat{\mathbf{M}}\hat{\mathbf{M}})\hat{\mathcal{M}}\hat{Q}$ $(\hat{\mathbf{M}}\hat{\mathbf{M}})\hat{Q}\hat{Q}]$ ELECTROCHROMISM (Kerr effect)	$(\hat{\mathbf{M}}\hat{\mathbf{M}})\hat{\mathbf{M}}.\hat{\mathcal{M}}^{\dagger\ddagger}$ $(\hat{\mathbf{M}}\hat{\mathbf{M}})\hat{\mathbf{M}}\hat{Q}^{\dagger\ddagger}$ QUADRATIC ELECTRIC CIRCULAR DICHROISM
E,B	$\times(-1)^c$ $\times(-1)^c$	$(\hat{\mathbf{M}}.\hat{\mathcal{M}})\hat{\mathbf{M}}\hat{\mathbf{M}}^{\dagger\ddagger}$ $[(\hat{\mathbf{M}}.\hat{\mathcal{M}})\hat{\mathcal{M}}\hat{\mathcal{M}}^{\ddagger\ddagger}$ $(\hat{\mathbf{M}}.\hat{\mathcal{M}})\hat{\mathcal{M}}\hat{Q}^{\dagger\ddagger}$ $(\hat{\mathbf{M}}.\hat{\mathcal{M}})\hat{Q}\hat{Q}^{\dagger\ddagger}]$ MAGNETOELECTRIC CIRCULAR DICHROISM	$(\hat{\mathbf{M}}.\hat{\mathcal{M}})\hat{\mathbf{M}}.\hat{\mathcal{M}}$ $(\hat{\mathbf{M}}.\hat{\mathcal{M}})\hat{\mathbf{M}}\hat{Q}$ MAGNETOELECTROCHROMISM

[a] Matrix elements of the operators in parentheses originate in perturbation by the field and enter divided by the corresponding energy differences. The effect names in parentheses refer to the corresponding dispersion phenomena. The effects labeled with a dagger (†) are differential between left-handed and right-handed circularly polarized light. They do not appear for linearly polarized light or unpolarized light. The effects labeled with a double dagger (‡) are chiral. They do not appear unless the sample is chiral, and enantiomers show opposite effects.

[b] Also called magnetic-field-induced absorption difference (MIAD) or magnetochiral dichroism (MCHD).

[c] Sign change for reversal of either field.

other effects listed in Table 1.1 can be handled similarly. Finally, in Chapter 7 we shall address effects which are irreducibly of rank 4, such as two-photon absorption and the Raman effect. These are outside the scope of Table 1.1, which deals only with one-photon phenomena.

1.3.3 Chiroptical Spectroscopy

Complex Rotation. Linear and Circular Effects. Light propagating through matter generally changes both its intensity and its state of polarization in a complicated manner (Appendix III). Incident linearly polarized light in general emerges from a spectroscopic sample with elliptical polarization. The major axis of the ellipse is rotated by $\phi(\tilde{\nu})$ radians relative to the original direction of linear polarization (measured clockwise when viewing the light source). The ellipticity is $\theta(\tilde{\nu})$ radians, i.e., the ratio of the minor to the major axis of the ellipse is $\tan\theta(\tilde{\nu})$. It is defined as positive if the emerging light has right-handed elliptical polarization and negative if it has left-handed elliptical polarization.

Elliptical light can be viewed as a sum of a right-handed circular component of amplitude $|\mathbf{E}_0^{(R)}|$ and a left-handed circular component of amplitude $|\mathbf{E}_0^{(L)}|$. Right-handed elliptical polarization corresponds to $|\mathbf{E}_0^{(R)}| > |\mathbf{E}_0^{(L)}|$, left-handed elliptical polarization to $|\mathbf{E}_0^{(R)}| < |\mathbf{E}_0^{(L)}|$. The value of $\tan\theta$ is given by $(|\mathbf{E}_0^{(R)}| - |\mathbf{E}_0^{(L)}|)/(|\mathbf{E}_0^{(R)}| + |\mathbf{E}_0^{(L)}|)$.

In analogy to the complex refractive index $\mathrm{N}(\tilde{\nu})$ defined in (1.22), it is useful to define the complex rotation $\Phi(\tilde{\nu})$ of a sample:

$$\Phi(\tilde{\nu}) = \phi(\tilde{\nu}) - i\theta(\tilde{\nu}) = [\phi_{\mathrm{LD}}(\tilde{\nu}) + \phi_{\mathrm{CD}}(\tilde{\nu})] - i[\theta_{\mathrm{LD}}(\tilde{\nu}) + \theta_{\mathrm{CD}}(\tilde{\nu})] \qquad (1.80)$$

The minus sign originates in the convention adopted in Section 1.1.1. Strictly speaking, we define the imaginary part of $\Phi(\tilde{\nu})$ as $\tanh^{-1}(\tan\theta)$, but for the small values of θ of interest here, this can be replaced by θ.

As indicated in (1.80), in cases of interest in the present context [small $\phi(\tilde{\nu})$, small $\theta(\tilde{\nu})$], it is possible to view $\phi(\tilde{\nu})$ and $\theta(\tilde{\nu})$ as superpositions of linearly additive linear (subscript LD) and circular (subscript CD) effects. The latter are known as optical activity and form the subject of chiroptical spectroscopy. In the absence of a magnetic field, only chiral molecular assemblies (individually chiral molecules or chiral orientation distributions of achiral molecules) can contribute to optical activity. A chiral object is one which is not superimposable onto its mirror image. In the presence of a component of a magnetic field parallel to the light propagation direction, all molecules will contribute to optical activity.

At first, we shall consider linear and circular effects jointly, for two reasons. First, in measurements on partially oriented samples, they occur intimately intertwined. Second, magnetically induced circular effects actually originate in a magnetic perturbation of the ordinary linear effects (Table 1.1), strange though it may seem at first sight.

The first linear effect is the rotation $\phi_{\mathrm{LD}}(\tilde{\nu})$ due to *linear dichroism.* In a uniaxial sample, this measures the difference in the absorption coefficients for light linearly polarized in the direction of the unique axis, $\varepsilon_Z(\tilde{\nu})$, and for light linearly polarized perpendicular to it, $\varepsilon_X(\tilde{\nu}) = \varepsilon_Y(\tilde{\nu})$:

$$\phi_{LD}(\tilde{\nu}) = (1/4)l[\alpha_Z(\tilde{\nu}) - \alpha_Y(\tilde{\nu})] = (2.303/4)lc[\varepsilon_Z(\tilde{\nu}) - \varepsilon_Y(\tilde{\nu})] \qquad (1.81)$$

Here and in the following, l is the length of the sample in cm and c is the molar concentration of the solute. In spectral regions where the solvent is transparent, $\phi_{LD}(\tilde{\nu})$ is dominated by the solute.

The second linear effect is the ellipticity $\theta_{LD}(\tilde{\nu})$ due to *linear birefringence.* In a uniaxial sample, this measures the difference in the refractive indices for light linearly polarized along Z, $n_Z(\tilde{\nu})$, and for light linearly polarized perpendicular to Z, $n_X(\tilde{\nu}) = n_Y(\tilde{\nu})$:

$$\theta_{LD}(\tilde{\nu}) = -\pi\tilde{\nu}l[n_Z(\tilde{\nu}) - n_Y(\tilde{\nu})] \qquad (1.82)$$

This quantity is usually dominated by the properties of the anisotropic solvent even in the spectral regions where it is transparent. In samples uniaxially oriented by an electric field the appearance of a nonzero $\theta_{LD}(\tilde{\nu})$ is known as the Kerr effect; in those uniaxially oriented by a magnetic field, as the Cotton-Mouton effect.

The quantities which enter into the expressions for ϕ_{LD} and θ_{LD}, namely, $\varepsilon_U(\tilde{\nu})$ and $n_U(\tilde{\nu})$, are already familiar from the preceding section, 1.3.2, where their relation to molecular quantities has been discussed in some detail.

In samples of lower than uniaxial symmetry, it is necessary to consider two additional linear effects, namely, diagonal linear dichroism and birefringence, which are defined with respect to directions bisecting the Z and Y axes (Appendix III). We shall not deal with this case here.

The first circular effect is the rotation $\phi_{CD}(\tilde{\nu})$ due to *circular birefringence.* This measures the difference in the refractive indices for left-handed and right-handed circularly polarized light, $n_L(\tilde{\nu}) - n_R(\tilde{\nu})$:

$$\phi_{CD}(\tilde{\nu}) = \pi l\tilde{\nu}[n_L(\tilde{\nu}) - n_R(\tilde{\nu})] \qquad (1.83)$$

In the absence of an outside magnetic field and in an optically inactive solvent, an optically active solute will be the sole contributor to this effect (optical rotatory dispersion, ORD). In the presence of an outside magnetic field with a component parallel to the light propagation direction, the solvent will tend to dominate the effect (magnetic optical rotatory dispersion, MORD).

The second circular effect is the ellipticity $\theta_{CD}(\tilde{\nu})$ due to *circular dichroism.* This measures the difference in the absorption coefficients for left-handed and right-handed circularly polarized light, $\varepsilon_L(\tilde{\nu}) - \varepsilon_R(\tilde{\nu})$:

$$\theta_{CD}(\tilde{\nu}) = (1/4)l[\alpha_L(\tilde{\nu}) - \alpha_R(\tilde{\nu})] = (2.303/4)lc[\varepsilon_L(\tilde{\nu}) - \varepsilon_R(\tilde{\nu})] \qquad (1.84)$$

In spectral regions where the solvent is transparent, $\theta_{CD}(\tilde{\nu})$ is dominated by the solute.

In view of (1.81)–(1.84), the complex rotation $\Phi(\tilde{\nu})$ defined in (1.80) can also be written as

$$\begin{aligned}\Phi(\tilde{\nu}) &= \Phi_{LD}(\tilde{\nu}) + \Phi_{CD}(\tilde{\nu}) \\ &= [\phi_{LD}(\tilde{\nu}) - i\theta_{LD}(\tilde{\nu})] + [\phi_{CD}(\tilde{\nu}) - i\theta_{CD}(\tilde{\nu})] \\ &= \pi l\tilde{\nu}\{[N_L(\tilde{\nu}) - N_R(\tilde{\nu})] + i[N_Z(\tilde{\nu}) - N_Y(\tilde{\nu})]\} \qquad (1.85)\end{aligned}$$

This form makes it clear that $\Phi(\tilde{\nu})$ summarizes the information on the linear and circular differential effects of a sample, $N_Z - N_Y$ and $N_L - N_R$, respectively. The overall attenuation of light intensity by the sample is not addressed by these differential quantities alone but, rather, depends on the complex refractive indices $N_Z(\tilde{\nu})$, $N_Y(\tilde{\nu})$, $N_L(\tilde{\nu})$, and $N_R(\tilde{\nu})$ individually or in combinations as required by the propagation direction and state of polarization of the light beam used (see Section 1.3.2). The preferred approach to the experimental determination of N_Z and N_Y is to perform measurements of absorbance and refractive index using light linearly polarized along Z and along Y, which propagates through an achiral uniaxial sample without a change in polarization. The evaluation is then simple and avoids the issue of rotation altogether.

It needs to be emphasized that $\Phi(\tilde{\nu})$ is a tensorial quantity whose value depends on the light propagation direction unless the sample is isotropic. In the latter case, and in the case of light propagation along the unique axis Z of a uniaxial sample, only the circular effects remain:

$$\Phi(\tilde{\nu}) = \pi l \tilde{\nu}[N_L(\tilde{\nu}) - N_R(\tilde{\nu})] \tag{1.86}$$

These are the two situations in which the measurement of the generally much weaker circular effects is relatively easy, and they are to be preferred by far in actual practice. They are the only two cases which we shall consider in detail in Chapter 6.

Experimental Quantities in Chiroptical Spectroscopy and Their Usual Units. So far, all formulas for rotation and ellipticity were expressed in radians. It is more common to work with degrees, using $\phi_{CD}^{deg}(\tilde{\nu})$. The number of degrees per radian is $180/\pi$, so that $\phi_{CD}^{deg}(\tilde{\nu}) = (180/\pi)\phi_{CD}(\tilde{\nu})$.

For historical reasons, the measured rotation is usually expressed in units of degrees per decimeter. From (1.83),

$$\phi_{CD}^{deg}(\tilde{\nu})/l = 1800\tilde{\nu}[n_L(\tilde{\nu}) - n_R(\tilde{\nu})] \tag{1.87}$$

where l is now in units of dm = 10 cm. The rotation is taken as positive if the polarization direction of linearly polarized light is rotated clockwise by the sample when viewed against the light source. The sample is then called dextrorotatory. Samples with negative rotation are called levorotatory.

Specific rotation $[\phi(\tilde{\nu})]$ in units of deg dm^{-1} cm^3 g^{-1} is defined as rotation in degrees per decimeter of path length observed on an isotropic solution containing 1 g of solute per cubic centimeter:

$$[\phi(\tilde{\nu})] = \phi_{CD}^{deg}(\tilde{\nu})/c'l = (1800\tilde{\nu}/c')[n_L(\tilde{\nu}) - n_R(\tilde{\nu})] \tag{1.88}$$

where l is the pathlength in dm and c′ is the concentration in g cm^{-3}.

Molar rotation $[\phi(\tilde{\nu})]_M$ is defined as the product of the specific rotation and the molecular mass M (g mol^{-1}), "for brevity" divided by 100,

$$[\phi(\tilde{\nu})]_M = (M/100)[\phi(\tilde{\nu})] \tag{1.89}$$

Since $c' = 10^{-3}$ Mc, where c is the molar concentration in mol L^{-1}, molar rotation is given by

$$[\phi(\tilde{\nu})]_M = M\phi_{CD}^{deg}(\tilde{\nu})/100c'l = 10\phi_{CD}^{deg}(\tilde{\nu})/cl \tag{1.90}$$

where l is in units of dm, so that it is equal to the rotation in degrees observed on a molar solution of a path length of one meter. The usual units of $[\phi(\tilde{\nu})]_M$ are deg L cm^{-1} mol^{-1}. In these units,

$$[\phi(\tilde{\nu})]_M = 100\phi_{CD}^{deg}(\tilde{\nu})/cl = (18\,000/\pi)[\phi_{CD}(\tilde{\nu})/cl] \tag{1.91}$$

where $\phi_{CD}^{deg}(\tilde{\nu})$ is the rotation in degrees, $\phi_{CD}(\tilde{\nu})$ is the rotation in radians, c is the molar concentration, and l is the path length in cm^{-1}. Equation (1.91) is important in that it relates the molar rotation both to the rotation $\phi_{CD}^{deg}(\tilde{\nu})$ observed on a standard commercial polarimeter and to the quantity $\phi_{CD}(\tilde{\nu})$ which appears in (1.80), (1.83), and (1.85).

Molar ellipticity $[\theta(\tilde{\nu})]_M$ is defined in direct analogy to molar rotation (1.91) as

$$[\theta(\tilde{\nu})]_M = 100\theta_{CD}^{deg}(\tilde{\nu})/cl = (18\,000/\pi)[\theta_{CD}(\tilde{\nu})/cl] \tag{1.92}$$

where $\theta_{CD}^{deg}(\tilde{\nu})$ is the ellipticity in degrees, $\theta_{CD}(\tilde{\nu})$ is the ellipticity in radians, c is the molar concentration, and l is the path length in cm. This equation is important in that it relates the molar ellipticity $[\theta(\tilde{\nu})]_M$ both to the measured ellipticity $\theta_{CD}^{deg}(\tilde{\nu})$ and to the quantity $\theta_{CD}(\tilde{\nu})$ which appears in (1.80), (1.84), and (1.85). Using (1.92) and (1.84), we obtain a relation between molar ellipticity and the difference of molar extinction coefficients for left-handed and right-handed circularly polarized light:

$$\begin{aligned}[\theta(\tilde{\nu})]_M &= (18\,000/\pi)(2.303/4)[\varepsilon_L(\tilde{\nu}) - \varepsilon_R(\tilde{\nu})] \\ &= 3298.2[\varepsilon_L(\tilde{\nu}) - \varepsilon_R(\tilde{\nu})] \\ &= 3298.2[E_L(\tilde{\nu}) - E_R(\tilde{\nu})]/cl \end{aligned} \tag{1.93}$$

where l is in cm. The relation of the measured ellipticity $\theta_{CD}^{deg}(\tilde{\nu})$ recorded on standard commercial instruments to the difference of optical densities $E_L(\tilde{\nu}) - E_R(\tilde{\nu})$ is

$$\theta_{CD}^{deg}(\tilde{\nu}) = [\theta(\tilde{\nu})]_M cl/100 = 33[E_L(\tilde{\nu}) - E_R(\tilde{\nu})] \tag{1.94}$$

and we see that one millidegree corresponds to 3.032×10^{-5} units of absorbance (optical density).

A useful dimensionless quantity is Kuhn's dissymmetry factor $g(\tilde{\nu})$, defined as $[\varepsilon_L(\tilde{\nu}) - \varepsilon_R(\tilde{\nu})]/\varepsilon(\tilde{\nu})$.

Complex Rotation: The Line Shapes. Like the complex refractive index $N(\tilde{\nu})$, the complex rotation $\Phi(\tilde{\nu})$ is an analytic complex function of $\tilde{\nu}$, and its real and imaginary parts are not independent but are related by the Kramers-Kronig transforms (1.65). We have seen in Section 1.3.2 that the contribution of a transition to the real part of $N(\tilde{\nu})$ had the "dispersion line shape" $g''(\tilde{\nu})$, while its contribution to the imaginary part of $N(\tilde{\nu})$ had the "absorption line shape" $g'(\tilde{\nu})$. Examples of these shapes for Lorentzian and Gaussian lines were given in Figure 1.12.

The association of $g''(\tilde{\nu})$ with the real part of $N(\tilde{\nu})$, i.e., with the ordinary refractive index, and the association of $g'(\tilde{\nu})$ with the imaginary part of $N(\tilde{\nu})$, i.e., with the absorption coefficient, is immutable. This is apparent in (1.73) since the multiplicative factor in front of $[g_f''(\tilde{\nu}) - ig_f'(\tilde{\nu})]$ is real. Even in a more general case of a partially oriented sample this will be so, as is clear from the presence of the absolute value signs in (1.60). This unique association makes it possible to refer to $g''(\tilde{\nu})$ as the dispersion line shape and to $g'(\tilde{\nu})$ as the absorption line shape.

With complex rotation, the situation is quite different, and both shapes can contribute to the real part $\phi(\tilde{\nu})$ as well as the imaginary part $\theta(\tilde{\nu})$. This is best seen from the form of a general expression for $\Phi(\tilde{\nu})$ in terms of molecular quantities. This can be derived by means similar to those which led to (1.62) and (1.73). Including the series expansion (1.42) of an exponential in expressions of the type $\exp(i\mathbf{K}\cdot\mathbf{r})(\boldsymbol{\varepsilon}_U\cdot\mathbf{p})$, as encountered in (1.41), truncated after the term $i\mathbf{K}\cdot\mathbf{r}$, one obtains

$$\Phi(\tilde{\nu}) = \text{const} \times i\sum_{f}\{-i\boldsymbol{\varepsilon}_U\cdot\mathbf{M}(0f) - (\boldsymbol{\varepsilon}_K \times \boldsymbol{\varepsilon}_U)\cdot[\mathcal{M}(0f)/i] - \pi\tilde{\nu}_f\boldsymbol{\varepsilon}_K\mathrm{Q}(0f)\boldsymbol{\varepsilon}_U\}$$
$$\times\{i\boldsymbol{\varepsilon}_V\cdot\mathbf{M}(f0) + (\boldsymbol{\varepsilon}_K \times \boldsymbol{\varepsilon}_V)\cdot[\mathcal{M}(f0)/i] - \pi\tilde{\nu}_f\boldsymbol{\varepsilon}_K\mathrm{Q}(f0)\boldsymbol{\varepsilon}_V\}[g_f'(\tilde{\nu}) - ig_f''(\tilde{\nu})] \tag{1.95}$$

where directions U, V, and K form a right-handed system of coordinates.

Like (1.62), this expression contains a product of two terms, each containing a sum of contributions from electric dipole, magnetic dipole, and electric quadrupole transition moments. In (1.62), the two terms in the product were complex conjugates of each other so that their product was real. In (1.95), they also differ by interchange of the vectors $\boldsymbol{\varepsilon}_U$ and $\boldsymbol{\varepsilon}_V$. This has profound consequences: Since they are no longer related by complex conjugation alone, their product does not need to be real. The occurrence of both directions perpendicular to the light propagation direction, $\boldsymbol{\varepsilon}_U$ and $\boldsymbol{\varepsilon}_V$, in the expression for complex rotation makes sense qualitatively, just as the presence of the single direction $\boldsymbol{\varepsilon}_U$ did in the expression for the complex index of refraction.

When the imaginary unit in front of the bracket in (1.95) is incorporated into the last bracket, $[ig''(\tilde{\nu}) + g'(\tilde{\nu})]$ results. If the product of the expressions in the folded brackets is real, the imaginary part of Φ (ellipticity) will thus have the dispersion line shape $g''(\tilde{\nu})$ and the real part of Φ (rotation) will have the absorption line shape $g'(\tilde{\nu})$. This occurs in samples which are not optically active, i.e., which show only linear and no circular dichroism and birefringence. If the product of the two expressions in folded brackets is purely imaginary, ellipticity will have the absorption line shape $g'(\tilde{\nu})$ and rotation will follow the dispersion line shape $g''(\tilde{\nu})$. This occurs in isotropic samples which are optically active, i.e., in samples which show only circular and no linear dichroism and birefringence. In the general case, the product of the two folded brackets will have a nonvanishing real part as well as a nonvanishing imaginary part. Then both $g'(\tilde{\nu})$ and $g''(\tilde{\nu})$ will contribute to both the real and the imaginary parts of $\Phi(\tilde{\nu})$. This happens in samples which exhibit both linear and circular dichroism and birefringence. As already indicated, it will be desirable to find a direction of light propagation in which the linear effects disappear, greatly simplifying the interpretation of the results.

Linear Dichroism and Birefringence: Mostly a Nuisance in Chiroptical Spectroscopy. The largest contribution on the right-hand side of equation (1.95) will normally be proportional to $|\mathbf{M}(0f)|^2$, for the same reasons which make electric dipole contributions dominate the right-hand side of equation (1.62). This contribution will depend on the product of the U coordinate of the transition moment vector **M**(0f) and the V coordinate of the transition moment vector **M**(f0). The two vectors are identical in the absence of a magnetic field. As noted above, the real part of this contribution to $\Phi(\tilde{\nu})$ will have the shape $g'(\tilde{\nu})$ of an absorption line, whereas its imaginary part will have the shape $g''(\tilde{\nu})$ of a dispersion curve. Thus this contribution will cause the plane of linearly polarized light to rotate if its wavelength is such that it is absorbed by the sample, and it will induce ellipticity in a large range of wavelengths, including the regions in which the sample is transparent. This behavior is just the opposite of that expected for optical activity.

The effect will be largest if the direction of the molecular transition moment **M**(0f) bisects the angle between the unit vectors ε_U and ε_V. It will disappear if the molecular transition moment is aligned with one or the other of these vectors. This behavior is that of a well-known optical element, a retardation plate with axes along U and V, for instance, the quarter-wave plate (mentioned in Section 1.1). If plane-polarized light is incident upon such a plate at an angle of 45° with respect to both U and V, it can be thought of as two beams linearly polarized along U and V propagating coherently. If the wavelengths lie within the region of absorption, one of these will be absorbed more than the other, resulting not only in an overall decrease of light intensity but also in rotation of its plane of polarization (Figure 1.14). Moreover, even in a region of transparency, one of the two components into which the light wave was decomposed will propagate at a different speed from the other, introducing a phase difference and thus ellipticity. The thickness of a quarter-wave plate is such as to induce

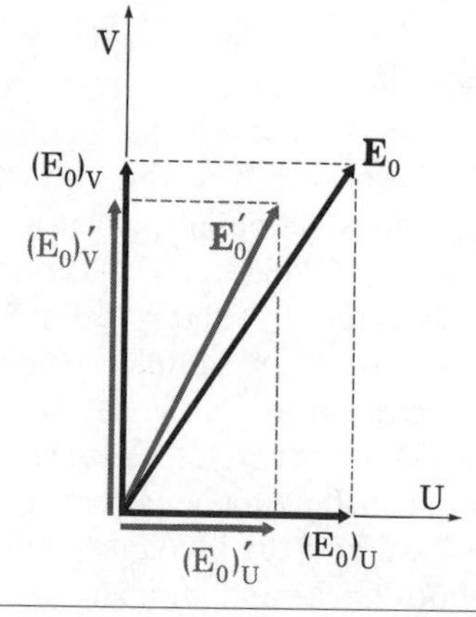

Figure 1.14 Rotation of the plane of polarized light by dichroic absorption. The direction of the electric vector $\mathbf{E}_0$ of the incident light is rotated to $\mathbf{E}'_0$ as its two components are attenuated to different degrees.

a phase difference of $\lambda/4$, and the light produced is circularly polarized (Figure 1.5).

The effect disappears by averaging over all orientations if the sample is isotropic. It also disappears in other instances in which the molecular transition moment vector **M**(0f) of the solute molecules is distributed uniformly with respect to the light propagation direction ε_K, so that the directions U and V are equivalent. An important example is the case of light propagating along the unique axis Z of a uniaxially oriented sample. Although the effect has been exploited for the measurement of very weak linear dichroism, its existence is mostly detrimental: It makes it extremely difficult to measure the circular dichroism and birefringence of most partially oriented samples.

Optical Activity—A Qualitative View. In isotropic samples and in other samples in which all directions perpendicular to the light propagation direction are exactly equivalent, $M_U(0f)M_V(f0)$ averages to zero, and it is then possible to detect relatively easily the presence of higher order cross-terms in (1.95). These will be present in optically active samples. As noted above, in optically active samples which show no linear dichroism, rotation of plane-polarized light is associated with a curve of the dispersion type, and production of ellipticity is associated with an absorption-like shape; in other words, the sample absorbs left-handed and right-handed circularly polarized light to different degrees and the two kinds of light propagate with different velocities. In order to identify those contributions which exhibit this type of behavior in our equation for the complex rotatory power $\Phi(\tilde{\nu})$, (1.95), we need to find such cross-terms in the product of the three brackets whose real part contains $g''(\tilde{\nu})$ and whose imaginary part is associated with $g'(\tilde{\nu})$. Such cross-terms can be of two possible origins, leading to natural and magnetic circular dichroism, respectively.

Natural Optical Activity. First, as we go beyond the electric dipole term in the expansion of $e^{i\mathbf{K}\cdot\mathbf{r}}$, we introduce the terms with $\mathscr{M}(0f)/i$ and Q(0f) shown in (1.95), both of which are real if the wavefunctions $|0\rangle$ and $|f\rangle$ are real (remember that $\hat{\mathscr{M}}$ is a purely imaginary operator). Multiplying either one by the leading first term from the other folded bracket, such as $i\varepsilon_V \cdot \mathbf{M}(f0)$, yields a purely imaginary number, whereas the previously considered product of two leading terms, $-i\varepsilon_U \cdot \mathbf{M}(0f)$ and $i\varepsilon_V \cdot \mathbf{M}(f0)$, was real. This will cause a reversal of the roles of $g_f'(\tilde{\nu})$ and $g_f''(\tilde{\nu})$.

We conclude that circular dichroism and circular birefringence will result and that their magnitude will be related to products of components of the vector **M**(0f) on the one hand and either of the vector $\mathscr{M}$(f0) or of the tensor Q(f0) on the other hand. Upon averaging over all molecular orientations encountered in an isotropic solution, one finds that the contribution from the **M**(0f), Q(f0) combination averages to zero, just as was the case for the **M**(0f), **M**(f0) contribution investigated in the previous paragraph. However, the **M**(0f), $\mathscr{M}$(f0) contribution does not vanish upon averaging and is alone responsible for circular dichroism and rotation of plane-polarized light by isotropic solutions of naturally optically active (chiral) molecules.

Magnetic Optical Activity. There is a second way in which the reversal of the role of the line shapes $g_f'(\tilde{\nu})$ and $g_f''(\tilde{\nu})$ relative to that observed for the linear effects can be accomplished and optical activity produced. All that needs to be done is to somehow introduce the imaginary unit i into the **M**(0f), **M**(f0) term in equation (1.95), i.e., to make the matrix element **M**(0f) complex. As already dicussed in Section 1.3.2, this can be achieved by perturbing the molecular wavefunctions by the application of an outside magnetic field. Such a perturbing field can be thought of as perturbing the ground and final states of the $0 \rightarrow f$ transition by mixing all other states of the molecule into the zero-field ground state $\Psi_0(\mathbf{q},\mathbf{Q})$ and the zero-field final state $\Psi_f(\mathbf{q},\mathbf{Q})$. Since the perturbing operator is $-\hat{\mathcal{M}} \cdot \mathbf{B}$ and thus purely imaginary, the first-order changes which it will cause in the initially real wavefunctions of the ground and final states will also be purely imaginary. This will cause the transition moment **M**(0f) from the ground state $\Psi_0(\mathbf{q},\mathbf{Q})$ to the final state $\Psi_f(\mathbf{q},\mathbf{Q})$ to become a complex vector (until now we have assumed that it is real). It can be written as a series in the powers of **B**. Those terms which are of odd power in the magnetic field **B** will be purely imaginary; those of even power in **B** will be real. Now, returning to equation (1.95), we see that the product of the real part of **M**(0f) with the real part of **M**(f0) from the other bracket will still behave as before, producing no optical activity and averaging to zero in isotropic solutions and for observation along the unique axis of a uniaxial sample. However, the product of the small imaginary term in **M**(0f) which is proportional to the first power of the perturbing magnetic field, taken from one of the brackets, with the large real part of **M**(f0), taken from the other bracket, will be imaginary. Thus, it will produce the sought reversal of the behavior of $g_f'(\tilde{\nu})$ and $g_f''(\tilde{\nu})$ in the overall expression for $\Phi(\tilde{\nu})$. The circular dichroism and rotation of the plane of plane-polarized light produced in this fashion are referred to as magnetic optical activity. Just like the **M**(0f), $\mathcal{M}$(f0) contribution to natural optical activity, magnetic optical activity does not disappear upon isotropic averaging.

It is obvious that higher order effects will result from the first-order perturbing influence of the magnetic field on higher order terms in the product in (1.95) and from the consideration of higher order perturbation theory for the magnetic field, but these are of no practical importance at present.

Optical Activity—A Quantitative View. As already indicated, in most areas of molecular spectroscopy, the birefringence (dispersion) effects are generally of much smaller importance than absorption (attenuation) effects. These two aspects of interaction of light with matter are of somewhat more comparable importance in chiroptical spectroscopy. The first, circular dichroism, is harder to measure but easier to interpret in terms of molecular structure. This is due to the narrower form of the absorption line shape $g'(\tilde{\nu})$ which limits the number of transitions which contribute at any one wavelength $\tilde{\nu}$. The second, optical rotation, is much easier to measure and for a long time dominated all discussion of optical activity. Unfortunately, it is much harder to interpret in terms of specific features of molecular structure, since the dispersion line shape $g''(\tilde{\nu})$ is very broad and very many excited states contribute in any observed spectral region (Figure 1.12).

We shall presently state some important results applicable to the measurement of complex rotation on samples in which linear dichroism and birefringence are absent, either for all directions of propagation, as in isotropic samples, or for a particular direction used, as in the measurement along the unique axis of a uniaxial sample. This will be the only situation which is treated in Chapter 6. The results are equally applicable to natural and magnetically induced optical activity.

We are now ready to consider the relation of the observable rotation and ellipticity to molecular wavefunctions. Because of the totally different physical origin of natural and magnetic optical activity (cf. Table 1.1), the two must be treated separately.

Natural Optical Activity. As long as solvent effects are neglected, the relation of molar ellipticity to molecular transition moments follows from (1.93) and (1.60), where we take K = Z, choose once $\boldsymbol{\varepsilon}_U = (\boldsymbol{\varepsilon}_X - i\boldsymbol{\varepsilon}_Y)/\sqrt{2}$ (left-handed circular polarization) and once $\boldsymbol{\varepsilon}_U = (\boldsymbol{\varepsilon}_X + i\boldsymbol{\varepsilon}_Y)/\sqrt{2}$ (right-handed circular polarization), and take the difference. Assuming the same line-shape function $g'(\tilde{\nu})$ for light of both polarizations, and a nondegenerate ground state,

$$
\begin{aligned}
\varepsilon_L(\tilde{\nu}) - \varepsilon_R(\tilde{\nu}) = \frac{-16\pi^3 N\tilde{\nu}}{2303hc} g'(\tilde{\nu}) \Big[\!\Big[& \operatorname{Im}[M_X(0f)\mathcal{M}_X(f0) + M_Y(0f)\mathcal{M}_Y(f0)] \\
& + \pi\tilde{\nu}\operatorname{Re}\{\mathbf{M}(0f) \times [\mathsf{Q}(f0)\boldsymbol{\varepsilon}_Z]\}_Z \\
& + \pi\tilde{\nu}\operatorname{Re}[\mathcal{M}_Y(0f)Q_{YZ}(f0) + \mathcal{M}_X(0f)Q_{XZ}(f0)] \\
& + \frac{i}{2}[\mathbf{M}(0f) \times \mathbf{M}(f0)]_Z + \frac{i}{2}[\mathcal{M}(0f) \times \mathcal{M}(f0)]_Z \\
& + \frac{i\pi^2\tilde{\nu}^2}{2}\{[\boldsymbol{\varepsilon}_Z\mathsf{Q}(0f)] \times [\mathsf{Q}(f0)\boldsymbol{\varepsilon}_Z]\}_Z \Big]\!\Big]
\end{aligned}
\qquad (1.96)
$$

Here, N is Avogadro's number, h is Planck's constant, and c is the speed of light. In the absence of an outside magnetic field, the initial and final wavefunctions can be chosen real, and we shall do so through the remainder of the section on natural optical activity. Then the matrix elements of $\hat{\mathbf{M}}$ and $\hat{\mathsf{Q}}$ are real and those of $\hat{\mathcal{M}}$ are purely imaginary, and $\mathbf{M}(0f) = \mathbf{M}(f0)$, $\mathsf{Q}(0f) = \mathsf{Q}(f0)$, $\mathcal{M}(0f) = -\mathcal{M}(f0)$. The vector product of two parallel or antiparallel vectors vanishes, so that the last three terms in (1.96) disappear. The sum in the square brackets in the third term is purely imaginary, so its real part vanishes also. This is all as expected; the right-hand side of (1.96) is equal to the difference of two absolute values squared and cannot contain imaginary terms. We are left with only the first two terms contributing to natural optical activity, and we obtain

$$
\begin{aligned}
\varepsilon_L(\tilde{\nu}) - \varepsilon_R(\tilde{\nu}) = \frac{16\pi^3 N\tilde{\nu}}{2303hc} g'(\tilde{\nu})\{ & \operatorname{Im}[M_X(f0)\mathcal{M}_X(0f) + M_Y(f0)\mathcal{M}_Y(0f)] \\
& + \pi\tilde{\nu}[M_Y(0f)Q_{XZ}(f0) - M_X(0f)Q_{YZ}(f0)]\}
\end{aligned}
\qquad (1.97)
$$

In an *isotropic solution*, the second term averages to zero. Isotropic averaging of the first term yields

$$
\varepsilon_L(\tilde{\nu}) - \varepsilon_R(\tilde{\nu}) = \frac{32\pi^3 N\tilde{\nu}}{2303 \times 3hc} g'(\tilde{\nu}) \operatorname{Im}[\mathbf{M}(f0) \cdot \mathcal{M}(0f)] \qquad (1.98)
$$

The computation makes use of the fact that isotropic averaging of the product $M_u \mathscr{M}_v$ yields $(\delta_{uv}/3)\mathbf{M} \cdot \mathscr{M}$. Equation (1.98) is known as the Rosenfeld formula. The expression $\mathrm{Im}[\mathbf{M}(\mathrm{f0}) \cdot \mathscr{M}(\mathrm{0f})]$ is referred to as the rotatory (rotational) strength of the transition $0 \rightarrow \mathrm{f}$, $\mathrm{R}^{\mathrm{iso}}(\mathrm{0f})$. It vanishes whenever the transition $0 \rightarrow \mathrm{f}$ is (i) electric-dipole forbidden, $\mathbf{M}(\mathrm{0f}) = 0$; or (ii) magnetic-dipole forbidden, $\mathscr{M}(\mathrm{0f}) = 0$; or (iii) such that the transition moments $\mathbf{M}(\mathrm{0f})$ and $\mathscr{M}(\mathrm{0f})/i$ are perpendicular. In any one of the three cases, the transition exhibits no optical activity. In molecules which belong to point symmetry groups with one or more improper rotation axes, $\mathbf{M}(\mathrm{0f})$ and $\mathscr{M}(\mathrm{0f})/i$ are always perpendicular or vanishing for any choice of f. Such molecules are called achiral and cannot exhibit optical activity in any of their transitions as long as they are in an isotropic medium and as long as there is no magnetic field in the direction of the light propagation. Even for a chiral molecule, it can be shown that the sum of the rotatory strengths of all transitions vanishes, $\sum_{\mathrm{f}} \mathrm{R}_{\mathrm{0f}}^{\mathrm{iso}} = 0$.

Dividing both sides of (1.98) by $\tilde{\nu}$ and integrating, we get

$$\begin{aligned}
\mathrm{R}^{\mathrm{iso}}(\mathrm{0f}) &= \mathrm{Im}[\mathbf{M}(\mathrm{f0}) \cdot \mathscr{M}(\mathrm{0f})] = \mathrm{Im}[\langle 0|\mathbf{M}|\mathrm{f}\rangle \cdot \langle \mathrm{f}|\mathscr{M}|0\rangle] \\
&= \frac{2303 \times 3hc}{32\pi^3 N} \int_{\mathrm{band}} \frac{\varepsilon_{\mathrm{L}}(\tilde{\nu}) - \varepsilon_{\mathrm{R}}(\tilde{\nu})}{\tilde{\nu}} \, \mathrm{d}\tilde{\nu} \\
&= 22.97 \times 10^{-40} \int_{\mathrm{band}} \frac{\varepsilon_{\mathrm{L}}(\tilde{\nu}) - \varepsilon_{\mathrm{R}}(\tilde{\nu})}{\tilde{\nu}} \, \mathrm{d}\tilde{\nu} \ \text{(cgs units: esu cm erg gauss}^{-1}\text{)} \\
&= 0.248 \int_{\mathrm{band}} \frac{\varepsilon_{\mathrm{L}}(\tilde{\nu}) - \varepsilon_{\mathrm{R}}(\tilde{\nu})}{\tilde{\nu}} \, \mathrm{d}\tilde{\nu} \quad \text{(Debye Bohr magneton)} \\
&= 7.51 \times 10^{-5} \int_{\mathrm{band}} \frac{[\theta(\tilde{\nu})]_{\mathrm{M}}}{\tilde{\nu}} \, \mathrm{d}\tilde{\nu} \quad \text{(Debye Bohr magneton)} \qquad (1.99)
\end{aligned}$$

where $\tilde{\nu}$ is in cm^{-1}, $\varepsilon(\tilde{\nu})$ is in units of $\mathrm{L\,mol^{-1}\,cm^{-1}}$, and where we used the relation 1 Debye Bohr magneton $= 0.92741 \times 10^{-38}$ cgs units. Both of these units are commonly encountered in the literature dealing with chiroptical spectroscopy. Very strongly optically active transitions have rotational strengths of the order of one Debye Bohr magneton.

The rotatory strength $\mathrm{R}^{\mathrm{iso}}(\mathrm{0f})$ characterizes the chiroptical strength of a transition and is somewhat analogous to the oscillator strength f defined in (1.24) to characterize the strength of an absorption band. The analogy is even closer to a less commonly used quantity, the dipole strength of an absorption band: $\mathrm{D(0f)} = |\mathbf{M}(\mathrm{0f})|^2$. This quantity has been mentioned briefly in connection with (1.61); it can be related to the measured absorption intensity of an isotropic sample by dividing both sides of (1.62) by $\tilde{\nu}$, using $\alpha(\tilde{\nu}) = 2303\, n'\varepsilon(\tilde{\nu})/N$, and integrating [isotropic averaging of $|\varepsilon_{\mathrm{U}}^* \cdot \mathbf{M}(\mathrm{0f})|^2$ yields $(1/3)|\mathbf{M}(\mathrm{0f})|^2$, and the magnetic dipole and electric quadrupole contributions are neglected]:

$$\begin{aligned}
\mathrm{D(0f)} = |\mathbf{M}(\mathrm{0f})|^2 &= \frac{2303 \times 3hc}{8\pi^3 N} \int_{\mathrm{band}} \frac{\varepsilon(\tilde{\nu})}{\tilde{\nu}} \, \mathrm{d}\tilde{\nu} \\
&= 9.188 \times 10^{-3} \int_{\mathrm{band}} \frac{\varepsilon(\tilde{\nu})}{\tilde{\nu}} \, \mathrm{d}\tilde{\nu} \qquad (1.100)
\end{aligned}$$

in units of Debye^2 (1 Debye = 10^{-18} esu cm). The units of $\varepsilon(\tilde{\nu})$ are L mol^{-1} cm^{-1} and $\tilde{\nu}$ is in cm^{-1}.

In studies on *partially oriented samples*, it is customary to define the rotatory strength tensor R(0f) of transition 0 → f. Its ZZ component is

$$\boldsymbol{\varepsilon}_Z \mathsf{R}(0f)\boldsymbol{\varepsilon}_Z = \tfrac{1}{2}\{\mathrm{Im}\,[\mathrm{M}_X(f0)\mathscr{M}_X(0f) + \mathrm{M}_Y(f0)\mathscr{M}_Y(0f)] + \pi\tilde{\nu}[\mathrm{M}_Y(0f)\mathrm{Q}_{XZ}(f0) - \mathrm{M}_X(0f)\mathrm{Q}_{YZ}(f0)]\} \quad (1.101)$$

so that (1.97), valid for light propagating along Z, can be written as

$$\Delta\varepsilon(\tilde{\nu}) = \frac{32\pi^3 N\tilde{\nu}}{2303hc}\, g'(\tilde{\nu})\boldsymbol{\varepsilon}_Z \mathsf{R}(0f)\boldsymbol{\varepsilon}_Z \quad (1.102)$$

and the element $[\mathsf{R}(0f)]_{ZZ}$ of the rotatory strength tensor can be obtained from a measurement on an oriented sample with light propagating along Z using

$$[\mathsf{R}(0f)]_{ZZ} = \boldsymbol{\varepsilon}_Z \mathsf{R}(0f)\boldsymbol{\varepsilon}_Z = \frac{2303hc}{32\pi^3 N}\int_{\text{band}} \frac{\varepsilon_L(\tilde{\nu}) - \varepsilon_R(\tilde{\nu})}{\tilde{\nu}}\, d\tilde{\nu} \quad (1.103)$$

For a general direction of light propagation K, the circular dichroism due to the f-th transition is given by (1.102) with Z replaced by K, and the general element of the tensor R(0f) is given by

$$[\mathsf{R}(0f)]_{UV} = \frac{1}{4}\sum_{U'V'} \mathbf{M}_{U'}(0f)[\varepsilon_{U'V'V}\mathrm{N}_{V'U}(0f) + \varepsilon_{U'V'U}\mathrm{N}_{V'V}(0f)] \quad (1.104)$$

where $\varepsilon_{TU'V'} = \boldsymbol{\varepsilon}_T \cdot \boldsymbol{\varepsilon}_{U'} \times \boldsymbol{\varepsilon}_{V'}$, so that $\varepsilon_{XYZ} = \varepsilon_{ZXY} = \varepsilon_{YZX} = 1$, $\varepsilon_{XZY} = \varepsilon_{YXZ} = \varepsilon_{ZYX} = -1$, and other $\varepsilon_{TU'V'}$'s vanish, and where

$$\mathrm{N}_{U'V'}(0f) = -i(\boldsymbol{\varepsilon}_{U'} \times \boldsymbol{\varepsilon}_{V'}) \cdot \mathscr{M}(0f) - \pi\tilde{\nu}\boldsymbol{\varepsilon}_{U'}\mathsf{Q}(0f)\boldsymbol{\varepsilon}_{V'} \quad (1.105)$$

The general element of R(0f), $[\mathsf{R}(0f)]_{KK}$, can be evaluated in principle using equation (1.103) and $\Delta\varepsilon(\tilde{\nu})$ measured with light propagating along direction K instead of Z, but as already mentioned, severe difficulties are encountered in the separation of the unwanted ellipticity θ_{LD} from the desired θ_{CD}, and such determinations are rarely performed in practice.

The assumption that the line-shape functions are the same for left-handed and right-handed circularly polarized light is good for transitions between individual well-resolved levels. When the absorption band consists of a number of overlapping vibrational components, the approximation need not hold, and the overall spectral band shape can be different in absorption and in CD spectra. This is of particular concern in the case of vibronic interactions (Chapter 2). The rotatory strengths $\mathrm{R}^{\mathrm{iso}}$ (0f) and R(0f) obtained from (1.99) and (1.102) can then still be used formally for the total integrated CD intensity of the band, but they have no simple physical meaning.

Solvent effects have been ignored so far. Corrections for them are rarely considered in practice. Considerations similar to those which led to (1.63) show that the right-hand side of (1.98) should be multiplied and the right-hand side of (1.99) should be divided by the effective polarizability of the solvent, $\alpha'(\tilde{\nu})$. In the approximation of Lorentz, (1.64), we have $\alpha'(\tilde{\nu}) = [n^2(\tilde{\nu}) + 2]/3$. In the case of (1.97), the first term in the folded bracket, Im [. . .], should be multiplied by $\alpha'(\tilde{\nu})$ and the second one by $\alpha'(\tilde{\nu})/\sqrt{n(\tilde{\nu})}$, and this makes solvent corrections in (1.101)–(1.104) similarly complicated.

In the presence of several transitions in the spectral region of interest, equation (1.98) for isotropic solutions acquires a form analogous to (1.62),

$$\Delta\varepsilon(\tilde{\nu}) = \varepsilon_L(\tilde{\nu}) - \varepsilon_R(\tilde{\nu}) = \frac{32\pi^3 N\tilde{\nu}}{2303hc}\sum_f g_f'(\tilde{\nu})\alpha'(\tilde{\nu})\,\mathrm{Im}\,[\mathbf{M}(f0)\cdot\mathcal{M}(0f)]$$

$$= \frac{32\pi^3 N\tilde{\nu}}{2303hc}\sum_f g_f'(\tilde{\nu})\alpha'(\tilde{\nu})R^{iso}(0f) \qquad (1.106)$$

and equation (1.97), applicable to oriented molecules, becomes

$$\Delta\varepsilon(\tilde{\nu}) = \frac{16\pi^3 N\tilde{\nu}}{2303hc}\sum_f g_f'(\tilde{\nu})\alpha'(\tilde{\nu})\,\{\mathrm{Im}\,[M_X(f0)\mathcal{M}_X(0f) + M_Y(f0)\mathcal{M}_Y(0f)]$$

$$+ \frac{\pi\tilde{\nu}}{\sqrt{n(\tilde{\nu})}}[M_Y(0f)Q_{XZ}(f0) - M_X(0f)Q_{YZ}(f0)]\} \qquad (1.107)$$

When the correction for the solvent effects is incorporated in the definition of $N_{UV}(0f)$ in (1.105), one obtains

$$N_{UV}(0f) = -i\alpha'(\tilde{\nu})(\varepsilon_U \times \varepsilon_V)\cdot\mathcal{M}(0f) - [\pi\tilde{\nu}/\sqrt{n(\tilde{\nu})}]\varepsilon_U Q(0f)\varepsilon_V \qquad (1.108)$$

The final expression for circular dichroism due to all overlapping transitions and measured in an arbitrary direction is

$$\Delta\varepsilon(\tilde{\nu}) = \frac{32\pi^3 N\tilde{\nu}}{2303hc}\sum_f g_f'(\tilde{\nu})\varepsilon_K R(0f)\varepsilon_K \qquad (1.109)$$

with the definitions given in (1.104) and (1.105).

Magnetic Optical Activity. Magnetic circular dichroism (MCD) is proportional to the first power of the longitudinal component of the magnetic field (Table 1.1). In order to appreciate its origin, we return to (1.96) and consider the perturbing effect of magnetic field whose component B_Z is directed along the light propagation direction Z. We have already noted that the last four terms in (1.96) vanish in the absence of a magnetic field since the initial and final wavefunctions can then be chosen real. Therefore, only the first two terms contribute to natural CD. We shall now assume that the sample is achiral or racemic so that natural CD vanishes. The MCD effect originates in the last four terms and thus has a totally distinct physical origin from that of natural CD and the two effects are additive. The first two terms in (1.96) are not affected by the magnetic field to first order (Table 1.1).

We shall adopt the customary neglect of all but the fourth term in (1.96), which contains only the matrix elements of the electric dipole operator $\hat{\mathbf{M}}$ and is generally much larger than the third, fifth, and sixth terms. As it stands, (1.96) does not allow explicitly for zero-field degeneracy of the initial or final states and must be modified. Let the degenerate components of the initial state be labeled by a subscript α, $\alpha = 1,2,\ldots,d$, and those of the final state be similarly labeled by subscript λ. Throughout the section on magnetic optical activity, we shall choose the wavefunctions $|0_\alpha\rangle$ and $|f_\lambda\rangle$ in their complex form in which $\hat{\mathcal{M}}_Z$ is diagonal within the ground and excited state manifolds. In the presence of a magnetic field, the several originally degenerate components of the ground state will have unequal energies and populations N_α. The total number of molecules in the ground state is $N = \sum_\alpha N_\alpha$. Equation (1.96) now acquires the form

$$\varepsilon_L(\tilde{\nu}) - \varepsilon_R(\tilde{\nu}) = \frac{-8\pi^3 N\tilde{\nu}}{2303hc} \sum_{\alpha\lambda} \frac{N_\alpha(B_Z)}{N} g'_{\alpha,\lambda}(\tilde{\nu},B_Z) i[\mathbf{M}(0_\alpha f_\lambda,B_Z) \times \mathbf{M}(f_\lambda 0_\alpha,B_Z)]_Z \tag{1.110}$$

Three factors in this expression depend on the magnetic field: (i) the fractional populations in the individual levels of the d-fold degenerate ground state N_α/N; (ii) the transition moment $\mathbf{M}(0_\alpha f_\lambda,B_Z)$ and its complex conjugate $\mathbf{M}(f_\lambda 0_\alpha,B_Z)$; and (iii) the transition energies whose Zeeman shifts we shall express through the line-shape functions $g'_{\alpha\lambda}(\tilde{\nu})$. The three factors give rise to the so-called C terms, B terms, and A terms, respectively.

Using degenerate perturbation theory to describe the effect of the magnetic field, we obtain for the energies of the various components of the initial and final states in the presence of the magnetic field

$$E_{0\alpha}(B_Z) = E_0 - \mathcal{M}_Z(0_\alpha 0_\alpha)B_Z \qquad E_{f\lambda}(B_Z) = E_f - \mathcal{M}_Z(f_\lambda f_\lambda)B_Z \tag{1.111}$$

where E_0 and E_f are the energies in the absence of the magnetic field. The energies of nondegenerate states remain unchanged in this approximation.

The Boltzmann population factor then becomes

$$N_\alpha(B_Z)/N = (1/d)[1 + \mathcal{M}_Z(0_\alpha 0_\alpha)B_Z/kT] \tag{1.112}$$

where k is the Boltzmann constant, T is absolute temperature, and where it is assumed that the Zeeman splitting is negligible relative to kT so that an exponential can be truncated after the linear term in its series expansion.

For the wavefunctions in a magnetic field, we obtain to first order in B_Z

$$|0,B_Z\rangle = |0\rangle - B_Z \sum_{\substack{j \\ (j\neq 0)}} \frac{\langle j|\hat{\mathcal{M}}_Z|0\rangle}{E_0 - E_j} |j\rangle$$

$$\langle f,B_Z| = \langle f| - B_Z \sum_{\substack{j \\ (j\neq f)}} \frac{\langle f|\hat{\mathcal{M}}_Z|j\rangle}{E_f - E_j} \langle j| \tag{1.113}$$

where the sums run over all electronic states except as indicated. For the transition moments, again to first order in B_Z, we obtain

$$\mathbf{M}(0_\alpha f_\lambda,B_Z) = \mathbf{M}(0_\alpha f_\lambda) + B_Z\left[\sum_{\substack{j \\ (j\neq f)}} \frac{\mathcal{M}_Z(jf_\lambda)}{E_j - E_f}\mathbf{M}(0_\alpha j) + \sum_{\substack{j \\ (j\neq 0)}} \frac{\mathcal{M}_Z(0_\alpha j)}{E_j - E_0}\mathbf{M}(jf_\lambda)\right] \tag{1.114}$$

Then, to first order in B_Z,

$$\mathbf{M}(0_\alpha f_\lambda, B_Z) \times \mathbf{M}(f_\lambda 0_\alpha, B_Z) = \mathbf{M}(0_\alpha f_\lambda) \times \mathbf{M}(f_\lambda 0_\alpha) + 2iB_Z \,\mathrm{Im}\left\{ \sum_{\substack{j \\ (j \neq f)}} \frac{\mathcal{M}_Z(f_\lambda j)}{E_j - E_f} \mathbf{M}(0_\alpha f_\lambda) \times \mathbf{M}(j0_\alpha) + \sum_{\substack{j \\ (j \neq 0)}} \frac{\mathcal{M}_Z(j0_\alpha)}{E_j - E_0} \mathbf{M}(0_\alpha f_\lambda) \times \mathbf{M}(f_\lambda j) \right\} \tag{1.115}$$

Finally, we need to express the line shape in the presence of the field, $g'_{\alpha\lambda}(\tilde{\nu}, B_Z)$, in terms of the zero-field line shape $g'(\tilde{\nu})$. We adopt the rigid-shift approximation: The Zeeman shift moves the spectral band without affecting its shape,

$$g'_{\alpha\lambda}(\tilde{\nu}, B_Z) = g'\{\tilde{\nu} + B_Z[\mathcal{M}_Z(f_\lambda f_\lambda) - \mathcal{M}_Z(0_\alpha 0_\alpha)]/hc\} \tag{1.116}$$

and the function g' is expanded in a Taylor series and truncated after the linear term, which is acceptable as long as the Zeeman shift is negligible relative to the bandwidth,

$$g'_{\alpha\lambda}(\tilde{\nu}, B_Z) = g'(\tilde{\nu}) + \frac{\partial g'(\tilde{\nu})}{\partial \tilde{\nu}} B_Z[\mathcal{M}_Z(f_\lambda f_\lambda) - \mathcal{M}_Z(0_\alpha 0_\alpha)]/hc \tag{1.117}$$

Now, we substitute (1.112), (1.115), and (1.117) into (1.110) and keep only terms linear in B_Z. Converting into our standard measure of energy, $\tilde{\nu}$, where appropriate, we obtain

$$\varepsilon_L(\nu) - \varepsilon_R(\nu) = \frac{-16\pi^3 N\tilde{\nu}}{2303 \times 3h^2c^2} B_Z \left\{ \frac{\partial g'(\tilde{\nu})}{\partial \tilde{\nu}} A(0f) + g'(\tilde{\nu})\left[B(0f) + C(0f)\frac{hc}{kT} \right] \right\} \tag{1.118}$$

where

$$A(0f) = \frac{3i}{2d} \sum_{\alpha\lambda} [\mathcal{M}_Z(f_\lambda f_\lambda) - \mathcal{M}_Z(0_\alpha 0_\alpha)][\mathbf{M}(0_\alpha f_\lambda) \times \mathbf{M}(f_\lambda 0_\alpha)]_Z \tag{1.119}$$

$$B(0f) = \frac{3}{d} \sum_{\alpha\lambda} \mathrm{Im}\left\{ \sum_{\substack{j \\ (j \neq 0)}} \frac{\mathcal{M}_Z(j0_\alpha)}{\tilde{\nu}_j} [\mathbf{M}(f_\lambda j) \times \mathbf{M}(0_\alpha f_\lambda)]_Z + \sum_{\substack{j \\ (j \neq f)}} \frac{\mathcal{M}_Z(f_\lambda j)}{\tilde{\nu}_j - \tilde{\nu}_f} \times [\mathbf{M}(j0_\alpha) \times \mathbf{M}(0_\alpha f_\lambda)]_Z \right\} \tag{1.120}$$

$$C(0f) = \frac{3i}{2d} \sum_{\alpha\lambda} \mathcal{M}_Z(0_\alpha 0_\alpha)[\mathbf{M}(0_\alpha f_\lambda) \times \mathbf{M}(f_\lambda 0_\alpha)]_Z \tag{1.121}$$

Equations (1.118)–(1.121) are the MCD analog of the CD expression (1.97). They are not directly useful to us in that (1.119)–(1.121) apply only for an assembly of identically oriented molecules, a situation of little interest in the present context. However, they serve as the starting point for the derivation of analogous expressions for the A, B, and C terms measured on isotropic or partially aligned samples. The numbers A(0f), B(0f), and C(0f) characterize the contribution of the transition $0 \rightarrow f$ to the MCD effect as measured on an assembly of identically oriented molecules and are referred to as the A, B, and C terms of the 0 → f transition in the oriented sample. In recent years, inorganic spectroscopists have adopted slightly modified definitions of these quantities, A(new) = $(2/3)\beta_e$A, B(new) = $-(2/3)\beta_e$B, and C(new) = $-(2/3)\beta_e$C, where β_e is the Bohr magneton. It is likely that the new definitions will eventually find universal acceptance, but for the moment essentially all of the published work on organic molecules appears to use the units that we adopt here. The reader should be warned that yet other definitions of the A, B, and C terms can also occasionally be found in the literature.

The line-shape function associated with the B and C terms has the usual absorption-like shape $g'(\tilde{\nu})$. That associated with the A term has the bisignate shape of a derivative, $\partial g'(\tilde{\nu})/\partial\tilde{\nu}$ (Figure 1.12). The temperature dependence of the C terms permits its experimental separation from the B term.

The C term originates in the magnetic-field-dependent differential population of the levels of the degenerate ground state [cf. (1.112)] and vanishes if the ground state is nondegenerate. The B term reflects the magnetic-field dependence of the transition moments [cf. (1.114)] and is always present if the transition is allowed. The A term reflects the shift of the transition energy due to the difference in the Zeeman shifts of the two levels connected by the transition [cf. (1.117). It vanishes if neither the ground state nor the excited state is degenerate.

In an isotropic solution, the results will still be given by (1.118), where we now define isotropic A, B, and C terms in a manner appropriate for ordinary solutions.

$$A^{iso}(0f) = \frac{i}{2d} \sum_{\substack{\alpha\lambda \\ \alpha'\lambda'}} [\mathcal{M}(f_{\lambda'}f_{\lambda})\delta_{\alpha\alpha'} - \mathcal{M}(0_\alpha 0_{\alpha'})\delta_{\lambda\lambda'}] \cdot \mathbf{M}(0_{\alpha'}f_{\lambda'}) \times \mathbf{M}(f_\lambda 0_\alpha) \qquad (1.122)$$

$$B^{iso}(0f) = \qquad (1.123)$$

$$\frac{1}{d}\sum_{\alpha\lambda} \mathrm{Im}\left\{ \sum_{\substack{j \\ (j\neq 0)}} \frac{\mathcal{M}(j0_\alpha)}{\tilde{\nu}_j} \cdot [\mathbf{M}(j0_\alpha) \times \mathbf{M}(0_\alpha f_\lambda)] + \sum_{\substack{j \\ (j\neq f)}} \frac{\mathcal{M}(f_\lambda j)}{\tilde{\nu}_j - \tilde{\nu}_f} \cdot \mathbf{M}(j0_\alpha) \times \mathbf{M}(0_\alpha f_\lambda) \right\}$$

$$C^{iso}(0f) = \frac{i}{2d} \sum_{\alpha\alpha'\lambda} \mathcal{M}(0_\alpha 0_{\alpha'}) \cdot \mathbf{M}(0_{\alpha'}f_\lambda) \times \mathbf{M}(f_\lambda 0_\alpha) \qquad (1.124)$$

In the computation, use is made of the fact that isotropic averaging of the product $\mathcal{M}_u M_v M'_w$, where u, v, and w are the three Cartesian axes in the molecular system, yields (1/6) $\mathcal{M} \cdot \mathbf{M} \times \mathbf{M}'$. If u, v, and w are not all different, the averaging yields zero.

It is customary to express the MCD results in terms of the ratios A/D, B/D, and C/D, where D is the dipole strength of the transition 0 → f [cf. (1.100)]. Allowing for the possible degeneracy of the ground state, we have

$$D(0f) = \frac{1}{d}\sum_{\alpha\lambda} |M(0_\alpha f_\lambda)|^2 = 9.188 \times 10^{-3} \int_{band} \frac{\varepsilon(\tilde{\nu})}{\tilde{\nu}}\, d\tilde{\nu} \qquad (1.125)$$

in units of Debye². The units of $\varepsilon(\tilde{\nu})$ are L mol^{-1} cm^{-1} and $\tilde{\nu}$ is in cm^{-1}.

The results (1.122)–(1.124) along with formula (1.118) are the MCD analog of the Rosenfeld formula for natural CD, (1.98). In order to derive the magnitude of the isotropic A, B, and C terms from the measured circular dichroism, the method of moments is normally used. One takes advantage of the fact that the line-shape function $g'(\tilde{\nu})$ is normalized and falls effectively to zero at both ends of the integration range, so that

$$\int_{\text{band}} \frac{\partial g'(\tilde{\nu})}{\partial \tilde{\nu}}\, d\tilde{\nu} = [g'(\tilde{\nu})]_{\text{band}} = 0 \tag{1.126}$$

$$\int_{\text{band}} \frac{\partial g'(\tilde{\nu})}{\partial \tilde{\nu}}\, \tilde{\nu}\, d\tilde{\nu} = [g(\tilde{\nu})\tilde{\nu}\, d\tilde{\nu}]_{\text{band}} - \int_{\text{band}} g(\tilde{\nu})\, d\tilde{\nu} = -1 \tag{1.127}$$

The center of the band $\tilde{\nu}_0$ is defined by

$$\tilde{\nu}_0 = \int_{\text{band}} g'(\tilde{\nu})\tilde{\nu}\, d\tilde{\nu} \tag{1.128}$$

Dividing both sides of (1.118) by $\tilde{\nu}$ and integrating, we obtain

$$B(0f) + C(0f)\frac{hc}{kT} = \frac{-2303 \times 3h^2c^2}{16\pi^3 N B_Z} \int_{\text{band}} \frac{\varepsilon_L(\tilde{\nu}) - \varepsilon_R(\tilde{\nu})}{\tilde{\nu}}\, d\tilde{\nu} \tag{1.129}$$

Using (1.93), substituting the values of the universal constants, converting to the units of Debye^2 and Bohr magneton, and continuing our standard practice of expressing all energy-related quantities in cm^{-1}, we obtain for unit magnetic field strength (1 gauss)

$$B(0f) + C(0f)\frac{hc}{kT} = -33.53^{-1} \int_{\text{band}} d\tilde{\nu}\, [\theta(\tilde{\nu})]_M/\tilde{\nu}$$

$$= -2.98 \times 10^{-2} \int_{\text{band}} d\tilde{\nu}\, [\theta(\tilde{\nu})]_M/\tilde{\nu} \tag{1.130}$$

The numerical values used are 1 Bohr magneton = 9.2741×10^{-21} erg gauss^{-1} [so that (1/hc) Bohr magneton = 4.6686×10^{-5} cm^{-1} gauss^{-1}] and 1 Debye = 10^{-18} esu cm.

Equation (1.130) permits the evaluation of the B and C terms from the observed ellipticity if the measurement is performed at several temperatures. For molecules with a nondegenerate ground state, C = 0, and a measurement at a single temperature suffices for the determination of B. Note that a positive MCD peak, $\varepsilon_L(\tilde{\nu}) > \varepsilon_R(\tilde{\nu})$, and $[\theta]_M > 0$, corresponds to negative B and C terms in the convention adopted here [(1.122)–(1.124)].

A direct integration of (1.118) followed by substitution for B(0f) + C(0f)/kT from (1.129) yields

$$A(0f) = \frac{2303 \times 3h^2c^2}{16\pi^3 N B_Z} \int_{\text{band}} \frac{\varepsilon_L(\tilde{\nu}) - \varepsilon_R(\tilde{\nu})}{\tilde{\nu}} (\tilde{\nu} - \tilde{\nu}_0)\, d\tilde{\nu}$$

$$= 33.53^{-1} \int_{\text{band}} d\tilde{\nu}\, [\theta(\tilde{\nu})]_M(\tilde{\nu} - \tilde{\nu}_0)/\tilde{\nu}$$

$$= 2.98 \times 10^{-2} \int_{\text{band}} d\tilde{\nu}\, [\theta(\tilde{\nu})]_M(\tilde{\nu} - \tilde{\nu}_0)/\tilde{\nu} \tag{1.131}$$

in units of Debye^2 Bohr magneton, with $[\theta]_M$ corresponding to a magnetic field of 1 gauss, and with $\tilde{\nu}$ in cm^{-1}, as in (1.130). Note that a positive A term results if the MCD plot of $[\theta]_M$ first dips to negative values and then grows to positive values as the wavenumber $\tilde{\nu}$ increases and passes through the band center $\tilde{\nu}_0$.

Expressions (1.129)–(1.131) for the A, B, and C terms are the MCD analog of the relation (1.99) for the rotatory strength $R^{iso}(0f)$ in natural CD. Like the latter, they can only be used if the band $0 \rightarrow f$ is well separated from other electronic bands. If this is not so, assumptions must be made about the nature of the overlap. It is usual to assume Gaussian band shapes for this purpose. In the following, we assume that a separation has been accomplished.

When equations (1.129)–(1.131) are applied to MCD curves measured on an isotropic sample, they produce the values $A^{iso}(0f)$, $B^{iso}(0f)$, and $C^{iso}(0f)$, and this is their usual application. In principle, they could be equally well used with MCD curves measured on molecules which are all equally oriented, and they would then produce the A(0f), B(0f), and C(0f) values defined in (1.119)–(1.121).

In the general case of MCD measurements on *partially oriented samples*, the use of (1.129)–(1.131) produces A, B, and C terms whose values depend on the orientation distribution. In close analogy to the rotatory strength tensor $\mathsf{R}(0f)$ discussed for natural optical activity [(1.101)], we define the magnetic rotatory strength tensors $\mathsf{A}(0f)$, $\mathsf{B}(0f)$, and $\mathsf{C}(0f)$ of transition $0 \rightarrow f$. The tensors are defined so that for a sample composed of equally oriented molecules their ZZ components are equal to one-third the oriented A, B, and C terms given in expressions (1.119)–(1.121), $[\mathsf{A}(0f)]_{ZZ} = \frac{1}{3}A(0f)$, $[\mathsf{B}(0f)]_{ZZ} = \frac{1}{3}B(0f)$, $[\mathsf{C}(0f)]_{ZZ} = \frac{1}{3}C(0f)$. Then, (1.118), valid for light propagating along Z, can be written as

$$\Delta\varepsilon(\tilde{\nu}) = \frac{-16\pi^3 N\tilde{\nu}}{2303h^2c^2} B_Z \left\{ \frac{\partial g'(\tilde{\nu})}{\partial\tilde{\nu}} \boldsymbol{\varepsilon}_Z \mathsf{A}(0f)\boldsymbol{\varepsilon}_Z + g'(\tilde{\nu})\boldsymbol{\varepsilon}_Z \left[\mathsf{B}(0f) + \mathsf{C}(0f)\frac{hc}{kT} \right] \boldsymbol{\varepsilon}_Z \right\} \tag{1.132}$$

which is analogous to (1.102) for natural CD.

The ZZ components are obtained from a measurement on an oriented sample with light propagating along Z using

$$[\mathsf{A}(0f)]_{ZZ} = \boldsymbol{\varepsilon}_Z \mathsf{A}(0f)\boldsymbol{\varepsilon}_Z = \frac{2303h^2c^2}{16\pi^3 NB_Z} \int_{band} \frac{\varepsilon_L(\tilde{\nu}) - \varepsilon_R(\tilde{\nu})}{\tilde{\nu}} (\tilde{\nu} - \tilde{\nu}_0)\, d\tilde{\nu}$$

$$= 0.994 \times 10^{-2} \int_{band} d\tilde{\nu}\, [\theta(\tilde{\nu})]_M (\tilde{\nu} - \tilde{\nu}_0)/\tilde{\nu} \tag{1.133}$$

$$[\mathsf{B}(0f)]_{ZZ} + \frac{hc}{kT}[\mathsf{C}(0f)]_{ZZ} = \frac{-2303h^2c^2}{16\pi^3 NB_Z} \int_{band} \frac{\varepsilon_L(\tilde{\nu}) - \varepsilon_R(\tilde{\nu})}{\tilde{\nu}}\, d\tilde{\nu}$$

$$= -0.994 \times 10^{-2} \int_{band} d\tilde{\nu}\, [\theta(\tilde{\nu})]_M/\tilde{\nu} \tag{1.134}$$

in units of Debye^2, Bohr magneton, and cm^{-1}. These expressions are analogous to (1.103) for natural CD.

In the case of a general direction of light propagation K, one can use equation (1.132), in which K has been substituted for Z. The general elements of the tensors A(0f), B(0f), and C(0f) are given by

$$[\mathsf{A}(0\mathrm{f})]_{\mathrm{UV}} = \frac{i}{2\mathrm{d}} \sum_{\alpha\lambda} [\mathscr{M}_{\mathrm{U}}(\mathrm{f}_\lambda \mathrm{f}_\lambda) - \mathscr{M}_{\mathrm{U}}(0_\alpha 0_\alpha)][\mathbf{M}(0_\alpha \mathrm{f}_\lambda) \times \mathbf{M}(\mathrm{f}_\lambda 0_\alpha)]_{\mathrm{V}} \tag{1.135}$$

$$[\mathrm{B0f})]_{\mathrm{UV}} = \frac{1}{\mathrm{d}} \sum_{\alpha\lambda} \mathrm{Im}\left\{ \sum_{\substack{\mathrm{j} \\ (\mathrm{j} \neq 0)}} \frac{\mathscr{M}_{\mathrm{U}}(\mathrm{j}0_\alpha)}{\tilde{\nu}_{\mathrm{j}}} [\mathbf{M}(\mathrm{f}_\lambda \mathrm{j}) \times \mathbf{M}(0_\alpha \mathrm{f}_\lambda)]_{\mathrm{V}} + \sum_{\substack{\mathrm{j} \\ (\mathrm{j} \neq \mathrm{f})}} \frac{\mathscr{M}_{\mathrm{U}}(\mathrm{f}_\lambda \mathrm{j})}{\tilde{\nu}_{\mathrm{j}} - \tilde{\nu}_{\mathrm{f}}} [\mathbf{M}(\mathrm{j}0_\alpha) \times \mathbf{M}(0_\alpha \mathrm{f}_\lambda)]_{\mathrm{V}} \right\} \tag{1.136}$$

$$[\mathsf{C}(0\mathrm{f})]_{\mathrm{UV}} = \frac{i}{2\mathrm{d}} \sum_{\alpha\lambda} \mathscr{M}_{\mathrm{U}}(0_\alpha 0_\alpha)[\mathbf{M}(0_\alpha \mathrm{f}_\lambda) \times \mathbf{M}(\mathrm{f}_\lambda 0_\alpha)]_{\mathrm{V}} \tag{1.137}$$

These expressions are the MCD analog of the natural CD expressions (1.104) and (1.105).

The values of these general elements can be obtained in principle from $[\theta(\tilde{\nu})]_{\mathrm{M}}$ measured with light propagating along direction K, using equations (1.132) and (1.134). As already mentioned in the discussion of natural CD, severe difficulties are encountered in the separation of the unwanted ellipticity θ_{LD} from the desired θ_{CD}, and such determinations are rarely if ever performed in practice.

Several complications remain to be mentioned. First, as in the case of natural CD, vibronic perturbations can complicate the analysis of MCD data, in particular the B terms. These can still be formally determined using (1.129) and (1.130), but their simple physical significance is lost. When the vibrational fine structure in MCD spectra is of interest, the rigid shift approximation adopted here generally needs to be abandoned and the perturbation expansion needs to be performed in vibronic rather than electronic states. The results are then much more complicated.

Second, we have ignored solvent effects so far. Considerations similar to those which led to (1.63) show that the observed dichroism will increase by the same factor $\alpha'(\tilde{\nu})^2/\mathrm{n}(\tilde{\nu})$ as the observed absorption. Here $\alpha'(\tilde{\nu})$ is the effective polarizability of the solvent and $\mathrm{n}(\tilde{\nu})$ its refractive index. Thus, the right-hand sides of (1.110), (1.118), (1.125), and (1.132) have to be multiplied by $\alpha'(\tilde{\nu})^2/\mathrm{n}(\tilde{\nu})$ and the right-hand sides of (1.129), (1.130), (1.131), (1.133), and (1.134) have to be divided by this factor. Since the MCD effect is increased by the same factor as the absorption coefficient, the two corrections cancel when the ratios A/D, B/D, and C/D are calculated. This is the usual practice in MCD spectroscopy.

Third, in the presence of several transitions in the spectral region of interest, a summation over f has to be incorporated in the formulas for $\Delta\varepsilon(\tilde{\nu})$ such as (1.132). The final expression for magnetic circular dichroism due to all overlapping transitions and measured in an arbitrary direction then is

$$\Delta\varepsilon(\tilde{\nu}) = \frac{-16\pi^3 N \tilde{\nu}}{2303\mathrm{h}^2 c^2} \frac{\alpha'(\tilde{\nu})^2}{\mathrm{n}(\tilde{\nu})} \times \mathrm{B_Z} \sum_{\mathrm{f}} \left\{ \frac{\partial \mathrm{g}_{\mathrm{f}}'(\tilde{\nu})}{\partial \tilde{\nu}} \boldsymbol{\varepsilon}_{\mathrm{K}} \mathsf{A}(0\mathrm{f}) \boldsymbol{\varepsilon}_{\mathrm{K}} + \mathrm{g}_{\mathrm{f}}'(\tilde{\nu}) \boldsymbol{\varepsilon}_{\mathrm{K}} \left[\mathsf{B}(0\mathrm{f}) + \frac{\mathrm{h}c}{k\mathrm{T}} \mathsf{C}(0\mathrm{f}) \right] \boldsymbol{\varepsilon}_{\mathrm{K}} \right\} \tag{1.138}$$

with the definitions given in (1.135)–(1.137).

Optical Activity in Light Scattering. In addition to the phenomena of natural and magnetic circular dichroism and optical rotatory dispersion, natural and induced molecular optical activity can also be observed in light scattering, both of the Rayleigh and Raman types. Two kinds of measurement are possible. First, natural or linearly polarized light scattered by chiral samples is elliptically polarized. The degree of ellipticity is normally very small and hard to measure. Second, equivalent information is contained in the somewhat easier measurement of circular intensity difference Δ, i.e., the difference in the scattered intensity for right (I_R) and left (I_L) circularly polarized incident light, $\Delta = (I_R - I_L)/(I_R + I_L)$. In magnetic Rayleigh and Raman optical activity, the magnetic field must be parallel to the scattered light beam to generate an ellipticity, and parallel to the incident light beam to generate a circular intensity difference. Electric Rayleigh and Raman optical activity is induced by an electric field perpendicular to both the incident and scattered light beams.

Although investigations of this type nicely complement the investigations of vibrational optical activity by infrared circular dichroism, and may ultimately acquire considerable importance, we have elected not to include the expressions for such measurements on partially oriented samples in the present treatment. Similarly, we are omitting from further consideration other higher-order phenomena listed in Table 1.1, such as magnetic differential absorption, the quadratic Faraday effect, and magnetoelectric circular dichroism.

1.3.4 Partially Oriented Samples

The consideration of spectroscopic intensities given so far in Section 1.3 was based on the assumption that the observed molecule is in a fixed spatial relation to the propagation direction and polarization vector of the light. Actually, we were considering a large number of noninteracting molecules, all oriented alike. This model is referred to as the "oriented gas model."

The expressions for polarized intensity had the general forms $\varepsilon_U O(0f)\varepsilon_U$ for processes of rank 2 and those reducible to rank 2 and $\varepsilon_U\varepsilon_V{}^{(4)}O(0f)\varepsilon_U\varepsilon_V$ for processes of rank 4, i.e., they were the general elements of second- or fourth-rank tensors $O(0f)$ and ${}^{(4)}O(0f)$. The nature and possible complex conjugation of the unit vectors ε_U and ε_V were given by the nature of the optical measurement in question and by propagation direction and polarization properties of the light. The nature of the tensors $O(0f)$ and ${}^{(4)}O(0f)$ was given by the observable under investigation.

On several occasions, we have indicated how the final results change when the observed molecules are in a randomly oriented assembly, and we have stated the results of isotropic averaging without proof. This is how (1.61) resulted from (1.60) (one-photon absorption and emission), (1.99) from (1.97) (natural CD), and (1.122)–(1.124) from (1.119)–(1.121)(MCD).

Since the subject of this book is the spectroscopy of partially aligned solutes, it is clear that a more complicated averaging over a nonrandom orientation distribution is needed in order to arrive at useful results. The above-listed results for isotropic solutions will then merely correspond to a limiting case of random orientation distribution.

At this point, we shall only sketch the nature of the averaging process and give the general form of the results. This will be needed as the starting point for Chapters 5–7, in which we consider the ways in which the observed intensities can be analyzed to provide information on molecular optical properties and molecular orientation. The quantities required for the description of the orientation distribution are discussed in detail in Chapter 4. The mathematics of the averaging procedure is illustrated on a detailed example worked out in Appendix II.

Consider a set of laboratory axes X, Y, Z firmly connected with the sample. We choose the Z axis to coincide with the unique axis of a uniaxial sample. As outlined above, measurements of polarized intensities provide information on the elements of the tensor $\mathsf{O}(0f)$ or ${}^{(4)}\mathsf{O}(0f)$ in the laboratory system of axes. A full knowledge of the nine elements of $\mathsf{O}(0f)$ or the 81 elements of ${}^{(4)}\mathsf{O}(0f)$ in the laboratory system of coordinates will be sufficient to obtain $\boldsymbol{\varepsilon}_U \mathsf{O}(0f) \boldsymbol{\varepsilon}_U$ or $\boldsymbol{\varepsilon}_U \boldsymbol{\varepsilon}_V {}^{(4)}\mathsf{O}(0f) \boldsymbol{\varepsilon}_U \boldsymbol{\varepsilon}_V$ for any choice of the unit vectors $\boldsymbol{\varepsilon}_U$ and $\boldsymbol{\varepsilon}_V$ and thus to describe the outcome of polarized intensity measurement for light of any direction and any linear polarization.

Now choose a set of axes x, y, z associated with the molecular framework. Assume that the "oriented gas" sample is such that the axis x of each molecule is parallel to the laboratory axis X, its y axis is parallel to Y, and its z axis is parallel to Z. Then the optical tensor of the sample $\mathsf{O}(0f)$ or ${}^{(4)}\mathsf{O}(0f)$ will have the same elements in the laboratory and in the molecular set of axes, $[\mathsf{O}(0f)]_{ZZ} = [\mathsf{O}(0f)]_{zz}$, etc. If the molecular set of axes is now rotated relative to the laboratory axes, this equality will be lost, and each element of the optical tensor in the laboratory coordinates will be a linear combination of the elements in the molecular axes:

$$\begin{aligned} [\mathsf{O}(0f)]_{UV} &= \sum_{uv} [\mathsf{D}]_{uv}^{UV} [\mathsf{O}(0f)]_{uv} \\ [{}^{(4)}\mathsf{O}(0f)]_{STUV} &= \sum_{stuv} [{}^{(4)}\mathsf{P}]_{stuv}^{STUV} [{}^{(4)}\mathsf{O}(0f)]_{stuv} \end{aligned} \qquad (1.139)$$

The transformation coefficients $[\mathsf{D}]_{uv}^{UV}$ and $[{}^{(4)}\mathsf{P}]_{stuv}^{STUV}$ will be given by products of direction cosines (Chapter 4 and Appendix II). If the orientation of the molecular axes x, y, z in the "oriented gas" sample relative to the laboratory axes X, Y, Z is known, the transformation coefficients will be known and the knowledge of all the elements of the optical tensor in the laboratory frame will be sufficient for the determination of all its elements in the molecular frame.

In a partially aligned sample, the contribution from each molecule will still be given by (1.139), but the transformation coefficients $[\mathsf{D}]_{uv}^{UV}$ and $[{}^{(4)}\mathsf{P}]_{stuv}^{STUV}$ will now be different for each molecule since each is oriented differently. Relation (1.139) will still hold, but the values of the transformation coefficients will now be determined by averaging over all molecules under observation. We shall denote them $\langle [\mathsf{D}]_{uv}^{UV} \rangle$ and $\langle [{}^{(4)}\mathsf{P}]_{stuv}^{STUV} \rangle$.

As described above in some detail, only measurements using light whose electric vector either lies along Z or is perpendicular to Z are easily performed on uniaxial samples in practice. This means that only certain components of the optical tensors $\mathsf{O}(0f)$ or $^{(4)}\mathsf{O}(0f)$ of the sample in the laboratory system of axes are accessible. Further, certain elements are related to others by symmetry. For instance, for uniaxial samples, the only easily accessible elements are $[\mathsf{O}(0f)]_{ZZ}$ and $[\mathsf{O}(0f)]_{XX} = [\mathsf{O}(0f)]_{YY}$ for processes of rank 2. We shall adopt an abbreviated notation for those transformation coefficients which are actually needed in the case of uniaxial samples:

$$\begin{aligned}
[\mathsf{D}_Z]_{uv} &= \langle [\mathsf{D}]_{uv}^{ZZ} \rangle \\
[\mathsf{D}_Y]_{uv} &= \langle [\mathsf{D}]_{uv}^{YY} \rangle \\
[^{(4)}\mathsf{P}_{ZZ}]_{stuv} &= \langle [^{(4)}\mathsf{P}]_{stuv}^{ZZZZ} \rangle \\
[^{(4)}\mathsf{P}_{YY}]_{stuv} &= [^{(4)}\mathsf{P}_{XX}]_{stuv} = \langle [^{(4)}\mathsf{P}]_{stuv}^{YYYY} \rangle \\
[^{(4)}\mathsf{P}_{ZY}]_{stuv} &= [^{(4)}\mathsf{P}_{ZX}]_{stuv} = \langle [^{(4)}\mathsf{P}]_{stuv}^{ZYZY} \rangle \\
[^{(4)}\mathsf{P}_{YZ}]_{stuv} &= [^{(4)}\mathsf{P}_{XZ}]_{stuv} = \langle [^{(4)}\mathsf{P}]_{stuv}^{YZYZ} \rangle \\
[^{(4)}\mathsf{P}_{XY}]_{stuv} &= [^{(4)}\mathsf{P}_{YX}]_{stuv} = \langle [^{(4)}\mathsf{P}]_{stuv}^{XYXY} \rangle
\end{aligned} \qquad (1.140)$$

In Chapter 7, a couple of elements needed for work with circularly polarized light will be added to this list.

The accessible elements of the optical tensor in laboratory coordinates will be abbreviated similarly:

$$[\mathsf{O}(0f)]_U = [\mathsf{O}(0f)]_{UU} \qquad [^{(4)}\mathsf{O}(0f)]_{UV} = [^{(4)}\mathsf{O}(0f)]_{UVUV} \qquad (1.141)$$

In this notation, the transformation between the measurable elements of the optical tensor in the laboratory axes system and its elements in the molecular axes system will be

$$[\mathsf{O}(0f)]_U = \sum_{uv} [\mathsf{D}_U]_{uv}[\mathsf{O}(0f)]_{uv} \qquad (1.142)$$

$$[^{(4)}\mathsf{O}(0f)]_{UV} = \sum [^{(4)}\mathsf{P}_{UV}]_{stuv}[^{(4)}\mathsf{O}(0f)]_{stuv} \qquad (1.143)$$

For uniaxial samples, the elements of the tensors D and $^{(4)}\mathsf{P}$ can be expressed once and for all in terms of a limited number of quantities characterizing the orientation distribution function (Chapters 5 and 7, Appendix II). There are five such independent quantities in the case of D and nine additional ones in the case of $^{(4)}\mathsf{P}$. They can be chosen in a variety of ways, three of which are described in Chapter 4. The simplest expressions result when the elements of D and $^{(4)}\mathsf{P}$ are written in terms of the orientation factors $K_{uv} = \langle \cos u \cos v \rangle$ and $L_{stuv} = \langle \cos s \cos t \cos u \cos v \rangle$, where the angled brackets indicate averaging over the molecular assembly and cos u stands for the direction cosine of the laboratory axis Z with respect to the u-th molecular axis. These orientation factors are used throughout this book.

Even if the orientation factors can be somehow determined so that the elements of D and $^{(4)}\mathsf{P}$ become known, the measurement of the accessible polarized intensities, which provides knowledge of the elements of the optical tensor

O(0f) or $^{(4)}$O(0f) in the laboratory system of axes, will not be sufficient for the determination of its elements in the molecular axes in the general case. In Chapters 5–7, we shall consider the situation in detail. We shall find that under many circumstances much useful information can still result.

1.4 Comments and References

The reader interested in the prewar publications on the linear dichroism and polarized emission of dyes in stretched polymers is referred to J. Preston, *J. Soc. Dyers Colourists* **47**, 312 (1931) and A. Jablonski, *Nature* **133**, 140 (1934); *Acta Phys. Polonica* **3**, 421 (1934). Land's work on sheet polarizers is summarized in E. H. Land and C. D. West, *Alexander's Colloid Chemistry*, Vol. VI, Reinhold, New York, 1946, p. 160.

Early postwar developments were due primarily to H. Kuhn [*J. Chem. Phys.* **17**, 1198 (1949)] and his collaborators: R. Eckert and H. Kuhn, *Z. Elektrochem.* **64**, 356 (1960) (UV); H. Jakobi and H. Kuhn, *Z. Elektrochem.* **66**, 46 (1962) (UV); and H. Jakobi, A. Novak, and H. Kuhn, *Z. Elektrochem.* **66**, 863 (1962) (IR).

Reviews dealing with optical spectroscopy of partially aligned samples can be found in F. Dörr, in *Creation and Detection of the Excited State*, Vol. 1A, A. A. Lamola, Ed., Marcel Dekker, New York, 1971, Chap. 2; A. C. Albrecht, *Progr. React. Kinet.* **5**, 301 (1970); G. Meier, E. Sackmann, and J. G. Grabmeier, *Applications of Liquid Crystals*, Springer-Verlag, New York, 1975; B. Nordén, Ed., *Proceedings of the Nobel Workshop on Molecular Optical Dichroism and Chemical Applications of Polarized Spectroscopy*, University of Lund, Sweden, 1977; B. Nordén, *Appl. Spectrosc. Rev.* **14**, 157 (1978); L. B.-Å. Johansson and G. Lindblom, *Quart. Rev. Biophys.* **13**, 63 (1980); E. W. Thulstrup, *Aspects of the Linear and Magnetic Circular Dichroism of Planar Organic Molecules*, Springer-Verlag, New York, 1980. For references to over 100 papers dealing with linear dichroism of dyed stretched polymers, see E. W. Thulstrup and J. Michl, *J. Phys. Chem.* **84**, 82 (1980).

Polarized Light. An excellent historical survey is provided by W. Swindel, *Polarized Light* (*Benchmark Papers in Optics*, Vol. 1), Dowden, Hutchinson and Ross, Inc., distributed by Halsted Press, 1975. A good elementary introduction is found in D. Clarke and J. F. Grainger, *Polarized Light and Optical Measurement*, Pergamon Press, Oxford, 1971. Advanced treatments of the optics of polarized light are provided by P. S. Theocaris and E. E. Gdoutos, *Matrix Theory of Photoelasticity*, Springer-Verlag, Berlin, 1979; R. M. A. Azzam and N. M. Bashara, *Ellipsometry and Polarized Light*, North-Holland Publishing Co., Amsterdam, 1977; and A. Gerrard and J. M. Burch, *Introduction to Matrix Methods in Optics*, Wiley, New York, 1975.

Group Theory. Introductory references are suggested at the end of Appendix I.

Absorption, Emission, and Scattering of Light. There are numerous excellent books on molecular spectroscopy. Only a few can be mentioned here: See J. I. Steinfeld, *Molecules and Radiation*, The MIT Press, Cambridge, MA., 2nd Ed.,

1985; C. J. H. Schutte, *The Theory of Molecular Spectroscopy*, Vol. 1, North-Holland Publishing Co., Amsterdam, 1976; J. Avery, *The Quantum Theory of Atoms, Molecules and Photons*, McGraw-Hill, London, 1972; R. E. Moss, *Advanced Molecular Quantum Mechanics*, Chapman and Hall, London, 1973; and R. Loudon, *The Quantum Theory of Light*, Clarendon Press, Oxford, 1973, for general theory; J. D. Macomber, *The Dynamics of Spectroscopic Transitions*, Wiley, New York, 1976, for the gyroscopic model; E. B. Wilson, Jr., J. C. Decius, and P. C. Cross, *Molecular Vibrations*, McGraw-Hill, New York, 1955; and D. A. Long, *Raman Spectroscopy*, McGraw-Hill, New York, 1977, for vibrational spectroscopy. The classical volumes on the subject are G. Herzberg, *Molecular Spectra and Molecular Structure*, Vols. 1–3, Van Nostrand, Princeton, NJ, 1966. For an introduction to non-linear spectroscopy, including two-photon absorption, see M. D. Levenson, *Introduction to Non-linear Laser Spectroscopy*, Academic Press, New York, 1982; for a more advanced treatment of various aspects of spectroscopy with coherent light, see J. I. Steinfeld, Ed., *Laser and Coherence Spectroscopy*, Plenum, New York, 1978; and for a discussion of the frontiers in the spectroscopy of molecules in solid media, see V. M. Agranovich and R. M. Hochstrasser, Eds., *Spectroscopy and Excitation Dynamics of Condensed Molecular Systems*, North-Holland, Amsterdam, 1983. In the development of the semiclassical description of the absorption and emission process we have followed the book by Piepho and Schatz quoted below. A thorough discussion of the nature of the interaction operator involved in optical spectroscopy, with particular attention to the relative merits of the dipole length and dipole velocity formulations, is given in D. Lee and A. C. Albrecht, *J. Chem. Phys.* **78**, 3382, 5373 (1983). The relation of scattering line shapes to molecular rotations is discussed in B. J. Berne and R. Pecora, *Dynamic Light Scattering*, Wiley, New York, 1976. Most aspects of molecular scattering are covered in the book by Barron mentioned below.

Chiroptical Spectroscopy. A rigorous and highly informative discussion of the conditions under which the linear and circular effects can be separated is given in J. Schellman and H. P. Jensen, Chem. Rev., in press (1987). Theoretical aspects of optical activity are covered in D. J. Caldwell and H. Eyring, *The Theory of Optical Activity*, Wiley-Interscience, New York, 1971. An excellent and thorough theoretical treatment of natural and magnetic optical activity is given by L. D. Barron, *Molecular Light Scattering and Optical Activity*, Cambridge University Press, Cambridge, 1982. This book is a must for any serious student of the theoretical aspects of these phenomena, but it does not go into specific applications at any great length. More applied aspects of natural optical activity are discussed in E. Charney, *The Molecular Basis of Optical Activity*, Wiley, New York, 1979, and S. F. Mason, *Molecular Optical Activity and the Chiral Discriminations*, Cambridge University Press, Cambridge, 1982. Theoretical aspects of magnetic optical activity as well as some applications in inorganic chemistry are discussed in another excellent monograph, S. B. Piepho and P. N. Schatz, *Group Theory in Spectroscopy*, Wiley, New York, 1983. References to the recent work extending CD and MCD spectroscopy to the IR region can be found in the book by Barron quoted above; see also S. F. Mason, in *Advances* *in IR and Raman Spectroscopy*, R. J. H. Clark and R. E. Hester, Eds., Heyden, London, 1981, Vol. 8, Chap. 5. A more accurate treatment of electronic MCD which avoids the Condon approximation is given in M. Z. Zgierski, J. Chem. Phys. **83**, 2170 (1985).

Static Field Effects. Electric dichroism and electrochromism are discussed in K. Yamaoka and E. Charney, *J. Am. Chem. Soc.* **94**, 8963 (1972); E. Fredericq and C. Houssier, *Electric Dichroism and Electric Birefringence*, Clarendon Press, Oxford, England, 1973; C. T. O'Konski, Ed., *Molecular Electro-Optics*, Marcel Dekker, New York, 1976, and S. Krause, Ed., *Molecular Electro-Optics*, Plenum, New York, 1981. Because of this extensive prior coverage, we say very little about this subject presently.

References to magnetic circular dichroism were given in the preceding paragraph. The theory of magnetic linear dichroism of paramagnetic molecules is treated by J. P. Riehl and F. S. Richardson, *J. Chem. Phys.* **68**, 4266 (1978). For a discussion of the theory of magnetic differential absorption (MDA), see G. Wagnière, *Z. Naturforsch* **39a**, 254 (1984); L. D. Barron and J. Vrbancich, *Mol. Phys.* **51**, 715 (1984). Magnetoelectric effects are discussed in S. Kielich, in *Molecular Electro-Optics*, C. T. O'Konski, Ed., Marcel Dekker, New York, 1976, Chapter 13; the quadratic Faraday effect in P. W. Atkins and M. H. Miller, *Mol. Phys.* **15**, 491, (1968); references to these and other higher-order effects are also found in the book by Barron referred to above.

A general analysis of symmetry restrictions for the appearance of natural and induced circular dichroism in oriented samples for both longitudinal and transverse fields is given in C. D. Churcher and G. E. Stedman, *J. Phys. C: Solid State Phys.* **15**, 5507 (1982); a very complete extension to a variety of natural and field-induced one-photon and multiphoton interactions is found in G. E. Stedman, Adv. Phys. **34**, 513 (1985).

2

Transition Moments

Transition moments are the key quantities of interest in this book. Although their role in optical spectroscopy of molecules has been briefly touched upon in Chapter 1, and although they have been formally defined as the matrix elements of transition operators between the initial and final states, at this point they are still likely to appear somewhat nebulous. In the present section, we first discuss in more detail the role which they play in determining the observed intensities and polarizations. Subsequently, we examine the ways in which transition moments are related to the geometric and electronic structure of molecules. We then take a look at the computational procedures which can be used to obtain them from molecular wavefunctions, using the Born-Oppenheimer approximation throughout.

Finally, we illustrate the abstract concepts on the example of the polarized spectra of partially aligned pyrene molecules contained in stretched polyethylene.

The crucial role played by transition moments in determining the directional properties of optical transitions is a reflection of the fact that these quantities are vectors or tensors and thus are intrinsically anisotropic. They correspond to transition amplitudes, are not observable as such, and can have complex values. Indeed, until the phases of both the initial and final wavefunction are fixed, they are only defined up to a multiplication by a complex unity. This leads to no ambiguities in the observable transition probabilities between stationary states, which contain products of transition moments in such a way that they are real and independent of the phase choices made. Since -1 is only a special case of complex unity, the sense of a transition vector moment vector is not uniquely defined although its direction is, and this is often indicated in the literature by the use of double-headed arrows.

2.1 Vibronic Transition Moments and Polarized Intensity Expressions

In Chapter 1, we referred repeatedly to the concept of normal coordinates. It is now appropriate to describe it in a little more detail even though the reader must be referred to specialized texts (Section 2.4) for a derivation of their properties.

The positions of the nuclei in a molecule are usually described in terms of three sets of coordinates, describing its translation, rotation, and internal motion (vibrations and internal rotations):

(i) The first set of coordinates are the three Cartesian coordinates of the molecular center of mass in a laboratory-fixed framework.

(ii) The second set are the three Euler angles describing the orientation of the principal axes of inertia, evaluated at the equilibrium geometry of the electronic state in question, relative to the laboratory system of axes. The equilibrium geometry can be described by the position vectors $\mathbf{R}_k^0$ of the N atoms in a molecular axes system (k = 1,2, . . . ,N). Only 3N − 6 of them are independent (if the equilibrium geometry is linear, 3N − 5), since their values must be such as to satisfy the conditions that the center of mass lies at the origin and the principal axes of inertia lie along the molecular axes.

(iii) The third set of coordinates describes the displacements $\Delta\mathbf{R}_k$ of the position vectors of the nuclei from their equilibrium positions $\mathbf{R}_k^0$. They are most conveniently described in terms of the 3N − 6 normal coordinates of the j-th electronic state: $\mathbf{Q}^{(j)} \equiv [Q_1^{(j)},Q_2^{(j)},Q_3^{(j)}, \ldots ,Q_{3N-6}^{(j)}]$. These have the property that at small displacements (in the harmonic approximation) they provide a separation of the overall molecular vibrational motion into the motion of a set of 3N − 6 uncoupled harmonic oscillators, each of which is described by motion along only one normal coordinate. At equilibrium in the j-th electronic state, all normal coordinates have the value zero, $\mathbf{Q} = \mathbf{0}^{(j)} \equiv [0,0,0, \ldots]$. We shall drop the superscript j when there is no danger of confusion, but in general both the equilibrium geometry and the normal coordinates of vibration change when the electronic state changes. While a detailed description of the relation between the normal coordinates and the displacements ΔR_k lies outside the scope of this presentation, a little more will be said about their properties in Section 2.2.1.

In the following, we shall express all position vectors in the molecular system of axes and shall ignore the translational and rotational degrees of freedom. Thus, we shall restrict our attention to internal motion of nuclei and electrons, i.e., to vibrational and electronic spectroscopy in fluid or solid solutions where the quantization of molecular rotation and translation is lost due to intermolecular interactions.

The transition moments which will be of interest to us presently are the electric dipole, magnetic dipole, and electric quadrupole transition moments. The transition tensors of interest are the two-photon absorption and Raman scattering tensors. We shall first consider the transition moments and then the transition tensors.

The electric dipole operator is given by the spin-free expression (1.46), which we shall repeat for convenience:

$$\hat{\mathbf{M}} = -|e| \sum_{l=1}^{n} \hat{\mathbf{r}}_l + |e| \sum_{k=1}^{N} Z_k \hat{\mathbf{R}}_k \qquad (2.1)$$

where e is the charge of the electron, Z_k is the atomic number of nucleus k, n is the number of electrons, N is the number of nuclei, and $\mathbf{r}_l$ is the position vector of the l-th electron.

The magnetic dipole moment operator contains a spin-free and a pure-spin part [cf. (1.54)]:

$$\hat{\mathscr{M}} = \frac{-|e|}{2mc} \sum_{l=1}^{n} (\hat{\mathbf{r}}_l \times \hat{\mathbf{p}}_l + 2\hat{\mathbf{s}}_l) + \sum_{k=1}^{N} \left[\frac{Z_k|e|}{2M_k c} (\hat{\mathbf{R}}_k \times \hat{\mathbf{P}}_k) + \gamma_k \hat{\boldsymbol{\sigma}}_k \right] \qquad (2.2)$$

where $\hat{\mathbf{r}}_l \times \hat{\mathbf{p}}_i$ is the orbital angular momentum operator (spin-free) and $\hat{\mathbf{s}}_l$ the spin angular momentum operator of the l-th electron, and where $\hat{\mathbf{R}}_k \times \hat{\mathbf{P}}_k$ is the spin-free angular momentum operator due to the motion of the k-th nucleus and $\hat{\boldsymbol{\sigma}}_k$ the spin angular momentum operator of the k-th nucleus. The linear momentum operator of the l-th electron is $\hat{\mathbf{p}}_l = -i\hbar\boldsymbol{\nabla}_l$, and that of the k-th nucleus is $\hat{\mathbf{P}}_k = -i\hbar\boldsymbol{\nabla}_k$, where $\boldsymbol{\nabla}$ is the gradient operator, m and M_k are the masses of the electron and the k-th nucleus, respectively, c is the speed of light in vacuum, and γ_k is the gyromagnetic ratio of the k-th nucleus.

The spin parts of $\hat{\mathscr{M}}$ are of paramount importance in magnetic resonance spectroscopy but only rarely play a role in optical spectroscopy which is of interest to us presently. However, whenever we use expression (2.2) for $\hat{\mathscr{M}}$, it is important to include nuclear spin parts of the wavefunction in the evaluation of matrix elements and to integrate over all nuclear spin coordinates.

The electric quadrupole moment operator $\hat{Q}$ is given by

$$\hat{Q} = -|e| \sum_{l=1}^{n} \left[\hat{\mathbf{r}}_l \hat{\mathbf{r}}_l - \frac{1(\hat{\mathbf{r}}_l \cdot \hat{\mathbf{r}}_l)}{3} \right] + |e| \sum_{k=1}^{N} Z_k \left[\hat{\mathbf{R}}_k \hat{\mathbf{R}}_k - \frac{1(\hat{\mathbf{R}}_k \cdot \hat{\mathbf{R}}_k)}{3} \right] \qquad (2.3)$$

where 1 is the unit tensor [cf. (1.58); quadrupolar nuclei are assumed to be absent].

The developments given in the following for absorption and emission intensities will be formulated in terms of the electric dipole moment operator $\hat{\mathbf{M}}$, which makes by far the largest contribution of the three. Subsequently, we shall discuss in much less detail how analogous reasoning applies for the operators $\hat{\mathscr{M}}$ and $\hat{Q}$, which play a subordinate role in absorption and emission intensities in the optical region but are important for optical activity.

2.1.1 Absorption and Emission

According to (1.60), the intensities observable in one-photon absorption and emission measurements are dictated by the matrix elements of the three operators $\hat{\mathbf{M}}$, $\hat{\mathscr{M}}$, and $\hat{Q}$ between the initial vibronic (vibrational-electronic) level and the final vibronic level, projected into laboratory directions as dictated by the polarization and propagation directions of the light, and then squared.

Electric Dipole Transition Moments. In the Born-Oppenheimer approximation, the expression for the matrix element of $\hat{\mathbf{M}}$ between two vibronic states is

$$\mathbf{M}(\mathrm{jv,j'v'}) = \langle \psi_{\mathrm{j'}}(\mathbf{q},\mathbf{Q})\chi_{\mathrm{j',v'}}(\mathbf{Q})|\hat{\mathbf{M}}|\psi_{\mathrm{j}}(\mathbf{q},\mathbf{Q})\chi_{\mathrm{j,v}}(\mathbf{Q})\rangle \tag{2.4}$$

where the position vectors of the electrons (in ordinary space and spin space) have been collected in the vector $\mathbf{q}$ for convenience. The quantum numbers j and j′ label electronic states; the quantum numbers v and v′ label vibrational states. In the harmonic approximation, v is actually a set of quantum numbers, each corresponding to the number of vibrational quanta in a particular normal mode, as is v′. Both v and v′ are therefore best viewed as vectors, e.g., $\mathbf{v} \equiv (\mathrm{v}_1,\mathrm{v}_2,\ldots,\mathrm{v}_{3N-6})$. The same set of internal coordinates $\mathbf{Q}$ is used throughout on the right-hand side of (2.4); these could be the normal coordinates of the j-th state or those of the j′-th state or yet another set if desired.

It is useful to separate the integration over the electronic ($\mathbf{q}$) and nuclear ($\mathbf{Q}$) coordinates explicitly and to introduce a shorthand notation for the result of the former. For the electric dipole moment involving the electronic states j and j′, evaluated at nuclear geometry $\mathbf{Q}$, we use the symbol $\mathbf{M}(\mathrm{jj'},\mathbf{Q})$:

$$\hat{\mathbf{M}}(\mathrm{jj'},\mathbf{Q}) = \langle \psi_{\mathrm{j'}}(\mathbf{q},\mathbf{Q})|\hat{\mathbf{M}}|\psi_{\mathrm{j}}(\mathbf{q},\mathbf{Q})\rangle_{\mathbf{q}} \tag{2.5}$$

Although the vector $\hat{\mathbf{M}}(\mathrm{jj'},\mathbf{Q})$ is an expectation value and is normally written without the operator sign (the caret ^), it simultaneously is an operator in the space of the nuclear coordinates and we shall indicate so explicitly. Taking the expectation value with respect to the nuclear degrees of freedom, we obtain

$$\mathbf{M}(\mathrm{jv,j'v'}) = \langle \chi_{\mathrm{j',v'}}(\mathbf{Q})|\hat{\mathbf{M}}(\mathrm{jj'},\mathbf{Q})|\chi_{\mathrm{j,v}}(\mathbf{Q})\rangle_{\mathbf{Q}} \tag{2.6}$$

Since the electronic wavefunctions $\psi_{\mathrm{j}}(\mathbf{q},\mathbf{Q})$ form an orthonormal set (1.13), Eq. (2.5) combined with (2.1) yields

$$\hat{\mathbf{M}}(\mathrm{jj'},\mathbf{Q}) = \langle \psi_{\mathrm{j'}}(\mathbf{q},\mathbf{Q})| - |e| \sum_{k=1}^{n} \hat{\mathbf{r}}_k|\psi_{\mathrm{j}}(\mathbf{q},\mathbf{Q})\rangle_{\mathbf{q}} + |e| \sum_{k=1}^{N} Z_k\hat{\mathbf{R}}_k\delta_{\mathrm{jj'}} \tag{2.7}$$

In principle, the evaluation of $\mathbf{M}(\mathrm{jv',j'v'})$ will begin by obtaining the vector $\hat{\mathbf{M}}(\mathrm{jj'},\mathbf{Q})$ for a large number of molecular geometries $\mathbf{Q}$. For each choice of $\mathbf{Q}$, five steps are needed: (i) choosing a nuclear geometry defined by $\mathbf{Q}$, (ii) calculating the second term in (2.7) using the known positions of nuclei, (iii) calculating the electronic wavefunctions ψ_{j} and $\psi_{\mathrm{j'}}$ for those same nuclear positions, (iv) evaluating the first term in (2.7) by integration over the electronic coordinates $\mathbf{q}$, and (v) adding the two terms to obtain $\hat{\mathbf{M}}(\mathrm{jj'},\mathbf{Q})$. Once $\hat{\mathbf{M}}(\mathrm{jj'},\mathbf{Q})$ is known for a sufficient number of nuclear geometries, it is possible to average $\hat{\mathbf{M}}(\mathrm{jj'},\mathbf{Q})$ over $\mathbf{Q}$ with the weight factors given by the vibrational wavefunctions $\chi_{\mathrm{j,v}}(\mathbf{Q})$ and $\chi_{\mathrm{j',v'}}(\mathbf{Q})$ as indicated in (2.6). In practice, matters can be frequently simplified, and one needs to distinguish three cases:

(i) Permanent Moments. If $\mathrm{j} = \mathrm{j'}$ and $\mathrm{v} = \mathrm{v'}$, the vector $\mathbf{M}(\mathrm{jv,jv})$ represents the static dipole moment of the molecule in the electronic state j and the vibrational state v. Although it is important in rotational (microwave) spectroscopy, it is not of interest in the present context. For many purposes, it can be approximated by $\mathbf{M}_j^0$, which is the vector $\hat{\mathbf{M}}(jj,\mathbf{Q})$ evaluated at the equilibrium geometry of the j-th state, $\mathbf{Q} = \mathbf{0}^{(j)} = \mathbf{0}$. A single calculation of the electronic wavefunction $\psi_j(\mathbf{q},\mathbf{Q})$ at the equilibrium geometry $\mathbf{Q} = \mathbf{0}$ is then sufficient. This is the procedure used in the majority of reported dipole moment calculations.

The second and third cases correspond to transition dipole moments and are of interest to us presently.

(ii) Vibrational Transition Moments. In the second case, $j = j'$ and $v \neq v'$. The quantity $\mathbf{M}(jv,jv')$ represents the electric dipole transition moment of the vibrational transition $v \rightarrow v'$ for a molecule in the electronic state j. Suppressing the index j, we can rewrite (2.6) in the form

$$\mathbf{M}(jv,jv') = \mathbf{M}(vv') = \langle \chi_{v'}(\mathbf{Q}) | \hat{\mathbf{M}}(\mathbf{Q}) | \chi_v(\mathbf{Q}) \rangle \tag{2.8}$$

We again need to know the dipole moment of the molecule in the electronic state j, $\hat{\mathbf{M}}(jj,\mathbf{Q}) \equiv \hat{\mathbf{M}}(\mathbf{Q})$, as a function of molecular geometry $\mathbf{Q}$. However, now we can no longer afford the luxury of approximating $\hat{\mathbf{M}}(\mathbf{Q})$ by its value at equilibrium, $\mathbf{M}^0$. In such a zero-order approximation, $\hat{\mathbf{M}}(\mathbf{Q})$ would be a constant independent of $\mathbf{Q}$; it could be taken out of the brackets in (2.8), and the orthogonality of the vibrational wavefunctions $\chi_{j,v}$ and $\chi_{j,v'}$ stated in (1.12) would then yield vanishing transition moments $\mathbf{M}(jv,jv')$ for all vibrations. Although this level of approximation accounts for the weakness of IR transitions relative to UV transitions in molecular spectra, it is clearly not useful for a discussion of the polarization of vibrational transitions.

It is thus seen that the vibrational transition moments are related to the change of the molecular dipole moment caused by the change in the molecular geometry during a vibration. It is usual to simplify matters by using a McLaurin series for the description of the dipole moment as a function of geometry and truncating after the linear term:

$$\hat{\mathbf{M}}(\mathbf{Q}) = \hat{\mathbf{M}}^0 + \sum_{k=1}^{3N-6} \left[\frac{\partial \mathbf{M}(\mathbf{Q})}{\partial Q_k} \right]_0 \hat{Q}_k \tag{2.9}$$

This is sometimes referred to as the assumption of electrical harmonicity. When (2.9) is substituted into (2.8), the constant term $\mathbf{M}^0$ does not contribute since $\chi_v(\mathbf{Q})$ and $\chi_{v'}(\mathbf{Q})$ are orthogonal as noted above, and one obtains

$$\mathbf{M}(vv') = \sum_{k=1}^{3N-6} \left[\frac{\partial \mathbf{M}(\mathbf{Q})}{\partial Q_k} \right]_0 \langle \chi_{v'}(\mathbf{Q}) | \hat{Q}_k | \chi_v(\mathbf{Q}) \rangle \tag{2.10}$$

Then the calculation of transition moments of vibrational transitions is reduced to the evaluation of the derivatives of the molecular dipole moment with respect to normal mode displacements and the evaluation of matrix elements of Q_k. If mechanical harmonicity is assumed, the vibrational wavefunctions become products of harmonic oscillator wavefunctions corresponding to each normal mode Q_k, and the matrix element of Q_k vanishes unless v and v′ differ by only one quantum excitation in a single normal mode. Thus, for the case of excitation of one quantum of the f-th normal mode Q_f from the vibrational ground state, $\mathbf{M}(0f)$ is proportional to the value of $\partial\mathbf{M}(\mathbf{Q})/\partial Q_f$ at the equilibrium geometry.

(iii) Electronic Transition Moments. In the third case, $j \neq j'$, and two sets of normal coordinates $\mathbf{Q}^{(j)}$ and $\mathbf{Q}^{(j')}$ are natural. The discussion simplifies if they coincide, and we shall assume that they do. In general, they do not (the Dushinsky effect), and the analysis is then more complicated. We do, however, permit the two equilibrium geometries $\mathbf{0}^{(j)}$ and $\mathbf{0}^{(j')}$ to differ.

The quantity $\mathbf{M}(jv,j'v')$ represents the electric dipole transition moment of an electronic transition, specifically of the vibronic transition $jv \rightarrow j'v'$. It is important to note that whenever $j \neq j'$, the usual concept of molecular dipole moment due to the charges of electrons and nuclei is no longer involved: In the evaluation of $\mathbf{M}(jv,j'v')$ from (2.6) and (2.7), the second term on the right-hand side of (2.7) is absent, since $\delta_{jj'} = 0$, so that the nuclear charges do not enter into the evaluation of $\hat{\mathbf{M}}(jj',\mathbf{Q})$. The first term in (2.7) still represents a dipole moment of a charge distribution at a given geometry $\mathbf{Q}$, and this charge distribution still contains both negative and positive charges, but the latter are not due to the nuclei, and the charge distribution does not correspond to an electron density distribution in any stationary electronic state. Rather, the positively and negatively charged regions in space correspond to regions in which $\psi_j(\mathbf{q})$ and $\psi_{j'}(\mathbf{q})$ have the same and the opposite signs, respectively. We shall illustrate this in more detail in Section 2.2.2.

It is now useful to distinguish two types of electronic transitions, allowed and forbidden.

Symmetry-Allowed Electronic Transitions. When the electronic transition moment calculated at equilibrium geometry does not vanish, $\mathbf{M}^0(jj') \neq 0$, it is usually possible to neglect the dependence of the transition moment $\hat{\mathbf{M}}(jj',\mathbf{0})$ on molecular geometry, as in case (i), static dipole moments. Then we approximate the electronic transition moment by its value at equilibrium geometry, $\mathbf{M}^0(jj')$:

$$\mathbf{M}^0(jj') = \mathbf{M}(jj',\mathbf{0}) \tag{2.11}$$

and equation (2.6) reduces to

$$\mathbf{M}(jv,j'v') = \langle \chi_{j',v'}(\mathbf{Q}) | \chi_{j,v}(\mathbf{Q}) \rangle \mathbf{M}^0(jj') \tag{2.12}$$

The vibrational overlap integral on the right-hand side of (2.12) is known as the Franck-Condon overlap integral. In the general case, it does not reduce to $\delta_{vv'}$, since the functions $\chi_{j,v}(\mathbf{Q})$ and $\chi_{j',v'}(\mathbf{Q})$ are eigenfunctions of different vibrational Hamiltonians, one containing the potential energy surface of state j, the other that of state j′. Only if the shapes of the two surfaces and the equilibrium geometries happen to be identical will the integral $\langle \chi_{j',v'}(\mathbf{Q}) | \chi_{j,v}(\mathbf{Q}) \rangle$ be equal to $\delta_{vv'}$. Then only the $\mathbf{M}(jv,j'v)$ transition moment will be nonzero (a Franck-Condon allowed electronic transition). As the two surfaces begin to differ, either in their shapes (force constants) or in their positions (equilibrium geometries) or both, other transition moments $\mathbf{M}(jv,j'v')$, $v \neq v'$, will also become different from zero. If the two surfaces differ greatly, the transition moment for the $jv \rightarrow j'v$ transition will have a much smaller value than some of the others (a strongly Franck-Condon-forbidden transition).

The Franck-Condon overlap integral vanishes by symmetry if $\chi_{j,v}(\mathbf{Q})$ and $\chi_{j',v'}(\mathbf{Q})$ belong to different irreducible representations of the molecular symmetry group. For instance, if the initial state jv is the lowest vibrational level of the ground state, only totally symmetric vibrational levels v′ of the final state j′ can appear in absorption. These can be constructed from any combination of single quanta of totally symmetric vibrations, plus such combinations of quanta of other vibrations which possess total symmetry, e.g., even numbers of quanta of antisymmetric vibrations. The totally symmetric modes which generally appear with the most intensity are those which describe the motion which brings the molecule from the equilibrium geometry in the initial state, $\mathbf{0}^{(j)}$, to the equilibrium geometry in the final state, $\mathbf{0}^{(j')}$. Modes which are not totally symmetric do not correspond to motion of this kind, i.e., they do not displace the equilibrium geometry. Even numbers of quanta of such modes will appear strongly only if they have very different frequencies in the initial and final states, and this is relatively rare.

The fact that the vectors $\mathbf{M}(jv,j'v')$ and $\mathbf{M}^0(jj')$ are collinear [equation (2.12)] under the assumption that $\hat{\mathbf{M}}(jj',\mathbf{Q})$ is independent of molecular geometry shows that all the allowed vibronic components of an electronic band have the same polarization. The calculation of the vibronic transition moments then requires the evaluation of overlap integrals between vibrational wavefunctions and evaluation of the electronic transition moment at either equilibrium geometry, $\mathbf{0}^{(j)}$ or $\mathbf{0}^{(j')}$. This matter is taken up in Section 2.2.2.

The oscillator strength of a transition is proportional to the square of the transition moment and to the transition energy [equation (1.61)]. For a given electronic transition $j \to j'$, all vibronic components differing in v and v′ usually have energies very similar to the average excitation energy in the $jv \to j'v'$ manifold. If this dependence is neglected, the total intensity of the electronic transition from any one vibrational level integrated over all its vibronic components is simply related to $|\mathbf{M}^0(jj')|^2$, since the sum of intensities of all lines originating in the v-th level of the $j \to j'$ transition is proportional to

$$\begin{aligned}\sum_{v'} |\mathbf{M}(jv,j'v')|^2 &= |\mathbf{M}^0(jj')|^2 \langle \chi_{j,v}(\mathbf{Q})|\sum_{v'}|\chi_{j',v'}(\mathbf{Q})\rangle\langle\chi_{j',v'}(\mathbf{Q})|\chi_{j,v}(\mathbf{Q})\rangle \\ &= |\mathbf{M}^0(jj')|^2 \langle \chi_{j,v}(\mathbf{Q})|\chi_{j,v}(\mathbf{Q})\rangle \\ &= |\mathbf{M}^0(jj')|^2 \qquad (2.13)\end{aligned}$$

where we used the fact that the summation runs over all members of a complete set, so that

$$\sum_{v'} |\chi_{j'v'}(\mathbf{Q})\rangle\langle\chi_{j'v'}(\mathbf{Q})| = \hat{1} \qquad (2.14)$$

It is common to refer to the geometry-independent quantity $M^0(jj')$ as the electronic transition moment of transition $j \to j'$ and to the squares of the Franck-Condon overlap integrals, which dictate the distribution of the total intensity of the transition into the individual vibronic lines, as the Franck-Condon factors.

When the molecules in the initial state are in thermal equilibrium, the population of v-th level is given by the Boltzmann factor $e^{-E_v/kT}$. At room temperature, the population of the lower excited vibrational levels of typical low-frequency vibrations is considerable. Transitions originating in such levels ($v \neq 0$) are responsible for "hot bands" in the spectra and cause a general broadening of the spectral features observed in solution, where they are usually not individually resolved. This can be removed by cooling the sample.

Symmetry-Forbidden Electronic Transitions. Transitions for which the transition moment at equilibrium geometry vanishes, $\mathbf{M}^0(jj') = 0$, are called electronically symmetry-forbidden. Here, we also need to consider the cases in which $\mathbf{M}^0(jj')$ is "accidentally" small but does not vanish exactly by symmetry. This can be due to the presence of approximate symmetries in the molecular Hamiltonian, e.g., alternant pairing for alternant hydrocarbons or electronic spin multiplicity. The latter case will be addressed in some detail at the end of Section 2.2.2.

The situation which results when $\mathbf{M}^0(jj')$ is small or vanishes altogether is similar to case (ii) above: It is no longer reasonable to neglect the dependence of $\hat{\mathbf{M}}(jj',\mathbf{Q})$ on the nuclear geometry. Then expression (2.12) cannot be used, and it is necessary to go back to expressions (2.6) and (2.7), which now acquire the form

$$\mathbf{M}(jv,j'v') = \langle \chi_{j',v'}(\mathbf{Q}) | \langle \psi_{j'}(\mathbf{q},\mathbf{Q}) | -|e| \sum_{k=1}^{n} \hat{\mathbf{r}}_k | \psi_j(\mathbf{q},\mathbf{Q}) \rangle_{\mathbf{q}} | \chi_{j,v}(\mathbf{Q}) \rangle_{\mathbf{Q}} \qquad (2.15)$$

The simplest procedure to follow at this point is known as the Herzberg-Teller treatment. In order to describe the dependence of $\hat{\mathbf{M}}(jj',\mathbf{Q})$ on $\mathbf{Q}$, we recognize that the electronic wavefunctions $\psi_j(\mathbf{q},\mathbf{Q})$ and $\psi_{j'}(\mathbf{q},\mathbf{Q})$ are obtained by diagonalization of the electronic Hamiltonian $\hat{H}_{el}(\mathbf{q},\mathbf{Q})$ and that the latter depends on the molecular geometry $\mathbf{Q}$. Expanding around the equilibrium geometry of, say, the l-th state, $\mathbf{0}^{(l)}$, whose normal coordinates are $Q_k^{(l)}$,

$$\hat{H}_{el}(\mathbf{q},\mathbf{Q}) = \hat{H}_{el}(\mathbf{q},\mathbf{0}^{(l)}) + \hat{H}_{ev}(\mathbf{q},\mathbf{Q}^{(l)}) \qquad (2.16)$$

The operator $\hat{H}_{ev}(\mathbf{q},\mathbf{Q}^{(l)})$ expresses the difference between the electronic Hamiltonian at the equilibrium geometry $\mathbf{0}^{(l)}$ and at the geometry described by $\mathbf{Q}^{(l)}$ using the normal coordinates of the l-th state. It is referred to as the vibronic coupling operator. For geometries $\mathbf{Q}^{(l)}$ which lie in the vicinity of $\mathbf{0}^{(l)}$, a McLaurin series expansion and truncation after the linear term yield

$$\hat{H}_{ev}(\mathbf{q},\mathbf{Q}^{(l)}) = \sum_{k=1}^{3N-6} \left[\frac{\partial \hat{H}_{el}(\mathbf{q},\mathbf{Q}^{(l)})}{\partial Q_k^{(l)}} \right]_0 \hat{Q}_k^{(l)} \qquad (2.17)$$

where the derivative is taken at the equilibrium geometry $\mathbf{0}^{(l)}$ and k runs over all normal modes of vibration.

Now, we use first-order perturbation theory to obtain the electronic wavefunction of the j-th electronic state at an arbitrary geometry $\mathbf{Q}^{(j)}$ in the vicinity of both equilibrium geometries, $\mathbf{0}^{(j)}$ and $\mathbf{0}^{(l)}$:

$$\psi_j(\mathbf{q},\mathbf{Q}^{(j)}) = N' \Bigg\{ \psi_j(\mathbf{q},\mathbf{0}^{(j)}) + \sum_{l \neq j} \psi_l(\mathbf{q},\mathbf{0}^{(l)}) \sum_{k=1}^{3N-6} \langle \psi_l(\mathbf{q},\mathbf{0}^{(l)}) \Bigg| \left[\frac{\partial \hat{H}_{el}(\mathbf{q},\mathbf{Q}^{(j)})}{\partial Q_k^{(j)}} \right]_0 \Bigg| \psi_j(\mathbf{q},\mathbf{0}^{(j)}) \rangle_{\mathbf{q}} \times Q_k^{(j)} (E_l - E_j)^{-1} \Bigg\} \qquad (2.18)$$

In the following, we assume that the $\mathbf{Q}$ dependence of $\psi_j(\mathbf{q},\mathbf{Q}^{(j)})$ is small so that the normalization factor N′ can be reasonably replaced by unity.

The energy denominator in (2.18) discriminates in favor of mixing among states which are close in energy. In particular, when j is the ground state, (j = 0), the mixing can be frequently neglected altogether, and we shall do so in the following. Further, when only one other state l has an energy close to that of the final state of the transition, j′, the dependence of the electronic wavefunction of the final state $\psi_{j'}(\mathbf{q},\mathbf{Q})$ on molecular geometry $\mathbf{Q}^{(j')}$ will be described adequately by restricting the summation in (2.18) to a single term l. With this assumption, the expression for the transition moment, (2.15), becomes

$$\mathbf{M}(0v,j'v') = \langle\chi_{j',v'}(\mathbf{Q}^{(j')})|\chi_{0,v}(\mathbf{Q}^{(j')})\rangle_{\mathbf{Q}}\mathbf{M}^0(0j')$$
$$+\sum_{k=1}^{3N-6}\langle\chi_{j',v'}(\mathbf{Q}^{(j')})|\langle\psi_1(\mathbf{q},\mathbf{0}^{(1)})|\left[\frac{\partial\hat{H}_{el}(\mathbf{q},\mathbf{Q}^{(j')})}{\partial\mathbf{Q}_k^{(j')}}\right]_0|\psi_{j'}(\mathbf{q},\mathbf{0}^{(j')})\rangle_{\mathbf{q}}$$
$$\times\,\hat{\mathbf{Q}}_k^{(j')}\,|\chi_{0,v}(\mathbf{Q}^{(j')})\rangle_{\mathbf{Q}}(E_l - E_{j'})^{-1}\mathbf{M}^0(0l) \qquad (2.19)$$

The first term in (2.19) is identical with that obtained in (2.12) for an allowed transition. It is presently assumed to be very small or exactly zero. The second term results from vibronic mixing and represents a first-order vibronic contribution to the transition moment. It is seen that in this description the forbidden transition 0 → j′ "steals" or "borrows" intensity from an allowed one, 0 → l.

In many cases of interest, $\mathbf{M}^0(0j')$ is small but not exactly zero. Then three cases can be distinguished.

First, when the zero-order transition moments $\mathbf{M}^0(0j')$ and $\mathbf{M}^0(0l)$ are parallel, all the various components of the absorption band which differ in the vibrational level v′ of the final state j′ will be polarized equally along the direction of $\mathbf{M}^0(0j')$. This will occur in molecules of high symmetry when the excited states j′ and l belong to the same irreducible representation. The presence of vibronic coupling will then not be apparent in polarized absorption (or emission) spectra, although it may still appear clearly in other types of spectra, for instance in magnetic circular dichroism.

Second, when the directions of $\mathbf{M}^0(0j')$ and $\mathbf{M}^0(0l)$ differ, the electronic absorption band 0 → j′ will in general consist of a series of components of different polarization. Transitions into those vibrational levels v′ of the j′-th state which are of the same symmetry as v, i.e., totally symmetric if v is the lowest vibrational level of the ground state (v = 0), will be polarized along $\mathbf{M}^0(0j')$. In addition, components polarized along $\mathbf{M}^0(0l)$ may also appear due to the presence of the second term in (2.19). These will excite vibrational levels v′ in the final state j′ which are of symmetry such that the matrix element in front of $\mathbf{M}^0(0l)$ in (2.19) does not vanish. The vibrations corresponding to these levels are referred to as the coupling vibrations. If the transition occurs from the lowest vibrational level of the ground state (v = 0), their symmetry will be given by the direct product of the symmetries of the electronic states j′ and l. This follows from the form of (2.19): We first note that the function of electronic (but not nuclear) coordinates, $[\partial\hat{H}_{el}(\mathbf{q},\mathbf{Q}^{(jz)})/\partial Q_k^{(jz)}]_0$, belongs to the same irreducible representation of the molecular symmetry group as $Q_k^{(j')}$, since the product of the two represents an additive contribution to the molecular Hamiltonian (2.17) and must therefore be totally symmetric. Now, the integration over electronic coordinates in the second term in (2.19) will yield a nonvanishing result only for those normal modes Q_k for which the direct product $\Gamma(\psi_l) \times \Gamma(Q_k) \times \Gamma(\psi_{j'})$ contains the totally symmetric irreducible representation. If subsequent integration over nuclear coordinates is to yield a nonvanishing result, the direct product $\Gamma(\chi_{0,v}) \times \Gamma(Q_k^{(j')}) \times \Gamma(\chi_{j',v'})$ must contain the totally symmetric irreducible representation. If $v = 0$, this requires that $\chi_{j',v'}$ transform like $\Gamma(Q_k)$ and therefore like $\Gamma(\psi_l) \times \Gamma(\psi_{j'})$.

Third, more than one state l may be involved in "lending" intensity to the electronic transition $0 \rightarrow j'$, some with the zero-order transition moment parallel to $\mathbf{M}^0(0j')$, others not. This third case is a simple superposition of the first two. Totally symmetric coupling vibrations will produce lines polarized like the origin, and vibrations of symmetry $\Gamma(\psi_0) \times \Gamma(\psi_{j'})$ will produce lines polarized parallel to $\mathbf{M}^0(0l)$.

If $\mathbf{M}^0(0j')$ vanishes altogether, the first term in (2.19) will not contribute, and all observed components of the electronic transition will be polarized along the direction of $\mathbf{M}^0(0l)$. The origin of the transition ($v = v' = 0$) will have zero intensity, and only vibrational levels of overall symmetry given by the direct product of symmetries of the j′-th and l-th states will appear.

The process of acquisition of transition intensity and polarization by one zero-order electronic transition at the expense of another is known as vibronic intensity borrowing or stealing. We shall encounter this concept again in Section 2.2.2 in the discussion of spin-orbit coupling and vibronically induced spin-orbit coupling, which has been postponed until after the description of the ways in which electronic wavefunctions are usually represented. Intensity borrowing in ordinary absorption and emission is only one of the many manifestations of vibronic coupling, which also plays an important role in natural and magnetic circular dichroism, in two-photon absorption and resonant Raman spectroscopy, in vibrational spectroscopy of molecules in excited states, and elsewhere.

Electric Quadrupole and Magnetic Dipole Transition Moments. Considerations very similar to those we have outlined for the electric dipole contribution to absorption or emission intensity and polarization apply also to electric quadrupole and magnetic dipole contributions, and it is only necessary to replace $\hat{\mathbf{M}}$ by $\hat{Q}$ or $\hat{\mathcal{M}}$ in (2.4) and to proceed as above to obtain the corresponding expressions for the matrix elements between two vibronic states.

For electric quadrupole moments, the three cases distinguished above correspond to (i) the permanent (static) electric quadrupole moment Q(jv,jv), generally approximated by its value at the equilibrium geometry, $Q^0(jj)$; (ii) the electric quadrupole transition moment of the vibrational transition $v \rightarrow v'$ for a molecule in the electronic state j, given by an expression analogous to (2.8) and related to the change of the molecular quadrupole moment caused by the change of the molecular geometry during a vibration; and (iii) the electric quadrupole transition moment of the vibronic transition $jv \rightarrow j'v'$, given by an expression analogous to (2.12) for an electric quadrupole allowed transition and by an expression analogous to (2.19) for an electric quadrupole forbidden transition. None of the three are of any great practical importance in molecular absorption or emission spectroscopy or in measurements of optical activity on isotropic solutions, but the electric quadrupole transition moments of vibrational and vibronic transitions are likely to play a significant role in optical activity of partially aligned molecules (Section 2.1.2).

For magnetic dipole moments, the three cases are complicated by the presence of the spin terms in the operator, $\hat{\mathcal{M}}$. This requires explicit recognition of the spin dependence of the total wavefunction, while so far we have simply assumed that the spin part of the wavefunction remains unchanged in a transition. We shall consider here only the level of approximation in which the electronic and nuclear wavefunctions are written as products of separable spin and space parts (pure-multiplicity state approximation, see Section 2.2.2). The first of the three cases corresponds to (i) $jv = j'v'$, where we need to distinguish two alternatives:

(a) If the electron and nuclear spin functions are the same for jv and j′v′, we obtain the static magnetic dipole moment of the molecule in the particular spin state defined by these functions. This is due to the sum of contributions from electron and nuclear spin and orbital motion. Orbital motion makes a nonzero contribution only in degenerate vibronic states, since $\hat{\mathbf{r}} \times \hat{\mathbf{p}}$ and $\hat{\mathbf{R}} \times \hat{\mathbf{P}}$ are pure imaginary operators. Such states occur only in molecules of very high symmetry (in the presence of a threefold or higher symmetry axis).

(b) If the spin functions are different for jv and j′v′, the spin-independent terms in (2.2) disappear after integration over spin. The remaining terms then represent a magnetic dipole transition moment (i) for an electron spin flip, which is responsible for transitions between individual levels of a multiplet and is important in electron paramagnetic resonance (EPR), or (ii) for a nuclear spin flip, which is responsible for transitions between hyperfine levels and is important in nuclear magnetic resonance (NMR), or (iii) possibly for simultaneous nuclear and electron flips, which are important for higher order effects in EPR.

Case (b) plays a certain role in optical spectroscopy inasmuch as electron spin flip transitions between different levels of a multiplet (say, a triplet) may affect spin-multiplicity forbidden transitions to a different electronic level (optically detected magnetic resonance, ODMR).

Next, we need to look at the case $j = j'$, $v \neq v'$ and the case $j \neq j'$. In these instances, the electron and nuclear spin functions need to be the same in the initial and final states if a nonvanishing contribution is to result at this level of approximation and only the spin-free part of $\hat{\mathcal{M}}$ actually contributes to the nonzero value.

Then (ii) if $j = j'$, $v \neq v'$, we obtain a magnetic dipole moment for a vibrational transition, and finally (iii) if $j \neq j'$, we obtain a magnetic dipole moment for a vibronic transition. In this last case the second term in (2.2) does not contribute. Once again, magnetic dipole allowed and forbidden transitions need to be considered and expressions similar to (2.12) and (2.19), respectively, apply.

Magnetic dipole transition moments, both vibrational (ii) and vibronic (iii), are essential for the understanding of optical activity in the respective spectral regions (Section 2.1.2) but in practice play almost no role in determining molecular absorption and emission intensities.

2.1.2 *Circular Dichroism*

As indicated by (1.102), (1.104), (1.105), (1.132), and (1.135)–(1.137), and as discussed more thoroughly in Chapter 6, the intensities observed in measurements of natural and magnetic circular dichroism of oriented molecules are dictated by products of the matrix elements of the electric dipole ($\hat{\mathbf{M}}$) and either the electric quadrupole ($\hat{\mathsf{Q}}$) or the magnetic dipole ($\hat{\mathcal{M}}$) operators, projected into laboratory directions. In natural circular dichroism, the matrix elements of all three operators between a vibronic level of the initial state (jv) and a vibronic level of the final state (j′v′) are needed. In the usual treatment of magnetic circular dichroism, only the matrix elements of the $\hat{\mathbf{M}}$ and $\hat{\mathcal{M}}$ operators are needed. However, they need to be evaluated for a very large number of vibronic states (in principle, infinite). Procedures similar to those just discussed in Section 2.1.1 for transition intensities can be used for the reduction of these various matrix elements over vibronic wavefunctions to elements over vibrational and electronic wavefunctions, and vibronic coupling can be introduced in a very similar manner. The results will not be described in detail here, and the interested reader will find suitable references in Section 2.4. Suffice it to say that vibrational transitions generally display only very weak optical activity both in infrared absorption and Raman scattering measurements, and therefore most interest has concentrated on electronic transitions so far.

2.1.3 *Two-Photon Absorption and Raman Scattering*

According to (1.18), the polarized absorption intensities (cross sections) in two-photon absorption experiments depend on the matrix elements of the tensor operator $\hat{\mathsf{T}}$ between the initial and final vibronic levels, projected into laboratory directions and squared. According to (1.19), Raman scattering intensities are similarly related to the matrix elements of the scattering tensor operator $\hat{\alpha}'$. There are considerable similarities between the two cases, the main practical difference being that the final state in two-photon absorption is normally electronically excited while in Raman scattering it is ordinarily vibrationally excited. Both $\hat{\mathsf{T}}$ and $\hat{\alpha}'$ can be expressed to a good approximation in terms of an infinite sum over products of matrix elements of the electric dipole operator $\hat{\mathbf{M}}$. For two-photon absorption by a molecule in the ground electronic state,

$$\hat{\mathsf{T}}(0\mathrm{f},\mathbf{Q},\tilde{\nu}_1) = \sum_{\mathrm{j}} \left[\frac{\hat{\mathbf{M}}(0\mathrm{j},\mathbf{Q})\hat{\mathbf{M}}(\mathrm{jf},\mathbf{Q})}{\tilde{\nu}_{\mathrm{j}} - \tilde{\nu}_1} + \frac{\hat{\mathbf{M}}(\mathrm{jf},\mathbf{Q})\hat{\mathbf{M}}(0\mathrm{j},\mathbf{Q})}{\tilde{\nu}_{\mathrm{j}} - \tilde{\nu}_2} \right] \tag{2.20}$$

where f is the final state and j runs over all electronic states of the molecule. The energies of the photons, $h\nu_1$ and $h\nu_2$, are such that

$$\tilde{\nu}_1 + \tilde{\nu}_2 = \tilde{\nu}_{\mathrm{f}} \tag{2.21}$$

and they are both smaller than the excitation energies of any of the allowed molecular states ψ_j, so that no one-photon absorption can occur [and the right-hand side of (2.20) is finite].

Procedures similar to those outlined in Section 2.1.1 for one-photon transition intensities can be used to reduce the elements of $\hat{\mathbf{M}}$ over vibronic wavefunctions to elements over vibrational or electronic wavefunctions and to introduce vibronic coupling if necessary. We shall not describe the results here.

For Raman scattering by a molecule in the j-th electronic state and lowest vibrational state, we have

$$\alpha'_j(\tilde{\nu}_1,\mathrm{f}) = \langle \chi_{j,f}(\mathbf{Q})|\hat{\alpha}_j(\tilde{\nu}_1,\mathbf{Q})|\chi_{j,0}(\mathbf{Q})\rangle \tag{2.23}$$

where $\hat{\alpha}_j(\tilde{\nu}_1,\mathbf{Q})$ is an operator acting only on the vibrational coordinates:

$$\hat{\alpha}_j(\tilde{\nu}_1,\mathbf{Q}) = \sum_{j'}\left[\frac{\hat{\mathbf{M}}(jj',\mathbf{Q})\hat{\mathbf{M}}(jj',\mathbf{Q})}{\tilde{\nu}_{j'} + \tilde{\nu}_1} + \frac{\hat{\mathbf{M}}(jj',\mathbf{Q})\hat{\mathbf{M}}(jj',\mathbf{Q})}{\tilde{\nu}_{j'} - \tilde{\nu}_1}\right] \tag{2.24}$$

where the summation runs over all electronic states of the molecule. This operator represents the frequency-dependent ($\tilde{\nu}_1$) electronic polarizability tensor of the molecule in its j-th electronic state as a function of molecular geometry **Q**. The two terms on the right of (2.24) have the same theoretical origin as the two terms on the right-hand side of (1.69). They make comparable contributions to nonresonant Raman intensity. The first one is sometimes referred to as antiresonant, the second as resonant. The need for the presence of the resonant term is intuitively obvious from the previously given simple arguments. The need for the presence of the antiresonant term is appreciated when it is realized that the initial state j can also be an electronically excited state. Then, the antiresonant term allows for resonances with electronic states lying at lower energies.

In the following, we shall describe in somewhat more detail the case of Raman scattering by a molecule in its electronic ground state, which is of the greatest significance in practice. We shall first consider *nonresonant Raman scattering*, in which the dependence on $\tilde{\nu}_1$ is of secondary importance. For the electronic ground state, we can drop the subscript j in (2.23) and replace it by 0 in the right-hand side of(2.24).

Since the vibrational eigenfunctions of a state form an orthonormal set (1.12), we note immediately that it is only the change in the polarizability tensor as a function of molecular geometry **Q** which contributes to $\alpha'(\nu_1,\mathrm{f})$ in equation (2.23). The value of the tensor at equilibrium geometry,

$$\hat{\alpha}(\tilde{\nu}_1,\mathbf{0}) = \hat{\alpha}^0(\tilde{\nu}_1) \tag{2.25}$$

can be taken outside the brackets, and the orthogonality of the vibrational wavefunctions $\chi_{0,f}$ and $\chi_{0,0}$ will produce a multiplicative factor of zero. Like the transition moment $\hat{\mathbf{M}}(jj',\mathbf{Q})$ discussed above, the polarizability tensor $\hat{\alpha}(\tilde{\nu}_1,\mathbf{Q})$ is a function of the normal coordinates Q_k ($k = 1,2,\ldots,3N-6$) and can be expanded in a McLaurin series around the equilibrium geometry $\mathbf{Q} = \mathbf{0}$ [cf. (2.9)]. If the expansion is truncated after the linear terms (assumption of electrical harmonicity), we have

$$\hat{\alpha}(\tilde{\nu}_1,\mathbf{Q}) = \hat{\alpha}^0(\tilde{\nu}_1) + \sum_{k=1}^{3N-6} \left[\frac{\partial\alpha(\tilde{\nu}_1,\mathbf{Q})}{\partial Q_k}\right]_0 \hat{Q}_k \qquad (2.26)$$

Substitution into (2.23) yields

$$\alpha'(\tilde{\nu}_1,f) = \sum_{k=1}^{3N-6} \left[\frac{\partial\alpha(\tilde{\nu}_1,\mathbf{Q})}{\partial Q_k}\right]_0 \langle\chi_{0,f}|\hat{Q}_k|\chi_{0,0}\rangle \qquad (2.27)$$

where we no longer indicate explicity the functional dependence of the vibrational wave functions on **Q**.

If mechanical harmonicity is also assumed, the vibrational wavefunctions become products of harmonic oscillator wavefunctions corresponding to each normal mode Q_k, and the matrix element of Q_k on the right-hand side of equation (2.27) vanishes unless $\chi_{0,f}$ and $\chi_{0,0}$ differ by only one quantum of excitation in one normal mode. Thus, for the case of excitation of one quantum of the f-th normal mode Q_f from the vibrational ground state, $\alpha'(\tilde{\nu}_1,f)$ is proportional to the value $[\partial\alpha(\tilde{\nu}_1,\mathbf{Q})/\partial Q_f]_0$ at the equilibrium geometry.

In *resonant Raman* experiments on ground-state molecules, $\tilde{\nu}_1$ is approximately equal to $\tilde{\nu}_{j'}$ for one or more of the excited states j′ in the sum in (2.24). The dependence of the scattering tensor on $\tilde{\nu}_1$ then acquires an overwhelming importance. A damping term analogous to $i\tilde{\gamma}$ in (1.69) can no longer be neglected and must be added to both denominators in (2.24). Since only the square of the absolute value of a projection of $\alpha'(\tilde{\nu}_1,f)$ into laboratory directions is observable, the presence of the imaginary unit causes no difficulties. As $\tilde{\nu}_1$ approaches $\tilde{\nu}_{j'}$, the resonant (second) term in (2.24) becomes dominant. The first term is then usually many orders of magnitude smaller and can be ignored altogether. Moreover, only one excited state normally need be considered in the sum on the right-hand side of (2.24): the one whose denominator is close to zero.

If the transition into this resonant j′-th state is allowed, it is reasonable to ignore the dependence of $\hat{\mathbf{M}}(0j',\mathbf{Q})$ on nuclear geometry **Q** and to replace this transition moment by $\mathbf{M}^0(0j')$, the value at equilibrium geometry. Next, an expansion of a unit operator in terms of the vibrational eigenfunctions of the j′-th state [cf. (2.14)] is inserted between the two vector operators $\mathbf{M}^0(0j')$ in (2.24) and the assumption is made that the vibrational motion in the j′-th state is harmonic. One then obtains

$$\alpha'(\tilde{\nu}_1,f) = \sum_v |\mathbf{M}^0(0j')|^2 \frac{\langle\chi_{0,f}|\chi_{j',v}\rangle\langle\chi_{j'v}|\chi_{0,0}\rangle}{\tilde{\nu}_{j',v} - \tilde{\nu}_1 + i\tilde{\gamma}_{j'}} \qquad (2.28)$$

where $\tilde{\nu}_{j',v}$ is the wavenumber of the transition from the vibrational level $\chi_{0,0}$ of the ground electronic state to the vibrational level $\chi_{j',v}$ of the resonant excited state, and where the expression in the numerator is recognized as a product of two Franck-Condon overlap integrals. Expression (2.28) represents the so-called A term in resonant Raman scattering.

When the resonant state j′ is forbidden or only weakly allowed in absorption, i.e., when $\hat{\mathbf{M}}(0j',\mathbf{Q})$ is zero or small, it is no longer reasonable to neglect its dependence on molecular geometry. Then the use of the McLaurin expansion (2.9) yields

$$\alpha'(\tilde{\nu}_1,\mathrm{f}) = \sum_{\mathrm{v}} \left\{ |\mathbf{M}^0(0\mathrm{j}')|^2 \frac{\langle \chi_{0,\mathrm{f}}|\chi_{\mathrm{j}',\mathrm{v}}\rangle\langle \chi_{\mathrm{j}',\mathrm{v}}|\chi_{0,0}\rangle}{\tilde{\nu}_{\mathrm{j,v}} - \tilde{\nu}_1 + i\tilde{\gamma}_{\mathrm{j}'}} + \mathbf{M}^0(0\mathrm{j}')\left[\frac{\partial \mathbf{M}(0\mathrm{j}',\mathbf{Q})}{\partial Q_\mathrm{f}}\right]_0 \right.$$
$$\left. \times \frac{\langle \chi_{0,\mathrm{f}}|\hat{Q}_\mathrm{f}|\chi_{\mathrm{j}',\mathrm{v}}\rangle\langle \chi_{\mathrm{j}',\mathrm{v}}|\chi_{0,0}\rangle + \langle \chi_{0,\mathrm{f}}|\chi_{\mathrm{j}',\mathrm{f}}\rangle\langle \chi_{\mathrm{j}',\mathrm{v}}|\hat{Q}_\mathrm{f}|\chi_{0,0}\rangle}{\tilde{\nu}_{\mathrm{j}'\mathrm{v}} - \tilde{\nu}_1 + i\tilde{\gamma}_{\mathrm{j}'}} \right\} \quad (2.29)$$

It is possible to carry this expression further along the lines discussed in Section 2.1.1 for Herzberg-Teller coupling, but we shall not do so here.

The first contribution on the right-hand side of (2.29) is recognized as the A term already described by (2.28); the second is known as the B term. A more complete treatment also yields a C term, which contains derivatives of transition moments along two normal modes, or twice the same normal mode. The A term is normally the leading contributor to scattering for totally symmetric modes f, since these are the only modes that produce nonzero Franck-Condon overlaps $\langle \chi_{0,\mathrm{f}}|\chi_{\mathrm{j}',\mathrm{v}}\rangle$ and $\langle \chi_{\mathrm{j}',\mathrm{v}}|\chi_{0,0}\rangle$ by displacing the equilibrium nuclear positions. Those modes f which describe the distortions experienced by the molecule upon going from the ground to the j′-th excited state generally appear with most intensity in the resonant Raman spectrum.

Non-totally symmetric modes f gain resonant Raman intensity via the B term. This is particularly important when the resonant state j′ is only weakly allowed but gains intensity by vibronic coupling to a nearby strong transition. The vibrational modes responsible for the vibronic mixing can then appear with considerable intensity.

2.2 Computation of Transition Moments from First Principles

We now turn our attention to ways in which molecular transition moments can be evaluated in an a priori fashion. The methods used for this purpose can be divided into two classes. Some begin with a quantum-mechanical calculation of the molecular wavefunctions, electronic and vibrational, while others start with a classical model which represents the molecule as a set of charged particles in a force field ("valence-optical model"). Only the former are useful for the calculation of electronic transition moments, while both can be used for vibrational transition moments. Here, we shall concentrate heavily on the less empirical methods of the former kind.

2.2.1 Computation of Vibrational Transition Moments from First Principles

The computation of the electric dipole transition moment $\mathbf{M}(\mathrm{vv}')$, the electric quadrupole transition tensor $\mathsf{Q}(\mathrm{vv}')$, and the magnetic dipole transition moment $\mathcal{M}(\mathrm{vv}')$ for the vibrational transition $\mathrm{v} \rightarrow \mathrm{v}'$ in a molecule in a given electronic state using expression (2.8) and its analogs for $\hat{Q}$ and $\hat{\mathcal{M}}$ requires knowledge of the vibrational wavefunctions of the initial state χ_v and the final state $\chi_{v'}$ and knowledge of the molecular electric dipole moment $\hat{\mathbf{M}}(\mathbf{Q})$, electric quadrupole moment $\hat{Q}(\mathbf{Q})$, or magnetic dipole moment $\hat{\mathcal{M}}(\mathbf{Q})$ as a function of molecular geometry. These types of information are generally not accurately known, even for relatively simple molecules, and approximations are necessary.

Under conditions of interest in the present context, the vibrational transitions will involve molecules which initially contain little or no vibrational excitation, and this simplifies matters considerably. In the limit of small vibrations in reasonably stiff molecules, it is reasonable to use the harmonic approximation for the potential energy which governs nuclear motions (mechanical harmonicity). It can be shown that in this approximation, the molecular vibrational motion can be characterized by a set of $3N - 6$ one-dimensional harmonic oscillators characterized by frequencies ν_k and normal mode coordinates Q_k ($k = 1,2, \ldots, 3N - 6$), which can be expressed as functions of the displacements ΔR_k of nuclei from their equilibrium positions. For molecules with linear equilibrium geometries, there are $3N - 5$ normal modes of vibration. The iterative procedures used to derive the explicit form of the Q_j's in terms of a set of internal atomic coordinates from the observed vibrational frequencies of a series of isotopically substituted analogs of a molecule are referred to as "normal mode analysis." Such results are available for a fairly large number of molecules. It is also possible to proceed in an a priori fashion: to compute the potential energy of the molecule (i.e., its total internal energy minus the internal kinetic energy of the nuclei) by a quantum-chemical method or a parameterized molecular mechanics method for a sufficiently large number of nuclear geometries, to fit the resulting potential energy surface to a harmonic surface (assumption of mechanical harmonicity), and to determine the normal modes and frequencies by solving the corresponding secular problem. In this case, only an approximate agreement with the actually observed vibrational frequencies can be obtained. Either way, approximate vibrational wavefunctions $\chi_v(\mathbf{Q})$ become available.

They have the form of the product of wavefunctions, one for each normal mode of vibration. For the k-th mode, the harmonic oscillator wavefunction is

$$\chi_{v_k}(Q_k) = N_{v_k} \exp(-\gamma_k Q_k^2/2) H_{v_k}(\gamma_k^{1/2} Q_k) \tag{2.30}$$

where N_{v_k} is the normalization factor, γ_k is given by

$$\gamma_k = 2\pi\nu_k/h = \nu_k/\hbar \tag{2.31}$$

and $H_{v_k}(\gamma_k^{1/2} Q_k)$ is the Hermite polynomial of degree v_k. The first three Hermite polynomials are

$$H_0(\alpha) = 1 \qquad H_1(\alpha) = 2\alpha \qquad H_2(\alpha) = 4\alpha^2 - 2 \tag{2.32}$$

For instance, the ground-state vibrational wavefunction of the molecule in the harmonic approximation is given by

$$\chi_0(\mathbf{Q}) = N \exp\left[-\sum_{k=1}^{3N-6} \frac{\gamma_k Q_k^2}{2}\right] \tag{2.33}$$

and a wavefunction in which one quantum of the f-th fundamental vibrational mode Q_f is excited is given by

$$\chi_{1(f)}(\mathbf{Q}) = N' Q_f \chi_0(\mathbf{Q}) \tag{2.34}$$

where N and N′ are normalization factors. Since the function $\chi_0(\mathbf{Q})$ is totally symmetrical, $\chi_{1(f)}(\mathbf{Q})$ has the symmetry of Q_f.

We can now address the second required quantity, the dipole moment $\hat{\mathbf{M}}(\mathbf{Q})$ as a function of **Q**. If the vibrational wavefunctions were obtained in an a priori fashion from a quantum-chemical calculation of molecular potential energy at a variety of geometries, then very little additional effort was required to also calculate the dipole moment at those same geometries. The dipole moment surface $\hat{\mathbf{M}}(\mathbf{Q})$ can then be fitted to a suitable functional form and the integration in (2.8) can be performed, yielding $\mathbf{M}(vv')$. Alternatively, electrical harmonicity can be assumed (2.9) and only the first derivatives $[\partial \mathbf{M}(Q)/\partial Q_k]_0$ computed and substituted into (2.10).

In the following, we assume both electrical and mechanical harmonicity and represent $\chi_v(\mathbf{Q})$ and $\chi_{v'}(\mathbf{Q})$ in the form of products of harmonic wavefunctions. The value of $\langle \chi_v(\mathbf{Q}) | Q_f | \chi_{v'}(\mathbf{Q}) \rangle$ for two such wavefunctions is zero unless v and v′ differ by exactly one quantum in the f-th normal mode and do not differ in any of the other normal modes. If $v' = v + 1$, the integral has the value $[(v + 1)/2\gamma_f]^{1/2}$.

Inspection of (2.10) now shows that there will be at most 3N − 6 nonzero transition moments from the vibrational ground state $\chi_0(\mathbf{Q})$ into excited vibrational states, each corresponding to the excitation of one quantum in one of the 3N − 6 normal modes (the fundamental vibrational absorptions). The transition moment for the excitation of f-th normal mode in the vibrational ground state will be given by

$$\langle \chi_0(\mathbf{Q}) | \hat{\mathbf{M}}(\mathbf{Q}) | \chi_{1(f)}(\mathbf{Q}) \rangle = (2\gamma_f)^{-1/2} \left[\frac{\partial \mathbf{M}(\mathbf{Q})}{\partial Q_f} \right]_0 \tag{2.35}$$

The dipole moment derivatives $[\partial \mathbf{M}(\mathbf{Q})/\partial Q_f]_0$ can be obtained if a scheme is available for evaluating the molecular dipole moment as a function of molecular geometry. As already mentioned above, this can be a quantum-chemical procedure or, more simply and less reliably, an empirical scheme based on bond moment additivities.

Relatively few vibrational transition moment calculations have been performed. Most often, only the symmetry properties of the vector $[\partial \mathbf{M}(Q)/\partial Q_f]_0$ are utilized. If the normal mode Q_f does not transform like any one of x, y, or z, the vector $[\partial \mathbf{M}(\mathbf{Q})/\partial Q_f]_0$ is zero and so is the vibrational transition moment. Thus, in general, only a fraction of the fundamental vibrational transitions actually have nonzero transition moments and IR intensities. The use of symmetry in determining the directions of transition moments from the properties of wavefunctions has been already discussed in Section 1.2.2.

If the assumptions of electrical and mechanical harmonicity are relaxed, additional vibrational transition moments become different from zero. When the McLaurin expansion (2.9) is continued to higher powers of Q_k, substitution into

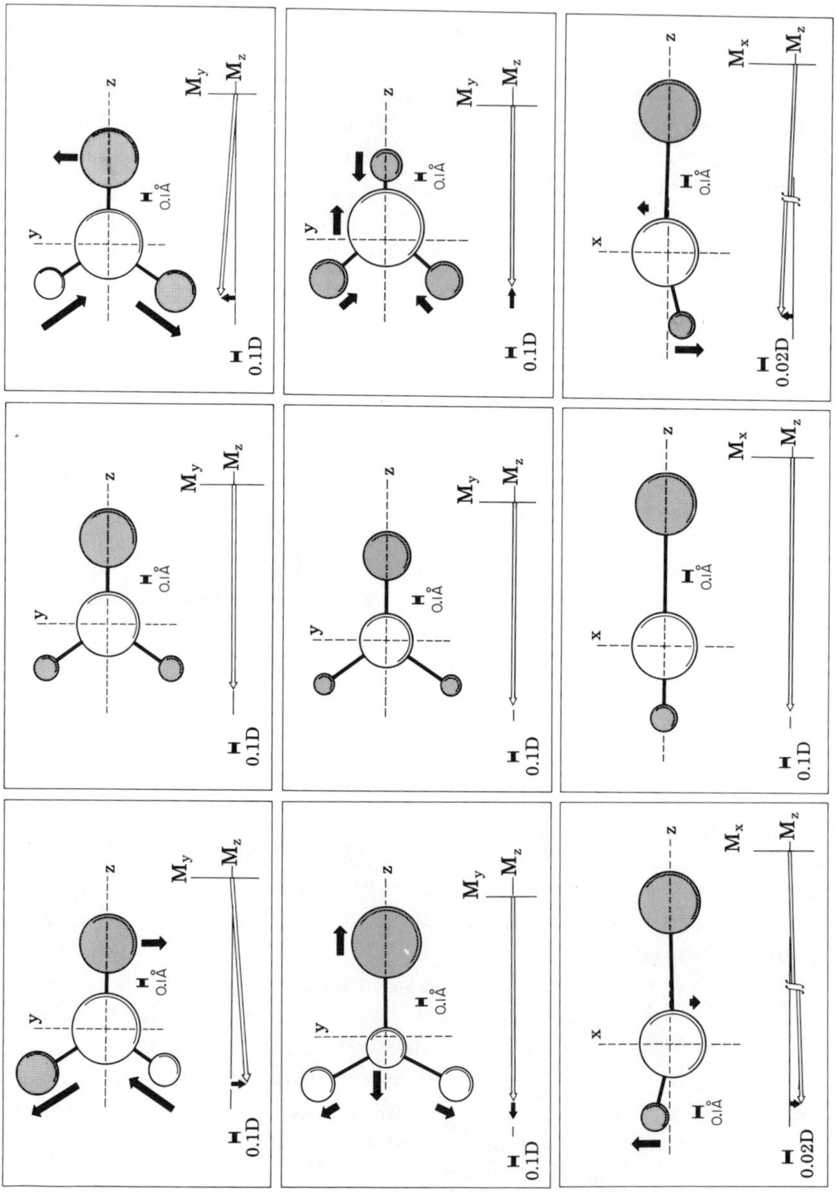
y
z
0.1Å
M_y
M_z
0.1D
x
M_x
0.02D

Figure 2.1 Schematic derivation of the vibrational transition moments of an x-polarized (bottom), a y-polarized (top), and a z-polarized (center) fundamental transition in formaldehyde as calculated by the MNDO method. The equilibrium geometry is shown in the center; geometries displaced by a small amount along a normal mode in two opposite directions are shown on the left and right, respectively. The displacement vectors are shown, and their scale in angstroms is indicated. The circles at the atoms show their total calculated net charge q (white, positive; black, negative). The radii of the circles are equal to $56|q| - 14.08$ on the C and O atoms and to $56|q| + 0.744$ on the H atoms for equilibrium geometries (central column). They are equal to $70|q| - 17.6$ for C and O and to $70|q| + 0.93$ for H in the left and right columns. The rehybridization contributions to the calculated dipole moments are not depicted. At the bottom of each part of the figure, the open arrows indicate the resulting total electric dipole moment of the molecule at each geometry, and the filled arrows give the difference with respect to the total dipole moment at equilibrium (the scale in Debye units is indicated).

(2.8) will yield a modified equation (2.10), which will now also contain terms of the types $\langle \chi_v(\mathbf{Q})|Q_k^p|\chi_{v'}(\mathbf{Q})\rangle$, $p = 2,3,\ldots$. Even when $\chi_v(\mathbf{Q})$ and $\chi_{v'}(\mathbf{Q})$ are taken to be harmonic wavefunctions, these terms will introduce nonvanishing dipole moments $\mathbf{M}(vv')$ for transitions between levels differing by more than one in the quantum number of the k-th mode. Transitions from the ground state $\chi_0(\mathbf{Q})$ into levels containing more than one quantum of excitation in one of the normal modes (overtone transitions) will now appear in the expected IR spectra.

When the assumption of mechanical harmonicity is relaxed and $\chi_v(\mathbf{Q})$ and $\chi_{v'}(\mathbf{Q})$ are no longer assumed to be products of harmonic oscillator wavefunctions, only the selection rules imposed by symmetry will remain. Symmetry permitting, all choices of v′ will give rise to nonvanishing transition moments from the vibrational ground state, so that not only fundamentals but also overtones and combination bands will appear in the expected IR spectra regardless of any assumptions concerning electric harmonicity (the upper state of combination bands corresponds approximately to simultaneous excitations of two or more normal modes). The very distinction between fundamentals, overtones, and combination bands will disappear if the harmonic approximation for the molecular potential energy surface and the separation of vibrational motions into individual normal modes are not even approximately valid.

The assumptions of electrical and mechanical harmonicity appear to be equally appropriate. It is not desirable to remove only one of them when a better approximation is desired.

Figure 2.1 provides an illustration of the origin of vibrational electric dipole transition moments. A quantum-mechanical procedure (MNDO) was used to determine the equilibrium geometry of formaldehyde and its normal modes and frequencies. For nuclear geometries along three of the modes, we show the molecular dipole moment and its change relative to the dipole moment at the equilibrium geometry. We also display the calculated net atomic charges, which help to understand the origin of the changes in the dipole moment. While the results are only approximate due to the approximate nature of the MNDO model, they serve to demonstrate the principles involved.

It is important to note that in all-valence electron models such as MNDO the calculated molecular dipole moment originates not only in the net charges on atoms, but also in atomic dipoles ("rehybridization charges"), due to one-center off-diagonal matrix elements of the electric dipole moment operator, such as those between a 2s and a 2p orbital on an atom. These contributions are not displayed in Figure 2.1.

It is of some interest to note that the dipole velocity formalism [see (1.43) and (1.44)] is not useful for the calculation of vibrational transition moments in the Born-Oppenheimer approximation. We shall meet this problem again in the discussion of magnetic dipole moments of vibrational transitions.

In addition to the a priori approach to vibrational electric dipole transition moments described so far, more empirical approaches have seen much use. In these, electro-optical parameters are assigned to bonds (bond moments and their derivatives) and are assumed to be transferable from one molecule to another. A detailed discussion lies outside the scope of this text (see Section 2.4).

Vibrational electric quadrupole transition moments can be calculated in a way entirely analogous to that used for electric dipole moments. The calculated net charges and rehybridization contributions are readily converted into the components of the molecular electric quadrupole moment, and their derivatives with respect to nuclear displacements along the normal modes, evaluated at the equilibrium geometry, provide the desired components of the transition moment.

The calculation of magnetic dipole transition moments for vibrational transitions is much more difficult. Assuming again that the normal modes have been obtained, we are faced with the problem of evaluating expressions analogous to (2.6) and (2.7):

$$\mathcal{M}(\mathrm{jv,jv'}) = \mathcal{M}(\mathrm{vv'}) = \langle\chi_{\mathrm{v}}(\mathbf{Q})|\hat{\mathcal{M}}(\mathbf{Q})|\chi_{\mathrm{v'}}(\mathbf{Q})\rangle_{\mathbf{Q}} \tag{2.36}$$

$$\hat{\mathcal{M}}(\mathbf{Q}) = \langle\psi_{\mathrm{j}}(\mathbf{q},\mathbf{Q})|\frac{-|e|}{2mc}\sum_{l=1}^{n}(\hat{\mathbf{r}}_l \times \hat{\mathbf{p}}_l + 2\hat{\mathbf{s}}_l)|\psi_{\mathrm{j}}(\mathbf{q},\mathbf{Q})\rangle_{\mathbf{q}} + \sum_{k=1}^{N}\left[\frac{Z_k|e|}{2M_k c}(\hat{\mathbf{R}}_k \times \hat{\mathbf{P}}_k) + \gamma_k\hat{\boldsymbol{\sigma}}_k\right] \tag{2.37}$$

where integration over $\mathbf{Q}$ in (2.36) now includes integration over nuclear spin. We first note that once again the right-hand side of (2.36) vanishes if $\hat{\mathcal{M}}(\mathbf{Q})$ is independent of $\mathbf{Q}$. Thus, all that matters is the change of the magnetic dipole moment of the molecule as its geometry is displaced from equilibrium. Inasmuch as nuclear and electron spin operators are independent of $\mathbf{Q}$, they will not contribute.

Next, we note a calamity: For any choice of the geometric parameter $\mathbf{Q}$, the first term on the right-hand side of (2.37) vanishes by integration over the electronic coordinates $\mathbf{q}$, since it represents the diagonal matrix element of a purely imaginary operator (remember that $\hat{\mathbf{p}} = -i\hbar\nabla$). In the Born-Oppenheimer approximation, the magnetic dipole moment of the vibrating molecule is therefore due solely to the motion of the unscreened nuclear charges. This is clearly absurd, and the approximation cannot be used to obtain useful results for vibrational magnetic dipole transition moments, or, for that matter, dipole velocity transition moments (i.e., matrix elements of $\hat{\mathbf{p}}$).

The proper way to proceed is to remove the approximation. However, non-Born-Oppenheimer calculations for large molecules or even small ones are very difficult, and the proper procedure therefore appears to be impractical.

The usual solution has been to set up a classical mechanical model of the molecule. The simplest such model is known as the "fixed partial charge model." The charge distribution is calculated at the equilibrium molecular geometry; electronic charges are allocated to individual atoms and allowed to collapse into the nuclei. The molecule is then represented as a set of charged points whose motion generates a current and thus a magnetic moment. Their masses are those of the nuclei, and their charges are in general fractional. Their motion along any normal mode is not allowed to cause any transfer of electron density from one atom to another, so that the fractional charges remain constant, and this explains the name of the model. It is clear that this type of model can be used for the calculation of electric dipole transition moments as well. Its performance is not very good, since the assumption that the charge distribution remains fixed during a vibration is unrealistic (cf. Figure 2.1). Relative to this problem, corrections for mechanical anharmonicity are insignificant.

A better classical model, which appears quite useful for the calculation of the matrix elements of both magnetic dipole and electric dipole transition moments of vibrational transitions (and probably will work for electric quadrupole moments as well), is the "charge flow model." In this model, all electron densities are still collapsed into their nearest nuclei as before, so that the molecule is still represented by a set of points with fractional charges executing normal mode motions. Now, the magnetic dipole moment produced by this motion is thought of as consisting of two parts. One is the same as that in the fixed partial charge model and is due to that part of the charges which stays at their respective nuclei; the other is due to the current corresponding to the flow of charge from one atom to another. The classical nature of the model becomes very clear here: We must specify the trajectory along which the negative fractional electron charge flows from one atom to another during the vibrational motion before we can compute the contribution of this charge flow to the magnetic dipole moment, and this violates the uncertainty principle. The assumption usually made is that the charge flows from one point representing an atom to the next point representing an atom along the line representing the bond between them. In spite of its fundamental conceptual shortcomings, the charge flow model produces acceptable results for magnetic dipole moments.

In the presence of lone pairs on one or more of the atoms, it usually becomes unrealistic to collapse all of the electronic density into the nuclei in constructing the model. This is seen readily when it is realized that the centroid of charge due to, say, the lone pair on nitrogen in ammonia does not coincide with the position of the nitrogen atom. In such cases the electrons of the lone pairs are represented as yet another point charge. The definition of the lone-pair electron density is obtained by localizing the molecular orbitals from their usual canonical form and identifying one of the localized orbitals as representing the lone pair.

An alternative that appears much superior to the classical models has been developed recently. In this approach, the magnetic moment of a vibrational transition is obtained from a quantum mechanical calculation on a molecule in outside magnetic field, introduced explicitly into the Hamiltonian (Section 2.4).

If one uses one of the classical models—the fixed partial charge model or the charge flow model—one has the choice of obtaining the fractional charges on the moving atoms in one of two ways. The fundamentally more satisfactory way is a quantum-chemical calculation by an ab initio or semiempirical method; the easier way is the use of one of the empirical schemes based on bond moments and their derivatives with respect to geometric changes, already briefly alluded to in the discussion of electric dipole transition moments. These matters are not further discussed here, and the reader is directed to the references in Section 2.4.

2.2.2 *Computation of Electronic Transition Moments*

The use of equation (2.7) for the calculation of $\hat{\mathbf{M}}(jj',\mathbf{Q})$ and the analogous equations for $\hat{\mathrm{Q}}(jj',\mathbf{Q})$ and $\hat{\mathscr{M}}(jj',\mathbf{Q})$ requires the knowledge of electronic wavefunctions of the ground and excited states at the desired geometry $\mathbf{Q}$. These are generally not known exactly. Approximate electronic wavefunctions are nowadays available for many molecules in the MO LCAO SCF CI form and are usually obtained as follows.

A basis set of one-electron atomic functions (AOs) λ_μ is used to construct molecular orbitals (MOs) ϕ_a in the form of linear combinations (LC) in a self-consistent-field (SCF) manner:

$$\phi_a = \sum_\mu c_{a\mu}\lambda_\mu \tag{2.38}$$

We shall number the MOs 1, . . . ,n/2, in the order of increasing energy. The procedure also produces a set of unoccupied (virtual) orbitals, labeled n/2 + 1, n/2 + 2, ... Here we assume that the number of electrons n is even, but very similar procedures are available if n is odd. Each orbital ϕ_a gives rise to two spinorbitals (products of spin and orbital functions), which we shall denote $\phi_a = \phi_a\alpha$ and $\bar{\phi}_a = \phi_a\beta$.

A specification of orbital occupancy represents an electron configuration. More explicitly, the one-electron MO functions ϕ_a and $\bar{\phi}_a$ are used to construct a series of spin- and symmetry-adapted many-electron configuration wavefunctions ${}^{2S+1}D_{(i)\kappa}$. Here, 2S + 1 is multiplicity, the subscript (i) indicates the projection of the total electron spin angular momentum into the direction of a quantizing magnetic field (z), and κ serves to distinguish the various possible configurations obtainable from the one-electron functions by specification of occupancies. Each spin- and symmetry-adapted configuration wavefunction consists of one or a combination of several Slater determinants obtained by antisymmetrizing a product of those molecular spinorbitals which are occupied by the n electrons in the particular configuration. The combinations are chosen so as to produce spin and symmetry eigenstates. A determinant is a sum of n! terms; each term can be written as a product of a function of space coordinates and a function of spin coordinates of all the electrons.

When the lowest-energy n/2 MOs are chosen for the construction of ${}^{2S+1}D_{(0),\kappa}$, the closed-shell ground singlet configuration ${}^1D_{(0),g}$ ordinarily results and can be written as a single Slater determinant:

$$\begin{aligned} {}^1D_{(0),g} &= \frac{1}{\sqrt{n!}}\sum_P (-1)^P \phi_1(1)\bar{\phi}_1(2)\phi_2(3)\bar{\phi}_2(4)\cdots\phi_{n/2}(n-1)\bar{\phi}_{n/2}(n) \\ &= |\phi_1\bar{\phi}_1\phi_2\bar{\phi}_2\cdots\phi_{n/2}\bar{\phi}_{n/2}| \end{aligned} \tag{2.39}$$

The symbol P stands for a permutation of the labels of electrons whose coordinates represent the independent variables on the right-hand side, so that the sum has n! terms. In the exponent, P stands for the parity of the permutation.

Other configurations are referred to as excited configurations. Singly excited configurations differ from ${}^1D_{(0),g}$ in that one electron is promoted from orbital a ($a \leq n/2$) to a higher energy orbital b ($b > n/2$). In doubly excited configurations, two electrons are so promoted, etc. In multiply excited configurations, more than one spin assignment to orbitals is frequently possible for each of the three components of a triplet, etc., and all independent possibilities must be considered.

Examples of singlet and triplet singly excited configurations are

$$ {}^1D_{(0),a\to b} = \frac{1}{\sqrt{2}}\{|\phi_1\bar{\phi}_1 \cdots \phi_a\bar{\phi}_b \cdots \phi_{n/2}\bar{\phi}_{n/2}| - |\phi_1\bar{\phi}_1 \cdots \bar{\phi}_a\phi_b \cdots \phi_{n/2}\bar{\phi}_{n/2}|\} \tag{2.40} $$

$$ {}^3D_{(+1),a\to b} = |\phi_1\bar{\phi}_1 \cdots \phi_a\phi_b \cdots \phi_{n/2}\bar{\phi}_{n/2}| \tag{2.41} $$

$$ {}^3D_{(0),a\to b} = \frac{1}{\sqrt{2}}\{|\phi_1\bar{\phi}_1 \cdots \phi_a\bar{\phi}_b \cdots \phi_{n/2}\bar{\phi}_{n/2}| + |\phi_1\bar{\phi}_1 \cdots \bar{\phi}_a\phi_b \cdots \phi_{n/2}\bar{\phi}_{n/2}|\} \tag{2.42} $$

$$ {}^3D_{(-1),a\to b} = |\phi_1\bar{\phi}_1 \cdots \bar{\phi}_a\bar{\phi}_b \cdots \phi_{n/2}\bar{\phi}_{n/2}| \tag{2.43} $$

Finally, pure multiplicity (pure spin) states are represented by linear combinations of configurations (configuration interaction, CI), with coefficients determined once again by solving a system of secular equations derived from the variational theorem:

$$ {}^{2S+1}\psi_{(i),f} = \sum_{\kappa} C_{\kappa f}\, {}^{2S+1}D_{(i),\kappa} \tag{2.44} $$

In general, a pure multiplicity state wavefunction ${}^{2S+1}\psi_{(i)}$ is then represented by a sum of terms, each of which has the form of a product of a space part and a spin part.

The various components (i) of a multiplet differ slightly in their energy. The terms in the Hamiltonian which are responsible for these differences can be collected in the so-called spin Hamiltonian. The most important among these are the Zeeman term, which dominates in strong outside magnetic fields and is absent in zero field, the electron spin dipole–spin dipole interaction, the spin-orbit coupling, and the coupling of electron to nuclear spins. For our purposes, the best choice of (i) is that which diagonalizes the spin Hamiltonian. The one we have used so far does so in the limit of strong magnetic field [the spin functions of the electrons in orbitals ϕ_a and ϕ_b are $\alpha(1)\alpha(2)$ for $i = +1$, $(1/\sqrt{2})[\alpha(1)\beta(2) + \alpha(2)\beta(1)]$ for $i = 0$, and $\beta(1)\beta(2)$ for $i = -1$].

Multiplicity-Allowed Electronic Transitions. Once the state wavefunctions ${}^{2S+1}\psi_{(i),j}(\mathbf{q},\mathbf{Q})$ and ${}^{2S'+1}\psi_{(i'),j'}(\mathbf{q},\mathbf{Q})$ have been obtained at a particular geometry $\mathbf{Q}$, the transition moments and tensors $\hat{\mathbf{M}}(jj',\mathbf{Q})$, $\hat{\mathbf{Q}}(jj',\mathbf{Q})$, and $\hat{\mathcal{M}}(jj',\mathbf{Q})$, $j \neq j'$, can be evaluated by integration over the coordinates of the electrons as shown in (2.5). Integration over the spin coordinates of the electrons will yield zero unless the transition is multiplicity-allowed (spin-allowed), $S = S'$ and $i = i'$, due to the orthogonality of spin functions which differ in S or i, and of electronic functions which differ in j. This is a result of the fact that $\hat{\mathbf{M}}$ and $\hat{Q}$ are spin-free operators, and $\hat{\mathscr{M}}$ contains a sum of spin-free and space-free parts (the latter does not contribute when $j \neq j'$, and we omit it in the following).

Thus, in the pure multiplicity approximation, only transitions between states of like multiplicity have nonzero moments. This approximation is adequate for a discussion of spin-allowed transitions (e.g., singlet → singlet or triplet → triplet) unless very heavy atoms are present in the molecule. It is not adequate for a discussion of multiplicity-forbidden transitions (e.g., singlet → triplet), and these will be taken up below.

Integration over the space coordinates of the electrons for multiplicity-allowed transitions is performed in several steps. First, substitution from (2.44) yields

$$\mathbf{M}(jj') = \sum_{\kappa}\sum_{\kappa'} C_{\kappa j} C_{\kappa' j'} \langle D_{\kappa'} | \hat{\mathbf{M}} | D_{\kappa} \rangle \tag{2.45}$$

where the argument $\mathbf{Q}$ and the labels S and i have been dropped, and similar results are obtained for $Q(jj')$ and $\mathscr{M}(jj')$. The integral on the right-hand side of (2.45) looks complicated since it consists of $(n!)^2$ terms when the determinants are expanded. However, $\hat{\mathbf{M}}$, $\hat{Q}$, and $\hat{\mathscr{M}}$ are equal to sums of the one-electron operators

$$\hat{\mathbf{m}}_l = -|e|\hat{\mathbf{r}}_l \tag{2.46}$$

$$\hat{q}_l = -|e|[\hat{\mathbf{r}}_l\hat{\mathbf{r}}_l - 1(\hat{\mathbf{r}}_l \cdot \hat{\mathbf{r}}_l)/3] \tag{2.47}$$

and

$$\hat{\boldsymbol{\mu}}_l = (-|e|/2mc)(\hat{\mathbf{r}}_l \times \hat{\mathbf{p}}_l) \tag{2.48}$$

respectively [cf. (2.1)–(2.3)], and this causes most of these $(n!)^2$ terms to vanish. The rest are all equal to one of only a small number of values. A detailed consideration produces the so-called Slater rules for the evaluation of such matrix elements. For instance, for singlet configurations and the operator $\hat{\mathbf{M}}$, one has

$$\langle {}^1D_g | \hat{\mathbf{M}} | {}^1D_g \rangle = \sum_a \langle \phi_a | \hat{\mathbf{m}} | \phi_a \rangle \tag{2.49}$$

$$\langle {}^1D_{a\to b} | \hat{\mathbf{M}} | {}^1D_g \rangle = \sqrt{2} \langle \phi_b | \hat{\mathbf{m}} | \phi_a \rangle \tag{2.50}$$

$$\langle {}^1D_{a'\to b'} | \hat{\mathbf{M}} | {}^1D_{a\to b} \rangle = \langle \phi_{b'} | \hat{\mathbf{m}} | \phi_b \rangle \delta_{aa'} - \langle \phi_{a'} | \hat{\mathbf{m}} | \phi_a \rangle \delta_{bb'} \tag{2.51}$$

and similar results for multiply excited configurations. If the two determinants D_κ and $D_{\kappa'}$ differ in the occupancy of more than one spinorbital in the configurations κ and κ', the matrix element vanishes.

The matrix elements $\langle \phi_{a'} | \hat{\mathbf{m}} | \phi_a \rangle$, $\langle \phi_{a'} | \hat{q} | \phi_a \rangle$, and $\langle \phi_{a'} | \hat{\boldsymbol{\mu}} | \phi_a \rangle$ are evaluated by substituting from (2.38) to obtain expressions such as

$$\langle \phi_{a'} | \hat{\mathbf{m}} | \phi_a \rangle = \sum_{\mu}\sum_{\nu} c_{a'\mu} c_{a\nu} \langle \lambda_\mu | \hat{\mathbf{m}} | \lambda_\nu \rangle \tag{2.52}$$

It is not possible to postpone the integration any longer, and at this point the calculations for $\hat{\mathbf{M}}$, $\hat{\mathrm{Q}}$, and $\hat{\mathcal{M}}$ diverge, since we need to substitute the explicit forms for $\hat{\mathbf{m}}$, $\hat{\mathrm{q}}$, and $\hat{\boldsymbol{\mu}}$. This leads to integrals of the types $\langle\lambda_\mu|\hat{\mathrm{r}}_\mathrm{u}|\lambda_\nu\rangle$, $\langle\lambda_\mu|\hat{\mathrm{r}}_\mathrm{u}\hat{\mathrm{r}}_\mathrm{v}|\lambda_\nu\rangle$, and $\langle\lambda_\mu|\hat{\mathrm{r}}_\mathrm{u}\hat{\mathrm{p}}_\mathrm{v} - \hat{\mathrm{r}}_\mathrm{v}\hat{\mathrm{p}}_\mathrm{u}|\lambda_\nu\rangle$ (u = x, y, or z). These can either be evaluated by substituting the explicit form of the atomic orbitals λ_μ and λ_ν and actually integrating the resulting expression, or approximated using suitable physical arguments. The former is done in the so-called ab initio methods of calculating molecular electronic structure; the latter is usually done in the so-called semiempirical methods. In the latter, the calculation of the wavefunction is normally based on the assumption of zero differential overlap, in which the set of orbitals λ_μ is orthonormal. In the calculation of matrix elements of operators such as $\hat{\mathbf{r}}$ and $\hat{\mathbf{p}}$, the fact that such orbitals are not completely localized on a single atomic center is sometimes ignored and the matrix elements are obtained as shown below for truly localized atomic orbitals (which are not mutually orthogonal when located at different atoms). Although this approximation frequently has little effect on the results, it is preferable to recognize explicitly that the orbitals λ_μ are obtained from a truly localized set of atomic orbitals λ'_μ by a Löwdin orthogonalization:

$$\lambda_\mu = \sum_\kappa (\mathsf{S}^{-1/2})_{\mu\kappa}\lambda'_\kappa \tag{2.53}$$

where S is the overlap matrix, i.e., $\mathrm{S}_{\mu\nu} = \langle\lambda'_\mu|\lambda'_\nu\rangle$.

Here we shall describe briefly the procedures used in most semiempirical methods, since they provide ready intuitive insight. For simplicity, we choose a π-electron system, since it is quite typical of the kinds of molecules to which the methods of polarized electronic spectroscopy have been applied most often so far and since we are primarily interested in the illustrative value.

The basis orbitals λ'_μ in a simple π-electron wavefunction are thought of as a set of normalized $\mathrm{p_z}$ orbitals, oriented with their z axis perpendicular to the molecular plane. Each atom which forms a part of the conjugated system carries one such orbital. Each has cylindrical symmetry around its own z axis, but its radial dependence is not explicitly specified. Let $\mathbf{R}_\mu$ be the position vector of the μ-th atomic center in the coordinate system chosen for the molecule. This center carries orbital λ'_μ. The vectors $\mathbf{R}_\mu$ are known if the geometry of the molecule is known. The integral $\langle\lambda'_\mu|\hat{\mathbf{r}}|\lambda'_\mu\rangle$ can then be expressed as

$$\langle\lambda'_\mu|\hat{\mathbf{r}}|\lambda'_\mu\rangle = \langle\lambda'_\mu|\hat{\mathbf{R}}_\mu|\lambda'_\mu\rangle + \langle\lambda'_\mu|\hat{\mathbf{r}} - \hat{\mathbf{R}}_\mu|\lambda'_\mu\rangle \tag{2.54}$$

Now, $\mathbf{R}_\mu$ is a constant vector given by the location of the μ-th nucleus and independent of the electron coordinates. It can be taken in front of the first bracket on the right-hand side, which then reduces to a normalization integral. The second bracket vanishes by symmetry, since $\mathbf{r} - \mathbf{R}_\mu$ is the position vector of the electron in a coordinate system whose origin is located at the μ-th atom, and λ'_μ is cylindrically symmetrical. This yields the very simple result

$$\langle\lambda'_\mu|\hat{\mathbf{r}}|\lambda'_\mu\rangle = \mathbf{R}_\mu \tag{2.55}$$

The off-diagonal elements $\langle\lambda'_\mu|\hat{\mathbf{r}}|\lambda'_\nu\rangle$, in which μ and ν are different centers, are evaluated similarly, using the Mulliken approximation for the overlap density of two atomic orbitals:

$$\lambda'_\mu \lambda'_\nu = S_{\mu\nu}(\lambda'_\mu \lambda'_\mu + \lambda'_\nu \lambda'_\nu)/2 \tag{2.56}$$

One obtains

$$\langle \lambda'_\mu | \hat{\mathbf{r}} | \lambda'_\nu \rangle = (S_{\mu\nu}/2)(\langle \lambda'_\mu | \hat{\mathbf{r}} | \lambda'_\mu \rangle + \langle \lambda'_\nu | \hat{\mathbf{r}} | \lambda'_\nu \rangle) \tag{2.57}$$

and finally,

$$\langle \lambda'_\mu | \hat{\mathbf{r}} | \lambda'_\nu \rangle = (S_{\mu\nu}/2)(\mathbf{R}_\mu + \mathbf{R}_\nu) \tag{2.58}$$

In all-valence-electron models, one also encounters the off-diagonal matrix elements of $\hat{\mathbf{r}}$ between different AOs ($\mu \neq \nu$) both located at the same center (e.g., a 2s and a 2p orbital on nitrogen). These are usually evaluated by direct quadrature, assuming a suitable explicit functional form for the AOs. They are responsible for the "rehybridization" contributions to dipole moments.

Returning now to the π-electron model, we note that integrals of the type $\langle \lambda'_\mu | \hat{\mathbf{r}}\hat{\mathbf{r}} | \lambda'_\nu \rangle$ can be similarly expressed as

$$\langle \lambda'_\mu | \hat{\mathbf{r}}\hat{\mathbf{r}} | \lambda'_\mu \rangle = \mathbf{R}_\mu \mathbf{R}_\mu \tag{2.59}$$

$$\langle \lambda'_\mu | \hat{\mathbf{r}}\hat{\mathbf{r}} | \lambda'_\nu \rangle = (S_{\mu\nu}/4)(\mathbf{R}_\mu + \mathbf{R}_\nu)(\mathbf{R}_\mu + \mathbf{R}_\nu) \tag{2.60}$$

so that, for instance, the xy element of the operator $\hat{\mathbf{r}}\hat{\mathbf{r}}$ between λ'_μ and λ'_ν is

$$\langle \lambda'_\mu | \hat{\mathrm{x}}\hat{\mathrm{y}} | \lambda'_\nu \rangle = (S_{\mu\nu}/4)[(\mathbf{R}_\mu)_x + (\mathbf{R}_\nu)_x][(\mathbf{R}_\mu)_y + (\mathbf{R}_\nu)_y] \tag{2.61}$$

where $(\mathbf{R}_\mu)_u$ stands for the u-th coordinate of vector $\mathbf{R}_\mu$.

The expressions for the integrals $\langle \lambda'_\mu | \hat{\mathbf{r}} \times \hat{\mathbf{p}} | \lambda'_\nu \rangle$ in π-electron models are more complicated. The diagonal elements vanish, since $\hat{\mathbf{p}}$ is purely imaginary, and expressions for the off-diagonal elements are obtained by the following series of steps. First, it is recognized that in a semiempirical model for a planar π system only the component of the matrix element taken in the direction perpendicular to the plane is different from zero, since all AOs available are antisymmetric relative to the molecular plane. Now,

$$\langle \lambda'_\mu | (\hat{\mathbf{r}} \times \hat{\mathbf{p}})_z | \lambda'_\nu \rangle = \langle \lambda'_\mu | [(\hat{\mathbf{r}} - \hat{\mathbf{R}}_\mu) \times \hat{\mathbf{p}}]_z | \lambda'_\nu \rangle + \langle \lambda'_\mu | (\hat{\mathbf{R}}_\mu \times \hat{\mathbf{p}})_z | \lambda'_\nu \rangle \tag{2.62}$$

The first bracket on the right-hand side vanishes since $[(\hat{\mathbf{r}} - \hat{\mathbf{R}}_\mu) \times \hat{\mathbf{p}}]_z$ is the operator for the z component of angular momentum relative to the origin located at the center μ, and the orbital λ'_μ, being of the p_z atomic type, has no angular dependence with respect to rotation around a line passing through its center and parallel to z. The second integral on the right-hand side contains the constant vector $\mathbf{R}_\mu$, which can be taken out in front. Then,

$$\langle \lambda'_\mu | (\hat{\mathbf{r}} \times \hat{\mathbf{p}})_z | \lambda'_\nu \rangle = (\mathbf{R}_\mu \times \langle \lambda'_\mu | \hat{\mathbf{p}} | \lambda'_\nu \rangle)_z \tag{2.63}$$

Now the matrix element of the linear momentum operator $\hat{\mathbf{p}} = -i\hbar\nabla$ can be evaluated by explicit integration if a suitable functional form is assumed for the atomic orbitals. It is more in keeping with the spirit of the semiempirical models to determine it from some other parameters already present in the model.

This can be done by demanding that the usual commutation relation between $\hat{\mathbf{p}}$ and $\hat{\mathbf{r}}$ be fulfilled for exact solutions of the π-electron model, and the result is

$$\langle\lambda'_\mu|\hat{\mathbf{p}}|\lambda'_\nu\rangle = (i\beta_{\mu\nu}\mathrm{m}/\hbar)(\mathbf{R}_\nu - \mathbf{R}_\mu) \tag{2.64}$$

where $\beta_{\mu\nu}$ is the resonance integral between AOs λ'_μ and λ'_ν, and m is electron mass. For the usual values of $\beta_{\mu\nu}$, the results obtained in this fashion are quite similar to those obtained by explicit integration over Slater orbitals with the usual exponents.

In the singlet ground state of ordinary organic molecules, ${}^1\mathrm{D}_{(0),g}$ predominates by far (C_{0j} is nearly unity). In the low-energy excited states of such molecules, both singlet and triplet, the main contributing configurations are of the singly excited type. However, it is not unusual for a low-energy excited state to contain significant contributions of the doubly excited type. In organic radicals, biradicals, and many transition metal complexes, the situation tends to be even more complicated.

It is fairly common for one or more of the low-energy singlet excited states of ordinary organic molecules to be represented nearly exclusively by only one singly excited configuration ${}^1\mathrm{D}_{k\to l}$. Often the excitation of an electron from the highest energy occupied MO (HOMO) to the lowest unoccupied MO (LUMO) of the ground state is of this kind (${}^1\mathrm{D}_{n/2\to n/2+1}$). Whenever this situation occurs, it is very easy to visualize the magnitude and orientation of the transition moment in terms of the MOs involved. This is likely to provide the reader with a better feeling for the relation of electric dipole transition moments to more familiar concepts. Our task will be simplified if we choose a $\pi\pi^*$ transition in a planar conjugated molecule, since we have already used this example in (2.54)–(2.64) and since a simple pictorial representation of π MOs is available. Similar considerations would apply in other instances but would be complicated by the presence of one-center ("rehybridization") contributions and would be harder to represent graphically.

In order to represent a π orbital ϕ_k, we view the plane of the molecule from above and draw circles at the individual centers in the conjugated system, such that their radius is proportional to the MO coefficient $c_{k\mu}$ at the center μ. The circle is white if $c_{k\mu}$ is positive and black if $c_{k\mu}$ is negative (see Figure 2.2). The contribution of an electron in MO ϕ_k to the total electron charge density in the molecule is given by $-|e|\phi_k^2$. Substitution for ϕ_k from (2.38) gives a double sum which contains diagonal terms, $-|e|c_{k\mu}^2\lambda_\mu^2$, and off-diagonal terms, $-|e|c_{k\mu}c_{k\nu}\lambda_\mu\lambda_\nu$. In the following, we simplify matters by displaying the contributions of the orthonormal orbitals λ_μ as if they were identical with the completely localized atomic orbitals λ'_μ (which are not orthogonal). The diagonal terms are easily visualized, since $c_{k\mu}^2$ is proportional to the area of the circle at center μ, where the AO λ_μ is located. In standard π-electron models of the zero-differential-overlap (ZDO) type, these diagonal terms play a key role in the evaluation of molecular properties. For instance, in such models the total electric dipole moment of the molecule is obtained by summing the negative charge contributions at each center provided by the electrons in all occupied MOs and considering the positive charge contribution of each nucleus in the molecule. The resulting dipole moment is given by the dipole moment of the classical charge distribution of positive point charges at the location of the nuclei and negative point charges due to electron density located at the individual centers of the conjugated system, with magnitudes given by $-|e|c_{\mu k}^2$.

A similar pictorial representation is possible within the π-electron model for electric dipole transition moments $\langle ^1D_0|\hat{\mathbf{M}}|^1D_{k\to l}\rangle$. In ZDO models, these are again given by the electric moments of a distribution of point charges on the centers of the conjugated system. Now, however, the magnitudes of these charges are not given by the distribution of electron charge density due to an electron in a single MO, neither in its initial orbital ϕ_k nor in the final orbital ϕ_l, but rather by the overlap charge density $\rho_{kl} = -|e|\phi_k\phi_l$, also referred to as the transition charge density. Note that it is positive in some parts of space and negative in others; its integral vanishes, since ϕ_k and ϕ_l are mutually orthogonal. This is an interference term between the one-electron wavefunctions ϕ_k and ϕ_l, and it is qualitatively reasonable that it should play a role in the absorption process where an outside perturbation causes a mixing of a stationary wavefunction containing ϕ_k with a stationary wavefunction containing ϕ_l.

Figure 2.2 shows in a stepwise fashion the derivation of the electric dipole transition moment for the HOMO → LUMO excitation in naphthalene, which corresponds to its ^1L_a transition, observed at about 36 000 cm^{-1}. The two orbitals involved are shown on the left, their overlap probability density in the center, and their overlap charge density and its dipole moment $\mathbf{M}^0(jj')$ on the right. A transition quadrupole moment would be derived similarly by consideration of the quadrupole moment of the classical charge distribution shown in the center.

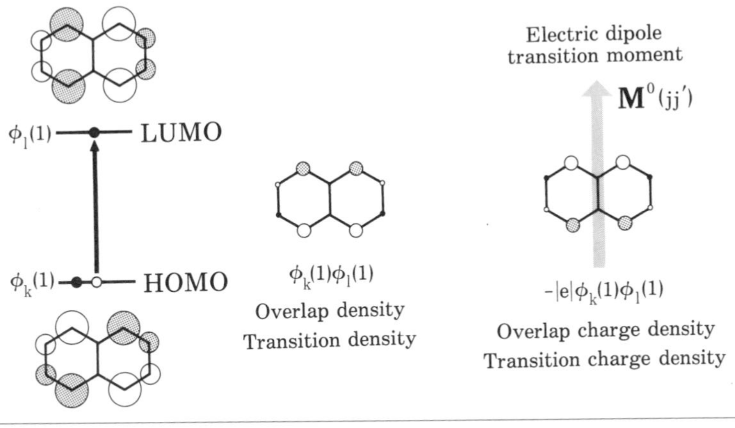

Figure 2.2 Schematic derivation of the electronic transition moment of the L_a transition in naphthalene. Left: coefficients of the highest occupied and lowest unoccupied MOs. Center: the transition density for the HOMO → LUMO promotion. Right: the transition charge density and its dipole moment (the transition moment). The circles indicate the sign (white, positive; dark, negative) and magnitude (proportional to radius) of contributions on the individual carbon p_z orbitals.

Multiplicity-Forbidden Electronic Transitions. In most organic and many inorganic molecules which do not contain very heavy atoms, the approximation of pure multiplicity states is adequate for the discussion of multiplicity-allowed transitions. As noted earlier, in this approximation the transition moment $\mathbf{M}^0(jj')$ vanishes if the multiplicities of the states j and j′ differ, and so does the moment of any other spin-free transition operator. For instance, in this approximation, ordinary organic molecules with singlet ground states will typically exhibit a variety of transitions into excited singlet states—some with large transition moments, some with small transition moments, and perhaps even some with vanishing transition moments—but all their transitions into excited triplet states will have zero transition moments. Similarly, transitions among the three components of a triplet state will have zero electric dipole and quadrupole moments. Magnetic dipole transition moments between different components (i) of an electronic state of a given pure multiplicity need not vanish, since $\hat{\mathscr{M}}$ operates on both electron and nuclear spin coordinates, and is responsible for the transitions observed in magnetic resonance spectroscopy.

However, strictly speaking, in nonlinear molecules, the spin angular momentum and its z component are not constants of motion and the classification of states by multiplicity is only approximate. Although transition moments of spin-free operators between states of different S and S_z "quantum numbers" are very small, they do not vanish exactly, and the corresponding absorptions and emissions can be observed. We shall now take up the calculation of transition moments of such "multiplicity-forbidden" transitions.

Among the Hamiltonian terms which cause the mixing of zero-order pure multiplicity states, spin-orbit coupling is the most important, and we shall describe its effects briefly. For this purpose, we shall assume for simplicity that the pure multiplicity wavefunction ${}^{2S+1}\psi_{(i)}$ is represented by a single product of a space part ${}^{2S+1}\Phi$ and a spin part ${}^{2S+1}\Lambda_{(i)}$, where the label (i) numbers the spin functions chosen as eigenfunctions of the spin Hamiltonian. In a strong external magnetic field, i refers to the component of spin angular momentum along the field direction z in units of $\hbar$, since the Zeeman term dominates the spin Hamiltonian (for singlets, i = 0; for triplets, i = +1, 0, or −1). In zero magnetic field, (i) refers to molecular axes obtained by diagonalization of the zero-field spin Hamiltonian. For singlets, i = 0; for triplets, i = x, y, or z, and i refers to the axis with zero component of spin angular momentum. These triplet spin functions are

$$(1/\sqrt{2})[\alpha(1)\alpha(2) - \beta(1)\beta(2)] \qquad \text{for } i = x$$

$$(1/\sqrt{2})[\alpha(1)\alpha(2) + \beta(1)\beta(2)] \qquad \text{for } i = y$$

and

$$(1/\sqrt{2})[\alpha(1)\beta(2) + \alpha(2)\beta(1)] \qquad \text{for } i = z$$

In magnetic fields of intermediate strength, the situation is more complicated. In the following, we shall assume the zero-field case.

The interaction between the spin angular momentum of an electron and the orbital momentum of the same electron causes the presence of a minor term in the molecular Hamiltonian, which can be written as

$$\hat{H}_{SO}(\mathbf{q},\mathbf{Q}) = \frac{1}{2m^2c^2} \sum_{l=1}^{n} (\boldsymbol{\nabla}V \times \hat{\mathbf{p}}_l) \cdot \hat{\mathbf{s}}_l \tag{2.65}$$

where V is the electrostatic potential experienced by the electron, and the other symbols have the meaning introduced earlier. The gradient vector $\boldsymbol{\nabla}V$ is particularly large in the vicinity of a nucleus. Since electrons in molecules spend most of their time very close to nuclei, so much so that it is quite reasonable to approximate their wavefunctions in the LCAO fashion, it is also reasonable to approximate each term in the sum in (2.65) as a collection of N terms, each originating at one of the nuclei and each containing only the potential V due to the nucleus in question. The gradient of a potential due to the point charge $Z_k|e|$ on the k-th nucleus can be written as

$$\boldsymbol{\nabla}V_k = (Z_k e^2/|\mathbf{r}_k|^3)\mathbf{r}_k \tag{2.66}$$

where $\mathbf{r}_k = \mathbf{r} - \mathbf{R}_k$. Spin-orbit coupling increases in importance as the atomic number increases, and this is known as the "heavy atom effect."

The combination of (2.65) with (2.66) produces the usual form of $\hat{H}_{SO}$,

$$\hat{H}_{SO}(\mathbf{q},\mathbf{Q}) = \frac{e^2}{2m^2c^2} \sum_{l=1}^{n} \sum_{k=1}^{N} \frac{Z_k}{|\mathbf{r}_k|^3} \hat{\mathbf{l}}_l \cdot \hat{\mathbf{s}}_l \tag{2.67}$$

where $\hat{\mathbf{l}}_l = \hat{\mathbf{r}}_l \times \hat{\mathbf{p}}_l$ is the orbital angular momentum operator of the l-th electron. The presence in the Hamiltonian of the orbital and spin angular momentum operators combined into a single product $\hat{\mathbf{l}} \cdot \hat{\mathbf{s}}$ causes the mutual mixing of pure multiplicity states ${}^{2S+1}\psi_{(i)}$. The former acts only on the spatial part of the total electronic wavefunction of such a state, ${}^{2S+1}\Phi$, the latter only on its spin part, ${}^{2S+1}\Lambda_{(i)}$. The presence of an inseparable scalar product of the two, $\hat{\mathbf{l}} \cdot \hat{\mathbf{s}}$, in the total electronic Hamiltonian $\hat{H}_{el}$ will cause an interaction between various pure multiplicity wavefunctions of the type ${}^{2S+1}\psi_{(i)} = {}^{2S+1}\Phi^{2S+1}\Lambda_{(i)}$ which differ in S. The resulting eigenfunctions ψ of $\hat{H}_{el}$ will therefore be represented by mixtures of pure multiplicity functions which differ in multiplicity. Then the integral which defines the transition moment in (2.4) will no longer vanish exactly by integration over spin, even when the transition operator is spin-free. However, since $\hat{H}_{SO}$ is only a very small term in $\hat{H}_{el}$, the mixing is normally not very severe, and each final wavefunction ψ contains predominantly only a function ${}^{2S+1}\psi$ of one multiplicity. In common parlance, such "impure singlets," "impure triplets," etc., are still called simply singlets, triplets, etc., as we have been doing so far, but transitions between them are now weakly allowed.

Spin-orbit coupling contributes to the minute energy differences between the components of a triplet state, and higher multiplicity states, which differ in (i). However, in organic molecules most of this difference is due to another small term in the Hamiltonian, the dipole-dipole interaction between the magnetic dipole moments of different electrons. An investigation of the transitions between the levels which differ in (i) is the domain of ESR spectroscopy.

Since spin-orbit coupling is a very weak effect except in molecules containing very heavy atoms, it can be treated by first-order perturbation theory. This yields, for instance, for one of the three "impure triplet" wavefunctions $\psi_{j'(i)}$ of the lowest triplet state T_1,

$$\begin{aligned}\psi_{j'(i)} &= \psi_{T_1(i)} \\ &= N[{}^3\Phi_{j'}{}^3\Lambda_{(i)} + \sum_n {}^1\Phi_n{}^1\Lambda_{(0)}\langle {}^1\Phi_n{}^1\Lambda_{(0)}|\hat{H}_{SO}|{}^3\Phi_{j'}{}^3\Lambda_{(i)}\rangle(E_n - E_{j'})^{-1}] \end{aligned} \quad (2.68)$$

where the sum runs in principle over all components of all pure multiplicity states, but only the singlet states ${}^1\psi_n = {}^1\Phi_n\,{}^1\Lambda_{(0)}$ are shown since the others will not contribute in the following example. The integration is over all space and spin coordinates of the electrons.

In a similar fashion, the mixing causes the ground-state wavefunction to become an "impure singlet" ψ_j (which has only the component i = 0)

$$\begin{aligned}\psi_j &= \psi_{S_0} \\ &= N'[{}^1\Phi_j{}^1\Lambda_{(0)} + \sum_{m,i} {}^3\Phi_m{}^3\Lambda_{(i)}\langle {}^3\Phi_m{}^3\Lambda_{(i)}|\hat{H}_{SO}|{}^1\Phi_j{}^1\Lambda_{(0)}\rangle(E_m - E_j)^{-1}] \end{aligned} \quad (2.69)$$

Since the perturbation is so weak, the normalization factors N and N′ are almost exactly equal to 1 and shall be omitted in the following.

The electric dipole transition moment of an electronic transition from the ground "singlet" state $\psi_j = \psi_{S_0}$ to the i-th component of the lowest "triplet" state $\psi_{j'(i)} = \psi_{T_1(i)}$ now becomes

$$\begin{aligned}\mathbf{M}[jj'(i)] &= \mathbf{M}[S_0T_1(i)] \\ &= \sum_n \mathbf{M}(jn)\langle {}^1\Phi_n{}^1\Lambda_{(0)}|\hat{H}_{SO}|{}^3\Phi_{j'}{}^3\Lambda_{(i)}\rangle(E_n - E_{j'})^{-1} \\ &\quad + \sum_{m,(i)} \mathbf{M}(jm)\langle {}^3\Phi_m{}^3\Lambda_{(i)}|\hat{H}_{SO}|{}^1\Phi_j{}^1\Lambda_{(0)}\rangle(E_m - E_j)^{-1} \end{aligned} \quad (2.70)$$

where we used the fact that $\langle {}^{2S+1}\Lambda_{(i)}|{}^{2S'+1}\Lambda_{(i')}\rangle = \delta_{SS'}\delta_{ii'}$. Similar expressions are obtained for $\mathbf{Q}[S_0T_1(i)]$ and $\mathcal{M}[S_0T_1(i)]$. Thus, in this description, the $S_0 \rightarrow T_1(i)$ transition acquires nonvanishing transition moments by borrowing from "multiplicity-allowed" singlet-singlet and triplet-triplet transitions.

Application of group theory to (2.70) produces selection rules for singlet-triplet transitions. Since $\hat{H}_{SO}$ is an additive part of the total molecular Hamiltonian operator, it must belong to the totally symmetric irreducible representation. It will therefore mix a singlet wavefunction ${}^1\Phi_n{}^1\Lambda_{(0)}$ with a triplet wavefunction ${}^3\Phi_{j'}{}^3\Lambda_{(i)}$ only if their total (space and spin) symmetries are the same. The singlet spin function ${}^1\Lambda_{(0)}$ is totally symmetrical. In the absence of an external static magnetic field, the triplet spin eigenfunctions of the spin Hamiltonian, ${}^3\Lambda_{(i)}$, transform as the rotations $\mathcal{R}_i$ along the i-th molecular axis.

Thus, the overall symmetry of a singlet wavefunction ${}^1\Phi_n{}^1\Lambda_{(0)}$ is equal to that of its orbital part ${}^1\Phi_n$. On the other hand, if none of the $\mathcal{R}_i$'s form a basis for a degenerate irreducible representation, the three components of the triplet state j′ will have total symmetries given by $\Gamma(^3\Phi_{j'}) \times \Gamma(\mathcal{R}_x)$, $\Gamma(^3\Phi_{j'}) \times \Gamma(\mathcal{R}_y)$, and $\Gamma(^3\Phi_{j'}) \times \Gamma(\mathcal{R}_z)$, where the Γ's stand for the irreducible representations which transform like $^3\Phi_{j'}$ and $\mathcal{R}_i$ (i = x,y,z), respectively. Thus, the requirement of a nonvanishing spin-orbit coupling matrix element between the two states is that the direct product $\Gamma(^1\Phi_n) \times \Gamma(^3\Phi_{j'}) \times \Gamma(\mathcal{R}_i)$ must contain the totally symmetric irreducible representation for at least one of the three choices of i. It is readily verified that the matrix element rarely vanishes for all choices of i.

According to (2.70), the requirement for the μ-th component of **M**(jj′) not to vanish is that the direct product $\Gamma(^1\Phi_j) \times \Gamma(\mu) \times \Gamma(^3\Phi_{j'}) \times \Gamma(\mathcal{R}_i)$ should contain the totally symmetric representation for at least one choice of i. Which i this turns out to be determines which one of the three components $^3\Lambda_{(i)}$ of the triplet state is engaged in the transition. This is usually immaterial, since they lie so close in energy that they are not resolved in the spectrum and their populations are in rapid equilibrium at all but the lowest temperatures.

The ground state is usually totally symmetric, and the product then simplifies to $\Gamma(\mu) \times \Gamma(^3\Phi_{j'}) \times \Gamma(\mathcal{R}_i)$. Since i can have any one of three values, this product frequently contains the totally symmetric representation for more than one choice of μ, and mixed polarization results. On relatively rare occasions, the product does not contain the totally symmetric representations for any choice of μ. Then it becomes essential to consider vibronic coupling and the geometry dependence of $\hat{H}_{SO}$.

Symmetry- and Multiplicity-Forbidden Electronic Transitions. At times, it is not possible to neglect the dependence of the spin-orbit coupling on molecular geometry. The most common example of this situation is a transition which is not only multiplicity-forbidden but whose transition moment also vanishes by symmetry at the equilibrium geometry.

In analogy to (2.16), $\hat{H}_{SO}(\mathbf{q},\mathbf{Q})$ can be written as

$$\hat{H}_{SO}(\mathbf{q},\mathbf{Q}) = \hat{H}_{SO}(\mathbf{q},\mathbf{0}) + \hat{H}_{vSO}(\mathbf{q},\mathbf{Q}) \qquad (2.71)$$

Usually, $\hat{H}_{SO}(\mathbf{q},\mathbf{Q})$ is approximated by the linear terms in a McLaurin expansion analogous to (2.17):

$$\hat{H}_{vSO}(\mathbf{q},\mathbf{Q}) = \sum_{k=1}^{3N-6} \left[\frac{\partial \hat{H}_{SO}(\mathbf{q},\mathbf{Q})}{\partial Q_k}\right]_0 \hat{\mathbf{Q}}_k \qquad (2.72)$$

where k runs over all normal modes of vibration and the derivative is taken at the equilibrium geometry **0**. A simultaneous treatment of vibronic coupling and spin-orbit coupling then uses the perturbation operator

$$\hat{H}_{pert}(\mathbf{q},\mathbf{Q}) = \hat{H}_{ev}(\mathbf{q},\mathbf{Q}) + \hat{H}_{SO}(\mathbf{q},\mathbf{0}) + \hat{H}_{vSO}(\mathbf{q},\mathbf{Q}) \qquad (2.73)$$

Experimental evidence shows that $\hat{H}_{ev}(\mathbf{q},\mathbf{Q}) > \hat{H}_{SO}(\mathbf{q},\mathbf{0}) > \hat{H}_{vSO}(\mathbf{q},\mathbf{Q})$. Second-order perturbation theory with the use of a few reasonable approximations yields for the wavefunction of the triplet state at geometry **Q** located in the vicinity of **0**

$$\psi_{j'(i)}(\mathbf{q},\mathbf{Q}) = \psi_{T_1(i)}(\mathbf{q},\mathbf{Q})$$

$$= N'\Big\{ {}^3\Phi_{j'}(\mathbf{q},\mathbf{0})^3\Lambda_{(i)}(\mathbf{q})$$

$$+ \sum_n {}^1\Phi_n(\mathbf{q},\mathbf{0})^1\Lambda_{(0)}(\mathbf{q})\langle {}^1\Phi_n(\mathbf{q},\mathbf{0})^1\Lambda_{(0)}(\mathbf{q})|\hat{H}_{SO}(\mathbf{q},\mathbf{0})|{}^3\Phi_{j'}(\mathbf{q},\mathbf{0})^3\Lambda_{(i)}(\mathbf{q})\rangle_{\mathbf{q}}(E_n - E_{j'})^{-1}$$

$$+ \sum_n {}^1\Phi_n(\mathbf{q},\mathbf{0})^1\Lambda_{(0)}(\mathbf{q})\langle {}^1\Phi_n(\mathbf{q},\mathbf{0})^1\Lambda_{(0)}(\mathbf{q})|\hat{H}_{vSO}(\mathbf{q},\mathbf{Q})|{}^3\Phi_{jz}(\mathbf{q},\mathbf{0})^3\Lambda_{(i)}(\mathbf{q})\rangle_{\mathbf{q}}(E_n - E_{j'})^{-1}$$

$$+ \sum_n \sum_{l \neq j'} {}^1\Phi_n(\mathbf{q},\mathbf{0})^1\Lambda_{(0)}(\mathbf{q})\langle {}^1\Phi_n(\mathbf{q},\mathbf{0})^1\Lambda_{(0)}(\mathbf{q})|\hat{H}_{SO}(\mathbf{q},\mathbf{0})|{}^3\Phi_l(\mathbf{q},\mathbf{0})^3\Lambda_{(i)}(\mathbf{q})\rangle_{\mathbf{q}}$$

$$\times \langle {}^3\Phi_l(\mathbf{q},\mathbf{0})^3\Lambda_{(i)}(\mathbf{q})|\hat{H}_{ev}(\mathbf{q},\mathbf{0})|{}^3\Phi_{j'}(\mathbf{q},\mathbf{0})^3\Lambda_{(i)}(\mathbf{q})\rangle_{\mathbf{q}}(E_n - E_l)^{-1}(E_l - E_{j'})^{-1}$$

$$+ \sum_n \sum_{l \neq n} {}^1\Phi_n(\mathbf{q},\mathbf{0})^1\Lambda_{(0)}(\mathbf{q})\langle {}^1\Phi_l(\mathbf{q},\mathbf{0})^1\Lambda_{(0)}(\mathbf{q})|\hat{H}_{SO}(\mathbf{q},\mathbf{0})|{}^3\Phi_{j'}(\mathbf{q},\mathbf{0})^3\Lambda_{(i)}(\mathbf{q})\rangle_{\mathbf{q}}$$

$$\times \langle {}^1\Phi_n(\mathbf{q},\mathbf{0})^1\Lambda_{(0)}(\mathbf{q})|\hat{H}_{ev}(\mathbf{q},\mathbf{Q})|{}^1\Phi_l(\mathbf{q},\mathbf{0})^1\Lambda_{(0)}(\mathbf{q})\rangle_{\mathbf{q}}(E_n - E_l)^{-1}(E_l - E_{j'})^{-1}\Big\} \tag{2.74}$$

where n runs over all singlet states as in (2.68), and l runs over all states except as indicated. Since $\hat{H}_{ev}$ does not mix states of different multiplicity, this means in practice that nonzero contributions arise in the first double sum only when l refers to a triplet state (vibronic mixing of the T_1 state within the triplet manifold, spin-orbit coupling with the singlet manifold), while in the second double sum only those contributions for which l refers to a singlet state do not vanish (vibronic mixing within the singlet manifold, spin-orbit coupling to the T_1 state).

If $\psi_j(\mathbf{q},\mathbf{Q}) = \psi_{S_0}(\mathbf{q},\mathbf{Q})$ is assumed to be adequately described by the pure spin function ${}^1\Phi_j(\mathbf{q},\mathbf{Q})^1\Lambda_{(0)}(\mathbf{q})$ and the normalization factor N′ is set equal to 1, the expression for the electric dipole transition moment from the ground state to the (i) component of the triplet state becomes

$$\mathbf{M}[jv,j'(i)v']$$

$$= \langle \chi_{j,v}(Q)|\chi_{j'v'}(Q)\rangle_{\mathbf{Q}}$$

$$\times \sum_n \langle {}^1\Phi_n(\mathbf{q},\mathbf{0})^1\Lambda_{(0)}(\mathbf{q})|\hat{H}_{SO}(\mathbf{q},\mathbf{0})|{}^3\Phi_{j'}(\mathbf{q},\mathbf{0})^3\Lambda_{(i)}(\mathbf{q})\rangle_{\mathbf{q}}(E_n - E_{j'})^{-1}\mathbf{M}^0(jn)$$

$$+ \sum_n \sum_k^{3N-6} \langle \chi_{j,v}(\mathbf{Q})|\langle {}^1\Phi_n(\mathbf{q},\mathbf{0})^1\Lambda_{(0)}(\mathbf{q})|\left[\frac{\partial \hat{H}_{SO}(\mathbf{q},\mathbf{Q})}{\partial Q_k}\right]_0|{}^3\Phi_{j'}(\mathbf{q},\mathbf{0})^3\Lambda_{(h)}(\mathbf{q})\rangle_{\mathbf{q}}$$

$$\times Q_k|\chi_{j',v'}(\mathbf{Q})\rangle_{\mathbf{Q}}(E_n - E_{j'})\mathbf{M}^0(jn)$$

$$+ \sum_n \sum_{l \neq j'} \langle \Phi_m(\mathbf{q},\mathbf{0})^1\Lambda_{(0)}(\mathbf{q})|\hat{H}_{SO}(\mathbf{q},\mathbf{0})|{}^3\Phi_k(\mathbf{q},\mathbf{0})^3\Lambda_{(h)}(\mathbf{q})\rangle_{\mathbf{q}}$$

$$\times \sum_{k=1}^{3N-6} \langle \chi_{j,v}(\mathbf{Q})|\langle {}^3\Phi_l(\mathbf{q},\mathbf{0})^3\Lambda_{(i)}(\mathbf{q})|\left[\frac{\partial \hat{H}_{el}(\mathbf{q},\mathbf{Q})}{\partial Q_k}\right]_0|{}^3\Phi_{j'}(\mathbf{q},\mathbf{0})^3\Lambda_{(i)}(\mathbf{q})\rangle_{\mathbf{q}}$$

$$\times\ Q_k|\chi_{j'v'}(\mathbf{Q})\rangle_{\mathbf{Q}}(E_n - E_l)^{-1}(E_l - E_{j'})^{-1}\mathbf{M}^0(jn)$$

$$+\sum_n \sum_{l\neq n} \langle {}^1\Phi_l(\mathbf{q},0)^1\Lambda_{(0)}(\mathbf{q})|\hat{H}_{SO}(\mathbf{q},0)|{}^3\Phi_{j'}(\mathbf{q},0)^3\Lambda_{(i)}(\mathbf{q})\rangle_{\mathbf{q}}$$

$$\times \sum_{k=1}^{3N-6} \langle\chi_{j,v}(\mathbf{Q})|\langle {}^1\Phi_n(\mathbf{q},0)^1\Lambda_{(0)}(\mathbf{q})|\left[\frac{\partial \hat{H}_{el}(\mathbf{q},\mathbf{Q})}{\partial Q_k}\right]_0 |{}^1\Phi_l(\mathbf{q},\mathbf{Q})^1\Lambda_{(0)}(\mathbf{q})\rangle_{\mathbf{q}}$$

$$\times\ Q_k|\chi_{j'v'}(\mathbf{Q})\rangle_{\mathbf{Q}}(E_n - E_l)^{-1}(E_l - E_{j'})^{-1}\mathbf{M}^0(jn) \qquad (2.75)$$

where the above comments on the nature of the summations still apply and k runs over all normal modes of vibration. Expression (2.75) represents a generalization of the first term in (2.70). When the mixing of triplet character into the ground-state singlet wavefunction ψ_{S_0} is also considered, a term representing a similar generalization of the second term in (2.70) must be added.

The use of group theory with (2.75) produces selection rules for phosphorescence polarization and S − T absorption as was the case for (2.70). We assume for simplicity that the electronic wavefunction of the ground-state ${}^1\Phi_j$ is totally symmetric and that the emission occurs from the lowest vibrational level of the triplet, so that $\chi_{j',v'}$ is also totally symmetric. The single sum in (2.75) corresponds to the contribution of ordinary spin-orbit coupling which has already been discussed. The μ-th component of the first term vanishes for those choices of (i) for which the direct product $\Gamma(\mu) \times \Gamma(\mathscr{R}_i) \times \Gamma({}^3\Phi_{j'})$ does not contain the totally symmetric representation. This component only provides phosphorescent intensity for transitions to the totally symmetric vibrational levels of the ground state.

The remaining terms in (2.75) correspond to vibronically induced spin-orbit coupling. The second sum reflects the dependence of the spin-orbit coupling operator on molecular geometry. Like the first term, it is due to a direct coupling of a perturbing singlet state ${}^1\Phi_n{}^1\Lambda_{(0)}$ to the (i) component of the initial triplet state of the phosphorescent transition, ${}^3\Phi_{j'}{}^3\Lambda_{(i)}$. Such a perturbing singlet state ${}^1\Phi_n{}^1\Lambda$ will contribute to the μ-th component of $\mathbf{M}[jv,j'(i)v']$ when it transforms according to the same irreducible representation as μ and then only through those normal modes Q_k for which $\Gamma({}^1\Phi_n) \times \Gamma(Q_k) \times \Gamma({}^3\Phi_{j'}) \times \Gamma(\mathscr{R}_i)$ contains the totally symmetric representation, since for others the integration over the electronic coordinates $\mathbf{q}$ yields zero. The subsequent integration over the nuclear coordinates $\mathbf{Q}$ yields zero unless the direct product $\Gamma(\chi_{j,v}) \times \Gamma(Q_k)$ contains the totally symmetric representation. Thus, the ground-state vibrational levels which will be accessible by phosphorescent emission from the (i) component of the triplet state as a result of the presence of the second term in (2.75) will be of symmetry $\Gamma(\mu) \times \Gamma({}^3\Phi_{j'}) \times \Gamma(\mathscr{R}_i)$.

The third term in (2.75) represents a spin-orbit coupling of the perturbing singlet state ${}^1\Phi_n$ to an intermediate triplet state ${}^3\Phi_l{}^3\Lambda_{(i)}$, which is then vibronically coupled to the phosphorescent triplet ${}^3\Phi_l{}^3\Lambda_{(i)}$. Once again, if ${}^1\Phi_n$ is to contribute to the μ-th component of $\mathbf{M}[jv,j'(i)v']$, it must transform like μ. The space part of the intermediate triplet ${}^3\Phi_l$ must be of symmetry $\Gamma(\mu) \times \Gamma(\mathscr{R}_i)$ in order for the matrix element of $\hat{H}_{SO}$ not to vanish. The contributing normal modes Q_k will once again be those of symmetry $\Gamma(\mu) \times \Gamma({}^3\Phi_{j'}) \times$

$\Gamma(\mathscr{R}_i)$; this will be the symmetry of the vibrational levels of the ground state reached in the phosphorescent transition.

Finally, the fourth term in (2.75) represents vibronic coupling of the perturbing singlet state $^1\Phi_n$ to an intermediate singlet state $^1\Phi_l$, which is then spin-orbit coupled to the phosphorescent triplet $^3\Phi_{j'}{}^3\Lambda_{(i)}$. Here again, if $^1\Phi_n$ is to contribute to the μ-th component of M[jv,j′(i)v′], it must transform like μ. Arguments similar to those above again show that the vibrational levels of the ground state which will be reached by phosphorescence from the (i) component of T_1 are those transforming as $\Gamma(\mu) \times \Gamma(^3\Phi_{j'}) \times \Gamma(\mathscr{R}_i)$.

Since three components (i) of T_1 are available, these selection rules are not very strict, and at least some vibrational symmetries are usually accessible with more than one polarization μ so that mixed polarization results. Very similar rules apply to singlet-triplet absorption starting at the vibrationless level of the S_0 ground state. The allowed vibrational symmetries then refer to the vibrational levels of the triplet state.

A discussion of the procedures for numerical evaluation of the matrix elements of $\hat{H}_{ev}$, $\hat{H}_{SO}$, and $\hat{H}_{vSO}$ lies outside the scope of the present work.

2.2.3 Computation of Two-Photon Absorption and Raman Scattering Tensors

The use of formula (2.20) for the elements of the two-photon absorption tensor T requires only the knowledge of quantities whose computation has been already discussed: electric dipole transition moments and state energies. However, it is rendered difficult by the presence of an infinite sum and by the need to compute large numbers of transition moments, many of which are between two excited states.

This is a somewhat specialized topic, to which relatively little attention has been paid so far. The best present-day procedures use the semiempirical CNDO/S model to obtain excited state energies and wavefunctions needed for the evaluation of electric dipole transition moments. Fairly extensive configuration interaction, including doubly excited configurations, is needed to obtain reasonable results (several hundred configurations for a molecule the size of naphthalene). The truncation of the CI expansion also automatically truncates the infinite sum in (2.20) and makes the computation feasible.

Little is known about the quality of the calculated absolute cross sections, since experimental data for comparison are scarce, but relative values for different transitions within one molecule are quite satisfactory. Leading references can be found in Section 2.4.

The expression for the elements of the Raman scattering tensor α' for the nonresonant case follows from (2.24) and (2.27). It resembles the case of the two-photon tensor T in that it contains an infinite sum, excited state energies, and electric dipole transition moments. The latter are of the form $\mathbf{M}(0j',\mathbf{Q})$ and now all involve the ground state, which simplifies matters. On the other hand, it is their dependence on molecular geometry $\mathbf{Q}$ which really counts in view of (2.27), and this complicates matters substantially. This approach has seen only limited use.

Some of the quantum-chemical calculations of nonresonant Raman intensities which have been performed so far have replaced the alternating electric field of the electromagnetic wave with a static electric field of comparable amplitude. In this fashion, the dependence of the Raman intensity on the wavenumber of the exciting light is lost, but the calculation is greatly simplified. The elements of the molecular static polarizability tensor,

$$\alpha_{uv} = (\partial/\partial E_u)\langle\psi_j(\mathbf{q},\mathbf{Q})|\hat{M}_v|\psi_j(\mathbf{q},\mathbf{Q})\rangle \tag{2.76}$$

where **E** is the static electric field, are calculated from wavefunctions of the j-th state (usually, the ground state, $j = 0$) for a sufficient number of geometries **Q** to permit numerical differentiation with respect to Q_k, which then yields the desired element of α'_{uv} as needed in (2.27). For medium-sized and large molecules, CNDO wavefunctions are usually used for $\psi_j(\mathbf{q},\mathbf{Q})$. One of the difficulties of this and other approximate methods is the use of a minimum basis set, which underestimates some of the polarizabilities severely. For instance, H_2 is calculated to have zero polarizability in the direction perpendicular to the bond, whereas in reality its value is 70% of the parallel polarizability. An improvement in the calculations is obtained when 2p functions on hydrogen atoms are added to the basis set. Leading references to this type of work can be found in Section 2.4.

Even less has been done so far in the way of quantum-mechanical calculations of resonant Raman intensities. Formula (2.29) provides the general expression, and further progress certainly seems likely.

Most of the analyses of Raman intensities published so far have been more empirical in nature and have employed electro-optical parameters describing bond polarizabilities, their derivatives, and their anisotropies. A discussion of these procedures lies outside the scope of this book (see Section 2.4).

2.3 Illustration: Spectra of Partially Aligned Pyrene

In this section, the use of some of the rather abstract concepts presented in previous sections will be illustrated on the example of pyrene. First we shall describe the expectations which one has for the polarized vibrational and electronic spectra of pyrene based on symmetry and other tools described in this chapter so far. Then we shall confront the expectations with experimental data obtained on partially aligned solutions of pyrene. Detailed analysis and the consideration of less common types of measurements are postponed until Chapter 8; here, we only intend to demonstrate some of the fundamental aspects of the interplay between theory and experiment.

2.3.1 Symmetry Analysis

The formula of pyrene is shown in Figure 2.3. It is a planar molecule of D_{2h} symmetry. We choose the x axis to lie perpendicular to the plane and show the y and z axes in the figure. The character table (Appendix I) reveals the existence of four irreducible representations of the gerade type—A_g, B_{1g}, B_{2g}, and B_{3g}—and four of the ungerade type—A_u, B_{1u}, B_{2u}, and B_{3u}. The former are symmetric, the latter antisymmetric with respect to inversion through the

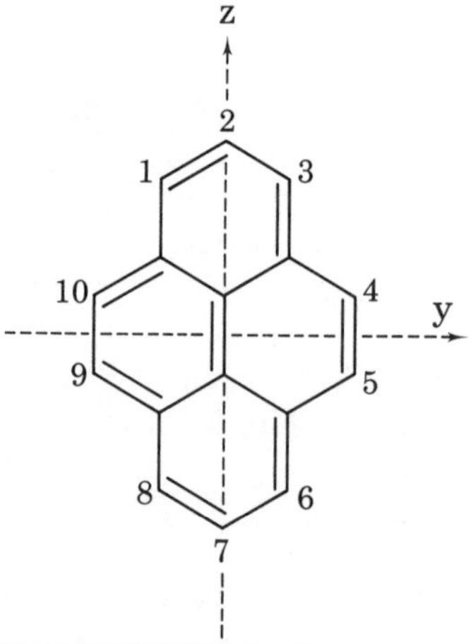

Figure 2.3 The numbering of carbon atoms in pyrene and choice of axes.

center of symmetry. The table further shows the transformation properties of operators important for spectroscopic transitions. The components of the electric dipole operator $\hat{\mathbf{M}}$ transform like x, y, and z ($\hat{M}_x$: B_{3u}; $\hat{M}_y$: B_{2u}; $\hat{M}_z$: B_{1u}). The components of the magnetic dipole operator $\hat{\mathcal{M}}$ and the triplet spin functions $^3\Lambda_{(i)}$ transform like rotations along x, y, and z ($\hat{\mathcal{M}}_x$ and $^3\Lambda_{(x)}$: B_{3g}; $\hat{\mathcal{M}}_y$ and $^3\Lambda_{(y)}$: B_{2g}; $\hat{\mathcal{M}}_z$ and $^3\Lambda_{(z)}$: B_{1g}). The components of the operators for the electric quadrupole tensor $\hat{Q}$, the two-photon absorption tensor $\hat{T}$, and the Raman scattering tensor $\hat{\alpha}'$ transform like the coordinate products uv (u,v = x,y,z) ($\hat{Q}_{yz}$, $\hat{T}_{yz}$, and $\hat{\alpha}'_{yz}$: B_{3g}; $\hat{Q}_{zx}$, $\hat{T}_{zx}$, and $\hat{\alpha}'_{zx}$: B_{2g}; $\hat{Q}_{xy}$, $\hat{T}_{xy}$, and $\hat{\alpha}'_{xy}$: B_{1g}; $\hat{Q}_{xx}$, $\hat{Q}_{yy}$, $\hat{Q}_{zz}$, $\hat{T}_{xx}$, $\hat{T}_{yy}$, $\hat{T}_{zz}$, $\hat{\alpha}'_{xx}$, $\hat{\alpha}'_{yy}$, and $\hat{\alpha}'_{zz}$: A_g).

Vibrational Spectra. The presence of 16 carbon and 10 hydrogen atoms implies the presence of $3 \times 26 - 6 = 72$ normal modes of vibration. Using the method outlined in Appendix I, these can be shown to fall into the following symmetry categories: $13a_g + 4b_{1g} + 7b_{2g} + 12b_{3g} + 5a_u + 12b_{1u} + 12b_{2u} + 7b_{3u}$. The first four groups among these, 36 normal modes altogether, should be Raman-active as fundamentals; the last three, 31 modes altogether, should be IR-active as fundamentals in the electric dipole approximation (12 each of y and z polarization, 7 of x polarization). In addition, the spectra should contain the usually less intense combination and overtone bands. For instance, an inspection of the table of products of irreducible representations in Appendix I shows that the following combinations of two vibrations can appear in the IR spectrum: z-polarized (b_{1u}): b_{1u} with a_g, a_u with b_{1g}, b_{2u} with b_{3g}, b_{3u} with b_{2g}; y-polarized (b_{2u}): b_{2u} with a_g, a_u with b_{2g}, b_{1u} with b_{3g}, b_{3u} with b_{1g}; x-polarized (b_{3u}): b_{3u} with a_g, a_u with b_{3g}, b_{1u} with b_{2g}, b_{2u} with b_{1g}. Similar statements can be made concerning the appearance of overtones and combination bands in the Raman spectrum.

Singlet-Singlet Electronic Transitions. The 16 $2p_z$ valence shell orbitals on the carbon atoms which are of π symmetry give rise to eight "bonding" and eight "antibonding" molecular orbitals of π symmetry. The methods of Appendix I show that their symmetries are $3a_u + 3b_{1g} + 5b_{2g} + 5b_{3u}$. The bonding orbitals are doubly occupied in the ground configuration (A_g), and the antibonding orbitals are vacant. Low-energy singly excited configurations are those originating in the promotion of an electron from the highest occupied MO (b_{1g}) or the one immediately below (b_{2g}) to the lowest unoccupied MO (a_u) or the one immediately above (b_{3u}). Among these are two excited singlet configurations of symmetry B_{1u} ($^1D_{b_{1g}\to a_u}$ and $^1D_{b_{2g}\to b_{3u}}$) and two of symmetry B_{2u} ($^1D_{b_{1g}\to b_{3u}}$ and $^1D_{b_{2g}\to a_u}$). These are the symmetries of the space part of the wavefunction; however, for singlet states the symmetry of the total wavefunction is equal to that of the space part, since the singlet spin wavefunction is totally symmetric. The mixing of the two B_{1u} configurations will produce two excited singlet states of symmetry B_{1u}; in the order of increasing energy these are known as the ^1L_a (mostly $^1D_{b_{1g}\to a_u}$) and ^1B_a (mostly $^1D_{b_{2g}\to b_{3u}}$) states. The two B_{2u} configurations have very similar energies and mix very strongly in an approximately 1:1 ratio to yield two excited singlet states of B_{2u} symmetry. The in-phase combination is known as ^1B_b state, and the out-of-phase combination is known as the ^1L_b state. The fact that the mixing is nearly exactly 1:1 is a consequence of the approximate validity of the alternant pairing theorem (see Section 2.4).

All four excited singlet states under consideration are seen to be electric dipole allowed for one-photon transitions from the A_g ground state, two with z polarization (B_{1u}) and two with y polarization (B_{2u}). Calculation of the transition moments by the methods outlined in Section 2.2.2 reveals an interesting result which is a consequence of the approximately 1:1 mixing of the two B_{2u} configurations but does not follow from geometrical symmetry alone: The y-polarized transition from the A_g ground state into the L_b state has a very small transition moment, that into the B_b state a very large one. The z-polarized transitions into the L_a and B_a states both have fairly large transition moments. A calculation of state energies yields L_b, L_a, B_b, B_a in the order of increasing energy, with L_b and L_a close together.

Under these circumstances, it is likely that the nearly forbidden L_b transition will derive a significant fraction of its total intensity by using the Herzberg-Teller vibronic coupling mechanism discussed in Section 2.1.1. The symmetry of the ^1L_b state is B_{2u}; that of the states from which intensity can be borrowed is B_{1u} for L_a and B_a and B_{2u} for B_b. The borrowing from L_a is favored by the small energy difference $L_b - L_a$. Inspection of the appropriate entries in Appendix I shows that the vibrational excitation which needs to be added onto the origin (0–0 band) of the L_b transition is either one or more quanta combining into overall symmetry species a_g (borrowing from B_b) or a combination of vibrations whose overall symmetry is b_{3g} (borrowing from L_a and B_a). The simplest way to produce the latter is to take one quantum of a normal mode of b_{3g} symmetry, but combinations are also possible, such as a_g with b_{3g}.

So far, we have been discussing absorption, but similar considerations apply to fluorescence. For instance, emission from the lowest vibrational level of the L_b state into the ground state will have y-polarized components which terminate in one of the a_g vibrational levels of the ground electronic state and z-polarized components which terminate in one of its b_{3g} vibrational levels.

Overall, then, we expect three relatively intense one-photon electronic transitions of the $\pi\pi^*$ type—L_a, B_b, and B_a, with vibrational fine structure containing a_g vibrations, polarized along z, y, and z, respectively—and a very weak L_b transition whose weak origin is polarized along y and whose vibrational fine structure contains y-polarized peaks for vibrations of symmetry a_g and z-polarized peaks for vibrations of symmetry b_{3g}.

Electronic transitions of the $\pi\pi^*$ type that should be observable by two-photon absorption from the A_g ground state must terminate in either A_g or B_{3g} excited states. π-Electron calculations place such states in the vicinity of the B_b band and at higher energies.

In principle, electronic transitions of types other than $\pi\pi^*$ must also be possible. Transitions such as $\pi\sigma^*$ and $\sigma\pi^*$ are expected to have low intensities and to lie at relatively high energies, where they will be buried under the much more intense $\pi\pi^*$ absorption. Excited states of B_{3u} symmetry are allowed in one-photon absorption, B_{1g} and B_{2g} in two-photon absorption. They have not yet been detected in spite of their distinctive polarization properties. Configurations of the $\sigma\sigma^*$ type are of the same symmetries as $\pi\pi^*$ configurations, and the two therefore mix when state wavefunctions are calculated. Since the $\sigma\sigma^*$ excitations are of much higher energy, their contribution to the properties of low-energy "$\pi\pi^*$" transitions is usually neglected. They undoubtedly play a role in the vacuum UV region. There, Rydberg states will also be important.

Singlet-Triplet Electronic Transitions. The triplet analogs of the four singlet singly excited configurations considered above are $^3D_{b_{1g}\rightarrow a_u}$ and $^3D_{b_{2g}\rightarrow b_{3u}}$ (with B_{1u} symmetry for the space part of the wavefunction) and $^3D_{b_{1g}\rightarrow b_{3u}}$ and $^3D_{b_{2g}\rightarrow a_u}$ (with B_{2u} symmetry for the space part). As before, configuration mixing will produce four excited states. Those with a B_{1u} space part are labeled ^3L_a and ^3B_a, while those with a B_{2u} space part are labeled ^3L_b and ^3B_b.

The three spin functions of the triplet state transform like rotations: B_{1g}, B_{2g}, and B_{3g}. Therefore, the total symmetries of the three components of each of the four triplet excited states are given by the direct products (Appendix I) $B_{1u} \times (B_{1g},B_{2g},B_{3g}) = (A_u,B_{3u},B_{2u})$ for ^3L_a and ^3B_a, and $B_{2u} \times (B_{1g},B_{2g},B_{3g}) = (B_{3u},A_u,B_{1u})$ for ^3L_b and ^3B_b. Two of the three components of each triplet state are thus electric dipole allowed for transitions from the ^1A_g ground state. For ^3L_a and ^3B_a, the contributing transitions are polarized along x and y; for ^3L_b and ^3B_b, along x and z. In each case, mixed polarization must therefore be expected in principle. Symmetry arguments alone do not indicate which of the two contributing components dominates for any one of the four states, but closer investigation suggests that it is the x-polarized one.

The lowest triplet state in pyrene turns out to be the ^3L_a state, and its three components have total symmetries A_u, B_{3u}, and B_{2u}. The transition from this state to the ground state is observed as phosphorescence. Its 0–0 transition, usually referred to as its origin, is expected to exhibit mixed x,y polarization. Transitions from the lowest vibrational level of the ^3L_a triplet into those vibrational levels of the ground state which are of a_g vibrational symmetry and therefore overall A_g symmetry will also show x,y polarization. Non-totally symmetric vibrational levels of the electronic ground state can be reached by vibronic coupling: consideration of the table of direct products in Appendix I shows that in addition to a_g levels, electric dipole phosphorescent emission is also allowed into ground state vibrational levels of symmetries b_{1g}, b_{2g}, and b_{3g}. The emissions will again be of mixed polarizations. Emission into b_{1g} vibrational levels will have contributions x-polarized from the B_{2u}, y-polarized from the B_{3u}, and z-polarized from the A_u component of the triplet state. Emission into b_{2g} vibrational levels will have contributions y-polarized from the A_u component of the triplet state and z-polarized from the B_{3u} component. Finally, emission into b_{3g} vibrational levels will have contributions x-polarized from the A_u and z-polarized from the B_{2u} component of the triplet state. Many of these contributions may well have negligible intensities; a calculation would be required in order to make detailed predictions.

2.3.2 Examples of Spectra and Qualitative Interpretations

Partial alignment of molecules can be achieved by imposition of a suitable force field. This field can be electric, magnetic, or "mechanical" in nature. Relatively high degrees of orientation are most easily achieved by the use of mechanical force fields, that is, by placing a solute in an anisotropic solvent. Intermolecular attractive and repulsive forces can then transfer orientational preferences from the solvent onto the solute. The solvent can be either liquid, such as a nematic liquid crystal, or solid, such as a stretched polymer or a crystal. In the following, we provide a few examples of polarized spectra obtained on pyrene contained inside stretched polyethylene.

Polarized Infrared Absorption Spectra. Our first example (Figures 2.4 and 2.5) shows the room-temperature linear dichroic absorption spectra $E_Z(\tilde{\nu})$ and $E_Y(\tilde{\nu})$ of pyrene in stretched polyethylene in the infrared region. $E_Z(\tilde{\nu})$ has been measured with the electric vector of the light parallel to Z, the stretching direction of the polymer; $E_Y(\tilde{\nu})$, with the electric vector along Y, which lies perpendicular to the stretching direction. All directions perpendicular to Z were equivalent within the experimental accuracy.

The spectrum contains many sharp peaks which do not overlap with others and some that do. For all of the former, the dichroic ratio $d(\tilde{\nu}) = E_Z(\tilde{\nu})/E_Y(\tilde{\nu})$ is equal to one of only three numbers: 2.17 ± 0.17, 1.03 ± 0.05 and 0.27 ± 0.03. These cases are labeled by a circle, a dot, and an arrow, respectively, in Figure 2.4. A particularly clear illustration is provided by peaks in the 470–570 cm^{-1} region shown expanded in Figure 2.5.

The exact values of the dichroic ratios change as a function of the nature of the polymer used and other factors such as its degree of stretching and temperature. For instance, at the temperature of liquid nitrogen, the three values for the same sample become 2.55 ± 0.21, 1.00 ± 0.03, and 0.22 ± 0.03. However, in all uniaxial samples only three distinct narrow ranges of values of $d(\tilde{\nu})$ are observed for nonoverlapping peaks.

The symmetry analysis outlined above showed that only those vibrationally excited states which are of symmetry B_{1u}, B_{2u}, or B_{3u} are accessible by electric dipole radiation and are polarized along z, y, or x, respectively (Figure 1.9). In the following, we shall assume that the perturbation of molecular symmetry by

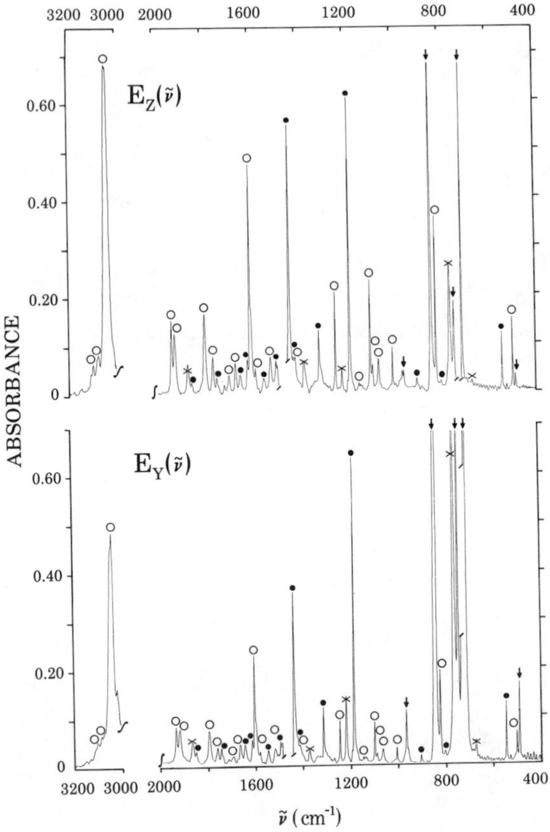

Figure 2.4 Polarized IR absorption spectra $E_Z(\tilde{\nu})$ and $E_Y(\tilde{\nu})$ of pyrene in stretched polyethylene (room temperature). Baseline subtraction has distorted the intensity of the strong dotted peak near 1420 cm^{-1}. Circles, dots, and arrows: see text. Reproduced by permission from J. G. Radziszewski and J. Michl, *J. Phys. Chem.* **85**, 2934 (1981).

the polyethylene environment is negligible and that the vibrational transitions in the actually observed molecules are all purely polarized along one of the molecular symmetry axes.

For each molecule, the absorption probability of a transition is proportional to the square of the projection of its electric dipole transition moment into the direction of the electric vector of the light, $|\boldsymbol{\varepsilon}_Z \cdot \mathbf{M}(0f)|^2$ or $|\boldsymbol{\varepsilon}_Y \cdot \mathbf{M}(0f)|^2$, when electric quadrupole and magnetic dipole contributions are neglected [equation (1.62)]. The contributions from all molecules must be summed to describe the properties of the whole sample. For instance, if all of them were perfectly aligned with their z axis along the Z direction of the sample, their x and y axes would lie in the XY plane. Then z-polarized transitions would appear in $E_Z(\tilde{\nu})$ but not at all in $E_Y(\tilde{\nu})$. On the other hand, x-polarized and y-polarized transitions would be absent in $E_Z(\tilde{\nu})$ but present in $E_Y(\tilde{\nu})$. The dichroic ratios would be $d_z = \infty$

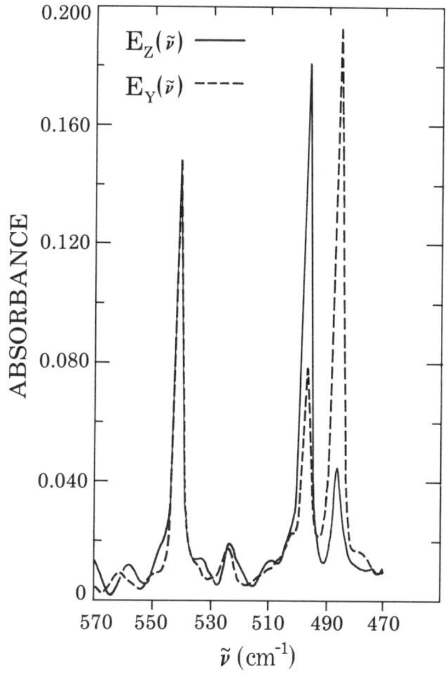

Figure 2.5 Polarized IR absorption spectra $E_Z(\tilde{\nu})$ and $E_Y(\tilde{\nu})$ of pyrene in stretched polyethylene (room temperature, baseline subtracted). Reproduced by permission from J. G. Radziszewski and J. Michl, *J. Phys. Chem.* **85**, 2934 (1981).

for z-polarized transitions and $d_x = d_y = 0$ for x- and y-polarized transitions. Another extreme situation would be a totally random distribution of molecular orientations. Then the average value of $|\boldsymbol{\varepsilon}_U \cdot \mathbf{M}(0f)|^2$ would have to be the same for U = X, U = Y, and U = Z, so that d = 1 for all transitions. Qualitatively, then, we see that the value of d_u increases as the average alignment of the u-th molecular axis with the sample Z axis increases.

The three unequal d_u values observed in Figures 2.4 and 2.5 for the IR transitions of pyrene imply that the three molecular axes x, y, and z are aligned to unequal degrees with the Z axis, with all identically polarized nonoverlapping transitions sharing a common d_u value.

Clearly, one of the molecular axes of pyrene lines up much better with Z than others, and we shall label it z′, while one lines up much worse than others, and we shall label it x′. The remaining axis is labeled y′. A more detailed analysis, described in Chapter 5, provides quantitative information on the degree of alignment.

So far, we have assigned all observed transitions as polarized along x′, y′, or z′ molecular axes, but we have no certainty as to how these axes are to be assigned to the three symmetry axes of pyrene x, y, and z, which we have introduced in Figure 2.3. Notice, however, that it would be sufficient to find out from other sources the absolute polarizations of only two differently polarized peaks among the many present to resolve the problem. Such information is indeed available from previous IR work on pyrene crystals and permits the assignment of x′ as identical with x (the out-of-plane axis), y′ as identical with y (the short in-plane axis), and z′ as identical with z (the long in-plane axis). The correlation of the average degree of alignment of the axes with molecular shape is not accidental; we shall see in Section 4.6.1 that the alignment of aromatic molecules in stretched polyethylene is quite closely related to their shape, so that the absolute assignment in pyrene would have been possible even if no crystal measurements had been made.

Polarized Ultraviolet Absorption Spectra. Our second example (Figure 2.6) shows the UV spectrum of pyrene contained in stretched polyethylene, taken at the temperature of liquid nitrogen. Only singlet-singlet $\pi\pi^*$ transitions are observed, and the weak L_b transition is actually not seen at this concentration. Here again, two independent spectra can be measured, $E_Z(\tilde{\nu})$ and $E_Y(\tilde{\nu})$, with the electric vector of the light parallel and perpendicular, respectively, to the stretching direction of the polymer. The presence of overlap of differently polarized transitions may not be apparent at first sight from the measured curves but is obvious when one plots the dichroic ratio $d(\tilde{\nu}) = E_Z(\tilde{\nu})/E_Y(\tilde{\nu})$ as a function of $\tilde{\nu}$. The ratio does not acquire a small number of discrete values, as was the case in the IR spectrum, but rather is a continuous function of $\tilde{\nu}$ (Figure 2.6). Making the usual assumption that solvent effects are negligible, symmetry dictates that the observed $\pi\pi^*$ transitions are polarized along either the molecular long axis z or the in-plane short axis y. From the IR spectrum shown previously and its accepted interpretation, it is clear that the z axis aligns better than the y axis with the stretching direction Z.

The $\pi\pi^*$ absorption of pyrene is likely to contain some transitions polarized along each of the two in-plane axes and is therefore likely to be given by a superposition of a contribution $A_z(\tilde{\nu})$, polarized along the z axis, and $A_y(\tilde{\nu})$, polarized along the y axis. We can think of these two contributions to the observed spectra just as we would of contributions of the spectra of two components in a mixture to the spectrum of the mixture. In our case, the contribution from $A_z(\tilde{\nu})$ will be enhanced relative to that from $A_y(\tilde{\nu})$ when the observation is done with light having its electric vector parallel to the stretching direction, that is, in $E_Z(\tilde{\nu})$, since the z axis is the one that aligns better with the stretching direction. On the contrary, in $E_Y(\tilde{\nu})$ the contribution $A_z(\tilde{\nu})$ should be less enhanced or perhaps even deemphasized relative to the contribution of $A_y(\tilde{\nu})$. This is exactly analogous to what we have already seen in the IR spectrum. A second glance at the curves $E_Z(\tilde{\nu})$ and $E_Y(\tilde{\nu})$ shows us now which transition is polarized in which direction: The first (L_a) and third (B_a) observed transitions are z-polarized; the second (B_b) is y-polarized.

However, one can ask for more: What would the purely z-axis polarized absorption $A_z(\tilde{\nu})$ look like, and what would $A_y(\tilde{\nu})$ look like? These questions normally do not come up in the analysis of mixtures because the pure components can be separated and measured independently. Even when that is not possible, however, one can still determine the spectra of the pure components of the mixtures if the number of different mixtures available is at least equal to the number of components contained in them. This requires that the spectra of the individual components be sufficiently linearly independent, i.e., that certain

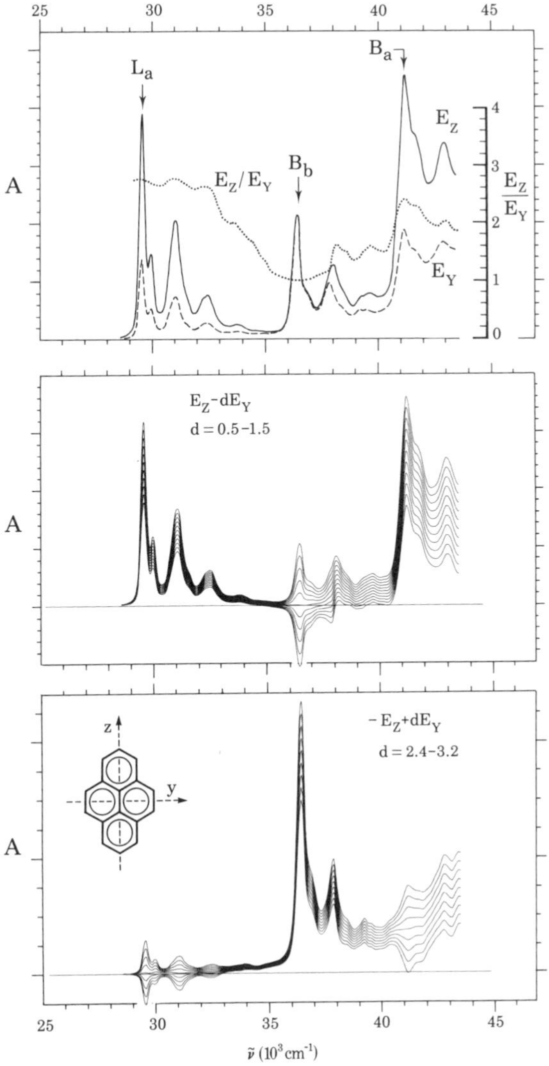

Figure 2.6 Pyrene in stretched polyethylene at 77 K. Absorbance in arbitrary units. Top: baseline-corrected polarized UV absorption spectra and the dichroic ratio E_Z/E_Y. Adapted by permission from F. W. Langkilde, M. Gisin, E. W. Thulstrup, and J. Michl, *J. Phys. Chem.* **87**, 2901 (1983). Center and bottom: linear combinations of the polarized spectra E_Z and E_Y.

spectral features of the mixture be associated with only one of its components and others only with the others. If we can make the same assumption here—that is, if we identify "long-axis polarized" peaks or shoulders which occur only in $A_z(\tilde{\nu})$ and others, "short-in-plane-axis polarized," which occur only in $A_y(\tilde{\nu})$—we could obtain the purely polarized "reduced" spectra.

In a molecule of high symmetry, such as pyrene, individual transitions have to be purely polarized along one of the three molecular axes in the absence of significant solvent effects, so the question is only: Do they overlap so much with other differently polarized transitions that we cannot distinguish them in the spectra? This clearly was not the case in the IR spectrum, but it is much more likely in the UV spectrum with its relatively broad bands. Since pyrene and similar molecules do not have intramolecular vibrational modes of frequencies lower than about 100 cm^{-1}, it is very likely that contributions due to peaks and shoulders assignable to 0–0 transitions are indeed purely polarized, but they may still overlap with contributions from other differently polarized transitions. Then it becomes a simple matter to find the reduced spectra $A_z(\tilde{\nu})$ and $A_y(\tilde{\nu})$ just as one would find the spectra of the two pure components from the spectra of two different mixtures. It is only necessary to find that linear combination of the observed spectra $E_Z(\tilde{\nu})$ and $E_Y(\tilde{\nu})$ which does not contain any of the long-axis polarized spectral features, and this will give the desired in-plane short-axis polarized reduced spectrum $A_y(\tilde{\nu})$. A similar removal of all short-axis polarized spectral features will yield the long-axis polarized reduced spectrum $A_z(\tilde{\nu})$. It is important to note that all the z-polarized features in the first case, and all the y-polarized features in the second case, disappear in the same linear combination (Figure 2.6). This is a good indication that the initial assumption that they are purely polarized was correct.

The resulting reduced spectra,

$$A_z(\tilde{\nu}) = E_Z(\tilde{\nu}) - 1.0E_Y(\tilde{\nu}) \tag{2.77}$$

$$A_y(\tilde{\nu}) = 1.75[E_Y(\tilde{\nu}) - 0.36E_Z(\tilde{\nu})] \tag{2.78}$$

are shown in Figure 2.7. The numerical factor in front of the bracket in equation (2.78) was chosen so as to put $A_z(\tilde{\nu})$ and $A_y(\tilde{\nu})$ on the same scale. Its magnitude was derived by procedures outlined in Chapter 5. As will be seen better there, in the present case a trial-and-error search for the proper linear combinations was actually not necessary, since sufficient information on the molecular

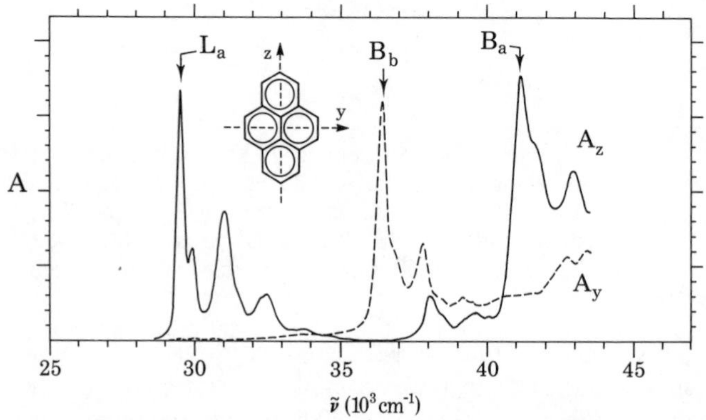

Figure 2.7 Reduced UV absorption spectra of pyrene in stretched polyethylene at 77 K, $A_z = E_Z - 1.0E_Y$ and $A_y = 1.75[E_Y(\tilde{\nu}) - 0.36E_Z(\tilde{\nu})]$. The absorbance scale is in arbitrary units.

orientation was available from the d_z and d_y values obtained in the IR measurements at liquid nitrogen temperature: Within experimental error, the coefficient 1.0 at $E_Y(\tilde{\nu})$ in (2.77) is equal to the measured d_y, and the coefficient 0.36 at $E_Z(\tilde{\nu})$ in (2.78) is equal to $1/d_z$.

It is interesting to note that the long-axis polarized spectrum (Figure 2.7) contains some peaks near 38 000 cm^{-1}, which cause a bump in the dichroic ratio in that region in Figure 2.6. The presence of these long-axis polarized vibronic peaks was not immediately obvious from the experimental spectra $E_Z(\tilde{\nu})$ and $E_Y(\tilde{\nu})$.

Polarized Fluorescence Spectra. Our third example is the polarized fluorescence spectrum obtained upon excitation into the z-polarized L_a band. Figure 2.8 shows the results obtained for pyrene dissolved in a 3-methylpentane glass at 77 K. This medium is rigid and does not permit any significant molecular rotation between the time the molecule absorbs a photon and the time when it subsequently emits one. The four spectra shown were obtained with the

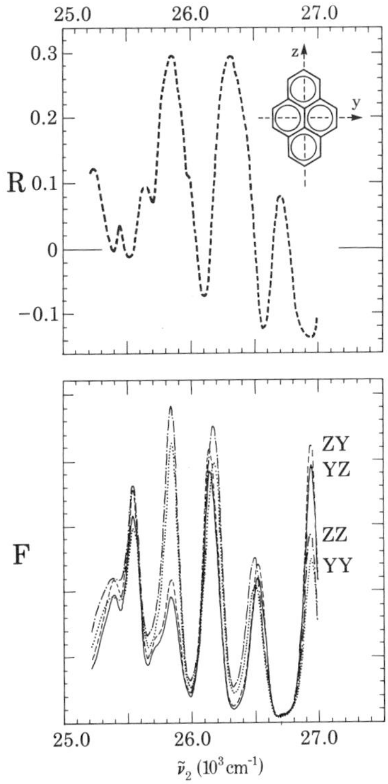

Figure 2.8 Polarized fluorescence of pyrene in glassy 3-methylpentane at 77 K, excited at the L_a origin. Bottom: the four polarized emission curves: full line, YZ; dashed line, ZY; dotted line, YY; dash-dot line, ZZ. The curves are not corrected for instrumental polarization biases nor for the 14.5° angle between the exciting and observed beam directions. Top: emission anisotropy. F. W. Langkilde and J. Michl, unpublished results.

exciting light polarized vertically (Z) or horizontally (Y) while the vertically (Z) or horizontally (Y) polarized emission intensity is being recorded. The direction of the observed beam was almost antiparallel to that of the exciting beam, so that one would expect the YY and ZZ intensities to be mutually equal, $F_{YY} = F_{ZZ}$, and the YZ and ZY intensities to be mutually equal, $F_{YZ} = F_{ZY}$, in the absence of instrumental distortions (the first letter refers to the polarization of the exciting light at $\tilde{\nu}_1$, the second one to that of the emitted light at $\tilde{\nu}_2$). This expectation is approximately fulfilled; the observed discrepancy can be used to correct for instrumental biases as described in Chapter 3. After such a correction, the so-called emission anisotropy $R = (F_{\parallel} - F_{\perp})/(F_{\parallel} + 2F_{\perp})$ has been computed and plotted in the upper part of Figure 2.8 ($F_{\parallel}$ is the corrected value of F_{ZZ} or F_{YY}, and $F_{\perp}$ is the corrected value of F_{YZ} or F_{ZY}).

The course of the R curve confirms what is already obvious from the four recorded spectra: The origin of the fluorescence near 26 900 cm^{-1} is represented predominantly by photons emitted with a polarization opposite to that of the photons absorbed ($F_{\parallel} < F_{\perp}$); if those absorbed had their electric vector along Z (Y), those emitted contribute mostly to intensity polarized along Y (Z).

Qualitatively, the significance of this observation is grasped readily. The transition moment of the absorbing L_a transition is oriented along the molecular axis z. The pyrene molecules are randomly oriented in the rigid glass, i.e., their z axis is equally likely to point in any direction. However, they are not all equally likely to absorb the exciting light. As we have just discussed in the interpretation of the IR and UV spectra, the absorption probability for a molecule is proportional to the square of the projection of the absorbing transition moment **M**(0f) into the direction of the electric vector of the exciting light, $|\boldsymbol{\varepsilon}_U \cdot \mathbf{M}(0f)|^2$. In our experiment, the absorbing transition moment is directed along the molecular axis z; $\mathbf{M}(0f) = \boldsymbol{\varepsilon}_z|\mathbf{M}(0f)|$, and the absorption probability is proportional to $|\boldsymbol{\varepsilon}_U \cdot \boldsymbol{\varepsilon}_z|^2$. Since the sample is isotropic, the number of molecules excited will be independent of the choice of U (no linear dichroism is observed). However, the ensemble of molecules which have been excited is not oriented randomly: those for which $|\boldsymbol{\varepsilon}_U \cdot \boldsymbol{\varepsilon}_z|^2$ is large, i.e., those whose z axis is aligned or nearly aligned with the laboratory axis U, are more susceptible to excitation than those for which $|\boldsymbol{\varepsilon}_U \cdot \boldsymbol{\varepsilon}_z|^2$ is small, and those whose z axis happens to lie perpendicular to U cannot be excited at all. We call the angle between the molecular z axis and the U direction z and say that the excitation probability distribution shows a $\cos^2 z$ dependence. This selection of a partially aligned assembly from a random assembly by the action of light is known as "photoselection." The resulting partially oriented assembly is quite analogous to the assembly of pyrene molecules partially aligned in stretched polyethylene which we have just discussed in the IR and UV absorption experiments. An important difference is that we know the orientation distribution function of the photoselected assembly if we know the direction of the absorbing transition moment in the molecular framework and if certain other conditions are fulfilled (Section 4.6.3).

If we now probe the photoselected assembly, say by allowing its molecules to emit light and observing its polarization, we obtain results similar to those mentioned in the IR and UV experiments. If the emitting transition moment is also directed along z, it will emit light polarized primarily along z; the intensity emitted with polarization perpendicular to z will be vanishingly small. Since the emitting molecules have their axes partially lined up with U, we must expect most of the emitted intensity to be polarized along U. If the emitting transition moment is directed perpendicular to z, we expect the emitted intensity to be polarized primarily perpendicular to U.

We conclude that the origin (0–0 component) of pyrene fluorescence is polarized along the x or y direction, and since the L_b transition is well established to be of the $\pi\pi^*$ type, the y direction must be the correct choice. In the rest of the fluorescence spectrum, peaks of both polarizations overlap, and this is responsible for the differences in the peak positions between the ZZ and YY spectra on the one hand and the ZY and YZ spectra on the other hand. Thus, the fourth peak, near 25 800 cm^{-1}, is primarily z-polarized, as is the high-energy edge of the second peak and the low-energy edge of the third peak. Such strongly mixed polarization within the L_b emission is undoubtedly due to vibronic coupling with the nearby z-polarized L_a transition. As in the treatment of UV data, it is again possible to separate the y-polarized and z-polarized fractions of the intensity. This task is now simplified by the fact that the orientation distribution function of the photoselected emitting assembly is known. As will be seen in more detail in Chapters 4 and 7, as long as the absorbing and emitting spectral features can be assumed to be purely polarized, as would be the case in a molecule of high symmetry, the reduction does not require a trial-and-error procedure. (We shall see in Chapter 8 that the intrinsically highly symmetrical pyrene molecules are sufficiently sensitive to perturbations by the solvent, even a nonpolar one such as 3-methylpentane, that they deviate from the expected behavior.)

If the z polarization of the absorbing L_a transition moment were not known, Figure 2.8 would still make it clear that it is perpendicular to the origin of the L_b band, i.e., it would establish the relative polarization of the two transitions but not the absolute polarization. Therefore, in itself, the measurement of polarized fluorescence in an isotropic rigid solution represents a relative polarization measurement method.

It is obvious that a repetition of the experiment for a variety of excitation wavelengths will produce information on the relative orientation of the various absorbing transition moments. Clearly, the polarized fluorescence experiment yields not only the polarizations of the individual vibrational components of the fluorescent emission, but also the polarization of the absorption spectrum, albeit only in a relative sense. A measurement of polarized phosphorescence can provide additional useful data.

Now, what happens when the pyrene sample is not isotropic to start with, but rather, partially uniaxially aligned? The effects of photoselection should then be superimposed on those of the originally present partial alignment. We shall see in Chapter 7 how they can be disentangled and how absolute polarizations of the absorbing and emitting dipoles can be obtained. At the same time, the experiment yields new information on the nature of the orientation distribution in the original partially aligned sample, not derivable from measurements of IR or UV dichroism alone. Here we shall provide only a brief qualitative description.

Figure 2.9 shows the polarized fluorescence spectra of pyrene contained in stretched polyethylene, again excited into the L_a band and taken at a temperature (77 K) where molecular rotation in this medium is negligible on the fluorescence time scale. The degree of anisotropy R is no longer unambiguously defined, and three curves representing simple intensity ratios $F_{UV}/F_{U'V'}$ are shown. Looking at the fluorescence origin (α) first, and comparing with Figure 2.8, we note that the intensity of the ZZ and ZY fluorescence curves is now enhanced relative to the others. This agrees with the assignment of the absorbing L_a transition moment to the z molecular direction, which is preferentially lined up with the polymer stretching direction Z.

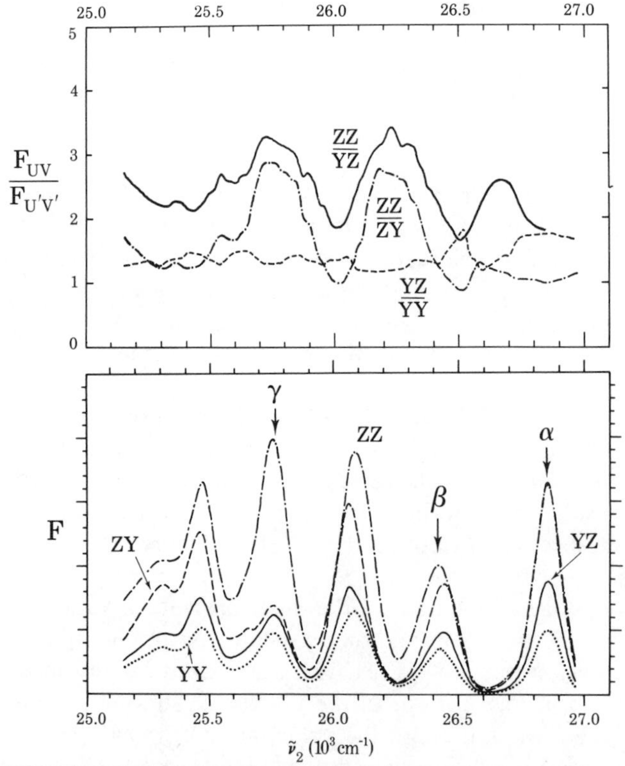

Figure 2.9 Polarized fluorescence of pyrene in stretched polyethylene at 77 K, excited at the L_a origin. Bottom, the four polarized emission curves: full line, YZ; dashed line, ZY; dotted line, YY; dash-dot line, ZZ. The curves are corrected for instrumental polarization biases but not for the 14.5° angle between the exciting and observed beam directions. Reproduced by permission from F. W. Langkilde, M. Gisin, E. W. Thulstrup, and J. Michl, *J. Phys. Chem.* **87**, 2901 (1983). Top, the polarization ratios: full line, F_{ZZ}/F_{YZ}; dashed line, F_{YZ}/F_{YY}; dash-dot line, F_{ZZ}/F_{ZY}.

When we shift our attention to the fourth peak (γ) in the fluorescence spectrum, located near 25 800 cm^{-1}, we note that the ZZ curve shows far more intensity than any of the others, while the YY curve has been suppressed tremendously relative to that shown in Figure 2.8. This is in good agreement with simple qualitative notions: The absorption as well as the emission are now z-polarized. Since the absorbing as well as the emitting transition moments are now approximately lined up with the Z direction and away from the Y direction, the intensity observed in the ZZ mode should be high and that in the YY mode should be small. Yet in the 3-methylpentane glass measurement, both were nearly equal (and after the introduction of corrections for experimental imperfections, they were exactly equal).

While such qualitative arguments are perhaps sufficient to confirm one's notions concerning the absolute polarizations of the electronic transitions and thus represent an advance relative to the 3-methylpentane glass measurements which yield only relative polarizations, much more information can be obtained using a more quantitative treatment, as we shall see in Chapters 7 and 8.

Polarized Raman Spectra. Our fourth and last example is that of the polarized Raman spectra. In this case also, two photons are involved, but there is no time lag between the destruction of the incident photon and the creation of a scattered photon, so that one need not be concerned with suppressing molecular rotation. The polarized Raman scattering spectrum of pyrene is easy to obtain in fluid solution, but so far we have found it impossible to obtain a quantitatively reliable spectrum in stretched polyethylene. Since we still wish to illustrate the effect of partial solute orientation on polarized Raman spectra, we have chosen a molecule for which we were more successful. Figure 2.10 presents a comparison of the polarized nonresonant Raman spectra of 2,3-dimethylnaphthalene obtained in an isotropic fluid solution with those obtained in stretched polyethylene. Judging by its polarized IR and UV spectra, 2,3-dimethylnaphthalene orients in the latter medium somewhat better than pyrene. The long axis labeled z in Figure 2.10 lines up the best with the stretching direction, and the out-of-plane axis x the worst.

The segment of the spectrum shown in Figure 2.10 contains two vibrations. Standard rules for the depolarization ratio in ordinary isotropic solutions can be applied to recognize the 417 cm^{-1} vibration as non-totally symmetric (depolarization ratio 0.75) and the 449 cm^{-1} vibration as totally symmetric (depolarization ratio 0.12). The spectra of the partially oriented sample contain information on the symmetry of the 417 cm^{-1} vibration and on the anisotropy of the diagonal scattering tensor of the a_1 symmetry 449 cm^{-1} vibration.

The 417 cm^{-1} vibration could belong to the a_2 symmetry species (nonvanishing Raman tensor elements $\alpha'_{xy} = \alpha'_{yx}$), the b_1 species (tensor elements $\alpha'_{xz} = \alpha'_{zx}$), or the b_2 species (tensor elements $\alpha'_{yz} = \alpha'_{zy}$) of the C_{2v} group. A decision between these possibilities cannot be reached from the isotropic spectrum. Already a qualitative consideration of the intensities of the 417 cm^{-1} peak in the four polarized Raman spectra in Figure 2.10 suggests that the nonvanishing tensor element has z as one of its subscripts, since the intensities I_{ZZ} and I_{ZY} are distinctly higher than I_{YY} and I_{XY} and we know that the z axis is much

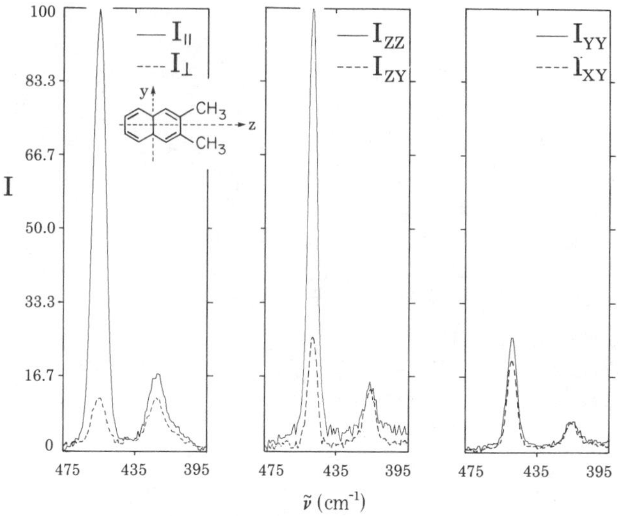

Figure 2.10 Polarized room-temperature Raman spectra of 2,3-dimethylnapthalene in fluid solution in benzene ($I_{\parallel}$ and $I_{\perp}$) and in stretched polyethylene [$I_{ZZ}(\tilde{\nu})$, $I_{ZY}(\tilde{\nu}) = I_{YZ}(\tilde{\nu})$, I_{YY}, and $I_{XY}(\tilde{\nu})$]. Excited with the 488 nm line of an Ar ion laser (600 mW). J. G. Radziszewski and J. Michl, unpublished results.

better aligned with Z than the x and y axes are. A decision between b_1 and b_2 will be possible only if the orientation properties of the molecular x and y axes are sufficiently different. This will clearly require a more quantitative analysis.

The 499 cm^{-1} vibration belongs to the a_1 species, so its nonvanishing matrix elements are α'_{xx}, α'_{yy}, and α'_{zz}. In this case, the question is, what are their relative magnitudes? The information contained in the isotropic depolarization ratio goes part way toward answering this question but provides no clue as to which one of the three elements is the smallest and which the largest. A qualitative consideration of the intensities of the 449 cm^{-1} peak in the four polarized spectra in Figure 2.10 suggests a partial answer: The much higher intensity I_{ZZ} relative to I_{YY} leaves little doubt that α'_{zz} is the largest tensor element, since the z axis is lined up with Z considerably better than the x or y axes are. Once again, a decision concerning the relative magnitude of α'_{yy} and α'_{xx} depends critically on the difference in the degree of alignment of the x and y axes and clearly calls for a more quantitative evaluation.

2.4 Comments and References

A classic introduction to many of the subjects covered in Chapter 2 is provided by H. Eyring, J. Walter, and G. E. Kimball, *Quantum Chemistry*, Wiley, New York, 1944. An excellent treatment of molecular vibrations, normal coordinates, and vibrational transition moments can be found in the book by Wilson, Decius, and Cross quoted in Section 1.4; another classic treatment also quoted there is that of Herzberg. The empirical valence-optical model based on the use of electro-optical parameters for the estimation of vibrational transition moments is described in L. M. Sverdlov, M. A. Kovner, and E. P. Krainov, *Vibrational Spectra of Polyatomic Molecules*, Wiley, New York, 1974. An excellent recent treatment of both the empirical and quantum-mechanical methods for the calculation of vibrational intensities is provided in W. B. Person and G. Zerbi, Eds., *Vibrational Intensities in Infrared and Raman Spectroscopy*, Elsevier Amsterdam, 1982.

The theory of π-electron systems, including the pairing theorem for alternant hydrocarbons, is treated by L. Salem in *The Molecular Orbital Theory of Conjugated Systems*, Benjamin, New York, 1966. Electronic transition moments are discussed, for instance, in R. L. Ellis and H. H. Jaffé, in *Semiempirical Methods of Electronic Structure Calculations*, G. A. Segal, Ed., Plenum, New York, 1977, Part B, Chapter 2. Good treatments of vibronic coupling and spin-orbit coupling are found there and also in F. Dörr, in *Creation and Detection of the Excited State*, Vol. 1A, A. A. Lamola, Ed., Marcel Dekker, New York, 1971, p. 53, and in S. P. McGlynn, T. Azumi, and M. Kinoshita, *Molecular Spectroscopy of the Triplet State*, Prentice-Hall, Englewood Cliffs, NJ, 1969. A book dedicated to vibronic coupling has appeared recently: G. Fisher, *Vibronic Coupling*, Academic Press, New York, 1984. Vibronic coupling in natural CD spectra is treated in the books by Caldwell and Eyring and by Charney quoted in Section 1.4. Transition moments for natural CD spectroscopy are discussed at length in A. E. Hansen and T. D. Bouman, *Adv. Chem. Phys.* **44**, 545 (1980) and in N. Harada and K. Nakanishi, *Circular Dichroic Spectroscopy*, University Sciences Books, Mill Valley, CA, 1983. Vibronic coupling in MCD is treated in the book by Piepho and Schatz quoted in Section 1.4 and in M. Z. Zgierski, *J. Chem. Phys.*, in press (1986). A good starting point for theoretical aspects of vibrational CD is the paper by S. Abbate, L. Laux, J. Overend, and A. Moscowitz, *J. Chem. Phys.* **75**, 3161 (1981), which describes the "charge flow model" [a factor e is missing in front of the last term in equation (10) of this paper]. An application of the charge flow model to an oriented sample is described in S. Abbate, L. Laux, V. Pultz, H. A. Havel, J. Overend, and A. Moscowitz, *Chem. Phys. Lett.* **113**, 202 (1985). A theoretical development that circumvents the usual difficulties in the calculation of vibrational magnetic dipole transition moments is described in P. J. Stephens, *J. Phys. Chem.* **89**, 748 (1985), and applied in M. A. Lowe, G. A. Segal, and P. J. Stephens, *J. Am. Chem. Soc.* **108**, 248 (1986). A review of VCD theory is found in P. J. Stephens and M. A. Lowe, *Annu. Rev. Phys. Chem.* **36**, 231 (1985). For theory of vibrational MCD, see L. Laux, V. Pultz, C. Marcott, J. Overend, and A. Moscowitz, *J. Chem. Phys.* **78**, 4096 (1983). The problems associated with the use of the Born-Oppenheimer approximation in the dipole velocity formalism were first pointed out by C. A. Mead and A. Moscowitz, *Int. J. Quant. Chem.* **1**, 243 (1967). The theory of vibrational structure in electronic MCD spectra is described by M. Z. Zgierski, *J. Chem. Phys.* **83**, 2170 (1985).

A good introduction to two-photon absorption tensor calculations is the paper by G. Hohlneicher and B. Dick, *J. Chem. Phys.* **70**, 5427 (1979). The book by Long quoted in Section 1.4 provides an excellent introduction to the concept of the Raman scattering tensor. For quantum-mechanical calculations of the Raman tensor, see, e.g., R. E. Bruns, in *Vibrational Intensities in Infrared and Raman Spectroscopy*, W. B. Person and G. Zerbi, Eds., Elsevier, Amsterdam, 1982, Chap. 7; and M. Spiekermann, D. Bougeard, H.-J. Oelichmann, and B. Schrader, *Theoret. Chim. Acta* **54**, 301 (1980). Empirical procedures of calculation, based on the use of electrooptical parameters, are covered by J. Gussoni in Chapter 11 of the book edited by Person and Zerbi. Good introductions to the theory of the resonant Raman effect are found in T. G. Spiro and P. Stein, *Ann. Rev. Phys. Chem.* **28**, 501 (1977) and W. Siebrand and M. Z. Zgierski, in *Excited States*, Vol. 4, E. C. Lim, Ed., Academic Press, New York, 1979, p. 1. More recent references are found in P. M. Champion and A. C. Albrecht, *Ann. Rev. Phys. Chem.* **33**, 353 (1982), and B. Hudson, P. B. Kelly, L. D. Ziegler, R. A. Desiderio, D. P. Gerrity, W. Hess, and R. Bates, *Adv. Laser Spectrosc.*, B. A. Garetz and J. R. Lombardi, Eds., Hayden Press, London, Vol. 3 (1986).

3

Techniques of Alignment and Measurement

The experimental aspects of optical measurements on partially aligned molecules are discussed in this chapter. We begin by describing the various ways in which molecules can be aligned. Subsequently, we outline the procedures used in recording polarized optical spectra of partially aligned samples.

We have attempted to provide sufficient detail to make the descriptions practically useful, but we still recommend that the original literature be consulted by those wishing to perform these kinds of experiments. In only one instance did we make a conscious effort to provide complete guidance to an experiment. This is the case of measurement of linear dichroism on stretched polyethylene. In our opinion, this is the easiest, simplest, and least expensive method of all and can be introduced in any laboratory with an ordinary UV-visible or Fourier-transform IR spectrometer within a matter of hours after the purchase, loan, or theft of a polarizer or, perhaps surprisingly, even without a polarizer. The degree of alignment obtained in stretched polyethylene for a given solute is among the highest obtainable in any anisotropic solvent or, for that matter, any other method of alignment in solid or fluid solution.

3.1 Molecular Alignment

Some naturally occurring samples, mostly crystals and biological specimens, contain fully or partially aligned molecules. Frequently, however, it is desired to impart orientation to molecules in the laboratory, starting with an unoriented sample, and this can be accomplished in a variety of ways.

Single Crystals. Growing single crystals of the molecules to be studied is a time-honored procedure which, at least in principle, is capable of providing absolute polarization assignments if the crystal structure is known and suitable. In practice, difficulties are frequently encountered both in the measurement and in the interpretation. Extremely thin single crystals are often required for absorption studies, particularly in the UV-visible region. This requirement can be circumvented by measuring reflection spectra and performing a Kramers-Kronig transform [see (1.65)], but this is not a simple procedure. Recent progress in photoacoustic spectroscopy suggests that it may be a useful technique in this regard. The interpretation of crystal spectra in terms of molecular properties is often complicated by large intermolecular interactions (Davydov splitting), particularly important for strong transitions in the UV-visible region of the spectrum.

The difficulties of work with neat single crystals are removed if the molecules to be studied are dissolved in a single crystal of a suitable host material that is transparent in the region of interest. The solute molecules must fit nearly perfectly into the lattice sites of the host, and it is assumed that they acquire the orientation of the host molecules they have replaced. Since the solid solution is dilute, very thin crystals are no longer required. Also, intermolecular interactions are limited to solute-host interactions, comparable in magnitude to the usual solute-solvent interactions. Unfortunately, it is frequently difficult to find a suitable host crystal into which the solute of interest enters in an isomorphous fashion and which is transparent throughout the spectral region of interest. Also, there may be some uncertainty about the exact orientation of the solute molecules relative to the lattice of the host, unless the fit is perfect. In such cases, methods described in this book for partially oriented samples may be used to obtain information on the orientation distribution.

The problems associated with growing suitable single crystals and the various difficulties associated with the single-crystal methods make their use tedious and have contributed to the reputation of spectroscopy with polarized light as being a difficult technique accessible only to specialists. This reputation is undeserved, since today less demanding methods, using only partially aligned samples, are available for the determination of absolute polarizations. Here, we refer in particular to the measurement of linear dichroism obtained with liquid crystals or stretched polymers as orienting solvents. It must be acknowledged, however, that many of the polarization assignments based on these newer methods can be considered absolute only because it was possible to compare results of a sufficient number of selected cases with the results of fundamental studies on single crystals.

Anisotropic Solvents. Partial orientation can be imparted to molecules in a variety of ways. Most of them produce uniaxial orientation, which we shall refer to as alignment. Uniaxially aligned samples possess a unique direction (optical axis, unique sample axis), with respect to which the solute molecules are nonrandomly oriented; all directions perpendicular to this axis are equivalent. Because of the overwhelming practical importance of uniaxial alignment in partially oriented assemblies, this book concentrates on it almost exclusively.

The use of anisotropic solvents is the most important method for achieving partial alignment of molecules. Among these, thermotropic liquid crystals, lyotropic liquid crystals containing lipid layers or bilayers (membranes) in various forms, and stretched polymers have found the most extensive use. They are discussed in Sections 3.1.1, 3.1.2, and 3.1.3, respectively.

Electric Field. Another method is alignment by strong electric fields, which produces electric dichroism, discussed in Section 3.2.3. For small molecules, only a small degree of alignment can be achieved in this fashion, but certain polymers such as DNA can be oriented very well. The birefringence associated with the partial alignment by the electric field is known as the Kerr effect.

The strong electric field of a very intense laser pulse is capable of aligning molecules in liquids. This phenomenon, known as the optical Kerr effect, is useful in the study of molecular rotational motion. This type of alignment has also been used to obtain Raman spectra of partially oriented molecules.

Magnetic Field. Like an electric field, a magnetic field can also have an orienting effect on molecules, but the orienting action is generally much weaker. The molecules to be oriented must exhibit either diamagnetic anisotropy or a permanent magnetic moment strongly coupled to the molecular framework. The birefringence associated with the partial alignment by the magnetic field is known as the Cotton-Mouton effect. Because of its weakness, the effect has found relatively little use. However, at times very large entities (domains) composed of many molecules can be aligned sufficiently, and magnetic fields can be used to achieve a uniform alignment of many liquid crystals.

Flow Field. The hydrodynamic shear associated with liquid flow can also align molecules. The orientation is given by the hydrodynamic properties of the molecules and is conveniently measurable only for macromolecules. It is detectable as the so-called flow dichroism and the associated birefringence (the Maxwell effect). The usual way of providing the necessary hydrodynamic shear is to use a Couette cell, in which a liquid experiences a flow gradient across the annular gap between a rotating cylinder and a fixed cylinder.

Photoselection. A uniquely different method is photoselection, in which no attempt is made to physically align all the molecules of the sample. Rather, a partially oriented subset of the total random molecular assembly is selected by absorption of photons from a light beam. The anisotropy of the selected subassembly is then probed by means of its absorption or emission properties. The experimental methods are the same as those described in Sections 3.2.6 and 3.2.7. The interpretation is based on methods discussed in Chapter 7 (cf. Section 7.3.5), which, however, deals primarily with the more complicated case of photoselection in an already preoriented molecular assembly.

3.1.1 Anisotropic Solvents: Thermotropic Liquid Crystals

Thermotropic liquid crystals are melts which are capable of existence in a temperature range between the melting point and the isotropic transition point. They ordinarily consist of a large number of microscopic domains in each of which the constituent molecules exhibit partial order. Under suitable conditions (proximity of a properly treated surface, application of electric or magnetic fields), it is possible to line up the axes of all the domains and produce a macroscopic volume of clear, partially aligned uniaxial (or biaxial) liquid suitable for optical measurements.

There are several types of liquid crystals which differ in the details of molecular order (Figure 3.1). Smectic liquid crystals are formed by rod-shaped molecules which aggregate side by side, forming layers which slide more or less freely against each other. Nematic liquid crystals are formed by rod-shaped molecules which line up approximately parallel and slide more or less freely past each other in the direction of alignment (Z). Cholesteric liquid crystals are formed by elongated optically active molecules and have helical structure. They can be thought of as consisting of a series of adjacent layers of the nematic phase, with the alignment axis Z of each layer parallel to the same plane. As one moves in the direction perpendicular to this plane, which is the axis of the cholesteric crystal (X), one notes that the directions of the alignment axes of every two neighboring layers differ by a constant angle. The repeat distance of the alignment axis direction is known as the pitch. The length of the pitch depends on the composition of the liquid crystal, and it is possible to adjust it so that it becomes infinitely long, transforming the cholesteric into a nematic liquid crystal ("compensated nematic mixture"). Most studies of interest in the present context have been performed on solutions in nematic liquid crystals, and a few on solutions in smetic liquid crystals.

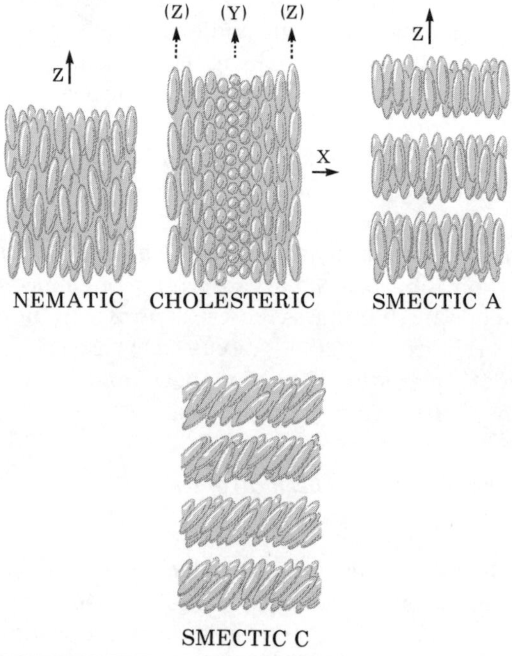

Figure 3.1 Internal structure of representative liquid crystals.

The anisotropic intermolecular potential in the liquid crystal is a function of its order. It causes a partial uniaxial alignment of the solute molecules relative to the unique axis ("director") of the liquid crystal. The degree of alignment depends on the nature and concentration of the solute and on the nature and temperature (T) of the sample. It is common to introduce the concept of "reduced temperature" T/T_{ni}, where T_{ni} is the clearing point of the nematic phase in the presence of the solute. This provides a suitable scale for a comparison of the degree of orientation as a function of the concentration of the solute. A striking example of an effect of a solute on the order in the liquid crystal is provided by the transformation of nematic crystals into cholesteric by addition of a chiral solute.

The macroscopic alignment of the liquid crystal sample is affected by fields applied from the outside. For instance, a strong electric field can increase the pitch of a cholestric liquid crystal from large to infinite and make the liquid crystal nematic.

Outside fields are commonly used to produce macroscopic domains in the liquid crystal, the other usual method being the use of surface forces. Nematic liquid crystals exhibit two important types of orientation relative to the surface with which they are in contact. In homeotropic alignment, the axis of the crystal is perpendicular to the surface. In homogeneous alignment, the axis of the crystal lies parallel to the surface, along a direction usually determined by rubbing the surface. One or the other alignment can often be achieved by a suitable choice of surface treatment and of an orienting electric or magnetic field.

An important consideration in the choice of a solvent for spectroscopic measurements is its region of transparency. Most liquid crystals contain aromatic chromophores. They offer limited spectral windows in the IR region, useful only for special purposes such as the study of C—O stretching vibrations in metal carbonyls, and are not suitable for UV measurements or even measurements over much of the visible region. However, there are two types of nematic thermotropic liquid crystals which are transparent throughout most of the UV-visible region. These are compensated nematic mixtures of suitable cholesteryl derivatives which transmit down to about 240 nm and the more recently discovered nematic mixtures of certain bicyclohexyl derivatives which are transparent to 200 nm, show only very weak birefringence, and have high dielectric anisotropy, so that they are well suited for orientation by electric field.

On quartz, the compensated nematic crystals tend to orient in a homogeneous fashion; if the surface is treated with egg lecithin, they tend to orient homeotropically. In either case, strong electric fields are usually used to ensure good uniform alignment. It is particularly difficult to achieve the perfect homeotropic alignment needed for studies of the optical activity of oriented molecules. Many surfaces are very sensitive and will respond even to a very minute unsymmetrical influence, such as the direction of flow of isotropic liquid during the filling of an optical cell. Thereafter, contact with such a surface will cause imperfections in the uniaxiality of the alignment. The use of compensated nematic crystals is essential in the studies of circular dichroism in liquid crystals, since the tendency of a chiral solute to induce a transition from nematic to cholesteric phase can be counteracted by adjusting the composition and the temperature of the liquid crystal mixture so that it is again compensated after the solute has been added.

The important nematic crystals of the bicyclohexyl type can be made to orient homogeneously or homeotropically on quartz, depending on the treatment of the quartz surface. A treatment with *N*-methyl-3-aminopropyltrimethoxysilane or coating with poly(vinyl alcohol) followed by surface rubbing with tissue paper is sufficient to produce excellent homogeneous alignment of a thin layer of the liquid crystal (~ 50 μm). A treatment with a detergent such as sodium *p*-*n*-octadecylbenzenesulfonate or with lecithin produces homeotropic alignment.

Finally, one type of orientation of cholesteric liquid crystals needs to be mentioned. This is the "Grandjean" texture obtained by applying a small pressure to a thin layer of a cholesteric liquid crystal held between two glass plates. In this texture the nematic layers of the cholesteric phase lie parallel to the surfaces, and the optical axis X (axis of the helix) lies normal to the surfaces. The director of each nematic layer will be called Z, and the direction perpendicular to it and to the axis of the helix will be called Y. As one proceeds down the axis of the helix, Z and Y rotate and cover 2π in one pitch length (Figure 3.1).

In work with liquid crystals as anisotropic solvents, temperature control is important. The nematic range of the liquid crystals is limited, and the nematic compensation of the cholesteric crystals as well as the degree of alignment of the solute is temperature-dependent. Therefore, sample cells thermostated to a fraction of a degree are commonly used. Compensated nematic liquid crystals can be supercooled. When cooled rapidly to very low temperatures (liquid nitrogen), they form rigid glasses in which alignment is preserved.

3.1.2 Anisotropic Solvents: Lyotropic Liquid Crystals, Membranes

Lyotropic nematic liquid crystals are solutions containing a large number of microscopic domains whose molecules exhibit partial order. The axes of these domains can be aligned by external forces such as a magnetic field, hydromechanical shear, or surface effects, producing macroscopic volumes of clear anisotropic liquid suitable for optical measurements. Most of the known systems contain water and an amphiphilic compound (lipid, detergent) as the main constituents. In these solutions, pieces of lipid bilayer membranes of various geometrical shapes are present depending on the type of the liquid crystal. Lamellar lyotropic liquid crystals containing more or less flat pieces of membrane and, to a lesser degree, nematic lyotropic liquid crystals believed to contain membranes in a rodlike shape, have been used most often as uniaxial media in optical studies. In many studies, the membranes were of natural origin, but artificial bilayers have also been used.

The lyotropic liquid crystals differ from thermotropic liquid crystals in that they accept both nonpolar and polar molecules. Sometimes, however, a polar solute will exhibit a very low degree of orientation, presumably when it resides primarily in the aqueous environment outside of the bilayers. The transparency of these liquid crystals is limited only by the properties of the amphiphile used and can be very good down to 250–300 nm.

In principle, monolayers adsorbed on surfaces can also be used as uniaxial orienting media, but highly sensitive spectroscopic methods are then required because of their small thickness.

3.1.3 Anisotropic Solvents: Stretched Polymers

General. Many linear polymers can be partially oriented by drawing or extruding at temperatures above their glass transition temperature. Under many circumstances, the orientation is uniaxial. This is found not only when rods are drawn or extruded, but also when relatively thin sheets are drawn, even though the decrease in sheet thickness upon stretching is much larger than the decrease in its width. However, it is also possible to generate more general types of orientation, for instance, by subsequent squashing in a direction perpendicular to the draw direction.

When the polymer is doped with a solute, anisotropic intermolecular forces impart partial orientation to the solute molecules. The degree of orientation of a given solute in a given polymer generally depends somewhat on the processing history of the polymer. For this reason, the orientation factors shown in Section 4.6 should be taken only as a general guide. For a given polymer and solute, the degree of orientation depends primarily on the stretching ratio. In some cases, a distinct change in the degree of orientation has been noted upon cooling to liquid nitrogen temperatures.

The most commonly used form of the polymer samples is a sheet whose plane contains the stretching direction. Attempts to produce thin polymer samples by cutting a stretched rod perpendicular to the stretching direction and polishing have so far produced samples suitable for measurement of absorption spectra but no specimens of optical quality useful for measurements of optical activity. The thickness of the polymer sample can be measured by a micrometer or, for thin sheets, more accurately by counting interference fringes in the infrared.

Commercial polymer sheets frequently contain additives and impurities which should be removed before use by extraction with a solvent which swells the polymer. It is relatively simple to prepare one's own polymer sheets from commercial polymer powder or pellets by melting and pressing or by casting from a solution.

Evaporation of volatile solutes from the polymer is sometimes a problem. Immersion of the sheet in liquid nitrogen, which is frequently done in order to produce low-temperature spectra, will also prevent this. Evaporation can also be prevented by placing the sheet between two quartz plates at room temperature, perhaps adding a drop of an inert liquid such as propylene glycol. In either case, improved optical contact is a side benefit.

Stretching. The stretching of polymers is usually done on simple mechanical devices in which a polymer film is clamped between two movable pairs of bars on each end (Figure 3.2). These pairs of bars should have no sharp points or

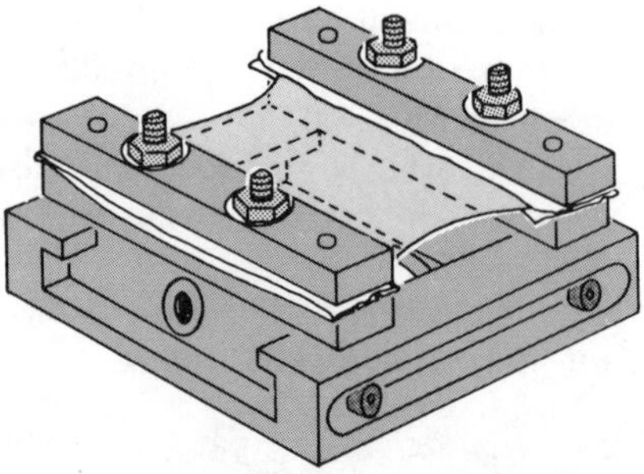

Figure 3.2 A stretcher for polymer sheets.

edges, which tend to cause tearing of the sheets. For instance, it is useful to insulate them from the polyethylene with a layer of Teflon. Commercial sheets can usually be stretched in only one direction and break if stretching in a perpendicular direction is attempted. This structural anisotropy is due to the nature of the manufacturing process.

The stretching of some polymers requires no special precautions and can be done at room temperature (polyethylene). Others require careful temperature control.

Doping. The substrate can be introduced into the polymer in a variety of ways, all of which usually produce a monomolecular dispersion at all practically attainable concentrations. The most common procedures are (i) casting a film of the polymer from a solution to which the substrate has been added, (ii) swelling the polymer with a solution containing the substrate and evaporating the solvent, (iii) allowing the vapor of the substrate to diffuse into the polymer, and (iv) dissolving the substrate in a molten polymer. Different ways of introducing the substrate may lead to location of the substrate molecules at morphologically different sites in the polymer and therefore to different degrees of orientation. For instance, the solute molecules may act as seeds in the crystallization of a molten polymer, while the swelling solvent usually does not penetrate the crystalline regions. The stretching of the polymer can always be performed after the doping, but for techniques (ii) and (iii), which depend on diffusion, it can also be done beforehand. This leads to an approximately equal degree of orientation and usually has the advantage that the stretched polymer is thinner and the doping proceeds faster. The required doping time is a function of the rate of diffusion through the sample and can be as short as seconds or as long as days. It is shortened substantially by the use of elevated temperatures. It can also be shortened dramatically by placing the swelling solvent in an ultrasound bath, but this technique must be used with great caution since it may produce cavities in the polymer, which then fill up with microcrystals of the substrate. When a high concentration of the substrate in the polymer is desired, it is essential to use a saturated solution of the substrate in the swelling solvent. When using procedures (i) and (ii), it is important to allow the solvent to evaporate completely if reproducible orientation is desired, since solvent residues decrease the degree of solute orientation. With procedure (ii), it is imperative to remove carefully all crystalline residue from the surface of the sample. This is best done with a solvent which does not swell the polymer.

Not all polymers will accept all substrates. In general, nonpolar polymers, which have the spectroscopic advantage of interacting only weakly with the solute, are less well suited or even totally unusable for highly polar substrates. On the other hand, the latter can be studied conveniently in polar polymers. It is already clear from the feasibility of the doping technique (iii) that the solute molecules can diffuse quite rapidly through the polymer; those with high vapor pressure, such as naphthalene, will sublime out of a sheet rapidly. A case is known in which nonpolar polymer sheets containing free molecules of a polar substrate when freshly doped contained primarily microcrystalline aggregates after a few days.

The determination of the amount of solute contained in the sample poses no problem when procedures (i) and (iv) are used. With (ii) and (iii), it is harder. When all the solute can be extracted into a suitable solvent or the polymer can be dissolved, a spectrophotometric determination is straightforward. Frequently, it is assumed that the extinction coefficient is the same in the polymer as in a liquid solvent of similar composition. The amount of solute in the polymer can then be determined from its polarized absorption spectra by methods described in Chapter 5. Often, the knowledge of the absolute amount is actually not needed.

Individual Polymers. Many linear polymers are transparent throughout the UV-visible region, and some even through most of the IR region. Those enjoying the highest popularity are polyethylene and poly(vinyl alcohol), which produce very high degrees of orientation. Measurements have also been reported in other polymers such as polypropylene and poly(vinyl chloride), which, however, orient solutes less well.

Polyethylene is best suited for nonpolar substrates. It is transparent both in the UV-visible region at least to 200 nm and in the usual IR region except for three blind spots: 720–750 cm^{-1}, 1440–1480 cm^{-1}, and 2800–3000 cm^{-1}. Perdeuterated polyethylene is commercially available and can be used to cover these three regions. It is expensive, but the polymer sample can be regenerated by extraction of the solute and reused. With the polyethylene–perdeuterated polyethylene combination, complete coverage is possible from the far-IR to the onset of vacuum UV. The degree of orientation imparted to a substrate seems to depend remarkably little on its concentration.

Either low-density or high-density polyethylene can be used, but the former is normally preferable since it accepts larger amounts of substrates. Also various copolymers of ethylene with poly(vinyl acetate) have been used successfully. They offer the advantage of increased polarity and are useful for the more polar substrates. In order to increase the observed optical density, one can use a stack of several polyethylene sheets. The larger number of surfaces tends to cause significant light losses by scattering, even when a refractive-index-matching fluid is used. This can be avoided by pressing substrate-containing sheets together at about 100°C to produce a single sheet up to several millimeters thick which is then stretched. Of course, thermally sensitive substrates should then not be used. In principle, one can also start with a single thick sheet, but diffusion of the substrate into it may then take an excessive amount of time. In such thick sheets, light depolarization by scattering is significant, and they should not be used for quantitative determination of the degree of orientation.

Polyethylene is extremely easy to stretch even at room temperature. The degree of orientation at first increases rapidly with the stretching ratio, but after about a fourfold stretch it remains almost constant. Thin-sheet strips can easily be stretched in the hands, as anybody who has played with a sandwich bag knows. After a stretched polyethylene sheet is released, it retracts a little and then keeps its stretched shape. This makes it easy to mount the sheets in a spectrometer. The usual way of introducing the substrate into polyethylene is by swelling [procedure (ii)], the solvent of choice being chloroform. Toluene can also be used but is harder to evaporate from the sheet. The best solvents for washing off residual crystals from the surface are methanol and ethanol. Other methods, (iii) and (iv), have been used occasionally. Even method (i) can be used, with hot *p*-xylene as the solvent.

Poly(vinyl alcohol) equals polyethylene in its transparency in the UV-visible region but is much less useful in the IR region because of its considerable absorption. It can be used for polar as well as nonpolar substrates. In equilibrium with a humid atmosphere, it is an excellent host for orienting DNA.

Samples for measurement are usually prepared by procedure (i). Commercial poly(vinyl alcohol) powder and the substrate are dissolved in water and heated to produce a thick homogeneous solution containing 5–10% of the polymer. This is then cast on a glass plate, and the solvent is allowed to evaporate in a dust-free atmosphere over a period of several days. The sheet is peeled off and stretched at an elevated temperature (a stream of hot air from a hair dryer is adequate). The degree of solute orientation increases with the stretching ratio. The increase is slower but still nonnegligible at high stretching ratios. It is also possible to swell a clean poly(vinyl alcohol) sheet with a solution of the substrate in methanol containing a little water [procedure (ii)].

3.2 Optical Measurements

Spectrometers and Polarizers. Most of the measurements which fall within the scope of this book can be performed on commercially available spectrometers. The only modification which may be required is the addition of one or two linear polarizers, and even this is actually not absolutely necessary, as will be seen below. In the UV-visible region, calcite Glan prisms are the polarizing elements of choice down to about 220 nm. The much cheaper polarizing sheets of the Polaroid sunglass type have lower transmission, particularly in the UV region, and cut off at longer wavelengths. Quartz or magnesium fluoride polarizers must be used below 220 nm. These are of the Rochon or Wollaston type, and the spatial separation of the two emerging beams of mutually perpendicular polarization is generally inadequate for a straightforward incorporation into the optical trains of commercial spectrometers if the polarizer is to be rotated. Aluminum or gold grid polarizers are ideal for the IR region. Several are needed to cover the whole IR range accessible to a modern Fourier transform IR instrument.

The use of instruments of high luminosity, and in particular of Fourier transform instruments in the IR, is highly advisable to optimize the signal-to-noise ratio, since polarizing elements necessarily cut the available light intensity. The measurements of dichroism in the UV-visible region are the easiest in this regard and can be performed even on quite cheap instruments.

The transmission of a grating monochromator is different for light polarized parallel to the groove direction than for light polarized perpendicular to that direction, to a degree which depends on the wavelength. If the polarizer is to be mounted in the beam in an immovable fashion, one orientation is clearly preferable to all others. In such a case, the sample must usually be rotated. There are some inherent dangers in this unless the sample is perfectly homogeneous, since different sections will be illuminated as it is rotated. Perfect homogeneity is easier to achieve on liquid crystal samples than on stretched polymers.

If the linear polarizer is to be rotated, the sample can usually be kept mounted statically. It is important to align the axis of the polarizer parallel to the light beam in order to minimize the displacement of the transmitted beam. Because of the polarization biases introduced by monochromators and other optical elements, the rotation of the polarizer will generally affect the intensity of the transmitted beam. This is usually undesirable: For instance, many commercial instruments will adjust the slit width so as to keep the light intensity reaching the detector at optimum; then differently polarized spectra will actually be recorded with different resolution. In order to avoid the variation of transmitted intensity with the polarizer position, one can mount a depolarizing element, such as a wedge-type depolarizer, in front of the polarizer. Sometimes it is possible to perform the measurements with two arbitrary polarizer positions as long as they are mutually orthogonal. These can be chosen at $+45^\circ$ and -45° with respect to the preferred polarization direction of the monochromator, assuring equal intensity in both positions. This arrangement is usually slightly imperfect since the degree to which the light beam has a preferred polarization direction may vary weakly with wavelength.

The polarizers should be the last optical elements before the sample and/or the first after the sample in order to avoid depolarization by other elements. This rule cannot always be followed, since the sample may have to be contained within walls, for instance, when measurements are made on a liquid crystal or in a dewar vessel. In such cases, the beam should pass normal to these surfaces. When this cannot be achieved, it is advisable to use Fresnel's relations to compute the resulting bias between the differently polarized intensities. The surfaces should be of good optical quality and free of dust, fingerprints, etc. This is of particular importance in the UV region and less so in IR measurements.

Light Scattering. Scattering and depolarization by the uniaxial sample itself is frequently the largest source of concern in UV and, to a lesser degree, visible spectroscopy. It usually causes no trouble in the IR region. Liquid crystals and stretched polymers both scatter and depolarize light more than isotropic liquids do, and it is desirable to work with thin samples and to keep the sample as close to the photodetector as possible to reduce the baseline. For instance, passage through several millimeters of stretched polyethylene will completely depolarize a laser beam of visible light. The depolarization of the sample, and its homogeneity, can be checked by inserting it between crossed polarizers and looking for transmitted light. Often the scattering occurs at the surface of the sample and can be reduced by having the sample in optical contact with a suitable liquid. This is a side benefit in measuring spectra at 77 K on polymer sheets immersed in liquid nitrogen.

Corrections for Sample Birefringence. Uniaxial samples are birefringent, $n_{||} \neq n_{\perp}$. As a result, only light linearly polarized along Z or perpendicular to Z, and light of any polarization propagating along Z does not suffer a change in its polarization state as it passes through the sample. All of the optical experiments described in the following have been designed with this fact in mind and avoid the complications which would result for other types of light.

The anisotropic nature of the sample has yet another more subtle effect on optical measurements. We have seen in Chapter 1 that the solvent affects the observed optical properties by making the effective electric field experienced by the molecule different from the field applied from the outside, $\mathbf{E}_{\text{eff}}(\tilde{\nu}) = \alpha'(\tilde{\nu})\mathbf{E}(\tilde{\nu})$, and by reducing the speed of light to $c/n(\tilde{\nu})$. In the Lorentz approximation, the effective polarizability of the solvent $\alpha'(\tilde{\nu})$ is related to its refractive index $n(\tilde{\nu})$ by $\alpha'(\tilde{\nu}) = [n^2(\tilde{\nu}) + 2]/3$. When working with a uniaxial sample, it is appropriate to use $n_{||}(\tilde{\nu})$ when making the solvent correction for a measurement with light polarized along Z, and to use $n_{\perp}(\tilde{\nu})$ for light polarized perpendicular to Z. For instance, in measurement of electric dipole absorption, the measured $E_Z(\tilde{\nu})$ will be too large by the factor $[n_{||}^2(\tilde{\nu}) + 2]^2/9n_{||}(\tilde{\nu})$, and the measured $E_Y(\tilde{\nu})$ will be too large by the factor $[n_{\perp}^2(\tilde{\nu}) + 2]^2/9n_{\perp}(\tilde{\nu})$. When the dichroic ratio $E_Z(\tilde{\nu})/E_Y(\tilde{\nu})$ is evaluated, these factors will not cancel exactly. It is common to ignore this when dealing with weakly birefringent materials such as stretched polyethylene, but corrections become unavoidable when working with strongly birefringent solvents such as certain liquid crystals. It is possible that the Lorentz approximation is not adequate in some of these cases, but little else is available at present (see Section 3.3).

3.2.1 Linear Dichroism: Static Polarization

In all measurements of linear dichroism, it is desirable to obtain the largest possible number of linearly independent spectra of the partially oriented sample. For uniaxial samples, the number is two. Several choices exist.

(i) Measurement of E_Y and E_Z. The most straightforward way to proceed is to measure the absorption spectrum twice. One measurement is made with a light beam whose electric vector is directed along the unique (Z) axis of the uniaxial sample, which is given by the director of a liquid crystal or the stretching direction of a polymer ["parallel spectrum," $E_Z'(\tilde{\nu})$]. The other measurement is made with the electric vector directed perpendicular to the unique axis ["perpendicular spectrum," $E'_Y(\tilde{\nu})$]. The absorbances corresponding to the baselines, $E^0_Z(\tilde{\nu})$ and $E^0_Y(\tilde{\nu})$, respectively, must be recorded separately for each polarizer orientation and subtracted from the measured spectra to obtain the desired results:

$$E_U(\tilde{\nu}) = E'_U(\tilde{\nu}) - E^0_U(\tilde{\nu}) \tag{3.1}$$

where U = Z or Y. Although this can be done readily by hand, the use of a computer is most convenient and provides the curves $E_Z(\tilde{\nu})$ and $E_Y(\tilde{\nu})$ in a form ready for the formation of linear combinations needed in the reduction procedure described in Chapter 5. The measurement of the baselines is performed on the pure solvent. This is straightforward for liquid crystals but may present some difficulties with those stretched polymers which do not always yield samples of uniform quality. Then the solute may be extracted from the polymer using a solvent which swells it; or a baseline may be recorded before doping the stretched polymer.

The apparent molar extinction coefficients $\varepsilon_U(\tilde{\nu})$ are related to the measured absorbance $E_U(\tilde{\nu})$ by $\varepsilon_U(\tilde{\nu}) = E_U(\tilde{\nu})/cl$, where c is the molar concentration of the solute and l the path length. Frequently, c and l need not be determined, since they cancel when ratios are taken.

Since the polarizer has a relatively low transmittance, it produces a considerable contribution to the baseline, which already inevitably contains contributions due to the light scattering, reflection, and absorption by the pure solvent sample. In a double-beam instrument, the baseline can be reduced considerably if another polarizer is placed into the reference beam, possibly along with another pure solvent sample (Figure 3.3). Since it is nearly impossible to find two polarizers with identical transmission properties, the baseline then still reflects the difference between the polarizers. Only rarely will it be possible to avoid the baseline correction altogether. In a double-beam instrument, it is also possible to measure the difference $E'_Z(\tilde{\nu}) - E'_Y(\tilde{\nu})$ directly by using identical samples but opposite polarizer orientations in the two beams. A baseline correction is usually still required.

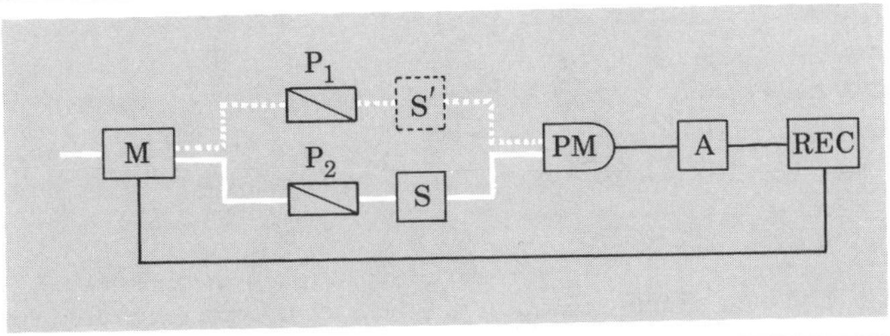

Figure 3.3 A schematic representation of a double-beam spectrophotometer for polarized spectroscopy. M, monochromator; P, polarizer; S, sample; PM, photomultiplier; A, amplifier; REC, recorder. An optional reference sample S′ is also shown.

In a Fourier transform IR measurement (Figure 3.4), the baseline is typically stored in the computer and is easy to subtract digitally, except for the regions of strong solvent absorption, for which no useful information can be obtained. It is very difficult to obtain good IR spectra of solutes in anisotropic samples on dispersive IR instruments, since IR absorptions are usually weak at the attainable solute concentrations. Often, nevertheless, a few of the strongest bands can be detected. A problem which may be encountered in IR measurements on thin polymer sheets is the occurrence of interference fringes. A simple way of minimizing their intensity is to scratch the polymer surface with fine sandpaper. Measurement at the Brewster angle is a more complicated possibility.

(ii) Measurement of E_Y or E_Z and E^{iso}. While the measurement of $E_Z(\tilde{\nu})$ and $E_Y(\tilde{\nu})$ ensures the largest possible overall difference between the two desired linearly independent spectra, it does not represent the only possible choice. A simple alternative is the measurement of either $E_Z(\tilde{\nu})$ or $E_Y(\tilde{\nu})$ combined with the measurement of the absorption spectrum of an isotropic sample containing the same number of solvent molecules in the light path, $E^{iso}(\tilde{\nu}) = [E_Z(\tilde{\nu}) + 2E_Y(\tilde{\nu})]/3$. This is relatively easy to obtain in liquid crystal solutions, which can be made isotropic by raising the temperature above the isotropic melting point. If the path length is kept constant, a correction for the change in density is necessary. It is more difficult to obtain $E^{iso}(\tilde{\nu})$ with stretched polymer samples (elution is usually necessary).

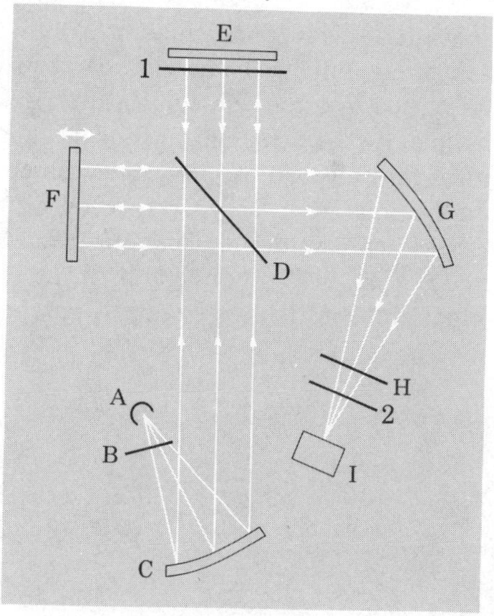

Figure 3.4 A schematic representation of a Fourier-transform infrared spectrometer. A, source; B, chopper; C, collimator; D, beam splitter; E, fixed mirror; F, movable mirror; G, focusing mirror; H, spectral filters; I, detector. The position of the sample is in the combined beams for ordinary FT spectroscopy (2) and in one arm of the interferometer for amplitude FT spectroscopy (1).

A measurement of this type can also be carried out in a differential fashion, with one sample in the sample beam and the other in the reference beam, and can be performed on both nematic and smectic liquid crystals.

If only the shapes of the reduced spectra or the polarizations of the individual transitions of a symmetrical molecule are of interest and the orientation factors need not be determined, the condition that the same number of molecules be observed in both spectra can be dropped, and $E^{iso}(\tilde{\nu})$ can then be measured in an ordinary solution as long as the absorbances in the two spectra are comparable (see Chapter 5). Under these circumstances, it is also possible to work with the $E_Z(\tilde{\nu})$ and $E_Y(\tilde{\nu})$ curves obtained by rotation of the sample as described under **(i)** even if the solute is distributed inhomogeneously.

The danger in combining two absorbance curves measured in two different solvents or at two different temperatures is that the intrinsic shapes of the spectra may not correspond to each other. This problem can be minimized by choosing solvents of very similar structure and by minimizing the temperature difference.

(iii) Measurement of E_ω and E_Y. In some instances, the sample geometry makes the measurement of $E_Z(\tilde{\nu})$ impractical, and yet an isotropic sample may not be available or its use may be undesirable. This situation is encountered in the studies of membranes in lamellar lyotropic liquid crystals, particularly those of natural origin, where it is undesirable to disrupt the lipid bilayers and the only aligned samples available are in the shape of thin layers, usually between two quartz plates. The alignment is such that the unique sample axis Z is normal to the surface. An absorption spectrum taken with light of any polarization propagating along Z then yields $E_Y(\tilde{\nu})$. A second linearly independent polarized spectrum must be taken with an incidence other than normal, preferably with the plane of polarization chosen so as to contain the Z axis (the "tilted plate method"). With this choice, the light remains linearly polarized as it propagates through the sample. The spectrum $E'_\omega(\tilde{\nu})$ recorded with the incidence angle $\pi/2 - \omega$ is related to the desired spectra $E_Z(\tilde{\nu})$ and $E_Y(\tilde{\nu})$ through equation (5.113), where εZ inside the sample is to be taken equal to $\arcsin[(\sin\omega)/n(\tilde{\nu})]$ on account of Snell's law [$n(\tilde{\nu})$ is the refractive index of the sample]. For $E'_Z(\tilde{\nu})$ and $E'_Y(\tilde{\nu})$, εZ is equal to 0 and $\pi/2$, respectively. However, in the derivation of $E'_\omega(\tilde{\nu})$ from the observed transmittance it must also be recognized that the path length l inside the sample increases as ω decreases, since the beam passes the sample cell at a skewed angle. The increase depends on $n(\tilde{\nu})$ as follows:

$$l_\omega/l = [1 - (\cos^2\omega)/n^2(\tilde{\nu})]^{-1/2} \qquad (3.2)$$

This information is not needed unless $E_Y(\tilde{\nu})$ is measured at a different angle of incidence than $E_\omega(\tilde{\nu})$. However, in accurate work, other corrections are needed for the effects of beam divergence, the birefringence of the sample, and multiple reflections at the four boundaries of the sample. These are relatively minor and need be considered only for weakly dichroic samples, which are normally investigated by polarization modulation methods.

The determination of the baselines $E^0_\omega(\tilde{\nu})$ and $E^0_Y(\tilde{\nu})$ becomes nontrivial for very thin samples because of interference effects. It is also difficult for optically thick samples. This is due to the fact that the difference of the two baselines is dominated by the difference in the reflection of light of the two polarizations at the sample surfaces at oblique incidence. This is determined by the Fresnel equations and is of fairly complex form, particularly when the effect of multiple polarized reflections is to be considered. The equations contain the refractive index of the uniaxial medium, which is affected by the presence of the solute, so that the baseline cannot be determined simply by measuring the transmission in the absence of the solute. In practice, mechanically thick and optically thin samples are used and the baseline correction is usually estimated by extrapolation from neighboring regions of transparence.

Those interested in further detail can consult the leading references given in Section 3.3.

(iv) Measurement of E_Y or E_Z and D. A combination which offers increased accuracy since it avoids the rotation of either the sample or the polarizer, so that a Rochon prism can be used, and yet permits both measurements to be performed on the same sample at the price of some complexity in data evaluation and an increased linear dependence of the two spectra obtained is the PNP method. The idea is to measure either $E_Z(\tilde{\nu})$ or $E_Y(\tilde{\nu})$ and to combine the result with the unpolarized absorbance $D(\tilde{\nu})$ measured with a depolarizer in front of the sample and the light beam direction perpendicular to the Z axis. The measurement is best performed on a dual-beam spectrometer which records $E_Z(\tilde{\nu}) = E'_Z(\tilde{\nu}) - E^0_Z(\tilde{\nu})$ [or $E_Y(\tilde{\nu}) = E'_Y(\tilde{\nu}) - E^0_Y(\tilde{\nu})$] and $D(\tilde{\nu}) = D'(\tilde{\nu}) - D^0(\tilde{\nu})$ when the reference beam contains a sample identical with that in the sample beam except for the absence of the solute. The evaluation requires also the knowledge of the background correction $\Delta(\tilde{\nu})$

$$\Delta(\tilde{\nu}) = E^0_Z(\tilde{\nu}) - E^0_Y(\tilde{\nu}) \tag{3.3}$$

measured by placing identical solute-free samples in both beams, with a polarizer aligned along Z in the sample beam and a polarizer aligned along Y in the reference beam.

If $E_Z(\tilde{\nu})$ and $D(\tilde{\nu})$ have been measured and the correction curve $\Delta(\tilde{\nu})$ is known, $E_Y(\tilde{\nu})$ can be obtained from

$$E_Y(\tilde{\nu}) = D(\tilde{\nu}) - \log\left[1 + 10^{-\Delta(\tilde{\nu})} - 10^{D(\tilde{\nu}) - E_Z(\tilde{\nu}) - \Delta(\tilde{\nu})}\right] \tag{3.4}$$

If $E_Y(\tilde{\nu})$ and $D(\tilde{\nu})$ have been measured, $E_Z(\tilde{\nu})$ can be obtained similarly using

$$E_Z(\tilde{\nu}) = D(\tilde{\nu}) - \log\left[1 + 10^{\Delta(\tilde{\nu})} - 10^{D(\tilde{\nu}) - E_Y(\tilde{\nu}) + \Delta(\tilde{\nu})}\right] \tag{3.5}$$

A derivation of expression (3.5) will now be given to provide an example of manipulations using the Lambert-Beer law [eq. (1.20)]. The light incident on the sample is attenuated from initial intensity I_0 to I', while that incident on the reference is attenuated from I_0 to I^0. For a light beam directed along X, we can write in general

$$I(\tilde{\nu}) = I_Z(\tilde{\nu}) + I_Y(\tilde{\nu}) \tag{3.6}$$

Using (1.20) for light of each polarization separately, we obtain for the observed ratio T of the intensity of the transmitted sample beam $I'(\tilde{\nu})$ to that of the transmitted reference beam $I^0(\tilde{\nu})$

$$T(\tilde{\nu}) = \frac{I'(\tilde{\nu})}{I^0(\tilde{\nu})} = \frac{I'_Z(\tilde{\nu}) + I'_Y(\tilde{\nu})}{I^0_Z(\tilde{\nu}) + I^0_Y(\tilde{\nu})}$$

$$= \frac{I_{0Z}(\tilde{\nu}) \times 10^{-E'_Z(\tilde{\nu})} + I_{0Y}(\tilde{\nu}) \times 10^{-E'_Y(\tilde{\nu})}}{I_{0Z}(\tilde{\nu}) \times 10^{-E^0_Z(\tilde{\nu})} + I_{0Y}(\tilde{\nu}) \times 10^{-E^0_Y(\tilde{\nu})}} \tag{3.7}$$

In the measurement with Z-polarized light in both beams, we have $I_{0Y}(\tilde{\nu}) = 0$ and

$$T_Z(\tilde{\nu}) = \frac{I'_Z(\tilde{\nu})}{I^0_Z(\tilde{\nu})} = 10^{-[E'_Z(\tilde{\nu}) - E^0_Z(\tilde{\nu})]} = 10^{-E_Z(\tilde{\nu})} \tag{3.8}$$

In the measurement with depolarized light in both beams, we have $I_{0Y} = I_{0Z} = I_0/2$ and

$$T_{Y,Z}(\tilde{\nu}) = \frac{10^{-E'_Z(\tilde{\nu})} + 10^{-E'_Y(\tilde{\nu})}}{10^{-E^0_Z(\tilde{\nu})} + 10^{-E^0_Y(\tilde{\nu})}} \tag{3.9}$$

The absorbances recorded in the two measurements will be

$$-\log T_Z = E_Z(\tilde{\nu}) \tag{3.10}$$

$$D(\tilde{\nu}) = -\log T_{Y,Z} = \log \frac{10^{-E^0_Z(\tilde{\nu})} + 10^{-E^0_Y(\tilde{\nu})}}{10^{-[E_Z(\tilde{\nu}) + E^0_Z(\tilde{\nu})]} + 10^{-[E_Y(\tilde{\nu}) + E^0_Y(\tilde{\nu})]}} \tag{3.11}$$

$$D(\tilde{\nu}) = \log \frac{1 + 10^{\Delta(\tilde{\nu})}}{10^{-E_Z(\tilde{\nu})} + 10^{-E_Y(\tilde{\nu}) + \Delta(\tilde{\nu})}} \tag{3.12}$$

where (3.1) and (3.3) were used. A rearrangement of this result yields equation (3.5).

It should be noted that not only $D(\tilde{\nu})$ but also $E_Y(\tilde{\nu})$ can in principle be obtained from a measurement using unpolarized light, by recording absorbance for a light beam propagating along Z and correcting for the baseline. Once the correction curve $\Delta(\tilde{\nu})$ is known for a given anisotropic solvent, it is therefore possible to obtain both $E_Y(\tilde{\nu})$ and $E_Z(\tilde{\nu})$ of a solute from two measurements on the same sample, entirely without the use of polarizers. In practice, this is hindered by the usual sample shape, which makes either the observation along Z or the observation along X impractical.

(v) Measurements without Polarizers. If one is willing to use two different samples containing the same solute but oriented differently or else differing in the degree of solute orientation, measurements using only unpolarized light become practical. The obvious precautions to be taken with such procedures, i.e., keeping the number of molecules observed the same and minimizing solvent effects on spectral shapes, have already been mentioned under **(ii)**. Clearly, the use of $E^{iso}(\tilde{\nu})$ and $E_Y(\tilde{\nu})$ outlined there does not require the use of polarizers.

Other possibilities for work with unpolarized light exist. First, a combination of $E^{iso}(\tilde{\nu})$ obtained as in **(ii)** with $D(\tilde{\nu})$ obtained as in **(iv)**, provided that the sample is only weakly dichroic and $\Delta(\tilde{\nu})$ is negligible so that equation (3.12) can be used to convert $D(\tilde{\nu})$ into $E_Y(\tilde{\nu}) + E_Z(\tilde{\nu})$. Second, the use of two dichroic samples both placed into the same beam, once with parallel and once with perpendicular mutual orientation (the self-polarized or SP method). The appropriate equations can be found in the reference given in Section 3.3. In a third method, an unpolarized absorption spectrum is measured twice, each time on a differently aligned sample. One spectrum is obtained on a cholesteric liquid crystal in Grandjean texture with light propagating along the optical axis (and the Z axes of the local nematic layers perpendicular to it). The other is obtained in the same geometry and with the same sample rendered nematic by application of a strong electric field in the light propagation direction (and the Z axis now parallel to it). If the solute alignment in the individual parallel nematic layers of which the cholesteric liquid crystal is composed is the same as the alignment in the nematic phase induced by the electric field, the two measurements yield $[E_Y(\tilde{\nu}) + E_Z(\tilde{\nu})]/2$ and $E_Y(\tilde{\nu})$, respectively. This is seen by considering that the local Y and Z axes of each nematic layer in the cholesteric liquid crystal are always perpendicular to the light propagation direction but, due to their helical arrangement, are uniformly distributed around the light propagation direction (Figure 3.1).

In most work on linear dichroism reported so far, the absorbance of the sample was determined by measuring the intensity of transmitted light. Other possibilities exist, however, and may become important in the future. Both photoacoustic and fluorescence spectroscopy are highly sensitive methods of detection of light absorption. A particular advantage of the photoacoustic method of detection is its insensitivity to light scattering. Wavelength-resolved fluorescence offers a similar advantage but is less sensitive than a measurement of total fluorescence intensity. When fluorescence is used to detect linear dichroism, it is important to avoid photoselection effects by spatial averaging. This is best done by using an integrating sphere.

3.2.2 Linear Dichroism: Modulated Polarization

General. The independently accumulated errors in the separate measurement of $E_Z(\tilde{\nu})$ and $E_Y(\tilde{\nu})$ or their various combinations as discussed in Section 3.2.1 are overwhelming when the degree of dichroism is very small. Since one is then interested in the difference between two nearly identical spectra, it becomes desirable to measure the difference directly. There is a complication in that the actually measured curves are $E'_Z(\tilde{\nu})$ and $E'_Y(\tilde{\nu})$ and if the two baselines $E^0_Z(\tilde{\nu})$ and $E^0_Y(\tilde{\nu})$ differ significantly, as is frequently the case in the short-wavelength range for stretched polymers, a direct measurement of the difference $LD'(\tilde{\nu})$, measured with the light beam propagating along X,

$$\begin{aligned} LD'(\tilde{\nu}) &= E'_Z(\tilde{\nu}) - E'_Y(\tilde{\nu}) \\ &= [E_Z(\tilde{\nu}) - E_Y(\tilde{\nu})] + [E^0_Z(\tilde{\nu}) - E^0_Y(\tilde{\nu})] = LD(\tilde{\nu}) + LD^0(\tilde{\nu}) \end{aligned} \qquad (3.13)$$

will be dominated by the baseline difference $LD^0(\tilde{\nu})$ rather than the desired difference $LD(\tilde{\nu})$. However, if the difference in the baselines is very well reproducible, as is usually the case for liquid crystals and often for stretched polymers, it can be subtracted with sufficient accuracy and the direct measurement will still extend the accessible range to far smaller degrees of dichroism.

Measurement of LD and $E_{Y,Z}$. A simultaneous direct measurement of $LD'(\tilde{\nu})$ and of $E'_{Y,Z}(\tilde{\nu})$, which is defined by

$$E'_{Y,Z}(\tilde{\nu}) = [E'_Z(\tilde{\nu}) + E'_Y(\tilde{\nu})]/2$$
$$= [E_Z(\tilde{\nu}) + E_Y(\tilde{\nu})]/2 + [E^0_Z(\tilde{\nu}) + E^0_Y(\tilde{\nu})]/2 = E_{Y,Z}(\tilde{\nu}) + E^0_{Y,Z}(\tilde{\nu}) \qquad (3.14)$$

where $E^0_{Y,Z}(\tilde{\nu})$ represents the baseline correction, can be performed on a modified commercial dichrograph. It contains a light source, a monochromator, a linear polarizer, and a polarization modulator oriented with its principal axis at 45° to that of the polarizer. The light beam passes through the sample and into a photomultiplier. In modern instruments, the polarization modulator is of the photoelastic type (Figure 3.5). It consists of a block of fused quartz or other suitable material which is subjected to a periodic stress by a piezoelectric transducer. It acts as a retarding element, but its birefringence varies periodically in time with frequency ω, so that the state of polarization of the beam passing through it is modulated in time. If the absorption of the light beam in the sample did not depend on the polarization of the light, the intensity of the transmitted beam would be constant. Since the detector does not respond to the state of polarization (a depolarizer may have to be placed in front of it to ensure that this is so), it would produce a dc signal. If the sample is dichroic and the polarization modulation suitable, the transmitted beam will vary in intensity and an ac signal will be superimposed on the original dc signal. The two signals can be separately recorded using a lock-in amplifier and, with suitable precautions, allow an evaluation of the average optical density of the sample $E'_{Y,Z}(\tilde{\nu})$ and its dichroism $LD'(\tilde{\nu})$. In practice, $E'_{Y,Z}(\tilde{\nu})$ is obtained from the dc current measured with the sample rotated around the beam axis to a position at which the ac signal disappears. The evaluation of $LD'(\tilde{\nu})$ from the

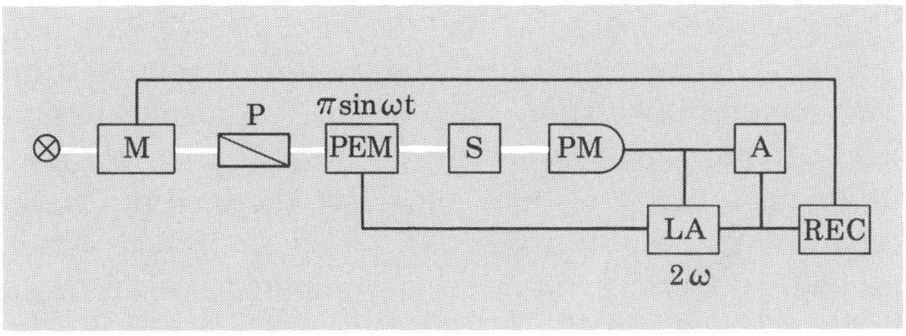

Figure 3.5 A schematic representation of a linear dichrograph of a newer type. M, monochromator; P, polarizer; PEM, photoelastic modulator producing time-dependent retardation $\pi \sin \omega t$; S, sample; PM, photomultiplier; LA, lock-in amplifier operating at frequency 2ω; A, dc amplifier; REC, recorder.

ac and dc signals is more complicated (see below). Correction for the baseline converts the results into $E_{Y,Z}(\tilde{\nu})$ and $LD(\tilde{\nu})$ as indicated in (3.13) and (3.14).

Measurement of LD and E^{iso} or D. If the modified dichrograph is not capable of measuring $E'_{Y,Z}(\tilde{\nu})$ and $E^0_{Y,Z}(\tilde{\nu})$ with sufficient accuracy, the $LD'(\tilde{\nu})$ curve must be combined with some other linearly independent spectral curve measured with a different instrumental setup. This tends to incur a penalty in lost accuracy. The simplest way to proceed in this situation would be to measure $E_Z(\tilde{\nu}) = E'_Z(\tilde{\nu}) - E^0_Z(\tilde{\nu})$ or $E_Y(\tilde{\nu}) = E'_Y(\tilde{\nu}) - E^0_Y(\tilde{\nu})$, using the same samples on which $LD(\tilde{\nu}) = LD'(\tilde{\nu}) - LD^0(\tilde{\nu})$ was determined. However, it appears to be more common to determine either $E^{iso}(\tilde{\nu}) = (E^{iso})'(\tilde{\nu}) - (E^{iso})^0(\tilde{\nu})$ by a measurement on an isotropic sample, with the concomitant uncertainties discussed under **(ii)** in Section 3.2.1, or $D(\tilde{\nu}) = D'(\tilde{\nu}) - D^0(\tilde{\nu})$ by measurement on the oriented sample with unpolarized light propagating in the X direction. The latter can be converted into $E_{Y,Z}(\tilde{\nu})$ if the correction curve $\Delta(\tilde{\nu}) = E^0_Z(\tilde{\nu}) - E^0_Y(\tilde{\nu})$ is known for the anisotropic solvent used. The requisite formula is obtained by rearranging equation (3.11) into the form

$$E_{Y,Z}(\tilde{\nu}) = D(\tilde{\nu}) + \log \frac{10^{[LD(\tilde{\nu})+\Delta(\tilde{\nu})]/2} + 10^{-[LD(\tilde{\nu})+\Delta(\tilde{\nu})]/2}}{10^{\Delta(\tilde{\nu})/2} + 10^{-\Delta(\tilde{\nu})/2}} \tag{3.15}$$

Conversion into $E^{iso}(\tilde{\nu})$ is accomplished using the relation

$$E^{iso}(\tilde{\nu}) = E_{Y,Z}(\tilde{\nu}) - LD(\tilde{\nu})/6 \tag{3.16}$$

which follows directly from the definition of the quantities involved.

When the primary emphasis is on the characterization of the solute alignment, some authors prefer to express their final results in the form of "reduced dichroism," defined as $LD(\tilde{\nu})/E^{iso}(\tilde{\nu})$. When the spectra of the solute are of intrinsic interest, it is advisable to process the data further as outlined in Chapter 5 ("stepwise reduction").

Dichrographs. The commercially available dichrographs are intended for the measurement of circular dichroism, and the modulated polarization varies between left-handed and right-handed. To accomplish this, the retardation introduced by the modulator varies between $-\pi/2$ and $+\pi/2$, so that the light polarization varies between left-handed and right-handed with the frequency of the oscillator, ω (for technical reasons, the variation of retardations actually used is larger). For the measurement of linear dichroism, the instrument must be modified so that the modulation is between two orthogonal forms of linearly polarized light, e.g., horizontally and vertically polarized. This requires the phase difference to sweep between $-\pi$ and $+\pi$. Then, the state of polarization will vary between horizontal and vertical at a frequency 2ω. The instrumental modification thus requires a change of the driving amplitude for the modulator and of the frequency detected by the lock-in amplifier. This is available as an option in some commercial instruments.

In older instruments, an electro-optical modulator was used in place of the currently favored photoelastic modulator. In that case, the above procedure is impractical in much of the wavelength region of interest, since excessive voltages would be required to drive the modulator from $-\pi$ to $+\pi$ retardation.

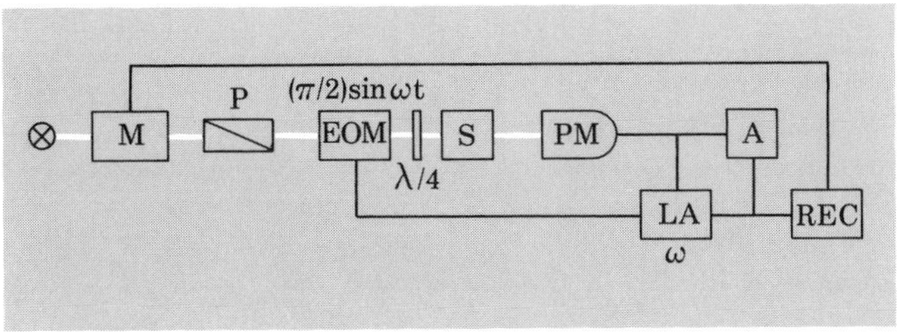

Figure 3.6 A schematic representation of a linear dichrograph of an older type. M, monochromator; P, polarizer; EOM, electro-optical modulator producing a time-dependent retardation $(\pi/2)\sin \omega t$; $\lambda/4$, a quarter-wave retarder; S, sample; PM, photomultiplier; LA, lock-in amplifier operating at frequency ω; A, dc amplifier; REC, recorder.

Then an alternative solution to the problem is preferable (Figure 3.6): the insertion of an additional quarter-wave retarding element oriented at 45° into the light path. This solution will work for any circular dichrograph, but the additional retarding element usually limits the useful UV range and has to be controlled to produce exactly the right retardation at all wavelengths. (An ordinary quarter-wave plate is too chromatic and would not work; a Fresnel rhomb is nearly achromatic but displaces the beam laterally; the more complex Oxley prism is the best choice.)

While the sensitivity of the modulated polarization measurement is orders of magnitude higher than of the static measurement described in Section 3.2.1, the quantity measured using a photoelastic modulator, $S(\tilde{\nu}) = i_{2\omega}(\tilde{\nu})/i_{dc}(\tilde{\nu})$, is a complicated function of the sought quantity, $E'_Z(\tilde{\nu}) - E'_Y(\tilde{\nu}) = LD'(\tilde{\nu})$. Here, $i_{2\omega}(\tilde{\nu})$ is the amplitude of the signal detected at frequency 2ω, and $i_{dc}(\tilde{\nu})$ is the intensity of the dc signal, both for light of wavenumber $\tilde{\nu}$. This function can be shown to be

$$LD'(\tilde{\nu}) = [2/(\ln 10)] \tanh^{-1} [C_1 S^{-1}(\tilde{\nu}) + C_2(\tilde{\nu})]^{-1} \qquad (3.17)$$

In this equation, C_1 contains several numerical factors such as $J_2(\delta_0)$, the Bessel function of the programmed maximum phase shift δ_0, which is ideally equal to $\pi/2$; and $C_2(\tilde{\nu})$ is given by $J_0(\delta_0)$ and is therefore approximately equal to -0.3. In practice, C_1 is a calibration constant and $C_2(\tilde{\nu})$ a calibration curve, to be determined experimentally. The calibration is based on comparison with a strongly dichroic standard sample whose $LD'(\tilde{\nu})$ has been measured by the static method. The constant C_1 needs to be determined only infrequently, using

$$C_1 = (S_+ - S_-)/2 \tanh[(LD'_+ \ln 10)/2] \qquad (3.18)$$

where the wavenumber at which S_+, S_-, and LD'_+ are measured corresponds to any strong peak in the standard sample. At this setting of $\tilde{\nu}$, the sample is first rotated in the dichrograph until the maximum positive signal S_+ is obtained. The linear dichroism measured at this orientation by the static method is LD'_+. Then, the sample is rotated in the dichrograph by 90°, and the maximum negative signal S_- is determined.

The value of $C_2(\tilde{\nu})$ must be determined at least for every peak in the spectrum, using

$$C_2(\tilde{\nu}) = -(C_1/2)[S_+(\tilde{\nu}) + S_-(\tilde{\nu})]/S_+(\tilde{\nu})S_-(\tilde{\nu}) \tag{3.19}$$

A procedure proposed for the correction of the analogous ratio $S(\tilde{\nu})$ recorded on a dichrograph with an electro-optical modulator to yield the true linear dichroism $LD'(\tilde{\nu})$ is the use of the expression

$$LD'(\tilde{\nu}) = \log_{10} \frac{C'_1(\tilde{\nu}) + S(\tilde{\nu})[1 - C'_2(\tilde{\nu})]}{C'_1(\tilde{\nu}) - S(\tilde{\nu})[1 + C'_2(\tilde{\nu})]} \tag{3.20}$$

where $C'_1(\tilde{\nu})$ and $C'_2(\tilde{\nu})$ are empirically determined calibration curves.

When the "tilted plate" method is used in measurement of linear dichroism by the polarization modulation method, the corrections already discussed in the static version of the method apply. The large difference between $E^0_Y(\tilde{\nu})$ and $E^0_\omega(\tilde{\nu})$ is a particular nuisance, since it produces a large baseline correction $LD^0_\omega(\tilde{\nu})$. Once again the reader is referred to the references given in Section 3.3 for details.

3.2.3 Electric Dichroism: Modulated Alignment

The measurements of linear dichroism resulting from electric orientation, known as electric dichroism, are relatively difficult for small molecules, since the degree of orientation achieved is quite small, even with the strongest electrostatic fields for which dielectric breakdown in the solvent can be avoided, up to about 10^5 V cm^{-1}. In order to enhance the sensitivity of the measurement, the orienting electric field is modulated and phase-sensitive detection is used to record the signal. The apparatus (Figure 3.7) consists of a light source, a monochromator, a linear polarizer, sample cell, and a detection system. The linearly polarized light beam passes between two parallel electrodes in the cell, and its electric vector forms an angle α with the direction Z of the electric field. One electrode is connected to an ac high-voltage power supply; the other may

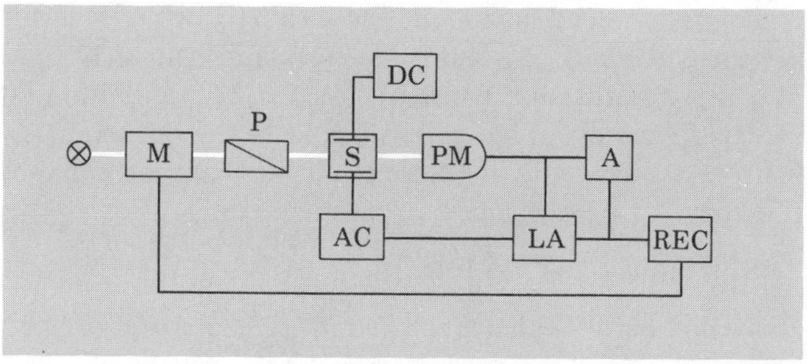

Figure 3.7 A schematic representation of an electrodichrograph. M, monochromator; P, polarizer; S, sample cell with electrodes; DC, source of dc voltage; AC, source of ac voltage; PM, photomultiplier; LA, lock-in amplifier; A, dc amplifier; REC, recorder.

be grounded or connected to a dc high-voltage power supply. Absorption of the light in the cell is modulated by the ac electric field applied across the sample. Modulation frequencies ω up to about 10^5 Hz are used commonly.

With a dc field strength E_{dc} and an ac field strength E_{ac}, the total applied field along Z is

$$E(t) = E_{dc} + E_{ac} \sin \omega t \tag{3.21}$$

Neglecting terms of fourth and higher order in E, the instantaneous relative change in the intensity $I_\alpha(\tilde{v},t)$ of the transmitted field due to the electric field is proportional to $E^2(t)$:

$$[I_\alpha^E(\tilde{v},t) - I_\alpha^0(\tilde{v})]/I_\alpha^0(\tilde{v}) = -2.303D(\tilde{v})L_\alpha(\tilde{v})E^2(t) \tag{3.22}$$

where the superscript on $I_\alpha(\tilde{v})$ refers to the presence or absence of the electric field, respectively, $\tilde{v}$ is the wavenumber, and $D(\tilde{v})$ is the optical density in the absence of the field. The quantity to be determined is $L_\alpha(\tilde{v})$, which plays a role somewhat analogous to the dichroic ratio of the preceding sections (cf. Section 5.3.2). It is usually quite small, with $L_\alpha(\tilde{v})E^2 \cong 10^{-5}$. Substituting from (3.21) into (3.22), one obtains

$$\begin{aligned} I_\alpha^E(\tilde{v},t) &= I_\alpha^0(\tilde{v}) - 2.303I_\alpha^0(\tilde{v})D(\tilde{v})L_\alpha(\tilde{v}) \\ &\quad \times [(E_{dc}^2 + \tfrac{1}{2}E_{ac}^2) + 2E_{dc}E_{ac} \sin \omega t - \tfrac{1}{2}E_{ac}^2 \cos 2\omega t] \end{aligned} \tag{3.23}$$

This result shows that the use of a lock-in amplifier will produce a total of three signals: a constant current $i_c(\tilde{v})$, a current $i_\omega(\tilde{v})$ at frequency ω, and a current $i_{2\omega}(\tilde{v})$ at frequency 2ω. Neglecting the field-dependent part of i_c, one obtains

$$|L_\alpha(\tilde{v})| = \frac{0.2171}{D(\tilde{v})E_{dc}E_{ac}} \frac{[i_\omega(\tilde{v})]_{max}}{i_c(\tilde{v})} = \frac{0.8684}{D(\tilde{v})E_{ac}^2} \frac{[i_{2\omega}(\tilde{v})]_{max}}{i_c(\tilde{v})} \tag{3.24}$$

where $[i_\omega(\tilde{v})]_{max}$ and $[i_{2\omega}(\tilde{v})]_{max}$ are the amplitudes of the currents $i_\omega(\tilde{v})$ and $i_{2\omega}(\tilde{v})$, respectively. The sign of $L_\alpha(\tilde{v})$ is obtained from the phases of $i_\omega(\tilde{v})$ or $i_{2\omega}(\tilde{v})$ relative to the phase of the ac field $E_{ac} \sin \omega t$. The dependence of $L_\alpha(\tilde{v})$ on α is trivial,

$$L_\alpha(\tilde{v}) = L_Z(\tilde{v}) \cos^2 \alpha + L_Y(\tilde{v}) \sin^2 \alpha \tag{3.25}$$

permitting the extraction of $L_Z(\tilde{v})$ and $L_Y(\tilde{v})$ or other suitable choices such as $L_{magic}(\tilde{v})$ with $\alpha = 54.7°$ (magic angle) by a suitable choice of α. In principle, any two linearly independent curves of this kind may be considered as the fundamental experimental data characteristic of electric dichroism. They depend on the nature of the molecular alignment (Section 4.6.2) and reflect the anisotropy of molecular optical properties (Section 5.1) and its response to the applied strong electric field, which complicates matters tremendously.

If the properties of the solute molecule were not affected by the electric field, the two usually measured curves would be related by

$$L_Z(\tilde{v}) = -2L_Y(\tilde{v}) \tag{3.26}$$

The validity of this relation can be checked by setting α equal to the magic angle, since then it follows from (3.25) that $L_{magic}(\tilde{\nu}) = 0$. In this simplified case, it is easy to visualize the physical significance of the two curves relative to the other dichroic methods discussed here, since

$$E_Z(\tilde{\nu}) = E^{iso}(\tilde{\nu})[1 + L_Z(\tilde{\nu})] \tag{3.27}$$

$$E_Y(\tilde{\nu}) = E^{iso}(\tilde{\nu})[1 + L_Y(\tilde{\nu})] \tag{3.28}$$

when normalized to unit strength of the electric field.

The change of unpolarized absorption intensity caused by application of electric field is known as electrochromism (see Section 1.3.2). Like electric dichroism, it is due both to molecular reorientation and to the modification of molecular properties by the perturbing effect of the electric field. It is not discussed here; the interested reader will find a leading reference in Section 3.3.

3.2.4 Liquid-Crystal-Induced Circular Dichroism

An intriguing procedure for the measurement of linear dichroism of a solute contained in a nematic layer of a liquid crystal utilizes the fact that a cholesteric liquid crystal, in its Grandjean texture, consists of a stack of nematic layers (Figure 3.1). Each layer is of monomolecular thickness and is oriented parallel to the cell surface, and its director is skewed by a small angle relative to those of its neighbors to form a helical array. The array is right-handed in a "right-handed" liquid crystal and left-handed in a "left-handed" liquid crystal. In each layer, achiral solute molecules are partially aligned with respect to the director Z of the layer and produce dichroic absorbances $E_Z(\tilde{\nu})$ and $E_Y(\tilde{\nu})$ for light polarized along the director (Z) and perpendicular to it (Y), respectively. Since the directors Z form a helix whose axis X is normal to the surface, this means that the average overall arrangement of the solute molecules in the sample is also helical.

The measurement consists of recording the circular dichroism $\Delta E(\tilde{\nu})$ of the cholesteric liquid crystal in the direction normal to the cell surface, i.e., along the optical axis X. To a good approximation,

$$\Delta E(\tilde{\nu}) = E_L(\tilde{\nu}) - E_R(\tilde{\nu}) = P\tilde{\nu}^3\,\Delta n\,[E_Z(\tilde{\nu}) - E_Y(\tilde{\nu})]/2(\tilde{\nu}^2 - \tilde{\nu}_0^2) \tag{3.29}$$

where P is the helical pitch length, positive for a right-handed liquid crystal and negative for a left-handed one. Further, $\Delta n = n_Z - n_Y$ is the linear birefringence of an individual nematic layer, with a mean refractive index n, and $\tilde{\nu}_0$ is the wavenumber of the cholesteric pitch band ($\tilde{\nu}_0 = 1/Pn$), at which right-handed (left-handed) circularly polarized light is reflected by the right-handed (left-handed) crystal. The values $E_Z(\tilde{\nu})$ and $E_Y(\tilde{\nu})$ contain the actual concentration and path length used in the measurement. For usual measurements in the UV region and with the usual choice of $\tilde{\nu}_0$ in the IR region, $\tilde{\nu}_0$ can be neglected compared to $\tilde{\nu}$, and equation (3.29) simplifies.

A minor contribution to the observed circular dichroism, neglected in (3.29), may arise from the differential radiation field strength in a chiral medium of left- and right-handed circularly polarized light which enters the medium with equal field amplitudes. This effect would be proportional to the circular birefringence of the polar solvent. Whether this correction is really needed still remains an open question, but at any rate it would be important only for nearly isotropic solutes $[E_Z(\tilde{\nu}) \cong E_Y(\tilde{\nu})]$.

In order to understand the physical origin of the process which converts the linear dichroism of the individual nematic layers, $E_Z(\tilde{\nu}) - E_Y(\tilde{\nu})$, into the circular dichroism observed in the direction perpendicular to these layers, $E_L(\tilde{\nu}) - E_R(\tilde{\nu})$, one needs to consider the nature of light propagation in a cholesteric liquid crystal. A planar wave entering in the direction parallel to the helix axis X splits up into two elliptically polarized components whose electric field vector endpoints are rotating in opposite directions with somewhat different velocities. Their eccentricities depend on the pitch of the helix. The major axis of one ellipse coincides with the minor axis of the other. The position of the axes changes gradually as the light proceeds along the helical axis in such a way that one axis stays parallel to the local nematic director while the other is perpendicular to it. Thus, the direction Z experiences a large electric field from one of the two elliptical components and a small one from the other. The direction Y perpendicular to the helical axis and the director experiences a small electric field from the former and a large one from the latter. If $E_Z(\tilde{\nu}) \neq E_Y(\tilde{\nu})$, the two contra-rotating waves are therefore attenuated to different degrees by the absorbing solute, and circular dichroism results.

For light in the region of the cholesteric pitch band, the two ellipses become circles and correspond to right-handed and left-handed circularly polarized light. In a right-handed (left-handed) cholesteric liquid crystal, the helix traced by the electric vector of the right-handed (left-handed) circularly polarized component is superimposable with the helix traced by the director Z of the crystal, and circularly polarized light of this handedness is then totally reflected, while light of the opposite handedness is transmitted essentially freely.

It is obvious from this description that the measured circular dichroism is a structural property of the system, reflecting the linear optical anisotropy of the helically arranged solute and not its chirality or interactions between individual solute molecules.

In concluding, it should be pointed out that all CD measurements on partially aligned samples are sensitive to experimental artifacts, calling for extreme care in the execution and interpretation of experiments (see Section 3.3).

3.2.5 Natural and Magnetic Circular Dichroism

Since the circular dichroism of ordinarily available samples is quite weak, it is always measured by the modulation method (Figure 3.8). Aligned samples of interest generally display strong linear dichroism and birefringence, making the separation of the much weaker circular dichroism extremely difficult if not impossible. Therefore, measurements are best done with the light beam propagating along the unique sample axis Z. If the sample is perfectly uniaxial, the light beam perfectly collimated, and the instrument performing ideally—in particular, without stray birefringence in the optical elements which precede the sample—circular dichroism will be the only effect measured. For this type

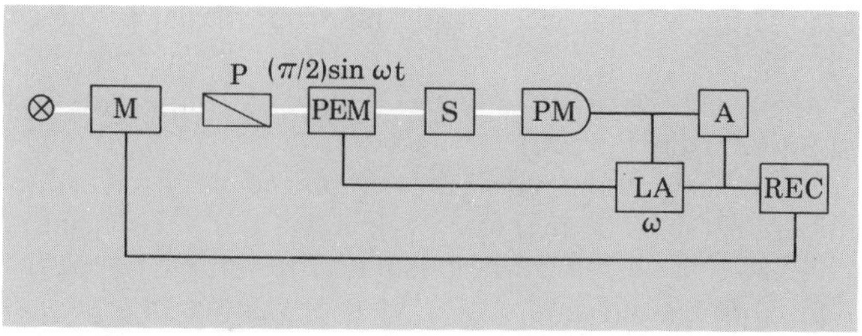

Figure 3.8 A schematic representation of a circular dichrograph. M, monochromator; P, polarizer; PEM, photoelastic or electro-optical modulator producing a time-dependent retardation $(\pi/2)\sin\omega t$; S, sample; PM, photomultiplier; LA, lock-in amplifier operating at frequency ω; A, dc amplifier, REC, recorder.

of measurement, no modification of the commercial circular dichrographs is necessary. Such instruments operate on the principle described in Section 3.2.2. The retardation actually used varies somewhat more than between $-\pi/2$ and $+\pi/2$, since this enhances the signal intensity, and the signal recorded is

$$S(\tilde{\nu}) = i_{\omega}(\tilde{\nu})/i_{dc}(\tilde{\nu}) \qquad (3.30)$$

where $i_{\omega}(\tilde{\nu})$ is the part of the detector current which appears at frequency ω, and $i_{dc}(\tilde{\nu})$ is the dc part. The relation between the desired difference, $\Delta E(\tilde{\nu}) = E_L(\tilde{\nu}) - E_R(\tilde{\nu})$ [cf. (3.29)], and the measured ratio $S(\tilde{\nu})$ is given by a relation similar to (3.17). However, since $\Delta E(\tilde{\nu})/E(\tilde{\nu})$ is always very small, it is normally adequate to approximate that relation by a simple proportionality, with the proportionality constant determined by calibration with a standard sample of known circular dichroism, such as *d*-camphorsulfonic acid.

As indicated in the above, measurements of circular dichroism on oriented samples are very difficult, and only a few have been reported to date. Attempts to obtain meaningful spectra have concentrated on the alignment and quality of the sample, in particular the removal of any residual linear dichroism and birefringence. To the first approximation, the measured circular dichroism differs from the "true" circular dichroism by a term proportional to the linear dichroism of the sample. The proportionality constant is $\sin\alpha\cos^2\beta$, where α is the static birefringence of the polarization modulator oriented at the angle β to the axis system of the induced birefringence. In principle, this term cancels when two measurements in which the sample position differs by rotation of 90° are summed. In practice, α and β may vary during the measurement time, making the compensation incomplete. Therefore, the averaging of several measurements at pairwise orthogonal orientations is essential. Measurements which yield an excessive baseline slope must be discarded.

In measurements of magnetic circular dichroism, whose sign depends on the relative directions of the magnetic field and the light beam, it is useful to subtract measurements recorded with opposed magnetic field directions. The proportionality of the recorded spectrum to the magnetic field provides a check for artifacts.

3.2.6 *Polarized Luminescence—Static*

General. A measurement of polarized emission and excitation spectra of a uniaxial sample, limited to light polarizations such that the electric vector is parallel to Z or perpendicular to it, requires the recording of five spectra, $E_{UV}(\tilde{\nu}_1,\tilde{\nu}_2)$. U describes the orientation of the polarizer P_1, located in the exciting beam of wavenumber $\tilde{\nu}_1$, and V gives that of the polarizer P_2, located in the emitted beam, which is observed at $\tilde{\nu}_2$. There are five independent possibilities for UV: ZZ, ZY, YZ, YY, and XY; the others yield no new information, since

$$E_{ZX} = E_{ZY}, \qquad E_{XZ} = E_{YZ}, \qquad E_{XX} = E_{YY}, \qquad E_{YX} = E_{XY} \tag{3.31}$$

It is ordinarily desirable to keep the angle θ between the exciting beam and the observed emitted beam constant. Then it is possible to record four of the five spectra by rotating the polarizers only. To obtain all five, the sample must be rotated as well, and at least one of the measurements must be performed with light propagating along the unique sample axis Z (for $\theta = 90°$, this is E_{YY}; for $\theta = 0°$ or $180°$, it is E_{XY}). Such measurements are difficult to do with stretched polymer sheets, and the use of thicker samples usually causes excessive depolarization. Fortunately, the interpretation rarely requires all five independent measurements.

Measurement. A static method of measurement in which the five spectra are recorded separately is a natural response to these requirements, but polarization modulation measurements have also been performed when only some of the spectra were desired, as is the case for unoriented samples. The experimental setup and signal evaluation for such modulated measurements are a generalization of the procedures described in Sections 3.2.2, 3.2.3, and 3.2.5 and will not be discussed further here.

An experimental arrangement suitable for the static method of measurement is shown in Figure 3.9. The depolarizers D_1 and D_2 at the slits of the monochromators M_1 and M_2 are essential to minimize the corrections needed to convert the observed intensities $E'_{UV}(\tilde{\nu}_1,\tilde{\nu}_2)$ to their true values $E_{UV}(\tilde{\nu}_1,\tilde{\nu}_2)$. Even with this precaution, polarization biases due to the sample walls, dewar windows, etc., enter into the measurement and must be corrected for. Suitable calibration is provided by a fluorescent sample which emits isotropically, such as a solution of a small fluorescent molecule in a solvent of low viscosity. A scattering surface, e.g., a magnesium oxide coating, can also be used for this purpose.

In measurements on isotropic highly viscous samples ("ordinary photoselection"), a simple correction method is available. In this case, only two of the five measurements are independent:

$$E_{ZZ} = E_{YY} \quad \text{and} \quad E_{XZ} = E_{ZX} = E_{XY} \tag{3.32}$$

and the four measurements available at $\theta = 90°$ without rotating the sample can be combined so that the polarization bias cancels. The ratio of the actual intensity of Z-polarized light incident along X on the sample to that of the Y-polarized light obtained after the polarizer P_1 is rotated is labeled *a*. The analogous bias favoring the detection of Z-polarized emitted light over X-polarized

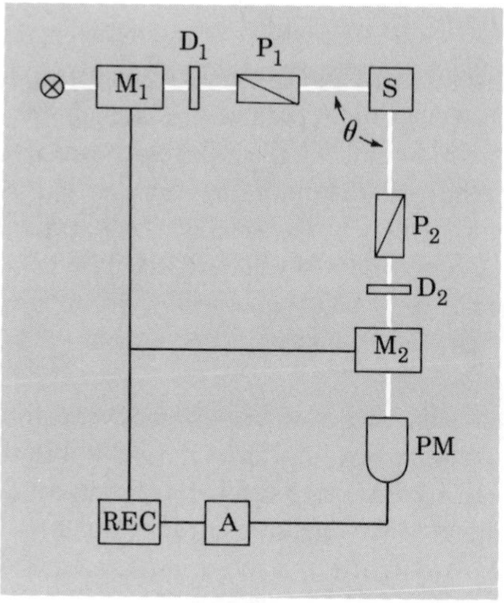

Figure 3.9 A schematic representation of a polarization spectrofluorimeter. M_1, exciting monochromator; D_1 and D_2, depolarizers; P_1 and P_2, polarizers; S, sample; M_2, analyzing monochromator; PM, photomultiplier; A, dc amplifier or photon counter; REC, recorder.

emitted light observed along the direction Y is labeled *b*. Then

$$E'_{ZZ} = abE_{ZZ} \qquad E'_{ZX} = aE_{ZX} \qquad E'_{YZ} = bE_{YZ} \qquad E'_{YX} = E_{YX} \tag{3.33}$$

and the ratio

$$(E'_{ZZ}/E'_{ZX})(E'_{YX}/E'_{YZ}) = (E_{ZZ}/E_{ZX})(E_{YX}/E_{YZ}) = E_{ZZ}/E_{ZX} \tag{3.34}$$

is independent of the polarization biases. A similar correction is possible for observation at angles θ different from 90°. Then a corrected ratio E_{ZZ}/E_{ZX} can be obtained from the four analogous measured spectra by finding the positive root of the quadratic equation (7.96).

The two-dimensional spectra $E_{UV}(\tilde{\nu}_1,\tilde{\nu}_2)$ have so far rarely been obtained. Commonly, either $\tilde{\nu}_2$ is held constant and $\tilde{\nu}_1$ is scanned, producing an excitation spectrum, or $\tilde{\nu}_1$ is held constant and $\tilde{\nu}_2$ is scanned, producing an emission spectrum. As in ordinary emission spectroscopy, both types of spectra still have to be corrected. The excitation spectrum requires a correction for the variation of the exciting light intensity with $\tilde{\nu}_1$, and the emission spectrum requires a correction for the variation of detection sensitivity with $\tilde{\nu}_2$. These corrections are not needed if only intensity ratios such as $E_{UV}/E_{U'V'}$ are desired, and the fully corrected excitation and emission spectra are rarely measured in polarization studies.

Precautions. In addition to paying attention to the standard sources of concern, such as scattering on surfaces, one must take several additional precautions.

For simply interpretable results to be obtained, the exciting light used in the measurement must be sufficiently weak that it does not depopulate the ground state significantly. This is usually not a danger except with long-lived phosphorescent excited states and/or intense laser sources. For similar reasons, corrections are necessary if one does not work with optically thin samples of absorbance less than 0.2–0.3.

Because phosphorescence lifetimes are often quite long, the measurement of the polarized phosphorescence excitation spectrum requires very slow scanning.

Depolarization. Because of the finite lifetime of the emitting excited state, the spectra $E_{UV}(\tilde{\nu}_1,\tilde{\nu}_2)$ are affected by rotational diffusion of the solute molecules in the sample. In the limit of rotation rapid relative to the emission lifetime, the emission is not polarized. In the limit of rotation very slow relative to the emission lifetime, the degree of polarization reaches its maximum and reflects intrinsic molecular properties alone. For the fluorescence of large molecules, this limit can be reached by the use of viscous solvents at or somewhat below room temperature, but the usual procedure, which is generally applicable, is to cool the sample to the temperature of liquid nitrogen or below, using a solvent which solidifies to a transparent rigid glass. In order to avoid strain birefringence and resultant depolarization, it is important to do this in a way which does not produce strain in the glass. This usually requires slow cooling and/or annealing. Suitable isotropic solvents are, for instance, 3-methylpentane, 2-methyltetrahydrofuran, and EPA (a 5:5:2 mixture of diethyl ether, isopentane, and ethyl alcohol). Suitable anisotropic solvents are supercooled compensated nematic liquid crystals and stretched polymers.

A phenomenon which is formally equivalent to rotational depolarization is concentration depolarization. This is caused by transfer of excitation energy between solute molecules of different orientation. To avoid interference from this source, it is usually necessary to use low solute concentrations, of the order of at most 10^{-3} *M*. It is always good practice to check that the results are concentration-independent.

3.2.7 Polarized Luminescence—Dynamic

Measurement of polarized luminescence on molecules which rotate on a time scale comparable with their emission lifetime provides information on the rotational motion. Similarly, such measurements on molecules which transfer their energy to others on this time scale provide information on energy transfer phenomena. Thus, rotational and concentration depolarization, which have so far been viewed only as a nuisance, are of considerable interest in their own right. There are three main ways of studying them: pulsed measurements, steady-state measurement, and modulation measurement.

Pulsed Measurements. In a pulsed experiment, the time change in the polarized intensity of the emission (or absorption, see Section 3.2.8) process involing the second photon, $E'_{UV}(\tilde{\nu}_1,\tilde{\nu}_2,\tau)$, is observed directly after excitation with a polarized very short pulse which delivers the first photon. Nitrogen or hydrogen flash lamps and mode-locked lasers are the common pulse sources. One can use an oscilloscope, a transient recorder, or single-photon counting electronics to measure the time-dependent response.

If the time profile of the pulse is given by $I(\tau)$, the time dependence of observed intensity $E''_{UV}(\tilde{\nu}_1,\tilde{\nu}_2,\tau)$ will be related to the intrinsic quantity $E'_{UV}(\tilde{\nu}_1,\tilde{\nu}_2,\tau)$ by the convolution integral

$$E''_{UV}(\tilde{\nu}_1,\tilde{\nu}_2,\tau) = \int_0^{\tau} I(\tau - \theta)E'_{UV}(\tilde{\nu}_1,\tilde{\nu}_2,\theta)\, d\theta \qquad (3.35)$$

If the duration of the exciting pulse is negligible relative to the rotation correlation time, $I(\tau - \theta)$ can be approximated by $\delta(\tau - \theta)$, and $E'_{UV}(\tilde{\nu}_1,\tilde{\nu}_2,\tau)$ is then observed directly.

Steady-State Measurements. In a continuous illumination experiment of the kind discussed in Section 3.2.6, the intensity of the exciting beam is kept constant. In this much simpler experimental arrangement, the signal observed once steady state has been reached is

$$E''_{UV}(\tilde{\nu}_1,\tilde{\nu}_2) = \int_0^{\infty} E'_{UV}(\tilde{\nu}_1,\tilde{\nu}_2,\tau)\, d\tau = C_0(\tilde{\nu}_1,\tilde{\nu}_2) \qquad (3.36)$$

so that much less information is obtained.

Modulation Measurements. In a phase-modulation experiment, the intensity of the exciting beam is modulated at a high frequency ω:

$$I(\theta) \propto 1 - \alpha \cos \omega t \qquad (3.37)$$

The intensity of the monitored beam is then also a periodic function of time,

$$E''_{UV}(\tilde{\nu}_1,\tilde{\nu}_2,\theta) \propto 1 - \beta \cos(\omega\tau + \phi) \qquad (3.38)$$

and it can be shown that the phase shift is given by

$$\tan \phi(\tilde{\nu}_1,\tilde{\nu}_2,\omega) = C_1(\tilde{\nu}_1,\tilde{\nu}_2,\omega)/C_2(\tilde{\nu}_1,\tilde{\nu}_2,\omega) \qquad (3.39)$$

and the degree of modulation is given by

$$\beta = \alpha[C_1^2(\tilde{\nu}_1,\tilde{\nu}_2,\omega) + C_2^2(\tilde{\nu}_1,\tilde{\nu}_2,\omega)]/C_0(\tilde{\nu}_1,\tilde{\nu}_2) \qquad (3.40)$$

where C_0 has been defined in (3.36) and

$$C_1(\tilde{\nu}_1,\tilde{\nu}_2,\omega) = \int_0^{\infty} E'_{UV}(\tilde{\nu}_1,\tilde{\nu}_2,\tau) \sin \omega\tau\, d\tau \qquad (3.41)$$

$$C_2(\tilde{\nu}_1,\tilde{\nu}_2,\omega) = \int_0^{\infty} E'_{UV}(\tilde{\nu}_1,\tilde{\nu}_2,\tau) \cos \omega\tau\, d\tau \qquad (3.42)$$

The phase-modulation method thus corresponds to a Fourier analysis of the effect of a δ-pulse signal and is physically equivalent to the pulsed excitation method as long as the measurement is performed at a series of frequencies.

After the usual correction for polarization biases which converts E'_{UV} to E_{UV}, all three methods provide information about the fundamental quantities of interest, $E_{UV}(\tilde{\nu}_1,\tilde{\nu}_2,\tau)$.

3.2.8 Photoinduced Linear Dichroism

Linear dichroism can be introduced into an isotropic sample or modified in a sample which is already anisotropic when solute molecules are excited with a beam of light (wavenumber $\tilde{\nu}_1$). The effect is greatest when the light is linearly polarized, and we shall call the polarization direction U. In order to observe the dichroic absorption spectrum of the excited or photoconverted solute molecules $E_{UV}(\tilde{\nu}_1,\tilde{\nu}_2)$, a second beam of wavenumber $\tilde{\nu}_2$ and polarization V is used. The required experimental arrangement will depend on factors such as the lifetime of the solute transformed by the excitation. If it is effectively infinite, i.e., if the solute undergoes a permanent photochemical conversion, it is possible to separate the excitation and the measurement far apart in time. The excitation can be performed with a suitable monochromatic light source and a linear polarizer, and the sample can subsequently be transferred into one of the instruments described in Sections 3.2.1 or 3.2.2 for the measurement of linear dichroism. It is important to remember that rotational diffusion may occur over a period of minutes or hours, even in apparently quite rigid samples such as organic glasses at liquid nitrogen temperatures, so that the time delay between excitation and measurement should not be excessive. The rate of rotational depolarization is a sensitive function of temperature, the size of the solute molecule, and the nature of the rigid solvent used.

For shorter excited state lifetimes, it becomes harder to produce adequate amounts of the excited species, and the intensity of the exciting beam becomes an important consideration. For lifetimes of the order of seconds to microseconds, intense flashlamps can be used. They are spatially extensive sources, and it is difficult or impossible to obtain much polarized monochromatic light from them. At times, however, even the use of broad-band and possibly even unpolarized exciting radiation will be useful for partially oriented samples, particularly if only qualitative conclusions are to be drawn about the degree of orientation. The pulsed nature of the excitation permits convenient observation in the dark periods and at the same time produces significant population in the excited state. In the measurement of IR spectra of the excited solute molecules, exciting visible or UV light does not interfere, and it is possible to use a continuous source. This facilitates work with monochromatic linearly polarized exciting radiation, using a Fourier transform IR spectrometer to produce the required high sensitivity.

Solutes with excited state lifetimes of the order of nanoseconds or shorter require the use of intense laser pulses to produce a sufficient concentration of the excited species. The apparatus suitable for this type of work is shown in Figure 3.10. If it is not already polarized, the laser pulse can be passed through a polarizer P_1 and its polarization can be rotated by a half-wave plate as necessary. The monitoring beam may be produced by a pulsed lamp or a laser. Much of the interest in this kind of experiment comes from studies in rotational diffusion, usually using picosecond laser pulses. A more detailed discussion of the signal-processing electronics lies outside the scope of this book.

The procedures described so far yield individual spectra $E'_{UV}(\tilde{\nu}_1,\tilde{\nu}_2)$ separately, but it is also possible to use polarization or orientation modulation techniques of the type described in Sections 3.2.2, 3.2.3, and 3.2.5. An interesting possibility available in the two-photon experiment is the use of excitation modulation, in which the exciting beam intensity is modulated at frequency ω and the ac component of the intensity of the second beam is detected using a lock-in amplifier.

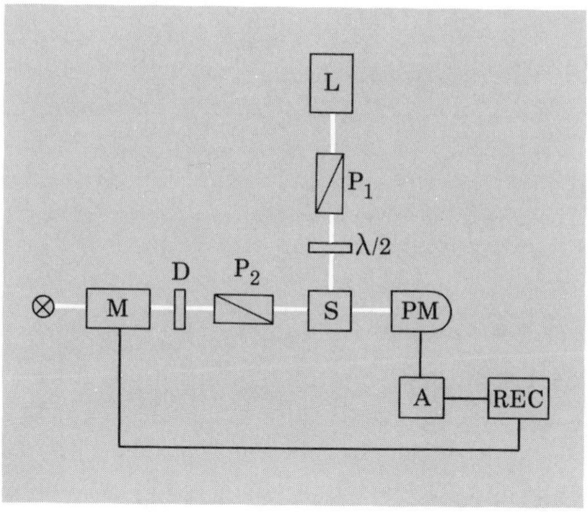

Figure 3.10 A schematic representation of a polarized laser flash photolysis instrument. L, pulsed laser; P_1 and P_2, polarizers; $\lambda/2$, half-wave retarder; M, monochromator; D, depolarizer; S, sample; PM, photomultiplier; A, data processor (such as transient recorder); REC, recorder.

This method is useful for excited states with lifetimes up to tens of milliseconds and has been used in both UV and IR regions.

Just as for polarized emission, only five independent spectra $E'_{UV}(\tilde{\nu}_1,\tilde{\nu}_2)$ are available in principle if only measurements with the electric vector of the light parallel to Z or perpendicular to Z are considered [(3.31)]. The analogy to emission studies is quite close, and similar precautions must be taken with respect to rotation depolarization, sample strain, and concentration depolarization. Because of the intrinsically lower sensitivity of absorption measurements, it is often impossible to avoid corrections for ground-state depletion in the interpretation of the data.

Even when the exciting light is unpolarized and/or when energy transfer randomizes the initial excitation, useful information on the polarization of transitions involving the metastable state can be obtained. An example in which the polarized IR spectrum of an excited triplet state was observed is presented in Section 8.2.4. The signal $I(\tilde{\nu})$ produced by the lock-in amplifier for absorption peaks of the metastable state is given by

$$I_{T_1}(\tilde{\nu}) = kI^0(\tilde{\nu})E_{T_1}(\tilde{\nu}) \tag{3.43}$$

and that for the absorption peaks of the ground state by

$$I_{S_0}(\tilde{\nu}) = -kI^0(\tilde{\nu})T_{S_0}(\tilde{\nu})E_{S_0}(\tilde{\nu}) \tag{3.44}$$

The proportionality constant k depends on the solute and the rate of excitation, $I^0(\tilde{\nu})$ is the monitoring beam intensity at the detector with no sample present, $T(\tilde{\nu})$ is transmittance, and $E(\tilde{\nu})$ is absorbance: $E_{T_1}(\tilde{\nu})$ is the absorbance of the steady-state triplet concentration that would be obtained if the excitation source were not chopped, and $E_{S_0}(\tilde{\nu})$ is the absorbance of the ground-state population with no excitation. The equations hold for nonoverlapping peaks. The difference in sign is due to the fact that (3.43) describes the consequences of metastable excited state population while (3.44) describes those of ground-state depletion.

For nonoverlapping peaks of the metastable state, the dichroic ratio is then given by

$$[E_{T_1}(\tilde{\nu})]_Z/[E_{T_1}(\tilde{\nu})]_Y = [I_{T_1}(\tilde{\nu})]_Z[I^0(\tilde{\nu})]_Y/[I_{T_1}(\tilde{\nu})]_Y[I^0(\tilde{\nu})]_Z \qquad (3.45)$$

where the subscripts Y and Z refer to the polarization of the monitoring beam.

3.2.9 Two-Photon Absorption

A laser experiment which does not appear to have been attempted so far on a partially aligned sample is the measurement of two-photon absorption cross-sections. Although the scattering inherent in such samples impedes experiments dependent on the coherent nature of laser light, it is likely that they could be performed.

Among the various methods useful for monitoring two-photon absorption by solutes contained in solids or liquids, detection of fluorescence has been the most popular. Figure 3.11 outlines the design of one of the most successful two-photon spectrofluorimeters. The light source is a dye laser pumped by a 1-MW nitrogen laser. Linearly polarized 5-ns pulses enter a double Fresnel rhomb, which acts as an achromatic half-wave plate. It rotates around its optical axis at 5 Hz. The light beam then passes through a fixed Fresnel rhomb and enters the sample. Its intensity is measured by a pyroelectric detector. As the double rhomb is rotated by angle ϕ from its reference position, the polarization state of the light entering the sample will be linear for $\phi = n\pi/4$ and circular for $\phi = (2n + 1)\pi/8$, where n is an integer. Sample fluorescence is analyzed by a monochromator and detected by a photomultiplier.

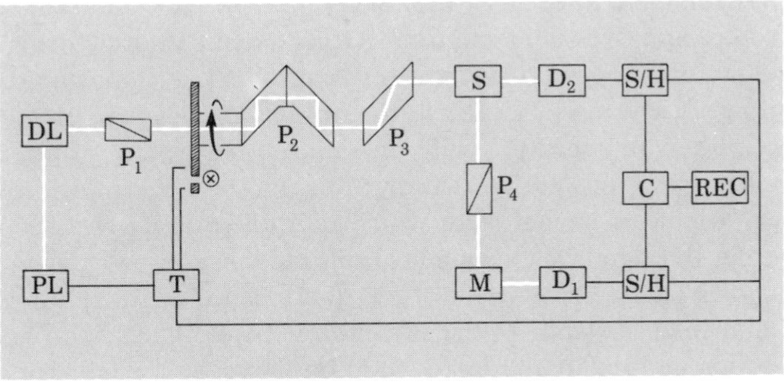

Figure 3.11 A schematic representation of a polarization two-photon spectrofluorimeter. PL, pump laser; DL, dye laser; P_1 and P_4, polarizers; P_2, double Fresnel rhomb; P_3, Fresnel rhomb; T, trigger; S, sample; M, monochromator; D_1 and D_2, detectors; S/H, sample-and-hold; C, computer; REC, plotter. Adapted by permission from B. Dick, H. Gonska, and G. Hohlneicher, *Ber. Bunsenges. Physik. Chem.* **85**, 746 (1981).

The measurement cycle is synchronized with the rotation of the double rhomb, which is attached to a disc which contains two regular rings of openings. The first set is spaced at $\pi/8$ and is used to trigger the nitrogen laser so as to produce an alternating sequence of exactly linearly polarized and exactly circularly polarized pulses as well as to initiate the sample-and-hold units connected to the two detectors and to control the operation of the data processing system. The second set of openings in the disc is spaced at $\pi/4$ and produces a trigger signal each time the laser pulse is circularly polarized. This permits the data processing system to produce two-photon excitation spectra and polarization curves.

The fluorescence intensity is proportional to the two-photon absorption cross section, the emission quantum yield, and the intensities of the two laser beams used. Most often, as in the instrument just described, both photons are taken from a single beam, and the fluorescence intensity is then proportional to the square of the beam intensity. The emission quantum yield may in general depend on the energy of the photons absorbed. In many aromatics, this dependence is only very weak. Moreover, it will cancel when ratios of fluorescence intensities are taken in a polarization study.

Several important precautions must be taken in this type of work. (i) The fluorescence must be analyzed by a monochromator or at least a narrow band-pass filter in order to remove Raleigh and Raman scattered light. (ii) The laser beam or beams should not be focused too tightly into the sample and their intensities should not be excessive in order to minimize the removal of excited molecules from the fluorescent pool by further photoprocesses. This is most easily tested by verifying that the signal dependence on beam intensity is purely quadratic. Pulsed dye lasers with peak powers of 5–50 kW appear to be optimal. (iii) Laser dye changes are unavoidable when complete spectra are taken, and it is not uncommon to use a dozen dyes for the spectrum of a single molecule. The wavelength regions of marginal dye performance near the lasing threshold require particular attention, and the matching of results obtained with sufficiently overlapping dyes is mandatory. Artifacts tend to occur here primarily due to spatial inhomogeneity of the laser beam. (iv) It is important to exclude participation of two-step processes in the production of the fluorescent signal. In practice, the main concern is with one-photon absorption from the ground to the triplet state followed by a second one-photon absorption producing a high-energy triplet or singlet state which may then decay to the fluorescent state. In order to exclude such complications, several approaches can be taken, such as studies of the dependence of the fluorescence intensity on laser pulse duration and intensity and on the presence of triplet quenchers. In isotropic media in which molecular rotation is rapid on the scale of the fluorescence lifetimes, fluorescence intensity will be independent of the state of polarization of the laser light (linear or circular) if the absorption process is two-step.

An example of two-photon absorption work, with reference to experimental details, is given in Section 8.3.7.

The use of fluorescence detection for monitoring two-photon absorption under conditions which do not allow for fast rotational averaging of molecular orientation is complicated by the fact that the observed fluorescence then is anisotropic and polarized and the overall process is of rank 6 (three photons are involved). The treatment of data of rank 6 for a general uniaxial initial orientation distribution would follow the lines sketched in Appendix II for rank 4. It will provide information on the sixth moments of the orientation distribution function. If such information is not desired and a study of molecular optical properties is the sole aim, it may be simplest to remove all dependence on the third photon by integrating the total fluorescence intensity, using, say, an integrating sphere and analyzing the data using the formulas of Section 7.5.1, appropriate for rank 4. These are also appropriate when altogether different means of detecting the two-photon absorption event are chosen, such as thermal blooming or photoacoustic detection.

3.2.10 Polarized Raman Scattering

Unlike the polarized emission and photoinduced transient dichroism experiments, which are most easily performed on home-built or at least considerably modified instruments, polarized Raman scattering measurements can be performed on standard commercial instruments. A schematic diagram of such an instrument is shown in Figure 3.12. If necessary, exciting laser light is linearly polarized by the polarizer P_1. Its polarization direction may be rotated by 90° using a half-wave plate. Raman scattered radiation emerging from the sample passes through the polarizer P_2, is subsequently depolarized, and then enters a monochromator and a detection system.

Polarized Raman experiments on isotropic samples are standard. The weakness of the effect requires the use of quite concentrated samples and relatively long path lengths (often several millimeters). The available anisotropic samples are usually quite dilute and depolarize light by scattering when their thickness exceeds a fraction of a millimeter. This makes it very difficult to obtain reliable

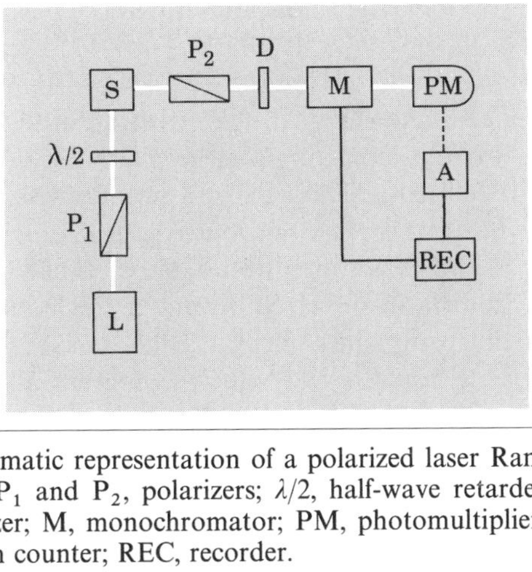

Figure 3.12 A schematic representation of a polarized laser Raman spectrometer. L, laser; P_1 and P_2, polarizers; $\lambda/2$, half-wave retarder; S, sample; D, depolarizer; M, monochromator; PM, photomultiplier; A, dc amplifier or photon counter; REC, recorder.

results for the relative intensities of a peak in the four independent spectra $E'_{UV}(\tilde{\nu}_1,\tilde{\nu}_2)$: E'_{ZZ}, $E'_{ZY} = E'_{YZ}$, E'_{YY}, and E'_{XY}. The intensity of non-totally symmetrical vibrations is generally low. The intensity of totally symmetric vibrations is quite high in the parallel arrangements E'_{ZZ} and E'_{YY} but very much smaller in the perpendicular arrangements E'_{YZ} and E'_{XY}, making quantitative evaluation difficult. The use of an optical multichannel analyzer with simultaneous recording of all lines would help to solve this problem.

A considerable enhancement of Raman intensity can be achieved by operation in the resonant mode, and successful experiments of this kind on a solute contained in a stretched polymer have been reported. However, there is a severe danger of damaging the sample by the intense laser radiation which is absorbed in this mode. In this case, the equality $E'_{ZY} = E'_{YZ}$ no longer holds.

3.3 Comments and References

Thermotropic Liquid Crystals. A highly readable introduction to liquid crystals can be found in G. H. Brown and P. P. Crooker, *Chem. Eng. News* **61**, 24 (1983). Information on the use of liquid crystals as uniaxial orienting media in optical spectroscopy can be found in Chapters 6 and 7 of H. Kelker and R. Hatz, *Handbook of Liquid Crystals*, Verlag Chemie, Weinheim, 1980 [a correction for the effects of solvent birefringence on the observed dichroic ratio is described on p. 277; see also H. S. Subramhanyam and D. Krishnamurti, *Mol. Cryst. Liq. Cryst.* **22**, 239 (1973) and A. Hauser, G. Pelzl, C. Selbmann, and D. Demus, *Mol. Cryst. Liq. Cryst.* **91**, 97 (1983)]. Useful hints for experimental work are given by W. H. de Jeu, *Physical Properties of Liquid Crystalline Materials*, Gordon and Breach, New York, 1980; and, for homeotropic alignment, by H. Kelker, R. Hatz, and G. Wirzing, *Z. Anal. Chem.* **267**, 161 (1973). Additional important sources are L. M. Blinov, *Electro-optical and Magneto-optical Properties of Liquid Crystals*, Wiley, New York, 1983; G. R. Luckhurst and G. W. Gray, Eds., *The Molecular Physics of Liquid Crystals*, Academic Press, New York, 1979; G. Meier, E. Sackmann, and J. G. Grabmaier, *Applications of Liquid Crystals*, Springer-Verlag, New York, 1975; and P. G. deGennes, *The Physics of Liquid Crystals*, Clarendon Press, Oxford, 1974.

There are many papers describing the preparation of thermotropic nematic liquid crystals in the homogeneous or planar alignment for optical studies, particularly in relation to its use in display technology. The Kelker and Hatz handbook is a good reference source. Orientation of compensated nematic mixtures by electric fields, including the preparation of supercooled oriented solids, has been described by E. Sackmann and H. Möhwald, *J. Chem. Phys.* **58**, 5407 (1973). In most of the studies, strong static electric fields are used to ensure a good uniform alignment of the liquid crystal. In G. Gottarelli, B. Samorì, and R. D. Peacock, *J. Chem. Soc. Perkin II* **1977**, 1208, the use of an alternating electric field to ensure a stable alignment is advocated. It is also possible to embed the liquid crystal in a microporous polypropylene film: E. J. Poziomek, T. J. Novak, and R. A. Mackay, in *Liquid Crystals and Ordered Fluids*, Vol. 3, J. F. Johnson and R. S. Porter, Eds., Plenum, New York, 1978, p. 259. The use of the liquid crystals of the bicyclohexyl type for studies of linear dichroism is described by H. Wedel and W. Haase, *Chem. Phys. Lett.* **55**, 96 (1978) and B. Samorì, P. Mariani, and G. P. Spada, *J. Chem. Soc. Perkins II* **1982**, 447. Their nematic phase may be aligned by coating the cell walls with SiO_n, according to J. L. Janning, *Appl. Phys. Lett.* **21**, 173 (1972); with silanes, according to T. Uchida, K. Ishikawa, and M. Wada, *Mol. Cryst. Liq. Cryst.* **60**, 37 (1980); or with poly(vinyl alcohol), according to T. R. N. Kutty and A. G. Fischer, *Mol. Cryst. Liq. Cryst.* **99**, 301 (1983). Homeotropic alignment may be obtained with an electric or magnetic field or by coating the cell walls with amphiphiles, according to K. Hiltrop and H. Stegemeyer, *Ber. Bunsenges. Phys. Chem.* **82**, 884 (1978).

The use of **lyotropic liquid crystals** is surveyed by F. Fuujiwara, L. W. Reeves, M. Suzuki, and J. A. Vanin, in *Solution Chemistry of Surfactants*, Vol. 1, K. Mittal, Ed., Plenum, New York, 1979, p. 63, and by G. J. T. Tiddy and M. F. Walsh in *Aggregation Processes in Solution*, E. Wyn-Jones and J. Gormally, Eds., Elsevier, New York, 1983. A review of optical spectroscopy of molecules aligned in membranes with emphasis on biological applications is found in L. B.-Å. Johansson and G. Lindblom, *Quart. Rev. Biophys.* **13**, 63 (1980). The preparation of highly transparent lyotropic nematic mesophase is described in L. B.-Å Johansson, O. Söderman, K. Fontell, and G. Lindblom, *J. Phys. Chem.* **85**, 3694 (1981). For the preparation of a macroscopically aligned lamellar mesophase, the reader is referred to L. B.-Å. Johansson, Å. Davidsson, G. Lindblom, and B. Nordén, *J. Phys. Chem.* **82**, 2604 (1978) and references therein.

Stretched Polymers. The procedures for the use of various stretched polymers as orienting media have been described more or less thoroughly in a large number of individual papers, but no book chapters or reviews devoted specifically to this topic seem to have appeared. For this reason, the subject has been covered in some detail in this volume. In addition to the long list of references available in E. W. Thulstrup and J. Michl, *J. Phys. Chem.* **84**, 82 (1980), the reader may consult B. Nordén, *Chem. Scripta*, **7**, 167 (1975); W. Hanle, H. Kleinpoppen, and A. Scharmann, *Z. Naturforsch.* **13a**, 64 (1958); Y. Tanizaki, *Bull. Chem. Soc. Japan* **33**, 979 (1960); K. Fiksiński and D. Frackowiak, *Spectrosc. Lett.* **13**, 873 (1980); and E. W. Thulstrup, J. Michl, and J. H. Eggers, *J. Phys. Chem.* **74**, 3868 (1970). Incorporation of dopants into polyethylene from a suspension in boiling water is described in G. Bauer, *Monatsh.* **102**, 1797 (1971), and incorporation from vapor phase is discussed in J. Konwerska-Hrabowska and J. H. Eggers, *Spectrosc. Lett.* **10**, 441 (1977). A fairly detailed description of several ways of doping polyethylene with solutes ranging from poorly soluble solids to simple gases is given in J. G. Radziszewski and J. Michl, *J. Am. Chem. Soc.*, **108**, 0000 (1986). The use of a layer of a polar substrate on the surface of a polyethylene sheet is described in the above-quoted paper by G. Bauer, and orientation of DNA in poly(vinyl alcohol) in Y. Matsuoka and B. Nordén, *Biopolymers* 22, 1731 (1983).

Optical properties of partially oriented polymers themselves have been the subject of intensive study. See, e.g., I. M. Ward, Ed., *Structure and Properties of Oriented Polymers*, Applied Science Publishers, London, 1975, and R. S. Stein and R. S. Finkelstein, *Ann. Rev. Phys. Chem.* **24**, 207 (1973). The subject is also discussed in books on polymer spectroscopy, such as J. Guillet, *Polymer Photophysics and Photochemistry*, Cambridge University Press, 1985; and W. Klöpffer, *Introduction to Polymer Spectroscopy*, Springer-Verlag, Berlin, 1984. The detailed mechanism by which orientation is transferred from the polymer to the solute has been the subject of intense speculation, but little is known with certainty. See, for instance, J. Konwerska-Hrabowska, G. W. Chantry, and E. A. Nicol, *Int. J. Infrared Millimeter Waves*, **2**, 1135 (1981) and references therein, and Y. T. Jang, P. J. Phillips, and E. W. Thulstrup, *Chem. Phys. Lett.* **93**, 66 (1982).

Stretching of polymers generally produces uniaxial alignment. On the other hand, unequal squeezing of a cubic block of polyacrylamide gel in two orthogonal directions is known to produce biaxial orientation of imbedded solute molecules. For an example and leading references, see A. O. Ganago, Yu. E. Erokhin, and Z. K. Makhneva, *Studia Biophysica* **80**, 193 (1980).

Other Means of Orienting Molecules. Leading references to other methods for inducing uniaxial orientation, such as liquid flow (cf. section 8.6.8), are given in the review by B. Nordén. *Appl. Spectrosc. Rev.* **14**, 157 (1978). Recently, alignment of DNA during electrophoretic migration has been detected: B. Åkerman, M. Jonsson, and B. Nordén, *J. Chem. Soc., Chem. Commun.* **1985**, 422. An interesting possibility is absorption into microscopic clay particles which can then be oriented macroscopically, e.g., by sedimentation. For examples of the use of montmorillonite, see V. C. Farmer and M. M. Mortland, *J. Chem. Soc.* (*A*) **1966**, 344 (pyridinium) and S. Yariv, J. D. Russell, and V. C. Farmer, *Israel J. Chem.* **4**, 201 (1966) (nitrobenzene).

Spectrometers. Linear Dichroism. No references need to be given for standard double-beam spectrophotometers. A useful source of information on absorption measurements is C. Burgess and A. Knowles, Eds., *Standards in Absorption Spectrometry*, Chapman and Hall, London, 1981. For use of a microspectrophotometer, see, e.g., E. Ohmes, J. Pauluhn, J.-U. Weidner, and H. W. Zimmermann, *Ber. Bunsenges. Physik. Chem.* **84**, 23 (1980). Information on polarizing elements for the short-wavelength region can be found in M. N. McDermott and R. Nowick, *J. Opt. Soc. Am.* **51**, 1008 (1961). A highly accurate oscillating beam spectrometer for differential measurement of very small optical densities is described in H. Seki and U. Itoh, *Rev. Sci. Instrum.* **51**, 22 (1980). The effects of scattering by polymer samples and a modification of a commercial spectrophotometer to minimize them are described in G. R. Kelly and T. Kurucsev, *Eur. Polym. J.*, **11**, 581 (1975). There is an extensive literature on light scattering by polymers; see, e.g., R. S. Stein, A. Misra, T. Yuasa, and F. Khambatta, *Pure Appl. Chem.* **49**, 915 (1977).

Fourier transform IR spectrometers are discussed in P. R. Griffiths, Ed., *Transform Techniques in Chemistry*, Plenum, New York, 1978. Photoacoustic spectroscopy is discussed in A. Rosencwaig, *Photoacoustics and Photoacoustic Spectroscopy*, Wiley, New York, 1980, and its application to the measurement of dichroism of a strongly absorbing sampling in D. Fournier, A. C. Boccara, and J. Badoz, *Appl. Phys. Lett.* **32**, 640 (1970); see also D. Frackowiak, S. Hotchandani, G. Bialek-Bylka, and R. M. Leblanc, *Photochem. Photobiol.* **42**, 567 (1985). Fluorescence-detected linear dichroism measurements are described in L. B.-Å. Johansson. G. Lindblom, and K. R. Naqvi, J. *Chem. Phys.* **74**, 3774 (1981).

Some of the less common methods of measurement of linear dichroism deserve mention here. The combination of measurements with unpolarized light along the orientation direction of a nematic liquid crystal followed by a similar measurement above the isotropic point of the crystal was introduced by H. Kelker, R. Hatz, and G. Wirzing, *Z. Anal. Chem.* **267**, 161 (1973), and used in a differential mode in J. R. Fernandes and S. Venugopalan, *Chem. Phys. Lett.* **53**, 407 (1978). The combination of measurement in a cholesteric liquid crystal with measurement in a nematic liquid crystal was described by F. D. Saeva and G. R. Olin, *Mol. Cryst. Liq. Cryst.* **35**, 319 (1976).

The "tilted plate" method is described in B. Nordén, G. Lindblom, and I. Jonás, *J. Phys. Chem.* **81**, 2086 (1977) and L. B.-Å. Johansson, Å. Davidsson, G. Lindblom, and B. Nordén, *J. Phys. Chem.* **82**, 2604 (1978). These papers discuss in detail the plethora of corrections which are needed for accurate work and give references to previous work. The theory of measurement on very thin layers is found in D. den Engelsen, *J. Phys. Chem.* **76**, 3390 (1972).

The PNP method is described in A. Yogev, L. Margulies, J. Sagiv, and Y. Mazur, *Rev. Sci. Instrum.*, **45**, 386 (1974), both the PNP and the SP methods are discussed in A. Yogev, L. Margulies, J. Sagiv, and Y. Mazur, *Chem. Phys. Lett.* **23**, 178 (1973).

The conversion of commercial dichrographs for the measurement of linear dichroism and the evaluation methods, including equations (3.17)–(3.19), are discussed in B. Samorì P. Mariani, and G. P. Spada, *J. Chem. Soc. Perkin II*, **1982**, 447. In the beginning of the "Results and Discussion" section of this paper, and also in B. Samorì, *J. Phys. Chem.* **83**, 375 (1979), the quantity $[E_Z(\tilde{\nu}) + E_Y(\tilde{\nu})]/2$ is called "isotropic absorption of the oriented sample," since it is measured with Z oriented at or near the magic angle relative to one of the planes of polarization, so that $LD'(\tilde{\nu})$ disappears. This is not the usual usage of this term, and it may be better to call this quantity "unpolarized absorption" or "average absorption"; cf. Samorì, *Mol. Cryst. Liq. Cryst.* **98**, 385 (1983). This paper also describes how the "average absorption" may be obtained from the detector dc current. Samorì and collaborators use an instrument with a photoelastic modulator and a commercial LD attachment. In Å. Davidsson and B. Nordén, *Chem. Scripta* **9**, 49 (1976), who proposed equation (3.20), as well as in G. Gottarelli, B. Samorì, and R. D. Peacock, *J. Chem. Soc. Perkin II*, **1977**, 1208, similar discussions are given for older instruments with an electro-optical modulator whose use requires the insertion of a quarter-wave plate. Additional detail on modulated polarization spectroscopy is found in K. W. Hipps and G. A. Crosby, *J. Phys. Chem.* **83**, 555 (1979); H. P. Jensen, J. A. Schellman, and T. Troxell, *Appl. Spectrosc.* **32**, 192 (1978); and in B. Nordén and S. Seth, *Appl. Spectrosc.* **39**, 647 (1985), who describe a device for absolute calibration. A simple description of the use of Muller calculus for the evaluation of measurements of linear dichroism is given in the appendix of B. Nordén,

Appl. Spectrosc. Rev. **14**, 157 (1978). A thorough and very instructive exposition of this subject and a detailed treatment of the theory of modulation spectroscopy with practical illustrations are found in J. Schellman, H. P. Jensen, and C. Foverskov, *Chem. Rev.*, in press (1987).

Electric dichroism and **electrochromism** are covered in great detail in the book by E. Fredericq and C. Houssier, *Electric Dichroism and Electric Birefringence*, Clarendon Press, Oxford, England, 1973, with particular emphasis on large molecules; in C. T. O'Konski, Ed., *Molecular Electro-Optics*, Marcel Dekker, New York, 1976; and in *Molecular Electro-Optics*, S. Krause, Ed., Plenum Press, New York, 1981. Electric dichroism and electrochromism of small molecules are covered in depth by W. Liptay, in *Excited States*, E. C. Lim, Ed., Academic Press, New York, 1974, p. 129.

Liquid-crystal-induced circular dichroism was first described by F. D. Saeva and J. J. Wysocki, *J. Am. Chem. Soc.* **93**, 5928 (1971) and has been used a few times since for the determination of polarization directions in the UV and IR regions. References can be found in W. G. Hill, S. F. Mason, and R. D. Peacock, *J. Chem. Soc. Perkin II*, **1977**, 1262; R. J. Dudley, S. F. Mason, and R. D. Peacock, *J. Chem. Soc., Faraday II*, **1975**, 997, from which equation (3.29) is taken; and F. D. Saeva, in *Liquid Crystals*, F. D. Saeva, Ed., Marcel Dekker, New York, 1979, Chapter 6. A detailed mathematical description of the propagation of light along the axis of a cholesteric crystal is found in S. Chandrasekhar, *Liquid Crystals*, Cambridge Univ. Press, 1977.

The measurement of **natural circular dichroism** is treated by L. Velluz, M. Legrand, and M. Grosjean, *Optical Circular Dichroism*, Academic Press, New York, 1965, who give references to older literature, and in the books by Mason and Charney quoted in Section 1.4. Special precautions required in measurements on aligned samples, including liquid-crystal-induced circular dichroism, have been discussed: see H.-G. Kuball and J. Altschuh, *Chem. Phys. Lett.* **87**, 599 (1982), Y. Shindo and Y. Ohmi, *J. Am. Chem. Soc.* **107**, 91 (1985), Y. Shindo, M. Nakagawa, and Y. Ohmi, *Appl. Spectrosc.* **39**, 860 (1985), and references therein, and Å. Davidsson, B. Nordén, and S. Seth, *Chem. Phys. Lett.* **70**, 313 (1980). The precautions will apply equally in the case of **magnetic circular dichroism**. An artifact specific to the latter and occurring even in measurements on isotropic solutions has been discussed in V. Schröder and M. V. Löwis of Menar, *Z. Phys. Chem.* (Leipzig), **264**, 721 (1983). A successful design of a sample cell suitable for oriented CD measurements in compensated liquid crystals is given in J. Altschuh, T. Karstens, and H.-G. Kuball, *J. Phys. E; Sci. Instrum.* **14**, 43 (1981). A miniaturized version has been used recently for the measurement of MCD of aligned anthracene: C. Puebla, *J. Mol. Struct.*, in press (1986). The use of a Fourier transform IR spectrometer for the measurement of CD in the IR region (VCD) is discussed by L. A. Nafie and D. W. Vidrine, in *Fourier Transform Infrared Spectroscopy*, Vol. 3, J. R. Ferraro and L. J. Basile, Eds., Academic Press, New York, 1982, Chapter 3; for a review of results see T. B. Freedman and L. A. Nafie, in *Topics in Stereochemistry*, E. L. Eliel and S. H. Wylen, Eds., Wiley, New York, Vol. 17, 1986. Artifacts in the measurment of circularly polarized luminescence are discussed in H. P. J. M. Dekkers, P. F. Moraal, J. M. Timpes, and J. P. Riehl, *Appl. Spectrosc.* **39**, 818 (1985).

Some of the classics in the area of application of **polarized luminescence and photodichroism** for polarization assignments on isotropic samples are G. J. Hoijtink and P. J. Zandstra, *Mol. Phys.* **3**, 371 (1960); H. Zimmermann and N. Joop, *Z. Elektrochem.* **64**, 1215 (1960); F. Dörr and M. Held, *Angew. Chem.* **72**, 287 (1960); A. C. Albrecht, *J. Mol. Spectrosc.* **6**, 84 (1961); *Progr. React. Kinet.* **5**, 301 (1970); and T. Azumi and S. P. McGlynn, *J. Chem. Phys.* **37**, 2413 (1963), who describe the basic experimental techniques and the corrections for instrumental polarization biases. A useful review is given by F. Dörr, in *Creation and Detection of the Excited State*, Vol. 1, A. A. Lamola, Ed., Marcel Dekker, New York, 1971, Chapter 2. The first chapter in this volume, by W. G. Herkstroeter, provides extensive tables of solvents suitable for use as low-temperature glasses; additional information on this subject is available on p. 102 of the chapter by Dörr and in G. Fischer and E. Fischer, *Mol. Photochem.* **8**, 279 (1977). A chapter on polarized fluorescence is found in J. R. Lakowicz, *Principles of Fluorescence Spectroscopy*, Plenum, New York, 1983. A discussion of the evaluation techniques in time-resolved polarized fluorescence work is found in A. J. Cross and G. R. Fleming, *Biophys. J.* **46**, 45 (1984). For multi-frequency phase-modulation fluorometry, see J. R. Lakowicz, H. Cherek, B. P. Maliwal, and E. Gratton, *Biochemistry* **24**, 376 (1985).

Techniques of polarized luminescence of partially aligned samples were described, e.g., by E. Sackmann and H. Möhwald, *J. Chem. Phys.* **58,** 5407 (1973), and L. L. Chapoy and D. B. DuPre, *J. Chem. Phys.* **70**, 2550 (1979), **73**, 3021 (1980), using liquid crystal samples. For references to studies of polymer samples, see Y. Nishijima. *Ber. Bunsenges. Phys. Chem.* **74**, 778 (1970), *J. Polym. Sci.* **C31**, 353 (1970); J. H. Nobbs, D. I. Bower, I. M. Ward, and D. Patterson, *Polymer* **15**, 287 (1974); D. I. Bower, in *Structure and Properties of Oriented Polymers*, I. M. Ward, Ed., Applied Science Publishers, London, 1975, Chapter 5; S. Hibi, M. Maeda, H. Kubota, and T. Miura, *Polymer* **18**, 143 (1977); J. J. Dekkers, G. Ph. Hoornweg, K. J. Terpstra, C. MacLean, and N. H. Velthorst, *Chem. Phys.* **34**, 253 (1978); L. Margulies and A. Yogev, *Chem. Phys.* **27**, 89 (1978); and M. Hennecke and J. Fuhrmann, *Colloid Polymer Sci.* **258**, 219 (1980), and references therein.

Studies of biological samples are reviewed by R. A. Badley, in *Modern Fluorescence Spectroscopy*, E. L. Wehey, Ed., Heyden, London, 1976, p. 91, and in G. S. Beddard and M. A. West, Eds., *Fluorescent Probes*, Academic Press, New York, 1981.

Useful sources of general information are J. N. Miller, Ed., *Standards in Fluorescence Spectrometry*, Chapman and Hall, London, 1981, and K. D. Mielenz, Editor, *Measurement of Photoluminescence*, Academic Press, New York, 1982 (Vol. 3 of "Optical Radiation Measurements"). A commercial instrument for the recording of two-dimensional spectra is described in T. H. Maugh, II, *Science* **203**, 1330 (1979). For a review of two-dimensional luminescence spectroscopy, see G. W. Suter, A. J. Kallir, and U. P. Wild, *Chimia* **37**, 413 (1983).

Permanent photodichroism can be induced by irradiation of isotropic samples in rare gas and other low-temperature matrices—see, e.g., J. K. Burdett, J. M. Grzybowski, R. N. Perutz, M. Poliakoff, J. J. Turner, and R. F. Turner, *Inorg. Chem.* **17**, 147 (1978); M. S. Baird, I. R. Dunkin, N. Hacker, M. Poliakoff, and J. J. Turner, *J. Am. Chem. Soc.* **103**, 5190 (1981)—and in glasses—see, e.g., J. Kolc, J. W. Downing, A. P. Manzara, and J. Michl, *J. Am. Chem. Soc.* **98**, 930 (1976). An example of a study of transient dichroism induced in isotropic samples by picosecond laser pulses is found in D. Waldeck, A. J. Cross, Jr., D. B. McDonald, and G. R. Fleming, *J. Chem. Phys.* **74**, 3381 (1981). A review of the early work on polarized triplet-triplet absorption can be found in H. Labhart and W. Heinzelmann, in *Organic Molecular Photophysics*, Vol. 1, J. B. Birks, Ed., Wiley, New York, 1973, p. 297. Modification of the dichroism of solutes in partially aligned samples by irradiation has also been studied. For polarized triplet-triplet absorption of such samples, see V. Zanker and J. Thies, *Z. Physik Chem. NF* **33**, 1 (1962) and T. Hoshi, M. Oka, M. Komuro, J. Yoshino, K. Ota, and K. Murofushi, *Z. Physik. Chem. NF* **112**, 129 (1978); for polarized IR absorption and the origin of (3.43)–(3.45), see M. B. Mitchell, W. A. Guillory, J. Michl, and J. Radziszewski, *Chem. Phys. Lett.* **96**, 413 (1983), and M. B. Mitchell, Ph.D. Dissertation, University of Utah, 1984; *Diss. Abstr.* **45**, 2567 (1985), no. 8424962.

In the optical Kerr effect, birefringence caused by alignment of solute molecules by an intense laser field is measured: K. Sala and M. C. Richardson, *Phys. Rev. A* **12**, 1036 (1975). Birefringence due to photoselection was investigated by H.-G. Kuball, W. Euing, and T. Karstens, *Ber. Bunsenges. Physik. Chem.* **74**, 316 (1970) and K. B. Eisenthal and K. E. Rieckhoff, *Phys. Rev. Lett.* **20**, 309 (1968).

A leading reference for **two-photon absorption spectroscopy** in solution is G. Hohlneicher and B. Dick, *Pure Appl. Chem.* **55**, 261 (1983).

Raman Scattering. Useful descriptions of the experimental precautions to be taken and corrections to be made in polarized Raman work can be found in a study of a doped liquid crystal: S. Jen, N. A. Clark, P. S. Pershan, and E. B. Priestley, *J. Chem. Phys.* **66**, 4635 (1977); P. S. Pershan, in *The Molecular Physics of Liquid Crystals*, G. R. Luckhurst and G. W. Gray, Eds., Academic Press, New York, 1979, Chapter 17; and in studies of neat polymers: J. Maxfield, R. S. Stein, and M. C. Chen, *J. Polym. Sci., Polym. Phys.* **16**, 37 (1978) and J. Purvis and D. I. Bower, *J. Polym. Sci., Polym. Phys. Ed.* **14**, 1461 (1976).

For Raman scattering studies of solutes in an isotropic fluid aligned by an intense laser pulse, see S. Kielich, Z. Ożgo, and L. Wolejko, *J. Raman Spectrosc.* **3**, 315 (1975).

4

Quantitative Aspects of Uniaxial Alignment

The purpose of this chapter is to provide the tools necessary for a quantitative analysis of the results of optical measurements with polarized light on partially aligned samples. We begin by introducing the concept of a molecular orientation distribution function and then proceed to outline three common ways in which it can be described. First, we outline the motivation for the introduction of the so-called orientation factors and use them to provide a definition of molecular orientation axes. Second, we introduce the Saupe orientation matrices, which offer a closely related equivalent notation. Third, we describe the theoretical physicist's approach, based on expansion in the Wigner rotation matrices. In this case, detailed understanding requires a background in spherical tensor calculus. Finally, we discuss the relation between molecular properties on the one hand, and the degree of alignment produced by the use of anisotropic solvents, an electric field, or photoselection on the other hand.

4.1 The Orientation Distribution Function

Let us choose a right-handed rectangular coordinate system defined by the unit vectors ε_X, ε_Y, ε_Z. This system is fixed with respect to the laboratory and will be used to describe the propagation direction and state of polarization of the light used in the experiments, as well as the macroscopic orientation of the sample. As noted in the introduction, we are interested primarily in uniaxial samples, i.e., those which have a unique axis such that all directions perpendicular to this axis are optically equivalent (rotation around the axis has no physical effect). We choose to orient the sample so that the unique axis coincides with the Z axis of the laboratory coordinate system. If the uniaxial alignment is polar, the unique sample axis has a positive and a negative end, given for instance by the direction of an orienting electric field. In that case, we choose the positive sense of the unique sample axis to coincide with the positive sense of the Z axis.

We assume that the sample consists of a very large number of identical rigid molecules of interest for the measurement, imbedded in an inert medium (solvent). Let us choose a right-handed rectangular coordinate system defined by unit vectors ε_x, ε_y, and ε_z to be connected rigidly with the molecular framework (Figure 4.1). The choice is the same with respect to the molecular framework in each one of the molecules but is otherwise arbitrary. We shall specify the orientation of the molecule by specifying the orientation of its coordinate system. In the case of flexible molecules which exist as a mixture of conformers, each conformer must be considered separately and the result averaged as required.

The coordinates of a vector **R** in the laboratory system of axes will be denoted R_X, R_Y, and R_Z. The coordinates in the molecular system of axes will be labeled R_x, R_y, and R_z.

In a general partially aligned sample, at any given instant no two molecules will be oriented exactly alike with respect to the laboratory system of coordinates XYZ. In view of their large number, a useful description of the partial alignment must be probabilistic in nature. We shall denote by $f(\Omega)\,d\Omega$ the probability that a given molecule finds itself in the region of orientations limited by Ω on one side and $\Omega + d\Omega$ on the other, where the symbol Ω stands for a suitable set of parameters defining the mutual orientation of the molecule-fixed unit vector set ε_x, ε_y, ε_z with respect to the laboratory-fixed set ε_X, ε_Y, ε_Z. Since every molecule must have some orientation, an integral over all possible values of Ω must be equal to 1, i.e., $f(\Omega)$ is normalized:

$$\int f(\Omega)\,d\Omega = 1 \tag{4.1}$$

Thus, the function $f(\Omega)$ has the meaning of probability density. It is commonly referred to as the molecular *orientation distribution function.* The knowledge of this function permits us to calculate averages $\langle a \rangle$ of an arbitrary angle-

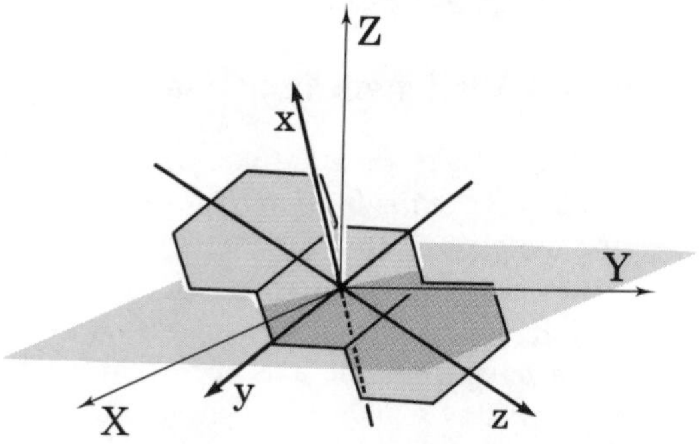

Figure 4.1 The laboratory (X,Y,Z) and molecular (x,y,z) systems of axes.

dependent quantity a(Ω) from

$$\langle a \rangle = \int a(\Omega) f(\Omega)\, d\Omega \tag{4.2}$$

Several practical realizations of Ω are common. We shall first describe its representation by the set of Euler angles α, β, γ. In order to bring a molecule into an orientation described by $\Omega \equiv (\alpha,\beta,\gamma)$, one first aligns it so that its axes coincide pairwise with the laboratory axes ($\boldsymbol{\varepsilon}_x = \boldsymbol{\varepsilon}_X$, $\boldsymbol{\varepsilon}_y = \boldsymbol{\varepsilon}_Y$, $\boldsymbol{\varepsilon}_z = \boldsymbol{\varepsilon}_Z$) and then performs three successive rotations on it as shown in Figure 4.2. The first rotation is by angle α around the axis z (which at this point is the same as Z). It brings the molecule from (A) to (B). The second rotation is by angle β around the axis y, which now in general no longer coincides with Y, and takes the molecule from (B) to (C). The third rotation is by angle γ around the axis z,

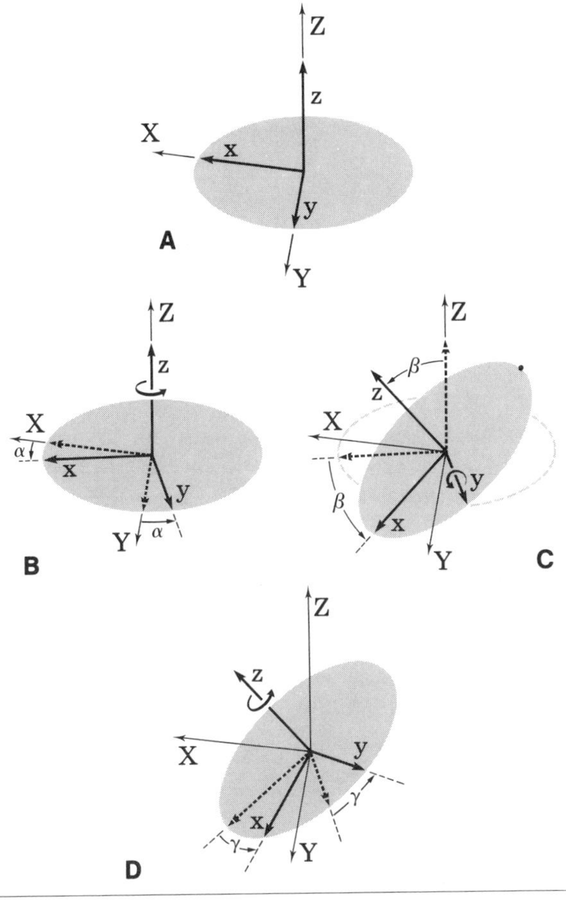

Figure 4.2 Rotation by the Euler angles α, β, and γ. See text. The shaded circle helps the visualization of the x,y plane but has no other significance.

which in the general case now no longer coincides with Z, and moves the molecule from (C) to (D). A rotation by a positive angle is taken to mean a counterclockwise rotation when viewed from the positive end of the rotation axis. Other definitions of Euler angles can be found in the literature, and some of the formulas based on these will differ from those given here.

With the above definition of Ω, the magnitude of $d\Omega$ is given by $(1/8\pi^2) \sin\beta \, d\alpha \, d\beta \, d\gamma$, and the normalization condition becomes

$$\int_{\alpha=0}^{2\pi} \int_{\beta=0}^{\pi} \int_{\gamma=0}^{2\pi} f(\alpha,\beta,\gamma)(1/8\pi^2) \sin\beta \, d\alpha \, d\beta \, d\gamma = 1 \tag{4.3}$$

In a uniaxial system, such as those of interest presently, all values of α must be equally probable. The distribution function is then constant with respect to α. It has the form $f(\beta,\gamma) = f(0,\beta,\gamma)$ and satisfies the normalization condition

$$\int_{\beta=0}^{\pi} \int_{\gamma=0}^{2\pi} f(\beta,\gamma)(1/4\pi) \sin\beta \, d\beta \, d\gamma = 1 \tag{4.4}$$

In molecules of special shape it may happen that all angles of rotation along the molecular z axis are equally probable, i.e., all values of γ are equally probable. The distribution function is then constant also with respect to γ, has the form $f(\beta) = f(0,\beta,0)$, and satisfies the normalization condition

$$\int_{\beta=0}^{\pi} f(\beta)(1/2) \sin\beta \, d\beta = 1 \tag{4.5}$$

Finally, in the absence of all anisotropic orientation, all values of the angle β are also equally probable, so that the distribution function is a constant independent of the Euler angles,

$$f(0,0,0) = 1 \tag{4.6}$$

Another set of angles which also interrelate $(\varepsilon_x,\varepsilon_y,\varepsilon_z)$ and $(\varepsilon_X,\varepsilon_Y,\varepsilon_Z)$, and to which we will also need to refer, are those between the molecular axes x, y, z and the laboratory-fixed axes X, Y, Z. Let u be one of the molecular axes and U one of the laboratory axes. We shall use the label uU for the angle between u and U; there are nine such angles (Figure 4.3). In uniaxial samples we need

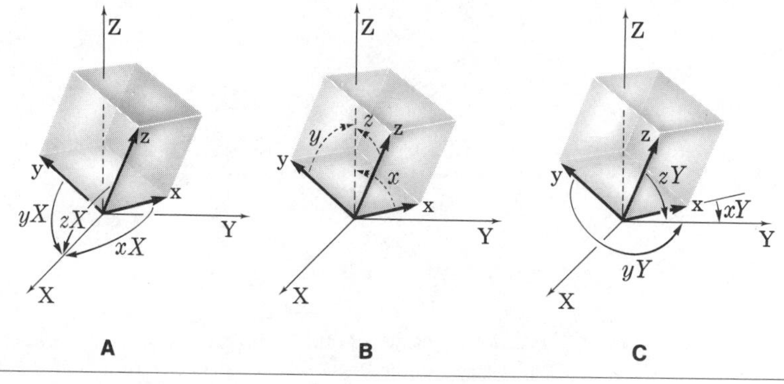

Figure 4.3 The angles between molecular and laboratory axes.

to use xZ, yZ, and zZ frequently, and we call them x, y, and z for short. The cosines of the angles uU are the direction cosines of the laboratory-fixed axes X, Y, Z in the molecular framework: $\cos x$, $\cos y$, and $\cos z$ are the direction cosines of the Z axis; $\cos xY$, $\cos yY$, and $\cos zY$ are those of the Y axis; and $\cos xX$, $\cos yZ$, and $\cos zX$ are those of the X axis. We note the existence of the relation

$$\cos^2 xU + \cos^2 yU + \cos^2 zU = 1 \tag{4.7}$$

where U = X, Y, or Z. The values of these direction cosines are related to the values of Euler angles:

$$\cos x = \cos xZ = -\sin\beta\cos\gamma \tag{4.8}$$

$$\cos y = \cos yZ = \sin\beta\sin\gamma \tag{4.9}$$

$$\cos z = \cos zZ = \cos\beta \tag{4.10}$$

$$\cos xY = \cos\alpha\sin\gamma + \sin\alpha\cos\beta\cos\gamma \tag{4.11}$$

$$\cos yY = \cos\alpha\cos\gamma - \sin\alpha\cos\beta\sin\gamma \tag{4.12}$$

$$\cos zY = \sin\alpha\sin\beta \tag{4.13}$$

$$\cos xX = \cos\alpha\cos\beta\cos\gamma - \sin\alpha\sin\gamma \tag{4.14}$$

$$\cos yX = -\sin\alpha\cos\gamma - \cos\alpha\cos\beta\sin\gamma \tag{4.15}$$

$$\cos zX = \cos\alpha\sin\beta \tag{4.16}$$

The special role played by the direction cosines of the Z axis in uniaxial systems is due to the fact that they do not change with the angle α of rotation around the unique axis Z, which has no physical effect. For such samples, a specification of the distribution function $f(x,y,z)$, combined with relation (4.7), is equivalent to a specification of the distribution function $f(\beta,\gamma)$, and either function is sufficient for a complete description of the uniaxial partial alignment. The fact that one of the angles x, y, z is redundant because of relation (4.7) may appear to be an inconvenience, but the use of the redundant set x, y, z permits a considerable simplification in the algebra relative to the nonredundant set β, γ. Among the descriptions of the uniaxial orientation distribution function which we shall now proceed to outline, the first two are based on the set x, y, z, and the third on the set β, γ.

4.2 The Orientation Factors

In order to outline the motivation for the use of the seemingly more cumbersome set of redundant angles x, y, z and to indicate the physical origin of the simplicity of the resulting expressions for a description of results of measurements using polarized light, we shall first briefly discuss such an experiment.

A more thorough treatment of the various types of experiments is found in the following chapters.

Let us assume that linearly polarized light with its electric vector parallel to the Z axis passes through the sample and interacts with an electric dipole transition moment **M** of a solute molecule whose orientation is characterized by the direction cosines cos x, cos y, and cos z.

$$\mathbf{M} = \varepsilon_x M_x + \varepsilon_y M_y + \varepsilon_z M_z \tag{4.17}$$

Let us now evaluate the probability of electric dipole light absorption, which is proportional to the square of the projection of the molecular transition dipole into the polarization direction Z of the light [(1.60), Figure 4.4]. Using (4.17), we have

$$(\varepsilon_Z \cdot \mathbf{M})^2 = (\varepsilon_Z \cdot \varepsilon_x M_x + \varepsilon_Z \cdot \varepsilon_y M_y + \varepsilon_Z \cdot \varepsilon_z M_z)^2 \tag{4.18}$$

$$(\varepsilon_Z \cdot \mathbf{M})^2 = (M_x \cos x + M_y \cos y + M_z \cos z)^2 \tag{4.19}$$

This expression is independent of the angle α of rotation of the molecule around the Z axis. If we now average over all molecular orientations Ω with their appropriate weights $f(\Omega)$, and indicate this by the use of pointed brackets, we obtain

$$\begin{aligned}\langle(\varepsilon_Z \cdot \mathbf{M})^2\rangle = {} & M_x^2\langle\cos^2 x\rangle + M_y^2\langle\cos^2 y\rangle + M_z^2\langle\cos^2 z\rangle \\ & + 2M_xM_y\langle\cos x \cos y\rangle + 2M_xM_z\langle\cos x \cos z\rangle \\ & + 2M_yM_z\langle\cos y \cos z\rangle\end{aligned} \tag{4.20}$$

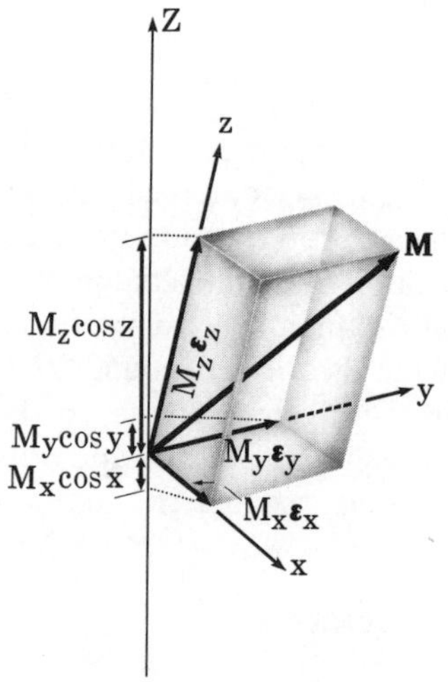

Figure 4.4 The projection of a transition moment **M** into the Z axis [equations (4.17)–(4.19)].

Clearly, averages of products of an even number of direction cosines, such as $\langle \cos u \cos y \rangle$, where u and v stand for x, y, or z, are likely to appear repeatedly in descriptions of the interaction of linearly polarized light with molecular transition moments, and it is useful to take them to be the primary parameters describing the molecular orientation. Equation (4.20) is simpler than the analogous expression based on the Euler angles (Chapter 5) and contains no strange numerical factors, and its individual terms have a simple and easily visualized physical meaning.

With the above as background, we shall now proceed to define the *orientation factors* K, L, M, etc:

$$K_{uv} = \langle \cos u \cos v \rangle \tag{4.21}$$

$$L_{stuv} = \langle \cos s \cos t \cos u \cos v \rangle \tag{4.22}$$

$$M_{qrstuv} = \langle \cos q \cos r \cos s \cos t \cos u \cos v \rangle \tag{4.23}$$

Throughout, each of the symbols q, r, s, t, u, v can be equal to any one of the angles x, y, z. Like the angles x, y, z themselves, the orientation factors K,L,M, . . . are redundant. Relations between them can be derived from the basic equation (4.7):

$$K_{xx} + K_{yy} + K_{zz} = 1 \tag{4.24}$$

$$L_{xxuv} + L_{yyuv} + L_{zzuv} = K_{uv} \tag{4.25}$$

$$M_{xxstuv} + M_{yystuv} + M_{zzstuv} = L_{stuv} \tag{4.26}$$

There are only $2j + 1$ independent orientation factors of order j, i.e., factors with j indices. Note that we do not define orientation factors of odd orders, since they are not needed for our discussion of optical experiments, but they could in principle be defined similarly.

The orientation factors are elements of symmetrical orientation tensors K (rank 2), ${}^{(4)}L$ (rank 4), ${}^{(6)}M$ (rank 6), etc. One can use the freedom available in the arbitrary selection of the x, y, z axis system with respect to the molecular framework to make the off-diagonal elements K_{uv}, $u \neq v$, vanish identically. This is achieved by finding the principal axes system of the orientation tensor K. Starting from an arbitrary axis system of axes, the principal axes are found by diagonalization of the 3×3 matrix K (For an example see Appendix II). The eigenvectors describe the transformation into a new system of axes; the eigenvalues are the averages of the squares of the direction cosines in the new system of axes, K_{zz}, K_{yy}, K_{xx}. These shall be labeled K_z, K_y, and K_x for short. We shall refer to the principal axes of the orientation tensor of rank 2, K, as *molecular orientation axes* and shall always refer the orientation factors to this system of axes unless otherwise noted. In order to remove the remaining ambiguity in the labeling of the axes, we specify $K_z \geqslant K_y \geqslant K_x$. The z axis will be referred to as the *effective orientation axis* or simply the *orientation axis*. It is sometimes called the "long" axis in the literature. By definition, this eigenvector of the orientation tensor K is the one among all possible molecular axes u for which $\langle \cos^2 u \rangle$ is the largest. The x axis is sometimes called the "short" axis and is the one for which $\langle \cos^2 u \rangle$ is the smallest.

If we take the value of $\cos^2 u$ as a measure of the degree to which an axis u lines up with the unique direction Z, we can say that the effective orientation axis z is that direction in the molecule which, on the average, is aligned the best. The y axis is that among all possible axes orthogonal to z which is aligned best, and the short axis x is that direction in the molecule which is aligned the worst. An additional discussion of the concept of the orientation axis can be found in Section 5.2.2.

In summary, we are left with the molecular orientation axes and with the following redundant set of nonvanishing K's and a relation between them:

$$1 \geqslant K_z = \langle \cos^2 z \rangle \geqslant K_y = \langle \cos^2 y \rangle \geqslant K_x = \langle \cos^2 x \rangle \geqslant 0 \qquad (4.27)$$

$$K_z + K_y + K_x = 1 \qquad (4.28)$$

Expression (4.20) for the absorption of a Z-polarized photon by an electric dipole transition moment now acquires the simple form

$$\langle (\boldsymbol{\varepsilon}_Z \cdot \mathbf{M})^2 \rangle = M_x^2 K_x + M_y^2 K_y + M_z^2 K_z \qquad (4.29)$$

Knowledge of the orientation factors K in itself provides only a very incomplete specification of the orientation distribution function $f(\Omega)$ but still describes it sufficiently for many experiments, as will be seen in the following. For experiments involving electric quadrupole absorption and those involving the interaction of more than one photon with a molecule, however, the higher order orientation factors L, M, etc., are also needed. In the general case, we have no freedom left to bring about reductions in their number. In the absence of symmetry the experimental data obtained from optical measurements rarely, if ever, yield enough information to completely determine all the higher order orientation factors. However, if the molecule possesses suitable elements of symmetry, the diagonalization (symmetry adaptation) of K produces simplifications in the higher tensors as well.

In the presence of a plane of symmetry containing the y and z axes, the orientation distribution function possesses the symmetry $f(\alpha,\beta,\gamma) = f(\alpha,\beta,\pi - \gamma)$. According to equations (4.8)–(4.10), operation $\gamma \to \pi - \gamma$ leaves $\cos y$ and $\cos z$ intact and converts $\cos x$ into $-\cos x$. Therefore, an average of a product of direction cosines in which x occurs an odd number of times will vanish due to integration over the angle γ. Thus, $K_{xy} = \langle \cos x \cos y \rangle = 0$ and $K_{xz} = \langle \cos x \cos z \rangle = 0$, but in general $K_{yz} \neq 0$. Similarly, $L_{xxxy} = \langle \cos^3 x \cos y \rangle = 0$, etc. If a plane of symmetry is located such that it contains the xy or xz axes instead, similar results are obtained by permutation of indices.

In general, the case which is of greatest practical importance is the one in which the molecule has two mutually perpendicular planes of symmetry. Taking both of these to lie parallel to the z axis, it is seen that the orientation distribution functions possess the additional symmetry property $f(\alpha,\beta,\gamma) = f(\alpha,\beta,\pi + \gamma)$. Equations (4.8)–(4.10) show that the operation $\gamma \rightarrow \pi + \gamma$ leaves only cos z intact and inverts the signs of both cos x and cos y. Therefore, to obtain a nonzero average, the total number of times cos x and cos y appear must also be even. Since the total number of indices is even, this is only possible if cos x, cos y, and cos z each appear an even number of times. In this case, the principal directions of K are determined by symmetry, as shown in detail on an example in Section 5.2.2 [eqs. (5.77)–(5.79)]. Only $1 + j/2$ independent orientation factors of order j remain: two K's, three L's, four M's, etc. Removing repeated indices just as we did for the K's, we are left with the following redundant set of nonvanishing higher-order orientation factors (u, v, and w are used to denote any one of the three subscripts x,y,z, such that $u \neq v$, $u \neq w$, $v \neq w$).

Choosing the three independent L's as L_z, L_y, and L_x,

$$1 \geqslant L_z = \langle \cos^4 z \rangle,\ L_y = \langle \cos^4 y \rangle,\ L_x = \langle \cos^4 x \rangle \geqslant 0 \tag{4.30}$$

and the linear dependency relationship for the remaining members of the redundant set is

$$L_{uv} = \langle \cos^2 u \cos^2 v \rangle = [(K_u - L_u) + (K_v - L_v) - (K_w - L_w)]/2 \tag{4.31}$$

Choosing the four independent M's as M_z, M_y, M_x, and M_{xyz}, we have

$$\begin{gathered} 1 \geqslant M_z = \langle \cos^6 z \rangle,\ M_y = \langle \cos^6 y \rangle,\ M_x = \langle \cos^6 x \rangle, \\ M_{xyz} = \langle \cos^2 x \cos^2 y \cos^2 z \rangle \geqslant 0 \end{gathered} \tag{4.32}$$

and the linear dependency relationships for

$$M_{uv} = \langle \cos^4 u \cos^2 v \rangle \tag{4.33}$$

are

$$M_{uv} + M_{uw} = L_u - M_u \tag{4.34}$$

$$M_{uv} + M_{vu} = L_{uv} - M_{xyz} \tag{4.35}$$

Note also the existence of the relations

$$L_z + L_y + L_x + 2(L_{xy} + L_{xz} + L_{yz}) = 1 \tag{4.36}$$

$$M_z + M_y + M_x + 6M_{xyz} + 3(M_{zy} + M_{yz} + M_{zx} + M_{xz} + M_{yx} + M_{xy}) = 1 \tag{4.37}$$

As the order of the orientation factors increases, so does the redundancy. The increase in the number of resulting relations between the orientation factors will undoubtedly reduce their practical utility. So far, the known applications involve only the factors K and L, where this is not a problem.

Table 4.1 provides a survey of the orientation factors K and L for some special orientation distributions.

Table 4.1
Orientation Factors for Special Uniaxial Orientation Distributions

	K_x	K_y	K_z	L_x	L_y	L_z	L_{xy}	L_{xz}	L_{yz}
Perfect alignment of z axis	0	0	1	0	0	1	0	0	0
Rodlike	$(1 - K_z)/2$	$(1 - K_z)/2$	K_z	$3(1 - 2K_z + L_z)/8$	$3(1 - 2K_z + L_z)/8$	L_z	$(1 - 2K_z + L_z)/8$	$(K_z - L_z)/2$	$(K_z - L_z)/2$
Photoselected with Z-polarized light, absorption z-polarized	1/5	1/5	3/5	3/35	3/35	3/7	1/35	3/35	3/35
Photoselected with natural light,[a] absorption x,y-polarized	3/10	3/10	2/5	6/35	6/35	9/35	2/35	1/14	1/14
Random	1/3	1/3	1/3	1/5	1/5	1/5	1/15	1/15	1/15
x, y, z Axes at magic angle (54.7°) to Z	1/3	1/3	1/3	1/9	1/9	1/9	1/9	1/9	1/9
x, y, or z Perfectly aligned with Z, with equal probability	1/3	1/3	1/3	1/3	1/3	1/3	0	0	0
Photoselected with natural light[a], absorption x-polarized[b]	1/5	2/5	2/5	3/35	9/35	9/35	2/35	2/35	3/35
Disc-like	K_x	$(1 - K_x)/2$	$(1 - K_x)/2$	L_x	$3(1 - 2K_x + L_x)/8$	$3(1 - 2K_x + L_x)/8$	$(K_x - L_x)/2$	$(K_x - L_x)/2$	$(1 - 2K_x + L_x)/8$
Perfect alignment of yz plane, y and z equivalent	0	1/2	1/2	0	3/8	3/8	0	0	1/8

[a]Propagating along Z. [b]Same as for Z-polarized light, absorption y,z-polarized.

4.3 Visualization of the Orientation Factors

Although it is not necessary for spectroscopic applications, it is still interesting to ask for a visualization of the physical significance of the orientation factors.

As mentioned above, K_u is a measure of the degree of alignment of the u-th molecular axis with the unique sample direction Z and provides us with an effective average alignment angle u_{av} with respect to the Z axis. This is defined as that acute angle which would produce the observed K_u value if all molecules were oriented so that their u axes lay at u_{av} or $\pi - u_{av}$ from the Z axis (both positions are equally probable in nonpolar distributions):

$$u_{av} = \cos^{-1} K_u^{1/2} \tag{4.38}$$

In an isotropic solution, $K_x = K_y = K_z = 1/3$, and the values of x_{av}, y_{av}, and z_{av} are all equal to the "magic angle," 54.7°. If the molecular long axis z is perfectly aligned in all molecules, the orientation of the y and x axes must be distributed randomly in the laboratory XY plane, since the sample is uniaxial, and we have $K_z = 1$, $K_x = K_y = 0$, so that the effective average angles of deviation are $z_{av} = 0°$ $y_{av} = x_{av} = 90°$ (Figure 4.5). If the molecular yz planes are perfectly aligned with the Z axis, with y and z equivalent, the molecular x axis must be distributed randomly in the laboratory XY plane. Then, $K_z = K_y = 1/2$, $K_x = 0$, and the effective average angles become $x_{av} = 90°$ and $z_{av} = y_{av} = 45°$ (Figure 4.6).

Partially aligned samples will be characterized by K_u values and average angles u_{av} between the limits $1/3 \leqslant K_z \leqslant 1$, $0 \leqslant K_y \leqslant 1/2$, $0 \leqslant K_x \leqslant 1/3$, and $0° \leqslant z_{av} \leqslant 54.7°$, $45° \leqslant y_{av} \leqslant 90°$, $54.7° \leqslant x_{av} \leqslant 90°$.

In general, the K_u values contain no information about the spread of the actual values of the angles u about their average values. This information is contained in the values of the higher order orientation factors. For a given K_u value, L_u provides a measure of the magnitude of the spread of the actual degree of alignment of the u-th molecular axis about its average value u_{av}. If $L_u = K_u^2$,

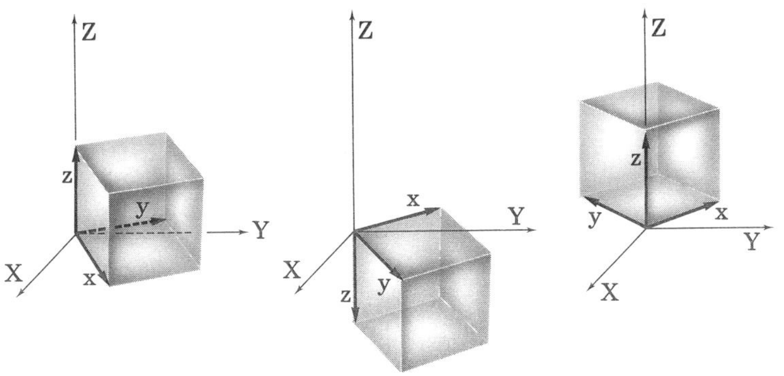

Figure 4.5 Three representative members of an orientation distribution with a perfect alignment of the z axis ($K_z = 1$, $K_x = K_y = 0$).

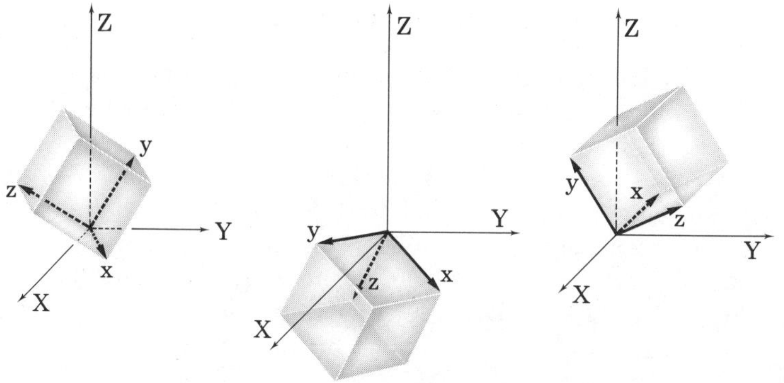

Figure 4.6 Three representative members of an orientation distribution with a perfect alignment of the yz plane with y and z equivalent ($K_z = K_y = 1/2$, $K_x = 0$).

there is no spread and the u axes of all molecules are inclined at the angle u_{av} or $\pi - u_{av}$ to the Z direction. If $L_u = K_u$, the spread is the largest possible for the given value of K_u. An example of two distributions with the same K_u value and different L_u values is shown in Figure 4.7. The values of the M's provide a further refinement in that they describe the mutual correlation of the three types of spread around the three effective average angles u_{av}, etc.

It is instructive to consider the possible ranges of values of the orientation factors and their interrelations, as well as to provide some idealized examples of orientation distributions and their orientation factors. It must be remembered that measurement of a finite number of orientation factors cannot provide a complete description of the orientation distribution function except in one purely hypothetical case mentioned later. On the other hand, if such a distri-

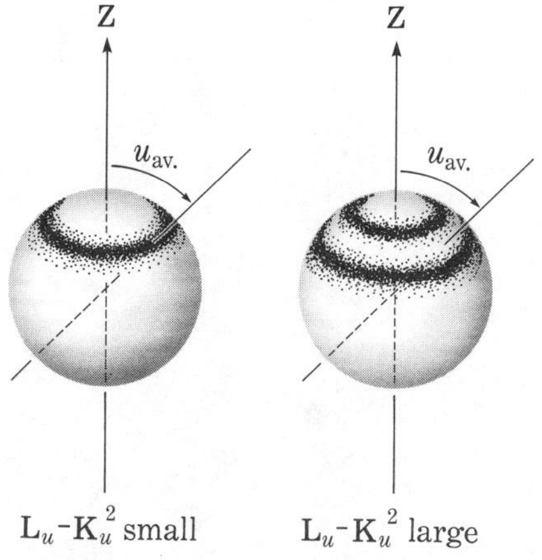

Figure 4.7 Two orientation distributions with the same value of K_u. The molecules are located at the center of the sphere; the points indicate the intersection of the u axis of each one with the surface of the sphere. Points on the southern hemisphere are not shown.

bution function is assumed, it is possible to calculate all desired orientation factors from it. In practice, this means that a finite number of measured values of K_u, L_u, etc., can always be fitted by an infinite number of different assumed distribution functions. Thus, it is in principle a hopeless task to try to derive the full orientation distribution function from optical experiments alone. On the other hand, such experiments can be used to show that a particular assumed form of orientation function is incorrect.

It is useful to view (K_x, K_y, K_z) as a point in three-dimensional space with coordinates K_x, K_y, and K_z along the x, y, and z axes, respectively. If we give up for the moment our requirement $K_z \geqslant K_y \geqslant K_x$, all possible points K lie on an equilateral triangle defined by points $Z^x \equiv (1,0,0)$, $Z^y \equiv (0,1,0)$, and $Z^z \equiv (0,0,1)$, since $\sum_u K_u = 1$ (Figure 4.8). These three points correspond to unique orientation distributions, one of which was alluded to above, namely, the perfect alignment of the molecular x, y, or z axis, respectively, with the sample unique axis Z, with the remaining two molecular axes aligned randomly in the plane perpendicular to Z. Every other point on the triangle corresponds to any of an infinite number of possible orientations. For instance:

(i) Points on the Z^u, Z^v side of the triangle correspond to those orientations in which $K_w = 0$, i.e., in which the w axis of every molecule is perpendicular to the Z axis of the sample.

(ii) For the completely random orientation, and for infinitely many others, $K_x = K_y = K_z = 1/3$. This point corresponds to the center of the triangle.

(iii) Points located on a line which connects the center of the triangle with the vertex Z^u, i.e., those for which $K_v = K_w$, correspond to all those distributions in which the molecular axes v and w are equivalent.

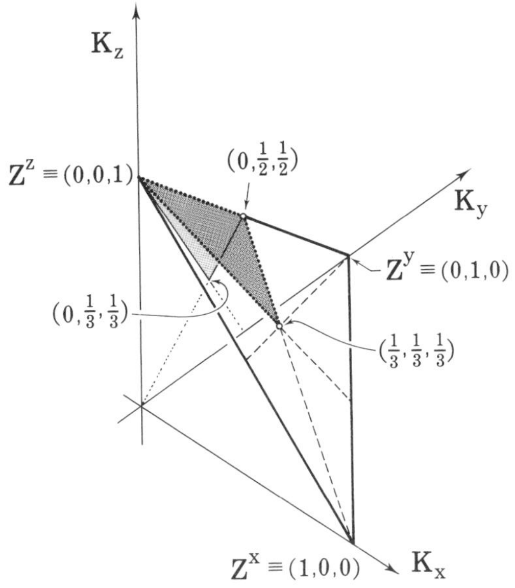

Figure 4.8 Points in the (K_x, K_y, K_z) space. See text.

If the molecular axes are labeled in the usual way so that $K_z \geqslant K_y \geqslant K_x$, all possible distributions will be imaged on points located on the dark triangle in Figure 4.8. A projection of this triangle onto the K_zK_y plane, more lightly shaded in Figure 4.8, will be referred to as the "orientation triangle" in which each orientation distribution is represented by a point with coordinates K_z and K_y. For points in the orientation triangle, $(1 - K_z)/2 \leqslant K_y \leqslant 1 - K_z$, $K_y \leqslant K_z$, and the vertical distance of the point from the line $K_z + K_y = 1$ is equal to the value of K_x.

Some particularly important distributions produce points which correspond to the three vertices of the orientation triangle (Figure 4.9). The totally random distribution is represented by the point $K_z = K_y = K_x = 1/3$. Perfect alignment of the yz planes of all molecules with the stretching direction, with total equivalence in the distribution of the directions of the y and z axes (cf. Figure 4.6), is represented by the point $K_z = K_y = 1/2$, $K_x = 0$. Perfect alignment of the z axes of all molecules with the stretching direction is represented by the point $K_z = 1$, $K_y = K_x = 0$ (cf. Figure 4.5). All points on the line $K_y = (1 - K_z)/2$, which limits the orientation triangle on the left in Figure 4.9, correspond to distributions in which the average alignment angles of the two short axes, x and y, are the same (e.g., rod-shaped molecules). The points on the line $K_z = K_y$, on the right in Figure 4.9, correspond to distributions in which the aver-

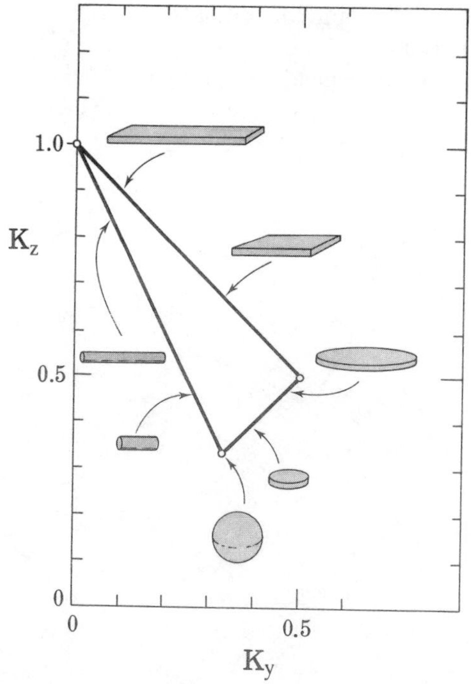

Figure 4.9 The orientation triangle with examples of solute shapes which suggest specific locations in the triangle.

age alignment angles of the two long axes, y and z, are identical (e.g., disc-shaped molecules). The points on the line $K_y = 1 - K_z$, uppermost in Figure 4.9, correspond to distributions in which all yz planes are parallel to the Z axis. The points inside the orientation triangle correspond to more general types of orientation distributions.

Any pair (K_z,K_y) can be represented as a linear combination of the pairs (1/3,1/3), (1/2,1/2), and (1,0) representing the three vertices:

$$(K_z,K_y) = 3K_x(1/3,1/3) + 2(K_y - K_x)(1/2,1/2) + (K_z - K_y)(1,0) \qquad (4.39)$$

Therefore, any general distribution can be thought of as a mixture of three extreme types of distributions represented by the vertices, for instance, those mentioned above, with weights given by $3K_x$, $2(K_y - K_x)$, and $(K_z - K_y)$. Although such representations have been considered helpful at times, their arbitrary nature is obvious, and the real orientation distribution will not in general be identical to any one described by the mixture of three vertex distributions.

The location of points corresponding to the actual orientation factors of numerous molecules in stretched polyethylene is discussed in Section 4.6.1. Such plots reveal the existence of a close relation between the location of a point in the orientation triangle and the molecular shape and can be used to interpolate the values of orientation factors for molecules which have not yet been studied. It is probable that similar relations also exist in other orienting media. Plots such as these provide an intuitive feeling for the nature of the orientation distribution. For instance, molecules whose points lie near the line of $K_y = K_x$ orient approximately like rods; those with points near the line $K_z = K_y$, approximately like discs; etc.

It is now instructive to go back to the larger triangle, $Z^xZ^yZ^z$ (Figure 4.8), and consider a similar representation of the fourth-order orientation parameters L. Given a point (K_x,K_y,K_z), all possible values of L_x, L_y, and L_z which might be associated with it in any real distribution must satisfy the relations $K_u^2 \leqslant L_u \leqslant K_u$. Now, $(\sqrt{L_x},\sqrt{L_y},\sqrt{L_z})$ can also be plotted as a point in the three-dimensional space shown in Figure 4.8. The three unique distributions which correspond to the points Z^x, Z^y, and Z^z when (K_x,K_y,K_z) is plotted also correspond to Z^x, Z^y, and Z^z when $(\sqrt{L_x},\sqrt{L_y},\sqrt{L_z})$ is plotted. In all other cases, the three coordinates of the point $(\sqrt{L_x},\sqrt{L_y},\sqrt{L_z})$ can in principle be chosen independently, and all possible points fill a three-dimensional object whose shape is dictated by the above inequalities: It is that part of the first octant which is limited by the triangle $Z^xZ^yZ^z$ on one side and by the unit sphere, which also passes through the points Z^x, Z^y, and Z^z, on the other (Figure 4.10). Once the point (K_x,K_y,K_z) is specified, the range of positions in which the point $(\sqrt{L_x},\sqrt{L_y},\sqrt{L_z})$ for any orientation distribution with the given K's can be located is much more severely restricted: It must be in a particular parallelepiped with edges parallel to the axes (Figure 4.11). The parallelepiped is defined by its body diagonal, which connects the point (K_x,K_y,K_z) on the triangle $Z^xZ^yZ^z$ with a point on the unit sphere whose position is given by $(\sqrt{K_x},\sqrt{K_y},\sqrt{K_z})$. If (K_x,K_y,K_z) lies on one of the sides of the $Z^xZ^yZ^z$ triangle, say the Z^uZ^v side,

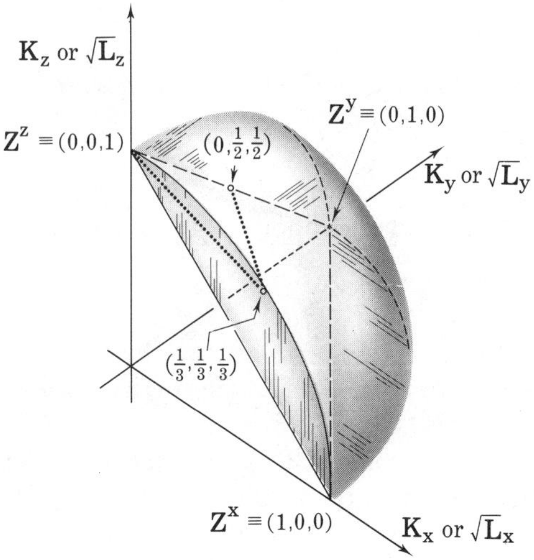

Figure 4.10 Points in the (K_x,K_y,K_z) and $(\sqrt{L_x},\sqrt{L_y},\sqrt{L_z})$ space. See text.

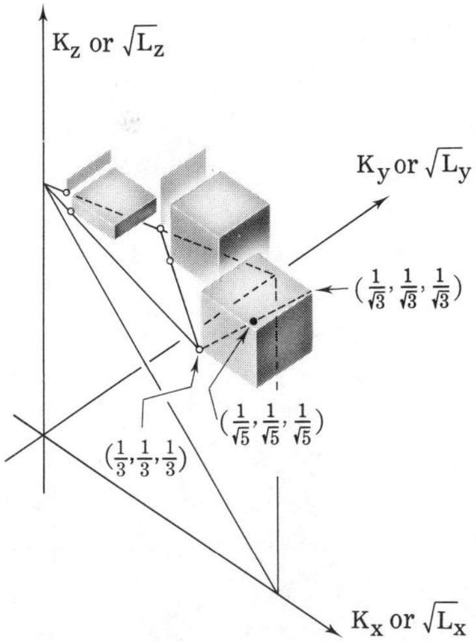

Figure 4.11 Regions in the $(\sqrt{L_x},\sqrt{L_y},\sqrt{L_z})$ space allowed for selected choices of (K_x,K_y,K_z), shown as white dots. See text.

the parallelepiped degenerates to a rectangle located in the K_uK_v plane. If (K_x,K_y,K_z) coincides with one of the vertices of the triangle, say Z^u, the parallelepiped shrinks into this point. This is in accordance with the unique nature of the three "vertex" distributions Z^u. For these three, the specification of (K_x,K_y,K_z) fully determines all higher order orientation factors, leaving no freedom for the choice of the L's.

There are two classes of particular orientation distributions which are fully characterized by five quantities, i.e., by a mere specification of (K_x, K_y, K_z) and $(\sqrt{L_x}, \sqrt{L_y}, \sqrt{L_z})$. These cases correspond to the points on the triangle and to those on the unit sphere. All other choices leave open an infinite number of possibilities for the selection of higher order orientation factors and thus correspond to an infinite number of possible orientation distributions.

First, points $(\sqrt{L_x}, \sqrt{L_y}, \sqrt{L_z})$ which lie on the triangle $Z^xZ^yZ^z$, i.e. for which $L_u = K_u^2$, are somewhat similar to the points at the vertices of the triangle in that they each describe a distribution in which any one of a small number (one, two, or four) of symmetry-related special directions in the molecule is equally likely to be perfectly aligned with the sample axis Z for all molecules in the assembly, while rotation around each of these special directions is completely random. The direction cosines $\cos \phi_u$ of these special directions in the molecule, taken with respect to the molecular framework x, y, z, are given by $\cos^2 \phi_u = L_u = K_u^2$. For instance, for the point at the center of the triangle, each of the possible four special directions forms the magic angle (54.7°) with each of the three molecular axes ($L_x = L_y = L_z = 1/9$). The multiplicity of the special directions is due to the fact that the specification of absolute values of the direction cosines describes more than one possible direction in the molecular framework. All of these enter with equal weights in the nonpolar uniaxial distributions considered here. "Sharp" distributions of this kind contain no scatter of the actual values of the squares of direction cosines of the molecular axes $\cos^2 u$ about their average values K_u. For them, the sum of the diagonal factors $L_x + L_y + L_z$ is minimized and the sum of the off-diagonal factors $L_{xy} + L_{xz} + L_{yz}$ is maximized for a given set of K_u's.

As the point $(\sqrt{L_x}, \sqrt{L_y}, \sqrt{L_z})$ moves away from the triangle $Z^xZ^yZ^z$ toward the surface of the unit sphere in Figures 4.10 and 4.11, keeping (K_x, K_y, K_z) constant, the scatter in the actual values of $\cos^2 u$ about their average value K_u increases until it reaches a maximum for points on the surface of the sphere, where $L_u = K_u$, so that $L_x + L_y + L_z = 1$, and $L_{xy} = L_{xz} = L_{yz} = 0$. Each point on the sphere corresponds uniquely to a mixture of the three distributions described by the corners of the triangle $Z^xZ^yZ^z$. To reach a distribution described by the point $(\sqrt{L_x}, \sqrt{L_y}, \sqrt{L_z})$ on the sphere, the distribution Z^x must enter with weight L_x, the distribution Z^y with weight L_y, and Z^z with weight L_z. It should be noted that among the distributions with $K_x = K_y = K_z = 1/3$, maximum scatter of the $\cos^2 u$ values about their average value K_u is not obtained with the random distribution ($L_x = L_y = L_z = 1/5$, $L_{xy} = L_{xz} = L_{yz} = 1/15$) but rather with the highly organized distribution containing equal weights of the three "vertex" distributions Z^x, Z^y and Z^z ($L_x = L_y = L_z = 1/3$, $L_{xy} = L_{xz} = L_{yz} = 0$).

If two of the molecular axes, say u and v, are equivalent by symmetry, the choices of permissible positions of $(\sqrt{L_x}, \sqrt{L_y}, \sqrt{L_z})$ for a given choice of K's are limited. Instead of being restricted to a parallelepiped, they are restricted to a rectangle obtained by cutting the parallelepiped diagonally by the plane u = v. If all three axes are equivalent, three such cutting planes exist and intersect in a line joining the center of the triangle with the center of the sphere octant. Then, $K_x = K_y = K_z = 1/3$ and $1/9 \leqslant L_x = L_y = L_z \leqslant 1/3$.

If we now reintroduce the axis-labeling convention $K_x \leqslant K_y \leqslant K_z$, the choice of points (K_x,K_y,K_z) is limited to the dark triangle in Figure 4.8, and the choice of the corresponding parallelepipeds in the plot of $(\sqrt{L_x},\sqrt{L_y},\sqrt{L_z})$ is limited accordingly. Note that the order of the magnitude of the L values is not necessarily $L_x \leqslant L_y \leqslant L_z$.

A type of orientation distribution which is sometimes assumed in the literature is a superposition of a perfect alignment of the z axis for a fraction f of the molecules, represented by Z^z, and of random distribution for the remaining fraction $1 - f$ (the Fraser-Beer model, see Section 5.2.1). All distributions of this kind yield $(\sqrt{L_x},\sqrt{L_y},\sqrt{L_z})$ points located on a segment of an ellipse which passes from the point Z^z ($f = 1$) to the random distribution point $(1/\sqrt{5},1/\sqrt{5},1/\sqrt{5})$ ($f = 0$). The ellipse can be obtained by constructing the circle in which the units sphere cuts the K_xK_z plane and projecting it into the $K_x = K_y$ plane (Figure 4.12). Of course, an infinite number of other distributions will also yield $(\sqrt{L_x},\sqrt{L_y},\sqrt{L_z})$ points on the same ellipse so that an experimental observation of a set of values $K_x = K_y = (1 - K_z)/2$, $L_x = L_y = (1 - L_z)/4$, which correspond to the ellipse, will not guarantee that the orientation distribution is of this type. So far, no real samples have been shown to produce such a set of L values. Not even molecules in which x and y axes are equivalent (rodlike molecules) need to produce such distributions, since the equivalence of the axes only limits the points to the $\sqrt{L_x} = \sqrt{L_y}$ plane and not exclusively to the ellipse described above. Even a more recent proposal that the actual orientation distributions of aromatic solutes in stretched polyethylene consist of a mixture of

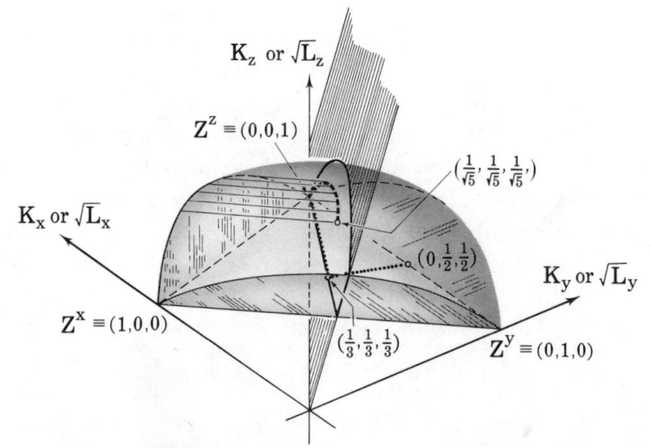

Figure 4.12 Points in the $(\sqrt{L_x},\sqrt{L_y},\sqrt{L_z})$ space. Orientation distributions composed of a perfectly aligned fraction and a random fraction are mapped on points on the thick dashed ellipse segment. The random distribution is mapped on the dot shown. A bottom view was used for clarity.

a randomly oriented fraction and a fraction oriented with the x axis perpendicular to the unique sample axis (points in the $\sqrt{L_x} = 0$ plane) is unnecessarily, and most likely unduly, restrictive.

The visualization of similar relations involving higher order orientation factors M, N, etc., is more difficult, since spaces of dimension 4, 5, etc., have to be considered. Given a point (K_x, K_y, K_z) and a point $(\sqrt{L_x}, \sqrt{L_y}, \sqrt{L_z})$ corresponding to some orientation distribution, all possible values of M_x, M_y, M_z, and M' $[M' = (M_{xxy} + M_{xyy} + M_{xxz} + M_{xzz} + M_{yyz} + M_{yzz})/6]$ which might be associated with this distribution must satisfy the relations $K_u L_u \leqslant M_u \leqslant L_u$. Now $(\sqrt[3]{M_x}, \sqrt[3]{M_y}, \sqrt[3]{M_z}, \sqrt[3]{M'})$ can be viewed as a point in four-dimensional space, etc.

4.4 The Saupe Orientation Matrices

The orientation factors K_{uv} defined by equation (4.21) lead to simple expressions for observables, as will be shown in Chapters 5, 6, and 7 and as is exemplified by equation (4.29). In the absence of any alignment they acquire the values $(1/3)\delta_{uv}$. It is possible to transform them simply into a set of equivalent quantities which have the nice property of being equal to zero in the absence of orientation. These are the elements of matrices defined by Saupe, which for the case of uniaxial orientation acquire the form

$$
\begin{aligned}
S_{uv} &= (1/2)\langle 3 \cos u \cos v - \delta_{uv}\rangle = (1/2)(3K_{uv} - \delta_{uv}) \\
S_{stuv} &= (1/8)[35\langle \cos s \cos t \cos u \cos v\rangle \\
&\quad - 5(\langle \cos s \cos t\rangle \delta_{uv} + \langle \cos s \cos u\rangle \delta_{tv} + \langle \cos s \cos v\rangle \delta_{tu} \\
&\quad + \langle \cos t \cos u\rangle \delta_{sv} + \langle \cos t \cos v\rangle \delta_{su} + \langle \cos u \cos v\rangle \delta_{st}) \\
&\quad + (\delta_{st}\delta_{uv} + \delta_{su}\delta_{tv} + \delta_{sv}\delta_{tu})]
\end{aligned} \tag{4.40}
$$

This set of quantities has been particularly popular with investigators working with nematic liquid crystals.

Using Saupe's matrix elements S_{uu} instead of the orientation factors K_u, the orientation triangle described in Section 4.3 acquires the form shown in Figure 4.13. The point which corresponds to isotropic distribution lies at the origin; the point which corresponds to a perfect alignment of the molecular z axis with the stretching direction corresponds to $S_{zz} = 1$, $S_{yy} = -1/2$; and the point which corresponds to a perfect alignment of the molecular plane with the stretching direction, with the molecular z and y axes being equivalent, corresponds to $S_{zz} = 1/4$, $S_{yy} = 1/4$. Rodlike distributions are represented by points on the line connecting the origin with the point $S_{zz} = 1$, $S_{yy} = -1/2$ and satisfy the relation

$$S_{xx} = S_{yy} = -(1/2)S_z \tag{4.41}$$

Disclike distributions are mapped on the line connecting the origin with the point $S_{zz} = 1/4$, $S_{yy} = 1/4$ and satisfy the relation

$$S_{xx} = -2S_{yy} = -2S_{zz} \tag{4.42}$$

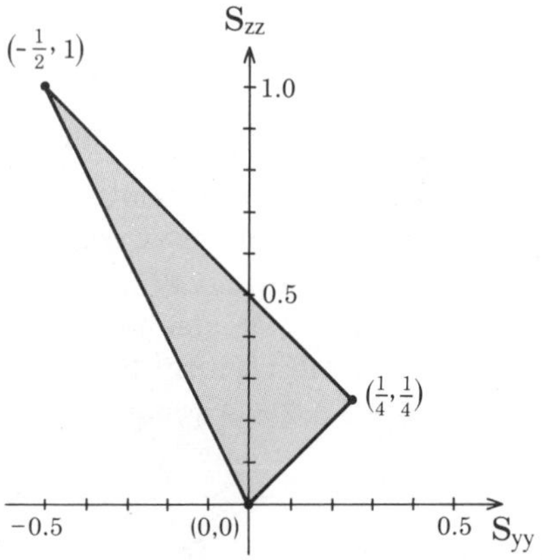

Figure 4.13 The orientation triangle in terms of the Saupe parameters.

Saupe's matrix elements are particularly suitable for the description of superstructures. Suppose that there is a set of axes Z′ aligned uniaxially relative to the Z axis. Let their degree of alignment be characterized by $S_{Z'Z'}$. Suppose that each of the Z′ axes represents a local direction for uniaxial alignment of axes z characterized by S'_{zz}. Then the overall uniaxial alignment of the axes z relative to Z, characterized by S_{zz}, is given by

$$S_{zz} = S'_{zz} S_{Z'Z'} \tag{4.43}$$

4.5 Order Parameters

The Wigner matrices $D^{(j)}_{mm'}(\alpha,\beta,\gamma)$ provide a convenient orthonormal complete basis set for the expansion of the distribution function $f(\Omega) = f(\alpha,\beta,\gamma)$:

$$f(\Omega) = \sum_{j=0}^{\infty} \sum_{m=-j}^{j} \sum_{m'=-j}^{j} \frac{2j+1}{8\pi^2} \langle D^{(j)*}_{m'm} \rangle D^{(j)}_{m'm}(\alpha,\beta,\gamma) \tag{4.44}$$

A specification of all of the infinite number of expansion coefficients $\langle D^{(j)*}_{m'm} \rangle$ would provide a complete specification of the orientation distribution function, just as a specification of the weights of the individual Fourier components specifies a periodic function. The coefficients $\langle D^{(j)*}_{m'm} \rangle$ provide a set of orientation order parameters.

In a uniaxial sample with a unique axis Z, all coefficients $\langle D^{(j)*}_{m'm} \rangle$ vanish except for those with $m' = 0$. Moreover, in a nonpolar sample, in which both directions of the Z axis are equivalent, all $\langle D^{(j)*}_{m'm} \rangle$'s vanish unless j is even. When it is further considered that $f(\Omega)$ is real, one can show that there are only $2j + 1$ real independent order parameters for each j. It is useful to take these to be the following combinations of the $\langle D^{(j)*}_{0m} \rangle$'s:

If m is odd:

$$A^{(j)}_{m} = (\langle D^{(j)*}_{0m} \rangle - \langle D^{(j)*}_{0-m} \rangle)/2 \tag{4.45a}$$

$$B_m^{(j)} = (\langle D_{0m}^{(j)*}\rangle + \langle D_{0-m}^{(j)*}\rangle)/2i \tag{4.45b}$$

If m is even:

$$A_m^{(j)} = (\langle D_{0m}^{(j)*}\rangle + \langle D_{0-m}^{(j)*}\rangle)/2 \tag{4.45c}$$

$$B_m^{(j)} = (\langle D_{0m}^{(j)*}\rangle - \langle D_{0-m}^{(j)*}\rangle)/2i \tag{4.45d}$$

It can be shown that $A_0^{(0)} = 1$ and that the other order parameters A and B acquire values between 0 and 1.

The explicit form of the Wigner matrices with $m' = 0$ is

$$D_{0m}^{(j)}(\alpha,\beta,\gamma) = \sqrt{\frac{4\pi}{2j+1}}\,(-1)^m Y_{jm}^*(\beta,\gamma) \tag{4.46}$$

where $Y_{jm}^*(\beta,\gamma)$ are the spherical harmonic functions. The explicit forms of the order parameters can be given by their relation to the previously described orientation factors:

$$
\begin{aligned}
A_0^{(0)} &= 1\\
A_0^{(2)} &= (1/2)(3K_z - 1)\\
A_1^{(2)} &= -\sqrt{3/2}K_{xz}\\
B_1^{(2)} &= \sqrt{3/2}K_{yz}\\
A_2^{(2)} &= (1/2)\sqrt{3/2}(K_x - K_y)\\
B_2^{(2)} &= -\sqrt{3/2}K_{xy}\\
A_0^{(4)} &= (1/8)(35L_z - 30K_z + 3)\\
A_1^{(4)} &= -(1/4)\sqrt{5}(7L_{xzzz} - 3K_{xz})\\
B_1^{(4)} &= (1/4)\sqrt{5}(7L_{yzzz} - 3K_{yz})\\
A_2^{(4)} &= (1/4)\sqrt{5/2}[7(L_y - L_x) - 6(K_y - K_x)]\\
B_2^{(4)} &= -(1/2)\sqrt{5/2}(7L_{xyzz} - K_{xy})\\
A_3^{(4)} &= (1/4)\sqrt{35}(3L_{xyyz} - L_{xxxz})\\
B_3^{(4)} &= (1/4)\sqrt{35}(3L_{xxyz} - L_{yyyz})\\
A_4^{(4)} &= (1/8)\sqrt{35/2}(L_x + L_y - 6L_{xy})\\
&= (1/8)\sqrt{35/2}[4(L_x + L_y) + 3(1 - L_z) - 6(K_x + K_y)]\\
B_4^{(4)} &= (1/2)\sqrt{35/2}(L_{xyyy} - L_{xxxy})
\end{aligned} \tag{4.47}
$$

It can be noted that $A_0^{(2)} = S_{zz}$. When expressed in terms of the order parameters $A_0^{(2)}$ and $A_2^{(2)}$, the orientation triangle acquires the form shown in Figure 4.14.

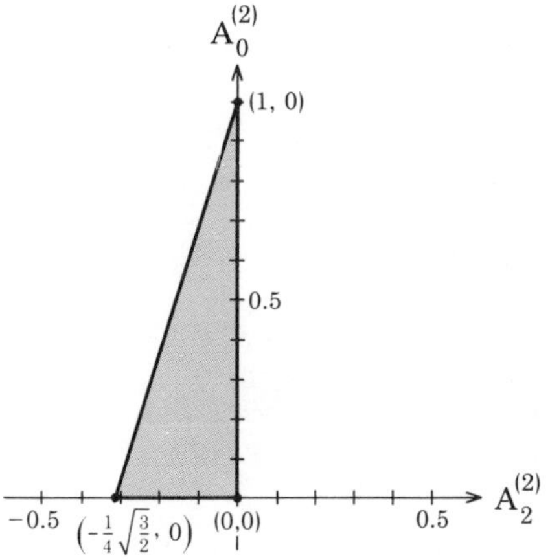

Figure 4.14 The orientation triangle in terms of the Wigner order parameters.

4.6 Alignment and Molecular Properties

In practical applications, it is important to have the ability to estimate the kind and degree of alignment molecules of a given solute will exhibit in response to a given orienting force. In this section we first discuss the relation between molecular properties, such as shape, dipole moment, and polarizability, and the orientation factors to be expected in anisotropic solvents or in an electric field.

The two situations are rather different. The forces acting on the solute molecules to produce orientation in a stretched polymer or a liquid crystal are exceedingly complex, and the relation of the orientation factors to molecular structure can at present be described only in a purely empirical way. On the other hand, the interaction energy between the electric field and a solute molecule can be written explicitly in terms of the molecular dipole moment and polarizability, so that explicit equations for the solute orientation distribution function can be written, albeit only in a way quite complicated by the effects of the solvent.

Subsequently, we discuss the production of aligned molecular assemblies by photoselection. This is not based on the use of any orienting forces, but rather on the use of the anisotropy of a beam of polarized or unpolarized light which effects a selection of molecules in a random assembly according to their orientation. In this instance, it is possible to derive relatively simple explicit equations for the orientation distribution function.

4.6.1 Anisotropic Solvents

Solute molecules contained in a uniaxially anisotropic solvent experience an orientation-dependent interaction potential due to the surrounding solvent molecules. This potential contains contributions from a wide array of intermolecular forces: London dispersion forces, van der Waals repulsion forces, solute dipole–solvent dipole forces, solute dipole–solvent-induced dipole forces, etc. Since this interaction potential is so complicated, it can at present be expressed only purely empirically as the effective time-averaged orientation-dependent interaction energy $W(\beta,\gamma)$. Formally, it is then possible to write an expression for the orientation distribution function $f(\beta,\gamma)$, since the α distribution is uniform:

$$f(\beta,\gamma) = \exp\left[\frac{-W(\beta,\gamma)}{kT}\right]\left\{\int_{\beta=0}^{\pi}\int_{\gamma=0}^{2\pi}\exp\left[\frac{-W(\beta,\gamma)}{kT}\right]\frac{1}{4\pi}\sin\beta\,d\beta\,d\gamma\right\}^{-1} \quad (4.48)$$

where k is the Boltzmann constant and T the absolute temperature.

For the orientation factors, we obtain

$$K_{uv} = \langle\cos u \cos v\rangle = \int_{\beta=0}^{\pi}\int_{\gamma=0}^{2\pi}\cos u \cos v\, f(\beta,\gamma)\frac{1}{4\pi}\sin\beta\,d\beta\,d\gamma \quad (4.49)$$

$$L_{stuv} = \langle\cos s \cos t \cos u \cos v\rangle$$
$$= \int_{\beta=0}^{\pi}\int_{\gamma=0}^{2\pi}\cos s \cos t \cos u \cos v\, f(\beta,\gamma)\frac{1}{4\pi}\sin\beta\,d\beta\,d\gamma \quad (4.50)$$

etc., where the direction cosines are given in terms of the angles β and γ by (4.8)–(4.10).

In order to predict the orientation factors or to interpret those observed, an explicit form of $f(\beta,\gamma)$ or $W(\beta,\gamma)$ is needed. If enough were known about the nature of the solute-solvent interaction, a physical model might be set up to derive such a form under guidance of the available experimental data. This appears more feasible for liquid crystals than for stretched polymers as solvents, since for the latter there is even uncertainty as to which region of the polymer the solute enters (the amorphous part, the amorphous-crystalline interface, or perhaps even the crystalline part).

One example of a functional form of $f(\beta,\gamma)$ which has found some popularity is based on the notion that the stretching of a polymer such as poly(vinyl-alcohol) or poly(acrylamide) distorts a spherical portion of the polymer into the shape of an ellipsoid of rotation, keeping the volume constant. The solute molecules are thought of as rods originally randomly oriented in the sphere, and it is assumed that their orientations follow the distortion of the sphere into an ellipsoid by following the position of the point on the sphere at which they were pointing originally. The assumed cylindrical symmetry of the molecules makes the distribution of the angle γ uniform. This type of orientation distribution would be expected for macroscopic rod-shaped objects such as nails imbedded in the polymer, which can then be viewed as a continuum. It may be valid for very large cylindrical macromolecules such as certain proteins. Its utility for smaller molecules is quite questionable.

This distribution is a function of only one variable, the degree of stretching R_s, and is independent of the thickness and length of the rod representing the molecule:

$$f(\beta,\gamma) = f(\beta) = R_s^2[1 + (R_s^2 - 1)\sin^2\beta]^{-3/2} \quad (4.51)$$

The degree of stretching is defined as the ratio of the long to the short axis of an ellipse into which a circle drawn on the surface of the polymer sheet distorts upon stretching. The function $f(\beta)$ is normalized according to (4.5). For the orientation factors it yields

$$K_z = [R_s^2/(R_s^2 - 1)]\{1 - [\pi/2 - \tan^{-1}(R_s^2 - 1)^{-1/2}](R_s^2 - 1)^{-1/2}\} \qquad (4.52)$$

$$K_x = K_y = (1 - K_z)/2 \qquad (4.53)$$

where z is the long axis of the rod.

An example of an explicit form for $W(\beta,\gamma)$ is that proposed some time ago for the case of liquid crystals, based on the use of dispersion forces alone:

$$W(\beta,\gamma) = Q(\alpha_{xx}\cos^2 x + \alpha_{yy}\cos^2 y + \alpha_{zz}\cos^2 z) \qquad (4.54)$$

where Q is an empirical constant and α_{uv} is the uv-th element of the molecular polarizability tensor. It is an open question as to how well it describes any real system.

An alternative is to expand $W(\beta,\gamma)$ in spherical harmonics and to fit as many parameters in the expansion as possible to experimental data, without attaching any great physical significance to them. Such fits have been found useful in investigations of the dependence of the order parameters on variables such as solute concentration or the temperature of the liquid crystal.

A more common approach is to give up the notion that an explicit form for $W(\beta,\gamma)$ can be found and merely use the general form of terms which are likely to appear in it to find inspiration concerning the type of physical quantities with which the orientation factors would be likely to correlate. Thus, one has been led to try $\Delta\alpha_u = 2\alpha_{uu} - \alpha_{vv} - \alpha_{ww}$ as an expression reflecting the action of the dispersion forces in a liquid crystal, and $\Delta l_u = 2l_u - l_v - l_w$, where l_t is the molecular dimension in direction t, has been proposed as a shape function which would presumably reflect the repulsive forces in the liquid crystal. Similar correlations have been attempted for stretched polymers. For many aromatic hydrocarbons, $\Delta\alpha_u$ and Δl_u are correlated, so that even a purely qualitative decision about the relative importance of the two kinds of forces cannot be reached from the experimental data. For more complicated molecules, $\Delta\alpha_u$ and Δl_u are less likely to correlate perfectly, but for only very few of them is $\Delta\alpha_u$ known with any certainty. The situation was explored in the case of several aromatic hydrocarbons. Their orientation factors in a compensated cholesteric liquid crystal correlated equally well with $\Delta\alpha_u$ and Δl_u. Those of some other aromatic molecules, for which $\Delta\alpha_u$ was not available, did not correlate too well with Δl_u. The conclusion that dispersion forces rather than repulsion forces dominate the solute-orienting behavior of liquid crystals would be rather premature. We consider the question open: If the $\Delta\alpha_u$ values for the other molecules had been available, their correlation with the experimental orientation factors might easily have been worse than the correlation of Δl_u.

Given the unsatisfactory present state of the theory of molecular alignment in anisotropic solvents, it is perhaps most instructive to show plots of K_z versus K_y for a variety of solutes in various orienting media if one is to convey a feeling for what to expect for new molecules. While the question of the orientation mechanisms remains unsettled, we suspect that among the various factors considered, molecular shape plays an important role. This belief is compatible with the data in Figures 4.15–4.17, which show the known orientation factors for three cases which have been studied extensively. Qualitatively, the data indicate that an increase in molecular dimension u leads to an increase in K_u relative to the other two orientation factors. However, molecular shape is not likely to be the sole determining factor, as indicated by the comparison of data for anthracene (**17**), acridine (**25**), and phenazine (**37**) in Figure 4.15.

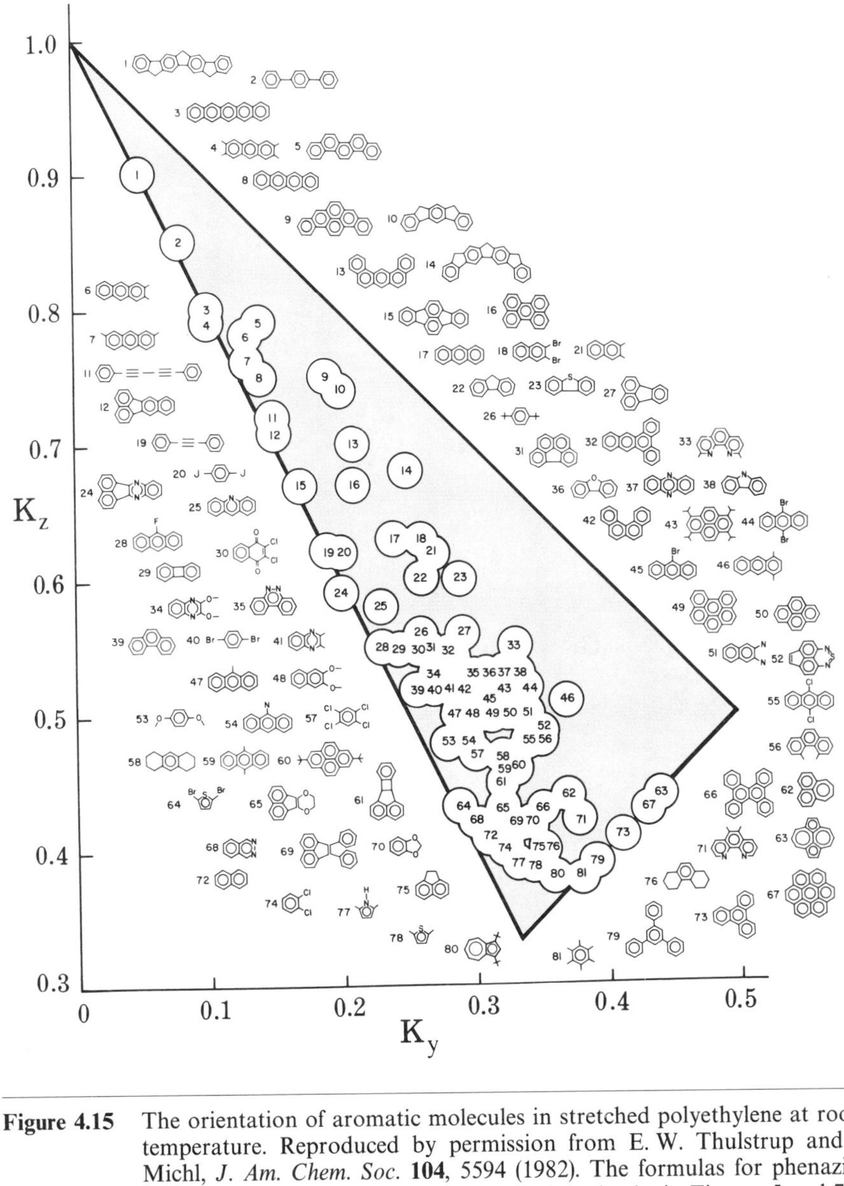

Figure 4.15 The orientation of aromatic molecules in stretched polyethylene at room temperature. Reproduced by permission from E. W. Thulstrup and J. Michl, *J. Am. Chem. Soc.* **104**, 5594 (1982). The formulas for phenazine (**37**) and acridine (**25**) were interchanged by mistake in Figures 5 and 7 of the original paper.

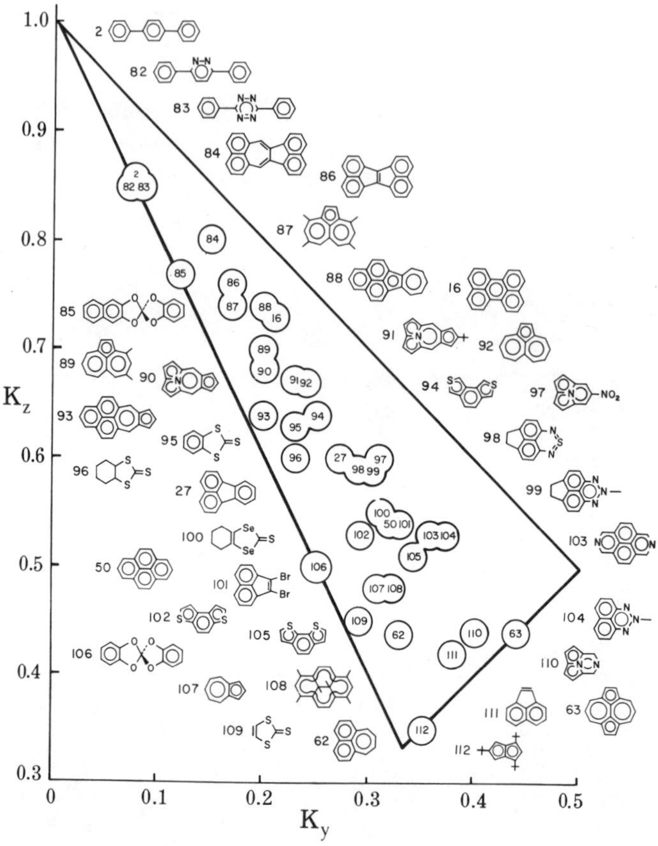

Figure 4.16 The orientation of aromatic molecules in stretched polyethylene at low temperature (77 K or below). Reproduced by permission from E. W. Thulstrup and J. Michl, *J. Am. Chem. Soc.* **104**, 5594 (1982).

4.6.2 Electric Field

When partial alignment of solute molecules is achieved by the application of an electric field, their orientation distribution function is predictable from first principles provided that enough is known about the molecules and that the actual electric field which is felt by a solute molecule can be derived from the knowledge of the outside applied field and the properties of the solvent and the solute. Assuming for the moment that the latter has been accomplished and that the interaction energy W of the solute molecule with the outside applied field **E** can be described by means of its effective dipole moment $\boldsymbol{\mu}_{\text{eff}}$ and its effective polarizability tensor α_{eff}, we have

$$W = -\boldsymbol{\mu}_{\text{eff}} \cdot \mathbf{E} - \tfrac{1}{2}\mathbf{E}\alpha_{\text{eff}}\mathbf{E} \qquad (4.55)$$

if hyperpolarizabilities are neglected.

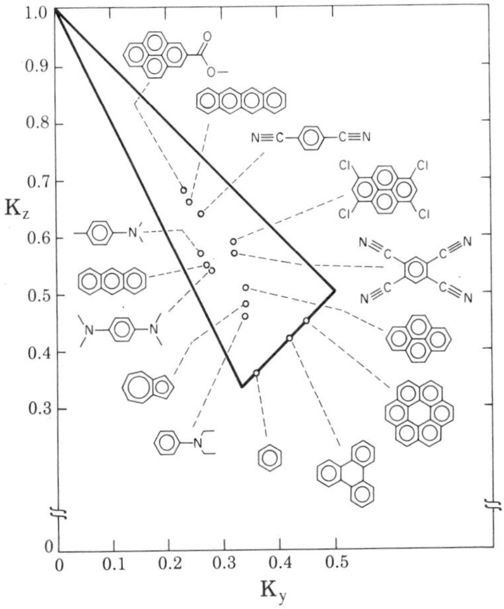

Figure 4.17 The orientation of aromatic molecules in a compensated nematic mixture of cholesteryl chloride and cholesteryl laurate (1.85:1 w/w, $T_{nem} = 30°C$). The data have been taken from E. Sackmann, P. Krebs, H. U. Rega, J. Voss, and H. Möhwald, *Mol. Cryst. Liq. Cryst.* **24**, 283 (1973).

At thermal equilibrium, the orientation distribution function $f(\Omega)$ is uniaxial and can be written as $f(\beta,\gamma)$, since the direction of **E** coincides with the Z axis ($|\mathbf{E}| = E_Z = E$). It is given by (4.48). The values of the orientation factors are given by (4.49) and (4.50). It is seen that the relation of the orientation factors K_{uv} to the effective molecular properties is complex but explicit.

Frequently, simplifying assumptions can be introduced. We shall illustrate this on an example of a molecule with effective polarizabilities $(\alpha_{eff})_{zz}$ along its axis z and $(\alpha_{eff})_{xx} = (\alpha_{eff})_{yy}$ along each of its other two axes, and with an effective permanent dipole moment $\boldsymbol{\mu}_{eff}$ directed along z, $|\boldsymbol{\mu}_{eff}| = (\mu_{eff})_z = \mu_{eff}$. We shall consider only the pure permanent-dipole orientation limit $(\alpha_{eff})_{zz} = (\alpha_{eff})_{yy} = (\alpha_{eff})_{xx}$ and the pure induced-dipole orientation limit $\boldsymbol{\mu}_{eff} = 0$.

For pure permanent-dipole orientation, the evaluation of the integrals yields

$$K_x(p) = K_y(p) = [\coth p - (1/p)]/p \tag{4.56}$$

$$K_z(p) = 1 - 2[\coth p - (1/p)]/p \tag{4.57}$$

where

$$p = \mu_{eff}E/kT \tag{4.58}$$

For pure induced-dipole orientation, we obtain

$$K_x(q) = K_y(q) = \frac{1}{2} - \frac{1}{4}\left[e^q\left(q^{1/2}\int_0^{q^{1/2}} e^{x^2}\,dx\right)^{-1} - \frac{1}{q}\right] \tag{4.59}$$

$$K_z(q) = \frac{1}{2}\left[e^q\left(q^{1/2}\int_0^{q^{1/2}} e^{x^2}\,dx\right)^{-1} - \frac{1}{q}\right] \tag{4.60}$$

where

$$q = [(\alpha_{eff})_{zz} - (\alpha_{eff})_{yy}]E^2/2kT \tag{4.61}$$

The relation of the effective quantities $\boldsymbol{\mu}_{eff}$ and α_{eff} to the intrinsic molecular dipole moment $\boldsymbol{\mu}$ and polarizability α can be shown to be

$$\boldsymbol{\mu}_{eff} = \mathsf{G}_1(1 - \mathsf{G}_2\alpha)^{-1}[\boldsymbol{\mu} + (1 - \mathsf{G}_3\alpha)\alpha\mathbf{E}_\Delta] \tag{4.62}$$

$$\alpha_{eff} = \mathsf{G}_1^2(1 - \mathsf{G}_2\alpha)^{-1}\alpha \tag{4.63}$$

where $\mathbf{E}_\Delta$ is the fluctuation of the local electric field, the tensor G_1 describes the modification of the applied electric field by the solvent, the tensor G_2 describes the reaction field caused by the polarization of the solvent by the total dipole moment of the solute, and the tensor G_3 is similar to G_2 except that it represents only the effects due to the electronic polarization of the medium. These quantities can be related to observable properties of the solvent, such as its dielectric constant and refractive index, using suitable approximate models.

4.6.3 *Photoselection: Linearly Polarized Light*

The selection of a subset of molecules from a more or less random assembly of rigidly held molecules by the process of interaction with the electromagnetic field of a collimated beam of light is known as photoselection. Ordinarily, it is the interaction of the electric field vector of the light with the electric dipole transition moment of the stationary solute molecules in the sample which is responsible for the selection process. In this context, "stationary" means nonrotating on the time scale of the irradiation. If a beam of light propagating along X is linearly polarized along Z, the electric vector is restricted to the Z direction and the photoselection is most effective. This is the case which shall be considered in some detail in this section. In Section 4.6.4, we shall list briefly some results for the less effective process of photoselection with unpolarized or circularly polarized light propagating along Z.

The expressions for the orientation distribution function of a photoselected molecular ensemble can be written explicitly. The probability of excitation of a molecule into the f-th excited state is proportional to the square of the projection of the molecular transition moment $\mathbf{M}(0f)$ into the direction ε_Z of the electric vector of the light [cf. (1.15), (1.47), and (1.60)]:

$$W_{el\,dip} \propto |\varepsilon_Z \cdot \mathbf{M}(0f)|^2 \tag{4.64}$$

In a random sample, those molecules which happen to be oriented with their transition moments $\mathbf{M}(0f)$ nearly parallel to the Z direction will be more likely to absorb the linearly polarized light than those whose transition moments lie nearly perpendicular to Z. After a period of irradiation, the set of molecules which were excited and the complementary set of molecules which were not excited will both represent molecular assemblies with nonrandom orientation. The orientation of both will be uniaxial, with Z as the unique axis.

If it is possible to perform optical observations on either set of molecules, those which were excited or those which were left unexcited, the objective of studying an aligned set of molecules has been achieved without actually using an aligning force. This is usually accomplished by observing the polarized emission of the excited molecules or the absorption of either set of molecules. This calls for the interaction of altogether two photons with the molecule and corresponds to an overall process of rank 4, discussed in Chapter 7. More complicated arrangements are also possible. For instance, the photoluminescence of an initial photoproduct can be excited and observed, corresponding to the use of three photons altogether. The overall process is then of rank 6, etc.

Negligible Depletion. The simplest expressions result for the orientation distribution of the set of molecules which have absorbed a photon if their number is negligible relative to the total number of molecules. In practice, this corresponds to a photochemical reaction carried out to a very small degree of conversion or to the population of a metastable excited state to a small stationary degree, i.e., under conditions far from saturation. In both instances, the photoselection occurs on an assembly which is so close to randomly oriented that the orientation distribution function, which will be of the type $f(\beta,\gamma)$ as used in (4.4), will take a form directly obtained from the interaction expressed in (4.19) and (4.64). Still, in the general case the existence of overlapping transitions and of the resulting mixed polarization must be considered.

We define $p_{\tilde{v}_0,Z}(\beta,\gamma)$ to be a measure of absorption probability as a function of molecular orientation (β,γ) for a Z-polarized photon at wavenumber $\tilde{v}_0$:

$$p_{\tilde{v}_0,Z}(\beta,\gamma) = \kappa(\tilde{v}) \sum_j g_j'(\tilde{v}_0)[M_x(0j)\cos x + M_y(0j)\cos y + M_z(0j)\cos z]^2$$

$$= \kappa(\tilde{v})\varepsilon_Z \cdot \left[\sum_j g_j'(\tilde{v}_0)\mathbf{M}(0j)\mathbf{M}(0j)\right] \cdot \varepsilon_Z \qquad (4.65)$$

Here $\kappa(\tilde{v})$ is a proportionality factor whose detailed nature has been discussed in Chapter 1 (see also Section 5.1), and $g_j'(\tilde{v})$ is the line-shape function of the j-th transition, whose value at the excitation wavenumber $\tilde{v}_0$ determines to what degree the j-th transisiton moment $\mathbf{M}(0j)$ will contribute to the photoselection process. The subscript Z indicates that the photoselecting light is linearly polarized along the Z laboratory axis. The sum runs over all transitions j, but typically at most only a few contribute to a detectable degree for any one choice of $\tilde{v}_0$. The direction cosines are related to the Euler angles through (4.8)–(4.10), and the equation applies to an arbitrary choice of the molecular axes system.

A measure of absorption probability averaged over a random assembly, for a photon of any polarization at $\tilde{v}_0$, is similarly provided by

$$p_{\tilde{v}_0}^{iso} = \kappa(\tilde{v}_0) \sum_j g_j'(\tilde{v}_0)|\mathbf{M}(0j)|^2/3 \qquad (4.66)$$

The orientation distribution function is then given by

$$f_{\tilde{v}_0,Z}(\beta,\gamma) = p_{\tilde{v}_0,Z}(\beta,\gamma)/p_{\tilde{v}_0}^{iso} \qquad (4.67)$$

since all orientations are initially equally probable. The presence of the denominator in (4.67) guarantees that $f_{\tilde{\nu}_0,Z}(\beta,\gamma)$ is properly normalized [cf. equation (4.4)].

Substitution into (4.2) produces the values of orientation factors. This can be illustrated by the calculation of K_y for the simple case in which only the j-th transition contributes to the absorption at $\tilde{\nu}_0$. Its line shape is then immaterial, and we obtain after integration over α

$$\begin{aligned} K_y(j,Z) &= \int_{\beta,\gamma} f_{j,Z}(\beta,\gamma)\cos^2 y\,(1/4\pi)\sin\beta\,d\beta\,d\gamma \\ &= \frac{3}{4\pi}\int_{\gamma=0}^{2\pi} d\gamma \int_{\beta=0}^{\pi} \sin\beta\,d\beta \\ &\quad \times \frac{[M_x(0j)(-\sin\beta\cos\gamma) + M_y(0j)\sin\beta\sin\gamma + M_z(0j)\cos\beta]^2 \sin^2\beta\sin^2\gamma}{|\mathbf{M}(0j)|^2} \end{aligned} \tag{4.68}$$

so that the result, in the arbitrary axes system chosen, is

$$K_y(j,Z) = [\tfrac{1}{5}M_x^2(0j) + \tfrac{3}{5}M_y^2(0j) + \tfrac{1}{5}M_z^2(0j)]/|\mathbf{M}(0j)|^2 \tag{4.69}$$

The general result for the orientation factors of molecules in the photoselected subset, assuming that its population is negligibly small, is obtained similarly. Referred to an arbitrary system of molecular axes, it is

$$K_{uv}(\tilde{\nu}_0,Z) = \frac{1}{5}\left[\delta_{uv} + (1+\delta_{uv})\frac{\sum_j g_j'(\tilde{\nu}_0)M_u(0j)M_v(0j)}{\sum_j g_j'(\tilde{\nu}_0)|\mathbf{M}(0j)|^2}\right] \tag{4.70}$$

Matters simplify if the choice of molecular axes is such as to diagonalize the orientation tensor $\mathsf{K}(\tilde{\nu}_0,Z)$, whose elements are defined by (4.70). According to this equation, this will happen when the total absorption tensor at $\tilde{\nu}_0$, given by the weighted sum of tensors contributed by the individual transitions $0 \to j$, $\kappa(\tilde{\nu})\sum_j g_j'(\tilde{\nu}_0)\mathbf{M}(0j)\mathbf{M}(0j)$, is diagonal. For this choice of molecular axes x, y, z, the expression for $p_{\tilde{\nu}_0,Z}(\beta,\gamma)$ simplifies:

$$\begin{aligned} p_{\tilde{\nu}_0,Z}(\beta,\gamma) &= \varepsilon_Z \mathsf{M}'(\tilde{\nu}_0)\varepsilon_Z \\ &= M'_{xx}(\tilde{\nu}_0)\cos^2 x + M'_{yy}(\tilde{\nu}_0)\cos^2 y + M'_{zz}(\tilde{\nu}_0)\cos^2 z \\ &= M'_{zz}\cos^2\beta + \tfrac{1}{2}[(M'_{xx} + M'_{yy}) + (M'_{xx} - M'_{yy})\cos 2\gamma]\sin^2\beta \end{aligned} \tag{4.71}$$

where M'_{xx}, M'_{yy}, and M'_{zz} are the diagonal elements of the diagonal tensor $\mathsf{M}'(\tilde{\nu}_0)$ and relations (4.8)–(4.10) were used.

In molecules of low symmetry, no a priori information is available for the directions of the principal molecular axes, and finding them may be very difficult; besides, note that they are a function of the choice of $\tilde{\nu}_0$. In molecules of high symmetry, the x, y, and z directions are determined by symmetry except for a possible permutation of labels, and this makes matters much simpler. In the following, we shall always use the principal axes system, in which $M'(\tilde{\nu}_0)$ and $K(\tilde{\nu}_0,Z)$ are diagonal, unless otherwise specified.

It is useful to introduce the concept of u-polarized fraction of absorption at $\tilde{\nu}_0$, $r_u(\tilde{\nu}_0)$. It is defined with reference to the principal orientation axes u = x, y, z as that fraction of the absorption intensity which is polarized along u:

$$r_u(\tilde{\nu}_0) = \frac{M'_{uu}(\tilde{\nu}_0)}{M'_{xx}(\tilde{\nu}_0) + M'_{yy}(\tilde{\nu}_0) + M'_{zz}(\tilde{\nu}_0)} = \frac{A_u(\tilde{\nu}_0)}{A_x(\tilde{\nu}_0) + A_y(\tilde{\nu}_0) + A_z(\tilde{\nu}_0)}$$

$$= \frac{A_u(\tilde{\nu}_0)}{3E^{iso}(\tilde{\nu}_0)} \qquad (4.72)$$

where $A_u(\tilde{\nu}_0)$ is the u-polarized part of the total absorbance at $\tilde{\nu}_0$ and $E^{iso}(\tilde{\nu}_0)$ is the absorbance of an isotropic sample at $\tilde{\nu}_0$ (see Chapter 5).

The polarized absorption fractions $r_u(\tilde{\nu}_0)$ are particularly useful when dealing with highly symmetrical molecules, where the axes u = x, y, z are known from symmetry.

In the principal axes system, expression (4.70) for the elements of the orientation tensor $K(\tilde{\nu}_0,Z)$ simplifies to

$$K_u(\tilde{\nu}_0,Z) = 1/5 + (2/15)A_u(\tilde{\nu}_0)/E^{iso}(\tilde{\nu}_0) \qquad (4.73)$$

when written in terms of the purely polarized absorbances $A_w(\tilde{\nu}_0)$. In terms of the polarized absorption fractions $r_w(\tilde{\nu}_0)$, the result is

$$K_u(\tilde{\nu}_0,Z) = [1 + 2r_u(\tilde{\nu}_0)]/5 \qquad (4.74)$$

The arguments $\tilde{\nu}_0$ and Z on the orientation factor serve to emphasize the fact that it is a function of $\tilde{\nu}_0$ and that the result is valid for photoselecting light linearly polarized along the Z axis.

As already pointed out, in low-symmetry molecules, the location of the principal axes will usually be difficult to establish; moreover, it changes continuously with the choice of $\tilde{\nu}_0$. Thus, the situation is complicated unless only one transition moment **M**(0j) contributes at $\tilde{\nu}_0$. However, in high-symmetry molecules, (4.74) is an extremely useful relation. From (4.73) and (4.74) it is seen that the orientation axis z will lie in the direction w for which $A_w(\tilde{\nu}_0)$ and $r_w(\tilde{\nu}_0)$ are the largest. If the other two fractions are equal, $r_u(\tilde{\nu}_0) = r_v(\tilde{\nu}_0)$, the distribution will be of the rodlike type, $K_y(\tilde{\nu}_0,Z) = K_x(\tilde{\nu}_0,Z)$. This will obviously always be true if the absorption at $\tilde{\nu}_0$ is purely polarized along w, even in a low-symmetry molecule.

We shall now consider the consequences of the general results (4.73) and (4.74), which are valid in the principal system of axes for molecules of any symmetry.

(i) If only one transition moment, **M**(0j), contributes at $\tilde{\nu}_0$ in a molecule of any symmetry, the orientation axis z lies in the direction of **M**(0j), and the orientation factors $K_u(z,Z)$ acquire particularly simple values:

$$K_z(z,Z) = 3/5 \qquad K_y(z,Z) = K_x(z,Z) = 1/5 \qquad (4.75)$$

The orientation tensor K(z,Z) is diagonal for any choice of the x and y axes, and the distribution is rodlike around the z axis.

(ii) If there are only two active mutually perpendicular transition moments contributing equally at $\tilde{\nu}_0$ (e.g., in the case of degenerate transitions in molecules of very high symmetry), the orientation factors $K_u(yz,Z)$ are again simple:

$$K_z(yz,Z) = K_y(yz,Z) = 2/5 \qquad K_x(yz,Z) = 1/5 \tag{4.76}$$

The orientation tensor K(yz,Z) is diagonal for any choice of the y and z axes in the yz plane, and the distribution is disclike around the x axis.

(iii) In the more general case of only two mutually perpendicular transition moments contributing unequally to the absorption, $r_z(\tilde{\nu}_0) > r_y(\tilde{\nu}_0)$, $r_x(\tilde{\nu}_0) = 0$, result (4.74) shows that the point representing the orientation factors $K(\tilde{\nu}_0,Z)$ in the orientation triangle (Figure 4.9) will lie on the line segment of constant $K_x = 0.2$. This line runs parallel to the upper edge of the triangle and connects the points K(z,Z) = (0.6,0.2) and K(yz,Z) = (0.4,0.4) defined for the two special cases [(4.75), (4.76)]. The location of the general point on this line segment is dictated by the relative magnitudes of the polarized fractions $r_z(\tilde{\nu}_0)$ and $r_y(\tilde{\nu}_0)$ which produce it: Its distance from the endpoint K(z,Z) is $(2\sqrt{2}/5)r_y(\tilde{\nu}_0)$; its distance from the endpoint K(yz,Z) is $\sqrt{2}[2r_z(\tilde{\nu}_0) - 1]/5$.

(iv) The situation which obtains in the completely general case is depicted in Figure 4.18, which shows the mapping of the set of points with coordinates $r_z(\tilde{\nu}_0)$, $r_y(\tilde{\nu}_0)$, $r_x(\tilde{\nu}_0)$ which characterize the polarization of the absorption at $\tilde{\nu}_0$ onto the set of points in the orientation triangle $K_z(\tilde{\nu}_0,Z)$, $K_y(\tilde{\nu}_0,Z)$, $K_x(\tilde{\nu}_0,Z)$ which represent the orientation factors of the resulting photoselected assembly in the principal system of axes, provided that the initial random assembly of ground states is only negligibly depleted. Since $\sum_u r_u(\tilde{\nu}_0) = \sum_u K_u(\tilde{\nu}_0,Z) = 1$, both the $r_u(\tilde{\nu}_0)$'s and the K_u's contain only two degrees of freedom. Moreover, the axes are labeled so that $K_z \geqslant K_y \geqslant K_x$ and $r_z(\tilde{\nu}_0) \geqslant r_y(\tilde{\nu}_0) \geqslant r_x(\tilde{\nu}_0)$, so that the mapping can be viewed as transforming the triangle of points $[r_z(\tilde{\nu}_0), r_y(\tilde{\nu}_0)]$, shown on the left in Figure 4.18, linearly into the 2.5-times smaller triangle of

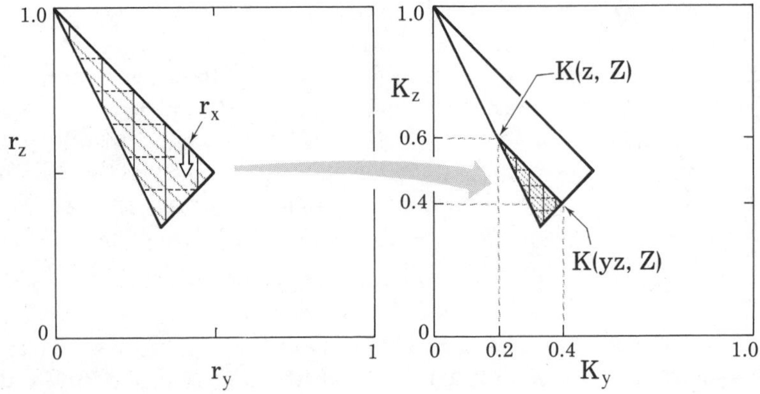

Figure 4.18 Mapping of points describing polarized absorption fractions $r_u(\tilde{\nu}_0)$ onto points describing the orientation factors $K_u(\tilde{\nu}_0,Z)$ of the initially photoselected assembly (linearly polarized light, electric vector along Z). See text.

points $[K_z(\tilde{\nu}_0,Z), K_y(\tilde{\nu}_0,Z)]$ shown on the right. The straight lines of constant $r_x(\tilde{\nu}_0)$, $r_y(\tilde{\nu}_0)$, or $r_z(\tilde{\nu}_0)$ in the triangle on the left transform into the corresponding straight lines of constant $K_x(\tilde{\nu}_0,Z)$, $K_y(\tilde{\nu}_0,Z)$, or $K_z(\tilde{\nu}_0,Z)$ in the triangle on the right. Figure 4.18 permits an instant visualization of the orientation factors $K_u(\tilde{\nu}_0,Z)$ to be expected for any choice of the $r_u(\tilde{\nu}_0)$'s and shows that the degree of orientation achieved for the transformed molecules in a single photoselection step is actually rather small even under the best of circumstances, since none of the points in the triangle on the right are very far from the origin, (1/3,1/3).

The elements of the orientation tensors of higher rank can also be written explicitly. For instance, a set of calculations similar to (4.66)–(4.69) shows that the orientation factors $L_{uv}(\tilde{\nu}_0,Z)$ which are sufficient for the discussion of any fourth-rank process in a photoselected set of symmetrical molecules, i.e., of a sixth-rank process on a random sample, are given in the principal axes system by

$$L_{uv}(\tilde{\nu}_0,Z) = [(1 + 2\delta_{uv})/105] \sum_w (1 + 2\delta_{uw} + 2\delta_{vw})A_w(\tilde{\nu}_0)/E^{iso}(\tilde{\nu}_0)$$

$$= [(1 + 2\delta_{uv})/105]\{3 + 2[A_u(\tilde{\nu}_0) + A_v(\tilde{\nu}_0)]/E^{iso}(\tilde{\nu}_0)\} \quad (4.77)$$

and in terms of the polarized absorption fractions by

$$L_{uv}(\tilde{\nu}_0,Z) = [(1 + 2\delta_{uv})/35][1 + 2r_u(\tilde{\nu}_0) + 2r_v(\tilde{\nu}_0)] \quad (4.78)$$

In the two particularly simple but important cases mentioned above—(i) a rodlike distribution resulting from a single significant absorbing transition moment **M**(0j) at $\tilde{\nu}_0$ (z-polarized) and (ii) a disclike distribution resulting from a single pair of mutually orthogonal equally absorbing moments (y- and z-polarized)—the orientation factors L(z,Z) and L(yz,Z) for the photoselected assembly acquire the values

$$L_z(z,Z) = 3/7$$

$$L_x(z,Z) = L_y(z,Z) = L_{xz}(z,Z) = L_{yz}(z,Z) = 3/35$$

$$L_{xy}(z,Z) = 1/35$$

$$L_z(yz,Z) = L_y(yz,Z) = 9/35$$

$$L_x(yz,Z) = L_{yz}(yz,Z) = 3/35$$

$$L_{xz}(yz,Z) = L_{xy}(yz,Z) = 2/35 \quad (4.79)$$

Substantial Depletion. Destructive Photoorientation. If the process of photochemical transformation of an initially random assembly of rigidly held molecules is continued until a significant fraction of the molecules have been converted to product, or if a stationary population of a metastable state represents a significant fraction of the total number of molecules, expressions (4.67)–(4.79) are no longer valid. The photoselection process is still based on the interaction embodied in (4.19) and (4.64), and the interaction takes the form expressed in (4.65), but after a finite radiation dose has been applied the photoselection no longer occurs from a random assembly, so (4.67) is invalid.

We shall now describe the results for this situation under the assumption that the sample is optically thin, i.e., that it absorbs only a negligible fraction of the incident light. In such a case, the resulting orientation distribution functions of both the starting and the transformed species are independent of position within the sample. We shall assume that the transformed species is not photochemically converted back into the starting material.

(i) The Remaining Molecules: The Orientation Distribution Function. We first define the irradiation dose $D(\tau)$ at time τ;

$$D(\tau) = \int_0^\tau I(t)\,dt \tag{4.80}$$

where $I(t)$ is the incident radiation intensity at time t. In the following, we shall usually drop the argument τ. Let $N(D)$ be the total number of molecules in the sample which remain untransformed after the sample has been exposed to the dose D. At that time, the orientation distribution function of these untransformed molecules will be $f'_{\tilde{v}_0,Z}(\beta,\gamma,D)$. An incremental radiation dose dD will cause the transformation of additional molecules to product or to a metastable state. This will occur at different rates for molecules at different orientations. For those with orientation within a volume element located at (β,γ), the conversion probability density will be proportional to their number, $N(D)f'_{\tilde{v}_0,Z}(\beta,\gamma,D)$, to the quantum yield of the transformation process, ϕ, and to the absorption probability $p_{\tilde{v}_0,Z}(\beta,\gamma)\,dD$, where D has been expressed in appropriate units (cf. Chapter 5) and $p_{\tilde{v}_0,Z}(\beta,\gamma)$, defined in (4.65), reflects the orientation dependence of the light-molecule interaction. In principle, ϕ is a function of ν_0, but we do not show this explicitly. We now simplify matters by again adopting the principal axes system, in which the absorption tensor $\mathsf{M}'(\tilde{v}_0)$ is diagonal and $p_{\tilde{v}_0,Z}(\beta,\gamma)$ acquires the form (4.71).

Thus, the change in the number of molecules oriented within a volume element at (β,γ) can be written as

$$-d[N(D)f'_{\tilde{v}_0,Z}(\beta,\gamma,D)] = N(D)f'_{\tilde{v}_0,Z}(\beta,\gamma,D)\phi p_{\tilde{v}_0,Z}(\beta,\gamma)\,dD \tag{4.81}$$

Integration from 0 to D yields

$$N(D)f'_{\tilde{v}_0,Z}(\beta,\gamma,D) = N(0)\exp[-\phi D p_{\tilde{v}_0,Z}(\beta,\gamma)] \tag{4.82}$$

where the random nature of the initial distribution has been recognized by setting $f'_{\tilde{v}_0,Z}(\beta,\gamma,0) = 1$.

Integration over β and γ yields an expression for the fraction of molecules remaining after dose D:

$$\frac{N(D)}{N(0)} = \frac{1}{4\pi}\int_0^\pi \sin\beta\,d\beta \int_0^{2\pi} d\gamma\,\exp[-\phi D p_{\tilde{v}_0,Z}(\beta,\gamma)] \tag{4.83}$$

Evaluation of the integral yields

$$\frac{N(D)}{N(0)} = e^{-a}\sum_{n=0}^{\infty}\frac{(2n)!}{(n!)^2}b^{2n}\sum_{k=0}^{2n}\frac{(-1)^k}{(2n-k)!k!}F_k(c) \tag{4.84}$$

where

$$a = [r_x(\tilde{\nu}_0) + r_y(\tilde{\nu}_0)]\eta \qquad (4.85)$$

$$b = [r_x(\tilde{\nu}_0) - r_y(\tilde{\nu}_0)]\eta/2 \qquad (4.86)$$

$$c = [2r_z(\tilde{\nu}_0) - r_x(\tilde{\nu}_0) - r_y(\tilde{\nu}_0)]\eta \qquad (4.87)$$

In these expressions, $r_u(\tilde{\nu}_0)$ is the already familiar u-polarized fraction of absorption and η is a parameter which measures the progress of the photochemical transformation, during which $\tilde{\nu}_0$ is assumed to remain constant. Like D, it starts at zero and grows to infinity:

$$\eta = \phi D[M'_{xx}(\tilde{\nu}_0) + M'_{yy}(\tilde{\nu}_0) + M'_{zz}(\tilde{\nu}_0)]/2 \qquad (4.88)$$

The function $F_0(c)$ is the error integral

$$F_0(c) = \int_0^1 \exp(-ct^2)\,dt = \frac{1}{2}\sqrt{\frac{\pi}{c}}\,\mathrm{erf}\,\sqrt{c} \qquad (4.89)$$

and the function $F_k(c)$ is its k-th derivative

$$F_k(c) = \frac{\partial^k F_0(c)}{\partial c^k} = \int_0^1 t^{2k} \exp(-ct^2)\,dt \qquad (4.90)$$

The values of these functions can be found in tables or evaluated numerically.

In the limit $[r_x(\nu_0) - r_y(\nu_0)] \to 0$, i.e., equal absorption properties in the x and y directions, we have $b \to 0$. Formula (4.84) remains valid if we set $b^0 \to 1$, and $b^n \to 0$ if $n \neq 0$, and simplifies to

$$\frac{N(D)}{N(0)} = e^{-a}F_0(c) \qquad \text{if} \quad r_x(\tilde{\nu}_0) = r_y(\tilde{\nu}_0) \qquad (4.91)$$

From (4.82), the orientation distribution function of the remaining molecules after dose D is now

$$f'_{\tilde{\nu}_0,Z}(\beta,\gamma,D) = \frac{\exp[-\phi D p_{\tilde{\nu}_0,Z}(\beta,\gamma)]}{N(D)/N(0)} \qquad (4.92)$$

with N(D)/N(0) given by (4.84).

(ii) The Remaining Molecules: Orientation Factors. The orientation distribution function $f'_{\tilde{\nu}_0,Z}(\beta,\gamma,D)$ given in (4.92) can now be used to obtain the orientation factors for the assembly of remaining molecules. The integration is more involved here than in the simple analogous case (4.68). It yields

$$K'_Z(\tilde{\nu}_0,Z,D) = \frac{\sum\limits_{n=0}^{\infty} \frac{(2n)!}{(n!)^2} b^{2n} \sum\limits_{k=0}^{2n} \frac{(-1)^k}{(2n-k)!k!} F_{k+1}(c)}{\sum\limits_{n=0}^{\infty} \frac{(2n)!}{(n!)^2} b^{2n} \sum\limits_{k=0}^{2n} \frac{(-1)^k}{(2n-k)!k!} F_k(c)} \qquad (4.93)$$

$$K'_y(\tilde{\nu}_0,Z,D) = \sum_{n=0}^{\infty} \frac{(2n+1)!}{(n!)^2} b^{2n} \times \frac{\dfrac{b}{(n+1)(2n+1)!} F_{2n+2}(c) + \displaystyle\sum_{k=0}^{2n+1} \frac{(-1)^k(2n+2-k+2b)}{(2n+2-k)!k!} F_k(c)}{2\displaystyle\sum_{n=0}^{\infty} \frac{(2n)!}{(n!)^2} b^{2n} \sum_{k=0}^{2n} \frac{(-1)^k}{(2n-k)!k!} F_k(c)} \tag{4.94}$$

$$L'_z(\tilde{\nu}_0,Z,D) = \frac{\displaystyle\sum_{n=0}^{\infty} \frac{(2n)!}{(n!)^2} b^{2n} \sum_{k=0}^{2n} \frac{(-1)^k}{(2n-k)!k!} F_{k+2}(c)}{\displaystyle\sum_{n=0}^{\infty} \frac{(2n)!}{(n!)^2} b^{2n} \sum_{k=0}^{2n} \frac{(-1)^k}{(2n-k)!k!} F_k(c)} \tag{4.95}$$

$$L'_y(\tilde{\nu}_0,Z,D) = \sum_{n=0}^{\infty} \frac{(2n+1)!}{(n!)^2} b^{2n} \times \frac{\dfrac{-2b}{(n+1)(2n+1)!} F_{2n+3}(c) + \displaystyle\sum_{k=0}^{2n+2} (-1)^k \frac{(4n+3)(2n+3-k)+4b(2n+3)}{(2n+3-k)!k!} F_k(c)}{4\displaystyle\sum_{n=0}^{\infty} \frac{(2n)!}{(n!)^2} b^{2n} \sum_{k=0}^{2n} \frac{(-1)^k}{(2n-k)!k!} F_k(c)} \tag{4.96}$$

The expressions for $K'_x(\tilde{\nu}_0,Z,D)$ and $L'_x(\tilde{\nu}_0,Z,D)$ are obtained from those for $K'_y(\tilde{\nu}_0,Z,D)$ and $L'_y(\tilde{\nu}_0,Z,D)$, respectively, by the substitution $b \to -b$. They look a little friendlier when expressed in terms of the order parameters $A_m^{(j)}$ (Section 4.5), but not much.

As (4.86) and (4.87) show, if the polarized absorption fractions $r_u(\tilde{\nu}_0)$ of the starting molecules are known, the only unknown parameter in the variables b and c is the reaction progress parameter η. This can be determined from the experimentally accessible fraction of remaining molecules, N(D)/N(0), and the presumably known $r_u(\tilde{\nu}_0)$ values, using (4.84)–(4.87). It is in principle also accessible from the knowledge of the diagonal elements of the absorption tensor, the quantum yield, and the irradiation dose according to (4.88).

Equal absorption properties in x and y directions. Expressions (4.93)–(4.96) simplify considerably if the x- and y-polarized absorption fractions are equal, $r_x(\tilde{\nu}_0) = r_y(\tilde{\nu}_0) \neq 1/3$. Then $b = 0$, and the only nonvanishing term in all the infinite sums corresponds to $n = 0$. For this, $b^n = 1$ must be used. One then obtains for the orientation factors of the remaining molecules as a function of the reaction progress parameter η defined in (4.88),

$$K'_z(\tilde{\nu}_0,Z,D) = F_1\{[3r_z(\tilde{\nu}_0) - 1]\eta\}/F_0\{[3r_z(\tilde{\nu}_0) - 1]\eta\} \tag{4.97}$$

$$K'_x(\tilde{\nu}_0,Z,D) = K'_y(\tilde{\nu}_0,Z,D) = [1 - K_z(\tilde{\nu}_0,Z,D)]/2 \tag{4.98}$$

$$L'_z(\tilde{\nu}_0,Z,D) = F_2\{[3r_z(\tilde{\nu}_0) - 1]\eta\}/F_0\{[3r_z(\tilde{\nu}_0) - 1]\eta\} \quad (4.99)$$

$$L'_x(\tilde{\nu}_0,Z,D) = L'_y(\tilde{\nu}_0,Z,D) = 3[1 - 2K_z(\tilde{\nu}_0,Z,D) + L_z(\tilde{\nu}_0,Z,D)]/8 \quad (4.100)$$

which implies that the orientation distribution of the remaining molecules is disclike $[r_z(\tilde{\nu}_0) > 1/3]$ or rodlike $[r_z(\tilde{\nu}_0) < 1/3]$, with the z direction serving as the symmetry axis. This is physically reasonable, since nothing distinguishes the x and y directions in the photoselection process. If $r_z(\tilde{\nu}_0) = 1/3$, there is no photoselection, the orientation remains random, and $K'_u = 1/3$, $L'_u = 1/5$ at all times.

Taking the limit $\eta \to \infty$ in expressions (4.97) and (4.99), we obtain for the last fraction of remaining molecules $K'_z \to 0$ if $r_z(\tilde{\nu}_0) > 1/3$ and $K'_z \to 1$ if $r_z(\tilde{\nu}_0) < 1/3$.

Thus, for $r_z(\tilde{\nu}_0) > 1/3$, a limiting disclike distribution is approached:

$$K'_x(\tilde{\nu}_0,Z,\infty) = K'_y(\tilde{\nu}_0,Z,\infty) = 1/2, \qquad K'_z(\tilde{\nu}_0,Z,\infty) = 0$$

$$L'_x(\tilde{\nu}_0,Z,\infty) = L'_y(\tilde{\nu}_0,Z,\infty) = 3/8, \qquad L'_z(\tilde{\nu}_0,Z,\infty) = 0$$

$$L'_{xy}(\tilde{\nu}_0,Z,\infty) = 1/8, \qquad L'_{xz}(\tilde{\nu}_0,Z,\infty) = L'_{yz}(\tilde{\nu}_0,Z,\infty) = 0$$

For $r_z(\tilde{\nu}_0) < 1/3$, a limiting rodlike distribution is approached: $K'_z(\tilde{\nu}_0,Z,\infty) = 1$, $L'_z(\tilde{\nu}_0,Z,\infty) = 1$, and all other $K'(\tilde{\nu}_0,Z,\infty)$ and $L'(\tilde{\nu}_0,Z,\infty)$ values equal zero.

The gradual change of the orientation factor of the remaining molecules $K'_z(\tilde{\nu}_0,Z,D)$ from its initial value of 1/3 to the end value, 0 if $r_z(\tilde{\nu}_0) > 1/3$ or 1 if $r_z(\tilde{\nu}_0) < 1/3$, is depicted in Figure 4.19, which shows the plot of $F_1(t)/F_0(t)$ against t and against log t or log $(-t)$. According to (4.97), $K'_z(\tilde{\nu}_0,Z,D)$ is equal to the ratio $F_1(t)/F_0(t)$, where $t = [3r_z(\tilde{\nu}_0) - 1]\eta$. We shall refer to t as the conversion degree parameter. During the photochemical transformation, t changes from 0 to $+\infty$ if $r_z(\tilde{\nu}_0) > 1/3$ (disclike case) and it changes from 0 to $-\infty$ if $r_z(\tilde{\nu}_0) < 1/3$ (rodlike case).

Note that for t larger than about 3, the ratio $F_1(t)/F_0(t)$ is well approximated by 1/2t. For instance, for the ideal case of a purely z-polarized absorption, $r_z(\tilde{\nu}_0) = 1$, $t = 2\eta$, and $N(D)/N(0) = F_0(t)$, according to (4.85), (4.87), and (4.91). Then $t = 3$ corresponds to an about half-converted sample $[F_0(3) = 0.504]$ and thus to $K'_z(\tilde{\nu}_0,Z,D) \cong 1/6$. For this and any higher degree of conversion N(D)/N(0), erf $\sqrt{c} \cong 1$ so that $F_0(c)$ can be approximated by $(1/2)\sqrt{\pi/c}$ [cf. (4.89)], and $K'_z(\tilde{\nu}_0,Z,D)$ is then equal to $(2/\pi)[N(D)/N(0)]^2$ to acceptable accuracy. This function goes to zero rapidly with progressing photoconversion. In our idealized case, when 10 percent of the sample is still left ($t \cong 80$), the orientation factor $K'_z(\tilde{\nu}_0,Z,D)$ of the remaining molecules already is only 0.0064. Clearly, a very high degree of orientation can be achieved for the remaining molecules toward the end of the photochemical transformation, and this will be summarized later (Figure 4.21).

Since we normally wish to preserve the convention $K_z \geqslant K_y \geqslant K_x$, a relabeling of the axes may be necessary. This order is achieved if $r_x(\tilde{\nu}_0) \geqslant r_y(\tilde{\nu}_0) \geqslant r_z(\tilde{\nu}_0)$, which is not necessarily the easiest choice for the application of formulas (4.93)–(4.96). With the possible exception of such a relabeling, the principal axes

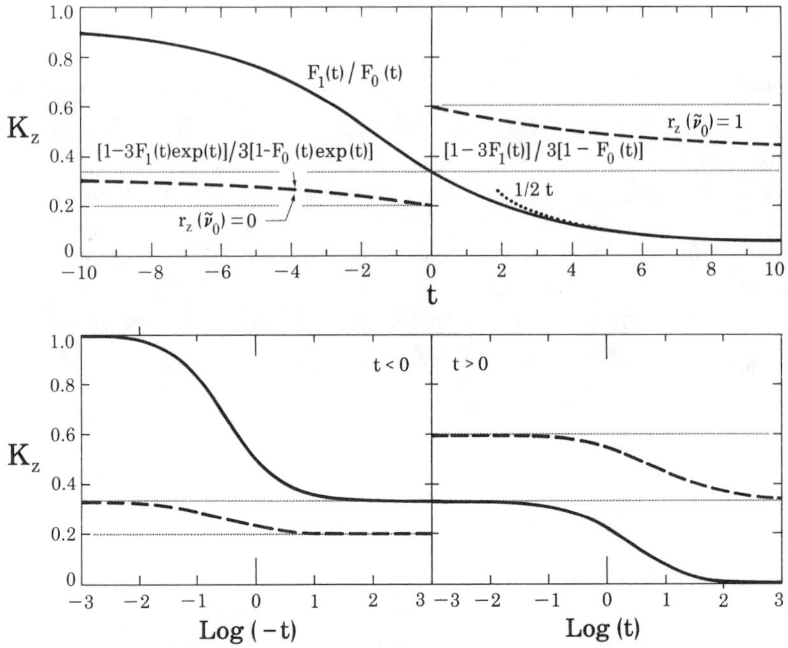

Figure 4.19 The orientation factors $K_z'(\tilde{\nu}_0,Z,D)$ of the remaining molecules (full lines) and the factor $K_z(\tilde{\nu}_0,Z,D)$ of the transformed molecules (dashed lines) for two examples $[r_z(\tilde{\nu}_0) = 1$ and $r_z(\tilde{\nu}_0) = 0]$ of an initially $(t = 0)$ random sample of molecules with equal absorption properties in the directions of molecular x and y axes. The light is linearly polarized along Z. The dotted curve represents the asymptotic limit 1/2t. See text [(4.97), (4.107)].

of the orientation tensor for the remaining molecules coincide with the principal axes of their absorption tensor $\mathsf{M}'(\tilde{\nu}_0)$.

(iii) The Transformed Molecules: The Orientation Distribution Function. To obtain the orientation distribution function $f_{\tilde{\nu}_0,Z}(\beta,\gamma,D)$ for the transformed molecules after dose D, we take advantage of the fact that the total molecular assembly of remaining and transformed molecules taken together is random. Its orientation distribution function is equal to unity, and we have

$$\frac{N(D)}{N(0)} f'_{\tilde{\nu}_0,Z}(\beta,\gamma,D) + \left[1 - \frac{N(D)}{N(0)}\right] f_{\tilde{\nu}_0,Z}(\beta,\gamma,D) = 1 \qquad (4.101)$$

Using (4.82) we obtain

$$f_{\tilde{\nu}_0,Z}(\beta,\gamma,D) = \frac{1 - \exp[-\phi D p_{\tilde{\nu}_0,Z}(\beta,\gamma)]}{1 - N(D)/N(0)} \qquad (4.102)$$

with N(D)/N(0) given by (4.84). In the limit of small D, this equation acquires the form (4.65), as it should. In the limit of very large irradiation doses D, the exponentials in (4.84) and (4.102) go to zero, except possibly for an infinitesimal fraction of molecules oriented so that $p_{\tilde{\nu}_0,Z}(\beta,\gamma) = 0$. As the photochemical transformation reaches completion, the orientation distribution function for the transformed molecules becomes

$$\lim_{D \to \infty} f_{\tilde{\nu}_0,Z}(\beta,\gamma,D) = 1 \qquad (4.103)$$

and corresponds to the random distribution, as expected on physical grounds. Thus, the maximum alignment of the transformed photoselected molecules is achieved at the beginning of the irradiation when D is small, and it then decreases as the irradiation proceeds.

(iv) The Transformed Molecules: Orientation Factors. The orientation factors K and L of the transformed molecules can be obtained from their orientation distribution function $f_{\tilde{\nu}_0,Z}(\beta,\gamma,D)$ given in (4.102). Multiplication by an appropriate product of direction cosines and integration yields

$$K_u(\tilde{\nu}_0,Z,D) = \frac{1/3 - [N(D)/N(0)]K'_u(\tilde{\nu}_0,Z,D)}{1 - N(D)/N(0)} \qquad (4.104)$$

$$L_{uv}(\tilde{\nu}_0,Z,D) = \frac{(1 + 2\delta_{uv})/15 - [N(D)/N(0)]L'_{uv}(\tilde{\nu}_0,Z,D)}{1 - N(D)/N(0)} \qquad (4.105)$$

where the fraction of the remaining molecules N(D)/N(0) is given by (4.84) and the orientation factors of the remaining molecules $K'_u(\tilde{\nu}_0,Z,D)$ and $L'_{uv}(\tilde{\nu}_0,Z,D)$ are determined from (4.93)–(4.96). In the limit $D \to \infty$, the orientation factors of the transformed molecules reach the isotropic values $K_u(\tilde{\nu}_0,Z,\infty) = 1/3$ and $L_{uv}(\tilde{\nu}_0,Z,\infty) = (1 + 2\delta_{uv})/15$. Initially, in the limit $D \to 0$, they are given by equations (4.74) and (4.78):

$$K_u(\tilde{\nu}_0,Z,0) = [1 + 2r_u(\tilde{\nu}_0)]/5$$

and

$$L_{uv}(\tilde{\nu}_0,Z,0) = [(1 + 2\delta_{uv})/35][1 + 2r_u(\tilde{\nu}_0) + 2r_v(\tilde{\nu}_0)]$$

It is seen that the degree of alignment of the transformed molecules is never very high.

Thus, for a given value of the reaction progress parameter η defined in (4.88), the orientation factors for both the transformed and the remaining molecular assemblies can be calculated if the polarized absorption fractions $r_u(\tilde{\nu}_0)$ of the starting molecules are known. The principal axes of the orientation tensors for both assemblies coincide with those of the absorption tensor. Relabeling of the axes may be necessary in order to satisfy the convention $K_z \geqslant K_y \geqslant K_x$.

In the large triangle of permitted K values, shown in Figure 4.8, the two points $K'(\tilde{\nu}_0,Z,D) \equiv [K'_x(\tilde{\nu}_0,Z,D), K'_y(\tilde{\nu}_0,Z,D), K'_z(\tilde{\nu}_0,Z,D)]$ and $K(\tilde{\nu}_0,Z,D) \equiv [K_x(\tilde{\nu}_0,Z,D), K_y(\tilde{\nu}_0,Z,D), K_z(\tilde{\nu}_0,Z,D)]$ define a straight line which passes through the center point of the triangle, (1/3,1/3,1/3). They lie on opposite sides of the center point. The ratio of the distance from K′ to the center point to the distance from K to the center point is equal to the ratio of the fraction of the transformed molecules to the fraction of remaining molecules,

$$\frac{1 - N(D)/N(0)}{N(D)/N(0)} = \frac{N(0)}{N(D)} - 1$$

Relations (4.104) and (4.105) provide a means of determining the fraction of remaining molecules N(D)/N(0) without disturbing the sample. Solving for N(D)/N(0), we have

$$\frac{N(D)}{N(0)} = \frac{1/3 - K_u(\tilde{\nu}_0,Z,D)}{K'_u(\tilde{\nu}_0,Z,D) - K_u(\tilde{\nu}_0,Z,D)} = \frac{(1 + 2\delta_{uv})/15 - L_{uv}(\tilde{\nu}_0,Z,D)}{L'_{uv}(\tilde{\nu}_0,Z,D) - L_{uv}(\tilde{\nu}_0,Z,D)} \qquad (4.106)$$

so that already a measurement of $K'_u(\tilde{\nu}_0,Z,D)$ and $K_u(\tilde{\nu}_0,Z,D)$ for a single choice of u provides the necessary information.

Equal absorption properties in x and y directions. In the special case $r_x(\tilde{\nu}_0) = r_y(\tilde{\nu}_0)$, for instance when the absorption is purely polarized along a line $[r_x(\tilde{\nu}_0) = r_y(\tilde{\nu}_0) = 0,\ r_z(\tilde{\nu}_0) = 1]$ or in a plane $[r_x(\tilde{\nu}_0) = r_y(\tilde{\nu}_0) = 1/2,\ r_z(\tilde{\nu}_0) = 0]$, expressions (4.104) and (4.105) acquire the form

$$K_z(\tilde{\nu}_0,Z,D) = \frac{1/3 - F_1\{[3r_z(\tilde{\nu}_0) - 1]\eta\}e^{-[1 - r_z(\tilde{\nu}_0)]\eta}}{1 - F_0\{[3r_z(\tilde{\nu}_0) - 1]\eta\}e^{-[1 - r_z(\tilde{\nu}_0)]\eta}} \qquad (4.107)$$

$$L_z(\tilde{\nu}_0,Z,D) = \frac{1/5 - F_2\{[3r_z(\tilde{\nu}_0) - 1]\eta\}e^{-[1 - r_z(\tilde{\nu}_0)]\eta}}{1 - F_0\{[3r_z(\tilde{\nu}_0) - 1]\eta\}e^{-[1 - r_z(\tilde{\nu}_0)]\eta}} \qquad (4.108)$$

$$K_x(\tilde{\nu}_0,Z,D) = K_y(\tilde{\nu}_0,Z,D) = [1 - K_z(\tilde{\nu}_0,Z,D)]/2 \qquad (4.109)$$

$$L_x(\tilde{\nu}_0,Z,D) = L_y(\tilde{\nu}_0,Z,D) = (3/8)[1 - 2K_z(\tilde{\nu}_0,Z,D) + L_z(\tilde{\nu}_0,Z,D)] \qquad (4.110)$$

The distribution of the transformed molecules is then rodlike if $r_z(\tilde{\nu}_0) > 1/3$, random if $r_z(\tilde{\nu}_0) = 1/3$, and disclike if $r_z(\tilde{\nu}_0) < 1/3$.

If the absorbing transition is purely polarized along the z axis, $r_z(\tilde{\nu}_0) = 1$, expression (4.107) for the orientation factor of the transformed molecules $K_z(\tilde{\nu}_0,Z,D)$ becomes $K_z(\tilde{\nu}_0,Z,D) = [1 - 3F_1(t)]/3[1 - F_0(t)]$, where t is defined as before, $t = [3r_z(\tilde{\nu}_0) - 1]\eta$, so that $t = 2\eta$. The initial value of t is 0, and during the phototransformation it increases to infinity. The plot of $K_z(\tilde{\nu}_0,Z,D)$ for this first idealized case against t and against log t is shown as a dashed line on the right-hand side of Figure 4.19.

If the absorbing transition is uniformly polarized in the xy plane, $r_z(\tilde{\nu}_0) = 0$, expression (4.107) for the orientation factor $K_z(\tilde{\nu}_0,Z,D)$ becomes $K_z(\tilde{\nu}_0,Z,D) = [1 - 3F_1(t)\exp(t)]/3[1 - F_0(t)\exp(t)]$, where t is defined as above, so that $t = -\eta$. The initial value of t is 0, and during the phototransformation it decreases to negative infinity. The plot of $K_z(\tilde{\nu}_0,Z,D)$ for this second idealized case against t and against log(−t) is shown as a dashed line on the left-hand side of Figure 4.19.

The general case, $0 < r_z(\tilde{\nu}_0) < 1$, requires the use of the full formula (4.107). For each value of $r_z(\tilde{\nu}_0)$ smaller than 1/3, t runs from zero to negative infinity, and the corresponding plot of $K_z(\tilde{\nu}_0,Z,D)$ in Figure 4.19 appears similar to that shown on the left as a dashed line for $r_z(\tilde{\nu}_0) = 0$, in that the limit of 1/3 is reached from below as t decreases. However, the starting point at t = 0 is not 0.2 but higher, as given by (4.74).

For each value of $r_z(\tilde{\nu}_0)$ larger than 1/3, t runs from zero to positive infinity and the plot of $K_z(\tilde{\nu}_0,Z,D)$ is similar to that shown on the right in Figure 4.19 in that the limit of 1/3 is reached from above as t increases. The starting point at $t = 0$ is not 0.6 but lower, as dictated by (4.74).

The relation between the orientation factors $K(\tilde{\nu}_0,Z,D)$ and $K'(\tilde{\nu}_0,Z,D)$ which characterize the transformed and the remaining molecules, respectively, in the photoselection process is displayed in yet another way in Figure 4.20 for the two limiting special cases, $r_z(\tilde{\nu}_0) = 1$ and $r_z(\tilde{\nu}_0) = 0$ [with $r_x(\tilde{\nu}_0) = r_y(\tilde{\nu}_0)$]. For the former case of a single absorbing transition moment $\mathbf{M}(0j)$ at $\tilde{\nu}_0$, polarized along z (bottom), and for the latter case of two mutually perpendicular equally

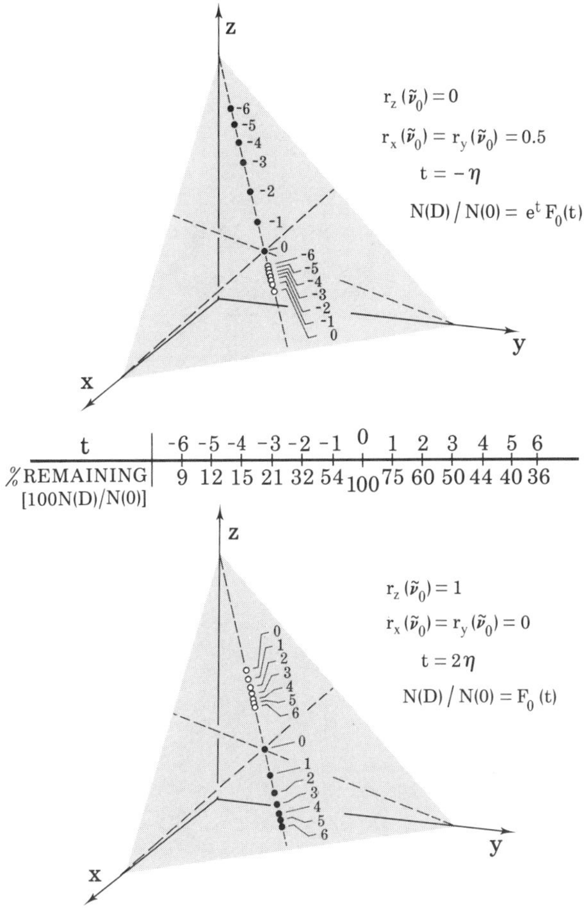

Figure 4.20 Two examples, $r_z(\tilde{\nu}_0) = 1$ and $r_z(\tilde{\nu}_0) = 0$, of the orientation of the assemblies of transformed (open circles) and remaining (filled circles) photoselected molecules with equal absorption properties in the x and y directions [$r_x(\tilde{\nu}_0) = r_y(\tilde{\nu}_0)$], as a function of the degree of conversion. The values of the conversion degree parameter t are shown. Excitation with Z-polarized light. See text.

absorbing moments at $\tilde{\nu}_0$ directed along x and y (top), the time sequence of points representing the orientation distribution in the (K_x,K_y,K_z) space (Figure 4.8) is shown and labeled by the corresponding t values. The points representing the orientation of the transformed molecules are empty; those representing the orientation of the remaining molecules are full. The points corresponding to the initial distributions (D = 0) correspond to t = 0. As the reaction progress parameter η, defined in (4.88), increases from its initial value of zero during the irradiation and eventually reaches infinity, the conversion degree parameter t changes as well. As discussed above, $t = 2\eta$ if $r_z(\tilde{\nu}_0) = 1$ (bottom), and $t = -\eta$ if $r_z(\tilde{\nu}_0) = 0$ (top). The relation of the t value to the percentage of molecules remaining, 100N(D)/N(0), is shown in the center of the figure for the two cases. A relabeling of the axes in the starting molecule and/or the photoproduct so as to assure $K_x \leqslant K_y \leqslant K_z$ would be necessary for the display of these results in the ordinary orientation triangle (Figure 4.9).

In Figure 4.21, the orientation factors $K_z(\tilde{\nu}_0,Z,D)$ of the transformed molecules and $K'_z(\tilde{\nu}_0,Z,D)$ of the remaining molecules are plotted against the fraction of molecules transformed by the photochemical process in the two limiting cases, $r_z(\tilde{\nu}_0) = 1$ and $r_z(\tilde{\nu}_0) = 0$. The points shown are labeled by the value of the conversion degree parameter t and correspond to the similarly labeled points in Figure 4.20.

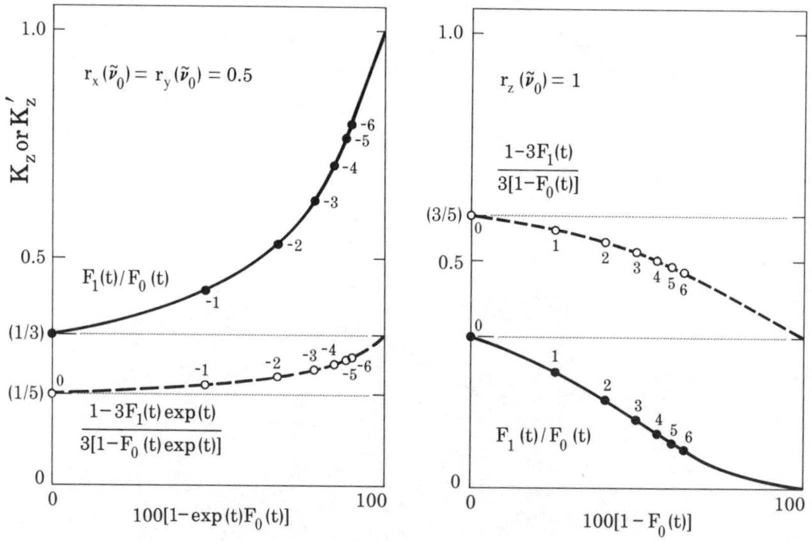

Figure 4.21 Photoselection with Z-polarized light for the cases $r_x(\tilde{\nu}_0) = r_y(\tilde{\nu}_0) = 0.5$ (left) and $r_z(\tilde{\nu}_0) = 1$ (right). The orientation factors $K_z(\tilde{\nu}_0,Z,D)$ (transformed molecules, dashed line) and $K'_z(\tilde{\nu}_0,Z,D)$ (remaining molecules, full line) as a function of the fraction of molecules transformed by the photochemical process. The values of the conversion degree parameter t at selected points are shown and correspond to those displayed in Figure 4.20. See text.

Nondestructive Photoorientation. The photoselection process may cause an alignment of an absorbing solute molecule even if the light absorption step does not convert it into another chemical species with different absorption properties. Such alignment, referred to as nondestructive orientation, can result from at least two primary mechanisms.

One mechanism, which can be termed *photoinduced rotational diffusion*, has apparently been looked for in vain so far but might be possible in principle. If a part of the energy of the absorbed photon is converted into heat, the viscosity of the rigid matrix may decrease sufficiently in the immediate vicinity of those molecules that were excited to permit some rotational diffusion before the heat is dissipated and molecular rotations cease. After a period of irradiation, this would lead to the depletion of those solute orientations in which the absorbing transition moment has a significant projection into the light polarization direction. If the transition moment is directed along the molecular axis x, after an infinite irradiation dose the solute orientation distribution would be that described by the bottom line in Table 4.1: perfect alignment of the yz plane with Z ($K_x = 0$, $K_y = K_z = 1/2$).

A bad omen for the occurrence of this mechanism is the fact that room-temperature investigations of rotational diffusion in ordinary solvents have failed to reveal a significant effect of local heating on the rotational diffusion constants (Section 8.9). On the other hand, cases are known in which the photoproduct from a matrix-isolated substrate molecule irradiated by polarized light emerges in a rotationally isotropic orientation (Section 8.2.9), and this suggests that further search may be worthwhile.

Another mechanism, which has already been found to operate, can be termed *photoinduced generalized pseudorotation*. This is applicable when the solute molecule undergoes a pseudorotation, i.e., a rearrangement to an identical chemical species in a different orientation, or when it undergoes an inversion to an enantiomer. In either case, the unpolarized absorption properties of the transformed solute are indistinguishable from those of the original solute. Assuming that the solute molecule cannot rotate as a whole, and assuming that the photoinduced pseudorotation or inversion process is not reversed thermally on the time-scale of the experiment, simple expressions can be derived for the degree of alignment in the photostationary state.

In the simplest case, each solute molecule can exist in two differently oriented but chemically identical or enantiomeric forms, related by mirroring in a plane by symmetry. When needed, we shall refer to the two forms of the molecule as singly primed and doubly primed, and shall label the quantities associated with them accordingly. We define a site-fixed common orthogonal system of axes for the two forms such that the mirror plane that interconverts them contains the x and z axes as shown on the top in Figure 4.21a.

The photoselecting excitation is assumed to be due to the j-th molecular transition. In one form of the molecule the absorbing electric dipole transition moment is $\mathbf{M}'(j)$, in the other it is its mirror image $\mathbf{M}''(j)$. The angle between $\mathbf{M}'(j)$ and $\mathbf{M}''(j)$ is labeled ω_j. The z axis is chosen parallel to the vector $\mathbf{M}'(j) + \mathbf{M}''(j)$ so that the vectors $\mathbf{M}'(j)$ and $\mathbf{M}''(j)$ lie in the yz plane. The sample

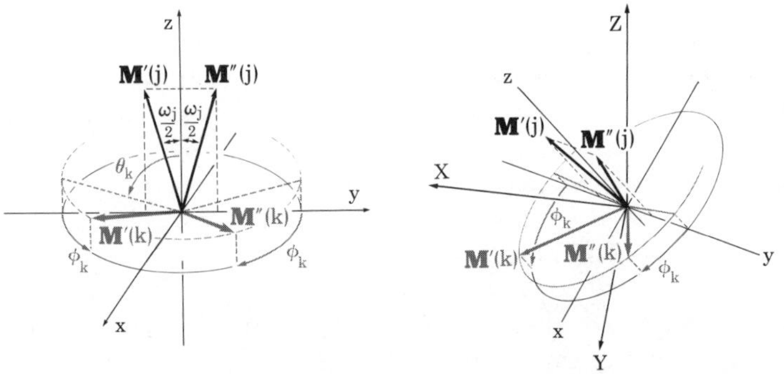

Figure 4.21a The molecular coordinate system x, y, z for a molecule subject to photo-induced generalized pseudorotation (left), and its general orientation in the laboratory system of axes X, Y, Z (right). See text. Reproduced by permission from J. G. Radziszewski, F. A. Burkhalter, and J. Michl, *J. Am. Chem. Soc.*, submitted for publication.

is initially racemic and isotropic, with the site x, y, and z axes pointed equally likely in any direction. Consider a subassembly of sites with x, y, and z axes oriented in the infinitesimal vicinity of some particular orientation in the laboratory, characterized by the Euler angles α, β, and γ (Figure 4.2 and the right-hand side of Figure 4.21a). Initially, the fractional populations of the two forms, $f'(\alpha,\beta,\gamma,\omega_j)$ and $f''(\alpha,\beta,\gamma,\omega_j)$, are both equal to one half. For most such subassemblies, the orientation is such that the probability of absorption of light linearly polarized along the laboratory axis Z is different for the two forms since the projections of **M**′(j) and of **M**″(j) into Z are different. Upon continued irradiation, the population of the two forms within the subassemblies will become unequal, the form with the larger projection being depleted, until a photostationary state is reached. The ratio of $f'(\alpha,\beta,\gamma,\omega_j)$ to $f''(\alpha,\beta,\gamma,\omega_j)$ in the photostationary state will be a function of the particular orientation chosen for the subassembly:

$$
\begin{aligned}
f'(\alpha,\beta,\gamma,\omega_j) &= (1/2)[\cos(\omega_j/2)\cos\beta \\
&\quad - \sin(\omega_j/2)\sin\beta\cos\gamma]^2/[\cos^2(\omega_j/2)\cos^2\beta + \sin^2(\omega_j/2)\sin^2\beta\cos^2\gamma] \\
f''(\alpha,\beta,\gamma,\omega_j) &= (1/2)[\cos(\omega_j/2)\cos\beta \\
&\quad + \sin(\omega_j/2)\sin\beta\cos\gamma]^2/[\cos^2(\omega_j/2)\cos^2\beta + \sin^2(\omega_j/2)\sin^2\beta\cos^2\gamma]
\end{aligned}
\qquad (4.111)
$$

When integrated over all subassemblies, the two forms are still equally abundant so that the rigid solution remains racemic if they are enantiomers. However, since the differently oriented subassemblies contribute differently to the absorption of linearly polarized light, $E_Z(j)$ will be different from $E_Y(j)$. The measured orientation factor of the j-th transition in the photooriented assembly in its photostationary state is independent of the length of the transition moments and depends only on the angle ω_j between them (note that the spectroscopic indistinguishability of the two forms means that the averaging is over both of them).

Integration yields the relation

$$K_j = (1/3)[1 - \sin^2 \omega_j/(1 + \sin \omega_j)] \qquad (4.112)$$

The minimum value of K_j is 1/6 and is reached at $\omega_j = \pi/2$. This is intuitively acceptable since then the act of photoconversion drives the transition moment as far from its original direction as is possible. On the other hand, at $\omega_j = 0$ or $\omega_j = \pi$, $K_j = 1/3$ and there is no photoorientation. The former happens when the transition moment lies in the xz plane, the latter when it is directed along y; in either case, the transition moment directions coincide in the two forms.

Once ω_j has been determined by measurement of the photoinduced dichroism at the wavelength of the exciting light, the orientation distribution of the molecular assembly is known. It can then be used to predict the orientation factor and thus the dichroism of any other transition moment in the molecule if its orientation in the x, y, z axes system is known. Let such a moment **M**′(k) in the singly primed form be located at angle θ_k from the molecular z axis and lie in a plane obtained by rotating the yz plane around the z axis by the angle ϕ_k: if **M**′(k) and **M**′(j) lie on the same side of the xz plane, $-\pi/2 \leq \phi_k \leq \pi/2$ and $\cos \phi_k \geq 0$; if they lie on opposite sides of the xz plane, $\pi/2 \leq \phi_k \leq 3\pi/2$ and $\cos \phi_k \leq 0$. By symmetry, in the doubly primed form the k-th transition moment **M**″(k) will still be located at angle θ_k from z but now will lie in a plane obtained by rotating the yz plane around z by the angle $\pi - \phi_k$. The result for the orientation factor K_k for the k-th transition is

$$K_k = (1/3)[1 - \cos \phi_k \sin 2\theta_k \sin \omega_j/(1 + \sin \omega_j)] \qquad (4.113)$$

This formula also shows the expected behavior. The k-th transition moment will not be aligned if $\omega_j = 0$ or $\omega_j = \pi$ since then there is no photoorientation to start with, and it will show a maximum effect if $\omega_j = \pi/2$ since then the photoorientation is most efficient. The dependence on θ_k is similar to that on ω_j: If $\theta_k = 0$ or $\theta_k = \pi/2$, **M**′(k) and **M**″(k) both lie in the z axis or both lie perpendicular to it, respectively. Then, their projections into the Z direction will be the same and there will be no dichroism for such a transition. Finally, the effect is largest when **M**′(k) lies in the yz plane. As long as $\cos \theta_k > 0$, the deviation of K_k from the isotropic value of 1/3 is toward lower values, in the same direction as for K_j. Indeed, if $\phi_k = 0$ and $2\theta_k = \omega_j$, the formulas (4.112) and (4.113) are identical. This is sensible, since the photoconversion will drive all those transition moment vectors that are located on the same side of the xz plane as **M**′(j) away from the Z direction as it drives the primed into the doubly primed form. However, if **M**′(k) is located on the opposite side of the xz plane from **M**′(j) so that $\cos \theta_k < 0$, light will flip it toward the Z axis as it drives the k-th transition moment away from this axis, and then an increase in K_k above 1/3 must be expected. When $\phi_k = \pm\pi/2$, there will be no effect since **M**′(k) and **M**″(k) will then both lie in the xz plane and will be collinear. The largest positive dichroism, $K_k = 1/2$, is expected when **M**′(k) coincides with **M**″(j) and is perpendicular to **M**′(j).

A measurement of the orientation factor K_k does not in itself permit the determination of the direction of the k-th transition moment but provides a relation between the angles ϕ_k and θ_k. Only one additional piece of information is then needed to determine the transition moment direction.

4.6.4 Photoselection: Unpolarized or Circularly Polarized Light

The principle of photoselection described in the beginning of Section 4.6.3 applies also for the action of a beam of spatially incoherent unpolarized or circularly polarized light. When its propagation direction is taken to be Z, the electric field vector takes all directions in the XY plane with equal probability. The ensemble of molecules excited by such light, as well as the complementary ensemble of the molecules left unexcited, will be uniaxially oriented, with Z as the unique axis. If circularly polarized light is used, the sample is assumed to be achiral so that it can exhibit no circular dichroism. The handedness of the light then plays no role.

The results also apply to photoselection with linearly polarized light propagating along Z, whose plane of polarization rotates in the XY plane or jumps between the various directions perpendicular to Z rapidly and uniformly on the time scale of the experiment, as in the measurement of photodichroism induced by a beam of light passing through a rapidly rotating polarizer.

In order to distinguish the two cases of photoselection, an argument or subscript Z was placed at the relevant quantities for photoselection by light linearly polarized along the laboratory direction Z in the preceding section, while the argument or subscript XY will be used now when dealing with photoselection by light whose electric vector lies uniformly in the laboratory plane XY.

The results for the latter case can be obtained most readily by recognizing that the squares of the direction cosines of a transition moment **M**(0j) with respect to the laboratory axes X, Y, and Z add up to 1, so that the equation for the orientation distribution function of the photoselected assembly at small degrees of conversion now is

$$\begin{aligned} f_{\tilde{\nu}_0,XY}(\beta,\gamma) &= \frac{\sum_j g'_j(\tilde{\nu}_0)(1/2)\{|M(0j)|^2 - [M_x(0j)\cos x + M_y(0j)\cos y + M_z(0j)\cos z]^2\}}{\sum_j g'_j(\tilde{\nu}_0)|M(0j)|^2/3} \\ &= [3 - f_{\tilde{\nu}_0,Z}(\beta,\gamma)]/2 \end{aligned} \tag{4.114}$$

with $f_{\tilde{\nu}_0,Z}(\beta,\gamma)$ given by (4.67).

The general results for the orientation factors K of the photoselected assembly at negligible degrees of conversion now become

$$K_{uv}(\tilde{\nu}_0,XY) = [3K_{uv}^{iso} - K_{uv}(\tilde{\nu}_0,Z)]/2 = [\delta_{uv} - K_{uv}(\tilde{\nu}_0,Z)]/2 \tag{4.115}$$

with $K_{uv}(\tilde{\nu}_0,Z)$ given by (4.70).

In the principal system of axes, the orientation tensor is diagonal, and its elements can be written in terms of the polarized absorption fractions in a form analogous to (4.74):

$$K_u(\tilde{\nu}_0,XY) = [2 - r_u(\tilde{\nu}_0)]/5 \tag{4.116}$$

As before, the values of the orientation factors and the assignment of the orientation axis to one of the three symmetry axes in general depend on the choice of $\tilde{\nu}_0$. This time, however, the orientation axis z will lie in the direction w for which $A_w(\tilde{\nu}_0)$ and $r_w(\tilde{\nu}_0)$ are the smallest, and the x axis in the direction where they are the largest. If the other two fractions are equal, $r_z(\tilde{\nu}_0) < r_y(\tilde{\nu}_0) = r_x(\tilde{\nu}_0)$, the distribution will be of the rodlike type, $K_y(\tilde{\nu}_0,XY) = K_x(\tilde{\nu}_0,XY)$. If the two smallest fractions are equal, either of the axes can be taken for the orientation axis z and the other for y, and the orientation is of the disclike type, $K_z(\tilde{\nu}_0,XY) = K_y(\tilde{\nu}_0,XY)$. If all three fractions are equal, no photoselection occurs.

In a molecule of any symmetry, the results simplify if only one transition moment **M**(0j) contributes at $\tilde{\nu}_0$. A second important special case occurs if only two transition moments contribute at $\tilde{\nu}_0$, provided that they are mutually orthogonal and contribute equally. This happens for doubly degenerate states in molecules of very high symmetry.

In the former case, (i), the direction of the single significant absorbing moment defines the principal x axis, and the y and z axes can be chosen in any mutually orthogonal pair of directions in the plane perpendicular to x. The photoselected orientation distribution is disclike, with orientation factors

$$K_x(x,XY) = 1/5 \qquad K_y(x,XY) = K_z(x,XY) = 2/5 \tag{4.117}$$

In the latter case, (ii), the directions of the two significant mutually perpendicular transition moments define the principal x and y axes (except that any set related by rotation around z is also acceptable), the photoselected orientation distribution is rodlike, and the orientation factors are

$$K_x(xy,XY) = K_y(xy,XY) = 3/10 \qquad K_z(xy,XY) = 2/5 \tag{4.118}$$

These results can be compared with (4.75) and (4.76). It will be noted that the tensors K(x,XY) and K(yz,Z) are identical. This is a consequence of the fact that the orientation distribution functions of the photoselected assembly in the two instances are identical, being both equal to $(3/2)\sin^2 x$. This result exemplifies the poorer photoselecting power of unpolarized (or circularly polarized) light relative to linearly polarized light: Photoselecting on a single direction with the former is only as efficient as photoselecting on a whole plane with the latter.

The mapping of the values of polarized absorption fractions at the wavenumber of the photoselecting light onto the orientation factors $K(\tilde{\nu}_0,XY)$ resulting for the photoselected assembly is shown in Figure 4.22. All possible points $r_x(\tilde{\nu}_0)$, $r_y(\tilde{\nu}_0)$, $r_z(\tilde{\nu}_0)$, with $r_x(\tilde{\nu}_0) \geqslant r_y(\tilde{\nu}_0) \geqslant r_z(\tilde{\nu}_0)$, are located in a triangle with vertices (1,0,0) at lower left (x), (1/3,1/3,1/3) on top and (1/2,1/2,0) on the lower right (xy) in the plot of $r_z(\tilde{\nu}_0)$ against $r_y(\tilde{\nu}_0)$, since $\sum_u r_u(\tilde{\nu}_0) = 1$. The mapping inverts this triangle into a five times smaller triangle of points $[K_x(\tilde{\nu}_0,XY), K_y(\tilde{\nu}_0,XY), K_z(\tilde{\nu}_0,XY)]$ in the bottom corner of the usual orientation triangle. The point x is mapped into K(x,XY), the point xy into K(xy,XY), the third vertex onto itself. The lines of constant $r_u(\tilde{\nu}_0)$ are mapped onto lines of constant $K_u(\tilde{\nu}_0,XY)$ as shown. The very small size of the $K(\tilde{\nu}_0,XY)$ triangle within the

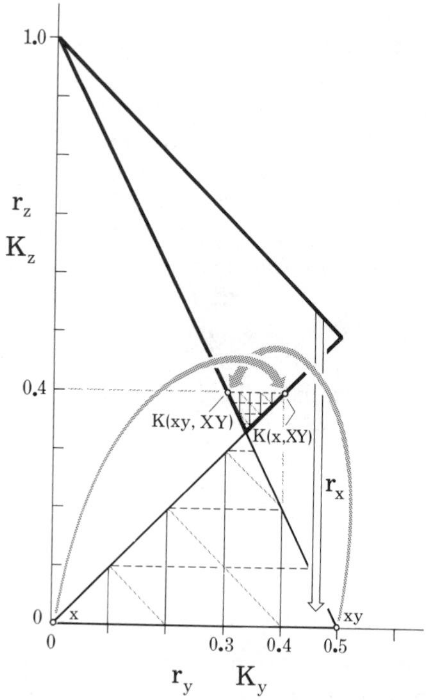

Figure 4.22 Mapping of points describing polarized absorption fractions $r_u(\tilde{\nu}_0)$ onto points describing the orientation factors $K_u(\tilde{\nu}_0,XY)$ of the initially photoselected assembly (unpolarized or circularly polarized light propagating along Z). See text.

usual orientation triangle provides a graphic illustration of the poor photoselecting power of the XY arrangement (cf. Figure 4.18).

Expressions for the orientation factors $L(\tilde{\nu}_0,XY)$ can also be obtained readily using (4.114):

$$L_{stuv}(\tilde{\nu}_0,XY) = [3L^{iso}_{stuv} - L_{stuv}(\tilde{\nu}_0,Z)]/2 \tag{4.119}$$

The orientation factors needed for the discussion of fourth-rank processes on photoselected high-symmetry molecules (sixth rank on a random sample) are obtained by substituting from (4.78) and using $L^{iso}_{uv} = (1 + 2\delta_{uv})/15$:

$$L_{uv}(\tilde{\nu}_0,XY) = [(1 + 2\delta_{uv})/35][3 - r_u(\tilde{\nu}_0) - r_v(\tilde{\nu}_0)] \tag{4.120}$$

In the two special cases mentioned above, (i) and (ii), these orientation factors acquire the values

$$L_{uv}(x,XY) = L_{uv}(yz,Z)$$
$$L_z(xy,XY) = 9/35 \qquad L_y(xy,XY) = L_x(xy,XY) = 6/35$$
$$L_{xz}(xy,XY) = L_{yz}(xy,XY) = 1/14 \qquad L_{xy}(xy,XY) = 2/35 \tag{4.121}$$

which can be compared with those for photoselection with linearly polarized light given in (4.79). The equality of $L_{uv}(yz,Z)$ and $L_{uv}(x,XY)$ results from the identity of the two orientation distribution functions already mentioned above.

If the degree of conversion is not negligible, expressions (4.114)–(4.121) are no longer valid. The treatment outlined in Section 4.6.3 remains in force, and expressions (4.80)–(4.83) and (4.101)–(4.106) are valid. However, all subscripts Z must now be replaced by subscripts XY so that (4.65) must be replaced by

$$p_{\nu 0,XY}(\beta,\gamma) = \frac{\kappa(\tilde{\nu})}{2} \sum_j g_j'(\tilde{\nu}_0)\{|M(0j)|^2 - [M_x(0j)\cos x + M_y(0j)\cos y + M_z(0j)\cos z]^2\} \qquad (4.122)$$

We shall not state the detailed results here.

A very high degree of orientation can again result for the last remaining small fraction of molecules. The best example is that of a single absorbing transition moment **M**(0j) at $\tilde{\nu}_0$. Its direction defines the z axis in the remaining molecules and leads to $p_{\tilde{\nu}0,XY}(\beta,\gamma) \propto \sin^2 z$ and thus to the same orientation distributions as in the case of two equivalent mutually perpendicular moments (x,y) in photoselection with linearly polarized light, since $p_{\tilde{\nu}0,Z}(\beta,\gamma) \propto \sin^2 z$ as well. The limiting distribution is again a perfect alignment of the z axis of the remaining molecules with the laboratory Z axis ($\cos z = 1$): $K_z'(z,XY) = L_z'(z,XY) = 1$, while all other orientation factors $K_u'(z,XY)$ and $L_{uv}'(z,XY)$ vanish. The equivalence of the case of a single absorbing transition moment (z) in XY photoselection and of an isotropically absorbing plane (xy) in Z photoselection implies that the filled and open points in the top triangle in Figure 4.20 also describe the time course of the gradual conversion of all molecules of the initial randomly oriented assembly in the case of XY photoselection.

Finally, we note that unpolarized or circularly polarized light should be very powerful in the so far hypothetical case of nondestructive photoorientation by photoinduced rotational diffusion. For the infinite dose limit, excitation of a z-polarized transition should produce a perfect alignment of the z axis along the light propagation direction Z (top line in Table 4.1, $K_x = K_y = 0$, $K_z = 1$).

4.7 A Note on More General Types of Orientation

Little experimental work has been done with samples oriented in a manner more general than uniaxial. However, it is of some interest to indicate briefly the types of generalization necessary for the description of the orientation distribution function in such samples.

In order to generalize the orientation factors and the Saupe matrices discussed so far, one needs to use direction cosines not only with respect to the Z axis, but also with respect to the other two sample axes X and Y. These are given in equations (4.11)–(4.16). One can then define

$$K_{UVuv} = \langle \cos uU \cos vV \rangle \qquad (4.123)$$

$$L_{STUVstuv} = \langle \cos sS \cos tT \cos uU \cos vV \rangle \qquad (4.124)$$

and similarly for higher order orientation factors. These are again redundant, and the description becomes quite complex in the general case. If measurements are performed only with the electric vector of the light directed along one of three orthogonal sample axes X, Y, or Z, and if they are limited to molecules of high symmetry, only some of the multitude of orientation factors will be needed, and it is then useful to introduce the notation

$$K_{Uu} = \langle \cos^2 uU \rangle \qquad (4.125)$$

Similarly, in the case of generally oriented samples, the Saupe matrices must be used in their general form

$$S_{UVuv} = (1/2)\langle 3 \cos uU \cos vV - \delta_{uv}\delta_{UV} \rangle \qquad (4.126)$$

The Wigner order parameters also become more complicated in that the condition $m' = 0$ is absent.

4.8 Comments and References

The definition of the Euler angles used in this book follows M. Tinkham, *Group Theory and Quantum Mechanics*, McGraw-Hill, New York, 1964. This reference provides the background necessary for the full appreciation of the spherical tensor method and the Wigner order parameters.

The use of orientation factors K to describe uniaxial orientation distributions has been described in E. W. Thulstrup, J. Michl, and J. H. Eggers, *J. Phys. Chem.* **74**, 3868 (1970); J. Michl, E. W. Thulstrup, and J. H. Eggers, *J. Phys. Chem.* **74**, 3878 (1970). The definition of the Euler angles used in these papers followed a different convention (H. Margenau and G. M. Murphy, *The Mathematics of Physics and Chemistry*, 2nd ed., Van Nostrand, Princeton, NJ, 1956, p. 286) than we do in the present text.

Further detail on the orientation factors K is found in J. Michl and E. W. Thulstrup, *Spectrosc. Lett.* **10**, 401 (1977); H.-G. Kuball, J. Altschuh, and A. Schönhofer, *Chem. Phys.* **49**, 247 (1980); E. W. Thulstrup and J. Michl, *J. Phys. Chem.* **84**, 82 (1980); and J. Michl and E. W. Thulstrup, *J. Chem. Phys.* **72**, 3999 (1980). The last-named paper also discusses the properties of the orientation factors L.

The Saupe orientation matrices are described in A. Saupe, *Mol. Cryst. Liq. Cryst.* **1**, 527 (1966). A brief description of the properties of the Wigner order parameters can be found in S. Jen, N. A. Clark, P. S. Pershan, and E. B. Priestley, *J. Chem. Phys.* **66**, 4635 (1977), but the formulas given there contain several misprints. A detailed treatment is available by C. Zannoni in *The Molecular Physics of Liquid Crystals*, G. R. Luckhurst and G. W. Gray, Eds., Academic Press, New York, 1979, Chapter 3.

Some of the explicitly assumed orientation distribution functions proposed in the literature can be found in O. Kratky, *Kolloid Z.* **64**, 213 (1933); W. Kuhn and F. Grün, *Kolloid Z.* **101**, 248 (1942) [rodlike orientation distributions, cf.

A. Yogev, L. Margulies, D. Amar, and Y. Mazur, *J. Am. Chem. Soc.* **91**, 4558 (1969)]; L. V. Smirnov, *Opt. Spektroskopiya* **3**, 123 (1957); Y. Tanizaki, *Bull. Chem. Soc. Japan* **38**, 1798 (1965); G. Bauer, *Monatsh.* **102**, 1789 (1971); K. R. Popov, *Opt. Spektrosc.* **38**, 102; **39**, 142, 368 (1975); B. Nordén, *J. Chem. Phys.* **72**, 5032 (1980) [these papers treat the "imbedded nails" model discussed in the text, cf. equations (4.51)–(4.53)]; L. Margulies and A. Yogev, *Chem. Phys.* **27**, 89 (1978) [a "planar-random" two-function model for orientation in stretched polyethylene]; L. Gårding and B. Nordén, *Chem. Phys.* **41**, 431 (1979) [the "ensemble method" for superstructures]; E. Sackmann, P. Krebs, H. U. Reja, J. Voss, and H. Möhwald, *Mol. Cryst. Liq. Cryst.* **24**, 283 (1973). The last-named reference introduced the notion that the uniaxial orienting potential experienced by a probe in a nematic liquid crystal could be expressed in terms of the polarizability anisotropy of the probe as in equation (4.54) and in the following text. This notion was extended to stretched polymers by M. Lamotte, *J. Chim. Phys.* **72**, 803 (1975) and by C. C. Bott and T. Kurucsev, *Chem. Phys. Lett.* **55**, 585 (1978). These three references contain collections of orientation factors for aromatic molecules; similar data are available for dyes in a liquid crystal [F. Jones and T. J. Reeve, *Mol. Cryst. Liq. Cryst.* **60**, 99 (1980)] and for a larger number of mostly aromatic solutes in stretched polyethylene; see legend to Figure 4.15 of J. G. Radziszewski and J. Michl, *J. Am. Chem. Soc.*, **108**, 0000 (1986).

Orientation by electric fields is treated in detail in the books listed under Electric Dichroism in Section 3.3.

The treatment of photoselection presented here is a generalization of the descriptions given by A. C. Albrecht, *J. Mol. Spectrosc.* **6**, 84 (1961); *Progr. React. Kinet.* **5**, 301 (1970); F. Dörr, in *Creation and Detection of the Excited State*, Vol. 1, A. A. Lamola, Ed., Marcel Dekker, New York, 1971, Chap. 2; and R. A. Deering, A. A. Kraus, Jr., D. M. Gray, and R. Kilkson, *J. Theor. Biol.* **13**, 295 (1966). Related results were derived for the anisotropy of hole-burning by W. Köhler, W. Breinl, and J. Friedrich, *J. Phys. Chem.* **89**, 2473 (1985), and by I. S. Osad'ko, S. L. Soldatov, and A. U. Jalmukhambetov, *Chem. Phys. Lett.* **118**, 97 (1985). The results for nondestructive photoorientation were taken from J. G. Radziszewski, F. A. Burkhalter and J. Michl, submitted for publication.

5

Processes of Rank Two

This chapter deals with electric dipole contributions to the absorption or emission of photons by partially aligned assemblies of molecules. The electric dipole moment operator $\hat{\mathbf{M}}$ is defined in (1.46). The quantity which will dominate the discussion of electric dipole absorption and emission is the electric dipole transition moment defined in (1.48),

$$\mathbf{M}(0f) = \langle \Psi_f | \hat{\mathbf{M}} | \Psi_0 \rangle \qquad (5.1)$$

Its a priori evaluation from molecular wavefunctions Ψ_0 and Ψ_f was discussed at length in Chapter 2.

We shall need to consider only unpolarized and linearly polarized light. Choosing both the initial and final wavefunctions to be real, and recognizing that $\hat{\mathbf{M}}$ is real as well, we see that we shall encounter only real transition moments $\mathbf{M}(0f)$,

$$\mathbf{M}(0f) = \mathbf{M}(f0) \qquad (5.2)$$

Since the electric dipole absorption and emission probabilities are related to the square of the transition moment, the quantity we must consider is actually a second-rank tensor $\mathsf{M}(0f)$ which is a direct product of the vector $\mathbf{M}(0f)$ with itself (Appendix II):

$$\mathsf{M}(0f) = \mathbf{M}(f0)\mathbf{M}(0f) \qquad (5.3)$$

and whose elements are therefore given by

$$[\mathsf{M}(0f)]_{uv} = M_u(f0)M_v(0f) \qquad (5.4)$$

We have already encountered it briefly in (1.15), in Section 1.3.4, and in the discussion of photoselection in Section 4.6.

Experimental measurements are usually made by observing the absorption of plane-polarized light by a partially oriented sample. As we have seen in Figures 2.4–2.6, the spectra observed with different alignments of the electric vector of the light with respect to the sample axes can differ from each other, and this phenomenon is called linear dichroism.

An analogous phenomenon in emission would result if molecules in a partially aligned sample emitted light due to some isotropic stimulation. This could be, for instance, spontaneous luminescence (thermal radiation), chemiluminescence, or totally orientationally relaxed photoluminescence (i.e., with the loss of any memory of excitation anisotropy). The measurement would then consist of analyzing the polarization of the emission. In the following discussion, we concentrate on absorption measurements, but the same formulas would apply to the case of emission.

It is possible to apply the formulas of this chapter even to the case of photoluminescence if the orientation factors K and L are taken to represent the orientation distribution of emitting molecules rather than the equilibrium orientation distribution of all solute molecules in the partially aligned sample. In doing this, one gives up an opportunity to obtain information on the equilibrium orientation distribution separately from the effects of photoselection, but it may still be possible to obtain some spectroscopic information about the solute molecules. It is important to note that this approach is valid only if the exciting light is polarized along the unique sample axis Z (or if the light propagates along Z but is natural or circularly polarized), since otherwise the photoselected orientation distribution is not uniaxial. At any rate, the approach has only limited appeal considering that a proper treatment separating the effects of the photoselection and of the originally present equilibrium orientation is feasible and yields more information, as discussed in Chapter 7.

Throughout this chapter, we assume that the electric dipole contribution to the transition intensity dominates and that magnetic dipole and electric quadrupole contributions can be safely neglected, as is normally the case.

The discussion begins with an outline of the general relations between the observable polarized intensities of absorption or emission by a uniaxial partially aligned assembly of solute molecules on the one hand, and the orientation distribution of the molecules in the sample, as well as the orientation of the molecular transition moment within the molecular framework, on the other hand. We then proceed to a more detailed discussion of the general case, in which the orientation distribution is not known a priori. In cases where this is known, the evaluation of the experimental data is greatly facilitated. This requires no separate discussion.

We first outline a practical method (referred to as the TEM model) which can be used to derive the orientation factors and transition moment directions from experimental data obtained on solutes contained in uniaxial media such as stretched polymers. This treatment has been used in much of the published work. In the subsequent section, three more severe approximations also commonly used in the analysis of experimental data will be described and analyzed.

These are the Fraser-Beer approximation, the Tanizaki approximation, and the Popov approximation.

Measurements of weak dichroism are best performed using light beams with polarization modulation, and the evaluation of this type of data is discussed next. We conclude the chapter with a brief description of a generalization of the treatment to orientations of a type more general than uniaxial.

5.1 Electric Dipole Absorption and Emission. Linear Dichroism in Absorption and in Spontaneous Luminescence

Formulation in Terms of Partial Absorbances. Let $\mathbf{M}(0f)$ be the electric dipole transition moment of the f-th molecular transition and $g'_f(\tilde{\nu})$ its line-shape function. Let ε_U be the unit vector in the direction of the electric field of the incident linearly polarized light of wavenumber $\tilde{\nu}$, U = X, Y, or Z. The transmittance $T'_U(\tilde{\nu})$, uncorrected for the baseline, is given in the electric dipole approximation by (1.20) and (1.63) as

$$T'_U(\tilde{\nu}) = 10^{-\{(4\pi^2\tilde{\nu}N/2303hc)[\alpha'^2_U(\tilde{\nu})/n_U(\tilde{\nu})]cl\,\Sigma_f\, g'_f(\tilde{\nu})\langle M^2_U(0f)\rangle + E^0_U(\tilde{\nu})\}} \qquad (5.5)$$

where the angled brackets ($\langle\ \rangle$) indicate averaging over all molecules in the interrogated sample. The summation over f runs over all transitions, coherence effects between overlapping transitions have been neglected, the molar concentration of the solute is c, the path length is l (cm), $E^0_U(\tilde{\nu})$ is the contribution of the baseline to the absorbance for light polarized along U, and the universal constants have the meaning defined in Chapter 1. The factor $\alpha'^2_U(\tilde{\nu})/n_U(\tilde{\nu})$ represents the modification of the absorption probability for light polarized along U by the presence of solvent molecules. As indicated by the subscript, it is a function of U, but the anisotropy in solvent effects will be neglected in the present treatment (see Section 5.5). As a result, we drop the subscript U, and the solvent correction will cancel in our final expressions for intensity ratios. In strongly birefringent orienting media, such as certain liquid crystals, this neglect is not warranted, and suitable estimates of this factor as a function of U must then be introduced (see Section 3.3).

Let the angles between $\mathbf{M}(0f)$ for an individual molecule and the laboratory axes be denoted fX, fY, and fZ. Then

$$\langle M^2_U(0f)\rangle = |\mathbf{M}(0f)|^2\langle\cos^2 fU\rangle \qquad (5.6)$$

We note that the squares of the direction cosines of a vector add up to 1: $\cos^2 fX + \cos^2 fY + \cos^2 fZ = 1$. Also, $\langle\cos^2 fX\rangle = \langle\cos^2 fY\rangle$, since the sample is uniaxial. We introduce the orientation factor K_f of the f-th transition:

$$\langle\cos^2 fZ\rangle = K_f \qquad \langle\cos^2 fX\rangle = \langle\cos^2 fY\rangle = (1 - K_f)/2 \qquad (5.7)$$

Let us now define the partial absorbance $A_f(\tilde{\nu})$ for the f-th transition (f = 1,2,...) as the contribution of the f-th transition to the absorbance measured on a sample of molecules all oriented with their f-th transition moment parallel to the electric vector of the light:

$$A_f(\tilde{\nu}) = (4\pi^2\tilde{\nu}N/2303hc)[\alpha'^2(\tilde{\nu})/n(\tilde{\nu})]cl g'_f(\tilde{\nu})|\mathbf{M}(0f)|^2 \qquad (5.8)$$

Then the results for polarized transmittance are

$$\begin{aligned} T'_X(\tilde{\nu}) &= T'_Y(\tilde{\nu}) = 10^{-\{\Sigma_f A_f(\tilde{\nu})(1-K_f)/2 + E^0_Y(\tilde{\nu})\}} \\ T'_Z(\tilde{\nu}) &= 10^{-\{\Sigma_f A_f(\tilde{\nu})K_f + E^0_Z(\tilde{\nu})\}} \end{aligned} \tag{5.9}$$

The measurement of the baselines yields

$$T^0_U(\tilde{\nu}) = 10^{-E^0_U(\tilde{\nu})} \tag{5.10}$$

Correcting the spectra for baselines, we obtain results for the absorbances $E_U(\tilde{\nu})$ from

$$E_U(\tilde{\nu}) = -\log\left[T'_U(\tilde{\nu})/T^0_U(\tilde{\nu})\right] \tag{5.11}$$

The results for all three $E_U(\tilde{\nu})$'s can be combined into a single matrix equation:

$$\begin{pmatrix} E_X(\tilde{\nu}) \\ E_Y(\tilde{\nu}) \\ E_Z(\tilde{\nu}) \end{pmatrix} = \sum_f A_f(\tilde{\nu}) \begin{pmatrix} (1-K_f)/2 \\ (1-K_f)/2 \\ K_f \end{pmatrix} \tag{5.12}$$

This important result will serve as a basis for the analysis of experimental data. Its validity is subject to the conditions stated in Section 5.5.

For an isotropic solution, averaging yields $K_f = 1/3$ and $E^0_Z(\tilde{\nu}) = E^0_Y(\tilde{\nu})$. Absorbance is then independent of U as expected:

$$E^{iso}_U(\tilde{\nu}) = E^{iso}(\tilde{\nu}) = \frac{1}{3}\sum_f A_f(\tilde{\nu}) \tag{5.13}$$

We see that $A_f(\tilde{\nu})$ also has another physical significance: It is three times the contribution of the f-th transition to the absorbance measured on an isotropic sample.

For very weak dichroism, the direct measurement of the $E_U(\tilde{\nu})$'s may not be possible with sufficient accuracy. Then the more sensitive methods which use light of modulated polarization are to be preferred. These methods produce linear combinations of the $E_U(\tilde{\nu})$'s whose evaluation is quite analogous (Section 5.3.1).

The number of different K_f values possible for a given solute and alignment depends on the symmetry of the solute. In Sections 5.1.1 and 5.1.2, we consider molecules of point group symmetry such that the molecular axes x, y, and z belong to different irreducible representations (two of them may belong to the same degenerate representation). We refer to these as "high-symmetry" molecules ("very high symmetry" if degeneracies are present). Examples are molecules which belong to the point groups C_{2v} and D_{2h} (high symmetry), or D_{6h} (very high symmetry). Applications to molecules of lower symmetry are discussed in Section 5.1.3.

In the high-symmetry case, (i) the molecular axes determined by symmetry, x, y, and z, coincide with the principal axes of the orientation tensor and (ii) the vector $\mathbf{M}(0f)$ can be directed only along x, y, or z (if 0 or f represents a component of a degenerate state, it should be chosen in an appropriate symmetry-adapted fashion).

Formulation in Terms of the Absorption Tensor. It is instructive to compare the expressions for the observed intensities $E_U(\tilde{\nu})$ using the tensorial formulation of Section 1.3.4 and the three most common ways of describing the second moments of the orientation distribution function. We introduce a molecule-fixed coordinate system of axes x,y,z and label the components of the transition moment **M** in this system M_x, M_y, M_z. Noting that $[\mathsf{M}(0f)]_{uv} = [\mathsf{M}(0f)]_{vu}$, we have for the average square of the projection of the f-th transition moment into the axis U

$$
\begin{aligned}
\langle M_U^2(0f)\rangle &= \left\langle \left[\sum_u M_u(0f)\cos uU\right]^2 \right\rangle \\
&= \left\langle \sum_{u,v} M_u(0f) M_v(0f) \cos uU \cos vU \right\rangle = \langle \boldsymbol{\varepsilon}_U \mathsf{M}(0f) \boldsymbol{\varepsilon}_U \rangle \\
&= \sum_{u,v} \langle \cos uU \cos vU \rangle M_u(0f) M_v(0f) \\
&= \sum_{u,v} (\mathsf{D}_U)_{uv} [\mathsf{M}(0f)]_{uv}
\end{aligned} \tag{5.14}
$$

where the direction cosines are those of (4.8)–(4.16). The left-hand side of this expression is equal to the U,U element of the absorption tensor $\mathsf{M}(0f)$ in the laboratory framework and is related to an observable quantity through (5.5). The result has the general form anticipated in (1.142). All information on the molecular orientation is contained in the second-rank tensor D_U with elements $\langle \cos uU \cos vU \rangle$, and all molecular information on the anisotropy of the f-th molecular dipole transition moment is contained in the elements of the second-rank tensor $\mathsf{M}(0f)$ in the molecular framework. We have already encountered this relation in (4.65) and (4.71), where we used the absorption tensor M to derive results for molecular orientation by photoselection.

Explicit expressions for the elements of D_U are obtained by substitution from (4.8)–(4.16). In terms of orientation factors, the elements of D_U are

$$(\mathsf{D}_Z)_{uv} = K_{uv} \tag{5.15}$$

$$(\mathsf{D}_Y)_{uv} = (\mathsf{D}_X)_{uv} = (\delta_{uv} - K_{uv})/2 \tag{5.16}$$

In terms of Saupe matrix elements, they are

$$(\mathsf{D}_Z)_{uv} = (2S_{uv} + \delta_{uv})/3 \tag{5.17}$$

$$(\mathsf{D}_Y)_{uv} = (\mathsf{D}_X)_{uv} = (\delta_{uv} - S_{uv})/3 \tag{5.18}$$

In terms of order parameters, the tensors D_U are

$$
\mathsf{D}_Z = \begin{pmatrix}
\frac{1}{3} - \frac{1}{3}A_0^{(2)} + \sqrt{\frac{2}{3}}A_2^{(2)} & -\sqrt{\frac{2}{3}}B_2^{(2)} & -\sqrt{\frac{2}{3}}A_1^{(2)} \\
-\sqrt{\frac{2}{3}}B_2^{(2)} & \frac{1}{3} - \frac{1}{3}A_0^{(2)} - \sqrt{\frac{2}{3}}A_2^{(2)} & \sqrt{\frac{2}{3}}B_1^{(2)} \\
-\sqrt{\frac{2}{3}}A_1^{(2)} & \sqrt{\frac{2}{3}}B_1^{(2)} & \frac{1}{3} + \frac{2}{3}A_0^{(2)}
\end{pmatrix} \tag{5.19}
$$

$$D_X = D_Y = \begin{pmatrix} \frac{1}{3} + \frac{1}{6}A_0^{(2)} - \sqrt{\frac{1}{6}}A_2^{(2)} & \sqrt{\frac{1}{6}}A_1^{(2)} & \sqrt{\frac{1}{6}}A_1^{(2)} \\ \sqrt{\frac{1}{6}}A_1^{(2)} & \frac{1}{3} + \frac{1}{6}A_0^{(2)} + \sqrt{\frac{1}{6}}A_2^{(2)} & -\sqrt{\frac{1}{6}}B_1^{(2)} \\ \sqrt{\frac{1}{6}}A_1^{(2)} & -\sqrt{\frac{1}{6}}B_1^{(2)} & \frac{1}{3} - \frac{1}{3}A_0^{(2)} \end{pmatrix} \quad (5.20)$$

Writing the result in terms of orientation factors explicitly, for Z-polarized light we obtain

$$\langle M_Z^2(0f)\rangle = \sum_{u,v} K_{uv}M_uM_v \quad (5.21)$$

and

$$\langle M_Z^2(0f)\rangle = K_zM_z^2 + K_yM_y^2 + K_xM_x^2 + 2(K_{yz}M_yM_z + K_{zx}M_zM_x + K_{xy}M_xM_y) \quad (5.22)$$

For light polarized in a direction perpendicular to Z, we have for a uniaxial sample

$$\langle M_X^2(0f)\rangle = \langle M_Y^2(0f)\rangle = \sum_{u,v} (1/2)(\delta_{uv} - K_{uv})M_uM_v \quad (5.23)$$

and

$$\langle M_Y^2(0f)\rangle = (1/2)(1 - K_z)M_z^2 + (1/2)(1 - K_y)M_y^2 + (1/2)(1 - K_x)M_x^2 - K_{yz}M_yM_z - K_{zx}M_zM_x - K_{xy}M_xM_y \quad (5.24)$$

If the molecule-fixed axes coincide with the molecular orientation axes, the terms with K_{yz}, K_{zx}, and K_{xy} vanish in (5.15), (5.16), (5.21)–(5.24). In this instance, the terms with S_{yz}, S_{zx}, and S_{xy} vanish in the expressions based on Saupe matrices, (5.17) and (5.18), and the terms with $A_1^{(2)}$, $B_2^{(2)}$, and $B_1^{(2)}$ vanish in the expressions based on order parameters, (5.19) and (5.20). In every case, considerable simplification results.

Two independent measurements, one with the electric vector of the light along the unique axis Z and one perpendicular to it, completely characterize the sample with respect to one-photon processes. The former requires light to propagate in a direction perpendicular to Z, and the latter is usually also performed in that geometry, but it could in principle also be measured with light of any polarization propagating along Z. Any other two linearly independent measurements could be used instead; one of these could be $E^{iso}(\tilde{\nu})$.

5.1.1 High Symmetry: Orientation Factors

For symmetrical molecules, we define reduced spectra in the directions x, y, and z in the molecular framework as the sums of partial absorbances due to all transitions polarized along x, y, or z, respectively:

$$A_u(\tilde{\nu}) = \sum_{f:u} A_f(\tilde{\nu}), \qquad u = x,y,z \quad (5.25)$$

where the summation is over all transitions f polarized along u. We have already encountered these quantities in our discussion of photoselection in Chapter 4 [cf. (4.72)].

The general expression for polarized absorbance now becomes

$$\begin{pmatrix} E_X(\tilde{\nu}) \\ E_Y(\tilde{\nu}) \\ E_Z(\tilde{\nu}) \end{pmatrix} = \begin{pmatrix} \frac{1}{2}(1-K_x) & \frac{1}{2}(1-K_y) & \frac{1}{2}(1-K_z) \\ \frac{1}{2}(1-K_x) & \frac{1}{2}(1-K_y) & \frac{1}{2}(1-K_z) \\ K_x & K_y & K_z \end{pmatrix} \begin{pmatrix} A_x(\tilde{\nu}) \\ A_y(\tilde{\nu}) \\ A_z(\tilde{\nu}) \end{pmatrix} \tag{5.26}$$

and again it is seen that $E_X(\tilde{\nu}) = E_Y(\tilde{\nu})$, so that only one of these two need be measured, say, $E_Y(\tilde{\nu})$.

Pure Polarization. In many cases, spectral features such as peaks or shoulders can be recognized in the accessible region of the absorption (emission) spectra. Usually at least some of them can be ascribed to only one of the three $A(\tilde{\nu})$ curves. In UV-visible spectra this separation is rarely so complete that all of the absorption at a particular wavelength is purely polarized along the u axis, since the individual differently polarized features partially overlap. Such complete separation is more common in IR spectra. Whenever it occurs, the evaluation of the orientation factors from (5.26) is particularly simple. In the region of wavenumbers $\tilde{\nu}_u$ where only u-polarized absorption contributes to the total absorption, $A_u(\tilde{\nu}_u) \neq 0$ and $A_v(\tilde{\nu}_u) = A_w(\tilde{\nu}_u) = 0$, and one has

$$E_Z(\tilde{\nu}_u) = K_u A_u(\tilde{\nu}_u) \tag{5.27}$$

$$E_Y(\tilde{\nu}_u) = \tfrac{1}{2}(1 - K_u)A_u(\tilde{\nu}_u) \tag{5.28}$$

Then the dichroic ratio $d_u = E_Z(\tilde{\nu}_u)/E_Y(\tilde{\nu}_u)$ measured at $\tilde{\nu}_u$ provides immediately the value of K_u:

$$K_u = d_u/(d_u + 2) \tag{5.29}$$

Also,

$$K_u = \frac{E_Z(\tilde{\nu}_u)}{E_Z(\tilde{\nu}_u) + 2E_Y(\tilde{\nu}_u)} = \frac{E_Z(\tilde{\nu}_u)}{3E^{iso}(\tilde{\nu}_u)} = \frac{3E^{iso}(\tilde{\nu}_u) - 2E_Y(\tilde{\nu}_u)}{3E^{iso}(\tilde{\nu}_u)} \tag{5.30}$$

providing an expression for K_u from the measurement of $E_Z(\tilde{\nu})$ and $E^{iso}(\tilde{\nu})$ or $E_Y(\tilde{\nu})$ and $E^{iso}(\tilde{\nu})$ as an alternative to the measurement of $E_Y(\tilde{\nu})$ and $E_Z(\tilde{\nu})$. If such a purely polarized region can also be found for another polarization direction v, $A_v(\tilde{\nu}_v) \neq 0$, $A_u(\tilde{\nu}_v) = A_w(\tilde{\nu}_v) = 0$, K_v is obtained similarly from $d_v = E_Z(\tilde{\nu}_v)/E_Y(\tilde{\nu}_v)$. Since $\sum K_u = 1$, if a third purely w-polarized region is also present its dichroic ratio d_w will be related simply to the already known ratios d_u and d_v:

$$d_u/(d_u + 2) + d_v/(d_v + 2) + d_w/(d_w + 2) = 1 \tag{5.31}$$

Mixed Polarization. Stepwise Reduction. In general, the spectrum will not contain two purely and differently polarized wavelength regions. It is then still possible to determine the K's if the mutually overlapping spectral features associated with the three $A(\tilde{\nu})$ curves are not shaped alike and are not located at exactly the same wavelengths. In such a case various linear combinations of the measured spectra, $E_Z(\tilde{\nu}) - cE_Y(\tilde{\nu})$, will contain the peaks and shoulders of different polarization to different degrees, and in some of them one or another set of spectral features will be absent altogether. Such linear combinations can be found by trial and error ("stepwise reduction").

If a set of peaks $A_u(\tilde{\nu})$ polarized along the axis u just disappears from the linear combination; i.e., if $E_Z(\tilde{\nu}) - cE_Y(\tilde{\nu})$ is a linear combination of only two of the three curves, $A_v(\tilde{\nu})$ and $A_w(\tilde{\nu})$, we have from (5.26)

$$E_Z(\tilde{\nu}) - cE_Y(\tilde{\nu}) = 0 \times A_u(\tilde{\nu}) + [-c/2 + K_v(1 + c/2)]A_v(\tilde{\nu}) + [-c/2 + K_w(1 + c/2)]A_w(\tilde{\nu}) \quad (5.32)$$

where the coefficient at $A_u(\tilde{\nu})$ vanishes, and this requires

$$c = d_u = 2K_u/(1 - K_u) \quad (5.33)$$

so that once again,

$$K_u = d_u/(d_u + 2) \quad (5.34)$$

Successive stepwise reduction of the spectral features due to u, v, and w thus successively provides all three K's. Of course, two stepwise reductions are sufficient, since $K_w = 1 - K_u - K_v$.

It is important to consider the likelihood of finding spectral features which can be ascribed exclusively to one of the polarization directions, overlapping or not. In practice, a check is frequently provided by the fact that more than one such feature can be found for each polarization direction, so that in the stepwise reduction procedure a whole set of peaks and shoulders, rather than just one of them, disappears simultaneously when one of the correct values for c is found. The peaks which are most likely to belong fully to only one of the three $A_u(\tilde{\nu})$'s are IR transitions or 0-0 components of electronic transitions. Barring an accidental exact superposition of the origins of two transitions, the latter will be purely polarized unless the interaction with the solvent is strong and destroys the local symmetry of the molecule, or unless vibronic interactions cause very low frequency nonsymmetrical vibrations to appear with considerable intensity. Neither of these situations is very likely to occur except for very weak transitions (see Section 8.1).

Figure 5.1 shows an example of the stepwise reduction procedure. The UV-visible absorption spectrum of benzo[*rst*]pentaphene contains a large number of overlapping electronic singlet-singlet transitions. It is clear from the figure that the 0-0 vibronic peak of the first of these, at 25 000 cm^{-1}, together with a number of other distinct spectral features, such as those at 33 000 and 41 000 cm^{-1}, disappears from the linear combination $E_Y(\tilde{\nu}) - (1/c)E_Z(\tilde{\nu})$ for $1/c \cong 0.2$, so that it would disappear from $E_Z(\tilde{\nu}) - cE_Y(\tilde{\nu})$ for $c = d_z \cong 5$. Similarly, the peaks near 30 000, 37 000, and 45 000 cm^{-1} all disappear from $E_Z(\tilde{\nu}) - cE_Y(\tilde{\nu})$ for $c = d_y \cong 0.4$. The fact that the members of the two groups of spectral features disappear simultaneously from the linear combinations shows that they have equivalent polarization directions; the only likely way in which this can happen is that they are purely polarized. It should be noted that the

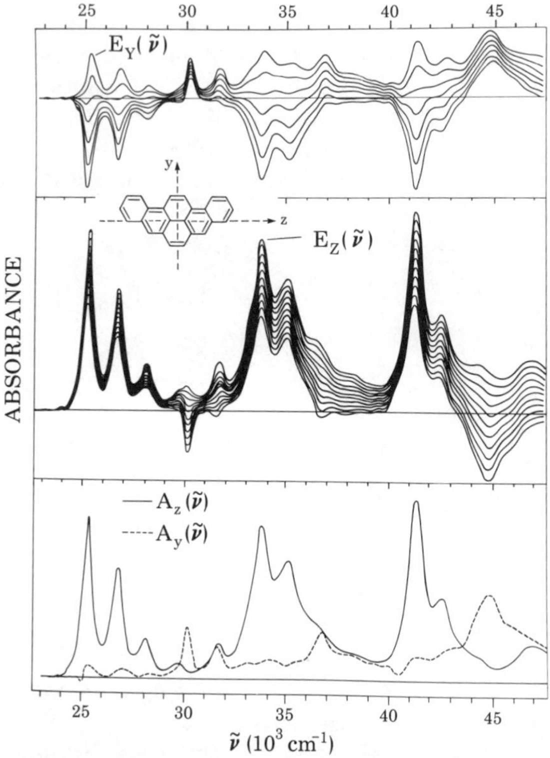

Figure 5.1 Stepwise reduction of the UV linear dichroic spectra of benzo[*rst*]pentaphene in stretched polyethylene. Absorbance is plotted vertically. Top: the family of curves $E_Y(\tilde{\nu}) - (1/c)E_Z(\tilde{\nu})$ with 1/c increasing from 0.0 (top) to 0.6 (bottom) in steps of 0.1. Center: the family of curves $E_Z(\tilde{\nu}) - cE_Y(\tilde{\nu})$ with c increasing from 0.0 (top) to 2.0 (bottom) in steps of 0.2. Bottom: the reduced spectra $E_Z(\tilde{\nu}) - 0.47E_Y(\tilde{\nu})$ (full line) and $E_Y(\tilde{\nu}) - 0.17E_Z(\tilde{\nu})$ (dashed line). Reproduced by permission from P. B. Pedersen, E. W. Thulstrup, and J. Michl, *Chem. Phys.* **60**, 187 (1981).

condition for a successful reduction of a spectral feature is that it be purely polarized. Whether or not it overlaps with other differently polarized spectral features is immaterial as long as they do not have exactly the same location and shape as the feature being reduced. Thus, the overall absorption curve in this region of the spectrum may well be of mixed polarization (cf. the y-polarized peak near 37 000 cm^{-1} in Figure 5.1). A closer inspection indicates that the proper values for the corresponding K's are $K = K_z = d_z/(d_z + 2) = 0.75$ and $K = K_y = d_y/(d_y + 2) = 0.19$; thus $K_x = 0.06$ according to the convention $K_z \geqslant K_y \geqslant K_x$ and since $K_x + K_y + K_z = 1$. The linear combinations $E_Z(\tilde{\nu}) - cE_Y(\tilde{\nu})$ and $E_Y(\tilde{\nu}) - (1/c)E_Z(\tilde{\nu})$ with the corresponding values of d_y and d_z substituted for c are shown again at the bottom of Figure 5.1.

Uneven Reduction—"Wiggles." It can be seen that even the strong peaks, which one would expect to be purely polarized, do not reduce to zero uniformly across their bandwidth (e.g., the peak at 25 000 cm^{-1}). The resulting "wiggles" have been observed in the reduction of the spectra of a variety of molecules in stretched polymers. They were observed even for the well-resolved origins of the lowest energy transitions, for which they cannot be attributed to vibronic mixing. We believe that one should resist the temptation to assign these weak spectral features to real transitions. Rather, we take the attitude that they provide a limitation on the amount of spectral detail which can be derived from the experimental spectra.

It is interesting to ask about the origin of this weak variation of the dichroic ratio observed even for what in all probability are intrinsically purely polarized transitions. There may be several contributing factors. Any displacement of the wavenumber scales in the measurement of $E_Z(\tilde{\nu})$ and $E_Y(\tilde{\nu})$ will produce wiggles in the shape of the derivatives of the real peaks. This need not be due just to an instrumental inaccuracy in wavelength setting (which would be absent in a polarization-modulated measurement). After all, the excitation energies of transitions are affected by the refractive index of a solvent and in a birefringent solvent will occur at slightly different wavenumbers for the two polarizations.

We believe that a more important effect is that of inhomogeneous broadening. The 0-0 transition peak in the orienting solvent is much broader than in an inert gas matrix or a Shpolskii matrix, since each molecule is exposed to a slightly different environment and will have a slightly different excitation energy. A group of environments characterized by a particular value of transition energy is referred to as a site. With narrow-band laser excitation, such sites can be excited selectively at very low temperature, producing very narrow emission lines. Each narrow line is accompanied by a "phonon side band," which corresponds to a simultaneous electronic excitation of a molecule and an excitation of intermolecular vibrations in its environment. Its intensity relative to the narrow "zero-phonon line" increases rapidly with increasing absolute temperature. As a result, the "site-selection" effect is not observed at ordinary temperatures.

We believe that orientation-dependent inhomogeneous broadening occurs for solutes in stretched polymers in a manner somewhat analogous to the above. A variety of different types of orientation exist, each characterized by a band of energies (at room temperature, much wider than the above sites) and by a distribution of orientations. Molecules contributing to the absorption at different wavenumbers within the 0-0 peak have slightly different average alignments with the Z axis. Moreover, in some cases (Section 8.1) even the direction of a transition moment within the molecular framework can be sensitive to the molecular environment. The result is a slightly wavenumber-dependent dichroic ratio and wiggles in the reduced spectra.

Absolute Assignment. One problem still left is the determination of the absolute polarization directions of the transitions. Arranging the three K's in the order $K_z \geqslant K_y \geqslant K_x$ automatically provides the labels z, y, x, but the identification of these labels with actual molecular axes is a different matter. If nothing is known about the nature of the orientation distribution except that it is uniaxial, or about the absolute polarization direction of any of the transitions, say from IR assignments, absolute identification is not possible. This is rarely if ever the case. If the orientation was achieved by electric field, the orientation distribution function is known fairly well. If it was achieved by the use of stretched polymers, enough information about the nature of the orientation of other molecules has already been accumulated that once again reliable absolute assignments are usually possible. As discussed in Chapter 3, a similar situation is found with solutions in liquid crystals. In these, a comparison with orientation factors obtained from NMR measurements is also possible.

In many molecules, comparison with other data, such as those obtained on single crystals, can be made for at least one transition, establishing a firm reference frame for all other transitions of like polarization. Also, comparison of UV-visible with IR transitions, which tend to overlap much less, can be of value. For many kinds of IR-active vibrational transitions, the transition moment direction is obvious (e.g., a C=O stretching vibration is polarized along the C=O bond direction, cf. Figure 2.1).

5.1.2 High Symmetry: Reduced Spectra

Having shown how the orientation factors K_u can be determined, we next address the problem of finding the reduced spectra $A_u(\tilde{\nu})$, $u = x,y,z$. This is essential for the analysis of UV-visible spectra in terms of the location and number of independent transitions, their polarization and fine structure, and can also be useful for IR spectra containing overlapping bands. Attempts to perform such analysis directly by inspection of the experimental $E(\tilde{\nu})$ curves or simple derived quantities such as the dichroic ratio can be highly misleading if the transitions overlap.

General. It is clear that it is impossible to determine three linearly independent curves $A_x(\tilde{\nu})$, $A_y(\tilde{\nu})$, and $A_z(\tilde{\nu})$ from the two independent experimental curves $E_Z(\tilde{\nu})$ and $E_Y(\tilde{\nu})$ in the absence of additional information. The most obvious remedy is to find a third spectrum, linearly independent of $E_Z(\tilde{\nu})$ and $E_Y(\tilde{\nu})$. Then, equation (5.26) will provide the three sought spectra $A_u(\tilde{\nu})$ once the matrix is inverted. The third curve cannot be any other spectrum measured on the same uniaxially stretched sheet or the isotropic solution spectrum, since these will not be linearly independent of the first two. In principle, it might be possible to use a solution in the same polymer uniaxially stretched to a different degree or, if solvent effects are negligible, a solution in a different uniaxially stretched polymer. If the orientation factors in the new oriented sample are $\{K_u^*\}$, the new spectra $E_Z^*(\tilde{\nu})$ and $E_Y^*(\tilde{\nu})$ will be linearly independent of the first two spectra $E_Z(\tilde{\nu})$ and $E_Y(\tilde{\nu})$, and therefore useful, only if

$$(3K_z^* - 1)/(3K_y^* - 1) \neq (3K_z - 1)/(3K_y - 1) \qquad (5.35)$$

A geometrical representation of this result can be provided: The straight line going through points (K_z,K_y) and (K_z^*,K_y^*) in the orientation triangle (Figure 4.9) must not pass through the point (1/3,1/3). The possibility that several different degrees of stretching of the same polymer could be used does not look very promising. Some results for polyethylene and poly(vinyl alcohol) are shown in

Figure 5.2, and it is clear that all points (K_z, K_y) lie near a line passing through the isotropic point, so that the two sides of inequality [(5.35)] are almost equal. If other molecules behave like those investigated so far, the determination of three independent $A(\tilde{\nu})$'s by this approach will generally be very inaccurate if not impossible.

Special Cases. There are three special cases in which reduced spectra can be obtained from only two measured curves $E_Z(\tilde{\nu})$ and $E_Y(\tilde{\nu})$. Fortunately, these occur very commonly in practice, thereby making the method widely applicable.

(i) The orienting properties of the z and y axes are similar in such a way that $K_z = K_y = (1 - K_x)/2$ (e.g., disc-shaped molecules). Then $A_z(\tilde{\nu}) + A_y(\tilde{\nu})$ and $A_x(\tilde{\nu})$ can be obtained separately:

$$\begin{aligned}[A_z(\tilde{\nu}) + A_y(\tilde{\nu})]/2 &= [K_z E_Z(\tilde{\nu}) - (1 - 2K_z)E_Y(\tilde{\nu})]/(3K_z - 1) \\ &= [(1 - K_x)E_Z(\tilde{\nu}) - 2K_x E_Y(\tilde{\nu})]/(1 - 3K_x) \end{aligned} \tag{5.36}$$

$$\begin{aligned} A_x(\tilde{\nu}) &= [2K_z E_Y(\tilde{\nu}) - (1 - K_z)E_Z(\tilde{\nu})]/(3K_z - 1) \\ &= [2(1 - K_x)E_Y(\tilde{\nu}) - (1 + K_x)E_Z(\tilde{\nu})]/(1 - 3K_x) \end{aligned} \tag{5.37}$$

(ii) The orienting properties of the y and x axes are similar in such a way that $K_z = (1 - K_y)/2$, $K_y = K_x$ (e.g., rod-shaped molecules). Then, $A_z(\tilde{\nu})$ and $[A_y(\tilde{\nu}) + A_x(\tilde{\nu})]$ can be obtained separately:

$$\begin{aligned} A_z(\tilde{\nu}) &= [(1 + K_z)E_Z(\tilde{\nu}) - 2(1 - K_z)E_Y(\tilde{\nu})]/(3K_z - 1) \\ &= [(3 - K_y)E_Z(\tilde{\nu}) - 2(1 + K_y)E_Y(\tilde{\nu})]/(1 - 3K_y) \end{aligned} \tag{5.38}$$

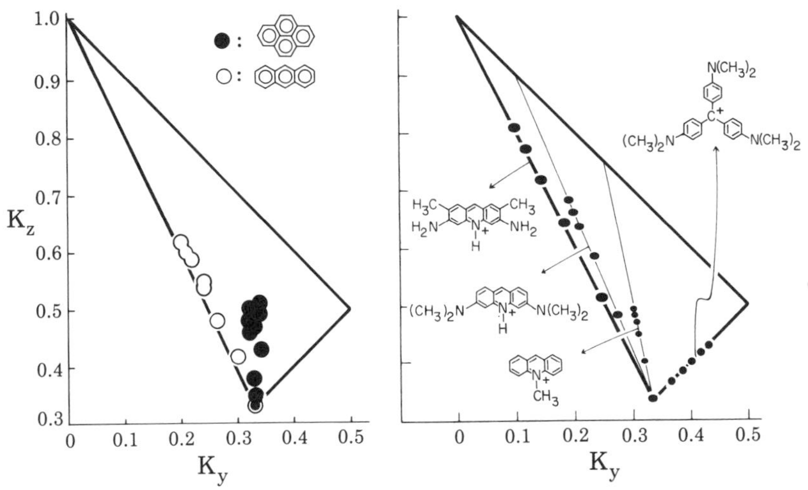

Figure 5.2 Orientation factors for selected aromatics in stretched polyethylene (left) and poly(vinyl alcohol) (right) for various degrees of stretching. Reproduced by permission from E. W. Thulstrup and J. Michl, *J. Am. Chem. Soc.* **104**, 5594 (1982), and Y. Matsuoka, *J. Phys. Chem.* **84**, 1361 (1980).

$$[A_y(\tilde{\nu}) + A_x(\tilde{\nu})]/2 = [2K_zE_Y(\tilde{\nu}) - (1 - K_z)E_Z(\tilde{\nu})]/(3K_z - 1)$$
$$= [2(1 - K_y)E_Y(\tilde{\nu}) - (1 + K_y)E_Z(\tilde{\nu})]/(1 - 3K_y) \qquad (5.39)$$

If the orienting properties of all three axes are equal ($K_z = K_y = K_x = 1/3$), no useful orientation is present, and the reduction cannot be performed.

(iii) A spectral region can be found in which one of the three components $A_z(\tilde{\nu})$, $A_y(\tilde{\nu})$, or $A_x(\tilde{\nu})$ is negligibly small. Then the other two components can be determined in this region. For instance, in many aromatic molecules, out-of-plane polarized intensity is negligible in the visible and near-UV regions, $A_x(\tilde{\nu}) = 0$. Then

$$A_z(\tilde{\nu}) = [(1 - K_y)E_Z(\tilde{\nu}) - 2K_yE_Y(\tilde{\nu})]/(K_z - K_y) \qquad (5.40)$$

$$A_y(\tilde{\nu}) = [2K_zE_Y(\tilde{\nu}) - (1 - K_z)E_Z(\tilde{\nu})]/(K_z - K_y) \qquad (5.41)$$

It is seen that the purely polarized spectra $A_z(\tilde{\nu})$ and $A_y(\tilde{\nu})$ are proportional to the curves given by the expression $E_Z(\tilde{\nu}) - [2K_u/(1 - K_u)]E_Y(\tilde{\nu})$, i.e., to the "correctly reduced" linear combinations of $E_Z(\tilde{\nu})$ and $E_Y(\tilde{\nu})$ from which the values of the K_u were obtained. These are shown in the bottom part of Figure 5.1. Note that the proportionality constants are not the same for $A_z(\tilde{\nu})$ and $A_y(\tilde{\nu})$.

Case (iii) is the most important in practice. Truly disc-shaped and truly rod-shaped molecules are relatively rare, but spectra of molecules of more general shape have often been analyzed as if they belonged to one of these limiting cases. This is a highly questionable procedure, and it usually produces incorrect results.

5.1.3 Low Symmetry

General. Molecules which do not possess at least two mutually perpendicular planes of symmetry or rotation axes shall be considered to be of low symmetry. In molecules without symmetry planes or rotation axes, electric dipole transition moments can be oriented in any direction with respect to the molecular framework.

The treatment is considerably simplified when the molecular orientation axes are used for the molecule-fixed axes system x,y,z. As noted in Section 4.2, in this system of axes, $K_{uv} = 0$ unless $u = v$. In low-symmetry molecules, the location of these molecular axes in the molecular framework does not follow from symmetry, and its determination is not trivial.

Let the f-th transition be characterized by the absorption curve $A_f(\tilde{\nu})$ due to a transition moment **M**(0f) with direction cosines

$$\cos \phi_u^f = M_u(0f)/|\mathbf{M}(0f)| \qquad (5.42)$$

in the molecular axes system (u = x,y,z). The transition moment **M**(0f) can be expressed in terms of its components in the molecular axes system according to (4.17):

$$\mathbf{M}(0f) = \sum_u \varepsilon_u M_u(0f) \qquad (5.43)$$

Equation (5.12) shows that for any particular choice of f the spectral features associated with $A_f(\tilde{\nu})$ will be eliminated in the following linear combination of $E_Z(\tilde{\nu})$ and $E_Y(\tilde{\nu})$:

$$E_Z(\tilde{\nu}) - [2K_f/(1 - K_f)]E_Y(\tilde{\nu}) \quad (5.44)$$

For a purely f-polarized spectral region $\tilde{\nu}_f$, we have

$$K_f = E_Z(\tilde{\nu}_f)/[E_Z(\tilde{\nu}_f) + 2E_Y(\tilde{\nu}_f)] = E_Z(\tilde{\nu}_f)/3E^{iso}(\tilde{\nu}_f) \quad (5.45)$$

in analogy to equation (5.30).

For overlapping transitions, the stepwise reduction procedure is used to determine a value of c such that the features contributed by the particular f-th transition chosen just disappear from the linear combination $E_Z(\tilde{\nu}) - cE_Y(\tilde{\nu})$. For this value, say $c = d_f$, this linear combination contains the term $A_f(\tilde{\nu})$ for this particular transition with zero weight. Substituting for $E_Z(\tilde{\nu})$ and $E_Y(\tilde{\nu})$ from (5.12), this implies

$$K_f = d_f/(d_f + 2) \quad (5.46)$$

in analogy to equation (5.34). In this fashion, a K_f value can be obtained for each observed transition f.

The experimentally accessible values K_f reflect the average alignment of the transition moments **M**(0f). They are not simply related to the average degree of alignment of the molecular axes x, y, and z, as was the case for high-symmetry molecules, since **M**(0f) no longer necessarily lies along one of these axes. Now, the K_f values also depend on the orientation of the transition moments **M**(0f) within the molecular framework.

Thus, our unknowns are the five independent K_{uv} values and, for each transition f, two unknown independent direction cosines $\cos \phi_u^f$ among the three in (5.42) describing the orientation of **M**(0f) with respect to the x,y,z axes.

In order to relate the experimentally determined K_f values to the unknowns, we write

$$K_f = \langle \cos^2 fZ \rangle = \left\langle \left[\sum_u \cos \phi_u^f \cos u \right] \left[\sum_v \cos \phi_v^f \cos v \right] \right\rangle$$
$$= \sum_{u,v} \cos \phi_u^f \, K_{uv} \cos \phi_v^f \quad (5.47)$$

If the K_f values have been determined for n transitions, there will be n equations of type (5.47) and 2n + 5 unknowns. Given the overwhelming number of unknowns, additional information will be needed to solve for any of them. This can have the form of symmetry relations if some symmetry is present, in the form of independent knowledge of certain transition moment directions, perhaps for vibrational transitions, etc.

If nothing at all were known about the K_{uv} values, one would need to know the transition moment directions for five differently polarized transitions in order to determine them. One could then write a system of five equations (5.47) with f = 1,2,3,4,5, in which the five independent K_{uv}'s are the only unknowns.

The equations can also be written in a way based on an alternative form of equation (5.47),

$$\sum_{u,v} \cos \phi_u^f (K_{uv} - \delta_{uv} K_f) \cos \phi_v^f = 0 \qquad (5.48)$$

After the K_{uv}'s have been determined, the orientation tensor can be diagonalized. The resulting eigenvectors define the location of the principal molecular orientation axes in the originally selected axes system, and the eigenvalues provide their orientation factors. Once the principal orientation axes are known, they serve as a new and more convenient set of molecular axes x,y,z, now with known orientation factors K_u. In this system of axes, equation (5.48) acquires the form

$$\sum_{u} (K_u - K_f) \cos^2 \phi_u^f = 0, \qquad u = x,y,z \qquad (5.49)$$

for each transition f. It contains two unknowns, say $\cos^2 \phi_y^f$ and $\cos^2 \phi_z^f$, since

$$\cos^2 \phi_z^f + \cos^2 \phi_y^f + \cos^2 \phi_x^f = 1 \qquad (5.50)$$

Note that even if the unknowns can be determined, knowledge of them will not fix the location of the transition moment direction in the molecular framework uniquely, since only the absolute values and not the signs of the direction cosines can be obtained.

In principle, the knowledge of the principal orientation axis directions and their associated orientation factors could come from a source other than the solution of the equation system (5.47), f = 1,2, ... For instance, with sufficient knowledge of the orientation mechanism, they could be predicted a priori. Although molecular shape alone is undoubtedly not the only determining factor for molecular orientation in anisotropic solvents, it is often possible to estimate the location of the principal orientation axes and the values of the orientation factors from molecular shape by comparison with symmetrical molecules of otherwise similar shape.

Regardless of how the location of the principal orientation axes and the orientation factors were obtained, it is still clear from equations (5.49) and (5.50) that a single measurement of K_f for an additional transition cannot permit the determination of the two unknown independent direction cosines which characterize its transition moment direction **M**(0f) in the molecular framework in the general case.

Special Cases. If the transition moment directions of more than five transitions are known a priori, the set of equations (5.49) is overdetermined and a least squares solution is likely to provide more reliable results. More often, however, fewer than five **M**(0f) directions will be known. Progress can then still be made if a suitable symmetry element is present in the molecule.

Consider, for instance, the effect of the presence of a plane of symmetry. Then one of the principal axes of the orientation tensor, say x, must be perpendicular to the plane, and $K_{xy} = K_{zx} = 0$. Only one off-diagonal orientation factor, K_{yz}, two diagonal orientation factors, K_y and K_z, and the location of the y and z axes remain unknown, since the value of K_x can be determined immediately: It will be the K_f value common to all transitions polarized perpendicular to the symmetry plane. Thus, it can be obtained from the observation of at least one among them, $K_x = K_f$.

In order to pursue this example further, let us take the location of the principal axis z within the molecular symmetry plane yz and its orientation factor K_z as our two unknowns. Once these are found, the location of the third principal axis y and the value of K_y will be obvious. In the principal axes system, the off-diagonal element K_{yz} will vanish.

A set of two equations of the type (5.49) will have to be solved for our two unknowns. To write the equations, two K_f values must be determined from measurements on two in-plane polarized transitions f which possess different directions of polarization. Let us call the larger of the two measured K_f values K_α and the smaller one K_β, and let α and β, respectively, be the angles which the transition moments make with the principal orientation axis z. Obviously, $|\alpha| \leqslant |\beta|$ and $K_z \geqslant K_\alpha \geqslant K_\beta \geqslant K_y$. From (5.49),

$$K_\alpha = K_z \cos^2 \alpha + K_y \sin^2 \alpha \qquad K_\beta = K_z \cos^2 \beta + K_y \sin^2 \beta \tag{5.51}$$

We then have for the unknown angles

$$\tan^2 \alpha = (K_z - K_\alpha)/(K_\alpha - K_y) \qquad \tan^2 \beta = (K_z - K_\beta)/(K_\beta - K_y) \tag{5.52}$$

Note that K_y is not an unknown independent of K_z since $K_y = 1 - K_x - K_z$.

Alternatively,

$$\begin{aligned} \cos^2 \alpha &= [2K_\alpha - (1 - K_x)]/[2K_z - (1 - K_x)] \\ \cos^2 \beta &= [2K_\beta - (1 - K_x)]/[2K_z - (1 - K_x)] \end{aligned} \tag{5.53}$$

showing how $|\alpha|$ and $|\beta|$ would be determined if the orientation factor K_z of the principal orientation axis were known. The values of these angles would then fix the position of the z axis in the molecular frame if the absolute transition moment direction of one or both transitions were known a priori. At present, however, we assume that K_z is not known and will have to be determined from (5.51) once α or β or both are known.

From (5.53),

$$\frac{\cos^2 \alpha}{\cos^2 \beta} = \frac{2K_\alpha - (1 - K_x)}{2K_\beta - (1 - K_x)} \tag{5.54}$$

so that the knowledge of one of the angles α, β is sufficient to determine the other. However, usually both angles are unknown, since the location of the orientation axis z in the molecular framework is unknown.

At times, the difference of the angles is known from molecular structure, and this is independent of the unknown orientation of the z axis. This can happen, for instance, in the case of two infrared transitions corresponding to well-localized vibrations whose transition moments have a known relation to molecular structure.

If $\Delta = \beta - \alpha$ is known, we rewrite (5.54) as

$$\tan 2\alpha \sin 2\Delta = \cos 2\Delta - \frac{2K_\beta - (1 - K_x)}{2K_\alpha - (1 - K_x)} \tag{5.55}$$

Now, if $|\Delta| \neq 0°, 90°$, the problem is solved. For α, we have

$$\tan 2\alpha = \cot 2\Delta - \frac{2K_\beta - (1 - K_x)}{[2K_\alpha - (1 - K_x)] \sin 2\Delta} \tag{5.56}$$

and K_z and $|\beta|$ can be obtained from (5.51)–(5.53).

Two special cases present problems. If $\Delta = 0°$, K_α and K_β coincide and do not represent two independent pieces of information, so that a solution cannot be obtained. If $|\Delta| = 90°$,

$$K_\alpha + K_\beta = 1 - K_x = K_y + K_z \tag{5.57}$$

This situation can be recognized if Δ is not known a priori, but it is not helpful for the determination of the unknowns.

In either case, a different set of transitions must be chosen for the solution of the problem.

A Geometrical Representation for the General Case. In order to appreciate just what information becomes available when K_f is measured for a transition in a molecule whose principal axes directions and orientation factors K_u are already known, a geometrical representation is useful. If we view $\cos \phi_u^f$ as the u-th coordinate of a point on a unit sphere located at the origin of the molecular coordinate system, all transition moment directions compatible with a given value of K_f are represented by arrows pointing from the center of the sphere to any one of the points lying on one of two closed curves on the surface of the sphere. The curves are defined as the locus of points satisfying (5.49) and (5.50) simultaneously, and their shape depends on the values K_x, K_y, K_z, and K_f (Figure 5.3). As K_f runs from its smallest possible value, K_x, to its largest possible value, K_z, the curves change in a continuous fashion which is easiest to describe by starting in the center of the range, with $K_f = K_y$ (center of Figure 5.3, recall that $0 \leqslant K_x \leqslant K_y \leqslant K_z \leqslant 1$). If $K_f = K_y$, equation (5.49) yields

$$\frac{\cos^2 \phi_z^f}{\cos^2 \phi_x^f} = \frac{K_y - K_x}{K_z - K_y} \tag{5.58}$$

and it is seen that the two curves are two great circles obtained by cutting the unit sphere by two planes containing the y axis and cutting the x, z plane at angles $\pm\alpha$ from the x axis, with

$$\tan \alpha = \sqrt{\frac{K_y - K_x}{K_z - K_y}} \tag{5.59}$$

In the case of rodlike orientation, $K_x = K_y$, and $\alpha = 0°$, so that both great circles are identical and lie in the xy plane (on the left in Figure 5.3). In the case of disclike orientation, $K_y = K_z$, and $\alpha = 90°$, so that both great circles are again identical but now lie in the yz plane (on the right in Figure 5.3).

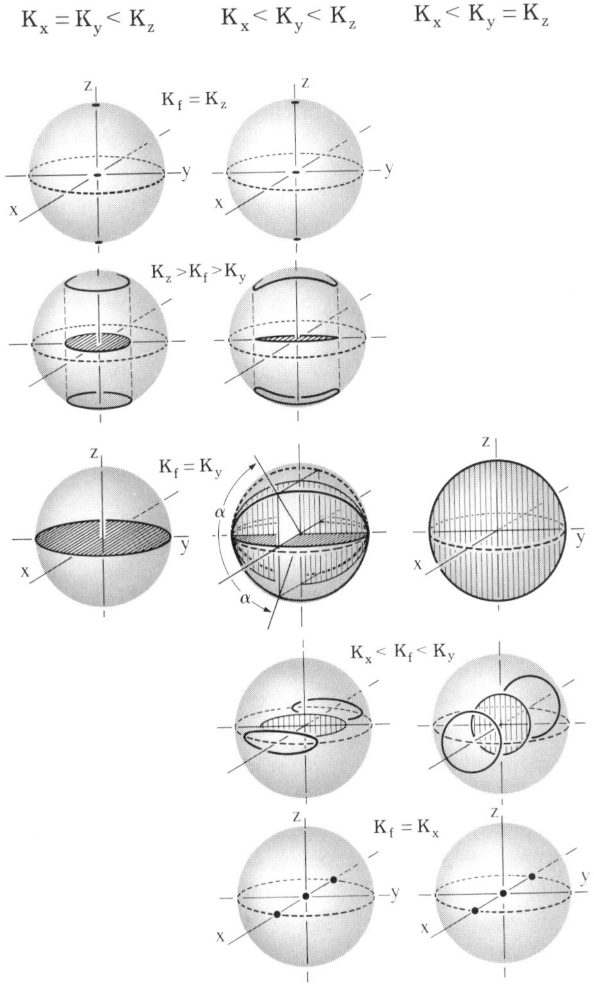

Figure 5.3 Relation of possible transition moment directions to an observed K_f value in a general low-symmetry molecule. See text.

Returning to the general orientation again (center of Figure 5.3), we note that the projection of the two great circles into the xy plane is an ellipse whose axes lie in the y and x directions. The length of the y half-axis is unity, and that of the x half-axis is $\sqrt{(K_z - K_y)/(K_z - K_x)}$. The projection of the great circles into the yz plane is an ellipse whose axes lie in the y and z directions. The length of the y half-axis is unity; that of the z half-axis is $\sqrt{(K_y - K_x)/(K_z - K_x)}$. Alternatively, of course, the great circles can be viewed as projections of the xy or yz ellipse onto the unit sphere along the z or x directions, respectively.

If $K_f \neq K_y$, the two closed curves can still be viewed as projections of an ellipse onto the unit sphere.

If $K_f > K_y$, equations (5.49) and (5.50) yield

$$(K_z - K_x)\cos^2\phi_x^f + (K_z - K_y)\cos^2\phi_y^f = K_z - K_f \tag{5.60}$$

Since $K_z - K_x$, $K_z - K_y$, and $K_z - K_f$ are all positive, this is an equation of an ellipse in the xy plane, with axes along x and y. The length of its shorter half-axis, directed along x, is $\sqrt{(K_z - K_f)/(K_z - K_x)}$; that of its longer half-axis, directed along y, is $\sqrt{(K_z - K_f)/(K_z - K_y)}$. In the limit $K_f = K_y$, this ellipse is identical with the xy ellipse described above. As K_f grows and approaches K_z, the ellipse shrinks, and in the limit $K_f = K_z$ it is reduced to a point at the origin. Its projections onto the unit sphere along the +z and −z directions, which define the two sought closed curves, change accordingly. Thus, if $K_f = K_z$, the transition moment **M**(0f) lies along the z axis (upper left corner in Figure 5.3).

If $K_f < K_y$, equations (5.49) and (5.50) yield

$$(K_y - K_x)\cos^2\phi_y^f + (K_z - K_x)\cos^2\phi_z^f = K_f - K_x \tag{5.61}$$

Now, $K_y - K_x$, $K_z - K_x$, and $K_f - K_x$ are all positive and (5.61) represents an equation of an ellipse in the yz plane, whose projections along the +x and −x directions onto the unit sphere produce the two sought closed curves. The longer half-axis of the ellipse is directed along y, and its length is $\sqrt{(K_f - K_x)/(K_y - K_x)}$; the shorter half-axis is directed along z, and its length is $\sqrt{(K_f - K_x)/(K_z - K_x)}$. In the limit $K_f = K_y$, the ellipse is identical with the yz ellipse already described. As K_f decreases and approaches K_x, the ellipse shrinks, and in the limit $K_f = K_x$ it is reduced to a point at the origin. Its projections onto the unit sphere change accordingly. If $K_f = K_x$, the transition moment **M**(0f) lies along the x axis (lower right corner in Figure 5.3).

We conclude that except for the special cases $K_f = K_x$ and $K_f = K_z$, a measurement of K_f will only permit a conclusion that **M**(0f) is directed to one of the infinite number of points on one of the two closed curves shown on each sphere in Figure 5.3.

The situation simplifies somewhat in three instances:

(i) The orientation is such that $K_y = K_z = (1 - K_x)/2$ (e.g., disc-shaped molecules), all of the admissible directions for **M**(0f) have the same value of $|\phi_x^f|$, and this value can be determined from equation (5.49):

$$\tan^2\phi_x^f = \frac{K_f - K_x}{K_z - K_f} \tag{5.62}$$

(ii) If the orientation is such that $K_y = K_x = (1 - K_z)/2$ (e.g., rod-shaped molecules), all of the possible **M**(0f) directions for a given K_f value have the same value of $|\phi_z^f|$, and equation (5.49) yields

$$\tan^2\phi_z^f = \frac{K_z - K_f}{K_f - K_y} \tag{5.63}$$

(iii) A single additional independent piece of information on the transition moment direction will permit its determination. Thus, if **M**(0f) is known to lie in the uv plane, we have $|\phi_w^f| = 90°$ (u, v, w stand for any combination of the three different molecular axes). Then the sum in (5.49) is reduced to two terms, only one unknown remains, and we obtain

$$\tan^2 \phi_v^f = \cot^2 \phi_u^f = \frac{K_v - K_f}{K_f - K_u} \qquad (5.64)$$

An example of this situation with u = y, v = z, w = x is the common case of $\pi\pi^*$ transitions in planar molecules of symmetry C_s or C_{2h}.

Concluding Comments. It should be emphasized again that only the numerical values and not the signs of the angles ϕ_u^f are accessible from the dichroic measurements. Sometimes the signs can be deduced from other data, such as polarized fluorescence. Such comparison may also permit a more precise determination of the transition moment direction for the transition responsible for the fluorescence. Results on symmetrical molecules contained in stretched polymers (Section 4.6.1) make it very likely that both the values of K_z and K_y and the location of the orientation axis can be estimated reasonably well from the molecular shape. If the shape can be changed in a way which does not affect transition moments appreciably, and the measurements are repeated, additional information on the location of the orientation axis as well as the signs of the ϕ_u^f's may be obtained.

At present, however, it appears that most often only molecules with at least one plane of symmetry can be handled satisfactorily and that a guess at the location of the orientation axis will frequently be the best one can do. The suggestion that the orientation axis lies in the direction perpendicular to the smallest molecular cross section is probably as good as any and is simple to use. The limited experience available suggests that this direction lies within perhaps 10° of the actual orientation axis. Of course, this possible error does not affect the accuracy of the relative magnitudes of the ϕ's. However, these magnitudes still depends on the correctness of the estimate of the K's.

In summary, in order to determine the absolute values of the direction cosines of transition moments in molecules of general symmetry with respect to the molecular framework from a measurement of dichroism, one must begin with five pieces of information in addition to the assumptions already invoked in the high-symmetry case. First, one needs the two independent values for the diagonal orientation factors K_u. Second, one needs the three angles which define the directions of the molecular orientation axes. Alternatively, one needs the five K_{uv}'s in an arbitrarily preselected system of axes. Given this information, a complete determination of the absolute values of at least some of the angles will only be possible in one of the following special cases: either the transition moment lies in the x or the z molecular orientation axis, or (i) the molecule orients so that $K_z = K_y$ (e.g., disc), (ii) the molecule orients so that $K_x = K_y$ (e.g., rod), or (iii) the molecule has another special property such as a plane of symmetry.

The five independent K_{uv} values which are needed can either be guessed from a comparison with related symmetrical molecules or be obtained from some independent measurement such as the determination of linear dichroism of transitions whose polarization directions are already known. A determination of the ϕ's, but not their signs, is in principle feasible in the general case if at least three linearly independent spectra, obtained from two different uniaxial samples, are available, assuming that the z, y, and x axes which diagonalize the orientation tensors are identical in the two samples. In some cases, considerable additional information on the ϕ's may be available from photoluminescence measurements on an isotropic or a partially aligned sample (see Chapter 7).

Before concluding Section 5.1, let us note that the discussion of the evaluation techniques presented here is all based on the assumption that the primary experimental data are the polarized spectra $E_Z(\tilde{\nu})$ and $E_Y(\tilde{\nu})$. These are the most linearly independent pair possible and are usually simple to obtain experimentally.

As mentioned in Section 3.2, at times it may be impossible to measure $E_Z(\tilde{\nu})$ due to the nature of the sample, or it may be more convenient or more accurate to start with a different pair of spectral curves, such as either $E_Z(\tilde{\nu})$ or $E_Y(\tilde{\nu})$ with $E^{iso}(\tilde{\nu})$, $E_\omega(\tilde{\nu})$ with $E_Y(\tilde{\nu})$, etc. In such cases, the evaluation of the spectral ratios d_u, directly for purely polarized transitions and by the stepwise reduction procedure for overlapping transitions, still proceeds exactly as shown. Now, however, the physical significance of the resulting d_u's and their relation to the orientation factors K_u in symmetrical molecules and to the factors K_f of transitions in low-symmetry molecules will be different. Relations (5.29) or (5.34) and (5.46) will have to be replaced by analogous relations similarly derived from (5.26) for symmetrical molecules and from (5.12) for low-symmetry molecules, with the right-hand sides modified appropriately and the requisite new linear combinations placed on the left-hand side of equation (5.32). The derivation of the reduced spectra of symmetrical molecules in the special cases amenable to solution will follow closely the lines given in (5.36)–(5.41). In each case, information exactly equivalent to that obtainable from the $E_Z(\tilde{\nu})$,$E_Y(\tilde{\nu})$ pair can be obtained, at least in principle. We shall not write the results of these derivations here, but a few examples of the procedure will be given in Section 5.3, dealing with modulation spectroscopy, in which case one of the spectral curves is of necessity different from both $E_Z(\tilde{\nu})$ and $E_Y(\tilde{\nu})$.

5.2 Approximate Methods

The analysis provided so far was based on a minimum of assumptions: uniaxial sample, isotropic solvent effects, recognizable contributions from purely polarized spectral features. These assumptions are all justified by experimental observations. Attempts to detect deviations from uniaxial symmetry in samples of stretched polymers of the kind used in these experiments have so far given negative results; deviations from isotropy of solvent effects have been detected, but only in special circumstances (Section 8.1), and single-crystal work on many molecules supports the notion that essentially purely polarized contributions are recognizable in observed UV and IR spectra. The analysis avoided unnecessary assumptions, such as those concerning the detailed nature of the orientation distribution function. Since the only properties of the orientation function which affect the outcome of experiments based on processes of rank 2 are its second moments, expressible, e.g., through the orientation factors K_u, it is appropriate to use such experiments to derive the K_u's, but it is not possible to use them to characterize the orientation distribution function in any greater detail. At any rate, additional information on the orientation distribution function would not make it possible to extract any additional spectroscopic information on the solute molecule as long as it has to do with a process of rank 2 (electric dipole or magnetic dipole one-photon transitions).

This philosophy has not been shared by all workers investigating linear dichroism of solutes in stretched polymers. Some prefer to hypothesize a detailed form of the distribution function, even if it cannot be verified, and work out its consequences for the observable second moments. Thus, in the Fraser-Beer model, it is hypothesized that a fraction f of the solute molecules are aligned perfectly with their z axis along Z and their x and y axes randomly distributed in the XY plane, while a fraction $1 - f$ are oriented totally randomly. In the Tanizaki model, it is assumed that there is a specific relation among the orientation distribution functions of solute molecules in poly(vinyl alcohol) which result for different degrees of stretching: At infinite stretching ratio all molecules are oriented alike, with a particular axis (fixed within the molecular framework) aligned along Z and with axes perpendicular to it aligned randomly in the XY plane. In the Popov model, solute molecules are assumed to be aligned with cylindrical symmetry around each polymer chain, and the second moment for the average chain alignment is specified.

We see two problems with such "detailed model" philosophy. One is of a fundamental nature: An illusion is created that more knowledge is available than is actually the case. It is conceivable that for some solutes the orientation distribution assumed in one or another of these detailed models is actually correct. However, in general it is highly unlikely, and it cannot be verified by optical measurements relying on processes of rank 2. Measurements utilizing processes of rank 4 (Chapter 7) can refute the model for a given solute if their results disagree with expectations derived from the detailed model, but even if they agree they do not prove the correctness of the model, since an infinite number of degrees of freedom still remain.

This is not to say that it would not be highly desirable to know the solute orientation distribution function in intimate detail. Such knowledge would permit a detailed understanding of the intermolecular interaction forces which are responsible for mechanical orientation and of the internal structure of the anisotropic orienting medium. However, this detailed information cannot come from studies of solute linear dichroism alone but can only be approached by a combination of a variety of experimental techniques.

The second problem with the "detailed model" philosophy is of a practical nature. The primary aim of the investigators using these models is to obtain spectroscopic information on solute molecules, such as $A_u(\tilde{\nu})$ or ϕ_u^f. Not only do the excessive assumptions contribute nothing to one's ability to extract such information from the measured spectra, they can actually detract from it to the extent that they provide a temptation to assume values for the second moments of the orientation distribution function which may be wrong for some solutes. For instance, the Fraser-Beer orientation distribution function is independent of the Euler angle γ and demands $K_x = K_y$ and thus a particular relation between K_y and K_z: $K_y = (1 - K_z)/2$. In a given polymer, this relation is fulfilled for some solutes and not for others. Applied indiscriminately, the model has led to wrong conclusions about $A_u(\tilde{\nu})$ and ϕ_u^f in many cases. The model is clearly not valid in general, but at least it appears possible to guess when it will give wrong results. The assumptions made in the Tanizaki and Popov models are less transparent, and it is harder to guess when they will produce incorrect conclusions about $A_u(\tilde{\nu})$ and ϕ_u^f.

A part of the motivation for the construction of detailed orientation models undoubtedly is the desire to understand the dependence of the measured spectra on the degree of stretching when a polymer is used as the anisotropic medium. So far, few detailed experimental investigations of this dependence have been published, and in our opinion none of them provide unequivocal support for any of the detailed models proposed.

Since the three detailed models mentioned above have found extensive use in the literature, we shall now proceed to analyze them in detail, translating them into the language developed in Section 5.1 as a common basis. We attempt to delineate their respective areas of applicability to help the reader evaluate the reliability of literature data.

5.2.1 *The Fraser-Beer Approximation*

Fraser and Beer described the orientation of polymer chains using the assumption that all angles of rotation of the chain along its own direction were equally probable. The formalism was then transferred to work on solutes contained in stretched polymers and exploited primarily by Mazur, Yogev, and collaborators. Here the assumption was that all angles of rotation of a solute molecule around its own Z axis are equally probable. This amounts to assuming a uniform distribution of the Euler angle γ, i.e., a distribution function of the type $f(0,\beta,0)$, with normalization condition given by (4.5). This is physically reasonable in some cases, for instance if the molecule is cylindrical. This type of orientation distribution is also encountered in some photoselected assemblies (Section 4.6). It is also reasonable for that infinitesimal fraction of molecules of any shape which happen to be perfectly aligned with their z axes along a polymer stretching direction, since the distribution is uniaxial. In general, however, the assumption is arbitrary, and there is no physical reason why it should be fulfilled. The arguments to the contrary which can be found in published literature are wrong and confuse the rotation around the stretching direction Z (Euler angle α) with rotation around the z axis of the molecule (Euler angle γ, cf. Figure 5.4).

In a more general sense, for processes of rank 2 the Fraser-Beer approximation amounts to postulating the effective orientational equivalence of two of the molecular axes. If these are x and y, this is identical to postulating $K_y = (1 - K_z)/2$. If they are y and z, this is identical to postulating $K_y = K_z$. With

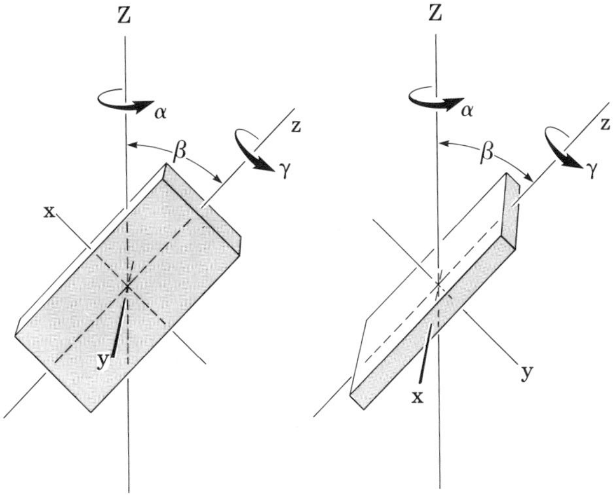

Figure 5.4 The Fraser-Beer model: two physically inequivalent molecular orientations assumed to be equally probable. The values of Euler angles α and β are equal for both, but $\gamma = 0$ on the left and $\gamma = \pi/2$ on the right.

either assumption, only one K needs to be measured to determine all three K's. This K value may be determined either by the stepwise reduction procedure or, in simple cases, from the dichroic ratio.

High-Symmetry Molecules. If the postulate is $K_y = (1 - K_z)/2$, it is common to define $f = K_z - K_y = (3K_z - 1)/2 = S_{zz}$. The reduced spectra $A_z(\tilde{\nu})$ and $A_x(\tilde{\nu}) + A_y(\tilde{\nu})$ may be calculated according to equations (5.38) and (5.39). If $A_x(\tilde{\nu}) = 0$, one obtains the usual form of the Fraser-Beer model

$$\frac{E_Z(\tilde{\nu})}{E_Y(\tilde{\nu})} = \frac{A_z(\tilde{\nu})f + (1/3)(1 - f)[A_y(\tilde{\nu}) + A_z(\tilde{\nu})]}{(1/2)A_y(\tilde{\nu})f + (1/3)(1 - f)[A_y(\tilde{\nu}) + A_z(\tilde{\nu})]} \tag{5.65}$$

It is not necessary to believe that f really represents a totally oriented fraction of molecules $[(K_z,K_y) = (1,0)]$ and $1 - f$ a totally random fraction of molecules $[(K_z,K_y) = (1/3,1/3)]$. Rather, it is possible to take the more realistic point of view that the orientation distribution function is not known and that f is an empirical measure of the degree of alignment. This is the attitude we shall adopt. Variation of f from 0 to 1 produces all (K_z,K_y) pairs lying on the $K_x = K_y$ side of the orientation triangle introduced in Chapter 4, i.e., all pairs possible within the Fraser-Beer model.

If the postulated relation is $K_z = K_y$, the reduced spectra are given by equations (5.36) and (5.37). This case was first analyzed by Nordén, who defined $f = 2(K_z - K_x) = 2(3K_z - 1) = 1 - 3K_x = 2S_{xx}$. Then

$$\frac{E_Z(\tilde{\nu})}{E_Y(\tilde{\nu})} = \frac{(2 + f)[A_z(\tilde{\nu}) + A_y(\tilde{\nu})] + 2(1 - f)A_x(\tilde{\nu})}{(1/2)(4 - f)[A_z(\tilde{\nu}) + A_y(\tilde{\nu})] + (2 + f)A_x(\tilde{\nu})} \tag{5.66}$$

As before, it is possible to think of f as either a hypothetical or a real fraction of molecules with y, z planes aligned perfectly with the stretching direction

$[(K_z,K_y) = (1/2,1/2)]$ and to think of $1 - f$ as a totally random fraction of molecules $[(K_z,K_y) = (1/3,1/3)]$. Variation of f from 0 to 1 then produces all (K_z,K_y) pairs lying on the $K_y = K_z$ side of the orientation triangle, i.e., all pairs possible within this model.

While it is clear from Figure 5.4 that there is no physical reason why the Fraser-Beer assumption should be generally valid for solutes other than those which are rodlike or disclike, it could be valid accidentally in practice, and it does not yet follow that it is wrong in general. Such a proof is, however, provided by experimental data for molecules of other than rodlike or disclike shapes. For instance, for molecules such as acenaphthylene in which not only z-polarized but also y-polarized transitions exhibit a dichroic ratio $E_Z(\tilde{\nu}_y)/E_Y(\tilde{\nu}_y)$ larger than 1, i.e., $K_z > K_y > 1/3$, the rodlike approximation, $K_x = K_y$, would require $K_x + K_y + K_z > 1$, which is impossible. The disclike approximation is also unacceptable, since K_y and K_z are clearly different. Thus, there is no doubt that neither the Fraser-Beer assumption of two equivalent short axes, which was most likely correct for the polymer itself and for some of the molecules to which it was applied, nor the assumption of two equivalent long axes can be used for a general solute molecule. Many examples of symmetrical molecules which probably are not sufficiently disc-shaped or rod-shaped in terms of orienting properties and for which questionable conclusions were reached by application of this model can be found in the literature.

Low-Symmetry Molecules. It has usually been assumed by the users of the Fraser-Beer approximation that nonoverlapping purely polarized transitions can be found in the spectra; then $d_f = E_Z(\tilde{\nu}_f)/E_Y(\tilde{\nu}_f)$. With the assumption $K_y = (1 - K_z)/2$, relation (5.58) for the direction cosine of the transition moment of such a purely polarized transition with respect to the orientation axis, ϕ_z^f, is identical with the expression used by Yogev and collaborators, who usually write

$$d_f = \frac{f \cos^2 \phi_z^f + (1 - f)/3}{(f/2) \sin^2 \phi_z^f + (1 - f)/3} \tag{5.67}$$

so that $\cos^2 \phi_z^f$ can be expressed in terms of the remaining orientation parameter K_z or f and the observed dichroism parameter K_f, d_f, or g_f, as preferred by various authors:

$$\begin{aligned} \cos^2 \phi_z^f &= [d_f/(2 + d_f) - (1 - f)/3]/f = [K_f - (1 - f)/3]/f \\ &= [d_f + 2(1 - K_z)(d_f - 1)/(3K_z - 1)]/(d_f + 2) \\ &= [(3K_z - 1) + g_f(3 - K_z)]/(3K_z - 1)(3 - g_f) \end{aligned} \tag{5.68}$$

where

$$f = K_z - K_y = (3K_z - 1)/2 \qquad K_f = d_f/(2 + d_f)$$

and

$$g_f = [E_Z(\tilde{\nu}_f) - E_Y(\tilde{\nu}_f)]/[E_Z(\tilde{\nu}_f) + E_Y(\tilde{\nu}_f)] = (d_f - 1)/(d_f + 1)$$

For many but not all of the molecules to which this procedure has been applied, the assumption of rodlike orientation appears to be acceptable.

With a correct choice of the orientation axis and K_z, an incorrect assumption of rodlike orientation will lead to too high values for ϕ_z^f for a transition polarized in the yz plane, while too low values of ϕ_z^f will be found for a transition in the xz plane. This is clear from equation (5.64) and from the fact that an incorrect rodlike assumption for a given K_z corresponds to an overestimate of K_x and underestimate of K_y.

In summary, the Fraser-Beer approximation is reasonable for molecules which orient like rods or discs. Since this is rarely known with certainty, and since in most cases the use of the approximation saves hardly any labor, it is best avoided. The published values of polarization angles and reduced spectra based on this approximation should be viewed critically, particularly for molecules which are clearly not rodlike or disclike.

5.2.2 The Tanizaki Approximation. Definitions of the Orientation Axis

General. One of the first quantitative models for the evaluation of linear dichroism of solutes in a stretched polymer, poly(vinyl alcohol), was proposed by Tanizaki. The fundamental assumption is that at infinite stretching all molecules would be aligned alike (Figure 5.5): One particular direction in the molecular framework, referred to as the "orientation axis," would be aligned exactly along the stretching direction, and all angles of rotation around this direction would be equally probable (as they must be if the sample is uniaxial). It should be noted that the term "orientation axis" has a different meaning in this model than in all the others; to denote this, we shall place it in quotation marks. Although originally the possibility was entertained that in planar molecules the "orientation axis" need not lie in the molecular plane, in recent work it has been generally assumed that it must. This is unfortunate, since the assumption implies the existence of a definite relation between K_z and K_y for any given stretch ratio.

It is further assumed that the dichroic ratio $d(\tilde{\nu})$, which Tanizaki labels $R_d(\tilde{\nu})$,

$$R_d(\tilde{\nu}) = E_Z(\tilde{\nu})/E_Y(\tilde{\nu}) \qquad (5.69)$$

measured for finite stretching ratios R_s can be extrapolated to obtain the dichroic ratios $R_{d\infty}(\tilde{\nu})$ which correspond to infinite stretching:

$$R_{d\infty}(\tilde{\nu}) = \frac{2(T-1) + (T+1)R_d(\tilde{\nu})}{2T + (T-1)R_d(\tilde{\nu})} \qquad (5.70)$$

where

$$T = \frac{R_s^2}{R_s^2 - 1}\left\{1 - \left[\frac{\pi}{2} - \tan^{-1}(R_s^2 - 1)^{-1/2}\right](R_s^2 - 1)^{-1/2}\right\} \qquad (5.71)$$

The stretch ratio R_s is defined in Section 4.6.1 as the ratio of the long to the short axis of an ellipse into which a circle drawn on the sheet before stretching

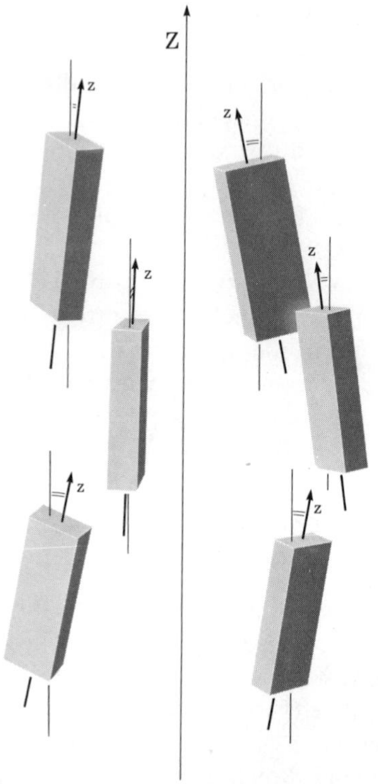

Figure 5.5 The Tanizaki model: the alignment of solute molecules in the limit of infinite stretching. The Tanizaki "orientation axis" of each molecule (thin line) is parallel to the stretching direction Z. Its angle with the molecular z axis (thick line) is indicated by a double arc and is the same for every molecule. The orientation distribution is uniform with respect to rotation around the molecular "orientation axis."

will distort as a result of the stretching. This formula corresponds to the orientation distribution (4.51) and was derived from the assumption that upon stretching, the contents of a sphere rearrange rigidly into the contents of an ellipsoid of rotation.

Next, $|\theta(\tilde{\nu})|$ is defined by

$$\cot^2 \theta(\tilde{\nu}) = R_{d\infty}(\tilde{\nu})/2 \tag{5.72}$$

In the papers using Tanizaki's method, wavelengths rather than wavenumbers are usually used, and a plot of $R_d(\lambda)$ against λ is usually shown along with the values of $|\theta(\lambda)|$ for selected values of λ.

If the absorption of a symmetrical molecule is purely polarized along the u-th axis at $\tilde{\nu}_u$, the physical meaning of $|\theta(\tilde{\nu}_u)| = |\theta_u|$ can be shown from the above assumptions to be the angle between the u-th axis and the molecular "orientation axis." Thus, $|\theta_u|$ can have only three possible values for purely polarized transitions: $|\theta_z| \leqslant |\theta_y| \leqslant |\theta_x|$. In those planar molecules which exhibit only $\pi\pi^*$ transitions, only $|\theta_z|$ and $|\theta_y|$ are observed.

In regions of overlapping transitions, intermediate values of $|\theta(\tilde{\nu})|$ can occur. Since the "orientation axis" is assumed to lie in the molecular plane, for planar molecules $|\theta_x| = 90°$ and $|\theta_z| + |\theta_y| = 90°$. Note that the labeling of axes used in this book (z orients best, x worst) is different from Tanizaki's (x orients best, z worst).

If it is now assumed that one of the observed absorption features is purely polarized along z or y, equations (5.70)–(5.72) and the relation $|\theta_z| + |\theta_y| = 90°$ can be used to obtain $|\theta_z|$ and $|\theta_y|$. If the out-of-plane polarized intensity $A_x(\tilde{\nu})$ is zero in the region of interest, the following expressions can be derived from the assumptions of the model for the absorption intensity $A_z(\tilde{\nu})$ polarized along z and the intensity $A_y(\tilde{\nu})$ polarized along y:

$$A_z(\tilde{\nu})/A_y(\tilde{\nu}) = [2 - R_{d\infty}(\tilde{\nu}) \cot^2 \theta_z]/[R_{d\infty}(\tilde{\nu}) - 2 \cot^2 \theta_z] = F(\tilde{\nu}) \quad (5.73)$$

$$E(\tilde{\nu}) = A_z(\tilde{\nu}) + A_y(\tilde{\nu}) \quad (5.74)$$

where $E(\tilde{\nu})$ is proportional to the baseline corrected absorbance of an unstretched dyed film. Then the spectrum $E(\tilde{\nu})$ can be combined with the ratio $F(\tilde{\nu})$, derived from $R_{d\infty}(\tilde{\nu})$ and $|\theta_z|$, to obtain the purely polarized spectra $A_z(\tilde{\nu})$ and $A_y(\tilde{\nu})$:

$$A_z(\tilde{\nu}) = \{F(\tilde{\nu})/[1 + F(\tilde{\nu})]\}E(\tilde{\nu}) \quad (5.75)$$

$$A_y(\tilde{\nu}) = E(\tilde{\nu}) - A_z(\tilde{\nu}) \quad (5.76)$$

These are the expressions used by Tanizaki et al., who, however, assume that they are valid even if $A_x(\tilde{\nu}) \neq 0$, in which case the out-of-plane polarized contributions $A_x(\tilde{\nu})$ will appear in the derived short-axis polarized $A_y(\tilde{\nu})$ curve and not in the long-axis polarized $A_z(\tilde{\nu})$ curve. This belief is incorrect unless the orientation is rodlike, as will be shown below.

In low-symmetry molecules, it is usually assumed by the users of Tanizaki's procedure that the transitions of interest are purely polarized. Then the angle $|\theta_f|$ which the f-th transition moment makes with the "orientation axis" is determined from T and $R_d(\tilde{\nu})$ by use of equations (5.70)–(5.72) as before. The location of the "orientation axis" cannot be obtained, but the results may be used for an estimate of relative angles between transition moments. The angle between transitions f and j is $\phi_{fj} = |\theta_f \pm \theta_j|$ if it is assumed that the orientation axis and the two transition moments are in the same plane. The problems involved when this assumption does not hold have been discussed by Fucaloro and Forster, who introduced a "cone analysis" for the description of such cases.

Definition of the Orientation Axis. A severe conceptual difficulty connected with the use of Tanizaki's model is due to the nature of the definition of the "orientation axis" and its possible nonuniqueness. In general, Tanizaki's "orientation axis" does not coincide with any of the symmetry axes of the molecule. In a sense, it corresponds to a "preferred orientation axis," i.e., a direction in the molecule which in a real sample often finds itself almost exactly aligned with the sample Z axis. In the case of mechanically orienting media such as liquid crystals and stretched polymers, it is common to identify such a direction either with the longest dimension of the molecule or with the direction

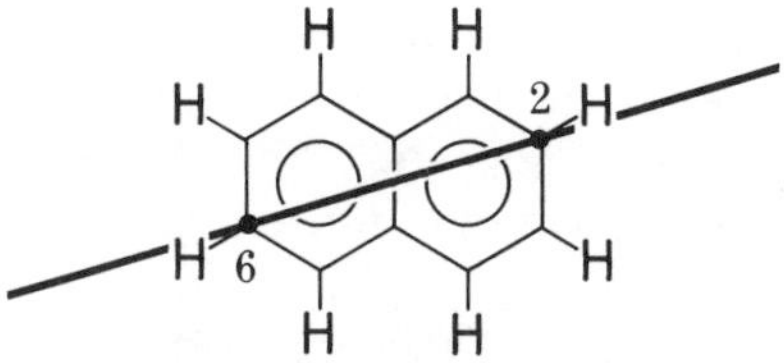

Figure 5.6 An assumed "preferred orientation axis" in naphthalene.

perpendicular to its smallest cross section. For instance, in naphthalene, such an axis might be believed to run approximately between positions 2 and 6 (Figure 5.6). Intuitively, it might be tempting to conclude that such an axis is the one among all possible directions in the molecule which aligns best, and this presumably provided Tanizaki's rationale for referring to it as the "orientation axis." It appears to us that the most useful quantitative rendition of the concept of alignment is provided if we define the direction which aligns best as the one which would yield the highest dichroic ratio $E_Z(\tilde{\nu}_f)/E_Y(\tilde{\nu}_f)$ if the f-th transition moment lay in it, i.e., the one for which $K_f = \langle \cos^2 fZ \rangle$ is largest. Other definitions are possible. For instance, the best aligned direction could be taken to be the one which yields the largest value for $|\cos fZ|$, but this does not change the following argument.

In symmetrical molecules, it is obvious that the above intuitive conclusion cannot possibly be correct unless Tanizaki's "orientation axis" coincides with one of the symmetry axes. This is clear from the fact that the largest (as well as the smallest) possible value of K_f must be identical with one of the eigenvalues of the orientation tensor, whose principal directions are dictated by molecular symmetry and must lie in the symmetry axes. Unfortunately, this is often misunderstood even in connection with orientation models different from that of Tanizaki. To see the point clearly, it is only necessary to realize that molecular axes are assumed to be rigidly fixed to the molecular framework and that in a symmetrical molecule any direction which does not coincide with one of the symmetry axes comes as a pair or possibly an even larger number of physically indistinguishable directions, so that the "preferred orientation axis" generally is not unique. In the above example of naphthalene, a C_2—C_6 direction is symmetry equivalent to the C_3—C_7 direction (Figure 5.7). We shall label α the smaller of the two angles formed by the intersection of these two directions. Any aligning force which forces the C_2—C_6 direction into alignment with

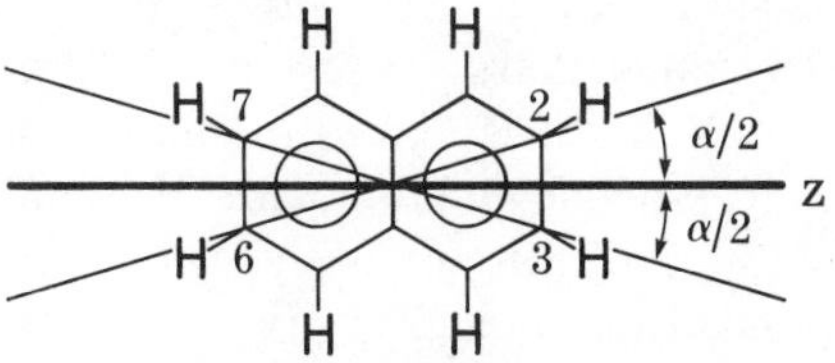

Figure 5.7 Symmetry-equivalent "preferred orientation axes" (thin lines) and the effective orientation axis z (thick line) in naphthalene.

the sample Z axis cannot fail to force the C_3—C_7 direction into alignment to exactly the same degree. Even if this force is infinitely strong, the "preferred orientation axis" C_2—C_6 will lie in the Z direction only in half the molecules. In the other half of the molecules, the C_3—C_7 direction will lie along Z and the C_2—C_6 direction will lie at an angle α from it. The K value for the C_2—C_6 direction will be

$$K_f = \langle \cos^2 fZ \rangle = (1/2)\cos^2 0 + (1/2)\cos^2 \alpha = (1 + \cos^2 \alpha)/2 \qquad (5.77)$$

In this same orientation distribution, the symmetry axis z which bisects the angle α and joins the midpoint of the C_2—C_6 bond with that of the C_3—C_7 bond always lies at the angle $\alpha/2$ with respect to the Z sample axis (Figure 5.7). Its K value is

$$K_{f'} = \langle \cos^2 f'Z \rangle = \cos^2(\alpha/2) = (1 + \cos\alpha)/2 \qquad (5.78)$$

Since $\alpha < \pi/2$,

$$K_{f'} - K_f = (1 - \cos\alpha)\cos\alpha/2 = \cos\alpha \sin^2(\alpha/2) > 0 \qquad (5.79)$$

We see that the symmetry axis direction z aligns better on the average than the C_2—C_6 direction in spite of the fact that no molecule is actually aligned with its symmetry axis z along the Z direction of the sample. In other words, although the C_2—C_6 and C_3—C_7 directions are the "preferred orientation axes" in our idealized orientation distribution, the long symmetry axis aligns best and is the "effective orientation axis," or orientation axis for short.

In view of the nonunique nature of Tanizaki's effective "orientation axis," we believe that it is better not to build a molecular axes system around it. The basic assumption of the Tanizaki model is clearly unphysical; even at infinite stretching ratio, all molecules cannot align with the same "orientation axis" along the stretching direction Z of the sample if symmetry demands that there be several indistinguishable such axes.

It is possible to avoid this problem for symmetrical molecules by assuming instead that at infinite stretching ratio, each of the equivalent "orientation axes" will have equal probability of perfect alignment. However, this will still be unsatisfactory for unsymmetrical molecules with several nearly equivalent preferred orientation axes. Imagine, for instance, that the naphthalene molecules of the preceding example have been converted into quinoline molecules by aza replacement in position 1. Now, the C_2—C_6 and C_3—C_7 directions are no longer equivalent, and the Tanizaki "orientation axis" will have to lie near one or the other. It is rather unphysical to imagine that as the stretching ratio is increased all the quinoline molecules will align themselves in only one of the two nearly equivalent ways, as demanded by the Tanizaki model, so that a transition polarized along C_2—C_6 will give a completely different dichroic ratio than one polarized along C_3—C_7.

In one case an experimental proof of the uniqueness of the "orientation axis" was claimed for an unsymmetrical molecule. In this investigation the authors showed that the conditions for the Fraser-Beer and Tanizaki assumptions lead to the following equation for the Fraser-Beer f:

$$f = (R_d - 1)(R_{d\infty} + 2)/(R_{d\infty} - 1)(R_d + 2) \qquad (5.80)$$

Experimentally, the expression for f was found to be constant for all wavelengths at a given stretch ratio, which was taken to prove the uniqueness of the "orientation axis." However, substituting into equation (5.80) from equation (5.70) we obtain $f = (3T - 1)/2$. Therefore, f is independent of R_d by definition, and all experimental spectral values cancel in its calculation. Since T is a constant determined solely by the stretch ratio, the fact that f is constant proves nothing about the solute, and Tanizaki's definition of the "orientation axis" remains as much of a problem as before.

A related complication in the method is that the direction of the "orientation axis" of this model for molecules of low symmetry is difficult to determine from the molecular structure. As discussed in Section 4.6, the usual definition of the orientation axis makes an estimate possible within reasonable error limits.

Critique of the Model. As pointed out by Tanizaki at the outset of his investigation and by other authors since, the whole concept of the simple orientation distribution assumed for an infinitely stretched poly(vinyl alcohol) sheet is rather unrealistic. Even in a very highly aligned polymer, there will undoubtedly be a continuous distribution of orientation for guest molecules around their (possibly numerous) most probable positions of alignment, and, most important, molecular planes will not be perfectly parallel to the stretching direction. Note, however, that even if the assumed nature of the orientation distribution in the hypothetical limit of an infinitely stretched polymer is incorrect, the procedure may in principle still lead to correct results.

In this respect, the first helpful factor is that only the second moments of the assumed distribution function (related to K_z, K_y, and K_x in our notation) affect the measured linear dichroism, and these can be correct even if the overall distribution function is incorrect. In the present case, however, the assumed distribution function is too restrictive. In order to see this, it is useful to first express dichroic ratios for purely polarized transitions at $\tilde{\nu}_x$, $\tilde{\nu}_y$, and $\tilde{\nu}_z$ by means of the orientation factors K_x, K_y, and K_z:

$$R_{d\infty}(\tilde{\nu}_u)/2 = \cot^2 \theta_u = \frac{K_u^\infty}{1 - K_u^\infty} = \frac{T - 1 + 2K_u}{2(T - K_u)}, \qquad u = x,y,z \qquad (5.81)$$

$$R_d(\tilde{\nu}_u)/2 = \frac{K_u}{1 - K_u}, \qquad u = x,y,z \qquad (5.82)$$

The quantities with the superscript ∞ correspond to the hypothetical limiting orientation distribution at infinite stretching. The assumption of perfect alignment of the molecular plane in this limit means $K_x^\infty = 0$ ($K_z^\infty + K_y^\infty = 1$), which, however, it is hard to imagine could be valid in general.

The second helpful factor built into the model is the extrapolation procedure by means of which the actually measured $R_d(\tilde{\nu})$ is converted to $R_{d\infty}(\tilde{\nu})$. The measured $R_d(\tilde{\nu})$ values reflect a certain set of orientation parameters K_z, K_y, K_x in the actual sample. The extrapolation to $R_{d\infty}(\tilde{\nu})$ corresponds to converting these into new values K_z^∞, K_y^∞, K_x^∞. It is not claimed in the model that the former set fits the requirements imposed by the form of the distribution function assumed for infinite stretching, but only that the latter does. It is therefore important to analyze what the requirement imposed on the latter ($K_x^\infty = 0$) implies for the former before drawing conclusions on the validity of the method.

The relation between $R_{d\infty}$ and R_d [equations (5.70), (5.71)] was originally derived by making certain specific assumptions about the nature of the stretching process. The physical meaning of T is the average value of the square of the direction cosine of "unit vectors" which are randomly distributed in an unstretched sheet, pointing with equal probability to any point on a unit sphere, and which gradually become aligned with the stretching direction during the stretching process which converts the unit sphere into an ellipsoid of rotation of equal volume. The ratio of the long axis to the short axis of the ellipsoid is equal to R_s. It has been known for some time that this form of the extrapolation does not fit some of the experimental data, but this is immaterial since the assumed distribution function is unrealistic anyway.

Equations (5.70) and (5.71) lead to the following relation between the orientation factors describing the actual distribution and those describing the hypothetical limiting distribution:

$$K_u^\infty = \frac{T - 1 + 2K_u}{3T - 1}, \qquad u = x,y,z \tag{5.83}$$

The assumed orientation distribution in the infinitely stretched sheet requires $K_z^\infty + K_y^\infty = 1$, $K_x^\infty = 0$. This will be fulfilled if and only if

$$K_z + K_y = (1 + T)/2 \qquad K_x = (1 - T)/2 \tag{5.84}$$

Thus, if the solute orientation distribution in the actually measured sheet happens to fulfill relation (5.84), Tanizaki's procedure will produce correct transition moment directions $|\theta_u|$; moreover, if $A_x(\tilde{\nu}) = 0$, it will also produce correct purely polarized spectra $A_z(\tilde{\nu})$ and $A_y(\tilde{\nu})$. In the orientation triangle, the K_z, K_y values for which Tanizaki's theory will produce correct results lie on a line parallel to the upper side of the triangle and separated from it by $(1 - T)/2\sqrt{2}$ as shown in Figure 5.8.

It has been stated repeatedly in the literature by authors other than Tanizaki that his model assumes a uniform distribution of the Euler angle γ which describes the rotation around the long axis of the molecule. This is rather misleading. What should actually be stated is that in the limit of infinite stretching, but not necessarily in the real sample, all angles of rotation around Tanizaki's "orientation axis" are equally likely. This is not an assumption if the infinitely stretched polymer is uniaxial and the axis is unique. If we recognize the existence of more than one equivalent "orientation axis" in symmetrical molecules, we must postulate that each has an equal probability of finding itself aligned with the sample Z axis in the infinite stretching limit and that all angles of rotation around the "orientation axis" which is so aligned are equally distributed. However, even if we were to choose one of these axes as the z axis of a molecular system used to define Euler angles, it would not be correct to state that the γ distribution is uniform in the infinite stretching limit. This would be true only

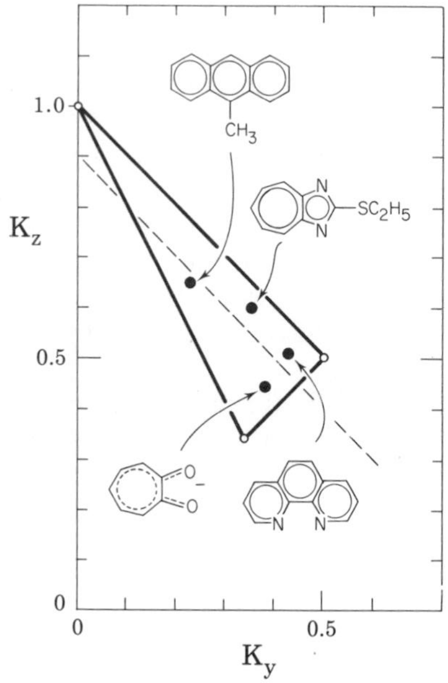

Figure 5.8 Orientation factors of a few solutes in stretched poly(vinyl alcohol), $R_s = 8$ ($T = 0.81$). In Tanizaki's model it is assumed that all points lie on the dashed line: $K_y + K_z = (T + 1)/2$. Reproduced by permission from E. W. Thulstrup and J. Michl, *J. Phys. Chem.* **84**, 82 (1980).

for that fraction of molecules which are aligned with the z axis along the Z direction, but not for those which are aligned so that one of their other equivalent "orientation axes" lies along Z. If we instead use the principal axes of the orientation tensor as the molecular axes, as we normally do, the requirement of a uniform γ distribution calls for $K_x = K_y$, and while this is not assumed in the Tanizaki model at all, neither for the real sample nor in the infinite stretching limit, it is also not excluded. The assumption that the "orientation axis" lies in the molecular plane amounts only to postulating relation (5.84).

Tanizaki's procedure will produce incorrect reduced spectra if $A_x(\lambda) \neq 0$ even if (5.84) holds, unless the orientation is rodlike. To see this, imagine that the function $F(\tilde{\nu})$

$$F(\tilde{\nu}) = \frac{2 - R_{d\infty}(\tilde{\nu}) \cot^2 \theta_z}{R_{d\infty}(\tilde{\nu}) - 2 \cot^2 \theta_z} \tag{5.85}$$

is evaluated from the measured spectra according to (5.73). Then, using $K_x^\infty = 0$, $K_y^\infty = 1 - K_z^\infty$,

$$R_{d\infty}(\tilde{\nu}) = \frac{K_z^\infty A_z(\tilde{\nu}) + (1 - K_z^\infty)A_y(\tilde{\nu}) + 0 \times A_x(\tilde{\nu})}{(1/2)(1 - K_z^\infty)A_z(\tilde{\nu}) + (1/2)K_z^\infty A_y(\tilde{\nu}) + (1/2)A_x(\tilde{\nu})} \tag{5.86}$$

If equations (5.75) and (5.76) are now used as one seeks to derive the reduced spectra from the experimental values of $F(\tilde{\nu})$ and $E(\tilde{\nu})$, the following results are obtained using (5.70) and (5.81), and assuming that (5.84) holds: The expression

$$\frac{F(\tilde{\nu})}{1+F(\tilde{\nu})}E(\tilde{\nu}) = A_z(\tilde{\nu}) - \frac{1}{\cot^2\theta_z - 1}A_x(\tilde{\nu}) \tag{5.87}$$

is obtained upon an attempt to obtain $A_z(\tilde{\nu})$, and

$$E(\tilde{\nu})\left[1 - \frac{F(\tilde{\nu})}{1+F(\tilde{\nu})}\right] = [A_x(\tilde{\nu}) + A_y(\tilde{\nu})] + \frac{1}{\cot^2\theta_z - 1}A_x(\tilde{\nu}) \tag{5.88}$$

is obtained in an attempt to obtain $A_y(\tilde{\nu})$ or $A_x(\tilde{\nu}) + A_y(\tilde{\nu})$. The desired results, i.e., $A_z(\tilde{\nu})$ and $A_x(\tilde{\nu}) + A_y(\tilde{\nu})$, are obtained only if $\theta_z = 0$, i.e., for rodlike orientations. In all other cases, and in particular all those studied experimentally so far, any $A_x(\tilde{\nu})$ contributions which may exist appear with negative weight in the curve claimed to be $A_z(\tilde{\nu})$, and their contribution to the curve claimed to be $A_x(\tilde{\nu}) + A_y(\tilde{\nu})$ is exaggerated.

In the practical application of the method, $|\theta_z|$ (or $|\theta_y|$) is first obtained from equations (5.70)–(5.72) for a transition assumed to be purely z- or y-polarized. This is equivalent to finding the K_z (or K_y) value for the actually measured distribution. Now, in many symmetrical molecules, purely or almost purely polarized transitions can be found not only along one but along both of the y and z directions. Then the above procedure can be used to find both $|\theta_y|$ and $|\theta_z|$ independently, and this is equivalent to finding K_y and K_z independently. If these values are in the relation demanded by equation (5.84), it will be found that $|\theta_y|$ and $|\theta_z|$ add up to 90° and the results will be correct. Frequently, however, the sum of the measured extreme values of $|\theta|$, $|\theta|_{min} + |\theta|_{max}$, was found to differ from 90°. Sometimes this could have been due to band overlap, but most often it is only a reflection of the fact that the fundamental assumption of the model is not fulfilled and that (5.84) does not hold.

The unrealistic nature of the restriction (5.84) is most clearly illustrated in the case of solutes shaped like prolate ellipsoids of rotation ("rod-shaped" molecules, e.g., acetylene) or oblate ellipsoids of rotation ("disc-shaped" molecules, e.g., coronene). By symmetry, $K_y = (1 - K_z)/2$ for the former ($K_z = T$, $|\theta_z| = 0°$) and $K_y = K_z$ [$K_z = (T + 1)/4$, $|\theta_z| = 45°$] for the latter. Thus, for a given stretch ratio R_s and thus a given constant T, equation (5.84) demands that all solutes of the former kind, whether javelin-like or nearly spherical, exhibit an equal degree of orientation and equal dichroic ratios [$2T/(1 - T)$ for z-polarized transitions and $2(1 - T)/(1 + T)$ for y-polarized transitions], while all solutes of the latter kind, whether discus-like or nearly spherical, exhibit another equal degree of orientation and equal dichroic ratios [$2(1 + T)/(3 - T)$ for in-plane polarized transitions and $2(1 - T)/(1 + T)$ for out-of-plane polarized transitions]. Yet spherical molecules, which differ only infinitesimally from either type of ellipsoid in the limit, must remain unoriented and have dichroic ratios equal to 1 for all transitions. This is clearly unphysical.

Combining equations (5.70) and (5.72), it can be shown that

$$|\theta_z| + |\theta_y| \gtreqless 90° \qquad \text{if } K_z + K_y \lesseqgtr \frac{1+T}{2} \tag{5.89}$$

Therefore, the sum $|\theta_z| + |\theta_y|$ will exceed 90° if the point representing the actual sample orientation lies below the dashed line in the orientation triangle of Figure 5.8, and it will be less than 90° if the point lies above it. The deviations from 90° are usually numerically small (the largest possible value is 109.5°, twice the magic angle) but correspond to considerable deviations of the K_z and K_y values from the line defined by equation (5.84), well beyond experimental error.

The above analysis applies to purely polarized spectral features in highly symmetrical molecules. If angles θ are derived from dichroic ratios measured in regions of mixed polarization, one or both of the extreme values $|\theta|_{min}$ and $|\theta|_{max}$ will fail to reach the values $|\theta_z|$ and $|\theta_y|$ which reflect the molecular orientation. If $|\theta|_{min}$ is obtained from a region of mixed polarization, $|\theta|_{min} + |\theta_y| > 90°$; if $|\theta|_{max}$ is obtained from such a region, $|\theta_z| + |\theta|_{max} < 90°$. Since the Tanizaki model frequently leads to incorrect conclusions for highly symmetrical molecules, its results for molecules of low symmetry are not likely to be useful.

In summary, the model is based on an approximation [eq. (5.84)] which has no sound physical basis. It is difficult to predict for which molecules it will be fulfilled and for which ones it will not; in this sense the approximation is much harder to defend than the Fraser-Beer approximation. A way to remove the worst practical problem of the Tanizaki approximation would be to permit the "orientation axis" to lie outside of the molecular plane. By permitting real angles of deviation, the (K_z, K_y) point in the orientation triangle would be allowed to move part way toward the point (1/3,1/3). Since it must occasionally lie in the opposite direction from the line defined by $K_y + K_z = (1 + T)/2$, it would also be necessary to permit complex angles between the "orientation axis" and the molecular plane. We believe that it would be best to abandon the model altogether, and would recommend a healthy dose of skepticism when using published results based on its use.

5.2.3 *The Popov Approximation*

Popov has introduced a model for stretched polymer work which represents simultaneously both a generalization of the description given in Section 5.1 in terms of the orientation factors K and a restriction in that an assumption is made concerning the nature of $f(\Omega)$. Popov uses a third parameter which characterizes the uniaxially stretched polymer: $A = \langle \cos^2 \theta \rangle$, where θ is the angle between the direction of a polymer chain and the macroscopic stretching direction. The value of A is independent of the solute used and depends only on the nature of the polymer and the degree of stretching. Unlike Tanizaki's T, which serves a similar purpose, A is not obtained from assumptions concerning the nature of the stretching process but is an empirical parameter to be evaluated from measurements on the polymer itself (e.g., its IR dichroism).

The remaining two parameters describe the average alignment of the solute molecules with respect to the polymer chain. The restriction on $f(\Omega)$ which is introduced is the assumption that the solute alignment relative to the polymer chain possesses cylindrical symmetry around the chain direction (Figure 5.9). The two parameters are analogous to the two orientation parameters K_z and

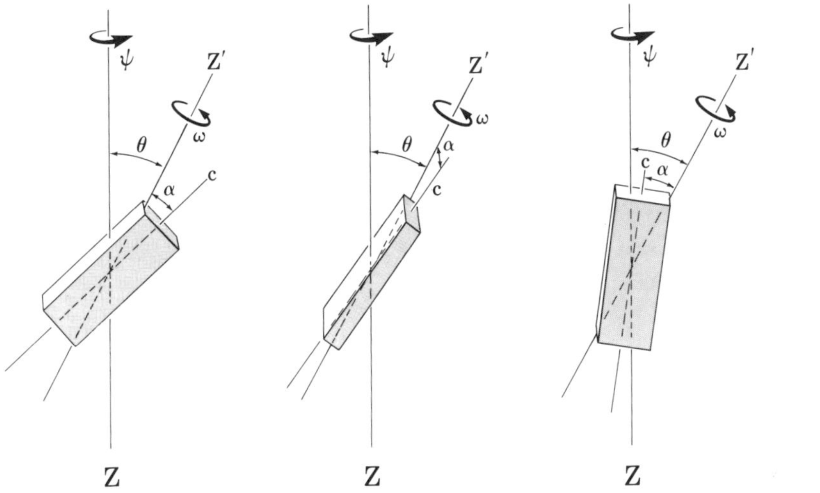

Figure 5.9 The Popov model: physically inequivalent molecular orientations assumed to be equally probable. The direction of a selected polymer chain is Z′. A molecule in its vicinity, with orientation axis z, has a position defined by angles α and δ (in the examples shown, $\delta = 0$). The three orientations shown differ only in the angle ω $(0, -\pi/2, \pi)$. The assumed orientation distribution is uniform relative to this angle of rotation. Reproduced by permission from E. W. Thulstrup and J. Michl, *J. Phys. Chem.* **84**, 82 (1980).

K_y, but in addition to being related to the chain direction Z′ rather than the stretching direction Z, they are defined somewhat differently: $B = \langle\cos^2\alpha\rangle$, $C = \langle\cos^2\delta\rangle$, where α is the angle between the molecular orientation axis and the projection of the polymer chain direction into the molecular plane, and δ is the angle between this projection direction and the polymer chain direction. The relation of these angles to the analogous Euler angles β' and γ' is

$$\alpha = \tan^{-1}(\cos\gamma' \tan\beta') \qquad \delta = \cos^{-1}(1 - \sin^2\beta' \sin^2\gamma')^{1/2} \tag{5.90}$$

where β' and γ' are referred to the local polymer chain direction Z′ rather than the stretching direction Z as β and γ normally are. It follows that the relation of the parameters B and C to the two parameters which describe the average alignment of the long axis and the short in-plane axis of the planar solute with the polymer chain direction, K'_z and K'_y, respectively, is

$$C = K'_z + K'_y \qquad BC = K'_z \tag{5.91}$$

Thus, our usual factors K_z and K_y, which describe the average alignment of the long axis and the short in-plane axis of the solute, respectively, with the stretching direction of the polymer Z, are related uniquely to the Popov parameters B and C once a particular value is selected for A:

$$B = \frac{2K_z + A - 1}{2(K_z + K_y + A - 1)} \tag{5.92}$$

$$C = 2\frac{K_z + K_y + A - 1}{3A - 1} \tag{5.93}$$

Assuming that all observed transitions of the solute are in-plane polarized, we have

$$g_f = \frac{E_Z(\tilde{\nu}_f) - E_Y(\tilde{\nu}_f)}{E_Z(\tilde{\nu}_f) + E_Y(\tilde{\nu}_f)}$$
$$= \frac{(3A - 1)\{[3C(1 - B) - 1] + 3C(2B - 1)\cos^2 \phi_z^f\}}{(3 - A) + C(3A - 1)(1 - B) + C(3A - 1)(2B - 1)\cos^2 \phi_z^f} \tag{5.94}$$

and

$$\cos^2 \phi_z^f = \frac{-(3A - 1)[3C(1 - B) - 1] + g_f[3 - A + C(3A - 1)(1 - B)]}{C(3A - 1)(2B - 1)(3 - g_f)} \tag{5.95}$$

The Popov model requires the values of A which might be assumed for any given polymer to be such that the K_z and K_y of all solute molecules lie within the segment of the orientation triangle defined by $K_z \geqslant K_y$, $K_y \geqslant (1 - K_z)/2$, $K_z + K_y \leqslant (1 + A)/2$. Once a value for A is assumed, this smaller "allowed" triangle is then mapped by the unique transformation $K_z, K_y \rightarrow B, C$ into a distorted triangle-like figure defined by the conditions $\frac{1}{2} \leqslant B \leqslant 2 - 1/C$, $C \leqslant 1$ (Figure 5.10). This "transformed orientation triangle" then contains points for all possible molecular orientations with respect to the polymer chain which

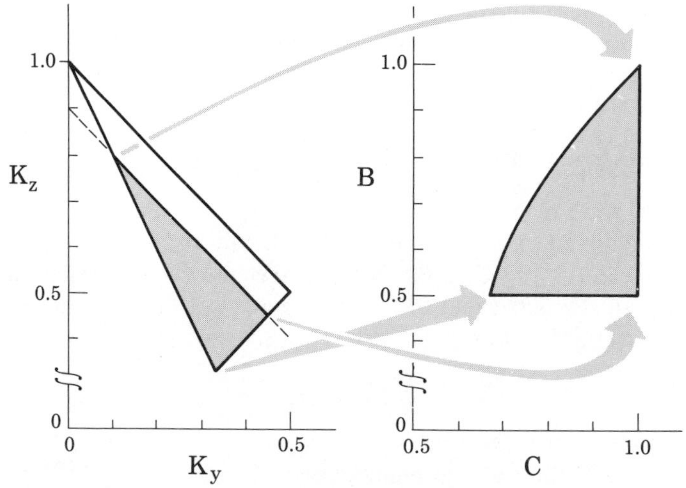

Figure 5.10 A mapping of the portion of the orientation triangle (K_y,K_z) allowed by the Popov model for the case $A = 0.8$ [the limiting line is $K_y + K_z = (1 + A)/2$] into the "transformed orientation triangle" (B,C). Reproduced by permission from E. W. Thulstrup and J. Michl, *J. Phys. Chem.* **84**, 82 (1980).

have the assumed cylindrical symmetry around that chain, as described by the values of B and C. According to (5.91), the point (B,C) = (1/2,2/3) corresponds to $K'_z = K'_y = 1/3$, the point (1/2,1) corresponds to $K'_z = K'_y = 1/2$, and the point (1,1) corresponds to $K'_z = 1$, $K'_y = 0$. The three vertices of the transformed orientation triangle are thus related to the three vertices of the original allowed orientation triangle. The lines joining the three new points have a meaning analogous to that of the sides of the original orientation triangle: The straight line B = 1/2 is the line of disclike orientation, the curve C = 1/(2 − B) is the line of rodlike orientation, and the straight line C = 1 is the line of perfect alignment of molecular planes, all with respect to the polymer chain direction Z′.

The assumption of cylindrical symmetry for the solute orientation distribution around the polymer chain is fundamental to the Popov model. The assumption limits the method to solutes whose out-of-plane axis aligns less well in the plane perpendicular to the stretching direction than does the short axis of the polymer chain. It would be justified if the individual polymer chains were far separated in space, but it is not clear why it should be satisfied in any real polymer. It implies, for instance, that the solute can never be aligned on the average to a higher degree than the polymer chain itself, and this is clearly wrong, since counterexamples are known (Section 8.8.1). This assumption is also inappropriate if the anisotropy of the solute orientation distribution is due to combined effects of photoselection and preexisting equilibrium orientation, as in the treatment of polarized photoluminescence by the methods of this chapter rather than those of Chapter 7.

In view of this difficulty and the considerable complication introduced by this elaboration of the simple model, it is important to ask whether there are any compensating advantages. One advantage might be the fact that in the Popov model the position of a point for a solute in the distorted orientation triangle is independent of the degree of stretching, so that there is only one such diagram for each type of polymer rather than a separate diagram for each degree of stretching. Such a condensation of data would be desirable. It still remains to be seen whether the experimental results really depend on the degree of stretching in the way predicted by this approximation.

5.3 Evaluation of Modulation Spectra

The discussion of the evaluation procedures given so far has been based on the assumption that $E_Z(\tilde{\nu})$ and $E_Y(\tilde{\nu})$ are the experimentally measured quantities. When dealing with weakly oriented samples, the relative magnitude of the errors inherent in the separate determination of $E_Z(\tilde{\nu})$, $E_Y(\tilde{\nu})$, or $E^{iso}(\tilde{\nu})$ increases, and it is then frequently desirable to use methods in which two quantities enter into the measurement simultaneously. This can be accomplished either by modulating the light beam between two states of polarization or by modulating the degree of alignment of the sample. The procedures which can be used to obtain orientation factors and transition moment directions from such differential measurements are quite similar to those already described in this chapter.

5.3.1 Polarization Modulation

Measurement of LD and $E_{Y,Z}$. When the state of polarization of the measuring light beam is modulated from being linearly polarized along the sample Z axis to linearly polarized along the sample Y or X axis, the quantities obtained after appropriate baseline corrections are $LD(\tilde{\nu}) = E_Z(\tilde{\nu}) - E_Y(\tilde{\nu})$ and $E_{Y,Z}(\tilde{\nu}) = [E_Z(\tilde{\nu}) + E_Y(\tilde{\nu})]/2$. The orientation factors K_u of symmetrical molecules can be found by the stepwise reduction procedure using equation (5.26), just as in (5.32)–(5.34):

$$LD(\tilde{\nu}) - cE_{Y,Z}(\tilde{\nu}) = \sum_{t=x,y,z} [K_t(1 - c/2) - (1/2)(1 - K_t)(1 + c/2)]A_t(\tilde{\nu}) \tag{5.96}$$

A u-polarized spectral feature will disappear in the trial-and-error search when c reaches the value

$$c = d_u = 2(3K_u - 1)/(K_u + 1) \tag{5.97}$$

and the desired value of K_u can then be obtained from

$$K_u = (2 + d_u)/(6 - d_u) \tag{5.98}$$

The reduced spectra can again be obtained only in special instances.

If $K_z = K_y = (1 - K_x)/2$, we have

$$[A_z(\tilde{\nu}) + A_y(\tilde{\nu})]/2 = E_{Y,Z}(\tilde{\nu}) + [(1/2)(1 - K_z)/(3K_z - 1)]LD(\tilde{\nu}) \tag{5.99}$$

$$A_x(\tilde{\nu}) = E_{Y,Z}(\tilde{\nu}) - [(1/2)(1 + K_z)/(3K_z - 1)]LD(\tilde{\nu}) \tag{5.100}$$

If $K_x = K_y = (1 - K_z)/2$, we have

$$A_z(\tilde{\nu}) = E_{Y,Z}(\tilde{\nu}) + (1/2)[(3 - K_z)/(3K_z - 1)]LD(\tilde{\nu}) \tag{5.101}$$

$$[A_x(\tilde{\nu}) + A_y(\tilde{\nu})]/2 = E_{Y,Z}(\tilde{\nu}) - [(1/2)(1 + K_z)/(3K_z - 1)]LD(\tilde{\nu}) \tag{5.102}$$

If one of the reduced spectra vanishes identically in the region of interest, say $A_x(\tilde{\nu}) = 0$, we obtain

$$A_z(\tilde{\nu}) = [(1 - 3K_y)E_{Y,Z}(\tilde{\nu}) + (1/2)(1 + K_y)LD(\tilde{\nu})]/(K_z - K_y) \tag{5.103}$$

$$A_y(\tilde{\nu}) = [(3K_z - 1)E_{Y,Z}(\tilde{\nu}) - (1/2)(1 + K_z)LD(\tilde{\nu})]/(K_z - K_y) \tag{5.104}$$

In low-symmetry molecules, the K_f values needed for equation (5.48) can now be similarly obtained from

$$K_f = (2 + c_f)/(6 - c_f) \tag{5.105}$$

where the c_f values result from the stepwise reduction procedure.

Measurement of LD and E_Y or E_Z. In some experimental arrangements, the quantities $LD(\tilde{\nu})$ and either $E_Z(\tilde{\nu})$ or $E_Y(\tilde{\nu})$ rather than $E_{Y,Z}(\tilde{\nu})$ are obtained. The orientation factors and reduced spectra of symmetrical molecules can then be obtained in an analogous fashion. For instance, using $E_Y(\tilde{\nu})$,

$$LD(\tilde{\nu}) - cE_Y(\tilde{\nu}) = \sum_{t=x,y,z} [K_t - \tfrac{1}{2}(1 - K_t)(1 + c)]A_t(\tilde{\nu}) \tag{5.106}$$

A u-polarized spectral feature will disappear when c reaches the value

$$c = d_u = (3K_u - 1)/(1 - K_u) \tag{5.107}$$

so that K_u can be obtained from

$$K_u = (1 + d_u)/(3 + d_u) \tag{5.108}$$

Expressions for the reduced spectra for symmetrical molecules are derived easily in analogy to (5.99)–(5.104), and the K_f values needed in equation (5.48) for low-symmetry molecules are given by

$$K_f = (1 + d_f)/(3 + d_f) \tag{5.109}$$

When $E_Z(\tilde{\nu})$ is used instead of $E_Y(\tilde{\nu})$, very similar formulas result in place of (5.106)–(5.108):

$$E_Z(\tilde{\nu}) - cLD(\tilde{\nu}) = \sum_{t=x,y,z} [K_t(1 - c) - c(K_t - 1)/2]A_t(\tilde{\nu}) \tag{5.110}$$

$$c = d_u = 2K_u/(3K_u - 1) \tag{5.111}$$

$$K_u = d_u/(3d_u - 2) \tag{5.112}$$

Measurement of LD and E^{iso}. A more complicated procedure occasionally encountered in the literature is to combine the $LD(\tilde{\nu})$ curve with the $E^{iso}(\tilde{\nu})$ curve obtained from $E^{iso}(\tilde{\nu}) = E_{Y,Z}(\tilde{\nu}) - (1/6)LD(\tilde{\nu})$ according to (3.16), where $E_{Y,Z}(\tilde{\nu})$ is obtained either directly or from $D(\tilde{\nu})$ according to (3.15). The $E^{iso}(\tilde{\nu})$ curve can also be measured separately on an unoriented sample. Procedures very similar to those just outlined can be used to obtain the orientation factors and reduced spectra.

Measurement of LD_ω and E_Y. In measurements using the "tilted plate method," useful for membranes, light strikes the sample at an external angle $\pi/2 - \omega$ with respect to its Z axis, i.e., at an internal angle $\pi/2 - \arcsin[(\sin\omega)/n]$. The observed linear dichroism corrected for baseline effects, $LD_\omega(\tilde{\nu})$, is related to the intrinsic sample properties $E_Z(\tilde{\nu})$ and $E_Y(\tilde{\nu})$ in a very complicated fashion. We shall treat only the simple case in which the sample is mechanically sufficiently thick for interference effects to be negligible but optically thin and has negligible birefringence, the linear dichroism is sufficiently large to warrant the neglect of multiple polarized reflections on the four sample boundaries, and the light beam is perfectly collimated. All of these assumptions appear to be satisfied to a sufficient degree under the experimental conditions ordinarily used.

For an optically thin sample of negligible birefringence, the optical paths of the beams of any polarization striking the sample at the incidence angle $\pi/2 - \omega$ will be the same. Given the above assumptions, (5.9) can be generalized, and after proper removal of the baseline effects one obtains

$$E_\varepsilon(\tilde{\nu}) = E_Z(\tilde{\nu})\cos^2 \varepsilon Z + E_Y(\tilde{\nu})\sin^2 \varepsilon Z \tag{5.113}$$

where εZ is the angle between the electric vector of the light and the sample

Z axis inside the sample. According to Snell's law of refraction, $\cos \varepsilon Z = (\cos \omega)/n(\tilde{\nu})$, so that the spectrum $E_\omega(\tilde{\nu})$ taken at the external angle of incidence $\pi/2 - \omega$ and corrected for baseline effects is given by

$$E_\omega(\tilde{\nu}) = E_Y(\tilde{\nu}) + [E_Z(\tilde{\nu}) - E_Y(\tilde{\nu})](\cos^2 \omega)/n^2(\tilde{\nu}) \qquad (5.114)$$

$$E_\omega(\tilde{\nu}) = E_Y(\tilde{\nu}) + LD(\tilde{\nu})(\cos^2 \omega)/n^2(\tilde{\nu}) \qquad (5.115)$$

Now, in the measurement of $LD_\omega(\tilde{\nu})$ the electric vector of the light oscillates between positions $\boldsymbol{\omega}$ and Y, and

$$LD_\omega(\tilde{\nu}) = E_\omega(\tilde{\nu}) - E_Y(\tilde{\nu}) = LD(\tilde{\nu})(\cos^2 \omega)/n^2(\tilde{\nu}) \qquad (5.116)$$

With the above assumptions, then, the linear dichroism $LD_\omega(\tilde{\nu})$ observed at the incidence angle $\pi/2 - \omega$ and corrected for baseline effects varies like $(\cos^2 \omega)/n^2(\tilde{\nu})$.

The spectrum $LD_\omega(\tilde{\nu})$ can now be combined with another linearly independent spectral curve obtained on the same sample. This can be done with the same optical path length in order to minimize corrections. The simplest procedure is to use either $E_Y(\tilde{\nu})$ or $E_\omega(\tilde{\nu})$. These are measured at the same angle of tilt ω as the $LD_\omega(\tilde{\nu})$ curve, and if the birefringence of the optically thin sample is indeed negligible, they all correspond to the same optical path length l_ω.

If one wishes to combine the $LD_\omega(\tilde{\nu})$ curve with a spectrum obtained on the same sample at another angle of incidence, say normal incidence, which minimizes the baseline correction, one must correct for the change in the optical path length as shown in equation (3.2).

We shall consider only the simple former cases in which $LD_\omega(\tilde{\nu})$ is reduced against $E_Y(\tilde{\nu})$ or $E_\omega(\tilde{\nu})$ measured at the same incidence angle $\pi/2 - \omega$. All baseline corrections, $LD^0_\omega(\tilde{\nu})$ and $E^0_Y(\tilde{\nu})$ or $E^0_\omega(\tilde{\nu})$, must also be obtained at the same angle of incidence.

The orientation factors K_u of symmetrical molecules can be found by the stepwise reduction procedure using equation (5.26), just as in (5.106)–(5.108):

$$LD_\omega(\tilde{\nu}) - cE_Y(\tilde{\nu}) = \sum_{t=x,y,z} [K_t(\cos^2 \omega)/n^2(\tilde{\nu}) - \tfrac{1}{2}(1 - K_t)(1 + c)]A_t(\tilde{\nu}) \qquad (5.117)$$

The value of $n^2(\tilde{\nu})$ to be used in this equation should be obtained at $\tilde{\nu}$ equal to the energy of the transition which is being reduced. In practice, $n^2(\tilde{\nu})$ has usually been assumed to be a constant independent of $\tilde{\nu}$.

A u-polarized spectral feature will disappear when c reaches the value

$$c = d_u = 2K_u(\cos^2 \omega)/n^2(\tilde{\nu})(1 - K_u) - 1 \qquad (5.118)$$

so that K_u can be obtained from

$$K_u = (1 + d_u)/[1 + d_u + 2(\cos^2 \omega)/n^2(\tilde{\nu})] \qquad (5.119)$$

Expressions for reduced spectra can be derived in a manner quite analogous to (5.99)–(5.104), as can an expression for K_f for a low-symmetry molecule.

Measurement of LD_ω and E_ω. If $E_\omega(\tilde{\nu})$ is used in the reduction procedure instead of $E_Y(\tilde{\nu})$, the resulting equations corresponding to (5.117), (5.118), and

(5.119) are

$$E_{\omega}(\tilde{\nu}) - cLD_{\omega}(\tilde{\nu}) = E_Z(\tilde{\nu})(1-c)(\cos^2\omega)/n^2(\tilde{\nu}) - E_Y[(1-c)(\cos^2\omega)/n^2(\tilde{\nu}) - 1]$$
$$= \sum_{t=x,y,z} \{K_t(1-c)(\cos^2\omega)/n^2(\tilde{\nu}) - (1-K_t)[(1-c)(\cos^2\omega)/n^2(\tilde{\nu})]/2\}A_t(\tilde{\nu}) \quad (5.120)$$

$$c = d_u = 1 + (1-K_u)/(3K_u - 1)\cos^2\omega \quad (5.121)$$

$$K_u = [1 + (d-1)(\cos^2\omega)/n^2(\tilde{\nu})]/[1 + 3(d-1)(\cos^2\omega)/n^2(\tilde{\nu})] \quad (5.122)$$

In concluding this section, we need to reemphasize that the stepwise reduction procedure is needed only for differently polarized overlapping transitions. For transitions known to be purely polarized, it is much simpler and quite equivalent to take the ratio of peak intensities in the two spectra which are being combined. As discussed in Section 5.1.1, this immediately produces the value of d_u.

5.3.2 *Orientation Modulation*

When the state of the sample is modulated from oriented to random by a periodic variation of the orienting electric field, the characteristic quantities of electric dichroism $L_Z(\tilde{\nu})$ and $L_Y(\tilde{\nu})$, related to $E_Z(\tilde{\nu}) - E_{iso}(\tilde{\nu})$ and $E_Y(\tilde{\nu}) - E_{iso}(\tilde{\nu})$, respectively, can be recorded.

As mentioned in Section 4.6.2, a more detailed interpretation of these quantities is complicated immensely by the modification of molecular properties such as transition energies and transition moments by the strong applied electric field. While this may be considered a disadvantage if only transition moment directions are sought, it represents an important opportunity to measure quantities which are difficult to obtain otherwise, such as dipole moments and polarizabilities of excited states and transition polarizabilities. Various simplified evaluation schemes have been proposed. However, a discussion of these matters lies outside the scope of this book, and the reader is directed to Section 5.5 for leading references.

We note in passing that when the effects of the electric field on molecular properties can be neglected, so that $L_{magic}(\tilde{\nu}) = 0$, the quantities $L_Z(\tilde{\nu})$ and $L_Y(\tilde{\nu})$ are given by

$$L_Z(\tilde{\nu}) = [E_Z(\tilde{\nu}) - E^{iso}(\tilde{\nu})]/E^{iso}(\tilde{\nu}) \quad (5.123)$$

$$L_Y(\tilde{\nu}) = [E_Y(\tilde{\nu}) - E^{iso}(\tilde{\nu})]/E^{iso}(\tilde{\nu}) \quad (5.124)$$

normalized to unit strength of the electric field. Then the orientation factors K_u defined in the principal orientation axes system of symmetrical molecules can be determined in the same way as in (5.32)–(5.34) by the stepwise reduction of either $E^{iso}(\tilde{\nu})L_Z(\tilde{\nu})$ or $E^{iso}(\tilde{\nu})L_Y(\tilde{\nu})$ against $E^{iso}(\tilde{\nu})$. In the former case,

$$E^{iso}(\tilde{\nu})L_Z(\tilde{\nu}) - cE^{iso}(\tilde{\nu}) = \sum_{t=x,y,z} [K_t - (c+1)/3]A_t(\tilde{\nu}) \quad (5.125)$$

so that all u-polarized spectral features will disappear in the trial-and-error procedure from the linear combination $E^{iso}(\tilde{\nu})L_Z(\tilde{\nu}) - cE^{iso}(\tilde{\nu})$ when c is chosen so that

$$c = d_u = 3K_u - 1 = 2S_{uu} \tag{5.126}$$

and as a result

$$K_u = (1 + d_u)/3 \tag{5.127}$$

and

$$S_{uu} = d_u/2 \tag{5.128}$$

In the latter case, the linear combination $E^{iso}(\tilde{\nu})L_Y(\tilde{\nu}) - cE^{iso}(\tilde{\nu})$ will lose u-polarized spectral features when c acquires the value

$$c = d_u = (1 - 3K_u)/2 = -S_{uu} \tag{5.129}$$

so that

$$K_u = (1 - 2d_u)/3 \tag{5.130}$$

and

$$S_{uu} = d_u \tag{5.131}$$

As usual, a reduction is unnecessary if a purely u-polarized transition is located at $\tilde{\nu}_u$. In this case one obtains

$$K_u = [1 + L_Z(\tilde{\nu}_u)]/3 \tag{5.132}$$

$$S_{uu} = -L_Y(\tilde{\nu}_u) \tag{5.133}$$

Reduced spectra can be obtained for symmetrical molecules under the usual conditions.

For low-symmetry molecules, one obtains similar relations from the K_f value of each transition. In the case of a purely polarized transition f, the result

$$K_f = [1 + L_Z(\tilde{\nu}_f)]/3 \tag{5.134}$$

is easily interpreted qualitatively. For a molecule of any symmetry, a purely polarized transition f will yield a positive value of $L_Z(\tilde{\nu})$ if its transition moment direction deviates from the molecular orientation axis z by less than the magic angle and a negative value of $L_Z(\tilde{\nu})$ if this deviation exceeds the magic angle. In favorable cases, the location of the z axis is known from molecular symmetry. In general, it is determined by a complex interplay of the molecular dipole moment and molecular polarizability.

5.4 More General Types of Orientation

The treatment of linear dichroism in partially oriented samples of lower than uniaxial symmetry is considerably more complicated than was the case for uniaxial samples (single-crystal work is not being considered in this book).

As noted in Section 4.7, many more independent orientation factors exist. Virtually no experimental work on this subject is available, although in principle more information about the solute molecules can be obtained. This dearth of experimental studies is undoubtedly also related to the current lack of a suitable method for reproducibly achieving a high degree of partial biaxial orientation on small molecules. Even with suitable samples, it would be a nontrivial task to perform measurements, since under most circumstances light will not propagate in such samples without having its state of polarization changed by the passage.

We shall consider briefly the simple case of a sample on which measurements are possible with the electric vector of light polarized along each of the three orthogonal X,Y,Z axes, leading to observed spectra $E_X(\tilde{\nu})$, $E_Y(\tilde{\nu})$, and $E_Z(\tilde{\nu})$. We shall further assume that the solute in question is of sufficiently high symmetry that only transitions along three mutually orthogonal axes x,y,z are possible, corresponding to reduced spectra $A_x(\tilde{\nu})$, $A_y(\tilde{\nu})$, and $A_z(\tilde{\nu})$, respectively. Using the notation for generalized orientation factors K_{Uu} introduced in Section 4.7, the two sets of spectra are related by the matrix equation

$$\begin{pmatrix} E_X(\tilde{\nu}) \\ E_Y(\tilde{\nu}) \\ E_Z(\tilde{\nu}) \end{pmatrix} = \begin{pmatrix} K_{Xx} & K_{Xy} & K_{Xz} \\ K_{Yx} & K_{Yy} & K_{Yz} \\ K_{Zx} & K_{Zy} & K_{Zz} \end{pmatrix} \begin{pmatrix} A_x(\tilde{\nu}) \\ A_y(\tilde{\nu}) \\ A_z(\tilde{\nu}) \end{pmatrix} \tag{5.135}$$

The reduction procedure described in Section 5.1.1 can be used in the usual way to find all the orientation factors K_{Uu} for which spectral features polarized along u can be found in the spectra. Removal of such u-polarized features from a linear combination of spectra $E_U(\tilde{\nu})$ and $E_V(\tilde{\nu})$ will be achieved in the linear combination

$$E_U(\tilde{\nu}) - d_u^{UV}E_V(\tilde{\nu}) = [K_{Ux} - d_u^{UV}K_{Vx}]A_x(\tilde{\nu}) + [K_{Uy} - d_u^{UV}K_{Vy}]A_y(\tilde{\nu}) + [K_{Uz} - d_u^{UV}K_{Vz}]A_z(\tilde{\nu}) \tag{5.136}$$

Once such a value of d_u^{UV} has been found by the usual trial-and-error procedure for which the difference $K_{Uu} - d_u^{UV}K_{Vu}$ vanishes, it provides a value for the ratio K_{uU}/K_{uV}. Then, using

$$K_{Uu} = d_u^{UV}K_{Vu} = d_u^{UW}K_{Wu} \tag{5.137}$$

and

$$K_{Uu} + K_{Vu} + K_{Wu} = 1 \tag{5.138}$$

all the unknown generalized orientation factors can be found from relations such as

$$K_{Uu} = \frac{1}{1 + d_u^{VU} + d_u^{WU}} \tag{5.139}$$

where U, V, and W are all different. Once the generalized orientation factors are determined, reduced spectra $A_u(\tilde{\nu})$ can in principle be obtained from the inversion of the matrix equation (5.135),

$$\begin{pmatrix} A_x(\tilde{\nu}) \\ A_y(\tilde{\nu}) \\ A_z(\tilde{\nu}) \end{pmatrix} = \frac{1}{D} \begin{pmatrix} K_{Yy}K_{Zz} - K_{Zy}K_{Yz} & K_{Zy}K_{Xz} - K_{Xy}K_{Zx} & K_{Xy}K_{Yz} - K_{Yy}K_{Xz} \\ K_{Zx}K_{Yz} - K_{Yx}K_{Zz} & K_{Xx}K_{Zz} - K_{Zx}K_{Xz} & K_{Xz}K_{Yx} - K_{Xx}K_{Yz} \\ K_{Yx}K_{Zy} - K_{Yy}K_{Zx} & K_{Xy}K_{Zx} - K_{Xx}K_{Zy} & K_{Xx}K_{Yy} - K_{Yx}K_{Xy} \end{pmatrix} \begin{pmatrix} E_X(\tilde{\nu}) \\ E_Y(\tilde{\nu}) \\ E_Z(\tilde{\nu}) \end{pmatrix} \tag{5.140}$$

where D is defined by

$$D = \begin{vmatrix} K_{Xx} & K_{Xy} & K_{Xz} \\ K_{Yx} & K_{Yy} & K_{Yz} \\ K_{Zx} & K_{Zy} & K_{Zz} \end{vmatrix} \tag{5.141}$$

If $D \neq 0$, and if the reductions are performed correctly, each reduced spectrum $A_u(\tilde{\nu})$ obtained in this fashion will contain only u-polarized features. In practice, it may be easier to determine the three reduced spectra $A_u(\tilde{\nu})$ by trial and error, combining the three linearly independent spectra E_X, E_Y, and E_Z until all contributions due to two of the A_u's disappear (cf. Section 7.3.2).

Only four of the nine generalized orientation factors of the kind K_{Uu} are independent, since they are connected by the six relations (5.138), u = x,y,z, and three additional relations,

$$K_{Ux} + K_{Uy} + K_{Uz} = 1, \qquad U = X,Y,Z \tag{5.142}$$

Five of these six equations are independent.

For aromatic molecules whose absorption is polarized along the axes y and z, it is practical to use the four generalized orientation factors, K_{Yy}, K_{Zy}, K_{Yz}, and K_{Zz} as the independent set of four.

5.5 Comments and References

Throughout this chapter as well as Chapters 6 and 7, we deal with an idealized system. We recognize that the solvent affects the energy differences between molecular states as well as the magnitudes and directions of transition moments (note that the immediate environment of a solute molecule is unsymmetrical); however, we assume that the sample is such that these effects are identical for all molecules of the solute independent of their orientation within the sample. Similarly, we recognize that the effective electromagnetic field interacting with the molecule is due not only to the field of the incident wave but also to the additional field caused by the polarization of the environment; however, we assume that the sample is such that the latter is the same for incident electric vectors directed along the X, Y, or Z axes. Effectively, then, we assume that the solvent is homogeneous and isotropic except for its possible ability to align the solute molecules.

In most instances, these approximations are quite acceptable. In Chapter 8, Section 8.1, we shall provide a rare example which violates detectably the first of the assumptions. The second assumption cannot be made if the sample has strong birefringence. Then it is possible to correct the experimental data for the difference in the effect of the Lorentz field for light polarized parallel and perpendicular to Z, respectively, by using the two different values of the refractive index appropriate for these two directions (see Section 3.3).

We assume throughout that the electric vector of the light is directed either along Z or perpendicular to it (X or Y) or else that the birefringence is negligible. If neither applies, the polarization of the light will be changed from linear to elliptical by the sample, and the simple equations given here will no longer hold.

Some leading references relevant to the neglect of the effects of solvent anisotropy adopted in this chapter are found in B. Samorì, *J. Phys. Chem.* **83**, 375 (1979); R. J. Dudley, S. F. Mason, and R. D. Peacock, *J. Chem. Soc. Faraday II* **1975**, 997; and S. Jen, N. A. Clark, P. S. Pershan, and E. B. Priestley, *J. Chem. Phys.* **66**, 4635 (1977). The effect of the shape of solute molecules on the solvent effects on transition moments is discussed in A. B. Myers and R. R. Birge, *J. Chem. Phys.* **73**, 5314 (1980).

The TEM model and the stepwise reduction procedure are described in E. W. Thulstrup and J. H. Eggers, *Chem. Phys. Lett.* **1**, 690 (1968); E. W. Thulstrup, J. Michl, and J. H. Eggers, *J. Phys. Chem.* **74**, 3868 (1970); J. Michl and E. W. Thulstrup, *Spectrosc. Lett.* **10**, 401 (1977); J. Michl, E. W. Thulstrup, and J. H. Eggers, *J. Phys. Chem.* **74**, 3878 (1970); E. W. Thulstrup, *Aspects of the Linear and Magnetic Circular Dichroism of Planar Organic Molecules*, Springer-Verlag, New York, 1980; and E. W. Thulstrup and J. Michl, *J. Am. Chem. Soc.* **104**, 5594 (1982). The last two of these discuss the use of the TEM model for low-symmetry molecules. The evaluation procedures for the "tilted plate" method are described in B. Nordén, G. Lindblom, and I. Jonáš, *J. Phys. Chem.* **81**, 2084 (1977) and L. B.-Å. Johansson, Å. Davidsson, G. Lindblom, and B. Nordén, *J. Phys. Chem.* **82**, 2604 (1978).

The wiggles observed in the region of zero-zero transitions in reduced spectra, due to differences in the positions of band maxima in $E_Z(\tilde{\nu})$ and $E_Y(\tilde{\nu})$, as discussed in Section 5.1.1, were reported repeatedly: e.g., J. Michl, E. W. Thulstrup, and J. H. Eggers, *Ber. Bunsenges. Physik. Chem.* **78**, 575 (1974); B. Nordén, *Spectrosc. Lett.* **10**, 455 (1977); and L. Margulies and A. Yogev, *Chem. Phys.* **27**, 89 (1978). The problem has been discussed at some length in J. Konwerska-Hrabowska, *J. Mol. Struct.* **45**, 95 (1978).

The Fraser-Beer model is based on the work of R. D. B. Fraser, *J. Chem. Phys.* **21**, 1511 (1953); **24**, 89 (1956); and M. Beer, *Proc. Roy. Soc. London, Ser. A* **236**, 136 (1956). Its application to the orientation of solutes has been primarily due to Yogev, Mazur, and collaborators. Their initial communications on the subject are A. Yogev, L. Margulies, D. Amar, and Y. Mazur, *J. Am. Chem. Soc.* **91**, 4558 (1969) and A. Yogev, J. Riboid, J. Marero, and Y. Mazur, *J. Am. Chem.*

Soc., **91**, 4559 (1969); many more followed. The extension of the Fraser-Beer model to disclike molecules ($K_y = K_z$) was first given by B. Nordén, *Chem. Scr.* **1**, 145 (1971). A modification of the model to accommodate nonuniform distributions of the Euler angle γ was described in L. Margulies and A. Yogev, *Chem. Phys.* **27**, 89 (1978).

The Tanizaki model was first described and applied in Y. Tanizaki and S.-I. Kubodera, *J. Mol. Spectrosc.* **24**, 1 (1967); see also H. Hiratsuka, Y. Tanizaki, and T. Hoshi, *Spectrochim. Acta, A* **28**, 2375 (1972). A long series of papers followed. The "cone analysis" is described in A. F. Fucaloro and L. S. Forster, *Spectrochim. Acta, A* **30**, 883 (1974).

The Popov model is outlined in K. R. Popov, *Opt. Spectrosc.* **39**, 142 (1975). Its use for the interpretation of fluorescence polarization spectra without separating the effects of photoselection and intrinsic orientation is summarized in J. J. Dekkers, W. P. Cofino, G. Ph. Hoornweg, C. Maclean, and N. H. Velthorst, *Chem. Phys.* **47**, 369 (1980).

A qualitative treatment of a partially oriented sample which is not uniaxial is given in A. O. Ganago, Yu. E. Erokhin, and Z. K. Makhneva, *Studia Biophysica* **80**, 193 (1980). Caution is required, since these authors occasionally refer to uniaxial samples with nonuniform distribution of the Euler angle γ as non-uniaxial.

Many additional references to the use and misuse of the approximate models can be found in E. W. Thulstrup and J. Michl, *J. Phys. Chem.* **84**, 82 (1980). A recent paper which emphasizes the inadequacies of the Fraser-Beer type approximation in work with nematic liquid crystals is E. H. Korte, *Mol. Cryst. Liq. Cryst.* **100**, 127 (1983).

Evaluation procedures for electrochromism are described in the books quoted under Electric Dichroism in Section 3.3.

6

Processes Reducible to Rank Two

In this chapter we treat phenomena which are formally of rank 3 or 4 but whose description requires only the second moments of the orientation distribution function, as given, e.g., by the orientation factors K. An example of limited practical importance is the magnetic dipole contribution to absorption or emission. More significant are natural circular dichroism (CD), observable only for chiral molecules, and magnetic circular dichroism (MCD), observable for all molecules. This chapter deals with optical activity as a molecular prop-- erty and is unrelated to Section 3.2.4, which describes liquid-crystal-induced circular dichroism.

A description of the interactions to be discussed in this chapter requires the use of the electric dipole moment operator $\hat{\mathbf{M}}$, already familiar from Chapter 5, as well as the use of the magnetic dipole moment operator $\hat{\mathscr{M}}$ defined in (1.54) and the electric quadrupole moment operator $\hat{\mathrm{Q}}$ defined in (1.58). The relevant quantities are the matrix elements of these operators, the electric dipole transition moment $\mathbf{M}(0\mathrm{f})$, the magnetic transition dipole moment $\mathscr{M}(0\mathrm{f})$, and the electric quadrupole transition moment $\mathrm{Q}(0\mathrm{f})$, all defined in a way entirely analogous to (5.1). The calculation of these quantities from first principles was discussed in Chapter 2.

While $\mathbf{M}(0\mathrm{f})$ is an ordinary (polar) vector, $\mathscr{M}(0\mathrm{f})$ is the vector product of two ordinary vectors (1.54) and properly speaking is an antisymmetric tensor of rank 2. Such a tensor has only three independent elements, behaves in many respects as a vector, and is usually called an axial vector or a pseudovector, as if it were of rank 1. On the other hand, $\mathrm{Q}(0\mathrm{f})$ is a traceless symmetric tensor of rank 2 (1.58). It has six independent elements and cannot be treated like a vector.

The operators $\hat{\mathbf{M}}$ and $\hat{\mathrm{Q}}$ are real, while $\hat{\mathscr{M}}$ is purely imaginary. When matrix elements are taken between real wavefunctions, the resulting transition moments

M(0f) and Q(0f) are real, while $\mathcal{M}$(0f) is purely imaginary. For the components of these transition moments along the laboratory axes X, Y, and Z appropriate for interaction with linearly polarized light, we have **M**(0f) = **M**(f0), Q(0f) = Q(f0), and $\mathcal{M}(0f) = -\mathcal{M}(f0)$. Unlike Chapter 5, the present chapter deals primarily with measurements using circularly polarized light propagating along Z, so that the components of the transition moments in the complex directions $X \pm iY$ will be required. Further, in the presence of a magnetic field, wavefunctions generally become complex. For all these reasons, it is important to keep track of the order of the symbols 0 and f.

The organization of Chapter 6 is as follows. We first briefly consider the magnetic dipole mechanism of absorption and then turn to optical activity. We introduce the subject by reference to Section 1.3.3 and make some general comments applicable both to CD and MCD measurements on partially aligned samples. We then proceed to a detailed discussion of CD spectroscopy of general uniaxial samples and then their MCD spectroscopy. Finally, we indicate how matters simplify in the special case of uniaxial orientation induced by photoselection on a random sample.

The CD and MCD results apply equally to the polarization of spontaneous luminescence (e.g., thermal luminescence or chemiluminescence) when it is remembered that the reversal of the direction of a circularly polarized photon as well as the switch from photon annihilation to photon creation correspond to the reversal of its handedness. We shall find that measurements of optical activity of partially oriented molecules are not particularly advantageous for the determination of the orientation factors, and in the general case we shall assume that these have already been determined from measurements of linear dichroism or otherwise. The interest in the measurement of optical activity of partially aligned molecules lies instead in the sometimes unique insight which the results provide into molecular properties.

6.1 Magnetic Dipole Absorption and Emission

Ordinarily, the magnetic dipole contributions to absorption and emission stand a chance of being detectable only when the electric dipole contributions vanish by symmetry. Magnetic dipole absorption is described by the tensor $\mathfrak{M}$(0f) which is the direct product of the magnetic dipole transition moment $\mathcal{M}$(0f) with its complex conjugate:

$$\mathfrak{M}(0f) = \mathcal{M}(f0)\mathcal{M}(0f) \tag{6.1}$$

Its components are

$$[\mathfrak{M}(0f)]_{uv} = \mathcal{M}_u(f0)\mathcal{M}_v(0f) \tag{6.2}$$

Although formally $\mathfrak{M}$(0f) is a tensor of rank 4, it behaves as a pseudotensor of rank 2, since $\mathcal{M}$(0f) is an antisymmetric tensor of rank 2, which behaves as a pseudovector.

For light propagating along K whose electric vector is directed along U, the magnetic dipole contribution to the absorption probability for the f-th transition is proportional to $\langle|(\varepsilon_K \times \varepsilon_U^*) \cdot \mathscr{M}(0f)|^2\rangle$ [see (1.60)]. We define $\varepsilon_W = \varepsilon_U \times \varepsilon_K$, and after averaging obtain for W = Z

$$\langle|\varepsilon_Z \cdot \mathscr{M}(0f)|^2\rangle = \sum_{u,v} [\mathsf{D}_X]_{uv}[\mathfrak{M}(0f)]_{uv} = \sum_{u,v} [\mathsf{D}_Y]_{uv}[\mathfrak{M}(0f)]_{uv} \tag{6.3}$$

For W = X or Y, we obtain

$$\langle|\varepsilon_X \cdot \mathscr{M}(0f)|^2\rangle = \langle|\varepsilon_Y \cdot \mathscr{M}(0f)|^2\rangle = \sum_{u,v} [\mathsf{D}_Z]_{uv}[\mathfrak{M}(0f)]_{uv} \tag{6.4}$$

The elements of the tensor D are given by (5.15) and (5.16). These results have the form anticipated in (1.142).

The analogy to the results for electric dipole absorption (5.14) is striking. The only difference is that the result for Z-polarized light is now obtained using D_Y, and the result for Y-polarized light is now obtained using D_Z. Intuitively, this makes sense, since the magnetic vector of the light is directed along a direction perpendicular to its electric vector. Thus, formulas (6.3) and (6.4) imply projecting the magnetic dipole transition moment into the direction of the magnetic vector of the light, taking a square, and averaging over the whole sample.

As in the case of electric dipole absorption, two independent measurements will completely characterize the sample with respect to one-photon processes.

6.2 Optical Activity of Aligned Molecules

Although the information in Table 1.1 would lead one to believe that the rank of the natural CD process is 3 (it is described by terms of the types $\hat{\mathbf{M}}\hat{\mathscr{M}}$ and $\hat{\mathbf{M}}\hat{\mathbf{Q}}$), and that the rank of the dominant contribution to the MCD process ($\hat{\mathscr{M}}\hat{\mathbf{M}}\hat{\mathbf{M}}$) is 4, it turns out that both can be described in terms of pseudotensors of rank 2. These are known as rotatory strength tensors. The simplification is due to the pseudovector nature of $\hat{\mathscr{M}}$ and to the fact that in the MCD measurement the direction of the outside magnetic field is kept parallel to the light propagation direction, so that it does not represent an independent variable (a perpendicular component of magnetic field is not active in inducing circular dichroism).

Just as in the measurement of dichroic absorption, two linearly independent measurements of optical activity will provide all the information on the orientation which can be obtained from a uniaxial sample by means of processes of rank 2. In absorption measurements, it is usual to choose $E_Z(\tilde{\nu})$ and $E_X(\tilde{\nu}) = E_Y(\tilde{\nu})$ and to measure both of them with the light propagation direction perpendicular to the sample axis Z. This provides the maximum difference between the observed dichroic spectra. In principle, of course, it is also possible to perform the measurements with light propagating along the unique Z axis of the sample, which then provides $E_X(\tilde{\nu}) = E_Y(\tilde{\nu})$ regardless of the state of polarization of the light, and to combine this with, say, a measurement on an unoriented sample $E^{iso}(\tilde{\nu})$. In the measurement of optical activity, the latter choice is far preferable, since severe experimental problems are encountered with linear dichroism masking the much weaker effects of circular dichroism when the measurements are performed with light propagating perpendicular to the unique optical axis Z (Section 3.2.5).

Without any loss of information, we assume in the following that the measurements available are the CD or MCD spectrum measured on the oriented sample with light propagating along the Z axis (Figure 6.1) and the CD or MCD spectrum measured on an isotropic sample containing the same number of solute molecules in the light path.

In general, the actually observed spectra will contain a sum of contributions from several overlapping transitions. In the discussion of dichroic absorption measurements in Chapter 5, we found that it was possible to determine the orientation factors K but that it was not possible to separate the two observed spectra into purely polarized components along the three molecular axes in the general case of overlapping transitions. Such separation became at least partly possible under certain fairly commonly satisfied conditions, such as high molecular symmetry. In the present case, strict separation into contributions from individual transitions is impossible if they overlap, even if molecular symmetry is high. In individual cases it might be possible to obtain such separation, at least approximately, if one can assume that the spectral line shape in circular dichroism is identical with the spectral line shape in absorption. However, it is well known that this assumption does not hold in general. Also, it may be possible to find essentially isolated transitions in the spectra, particularly among the first few at low energies.

In the following, we assume that the separation into contributions from individual transitions has somehow been accomplished and describe the theory of optical activity of partially oriented molecules in terms of such contributions. The contribution from the f-th transition is characterized by a line shape and by the ZZ element $[\mathsf{R}(0f)]_{ZZ}$ of the rotatory strength tensor $\mathsf{R}(0f)$ in the case of CD and the ZZ elements of the magnetic rotatory strength tensors $\mathsf{A}(0f)$, $\mathsf{B}(0f)$, and $\mathsf{C}(0f)$ in the case of MCD. The subscript ZZ refers to the laboratory system of axes. The ZZ elements of these tensors are obtained experimentally from the measured curves $\varepsilon_L(\tilde{\nu}) - \varepsilon_R(\tilde{\nu})$ as described in (1.103), (1.133), and (1.134). The integration is over the band due to the f-th transition.

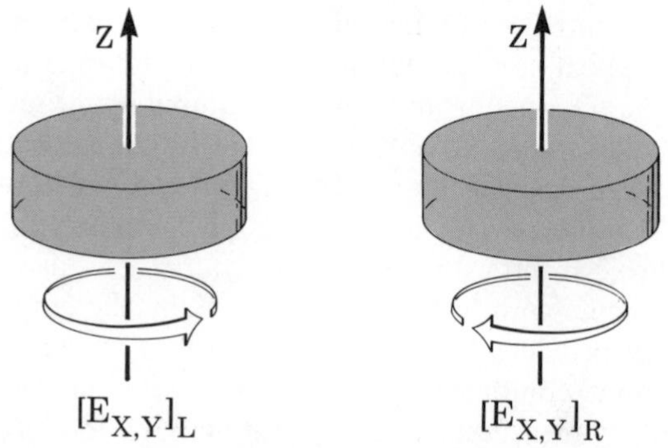

Figure 6.1 Sample orientation relative to the light propagation direction in the measurement of circular dichroism of a uniaxially aligned sample. The curved arrow indicates the sense of rotation of the endpoint of the electric vector of the light.

The rotatory tensors R(0f), A(0f), B(0f), and C(0f) are of rank 2 and are analogous to the absorption and emission tensors M(0f) and $\mathfrak{M}$(0f) which we have used to describe dichroic absorption and emission in Chapter 5 and in Section 6.1, respectively. All these tensors describe the properties of the macroscopic sample and thus depend both on the molecular optical anisotropies and on the molecular orientation distribution within the sample (cf. Section 1.3.4). The only reason we are now interested in only the ZZ elements while in the case of the absorption tensor we were also interested in the YY (or XX) elements has to do with practical aspects of the measurements as noted earlier. As pointed out already, it would have been possible to limit our attention to the ZZ elements in absorption also and to combine them with measurements on an isotropic sample. For the sake of consistency with the notation introduced in Section 1.3.4 [(1.141)] and used for the ZZ and YY elements of the absorption tensor, which are labeled $E_Z(\tilde{\nu})$ and $E_Y(\tilde{\nu})$ rather than $E_{ZZ}(\tilde{\nu})$ and $E_{YY}(\tilde{\nu})$, the ZZ elements of the rotatory tensors R, A, B, and C should also be labeled with only a single subscript, $[R]_Z$, $[A]_Z$, $[B]_Z$, and $[C]_Z$. However, this is not customary, and we shall not do so.

It is possible to separate the effects of the molecular optical anisotropy and those of the orientation distribution as anticipated in (1.142), by performing an explicit averaging of the former over the latter, using for instance the method outlined in Appendix II. This is quite analogous to the procedure outlined for absorption in Section 5.1, and once again only the orientation factors K are needed to describe the orientation distribution adequately. In the case of CD, the molecular property in question now is the molecular rotatory strength tensor R(0f) with elements $[R(0f)]_{uv}$, where u and v refer to the molecular axes system. For $[R(0f)]_{ZZ}$, we have

$$[R(0f)]_{ZZ} = \sum_{u,v} [D_Z]_{uv}[R(0f)]_{uv} = \sum_{u,v} K_{uv}R_{uv}(0f) \tag{6.5}$$

For $[A(0f)]_{ZZ}$, $[B(0f)]_{ZZ}$, and $[C(0f)]_{ZZ}$, we have similar results:

$$[A(0f)]_{ZZ} = \sum_{u,v} K_{uv}A_{uv}(0f) \tag{6.6}$$

$$[B(0f)]_{ZZ} = \sum_{u,v} K_{uv}B_{uv}(0f) \tag{6.7}$$

$$[C(0f)]_{ZZ} = \sum_{u,v} K_{uv}C_{uv}(0f) \tag{6.8}$$

Results for isotropic solutions are obtained by using $K_{uv} = (1/3)\delta_{uv}$:

$$[R^{iso}(0f)]_{ZZ} = (1/3) \sum_{u} R_{uu}(0f) = (1/3)R^{iso}(0f) \tag{6.9}$$

$$[A^{iso}(0f)]_{ZZ} = (1/3) \sum_{u} A_{uu}(0f) = (1/3)A^{iso}(0f) \tag{6.10}$$

$$[B^{iso}(0f)]_{ZZ} = (1/3) \sum_{u} B_{uu}(0f) = (1/3)B^{iso}(0f) \tag{6.11}$$

$$[C^{iso}(0f)]_{ZZ} = (1/3) \sum_{u} C_{uu}(0f) = (1/3)C^{iso}(0f) \tag{6.12}$$

The relation between $[\mathsf{R}^{iso}(0f)]_{ZZ}$ and the ordinary rotatory strength $R^{iso}(0f)$ follows from the comparison of (1.99) with (1.103). Similarly, (1.129) and (1.131) applied to an isotropic solution yield the ordinary quantities $A^{iso}(0f)$, $B^{iso}(0f)$, and $C^{iso}(0f)$, and comparison with (1.133) and (1.134) shows that the ZZ components of the three tensors measured on the same solution are three times smaller.

6.3 Natural Circular Dichroism

The elements of the rotatory strength tensor $\mathsf{R}(0f)$ in molecular frame, which appears in (6.5), are given by

$$R_{uv}(0f) = (1/4)\sum_{st} M_s(0f)[\varepsilon_{stv}N_{tu}(0f) + \varepsilon_{stu}N_{tv}(0f)] \tag{6.13}$$

where ε_{tuv} is equal to the triple product of unit vectors in directions t, u, and v, that is, $\varepsilon_{xyz} = \varepsilon_{zxy} = \varepsilon_{yzx} = 1$, $\varepsilon_{xzy} = \varepsilon_{yxz} = \varepsilon_{zyx} = -1$, and ε's with other combinations of subscripts vanish. This is directly comparable with the expression (1.104) written for the laboratory frame.

Equation (6.13) can be written out explicitly in the form

$$\begin{aligned} R_{uv}(0f) = {} & (1/4)M_x(0f)[\delta_{vz}N_{yu}(0f) + \delta_{uz}N_{yv}(0f) - \delta_{vy}N_{zu}(0f) - \delta_{uy}N_{zv}(0f)] \\ & + (1/4)M_y(0f)[-\delta_{vz}N_{xu}(0f) - \delta_{uz}N_{xv}(0f) + \delta_{vx}N_{zu}(0f) + \delta_{ux}N_{zv}(0f)] \\ & + (1/4)M_z(0f)[\delta_{vy}N_{xu}(0f) + \delta_{uy}N_{xv}(0f) - \delta_{vx}N_{yu}(0f) - \delta_{ux}N_{yv}(0f)] \end{aligned} \tag{6.14}$$

The symbol $N_{uv}(0f)$ contains contributions from two transition moments, the magnetic dipole transition moment $\mathcal{M}(0f)$ and the electric quadrupole transition moment $\mathsf{Q}(0f)$:

$$N_{uv}(0f) = -i\mathcal{M}_{u\times v}(0f) - \pi\tilde{\nu}_f Q_{uv}(0f) \tag{6.15}$$

where $\tilde{\nu}_f$ is the wavenumber of the f-th transition and the cross-product of two subscripts is defined in a manner similar to a vector product of basis vectors: $x \times x = 0$, $x \times y = z$, $x \times z = -y$, etc. A negative value of a subscript means that the subscripted quantity is to be multiplied by -1: $\mathcal{M}_{-z} = -\mathcal{M}_z$.

We will now show that only the magnetic dipole contributions enter the expression for $[\mathsf{R}^{iso}(0f)]_{ZZ}$ and thus for the ordinary rotatory strength $R^{iso}(0f)$, since the electric quadrupole contributions average out in a measurement on an isotropic solution. This can be seen by combining equations (6.5) and (6.13) to obtain

$$[\mathsf{R}^{iso}(0f)]_{ZZ} = (1/3)\sum_{s} R_{ss}(0f) = \frac{1}{6}\sum_{suv} \varepsilon_{uvs}M_u(0f)N_{vs}(0f) \tag{6.16}$$

Since ε_{uvs} is antisymmetric with respect to the exchange of the indices v and s, only that part of N_{vs} which is also antisymmetric with respect to that interchange will contribute, whereas the part which is symmetric will vanish. It is clear from equation (6.15) that the former is the magnetic dipole whereas the latter is the electric quadrupole contribution. The general expression for the ordinary (isotropic) rotational strength $R^{iso}(0f)$, equal to $3[R^{iso}(0f)]_{ZZ}$, is then given by

$$R^{iso}(0f) = \frac{-i}{2} \sum_{suv} \varepsilon_{uvs} M_u(0f) \mathscr{M}_{v \times s}(0f) = -i \sum_u M_u(0f) \mathscr{M}_u(0f)$$

$$= \mathbf{M}(0f) \cdot \mathscr{M}(0f)/i = \mathrm{Im}\,[\mathscr{M}(0f) \cdot \mathbf{M}(f0)] \qquad (6.17)$$

This result is known as the Rosenfeld equation [cf. (1.99)]. As noted above, we assume that the wavefunctions are taken real in the absence of an outside magnetic field.

The molecular properties which one might ultimately hope to obtain from rotatory strength tensors measured on oriented samples are the transition moments $\mathbf{M}(0f)$, $\mathscr{M}(0f)$, and $Q(0f)$. It can usually be assumed that the electric dipole transition moment $\mathbf{M}(0f)$ is already known from studies of linear dichroism. Even then, the number of remaining unknowns is formidable in the general case. Progress can be made, however, if the molecule or its orientation distribution function possess elements of symmetry. We shall consider these simplified cases in the following, assuming that the orientation factors K are known and that $[R(0f)]_{ZZ}$ and $[R^{iso}(0f)]_{ZZ}$ have been measured.

6.3.1 Molecules of D_2 Symmetry

Throughout this book we have reserved the label "high symmetry" for molecules which belong to point symmetry groups in which either (i) x, y, and z transform according to three different irreducible representations or (ii) two of them transform according to a doubly degenerate representation ("very high symmetry"). Among symmetry groups which permit chirality, only the D_2 point group belongs to type (i). Molecules which belong to this group have three mutually perpendicular twofold rotation axes and no other symmetry elements. Examples are a partially twisted biphenyl or ethylene.

In the symmetry-adapted molecular coordinate system x, y, z, the orientation tensor K is diagonal. Then expression (6.5) for the contributions provided by the f-th transition to the ZZ element of the rotatory strength tensor of the oriented sample observed along the Z axis, $[R(0f)]_{ZZ}$, as determined from (1.103), is

$$[R(0f)]_{ZZ} = K_x R_{xx}(0f) + K_y R_{yy}(0f) + K_z R_{zz}(0f) \qquad (6.18)$$

The analogous result measured on the isotropic sample is

$$[R^{iso}(0f)]_{ZZ} = [R_{xx}(0f) + R_{yy}(0f) + R_{zz}(0f)]/3 = R^{iso}(0f)/3 \qquad (6.19)$$

For a transition polarized along the molecular axis u, we have, from (6.13) and (6.15),

$$R_{vv}(0f) = \tfrac{1}{2} M_u(0f)[-i\mathscr{M}_u(0f) - \pi\tilde{\nu}_f Q_{v,v \times u}(0f)], \qquad v \neq u \qquad (6.20)$$

$$R_{uu}(0f) = 0 \qquad (6.21)$$

Equations (6.18) and (6.19) can be solved for the diagonal elements of the tensor R(0f), provided that $K_v \neq K_{|u \times v|}$:

$$R_{vv}(0f) = \{[R(0f)]_{ZZ} - 3K_{|u \times v|}[R^{iso}(0f)]_{ZZ}\}/(K_v - K_{|u \times v|}) \qquad (6.22)$$

where $v \neq u$. If the orientation is rodlike or disclike, the condition $K_v \neq K_{|u \times v|}$ need not be fulfilled.

Finally, we can combine equations (6.19)–(6.22) to obtain explicit expressions for the magnetic dipole and electric quadrupole contributions to the optical rotatory strength of u-polarized transitions:

$$-iM_u(0f)\mathscr{M}_u(0f) = 3[R^{iso}(0f)]_{ZZ} \qquad (6.23)$$

$$\begin{aligned}\pi\tilde{\nu}_f M_u(0f)Q_{st}(0f) &= \varepsilon_{ust}[R_{ss}(0f) - R_{tt}(0f)] \\ &= \varepsilon_{ust}\{2[R(0f)]_{ZZ} - 3(K_s + K_t)[R^{iso}(0f)]_{ZZ}\}/(K_s - K_t) \qquad (6.24)\end{aligned}$$

6.3.2 Other Chiral Molecules

In this section we consider molecules of symmetry D_n ($n \neq 2$) and C_n. Little can be done for molecules of C_1 symmetry without further special assumptions. In the other symmetry groups, there will be two kinds of allowed transitions: those polarized along the symmetry axis (excited state of A symmetry) and those polarized perpendicular to it. The C_2 symmetry group is a special case in which the two perpendicular directions are not equivalent and transitions polarized in a direction perpendicular to the symmetry axis produce excited states of B symmetry. In the other groups, the two perpendicular directions are equivalent by symmetry, and the excited state belongs to an E representation. In the absence of symmetry-lowering perturbations, the contributions to optical activity due to the two degenerate components of a transition cancel.

C_2 Symmetry. For molecules belonging to the C_2 symmetry group (one two-fold axis of symmetry), symmetry properties of the operators $\hat{\mathbf{M}}$, $\hat{\mathscr{M}}$, and $\hat{Q}$ are given in Table 6.1. Now, one of the principal orientation axes will coincide with the symmetry axis of the molecule, but the location of the two axes perpendicular to the symmetry axis may be difficult to establish and will have to be guessed from molecular shape or obtained from other available information such as IR linear dichroism of transitions with known polarization directions. In the principal orientation axes system, the observed circular dichroism will be given

Table 6.1
Nonvanishing Elements of $\hat{\mathbf{M}}$, $\hat{\mathscr{M}}$ and $\hat{Q}$ in the C_2 Symmetry Group

		$\hat{\mathbf{M}}, \hat{\mathscr{M}}$			$\hat{Q}$					
Sym.	Pol.	x	y	z	xx	yy	zz	yz	zx	xy
A	$\parallel$(z)	0	0	$M_z, \mathscr{M}_z$	Q_{xx}	Q_{yy}	Q_{zz}	0	0	Q_{xy}
B	$\perp$(x,y)	$M_x, \mathscr{M}_x$	$M_y, \mathscr{M}_y$	0	0	0	0	Q_{yz}	Q_{zx}	0

by the diagonal elements of the tensor $\mathsf{R}(0f)$ as described in equation (6.18). Equation (6.19) remains valid for the isotropic limit.

Transitions Polarized along the Symmetry Axis (Excited State of A Symmetry). From equations (6.5), (6.14), and (6.19), the information contained in Table 6.1 leads to a simple result:

$$R_{xx}(0f) = \{3K_y[\mathsf{R}^{iso}(0f)]_{ZZ} - [\mathsf{R}(0f)]_{ZZ}\}/(K_y - K_x) \tag{6.25}$$

$$R_{yy}(0f) = \{3K_x[\mathsf{R}^{iso}(0f)]_{ZZ} - [\mathsf{R}(0f)]_{ZZ}\}/(K_x - K_y) \tag{6.26}$$

$$R_{zz}(0f) = 0 \tag{6.27}$$

Here it was assumed that the symmetry axis coincides with the orientation axis of the molecule (the "long" axis). If this is not so and if we wish to maintain our standard convention $K_z \geqslant K_y \geqslant K_x$, it is necessary to relabel the axes in Table 6.1 and in equations (6.25)–(6.27) accordingly.

If the orientation distribution is such that $K_x = K_y$, formulas (6.25) and (6.26) are inapplicable and $R_{xx}(0f)$ and $R_{yy}(0f)$ cannot be determined separately. In such a case, the oriented and isotropic rotatory strengths are proportional to each other with a proportionality factor which is the same for all transitions polarized along the long axis and depends only on the orientation factors:

$$[\mathsf{R}(0f)]_{ZZ} = (3/2)(1 - K_z)[\mathsf{R}^{iso}(0f)]_{ZZ} \tag{6.28}$$

This result could conceivably be used in the determination of the orientation factors if they were not already known.

Returning now to the more general case of orientation ($K_x \neq K_y$), we note that the separate availability of $R_{xx}(0f)$ and $R_{yy}(0f)$ permits a determination of the individual contributions of magnetic dipole and electric quadrupole components of the optical activity [cf. equations (6.14) and (6.15), and Table 6.1]:

$$-iM_z(0f)\mathscr{M}_z(0f) = R_{xx}(0f) + R_{yy}(0f) = 3[\mathsf{R}^{iso}(0f)]_{ZZ} \tag{6.29}$$

$$\begin{aligned}\pi\tilde{\nu}_f M_z(0f)Q_{xy}(0f) &= \pi\tilde{\nu}_f M_z(0f)Q_{yx}(0f) = R_{xx}(0f) - R_{yy}(0f)\\ &= \{2[\mathsf{R}(0f)]_{ZZ} - 3(K_x + K_y)[\mathsf{R}^{iso}(0f)]_{ZZ}\}/(K_x - K_y)\end{aligned} \tag{6.30}$$

We note that these are just special cases of equations (6.23) and (6.24), which were valid for molecules of D_2 symmetry.

Perpendicularly Polarized Transitions (Excited State of B Symmetry). Here the situation is less favorable. A general transition of this kind will have nonvanishing components of the electric dipole transition moment **M**(0f) along both the x and y directions, which are defined as principal orientation axes, and it will not be possible to solve equations analogous to (6.18) and (6.19) for the three unknowns $R_{xx}(0f)$, $R_{yy}(0f)$, and $R_{zz}(0f)$. Only if it is known that a transition is polarized exactly along either the x or y principal orientation axis is it possible to make progress along lines quite analogous to those outlined in the section on D_2 symmetry molecules, in which such a condition is imposed by symmetry.

In the case of orientation distributions in which $K_x = K_y$, such as those of rodlike molecules, it is possible to solve for the sum $R_{xx}(0f) + R_{yy}(0f)$ and for

$R_{zz}(0f)$ separately:

$$R_{xx}(0f) + R_{yy}(0f) = [2/(3K_z - 1)]\{3K_z[\mathsf{R}^{iso}(0f)]_{ZZ} - [\mathsf{R}(0f)]_{ZZ}\} \tag{6.31}$$

$$R_{zz}(0f) = [1/(3K_z - 1)]\{2[\mathsf{R}(0f)]_{ZZ} - 3(1 - K_z)[\mathsf{R}^{iso}(0f)]_{ZZ}\} \tag{6.32}$$

Here again, it is assumed that the symmetry axis coincides with the orientation axis of the molecule, and it will be necessary to relabel the axes in Table 6.1 and in equations (6.31) and (6.32) if this is not so. For instance, in the case of disc-shaped molecules of C_2 symmetry in which the y and z axes are equivalent and x is the symmetry axis, if we use our standard requirement that $K_z \geqslant K_y \geqslant K_x$, the quantities which can be determined are $R_{yy}(0f) + R_{zz}(0f)$ and $R_{xx}(0f)$.

For perpendicularly polarized transitions in C_2 symmetry molecules, it is not possible to separate the contributions of magnetic dipole and electric quadrupole types, regardless of whether the orientation distribution is of a general type or of a rodlike or disclike type, unless it is known that the transition is polarized exactly along one of the principal orientation axes x or y. In this case the procedure is analogous to the case of molecules of D_2 symmetry.

C_n and D_n Symmetry, $n \geqslant 3$. Finally, we consider molecules of C_n and D_n symmetries in which $n \geqslant 3$. Here, the excited states of the parallel transitions have symmetry of type A and perpendicularly polarized transitions are degenerate (components f and f′), the excited state being of E symmetry. For these molecules, $K_x = K_y$ or $K_y = K_z$ by symmetry and electric quadrupole contributions to optical activity vanish. As a result, measurements of optical activity on oriented samples produce no information other than orientation factors. For parallel transitions, equations (6.28) and (6.29) hold. For perpendicularly polarized transitions the analogous equations (6.33) and (6.34) are valid, but it should be recalled that R(0f) and R(0f′) cancel each other in the absence of symmetry-lowering perturbations,

$$[\mathsf{R}(0f,0f')]_{ZZ} = (3/2)(1 - K_x)[\mathsf{R}^{iso}(0f,0f')]_{ZZ} \tag{6.33}$$

$$\begin{aligned} -iM_x(0f)\mathcal{M}_x(0f) &= -iM_x(0f')\mathcal{M}_x(0f') = -iM_y(0f)\mathcal{M}_y(0f) \\ &= -iM_y(0f')\mathcal{M}_y(0f') = (3/2)[\mathsf{R}^{iso}(0f,0f')]_{ZZ} \end{aligned} \tag{6.34}$$

In both cases it is assumed that the symmetry axis coincides with the orientation axis z, and the indices have to be relabeled appropriately if this is not the case. While these results can be used in principle to determine the orientation factors, this will be of little importance since they are usually more conveniently obtained from linear dichroism in absorption.

6.4 Magnetic Circular Dichroism

In the molecular frame, the elements of the magnetic rotatory tensors $\mathsf{A}(0f)$ and $\mathsf{C}(0f)$ which appear in (6.6) and (6.8) are given by

$$A_{uv}(0f) = \frac{i}{2d}\sum_{\alpha\lambda}[\mathcal{M}_u(f_\lambda f_\lambda) - \mathcal{M}_u(0_\alpha 0_\alpha)][\mathbf{M}(0_\alpha f_\lambda) \times \mathbf{M}(f_\lambda 0_\alpha)]_v \tag{6.35}$$

$$C_{uv}(0f) = \frac{i}{2d} \sum_{\alpha\lambda} \mathscr{M}_u(0_\alpha 0_\alpha)[\mathbf{M}(0_\alpha f_\lambda) \times \mathbf{M}(f_\lambda 0_\alpha)]_v \tag{6.36}$$

The tensor C(0f) vanishes if the ground state is not degenerate, and the tensor A(0f) vanishes if neither the ground nor the final state is degenerate.

The tensor B(0f), which is potentially nonzero for all allowed transitions in all molecules, is defined in the molecular frame by

$$B_{uv}(0f) = \frac{1}{d} \sum_{\alpha\lambda} \mathrm{Im} \left\{ \sum_{j \neq 0} \frac{\mathscr{M}_u(j0_\alpha)}{\tilde{\nu}_j} [\mathbf{M}(f_\lambda j) \times \mathbf{M}(0_\alpha f_\alpha)]_v + \sum_{j \neq f} \frac{\mathscr{M}_u(f_\lambda j)}{\tilde{\nu}_j - \tilde{\nu}_f} [\mathbf{M}(j0_\alpha) \times \mathbf{M}(0_\alpha f_\lambda)]_v \right\} \tag{6.37}$$

In these expressions, the subscripts α and λ denote the degenerate components of the ground and final states, d is the degeneracy of the ground state, and $\tilde{\nu}_j$ is the wavenumber of the j-th transition. The sum in (6.37) runs over all molecular states except as indicated.

Expressions (6.35)–(6.37) are directly comparable with the laboratory frame expressions (1.135)–(1.137).

We shall see in the following that the determination of the tensors A(0f) and C(0f) from measurements on partially oriented molecules provides no information beyond that accessible from measurements of dichroic absorption, and we shall mention them only briefly. On the other hand, determination of the tensor B(0f) from such measurements is capable of providing novel types of information about the molecule, as long as the molecule has at least some symmetry.

6.4.1 High-Symmetry Molecules

A and C Terms. In molecules with a threefold or higher-order symmetry axis, transitions involving excited doubly degenerate states exhibit A terms, and transitions involving a doubly degenerate ground state exhibit C terms in their MCD spectra (we do not consider molecules with several higher-order axes since they generally cannot be adequately oriented). Assuming for the moment that the symmetry axis coincides with the orientation axis z of the molecule, symmetry dictates $K_x = K_y = (1 - K_z)/2$.

We choose the components of the degenerate states in such a way that they diagonalize $\hat{\mathscr{M}}_z$. Only the z component of the magnetic moments $\mathscr{M}$(ff) and $\mathscr{M}$(00) can be different from zero. The expressions for the ZZ elements $[A(0f)]_{ZZ}$ and $[C(0f)]_{ZZ}$ obtained using (1.133) and (1.134) with light propagating along the Z axis of the sample then become

$$[A(0f)]_{ZZ} = K_z \frac{i}{2d} \sum_{\alpha\lambda} [\mathscr{M}_z(f_\lambda f_\lambda) - \mathscr{M}_z(0_\alpha 0_\alpha)][\mathbf{M}(0_\alpha f_\lambda) \times \mathbf{M}(f_\lambda 0_\alpha)]_z \tag{6.38}$$

$$[C(0f)]_{ZZ} = K_z \frac{i}{2d} \sum_{\alpha\lambda} \mathscr{M}_z(0_\alpha 0_\alpha)[\mathbf{M}(0_\alpha f_\lambda) \times \mathbf{M}(f_\lambda 0_\alpha)]_z \tag{6.39}$$

Both of these are simply expressed in terms of the ZZ elements $[A^{iso}(0f)]_{ZZ}$ and $[C^{iso}(0f)]_{ZZ}$ measured on the unoriented sample which are obtained when $K_z = 1/3$:

$$[A(0f)]_{ZZ}/[A^{iso}(0f)]_{ZZ} = [C(0f)]_{ZZ}/[C^{iso}(0f)]_{ZZ} = 3K_z \tag{6.40}$$

Clearly, no new information is obtained from oriented MCD measurements relative to oriented polarized absorption measurements, except that the separation of A or C terms from B terms may be facilitated. It is interesting to note that the orientational dependence of the signal observed for a transition polarized perpendicular to the symmetry axis follows the alignment of the symmetry axis, which is perpendicular to the transition moment. In this respect oriented MCD is very different from dichroic absorption.

As noted in Section 1.3.3, the quantity measured most commonly is not the value of the A term or the C term as such, but rather its ratio to the dipole strength D evaluated on the same sample, from a measurement of $E_{X,Y}(0f)$ or $E^{iso}(0f)$ and the use of (1.100). For either A/D or C/D, the dependence on the molecular alignment is given by

$$\frac{[A(0f)]_{ZZ}/[D(0f)]_{ZZ}}{[A^{iso}(0f)]_{ZZ}/[D^{iso}(0f)]_{ZZ}} = \frac{[C(0f)]_{ZZ}/[D(0f)]_{ZZ}}{[C^{iso}(0f)]_{ZZ}/[D^{iso}(0f)]_{ZZ}} = 4K_z(1 + K_z) \tag{6.41}$$

If the molecular symmetry axis coincides with the molecular x axis as defined by our convention $K_z \geqslant K_y \geqslant K_x$, similar formulas apply in which the subscript z is replaced by the subscript x. It is of some interest to note that the A/D and C/D values measured in the partially aligned solution increase over the values measured in an isotropic solution by up to a factor of 2 as the degree of alignment increases if the high-symmetry axis coincides with the molecular orientation axis, and decrease (and in the limit vanish) if the symmetry axis is perpendicular to the molecular orientation axis. The former instance is exemplified by rodlike molecules and the latter by disclike molecules.

B Terms. Many high-symmetry molecules as defined in the present text (not "very high symmetry") do not possess higher order symmetry axes and will therefore have no degenerate states and no A and C terms. In the remainder of this chapter, we shall consider molecules without degenerate states and limit our attention to B terms. In this case it is advantageous to work in terms of real wavefunctions. Then (6.37) can be written as

$$B_{uv}(0f) = \sum_{j \neq 0} \frac{\mathcal{M}_u(j0)/i}{\tilde{\nu}_j} [\mathbf{M}(fj) \times \mathbf{M}(0f)]_v + \sum_{j \neq f} \frac{\mathcal{M}_u(fj)/i}{\tilde{\nu}_j - \tilde{\nu}_f} [\mathbf{M}(j0) \times \mathbf{M}(0f)]_v \tag{6.42}$$

We shall now consider applications to molecules of high symmetry which belong to point symmetry groups in which no more than one of the functions x, y, and z belongs to any given one-dimensional irreducible representation. In these molecules the principal orientation axes coincide with the symmetry axes x, y, z. We shall assume that K_x, K_y, and K_z are known from measurements of linear dichroism and that $[\mathsf{B}(0f)]_{ZZ}$ and $[\mathsf{B}^{iso}(0f)]_{ZZ}$ have been measured. Applications to molecules of lower symmetry are considered in Section 6.4.2.

The ZZ elements of the tensor $\mathsf{B}(0f)$ observed on the oriented sample with light propagating along the Z axis, and on the isotropic sample, are given by (6.7) in the form

$$[\mathsf{B}(0f)]_{ZZ} = K_x B_{xx}(0f) + K_y B_{yy}(0f) + K_z B_{zz}(0f) \qquad (6.43)$$

$$[\mathsf{B}^{iso}(0f)]_{ZZ} = [B_{xx}(0f) + B_{yy}(0f) + B_{zz}(0f)]/3 \qquad (6.44)$$

For a transition f whose transition moment $\mathbf{M}(0f)$ is directed along the molecular axis u, B_{uu} vanishes since the vectors $\mathbf{M}(fj) \times \mathbf{M}(0f)$ and $\mathbf{M}(j0) \times \mathbf{M}(0f)$ in (6.42) are perpendicular to u. Equations (6.43) and (6.44) can therefore be solved for the remaining two diagonal elements of the tensor $\mathsf{B}(0f)$:

$$B_{vv}(0f) = \{[\mathsf{B}(0f)]_{ZZ} - 3K_{|u\times v|}[\mathsf{B}^{iso}(0f)]_{ZZ}\}/(K_v - K_{|u\times v|}) \qquad (6.45)$$

where $v \neq u$.

Equations (6.43)–(6.45) are completely analogous to equations (6.18), (6.19), and (6.22), valid for natural CD, with the magnetic rotatory tensor $\mathsf{B}(0f)$ now playing the role of the natural rotatory strength tensor $\mathsf{R}(0f)$. However, the physical significance of the two tensors is different. Whereas we have seen that $\mathsf{R}(0f)$ can be used to obtain information about magnetic dipole and electric quadrupole transition moments from the ground state to the excited state, the diagonal elements of $\mathsf{B}(0f)$ involve more complicated combinations of magnetic dipole and electric dipole transition moments and excitation energies. For a transition $0 \rightarrow f$ polarized along t, the diagonal elements are

$$B_{vv}(0f) = \varepsilon_{tuv} M_t(0f) \left\{ \sum_{\substack{j:v \\ \text{mag.dip.}}} M_u(fj) \frac{\mathscr{M}_v(0j)/i}{\tilde{\nu}_j} + \sum_{\substack{j:u \\ \text{el.dip.}}} M_u(0j) \frac{\mathscr{M}_v(fj)/i}{\tilde{\nu}_f - \tilde{\nu}_j} \right\} \qquad (6.46)$$

where the first sum runs over molecular states j which are magnetic dipole allowed for transition from the ground state with polarization along v, and the second sum runs over states which are electric dipole allowed for transition from the ground state with polarization along u.

A separate determination of the two diagonal elements of the B tensor for a t-polarized transition, $B_{vv}(0f)$ and $B_{uu}(0f)$, thus corresponds to a separation of the B term of the transition into two components. The first of these, $B_{vv}(0f)$, is due to the magnetic mixing of electric dipole u-polarized excited states into the final state and of magnetic dipole v-polarized excited states into the ground state. The second part, $B_{uu}(0f)$, corresponds to a magnetic mixing of v-polarized electric dipole excited states into the final state and magnetic dipole u-polarized states into the ground state. The situation is best illustrated on an example (Figure 6.2). We assume for simplicity that the only contributions to the B term are due to the second term on the right-hand side of (6.42), i.e., to the mutual mixing of excited states by the static magnetic field. This is not a bad approximation for $\pi\pi^*$ transitions of aromatic molecules such as quinoxaline, which we have chosen for our example.

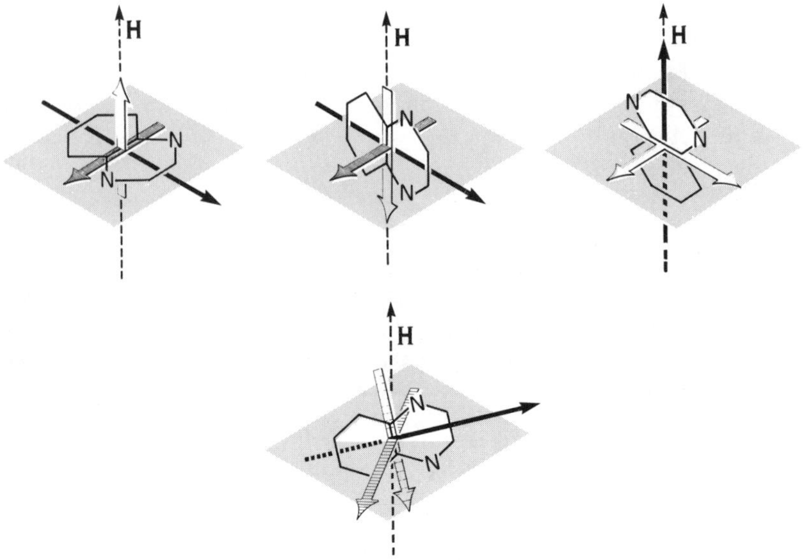

Figure 6.2 Decomposition of the B term in the MCD spectrum of a high-symmetry molecule (C_{2v}) into its B_{uu} and B_{vv} components. See text.

We simplify the example further by considering only three excited states, f, j_x, and j_y. The transition whose B term is being observed is $0 \to f$. We choose it to be long-axis (z) polarized and indicate the transition moment **M**(0f) by a long thin arrow in Figure 6.2. The two states which can be mixed into f by the magnetic field are j_x and j_y. They give rise to an out-of-plane polarized transition $0 \to j_x$ and short-in-plane-axis polarized transition $0 \to j_y$. Their respective transition moments $\mathbf{M}(0j_x)$ and $\mathbf{M}(0j_y)$ are indicated by fat arrows in Figure 6.2. Excited states j_x and j_y both contribute to the B term, the former through $B_{yy}(0f)$ and the latter through $B_{xx}(0f)$, as is clear from the consideration of the second term on the right-hand side of (6.46). However, the relative weights of their contributions provided by any particular molecule are a sensitive function of molecular orientation, as can be seen from the second term on the right-hand side of the general expression (1.136), with U = V = Z. This contains the term $[\mathbf{M}(0j) \times \mathbf{M}(0f)]_Z$. Since **M**(0f) lies along z, only the states corresponding to the shaded transition moments in Figure 6.2 yield nonvanishing contributions to the B term. If x is lined up with Z (top left), only j_y contributes; if y is lined up with Z (top center), only j_x contributes; and if z is lined up with Z (top right), neither contributes. In a general orientation (bottom), both j_x and j_y can contribute but usually to different degrees.

Clearly, in an isotropic sample, averaging will cause both j_x and j_y to have an equal opportunity to contribute to the observed B term, but this will not be so in a partially aligned sample.

General Uniaxial Orientation. In the case of a general uniaxial orientation ($K_x \neq K_y \neq K_z$), equation (6.45) permits one to determine both nonvanishing diagonal elements of the B tensor, $B_{vv}(0f)$ and $B_{uu}(0f)$, for a transition polarized along any one of the axes x, y, or z. If two of the orientation factors are equal, this is no longer possible. Difficulties occur for z-polarized transitions in molecules which orient like rods ($K_x = K_y$) and for x-polarized transitions in molecules which orient like discs ($K_y = K_z$).

Rodlike Orientation. For molecules which orient like rods, the results can be written as follows:

(i) For z-polarized transitions,

$$[B(0f)]_{ZZ}/[B^{iso}(0f)]_{ZZ} = 3K_x = 3K_y \tag{6.47}$$

so that no new information is obtained over what is usually already available from linear dichroism.

(ii) For x-polarized transitions and y-polarized transitions, B_{zz} can be determined;

$$B_{zz} = \frac{2[B(0f)]_{ZZ} - 3(1 - K_z)[B^{iso}(0f)]_{ZZ}}{3K_z - 1} \tag{6.48}$$

(iii) For x-polarized transitions, B_{yy} can be determined; for y-polarized transitions, B_{xx} can be determined. Their magnitude, B_{uu}, is given by

$$B_{uu} = 2\,\frac{3K_z[B^{iso}(0f)]_{ZZ} - [B(0f)]_{ZZ}}{3K_z - 1} \tag{6.49}$$

Disclike Orientation. In the case of molecules which orient like discs, we find:

(i) For x-polarized transitions, only the K's can be determined:

$$[B(0f)]_{ZZ}/[B^{iso}(0f)]_{ZZ} = 3K_y = 3K_z \tag{6.50}$$

(ii) For y-polarized transitions and z-polarized transitions, B_{xx} can be determined:

$$B_{xx} = \frac{3K_z[B^{iso}(0f)]_{ZZ} - [B(0f)]_{ZZ}}{3K_z - 1} \tag{6.51}$$

(iii) For y-polarized transitions, B_{zz} can be determined; for z-polarized transitions, B_{yy} can be determined. Their magnitude, B_{uu}, is given by

$$B_{uu} = \frac{2[B(0f)]_{ZZ} - 3(1 - K_z)[B^{iso}(0f)]_{ZZ}}{3K_z - 1} \tag{6.52}$$

6.4.2 Low-Symmetry Molecules

Little can be done for molecules of C_1 symmetry without further special assumptions.

If at least one symmetry element is present in the molecule, the situation simplifies considerably. We shall consider here in some detail the case of molecules of C_s symmetry, exemplified by aromatic molecules whose only symmetry operation is mirroring in the molecular plane. We assume as usual that the ground state is totally symmetric (A'). Transitions into excited states of A''

symmetry (e.g., $n\pi^*$ states) are electric dipole allowed with polarization x perpendicular to the molecular plane and magnetic dipole allowed with polarization in the molecular plane yz. Excited states of A′ symmetry (e.g., $\pi\pi^*$ states) are electric dipole allowed with polarization in the molecular plane yz and magnetic dipole allowed with polarization x perpendicular to the molecular plane.

For transitions which are electric dipole polarized along the x axis, $B_{xx} = 0$, and both nonvanishing diagonal elements of the B tensor can be found using equation (6.45), which is valid in the principal orientation axes system. The situation is less favorable for the usually more numerous in-plane polarized transitions to states of A′ symmetry. In the general case, B_{xx}, B_{yy}, and B_{zz} will all be nonzero and we shall not be able to determine them from the two measurements available. However, it should be noted that in the case that the A′ excited state as well as the ground state are magnetically mixed solely with other states of A′ symmetry, only the x components of $\mathbf{M}(j0) \times \mathbf{M}(0f)$ and $\mathbf{M}(fj) \times \mathbf{M}(0f)$ are nonzero [cf. (6.42)]. Further, only the x components of $\mathcal{M}(fj)/i$ and $\mathcal{M}(j0)/i$ are nonzero, so that B_{xx} is the sole nonvanishing element of B. In such a case, equations (6.43) and (6.44) combine to yield

$$[\mathsf{B}(0f)]_{ZZ}/[\mathsf{B}^{iso}(0f)]_{ZZ} = 3K_x \tag{6.53}$$

This result can now be used to check whether $\pi\pi^*$-$\pi\pi^*$ magnetic mixing is the only important mechanism which gives rise to the B term of a particular transition. For those transitions for which equation (6.53) is found to hold, this is likely to be true, unless $K_y = K_z$ and $B_{yy} = -B_{zz}$, in which case no conclusions can be drawn. This may be useful in investigations of the origin of the B terms in planar molecules in terms of the relative importance of $\pi\pi^*$-$\pi\pi^*$ and $\pi\pi^*$-$n\pi^*$ magnetic mixing.

6.5 Circular Dichroism of Photoselected Orientation Distributions

Photoselection with Natural or Linearly Polarized Light. *CPL and MCPL.* A special case of a uniaxially oriented molecular assembly is the one resulting from photoselection. As long as the depletion of the initial randomly oriented ground state molecular assembly is negligible, simple expressions can be written for the orientation factors of the photoselected assembly. For photoselecting light linearly polarized along Z, they are referred to as $K_{uv}(\tilde{\nu}_0,Z)$ and have been given in Section 4.6.3 [(4.70), (4.73), (4.74)]; for photoselecting light which is unpolarized and propagates along Z, they are referred to as $K_{uv}(\tilde{\nu}_0,XY)$ and have been given in Section 4.6.4 [(4.115), (4.116)]. Even when the degree of transformation to photoproduct is not negligible, explicit albeit complex expressions for the orientation factors have been given in (4.93) and (4.94) for Z-polarized photoselecting light. For unpolarized photoselecting light, similar expressions can be derived as indicated in Section 4.6.4. For molecules whose absorption properties are known, it is thus possible to prepare partially oriented assemblies with known orientation factors and to measure their natural or magnetic circular dichroism.

The measurement of the magnetic circular dichroism of the photoselected partially uniaxially oriented sample in emission (circularly polarized luminescence, CPL or MCPL, respectively) or in absorption is a process of rank 2 and can therefore be described using the methods of this chapter, with simple substitution of the now known values $K_{uv}(\tilde{\nu}_0,Z)$ or $K_{uv}(\tilde{\nu}_0,XY)$ into (6.5). From the point of view of the initially randomly oriented sample, the experiment actually corresponds to a process of overall rank 4, so that in a sense it does not belong in this chapter. As a matter of fact, the initial photoselection may already involve a two-photon process, such as successive or simultaneous absorption of two photons. Methods of Chapters 4 and 7 can then be combined to obtain the orientation factors of the photoselected assembly, and equation (6.5) can again be used to describe the results. Thus, the procedures given permit the description of a wide variety of processes of odd rank higher than 2, provided that the initial molecular assembly is randomly oriented.

Photoselection with Circularly Polarized Light. ***FDCD and Detection of Racemic Mixtures.*** It is also possible to use circularly polarized light in the initial photoselection step. As long as the starting molecules are achiral and no magnetic field along Z is present, the expressions of Section 4.6.4 for the orientation factors $K_{uv}(\tilde{\nu}_0,XY)$ apply. In the case of chiral molecules, it is necessary to consider the effect of the magnetic dipole and electric quadrupole interactions responsible for the differentiation between left-handed and right-handed circularly polarized light [cf. (6.13) and (6.15)]. These must be added to expression (4.64), which contains only the electric dipole interaction as it stands. This will modify the derivation given in Chapter 4 and introduce small correction terms into the expressions for the orientation factors resulting from the photoselection process. For a chiral substrate, the correction terms will be opposite in sign for left-handed and right-handed light and will produce different sets of orientation factors. The two kinds of light will also be absorbed at different rates, so that different photostationary populations will be achieved with a given light intensity. Therefore, measurement of total fluorescence intensity on a sample undergoing rapid orientation depolarization is already useful and produces the CD spectrum of the substrate (fluorescence-detected CD, or FDCD for short). The results for the other enantiomer will be similar except that the roles of left-handed and right-handed light will be reversed.

When the initially randomly oriented sample is a racemic mixture of two chiral enantiomers, photoselection with circularly polarized light will therefore produce (i) a different set of orientation factors and (ii) different total amounts of photoselection ("photoresolution") for the two enantiomers. One way to study these effects is to measure the circular dichroism of the photoselected molecules in absorption or in emission and to use (6.5) to analyze the results. This provides information on molecular chiroptic properties without an actual physical separation of the racemic mixture. The procedure can be used to prove that an outwardly optically inactive sample actually is a racemate. The effects are small, since both enantiomers are being observed at the same time and the difference in their orientation distribution factors and in the photoselected amounts is small. The process has an overall rank of 6, reducible to 4.

FDMCD. In the presence of a magnetic field in the +Z or −Z direction, the sample will be circularly dichroic, and photoselection with circularly polarized light will be affected so that the results of Section 4.6.4 for the orientation factors $K_{uv}(\tilde{\nu}_0,XY)$ will again have to be modified. Now, (4.64) has to be corrected by addition of the interactions described in Section 6.4. This will introduce additional terms into the derivation given in Chapter 4 and cause the appearance of small correction terms into the expressions for the orientation factors resulting from the photoselection process. The correction terms will be opposite in sign for left-handed and right-handed light and will produce different sets of orientation factors. The two kinds of light will also be absorbed at different rates, so that different photostationary populations will be achieved with a given light intensity. (The measurement of total fluorescence intensity on a sample undergoing rapid orientation depolarization produces the MCD spectrum of the starting material and is known as fluorescence-detected MCD, or FDMCD for short.) The results for the reversed direction of magnetic field (−Z instead of Z) will be similar, except that the roles of left-handed and right-handed light will be reversed. Magnetic CD cannot be used for photoresolution in the same sense that natural CD can, since there is nothing to resolve: The two members of a pair of enantiomers in CD are analogous to the two directions of magnetic field in MCD; a racemic mixture is analogous to the absence of magnetic field.

Although photoselection experiments appear suitable for the detection and study of the anisotropy of natural and magnetic CD, very little work has been done in this direction so far, and we shall not state the theoretical expressions for the various individual cases.

6.6 Comments and References

General references to the theory of natural and magnetic optical activity were given in Section 1.4.

One of the classics in the theory of optical activity of anisotropic solutions in N. Gō, *J. Chem. Phys.* **43**, 1275 (1965). Expressions for natural optical activity of uniaxially partially aligned solutes are given in H.-G. Kuball, T. Karstens, and A. Schönhofer, *Chem. Phys.* **12**, 1 (1976) and H.-G. Kuball, J. Altschuh, and A. Schönhofer, *Chem. Phys.* **43**, 67 (1979). Further references can be found there and in the historical review by J. K. O'Loane, *Chem. Rev.* **80**, 41 (1980).

The theory of natural and magnetically induced polarized emission (CPL, MCPL) can be found, for instance, in K. W. Hipps, *Chem. Phys.* **23**, 451 (1977) and J. P. Riehl and F. S. Richardson, *J. Chem. Phys.* **65**, 1988 (1976); see also F. S. Richardson and J. P. Riehl, *Chem. Rev.* **77**, 773 (1977), **86**, in press (1986). For the theory of photoselection with circularly polarized light (FDCD), see I. Tinoco, Jr., B. Ehrenberg, and I. Z. Steinberg, *J. Chem. Phys.* **66**, 916 (1977).

7

Processes of Rank Four

This chapter first mentions briefly the electric quadrupole contribution to the absorption or emission of light by partially aligned assemblies of molecules, which is described by the fourth-rank tensor ${}^{(4)}\mathsf{Q}(0f)$. This is a direct product of the electric quadrupole transition moment tensor $\mathsf{Q}(0f)$ with its complex conjugate:

$$ {}^{(4)}\mathsf{Q}(0f) = \mathsf{Q}(f0)\mathsf{Q}(0f) \qquad (7.1) $$

When molecular wavefunctions are chosen real, $\mathsf{Q}(0f)$ is real, $\mathsf{Q}(0f) = \mathsf{Q}(f0)$, and ${}^{(4)}\mathsf{Q}(0f)$ is real as well. Its elements are

$$ [{}^{(4)}\mathsf{Q}(0f)]_{stuv} = \mathsf{Q}_{st}(f0)\mathsf{Q}_{uv}(0f) \qquad (7.2) $$

The electric quadrupole transition moment $\mathsf{Q}(0f)$ is a symmetric tensor of rank 2, equal to the matrix element of the electric quadrupole moment operator defined in (1.58). The electric quadrupole contribution to absorption is of little importance in molecular spectroscopy.

The bulk of the chapter is devoted to processes based on the electric dipole interactions of two photons with a molecule. These are described by a fourth-rank tensor defined as a direct product of two tensors of rank 2:

$$ {}^{(4)}\mathsf{O}(0f) = \mathsf{O}(f0)\mathsf{O}(0f) \qquad (7.3) $$

The exact form of $\mathsf{O}(0f)$ depends on the particular two-photon process considered. For instance, in the case of photoluminescence,

$$ {}^{(4)}\mathsf{O}(0f) = {}^{(4)}\mathsf{M}(0f) = \mathsf{M}(f0)\,\mathsf{M}(0f) = \mathbf{M}(0f)\mathbf{M}(f0)\mathbf{M}(f0)\mathbf{M}(0f) \qquad (7.4) $$

where $\mathbf{M}$ is the electric dipole transition moment vector and M is the electric dipole absorption tensor, both of which are real if the molecular wavefunctions are chosen real. They are already familar from Chapter 5.

We first identify the four basic types of experiments which correspond to the two-photon processes. Then, following the general principles outlined in

Section 1.3.4, we describe the general relations between the observable polarized intensities obtained on a partially aligned assembly of solute molecules on the one hand, and the orientation distribution of the molecules, as well as the anisotropy of the molecular transition moments, on the other hand. Both the second moments (K) and the fourth moments (L) of the orientation distribution function enter into the description. We make the simplifying assumptions concerning the role of the solvent which were already discussed in Section 5.5.

Although in its most general form the theory is equally applicable to all four types of experiments, it is possible to specialize it further for those in which the two-photon molecular interactions occur simultaneously and those in which they occur successively. Practical methods which can be used to derive the orientation factors and transition moments from experimental data are then discussed in some detail for each group of experiments.

Measurements in which the state of polarization of the light does not change as it passes through the sample are preferable. In these, the light propagation direction, as well as the direction of linear polarization, coincide with one of the principal axes of the sample, X, Y, or Z, or circularly polarized light propagates along the unique Z axis. All of this chapter will be limited to experiments of this kind. The analysis of experiments of a more general nature is much more complicated.

7.1 Electric Quadrupole Absorption and Emission

The electric quadrupole contributions to absorption intensity stand a chance to be detectable only if the electric dipole contributions vanish by symmetry. For light propagating along K whose electric vector is directed along U, the electric quadrupole contribution to the absorption probability for the f-th transition is proportional to $\langle|\varepsilon_K \mathsf{Q}(0f)\varepsilon_U^*|^2\rangle$ [cf. (1.17) and (1.60)]. The electric quadrupole transition moment of the 0 → f transition, Q(0f), is quite analogous to the electric dipole transition moment **M**(0f) of Chapter 5, except that it is a tensor of rank 2 whereas **M**(0f) is a vector. Since Q(0f) is symmetric, quadrupole absorption intensity for light propagating along V through a uniaxial sample of partially aligned molecules and linearly polarized along U is the same as that for light polarized along V and propagating along U. As usual, the transformation between the quadrupole absorption tensor in the laboratory frame and the quadrupole absorption tensor in the molecular frame requires an averaging over the orientation distribution. It can be shown, for instance by the method outlined in Appendix II, that the result can be written in the form anticipated in (1.143):

$$\langle|\varepsilon_U \mathsf{Q}(0f)\varepsilon_V|^2\rangle = \sum_{stuv} [{}^{(4)}\mathsf{P}_{UV}]_{stuv}[{}^{(4)}\mathsf{Q}(0f)]_{stuv} \tag{7.5}$$

This equation is completely analogous to equation (5.14) for electric dipole absorption and equations (6.3) and (6.4) for magnetic dipole absorption. A similar expression can be written for the electric quadrupole contribution to emission.

All necessary information on the molecular orientation is contained in the fourth-rank tensor ${}^{(4)}\mathsf{P}_{UV}$, analogous to the second-rank tensor D_U introduced in Chapter 5 [(5.15)–(5.18)]. Since the directions U and V are perpendicular, only the two tensors $^{(4)}P_{XY}$ and $^{(4)}P_{XZ}$ are needed. Expressions for their matrix elements as a function of the orientation parameters K and L are given in the following [(7.9) and (7.11)].

7.2 Two-Photon Processes in Aligned Molecules—An Overview

Here we move on to the primary focus of this chapter and take up processes in which a molecule exchanges two polarized photons with the radiation field. Their polarizations relative to the laboratory axes U and V provide a handle for the exploration of the anisotropy of the molecular orientation distribution and the anisotropy of the molecular optical properties. However, unlike processes involving a single photon (Chapters 5 and 6), two-photon processes permit us to utilize the relative polarization of the two photons (i.e., U parallel to V, orthogonal to V, of like or opposite handedness as V, etc.) to probe the anisotropy of molecular optical properties even on isotropic samples (cf. Section 4.6). Although, strictly speaking, measurements on such samples lie outside the scope of this book, their understanding is necessary for proper appreciation of the measurements on aligned samples and they will be discussed briefly as well.

Two-Photon Processes. In all cases of practical interest, one of the photons ($\tilde{\nu}_1$) is taken from the incident beam and annihilated by interaction with the molecule. The other photon ($\tilde{\nu}_2$) can either be taken from the incident beam and annihilated or it can be created by the interaction with the molecule. The two exchanges of photons can be simultaneous (coherent) or separated in time (incoherent). If the second step is delayed, the intermediate state is given an opportunity to develop in time into a different intermediate state by photophysical or photochemical processes, and the molecule is also given an opportunity to rotate or to transfer energy to another molecule which may be differently oriented, with a similar final result. In this instance, the orientation factors describing the molecular orientation distribution become time-dependent, and the study of this time dependence provides information about molecular rotation and intermolecular energy transfer. Table 7.1 and Figures 7.1 and 7.2 summarize the common spectroscopic experiments which correspond to the

Table 7.1
Two-Photon Processes

Photon $\tilde{\nu}_1$, U Photon $\tilde{\nu}_2$, V	Annihilated Annihilated	Annihilated Created
Intensity	$\langle \lvert \varepsilon_U^* O \varepsilon_V^* \rvert^2 \rangle$	$\langle \lvert \varepsilon_U^* O \varepsilon_V \rvert^2 \rangle$
Successive events	Photoinduced dichroism	Photoluminescence
Tensor O	O = **M**(j)**M**(f)	O = **M**(j)**M**(f)
Simultaneous events	Two-photon absorption	Raman scattering
Tensor O	O = T(j,f)	O = α′(j,f)

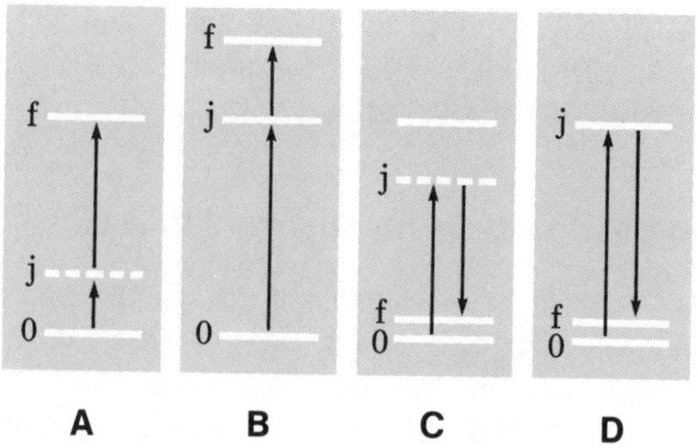

Figure 7.1 Simultaneous two-photon events. (A) Nonresonant two-photon absorption; (B) resonant two-photon absorption; (C) nonresonant Raman scattering; (D) resonant Raman scattering. Full horizontal lines represent energies of molecular eigenstates; broken ones, those of virtual states.

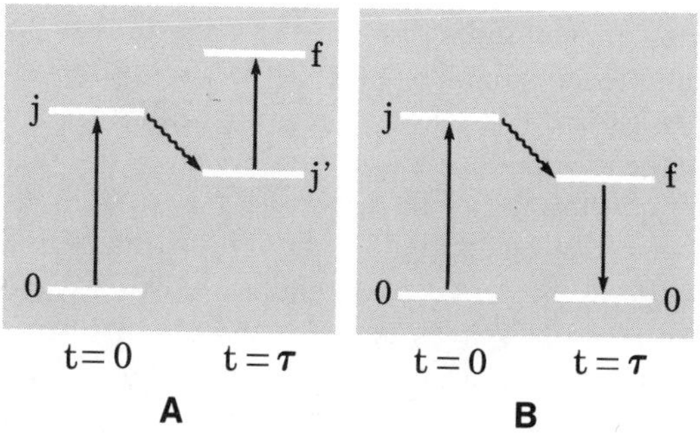

Figure 7.2 Successive two-photon events. (A) Photoinduced dichroism; (B) photoluminescence.

interaction of two photons with a molecule. Table 7.1 shows the quantities which must be averaged over the molecular orientation in order to obtain an expression for an elementary contribution to the observed intensity, as discussed in more detail in the following. The complex conjugation shown on ε_U^* and ε_V^* has no effect when the light is linearly polarized; it is discussed in Section 7.5.

In each case, measurements provide a spectral intensity as a function of two photon energies, $E_{UV}(\tilde{\nu}_1,\tilde{\nu}_2)$ ("two-dimensional" spectra). Usually, they are performed with either $\tilde{\nu}_1$ or $\tilde{\nu}_2$ held constant and the result is displayed as an ordinary "one-dimensional" spectrum, perhaps with several different choices of the wavenumber that is held constant. The spectra are given different names depending on whether $\tilde{\nu}_1$ or $\tilde{\nu}_2$ is held constant, as will be described in more detail below.

Notation. We shall generally treat the total observed intensity $E_{UV}(\tilde{\nu}_1,\tilde{\nu}_2)$ as an incoherent superposition of overlapping elementary contributions, weighted by the appropriate line-shape functions (overlapping spectral lines), in the same way as in the treatment of one-photon processes. This is almost always a reasonable approximation. In the description of the elementary two-photon processes, we shall assign the label j to the first step and f to the second step. Thus, the label (j,f) will specify the elementary two-photon interaction.

In Raman scattering and simultaneous two-photon absorption, $j = \tilde{\nu}_1$ contains information about the incident wavelength.

In Raman scattering, f contains information about the final vibrational level reached; $E_{UV}(j,f) = E_{UV}(\tilde{\nu}_1,f)$ refers to the intensity contributed by a particular vibration, and $E_{UV}(\tilde{\nu}_1,\tilde{\nu}_2)$ is the total intensity observed at a particular spectral position $\tilde{\nu}_2$ by all overlapping lines and involves a summation over f. Electronic Raman scattering is also known, but we shall not discuss it.

In simultaneous two-photon absorption, f contains information about the final electronic state reached and $E_{UV}(j,f) = E_{UV}(\tilde{\nu}_1,f)$ refers to the intensity provided by the f-th excited state. To obtain the observed intensity $E_{UV}(\tilde{\nu}_1,\tilde{\nu}_2)$, a summation over all contributing electronic states f is needed.

In photoluminescence we shall refer to absorption from the ground into the j-th electronic excited state (or a vibronic level thereof) and emission from the f-th excited state (usually the lowest vibrational level) into the ground state as yielding $E_{UV}(j,f)$. The label f fully characterizes the second step; it will have to contain the specification of the particular vibrational level of the ground state reached, if such detail is to be discussed. Emission into states other than the ground state is rarely observed in molecular spectroscopy. It would be handled similarly. Usually, several absorption lines (j), and at times several emission lines (f) also, will overlap; the total observed intensity $E_{UV}(\tilde{\nu}_1,\tilde{\nu}_2)$ will contain all of these contributions and is obtained by a double summation over j and f.

In photodichroism, the first photon excites the ground into the j-th electronic state, or a particular vibronic level thereof, which then may develop into the j′-th electronic state (usually the lowest vibrational level) and in turn absorb the second photon to produce the f-th state, or a particular vibronic level thereof if such detail is required. The elementary contribution to the intensity is $E_{UV}(j,f)$, and a double summation over j and f is needed to yield $E_{UV}(\tilde{\nu}_1,\tilde{\nu}_2)$.

Strictly speaking, the labels for the elementary contributions should be (0j,f0), (0j,j′f), etc., but in the following we shall only use (j,f), and the involvement of the other states will be understood.

In the general symbol for the contribution to the corrected observable intensity which is due to the (j,f)-th process, $E_{UV}(j,f)$, and in the symbol for the observed intensity, $E_{UV}(\tilde{\nu}_1,\tilde{\nu}_2)$, U is a unit vector in the direction of the electric vector of the first linearly polarized photon (incident light) and V that of the electric vector of the second photon. In the cases in which we wish to specialize, we shall use $F_{UV}(\tilde{\nu}_1,\tilde{\nu}_2)$ and $F_{UV}(j,f)$ for photoluminescent intensity or photo-induced optical density, and $I_{UV}(\tilde{\nu}_1,\tilde{\nu}_2)$ and $I_{UV}(\tilde{\nu}_1,f)$ for observed Raman scattering intensity or two-photon absorption cross section. For linearly polarized light, U and V are given by the direction of one of the laboratory axes X, Y, or Z. We shall also need to refer to circularly polarized light propagating along the positive or negative direction of the Z axis, with the electric vector described by $(X + iY)/\sqrt{2}$ or $(X - iY)/\sqrt{2}$. In such a case, U and V are indicated by plus and minus subscripts, respectively.

Finally, at times we shall also combine U and V into a single symbol, CON or DIS. Both refer to light of circular polarization propagating along the +Z or the −Z direction. The symbol CON (conrotatory) indicates that the sense of the rotation of the tip of the electric vector, viewed along the Z axis, is the same for both photons: Both are left-handed or both are right-handed if they propagate in the same direction; one is left-handed and one right-handed if they propagate in opposite directions. The symbol DIS (disrotatory) indicates that the sense of the rotation of the tip of the electric vector, viewed along the Z axis, is opposite for the two photons.

Expressions for Polarized Intensity. Just as was the case in ordinary absorption, it is useful to express the contributions $E_{UV}(j,f)$ in terms of the orientation factors on the one hand and of quantities describing the intrinsic molecular properties on the other hand. The orientation factors needed for processes of rank 4 will be both the K's and the L's. The quantities describing the molecular properties are tensors of rank 2, O, whose concrete realization depends on the particular experiment performed as discussed below.

The desired relation is obtained by averaging over the molecular orientation distribution (Section 1.3.4; Appendix II). The general result can be written in the form of a scalar product of two tensors of rank 4, as anticipated in (1.143):

$$E_{UV}(j,f) = \sum_{stuv} [{}^{(4)}\mathsf{P}_{UV}]_{stuv}[{}^{(4)}\mathsf{O}(j,f)]_{stuv} \qquad (7.6)$$

where the fourth-rank tensor ${}^{(4)}\mathsf{P}_{UV}$ depends on the choice of U and V and is only a function of the orientation distribution through the factors K and L. It is analogous to the second-rank tensor D_U of Section 5.1. The tensor ${}^{(4)}\mathsf{O}(j,f)$ is defined in equation (7.3) and is only a function of molecular properties and of the kind of measurement performed.

Simultaneous and Successive Two-Photon Processes. The power of formulation (7.6) is its generality: All processes of rank 4 are covered. The fundamental physical distinction between simultaneous two-photon processes, in which the two photons act coherently, and successive two-photon processes, in which coherence of the two one-photon events is lost, is not reflected in the form of (7.6) but enters through the choice of the molecular tensor O in (7.3). This is of rank 2 in any event. However, for simultaneous (coherent) two-photon processes, O cannot be reduced any further and is referred to as the Raman scattering tensor, $\mathsf{O} = \alpha'(\tilde{\nu}_1,f)$, or the two-photon absorption tensor, $\mathsf{O} = \mathsf{T}(\tilde{\nu}_1,f)$, while for successive (incoherent) two-photon processes, O can be written as a direct product of the molecular transition moment vectors **M**(j) and **M**(f) which refer to the two events separable in time, $\mathsf{O} = \mathbf{M}(j)\mathbf{M}(f)$. The transition moment **M**(j) is responsible for the initial excitation; the transition moment **M**(f) is responsible for the observed emission in the case of photoluminescence and for the observed photoinduced optical density in the case of photodichroism. Now, (7.3) dictates to the individual terms in the summation in (7.6) the form

$$[{}^{(4)}\mathsf{P}_{UV}]_{stuv}\alpha'_{st}(\tilde{\nu}_1,f)\alpha'^{*}_{uv}(\tilde{\nu}_1,f) \qquad \text{or} \qquad [{}^{(4)}\mathsf{P}_{UV}]_{stuv}T_{st}(\tilde{\nu}_1,f)T^{*}_{uv}(\tilde{\nu}_1,f)$$

in the case of simultaneous two-photon processes and the form

$$[{}^{(4)}\mathsf{P}_{UV}]_{stuv}M_s(j)M_t(f)M_u(j)M_v(f)$$

in the case of successive two-photon processes. (Remember that we always choose electric dipole transition moments in their real form, but the tensors α' and T can be complex.)

Thus, the physical coherence of the two one-photon interactions in the former case is reflected mathematically in the circumstance that the indices form the pairs (st), (uv); for instance, molecular symmetry may demand that certain $\alpha'_{st}(\tilde{\nu}_1,f)$'s vanish or be equal to others. The incoherence of the two interactions in the latter case is reflected in the fact that the indices form the pairs (su), (tv) which correspond to the two separate transition moments; thus, molecular symmetry may demand that certain products $\mathbf{M}_s(j)\mathbf{M}_u(j)$ vanish or be equal to others and that certain products $\mathbf{M}_t(f)\mathbf{M}_v(f)$ vanish or be equal to others. We shall see in the following how this difference leads to different expressions for the observed intensities $I_{UV}(\tilde{\nu}_1,\tilde{\nu}_2)$ for simultaneous and $F_{UV}(\tilde{\nu}_1,\tilde{\nu}_2)$ for successive two-photon processes.

The coherent nature of the two-photon interaction with the molecule in the case of the simultaneous two-photon processes is responsible for the fact that only a single summation over f is needed to convert the contributions $I_{UV}(\tilde{\nu}_1,f)$ into the observed intensities $I_{UV}(\tilde{\nu}_1,\tilde{\nu}_2)$, while a double summation over j and f is needed in the case of successive two-photon events to go from $F_{UV}(j,f)$ to $F_{UV}(\tilde{\nu}_1,\tilde{\nu}_2)$. In the former case, the summation over j was not really avoided but entered inseparably into the expressions for $\alpha'(\tilde{\nu}_1,f)$ or $T(\tilde{\nu}_1,f)$ as shown by (2.20) and (2.24). The presence of this second summation in the step from $F_{UV}(j,f)$ to $F_{UV}(\tilde{\nu}_1,\tilde{\nu}_2)$ complicates the analysis of the experimental data in photoluminescence and photodichroism experiments; its presence in the theoretical expressions for $\alpha(\tilde{\nu}_1,f)$ and $T(\tilde{\nu}_1,f)$ complicates their calculation from molecular wavefunctions. There is no general agreement as to which is worse.

The Tensor $^{(4)}\mathbf{P}_{UV}$. The values of the tensor elements $[^{(4)}P_{UV}]_{stuv}$ can be calculated, for instance by the method outlined in Appendix II, and tabulated once and for all. The results are

$$[^{(4)}P_{ZZ}]_{stuv} = L_{stuv} \tag{7.7}$$

$$\begin{aligned}[^{(4)}P_{XX}]_{stuv} = [^{(4)}P_{YY}]_{stuv} = (1/8)\{&\delta_{st}\delta_{uv}(2\delta_{su}+1)(1-K_{st}-K_{uv}) \\ &+(\delta_{st}-\delta_{uv})[\delta_{st}(2\delta_{s,|u\times v|}-3)K_{uv}-\delta_{uv}(2\delta_{|s\times t|,u}-3)K_{st}] \\ &+[2(\delta_{su}\delta_{tv}+\delta_{sv}\delta_{tu})-1]K_{|s\times t|,|u\times v|}+3L_{stuv}\}\end{aligned} \tag{7.8}$$

$$[^{(4)}P_{XZ}]_{stuv} = [^{(4)}P_{YZ}]_{stuv} = (1/2)(\delta_{su}K_{tv}-L_{stuv}) \tag{7.9}$$

$$[^{(4)}P_{ZX}]_{stuv} = [^{(4)}P_{ZY}]_{stuv} = (1/2)(\delta_{tv}K_{su}-L_{stuv}) \tag{7.10}$$

$$\begin{aligned}[^{(4)}P_{XY}]_{stuv} = [^{(4)}P_{YX}]_{stuv} = (1/8)\{&\delta_{st}\delta_{uv}(2\delta_{su}-1)(1-K_{st}-K_{uv}) \\ &+(\delta_{st}-\delta_{uv})[\delta_{st}(2\delta_{s,|u\times v|}-1)K_{uv}-\delta_{uv}(2\delta_{|s\times t|,u}-1)K_{st}] \\ &+[3(3\delta_{su}\delta_{tv}-\delta_{su}-\delta_{tv})-(3\delta_{sv}\delta_{tv}-\delta_{sv}-\delta_{tu})]K_{|s\times t|,|u\times v|}+L_{stuv}\}\end{aligned} \tag{7.11}$$

$$[{}^{(4)}\mathsf{P}_{++}]_{\text{stuv}} = [{}^{(4)}\mathsf{P}_{--}]_{\text{stuv}} = (1/4)\{\delta_{\text{st}}\delta_{\text{uv}}(2\delta_{\text{su}} - 1)(1 - \text{K}_{\text{st}} - \text{K}_{\text{uv}}) + (\delta_{\text{st}} - \delta_{\text{uv}})[\delta_{\text{st}}(2\delta_{\text{s},|\text{u}\times\text{v}|} - 1)\text{K}_{\text{uv}} - \delta_{\text{uv}}(2\delta_{|\text{s}\times\text{t}|,\text{u}} - 1)\text{K}_{\text{st}}] + [2(\delta_{\text{su}}\delta_{\text{tv}} + \delta_{\text{sv}}\delta_{\text{tu}}) - 1]\text{K}_{|\text{s}\times\text{t}|,|\text{u}\times\text{v}|} + \text{L}_{\text{stuv}}\} \tag{7.12}$$

$$[{}^{(4)}\mathsf{P}_{+-}]_{\text{stuv}} = [{}^{(4)}\mathsf{P}_{-+}]_{\text{stuv}} = (1/4)\{\delta_{\text{st}}\delta_{\text{uv}}(1 - \text{K}_{\text{st}} - \text{K}_{\text{uv}}) + (\delta_{\text{st}} - \delta_{\text{uv}})(\delta_{\text{uv}}\text{K}_{\text{st}} - \delta_{\text{st}}\text{K}_{\text{uv}}) + [2(\delta_{\text{su}}\delta_{\text{tv}} + \{\delta_{\text{sv}} - \delta_{\text{tu}}\}^2) - 1]\text{K}_{|\text{s}\times\text{t}|,|\text{u}\times\text{v}|} + \text{L}_{\text{stuv}}\} \tag{7.13}$$

$$[{}^{(4)}\mathsf{P}_{\text{UV}}]_{\text{stuv}} = [{}^{(4)}\mathsf{P}_{\text{UV}}]_{\text{uvst}} \tag{7.14}$$

Here the orientation factors K and L are those defined in Chapter 4, and the absolute value of the cross-product of two subscripts is defined as in Chapter 6 in a manner similar to the vector product of basis vectors:

$$|\text{x}\times\text{x}| = 0, \qquad |\text{x}\times\text{y}| = \text{z}, \qquad |\text{x}\times\text{z}| = \text{y}, \quad \text{etc.} \tag{7.15}$$

Any K with one or more indices equal to zero is itself equal to zero.

For linearly polarized light, the subscripts U and V define the polarization directions of the two photons involved in the interaction in a straightforward manner. The situation is more complicated for circularly polarized light: Whether a given sign refers to a left-handed or a right-handed photon depends on whether the photon is propagating along the +Z or the −Z laboratory direction and on whether it is being absorbed or created. The issue is addressed in more detail in Section 7.5, where measurements using circularly polarized light are discussed.

Although these results appear complicated relative to those for the elements of the analogous tensor D_{U} pertinent to one-photon processes [(5.15) and (5.16)], they are still simple compared with most alternative formulations, thanks to the use of the orientation factors K and L.

An explicit listing of the matrix elements defined by equations (7.7)–(7.13) is presented in (7.7a)–(7.13a).

$[{}^{(4)}\mathsf{P}_{\text{ZZ}}]_{\text{stuv}} =$

st \ uv	xx	yy	zz	yz	zx	xy	zy	xz	yx
xx	L_{xxxx}	L_{xxyy}	L_{xxzz}	L_{xxyz}	L_{xxzx}	L_{xxxy}	L_{xxzy}	L_{xxxz}	L_{xxyx}
yy	L_{yyxx}	L_{yyyy}	L_{yyzz}	L_{yyyz}	L_{yyzx}	L_{yyxy}	L_{yyzy}	L_{yyxz}	L_{yyyx}
zz	L_{zzxx}	L_{zzyy}	L_{zzzz}	L_{zzyz}	L_{zzzx}	L_{zzxy}	L_{zzzy}	L_{zzxz}	L_{zzyx}
yz	L_{yzxx}	L_{yzyy}	L_{yzzz}	L_{yzyz}	L_{yzzx}	L_{yzxy}	L_{yzzy}	L_{yzxz}	L_{yzyx}
zx	L_{zxxx}	L_{zxyy}	L_{zxzz}	L_{zxyz}	L_{zxzx}	L_{zxxy}	L_{zxzy}	L_{zxxz}	L_{zxyx}
xy	L_{xyxx}	L_{xyyy}	L_{xyzz}	L_{xyyz}	L_{xyzx}	L_{xyxy}	L_{xyzy}	L_{xyxz}	L_{xyyx}
zy	L_{zyxx}	L_{zyyy}	L_{zyzz}	L_{zyyz}	L_{zyzx}	L_{zyxy}	L_{zyzy}	L_{zyxz}	L_{zyyx}
xz	L_{xzxx}	L_{xzyy}	L_{xzzz}	L_{xzyz}	L_{xzzx}	L_{xzxy}	L_{xzzy}	L_{xzxz}	L_{xzyx}
yx	L_{yxxx}	L_{yxyy}	L_{yxzz}	L_{yxyz}	L_{yxzx}	L_{yxxy}	L_{yxzy}	L_{yxxz}	L_{yxyx}

(7.7a)

$8[{}^{(4)}P_{XX}]_{stuv} - 3L_{stuv} =$

st \ uv	xx	yy	zz	yz	zx	xy	zy	xz	yx
xx	$3(1-2K_{xx})$	$1-K_{xx}-K_{yy}$	$1-K_{xx}-K_{zz}$	$-K_{yz}$	$-3K_{xz}$	$-3K_{xy}$	$-K_{yz}$	$-3K_{xz}$	$-3K_{xy}$
yy	$1-K_{xx}-K_{yy}$	$3(1-2K_{yy})$	$1-K_{yy}-K_{zz}$	$-3K_{yz}$	$-K_{xz}$	$-3K_{xy}$	$-3K_{yz}$	$-K_{xz}$	$-3K_{xy}$
zz	$1-K_{xx}-K_{zz}$	$1-K_{yy}-K_{zz}$	$3(1-2K_{zz})$	$-3K_{yz}$	$-3K_{xz}$	$-K_{xy}$	$-3K_{yz}$	$-3K_{xz}$	$-K_{xy}$
yz	$-K_{yz}$	$-3K_{yz}$	$-3K_{yz}$	K_{xx}	$-K_{xy}$	$-K_{xz}$	K_{xx}	$-K_{xy}$	$-K_{xz}$
zx	$-3K_{xz}$	$-K_{xz}$	$-3K_{xz}$	$-K_{xy}$	K_{yy}	$-K_{yz}$	$-K_{xy}$	K_{yy}	$-K_{yz}$
xy	$-3K_{xy}$	$-3K_{xy}$	$-K_{xy}$	$-K_{xz}$	$-K_{yz}$	K_{zz}	$-K_{xz}$	$-K_{yz}$	K_{zz}
zy	$-K_{yz}$	$-3K_{yz}$	$-3K_{yz}$	K_{xx}	$-K_{xy}$	$-K_{xz}$	K_{xx}	$-K_{xy}$	$-K_{xz}$
xz	$-3K_{xz}$	$-K_{xz}$	$-3K_{xz}$	$-K_{xy}$	K_{yy}	$-K_{yz}$	$-K_{xy}$	K_{yy}	$-K_{yz}$
yx	$-3K_{xy}$	$-3K_{xy}$	$-K_{xy}$	$-K_{xz}$	$-K_{yz}$	K_{zz}	$-K_{xz}$	$-K_{yz}$	K_{zz}

(7.8a)

$2[{}^{(4)}P_{XZ}]_{stuv} + L_{stuv} =$

st \ uv	xx	yy	zz	yz	zx	xy	zy	xz	yx
xx	K_{xx}	0	0	0	0	K_{xy}	0	K_{xz}	0
yy	0	K_{yy}	0	K_{yz}	0	0	0	0	K_{xy}
zz	0	0	K_{zz}	0	K_{xz}	0	K_{yz}	0	0
yz	0	K_{yz}	0	K_{zz}	0	0	0	0	K_{xz}
zx	0	0	K_{xz}	0	K_{xx}	0	K_{xy}	0	0
xy	K_{xy}	0	0	0	0	K_{yy}	0	K_{yz}	0
zy	0	0	K_{yz}	0	K_{xy}	0	K_{yy}	0	0
xz	K_{xz}	0	0	0	0	K_{yz}	0	K_{zz}	0
yx	0	K_{xy}	0	K_{xz}	0	0	0	0	K_{xx}

(7.9a)

$2[^{(4)}P_{ZX}]_{stuv} + L_{stuv} =$

st \ uv	xx	yy	zz	yz	zx	xy	zy	xz	yx
xx	K_{xx}	0	0	0	K_{xz}	0	0	0	K_{xy}
yy	0	K_{yy}	0	0	0	K_{xy}	K_{yz}	0	0
zz	0	0	K_{zz}	K_{yz}	0	0	0	K_{xz}	0
yz	0	0	K_{yz}	K_{yy}	0	0	0	K_{xy}	0
zx	K_{xz}	0	0	0	K_{zz}	0	0	0	K_{yz}
xy	0	K_{xy}	0	0	0	K_{xx}	K_{xz}	0	0
zy	0	K_{yz}	0	0	0	K_{xz}	K_{zz}	0	0
xz	0	0	K_{xz}	K_{xy}	0	0	0	K_{xx}	0
yx	K_{xy}	0	0	0	K_{yz}	0	0	0	K_{yy}

(7.10a)

$8[^{(4)}P_{XY}]_{stuv} - L_{stuv} =$

st \ uv	xx	yy	zz	yz	zx	xy	zy	xz	yx
xx	$1 - 2K_{xx}$	$-(1 - K_{xx} - K_{yy})$	$-(1 - K_{xx} - K_{zz})$	K_{yz}	$-K_{xz}$	$-K_{xy}$	K_{yz}	$-K_{xz}$	$-K_{xy}$
yy	$-(1 - K_{xx} - K_{yy})$	$1 - 2K_{yy}$	$-(1 - K_{yy} - K_{zz})$	$-K_{yz}$	K_{xz}	$-K_{xy}$	$-K_{yz}$	K_{xz}	$-K_{xy}$
zz	$-(1 - K_{xx} - K_{zz})$	$-(1 - K_{yy} - K_{zz})$	$1 - 2K_{zz}$	$-K_{yz}$	$-K_{xz}$	K_{xy}	$-K_{yz}$	$-K_{xz}$	K_{xy}
yz	K_{yz}	$-K_{yz}$	$-K_{yz}$	$3K_{xx}$	$+K_{xy}$	$+K_{xz}$	$-K_{xx}$	$-3K_{xy}$	$-3K_{xz}$
zx	$-K_{xz}$	K_{xz}	$-K_{xz}$	$+K_{xy}$	$3K_{yy}$	$+K_{yz}$	$-3K_{xy}$	$-K_{yy}$	$-3K_{yz}$
xy	$-K_{xy}$	$-K_{xy}$	K_{xy}	$+K_{xz}$	$+K_{yz}$	$3K_{zz}$	$-3K_{xz}$	$-3K_{yz}$	$-K_{zz}$
zy	K_{yz}	$-K_{yz}$	$-K_{yz}$	$-K_{xx}$	$-3K_{xy}$	K_{xz}	$3K_{xx}$	$+K_{xy}$	$+K_{xz}$
xz	$-K_{xz}$	K_{xz}	$-K_{xz}$	$-3K_{xy}$	$-K_{yy}$	$-3K_{yz}$	$+K_{xy}$	$3K_{yy}$	$+K_{yz}$
yx	$-K_{xy}$	$-K_{xy}$	K_{xy}	$-3K_{xz}$	$-3K_{yz}$	$-K_{zz}$	$+K_{xz}$	$+K_{yz}$	$3K_{zz}$

(7.11a)

$4[{}^{(4)}P_{++}]_{stuv} - L_{stuv} =$

st \ uv	xx	yy	zz	yz	zx	xy	zy	xz	yx
xx	$1 - 2K_{xx}$	$-(1 - K_{xx} - K_{yy})$	$-(1 - K_{xx} - K_{zz})$	K_{yz}	$-K_{xz}$	$-K_{xy}$	K_{yz}	$-K_{xz}$	$-K_{xy}$
yy	$-(1 - K_{xx} - K_{yy})$	$1 - 2K_{yy}$	$-(1 - K_{yy} - K_{zz})$	$-K_{yz}$	K_{xz}	$-K_{xy}$	$-K_{yz}$	K_{xz}	$-K_{xy}$
zz	$-(1 - K_{xx} - K_{zz})$	$-(1 - K_{yy} - K_{zz})$	$1 - 2K_{zz}$	$-K_{yz}$	$-K_{xz}$	$-K_{xy}$	$-K_{yz}$	$-K_{xz}$	K_{xy}
yz	K_{yz}	$-K_{yz}$	$-K_{yz}$	K_{xx}	$-K_{xy}$	$-K_{xz}$	K_{xx}	$-K_{xy}$	$-K_{xz}$
zx	$-K_{xz}$	K_{xz}	$-K_{xz}$	$-K_{xy}$	K_{yy}	$-K_{yz}$	$-K_{xy}$	K_{yy}	$-K_{yz}$
xy	$-K_{xy}$	$-K_{xy}$	K_{xy}	$-K_{xz}$	$-K_{yz}$	K_{zz}	$-K_{xz}$	K_{yz}	K_{zz}
zy	K_{yz}	$-K_{yz}$	$-K_{yz}$	K_{xx}	$-K_{xy}$	$-K_{xz}$	K_{xx}	$-K_{xy}$	$-K_{xz}$
xz	$-K_{xz}$	K_{xz}	$-K_{xz}$	$-K_{xy}$	K_{yy}	$-K_{yz}$	$-K_{xy}$	K_{yy}	$-K_{yz}$
yx	$-K_{xy}$	$-K_{xy}$	K_{xy}	$-K_{xz}$	$-K_{yz}$	K_{zz}	$-K_{xz}$	$-K_{yz}$	K_{zz}

(7.12a)

$4[{}^{(4)}P_{+-}]_{stuv} - L_{stuv} =$

st \ uv	xx	yy	zz	yz	zx	xy	zy	xz	yx
xx	$1 - 2K_{xx}$	$1 - K_{xx} - K_{yy}$	$1 - K_{xx} - K_{zz}$	$-K_{yz}$	$-K_{xz}$	$-K_{xy}$	$-K_{yz}$	$-K_{xz}$	$-K_{xy}$
yy	$1 - K_{xx} - K_{yy}$	$1 - 2K_{yy}$	$1 - K_{yy} - K_{zz}$	$-K_{yz}$	$-K_{xz}$	$-K_{xy}$	$-K_{yz}$	$-K_{xz}$	$-K_{xy}$
zz	$1 - K_{xx} - K_{zz}$	$1 - K_{yy} - K_{zz}$	$1 - 2K_{zz}$	$-K_{yz}$	$-K_{xz}$	$-K_{xy}$	$-K_{yz}$	$-K_{xz}$	$-K_{xy}$
yz	$-K_{yz}$	$-K_{yz}$	$-K_{yz}$	K_{xx}	K_{xy}	K_{xz}	$-K_{xx}$	$-K_{xy}$	$-K_{xz}$
zx	$-K_{xz}$	$-K_{xz}$	$-K_{xz}$	K_{xy}	K_{yy}	K_{yz}	$-K_{xy}$	$-K_{yy}$	$-K_{yz}$
xy	$-K_{xy}$	$-K_{xy}$	$-K_{xy}$	K_{xz}	K_{yz}	K_{zz}	$-K_{xz}$	$-K_{yz}$	$-K_{zz}$
zy	$-K_{yz}$	$-K_{yz}$	$-K_{yz}$	$-K_{xx}$	$-K_{xy}$	$-K_{xz}$	K_{xx}	K_{xy}	K_{xz}
xz	$-K_{xz}$	$-K_{xz}$	$-K_{xz}$	$-K_{xy}$	$-K_{yy}$	$-K_{yz}$	K_{xy}	K_{yy}	K_{yz}
yx	$-K_{xy}$	$-K_{xy}$	$-K_{xy}$	$-K_{xz}$	$-K_{yz}$	$-K_{zz}$	K_{xz}	K_{yz}	K_{zz}

(7.13a)

In the general case, a measurement of intensities $E_{UV}(\tilde{\nu}_1,\tilde{\nu}_2)$ as a function of U, V, $\tilde{\nu}_1$, and $\tilde{\nu}_2$ may well be insufficient to determine the large number of unknown orientation factors and molecular tensor elements even if contributions from different (j,f)'s contained in each $E_{UV}(j,f)$ do not overlap or if they can be separated. The problem can still be manageable if it is possible to obtain a sufficient number of the unknowns, such as orientation factors, from other experiments on the same sample. A considerable simplification occurs if the molecular system of axes is chosen so as to diagonalize the matrix of the K_{st}'s. For this choice, the principal orientation axes, each K_{st} in the general formulas is replaced by $\delta_{st}K_s$. The simplified expressions are still perfectly general except that now the K's and L's refer to molecular axes systems whose location within the molecular framework may be unknown if the molecule has little or no symmetry. This is immaterial in some applications.

High-Symmetry Molecules. Some reasonable guesses at the location of the principal orientation axes in a molecule are usually possible from a consideration of its properties such as shape, but even if their location is known, two independent K's and nine independent L's remain as unknowns. It is therefore important that a significant simplification is avilable for molecules whose symmetry dictates the location of the principal orientation axes, up to a permutation, and reduces the number of L factors to be determined. Removal of the permutation ambiguity among the principal orientation axes, that is, an absolute polarization assignment, must be based on outside information but usually presents little difficulty, as discussed in Section 4.6. Using the principal axes system, the result for the fourth-rank tensor ${}^{(4)}\mathsf{P}$ for symmetrical molecules reduces from the general form given in (7.7)–(7.13) to a slightly less formidable set of expressions:

$$[{}^{(4)}\mathsf{P}_{ZZ}]_{stuv} = \delta_{|s\times t|,|u\times v|}L_{stuv} \tag{7.16}$$

$$[{}^{(4)}\mathsf{P}_{XX}]_{stuv} = (1/8)[\delta_{st}\delta_{uv}(2\delta_{su}+1)(1-K_{st}-K_{uv}) + (\delta_{su}\delta_{tv}+\delta_{sv}\delta_{tu})K_{|s\times t|,|u\times v|} + 3\delta_{|s\times t|,|u\times v|}L_{stuv}] \tag{7.17}$$

$$[{}^{(4)}\mathsf{P}_{XZ}]_{stuv} = (1/2)[\delta_{su}\delta_{tv}K_{tv} - \delta_{|s\times t|,|u\times v|}L_{stuv}] \tag{7.18}$$

$$[{}^{(4)}\mathsf{P}_{ZX}]_{stuv} = (1/2)[\delta_{su}\delta_{tv}K_{su} - \delta_{|s\times t|,|u\times v|}L_{stuv}] \tag{7.19}$$

$$[{}^{(4)}\mathsf{P}_{XY}]_{stuv} = (1/8)[\delta_{st}\delta_{uv}(2\delta_{su}-1)(1-K_{st}-K_{uv}) + (3\delta_{su}\delta_{tv}-\delta_{sv}\delta_{tu})K_{|s\times t|,|u\times v|} + \delta_{|s\times t|,|u\times v|}L_{stuv}] \tag{7.20}$$

$$[{}^{(4)}\mathsf{P}_{++}]_{stuv} = (1/4)[\delta_{st}\delta_{uv}(2\delta_{su}-1)(1-K_{st}-K_{uv}) + (\delta_{su}\delta_{tv}+\delta_{sv}\delta_{tu})K_{|s\times t|,|u\times v|} + \delta_{|s\times t|,|u\times v|}L_{stuv}] \tag{7.21}$$

$$[{}^{(4)}\mathsf{P}_{+-}]_{stuv} = (1/4)[\delta_{st}\delta_{uv}(1-K_{st}-K_{uv}) + (\delta_{su}\delta_{tv}-\delta_{sv}\delta_{tu})K_{|s\times t|,|u\times v|} + \delta_{|s\times t|,|u\times v|}L_{stuv}] \tag{7.22}$$

Further simplification of the explicit form of the general expression (7.6) for the observed intensity $E_{UV}(j,f)$ occurs when a specific form is taken for the tensor $^{(4)}O(j,f)$.

7.3 Successive Events: Photoluminescence and Photodichroism. The Static Case

In the present section, it is assumed throughout that the solute molecules cannot rotate on the time scale of the experiment and that there is no depolarization by energy transfer. In practice, these conditions can be guaranteed by the use of highly viscous or rigid solutions and low solute concentrations.

Photoluminescence. After the absorption of a photon, the excited molecule may emit a photon and return to a lower energy level. Such light emission is referred to as photoluminescence. The variables in this experiment are the energies and polarizations of the exciting and the observed emitted photons and, if the experiment is performed in a pulsed mode, the delay $\Delta\tau$ between the excitation event and the time at which the emission is observed. The emission does not need to originate from the originally reached excited state. Most commonly, intense emission is observed from only one or two excited states into which all other initially reached excited states convert rapidly. For singlet ground-state molecules, these emitting states are the lowest excited singlet state S_1 and the lowest triplet state T_1 (Figure 7.3). The emission from S_1 to the ground state S_0 usually has a lifetime of the order of nanoseconds and is referred to as fluorescence, whereas emission from T_1 to S_0, which is spin-forbidden, usually has a much longer lifetime, on the order of milliseconds or even seconds, and is referred to as phosphorescence. In a more general sense, fluorescence is emission of light which does not involve a change in spin multiplicity, whereas phosphorescence does. As mentioned above, the sample does not actually have to be oriented originally in order for the emitted light to have

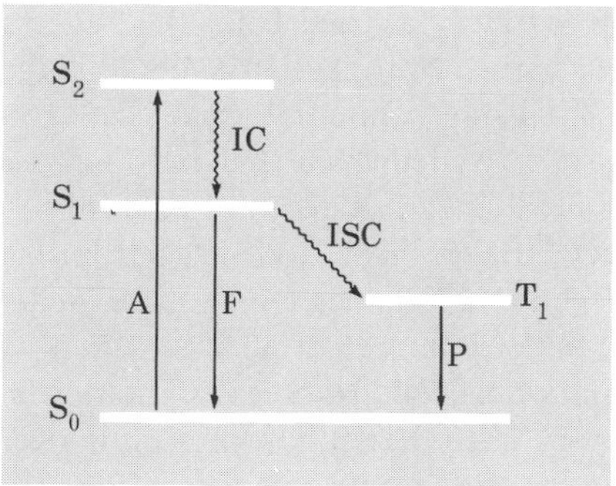

Figure 7.3 Jablonski diagram. Energies of singlet (S) and triplet (T) states and radiative (straight arrows) as well as nonradiative (wavy arrows) processes are shown. A, absorption; F, fluorescence; P, phosphorescence; IC, internal conversion; ISC, intersystem crossing.

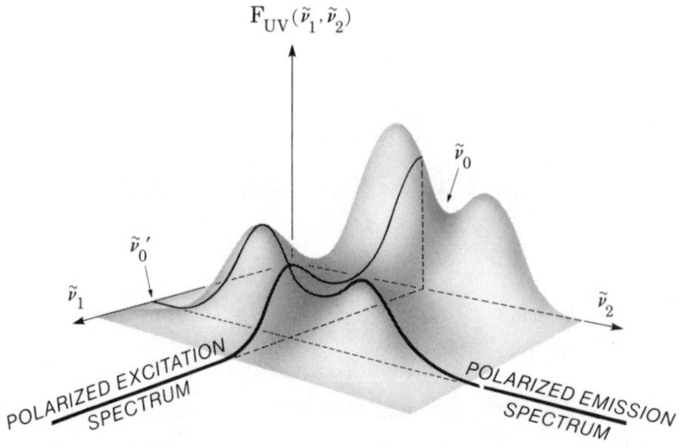

Figure 7.4 A two-dimensional photoluminescence spectrum.

an anisotropic polarization distribution. This effect has already been discussed in Section 4.6.3 and is known as photoselection.

Complete two-dimensional photoluminescence spectra $F_{UV}(\tilde{\nu}_1,\tilde{\nu}_2)$ are rarely measured (Figure 7.4). If $\tilde{\nu}_2$ is held constant at $\tilde{\nu}_0$, a one-dimensional polarized *excitation spectrum* $F_{UV}(\tilde{\nu}_1,\tilde{\nu}_0)$ is obtained. If $\tilde{\nu}_1$ is held constant at $\tilde{\nu}_0'$, a one-dimensional polarized *emission spectrum* $F_{UV}(\tilde{\nu}_0',\nu_2)$ is obtained. It is necessary to perform measurements at several suitable values of $\tilde{\nu}_0$ and $\tilde{\nu}_0'$ if a full analysis of an oriented sample is desired.

Photodichroism. Since not all molecules will luminesce, the studies of the polarization of emitted light are somewhat limited in scope. An experimentally more demanding but generally more applicable method of obtaining very similar information is to allow the initially excited molecule to absorb another polarized photon at time τ later (polarized flash photolysis). Once again, the sample does not have to be originally oriented for dichroism to be observed in the absorption of the second photon. It is therefore common to refer to this phenomenon as photoinduced linear dichroism. If the measurement is performed on a sample which is partially aligned from the beginning, as is the case of particular interest to us here, the proper name should perhaps be photomodified linear dichroism, but this is not commonly used. The choice of τ will determine the identity of the excited state which will be observed. If it is very short, absorption by the excited singlet of the molecule may be observed ($S_1 \rightarrow S_x$ spectrum). After longer times τ, the $T_1 \rightarrow T_x$ spectrum may be observable. If τ is chosen much longer than the lifetimes of any of the excited states, the spectrum observed will be only the ordinary dichroic spectrum of the initial sample, and all fourth-rank information will be lost unless a photochemical event took place and preserved a memory of the excitation. If a photochemical transformation occurred, the product species may be stable indefinitely, in which case arbitrarily large τ can be used in the experiment and there is no particular virtue in using a short pulse of light in the initial excitation. Steady-state illumination may be used even for studies of the $T_1 \rightarrow T_x$ spectrum if the triplet lifetime is sufficiently long, since detectable stationary state concentrations of the metastable triplet can then still be produced.

Just as in studies of photoluminescence, the complete two-dimensional spectra $F_{UV}(\tilde{\nu}_1,\tilde{\nu}_2)$ are rarely measured (Figure 7.5). When $\tilde{\nu}_2$ is held constant at $\tilde{\nu}_0$, the *action spectrum* $F_{UV}(\tilde{\nu}_1,\tilde{\nu}_0)$ results. When $\tilde{\nu}_1$ is held constant at $\tilde{\nu}'_0$, the *absorption spectrum* $F_{UV}(\tilde{\nu}'_0,\tilde{\nu}_2)$ of the photoproduced state is obtained. Once again, several $\tilde{\nu}_0$ and $\tilde{\nu}'_0$ values must be used for a complete study of a partially oriented sample.

Measurements. In order to obtain $F_{UV}(\tilde{\nu}_1,\tilde{\nu}_2)$ we are interested in experiments in which the exciting light is polarized along one of the principal axes of the sample, say U, and the intensity of the light emitted or absorbed by the photoselected excited molecules is decomposed into its components polarized along one or another of the principal sample axes, say V. In pulsed excitation experiments, $F_{UV}(\tilde{\nu}_1,\tilde{\nu}_2)$ will also depend on τ.

We shall consider five independent arrangements of the linear polarizations of the first and second photons, given by the combinations

$$F_{ZZ}, \quad F_{XX} = F_{YY}, \quad F_{XZ} = F_{YZ}, \quad F_{ZX} = F_{ZY}, \quad F_{XY} = F_{YX} \tag{7.23}$$

It is of no particular interest to use circularly polarized light in these experiments except when dealing with chiral molecules (Chapter 6). Inspection of the form of the tensors ${}^{(4)}\mathsf{P}_{++}$ and ${}^{(4)}\mathsf{P}_{+-}$ [(7.12a), (7.13a)] shows that the parts in which they differ do not contribute to photoluminescent intensity given by (7.6) when $[{}^{(4)}O(j,f)]_{stuv} = M_s(j)M_t(f)M_u(j)M_v(f)$, since $F_{++} = F_{+-}$. While measurements using one linearly polarized and one circularly polarized photon are also possible, they usually offer no advantages over those discussed here.

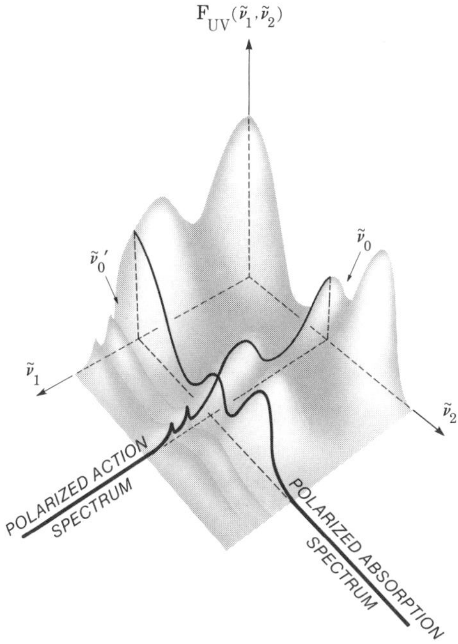

Figure 7.5 A two-dimensional photoinduced absorption spectrum.

The circumstance that for successive two-photon events $[^{(4)}O(j,f)]_{stuv}$ is a product of four numbers specifying the coordinates of the (real) transition moments **M**(j) and **M**(f) in the molecular system of axes can be used to simplify the sum of 81 terms implied in (7.6), since $M_s(j)M_u(j) = M_u(j)M_s(j)$ and $M_t(f)M_v(f) = M_v(f)M_t(f)$. For linear polarizations U and V, we have not only (7.14) but also $[^{(4)}P_{UV}]_{stuv} = [^{(4)}P_{UV}]_{utsv}$ [(7.7)–(7.11)]. As a result, the sum over all 81 possible allocations of the values x,y,z to the indices s, t, u, and v required in (7.6) can be reduced to a sum over 36 terms:

$$F_{UV}(j,f) = \sum_{(su)(tv)} [2(2 - \delta_{su} - \delta_{tv}) + \delta_{su}\delta_{tv}][^{(4)}P_{UV}]_{stuv} M_s(j)M_u(j)M_t(f)M_v(f) \tag{7.24}$$

where each of the unordered pairs (su) and (tv) can acquire any one of the six values (xx), (yy), (zz), (yz), (zx), and (xy).

The simple result (7.24) is valid only if it is assumed that the initial state is not depleted, i.e., that throughout the experiment the population of the excited or photochemically transformed state is only a negligible fraction of the population of the initial ground state. This condition is usually satisfied in the case of photoluminescence or transient photoinduced dichroism, since it is difficult to produce high populations of the excited species because of their rather short lifetimes. The condition is less likely to be valid in the case of permanent photodichroism, i.e., photodichroism due to a photochemical transformation, which in principle can be driven to completion. In the following, we shall assume negligible ground-state depletion unless specified otherwise (cf. Section 4.6.3). We shall also assume that the sample is optically thin at $\tilde{\nu}_1$, so that the intensity of the exciting light does not change significantly as it passes through the sample.

Since in general more than one absorbing transition moment and more than one emitting transition moment are involved due to spectral overlap, the observable intensity is given by

$$F_{UV}(\tilde{\nu}_1,\tilde{\nu}_2) = a(\tilde{\nu}_1)a'(\tilde{\nu}_2) \sum_{jf} g'_j(\tilde{\nu}_1)g'_f(\tilde{\nu}_2)F_{UV}(j,f) \tag{7.25}$$

where the quantity $a(\tilde{\nu}_1)$ depends on the intensity of the exciting beam and the number of solute molecules in the observed region, the quantity $a'(\tilde{\nu}_2)$ depends on the quantum yield for the production of the observed excited state and on the efficiency with which emitted photons are collected and detected, $g'_j(\tilde{\nu}_1)$ is the absorption line shape associated with the j-th transition, and $g'_f(\tilde{\nu}_2)$ is the emission or absorption line shape associated with the f-th transition.

7.3.1 High-Symmetry Molecules—Orientation Factors

Expressions for Polarized Intensity. In this section we consider molecules whose symmetry is such that the electric dipole transition moments **M**(j) and **M**(f) must be directed along one of the principal orientation axes x, y, or z. In this system of axes, only the nine diagonal terms, $\delta_{su} = \delta_{tv} = 1$, in the sum over the pairs (su) and (tv) in (7.24) can possibly make a nonvanishing contribution, while the 27 others vanish. Substitution of (7.24) into (7.25) now yields

$$F_{UV}(\tilde{\nu}_1,\tilde{\nu}_2) = a(\tilde{\nu}_1)a'(\tilde{\nu}_2)\sum_{jf}\sum_{st} g_j'(\tilde{\nu}_1)g_f'(\tilde{\nu}_2)M_s^2(j)M_t^2(f)[{}^{(4)}P_{UV}]_{stst} \tag{7.26}$$

where the summations over j and f run over all transition moments contributing to the initial excitation ($\tilde{\nu}_1$) and to the subsequent observation ($\tilde{\nu}_2$), respectively. The fact that each of these transition moments must lie along one of the axes x,y,z can be shown explicitly:

$$F_{UV}(\tilde{\nu}_1,\tilde{\nu}_2) = a(\tilde{\nu}_1)a'(\tilde{\nu}_2)\sum_{st} [{}^{(4)}P_{UV}]_{stst} \sum_{j:s} g_j'(\tilde{\nu}_1)M_s^2(j) \sum_{f:t} g_f'(\tilde{\nu}_2)M_t^2(f) \tag{7.27}$$

where s and t run over the indices x, y, and z; j runs over all transitions polarized along s; and f runs over all transitions polarized along t. We can now introduce "reduced" excitation spectra $A_s(\tilde{\nu}_1)$,

$$A_s(\tilde{\nu}_1) = \sum_{j:s} g_j'(\tilde{\nu}_1)M_s^2(j) \tag{7.28}$$

and "reduced" emission or photoinduced absorption spectra $B_t(\tilde{\nu}_2)$,

$$B_t(\tilde{\nu}_2) = \sum_{f:t} g_f'(\tilde{\nu}_2)M_t^2(f) \tag{7.29}$$

in complete analogy to the "reduced" absorption spectra introduced earlier in the introduction to linear dichroism in absorption in equation (5.25).

Only the summation over x, y, and z now remains explicitly in our expression for the observed two-dimensional spectral intensity:

$$F_{UV}(\tilde{\nu}_1,\tilde{\nu}_2) = a(\tilde{\nu}_1)a'(\tilde{\nu}_2)\sum_{st} [{}^{(4)}P_{UV}]_{stst}A_s(\tilde{\nu}_1)B_t(\tilde{\nu}_2) \tag{7.30}$$

We note that only a few elements of the fourth-rank tensor $[{}^{(4)}P_{UV}]_{stuv}$ are needed for this expression, namely those in which s = u and t = v. These elements, which depend on only two subscripts, s and t, can be thought of as forming a second-rank tensor S_{UV}. Its elements are

$$[S_{ZZ}]_{st} = L_{st} \tag{7.31}$$

$$[S_{XX}]_{st} = (1/8)[(1 + 2\delta_{st})(1 - K_s - K_t) + 3L_{st}] \tag{7.32}$$

$$[S_{XZ}]_{st} = (1/2)(K_t - L_{st}) \tag{7.33}$$

$$[S_{ZX}]_{st} = (1/2)(K_s - L_{st}) \tag{7.34}$$

$$[S_{XY}]_{st} = (1/8)[(3 - 2\delta_{st})[(1 - K_s - K_t) + L_{st}] \tag{7.35}$$

and our final expression for the observed intensity becomes

$$F_{UV}(\tilde{\nu}_1,\tilde{\nu}_2) = a(\tilde{\nu}_1)a'(\tilde{\nu}_2)\sum_{st} [S_{UV}]_{st}A_s(\tilde{\nu}_1)B_t(\tilde{\nu}_2) \tag{7.36}$$

Pure Polarization. Equations relating the observed two-dimensional spectra to the orientation factors K and L on the one hand and to the reduced spectra $A_s(\tilde{\nu}_1)$ and $B_t(\tilde{\nu}_2)$ on the other hand are now available. Our next task is to formulate a procedure for deriving the fundamental quantities describing the orientation and molecular properties from experiments. This will be completely analogous to the procedure used to analyze ordinary linear dichroism in absorption described in Section 5.1.1. First we concentrate on the derivation of the five unknown orientation factors K and L. This can be done if it is possible to recognize spectral features such as peaks and shoulders that are purely polarized along one molecular axis in absorption, say u, and one in emission, say v, in the two-dimensional spectra $F_{UV}(\tilde{\nu}_1,\tilde{\nu}_2)$. It does not matter if they overlap other such differently polarized features as long as the spectral shapes of the differently polarized features are not alike at all wavelengths. Typically, these features will be the 0-0 bands of intense transitions. Ideally, the fact that they are purely polarized will have been established independently, say by single-crystal studies. If these features do not overlap others of different polarization, the situation is particularly simple. In such a case, the double summation in (7.36) produces a single term in the wavelength region of this particular spectral feature indicated by the wavenumbers $\tilde{\nu}_u,\tilde{\nu}_v$:

$$F_{UV}(\tilde{\nu}_u,\tilde{\nu}_v) = a(\tilde{\nu}_1)a'(\tilde{\nu}_2)[S_{UV}]_{uv}A_u(\tilde{\nu}_u)B_v(\tilde{\nu}_v) \tag{7.37}$$

The four independent ratios of the observed intensities, taken between the quantities listed in (7.23), now provide equations for the unknown tensor elements $[S_{UV}]_{uv}$ and thus for the L's and K's:

$$\frac{F_{UV}(\tilde{\nu}_u,\tilde{\nu}_v)}{F_{U'V'}(\tilde{\nu}_u,\tilde{\nu}_v)} = \frac{[S_{UV}]_{uv}}{[S_{U'V'}]_{uv}} \tag{7.38}$$

Stepwise Reduction. If contributions from purely polarized spectral features of one polarization, say uv, overlap with those of other polarizations, ratios of observed intensities $F_{UV}(\tilde{\nu}_u,\tilde{\nu}_v)$ are no longer simply related to the ratios of the tensor elements $[S_{UV}]_{uv}$. Instead, it is necessary to use the stepwise reduction procedure already outlined for the simpler case of linear dichroism in absorption [cf. equation (5.32)]. Two two-dimensional spectra are chosen at a time, say $F_{UV}(\tilde{\nu}_1,\tilde{\nu}_2)$ and $F_{U'V'}(\tilde{\nu}_1,\tilde{\nu}_2)$, and are combined linearly with varying weights until a linear combination is found in which all uv-polarized spectral features just disappear. This combination is

$$F_{UV}(\tilde{\nu}_1,\tilde{\nu}_2) - ([S_{UV}]_{uv}/[S_{U'V'}]_{uv})F_{U'V'}(\tilde{\nu}_1,\tilde{\nu}_2) \tag{7.39}$$

The knowledge of the coefficient at $F_{U'V'}(\tilde{\nu}_1,\tilde{\nu}_2)$ in the sought linear combination thus fixes the ratio $[S_{UV}]_{uv}/[S_{U'V'}]_{uv}$. The procedure is repeated for all combinations of UV and U'V', but only four of the ratios can be experimentally independent.

The recording of two-dimensional spectra $F_{UV}(\tilde{\nu}_1,\tilde{\nu}_2)$ can be very tedious in practice unless an optical multichannel analyzer is available. If it is possible to identify a wavelength region of pure polarization in emission or in absorption, the reduction can be performed on one-dimensional cross sections of these spectra. If the emission is purely polarized along v at wavenumber $\tilde{\nu}_0$, the excitation spectra $F_{UV}(\tilde{\nu}_1,\tilde{\nu}_0)$ can be reduced against each other to yield the ratios $[S_{UV}]_{uv}/[S_{U'V'}]_{uv}$, where u can be x, y, and z if the excitation spectra contain features polarized along x, y, and z, respectively (Figure 7.6). If the

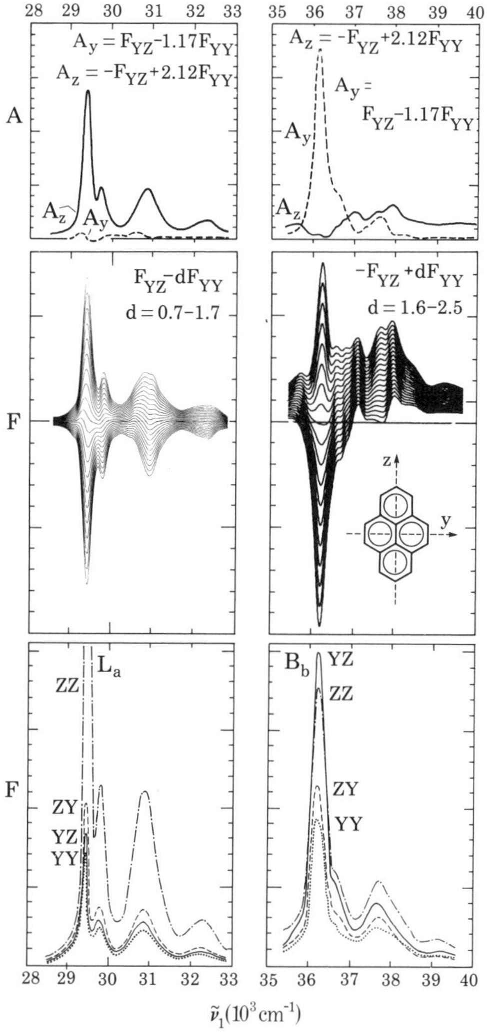

Figure 7.6 Stepwise reduction of polarized fluorescence excitation spectra of pyrene in stretched polyethylene at 77 K. The emission was monitored at the fourth fluorescence peak (z-polarized, γ in Figure 7.7). Left, the L_a absorption region; right, the B_b absorption region. Bottom, the four polarized excitation spectra: full line, YZ; dashed line, ZY; dotted line, YY; dash-dot line, ZZ. The curves are not corrected for instrumental polarization biases nor for the 14.5° angle between the exciting and observed beam directions. Center, families of linear combinations of the polarized excitation spectra F_{YZ} and F_{YY}. Top, the reduced excitation spectra: full line, $A_z = -F_{YZ} + 2.12F_{YY}$; dashed line, $A_y = F_{YZ} - 1.17F_{YY}$ (the two spectra are not on the same scale). After correction for polarization biases, the d values change from 2.12 to 2.23 and from 1.17 to 1.23. F. W. Langkilde and J. Michl, unpublished results.

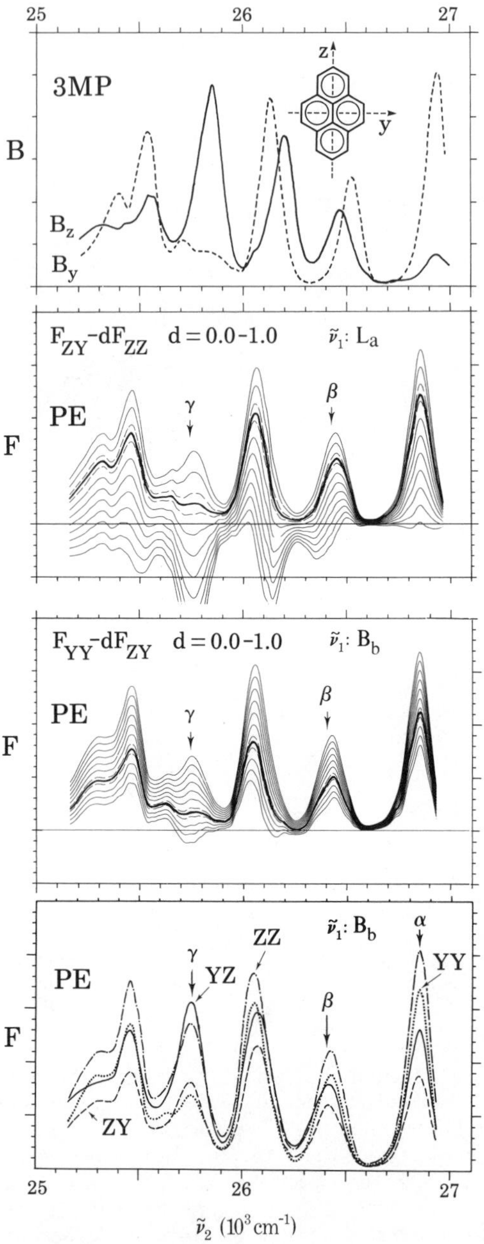

Figure 7.7 Stepwise reduction of polarized fluorescence emission spectra of pyrene in stretched polyethylene at 77 K. Bottom, the four polarized excitation spectra excited at the B_b origin: full line, YZ; dashed line, ZY; dotted line, YY; dash-dot line, ZZ. The curves are corrected for instrumental polarization biases but not for the 14.5° angle between the exciting and observed beam directions. Center, two families of linear combinations of the polarized fluorescence spectra: lower part, F_{YY} and F_{ZY}, excited at the

excitation spectrum is purely polarized along u at wavenumber ν'_0, the emission spectra $F_{UV}(\tilde{\nu}'_0,\tilde{\nu}_2)$ similarly yield the ratios $[S_{UV}]_{uv}/[S_{U'V'}]_{uv}$, where v can be x, y, or z if emission contains x-, y-, and z-polarized features, respectively (Figure 7.7).

By one or the other of the above procedures, we then end up with the four ratios:

$$[S_{ZZ}]_{uv}:[S_{XX}]_{uv}:[S_{XZ}]_{uv}:[S_{ZX}]_{uv}:[S_{XY}]_{uv}$$
$$= 8L_{uv}:[(1 + 2\delta_{uv})(1 - K_u - K_v) + 3L_{uv}]:4(K_v - L_{uv}):$$
$$4(K_u - L_{uv}):[(3 - 2\delta_{uv})(1 - K_u - K_v) + L_{uv}] \qquad (7.40)$$

for each combination of u and v for which purely polarized features can be identified in the spectra. In principle, the set of four ratios can be produced up to six times. The information obtained from uv-polarized features is the same as that obtained from vu-polarized features, serving as a check. Not all choices of u and v will provide four independent ratios. When u = v, $[S_{XZ}]_{uv} = [S_{ZX}]_{uv}$ and $[S_{XX}]_{uv} = 3[S_{XY}]_{uv}$, providing another check and leaving only two distinct ratios. Thus up to 18 different ratios can be obtained in principle, 12 for u ≠ v and 6 for u = v. This overdetermines the five independent orientation factors K and L.

7.3.2 High-Symmetry Molecules—Reduced Spectra

The reduced excitation spectra $A_s(\tilde{\nu}_1)$ and the reduced emission or photo-induced dichroism spectra $B_s(\tilde{\nu}_2)$ can be obtained by procedures similar to those described in Section 5.1.2. To obtain reduced excitation spectra, we rewrite equation (7.36) in the form

$$F_{UV}(\tilde{\nu}_1,\tilde{\nu}_2)/a(\tilde{\nu}_1)\tilde{a}'(\tilde{\nu}_2) = A_x(\tilde{\nu}_1)\sum_t [S_{UV}]_{xt}B_t(\tilde{\nu}_2) + A_y(\tilde{\nu}_1)\sum_t [S_{UV}]_{yt}B_t(\tilde{\nu}_2)$$
$$+ A_z(\tilde{\nu}_1)\sum_t [S_{UV}]_{zt}B_t(\tilde{\nu}_2) \qquad (7.41)$$

If the excitation spectrum is run with the value of $\tilde{\nu}_2$ held constant at $\tilde{\nu}_0$, it is given as a linear combination of the three sought reduced spectra $A_x(\tilde{\nu}_1)$, $A_y(\tilde{\nu}_1)$, and $A_z(\tilde{\nu}_1)$ with constant coefficients:

$$F_{UV}(\tilde{\nu}_1,\tilde{\nu}_2 = \tilde{\nu}_0)/a(\tilde{\nu}_1)a'(\tilde{\nu}_2) = A_x(\tilde{\nu}_1)C_x + A_y(\tilde{\nu}_1)C_y + A_z(\tilde{\nu}_1)C_z \qquad (7.42)$$

where the values of the constants C_x, C_y, and C_z depend on the choice of $\tilde{\nu}_0$. Five such spectra $F_{UV}(\tilde{\nu}_1,\tilde{\nu}_0)$ are available from the measurement. From these,

B_b origin (the fat line is the reduced y-polarized spectrum $F_{YY} - 0.65F_{ZY}$); upper part, F_{ZY} and F_{ZZ}, excited at the L_a origin (the fat line is the reduced y-polarized spectrum $F_{ZY} - 0.25F_{ZZ}$). Reproduced by permission from F. W. Langkilde, M. Gisin, E. W. Thulstrup, and J. Michl, *J. Phys. Chem.* **87**, 2901 (1983). Top: The reduced fluorescence spectra (full line, B_z; dashed line, B_y) of pyrene obtained in glassy 3-methylpentane (77 K, excited at the L_a origin) are shown for comparison [adapted by permission from F. W. Langkilde, E. W. Thulstrup, and J. Michl, *J. Chem. Phys.* **78**, 3372 (1983)].

we select the three which are the least linearly dependent. For instance, one of them could emphasize x-polarized features, another y-polarized features, and the third, z-polarized features. We then find three linear combinations of these spectra such that each contains only contributions from one polarization. This is easy if one of the contributions, say $A_x(\tilde{\nu}_1)$, vanishes throughout the region of interest, and is performed by exactly the same procedure as already discussed in Chapter 5. If none of the three contributions vanishes, an iterative solution of the problem can still be attempted. Thus, for instance, x-polarized features can be removed from, say, the UV spectrum by subtracting a suitable multiple of the U′V′ spectrum; then y-polarized features can be removed by subtracting a suitable multiple of the U″V″ spectrum, which, of course, will reintroduce some x-polarized features so that the multiplication constant for the U′V′ spectrum will have to be adjusted, and so on in an iterative fashion until both x- and y-polarized features are removed. A repetition of the procedure will produce all three reduced spectra if the five experimental spectra contain the necessary information, that is, are not excessively linearly dependent. If they are, the procedure will diverge (damping in the iterations may be helpful).

The procedure for finding reduced emission or photoproduced absorption spectra is similar and is based on rewriting equation (7.36) in the form

$$F_{UV}(\tilde{\nu}_1,\tilde{\nu}_2)/a(\tilde{\nu}_1)a'(\tilde{\nu}_2) = B_x(\tilde{\nu}_2)\sum_s [S_{UV}]_{sx}A_s(\tilde{\nu}_1) + B_y(\tilde{\nu}_2)\sum_s [S_{UV}]_{sy}A_s(\tilde{\nu}_1) + B_z(\tilde{\nu}_2)\sum_s [S_{UV}]_{sz}A_s(\tilde{\nu}_1) \qquad (7.43)$$

If the emission or photoinduced absorption spectra $F_{UV}(\tilde{\nu}_1,\tilde{\nu}_2)$ are now run with the excitation wavenumber $\tilde{\nu}_1$ held constant at $\tilde{\nu}_0'$ for the five possible orientations of the polarizers U,V, a set of linear combinations of the spectra $B_x(\tilde{\nu}_2)$, $B_y(\tilde{\nu}_2)$, and $B_z(\tilde{\nu}_2)$ will result, and these can be reduced exactly as described above for the excitation spectra.

The ease with which the reductions to obtain the spectra $A_s(\tilde{\nu}_1)$ and $B_t(\tilde{\nu}_1)$ can be performed depends significantly on the choice of wavenumber which is held constant: $\tilde{\nu}_2$ in the former case and $\tilde{\nu}_1$ in the latter. It is important that the measured spectra which are to be combined in the stepwise reduction procedure should be as linearly independent as possible. This is achieved by selecting the wavenumber to be held constant to correspond to a relatively purely polarized region of the spectrum. It is also possible to use several choices for this constant wavenumber, producing several spectra for a given polarizer orientation U,V and reducing these against each other.

7.3.3 Low-Symmetry Molecules

In molecules of low symmetry, we shall consider only the case of nonoverlapping transitions. Now, both the absorbing transition moment **M**(j) and the second transition moment **M**(f), which corresponds either to emission or to absorption by the photoproduced state, may have nonzero components in more than one direction of the principal axes system. In the absence of all symmetry, both of them may have nonvanishing components along each of the three axes, and as a result, all 36 terms of expression (7.24) for the observed contribution $F_{UV}(j,f)$ to the observed spectrum must be considered. Furthermore, there will now be either an uncertainty in the location of the principal axes within the molecule or, alternatively, an increase of the unknown orientation factors K from 2 to 5. There will also be nine independent orientation factors L as unknowns. As a result, a solution in the general case does not appear practical, even for nonoverlapping purely polarized transitions (i.e., no summations over j and f).

Simplifications occur if the molecules contain at least one plane of symmetry. This case is of interest since aromatic molecules are usually planar and also have suitable experimental properties for photoluminescence and photodichroism experiments. In the following, we shall consider two relatively simple cases as examples.

Example: In-Plane Absorption, Out-of-Plane Emission. In the first example, it is assumed that the initially excited transition moment **M**(j) lies in the yz plane of a molecule of C_s symmetry and that the observed transition moment **M**(f) is oriented along x, perpendicular to this plane. A practical realization of this case would be excitation of a $\pi\pi^*$ transition of an aromatic molecule of C_s symmetry, followed by purely out-of-plane polarized phosphorescent emission from its triplet state (Figure 7.8). We shall assume that the molecular axes x, y, z have been chosen so as to coincide with the principal orientation axes, and we shall not be concerned here about the problem of actually identifying the direction of the in-plane y and z axes with particular directions in the molecular framework, which has already been discussed in Section 5.1.3. In this system of axes, the only nonvanishing components of the two transition moments are $M_y(j)$, $M_z(j)$, and $M_x(f)$. The j-th transition is characterized by the excitation curve $A_j(\tilde{\nu}_1)$ whose transition moment **M**(j) lies at an angle ϕ_z^j with respect to the molecular z axis: $\cos\phi_z^j = M_z(j)/|\mathbf{M}(j)|$. The general equations (7.24) and (7.25) for the observed intensity now yield

$$\begin{aligned} F_{UV}(\tilde{\nu}_1,\tilde{\nu}_2) &= a(\tilde{\nu}_1)a'(\tilde{\nu}_2)g_j(\tilde{\nu}_1)g_f'(\tilde{\nu}_2)|\mathbf{M}(j)|^2|\mathbf{M}(f)|^2 \\ &\quad\times \sum_{(st)(uv)} [^{(4)}P_{UV}]_{stuv} \cos\phi_s^j \cos\phi_u^j \cos\phi_t^f \cos\phi_v^f \\ &= C(\tilde{\nu}_1,\tilde{\nu}_2) \sum_{(st)(uv)} [^{(4)}P_{UV}]_{stuv} \cos\phi_s^j \cos\phi_u^j \cos\phi_t^f \cos\phi_v^f \end{aligned} \tag{7.44}$$

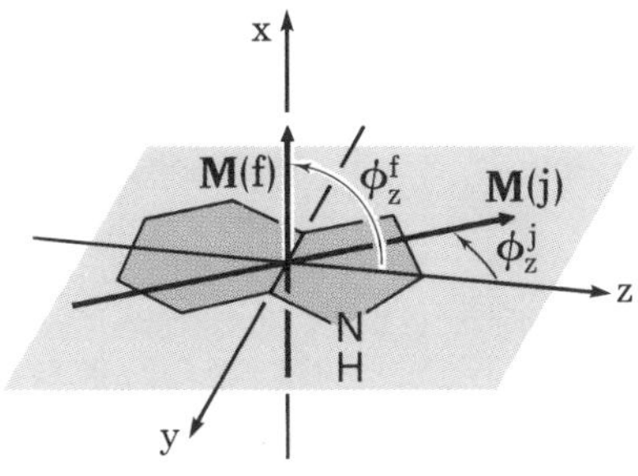

Figure 7.8 An example of transition moments responsible for $\pi\pi^*$ absorption (directions chosen arbitrarily) and out-of-plane polarized emission.

Now, for w = x,y,z, $\cos \phi^{f}_{w} = \delta_{wx}$, so that

$$F_{UV}(\tilde{\nu}_1,\tilde{\nu}_2) = C(\tilde{\nu}_1,\tilde{\nu}_2)\{[^{(4)}P_{UV}]_{yxyx} \cos^2 \phi^{j}_{y} + [^{(4)}P_{UV}]_{yxzx} \cos \phi^{j}_{y} \cos \phi^{j}_{z} + [^{(4)}P_{UV}]_{zxyx} \cos \phi^{j}_{z} \cos \phi^{j}_{y} + [^{(4)}P_{UV}]_{zxzx} \cos^2 \phi^{j}_{z}\} \quad (7.45)$$

Since the angles ϕ^{j}_{z} and ϕ^{j}_{y} add up to 90°, the results can be written in the form

$$F_{ZZ}(\nu_1,\nu_2) = C(\tilde{\nu}_1,\tilde{\nu}_2)(L_{xz} \cos^2 \phi^{j}_{z} + L_{xy} \sin^2 \phi^{j}_{z} + L_{xxyz} \sin 2\phi^{j}_{z}) \quad (7.46)$$

$$F_{XX}(\tilde{\nu}_1,\tilde{\nu}_2) = [C(\tilde{\nu}_1,\tilde{\nu}_2)/8] \times [(K_y + 3L_{xz}) \cos^2 \phi^{j}_{z} + (K_z + 3L_{xy}) \sin^2 \phi^{j}_{z} + 3L_{xxyz} \sin 2\phi^{j}_{z}] \quad (7.47)$$

$$F_{XZ}(\tilde{\nu}_1,\tilde{\nu}_2) = [C(\tilde{\nu}_1,\tilde{\nu}_2)/2] \times [K_x - L_{xz} \cos^2 \phi^{j}_{z} - L_{xy} \sin^2 \phi^{j}_{z} - L_{xxyz} \sin^2 \phi^{j}_{z}] \quad (7.48)$$

$$F_{ZX}(\tilde{\nu}_1,\tilde{\nu}_2) = [C(\tilde{\nu}_1,\tilde{\nu}_2)/2] \times [(K_z - L_{xz}) \cos^2 \phi^{j}_{z} + (K_y - L_{xy}) \sin^2 \phi^{j}_{z} - L_{xxyz} \sin^2 \phi^{j}_{z}] \quad (7.49)$$

$$F_{XY}(\tilde{\nu}_1,\tilde{\nu}_2) = [C(\tilde{\nu}_1,\tilde{\nu}_2)/8] \times [(3K_y + L_{xz}) \cos^2 \phi^{j}_{z} + (3K_z + L_{xy}) \sin^2 \phi^{j}_{z} + L_{xxyz} \sin^2 \phi^{j}_{z}] \quad (7.50)$$

This set of five equations is obtained for each transition j for which the five measurements can be performed, yielding a total of 4n intensity ratios for n such transitions. The equations contain five unknown orientation factors: two K's, L_{xy}, L_{xz}, and L_{xxyz}, some of which may be accessible from other measurements. In general, there will be 4n potentially linearly independent equations for n + 5 unknowns, and the minimum number of purely and differently polarized transitions j which need to be observed is two. A measurement of polarized photoluminescence on an isotropic sample does not provide further information on the unknown angles ϕ^{j}_{z} but serves only as a confirmation that the angle between the absorbing and emitting transition moment directions is 90°.

The presence of the molecular plane of symmetry containing the y and z axes does not guarantee that L_{xxyz} vanishes. Its nonzero value is related to the fact that neither the xz nor the xy plane are symmetry planes of the molecule. It is still possible, however, that the shape of the molecule is symmetrical or very nearly symmetrical with respect to at least one of these planes. Then if the orientation distribution of the molecules in the sample is dictated by their shape alone, L_{xxyz} will be equal to zero or nearly equal to zero. On the one hand, this will reduce the number of unknowns; but on the other hand, it will remove the possibility of determining the sign of the angles ϕ^{j}_{z}, which is otherwise in principle available from the measurements because of the presence of the term containing $\sin 2\phi_z^j$.

Example: In-Plane Absorption, In-Plane Emission. In the second example of a molecule of C_s symmetry, both the initially excited transition moment **M**(j) and the observed transition moment **M**(f) are assumed to lie in the yz symmetry plane of the molecule. A practical realization of this case could be an excitation of a $\pi\pi^*$ transition of an aromatic molecule of C_s symmetry, followed by fluorescence from one of its $\pi\pi^*$ excited states (Figure 7.9). Another example is excitation of a singlet $\pi\pi^*$ state of such a molecule followed by measurement of its polarized triplet-triplet $\pi\pi^*$ absorption.

Procedures similar to those just outlined lead to the expressions

$$F_{ZZ}(\tilde{\nu}_1,\tilde{\nu}_2) = C(\tilde{\nu}_1,\tilde{\nu}_2) \times [L_z a_1 + L_y a_2 + L_{yz}(a_3 + a_4 + a_5) + L_{yyyz}a_6 + L_{yzzz}a_7] \quad (7.51)$$

$$F_{XX}(\tilde{\nu}_1,\tilde{\nu}_2) = [C(\tilde{\nu}_1,\tilde{\nu}_2)/8][3(1 - 2K_z + L_z)a_1 + 3(1 - 2K_y + L_y)a_2 + (K_x + 3L_{yz})(a_3 + a_4 + a_5) + 3L_{yyyz}a_6 + 3L_{yzzz}a_7] \quad (7.52)$$

$$F_{XZ}(\tilde{\nu}_1,\tilde{\nu}_2) = [C(\tilde{\nu}_1,\tilde{\nu}_2)/2][(K_z - L_z)a_1 + (K_y - L_y)a_2 + (K_z - L_{yz})a_3 + (K_y - L_{yz})a_4 - L_{yz}a_5 - L_{yyyz}a_6 - L_{yzzz}a_7] \quad (7.53)$$

$$F_{ZX}(\tilde{\nu}_1,\tilde{\nu}_2) = [C(\tilde{\nu}_1,\tilde{\nu}_2)/2][(K_z - L_z)a_1 + (K_y - L_y)a_2 + (K_y - L_{yz})a_3 + (K_z - L_{yz})a_4 - L_{yz}a_5 - L_{yyyz}a_6 - L_{yzzz}a_7] \quad (7.54)$$

$$F_{XY}(\tilde{\nu}_1,\tilde{\nu}_2) = [C(\tilde{\nu}_1,\tilde{\nu}_2)/8][(1 - 2K_z + L_z)a_1 + (1 - 2K_y - L_y)a_2 + (3K_x + L_{yz})(a_3 + a_4) + (L_{yz} - K_x)a_5 + L_{yyyz}a_6 + L_{yzzz}a_7] \quad (7.55)$$

where

$$a_1 = \cos^2\phi_z^j \cos^2\phi_z^f \quad (7.56)$$

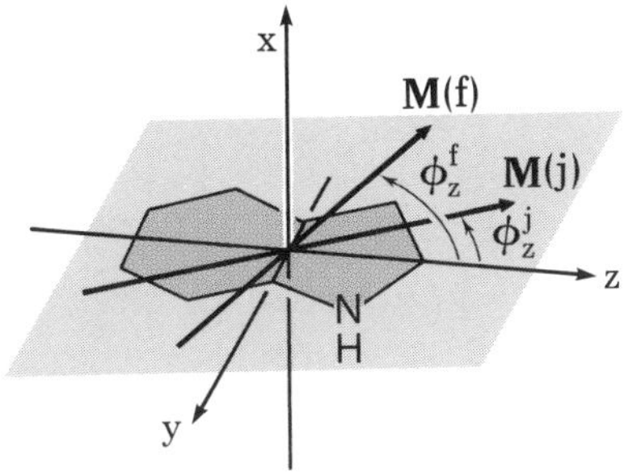

Figure 7.9 An example of transition moments responsible for $\pi\pi^*$ absorption and emission (directions chosen arbitrarily).

$$a_2 = \sin^2 \phi_z^j \sin^2 \phi_z^f \tag{7.57}$$

$$a_3 = \sin^2 \phi_z^j \cos^2 \phi_z^f \tag{7.58}$$

$$a_4 = \cos^2 \phi_z^j \sin^2 \phi_z^f \tag{7.59}$$

$$a_5 = \sin 2\phi_z^j \sin 2\phi_z^f \tag{7.60}$$

$$a_6 = \sin 2\phi_z^j \sin^2 \phi_z^f + \sin 2\phi_z^f \sin^2 \phi_z^j \tag{7.61}$$

$$a_7 = \sin 2\phi_z^j \cos^2 \phi_z^f + \sin 2\phi_z^f \cos^2 \phi_z^j \tag{7.62}$$

A similar set of expressions for the four intensity ratios can be obtained for each combination of j and f for which measurement is possible. The unknowns are two orientation factors K, five orientation factors L, and the angles ϕ_z^j and ϕ_z^f for each transition involved, except that $|\phi_z^j - \phi_z^f|$ can be obtained from independent measurements on an isotropic solution (Section 7.3.5). This is a rather formidable number of unknowns, even if information on some of these quantities is available from other types of measurements, and it may be necessary in practice to limit attention to instances in which L_{yyyz} and L_{yzzz} can be neglected. This will be justified if the orientation is dominated by molecular shape and if the shape is symmetrical or approximately symmetrical with respect to the xy or the xz plane of the molecule or both. With this neglect, the equations simplify to the form

$$\begin{aligned} F_{ZZ}(\tilde{\nu}_1,\tilde{\nu}_2) = C(\tilde{\nu}_1,\tilde{\nu}_2)[&L_z \cos^2 \phi_z^j \cos^2 \phi_z^f + 6L_{yz}(\cos \phi_z^j \cos \phi_z^f)(\sin \phi_z^j \sin \phi_z^f) \\ &+ L_y \sin^2 \phi_z^j \sin^2 \phi_z^f + L_{yz} \sin^2 (\phi_z^j - \phi_z^f)] \end{aligned} \tag{7.63}$$

$$\begin{aligned} F_{XX}(\tilde{\nu}_1,\tilde{\nu}_2) - (3/8)F_{ZZ}(\tilde{\nu}_1,\tilde{\nu}_2) = [C(\tilde{\nu}_1,\tilde{\nu}_2)/8]\{&K_x[1 + 2\cos^2 (\phi_z^j - \phi_z^f)] \\ &- 3(K_z - K_y) \cos (\phi_z^j - \phi_z^f) \cos (\phi_z^j + \phi_z^f)\} \end{aligned} \tag{7.64}$$

$$F_{XZ}(\tilde{\nu}_1,\tilde{\nu}_2) + (1/2)F_{ZZ}(\tilde{\nu}_1,\tilde{\nu}_2) = [C(\tilde{\nu}_1,\tilde{\nu}_2)/2](K_z \cos^2 \phi_z^f + K_y \sin^2 \phi_z^f) \tag{7.65}$$

$$F_{ZX}(\tilde{\nu}_1,\tilde{\nu}_2) + (1/2)F_{ZZ}(\tilde{\nu}_1,\tilde{\nu}_2) = [C(\tilde{\nu}_1,\tilde{\nu}_2)/2](K_z \cos^2 \phi_z^j + K_y \sin^2 \phi_z^j) \tag{7.66}$$

$$\begin{aligned} F_{XY}(\tilde{\nu}_1,\tilde{\nu}_2) - (1/8)F_{ZZ}(\tilde{\nu}_1,\tilde{\nu}_2) = [C(\tilde{\nu}_1,\tilde{\nu}_2)/8]\{&K_x[1 + 2\sin^2 (\phi_z^j - \phi_z^f)] \\ &- (K_z - K_y) \cos (\phi_z^j - \phi_z^f) \cos (\phi_z^j + \phi_z^f)\} \end{aligned} \tag{7.67}$$

and contain only five unknown orientation factors.

7.3.4 High-Symmetry Orientation Distributions

Several special cases of orientation distribution are of interest in the study of photoluminescence and photodichroism. One of these is the random distribution, to be discussed in Section 7.3.5; another is a distribution photoselected from the random distribution by a process of rank 2, to be discussed in Section 7.3.6. The one of present interest occurs when it is possible to assume that an anisotropic molecular orientation is similar to that of rods or discs. Historically, much of the work on photoluminescence of solutes contained in stretched polymers or liquid crystals was based on such an assumption. In some instances the assumption was justified, and it is of interest to describe the resulting simplifications, using the principal orientation axes system.

Rodlike Orientation. For the case of rodlike orientation, in which the two short molecular axes x and y are equivalent, all of the possibly nonvanishing orientation factors K and L can be expressed in terms of only two independent quantities. If we choose these as K_z and L_z, the other orientation factors (Table 4.1) are given by

$$K_x = K_y = (1/2)(1 - K_z) \tag{7.68}$$

$$L_x = L_y = (3/8)(1 - 2K_z + L_z) \tag{7.69}$$

$$L_{xy} = (1/8)(1 - 2K_z + L_z) \tag{7.70}$$

$$L_{xz} = L_{yz} = (1/2)(K_z - L_z) \tag{7.71}$$

If it is further assumed that molecular symmetry is such that transition moments are directed either along the long z axis of the molecule or perpendicular to it, and that differently polarized transitions do not overlap, very simple expressions result. For example, if the absorbing and emitting moments both lie in the z axis, the observed intensities are obtained from equation (7.36) as

$$F_{ZZ}(\tilde{v}_1,\tilde{v}_2) = a(\tilde{v}_1)a'(\tilde{v}_2)A_z(\tilde{v}_1)B_z(\tilde{v}_2)L_z = C(\tilde{v}_1,\tilde{v}_2)L_z \tag{7.72}$$

$$F_{XX}(\tilde{v}_1,\tilde{v}_2) = 3F_{XY}(\tilde{v}_1,\tilde{v}_2) = [3C(\tilde{v}_1,\tilde{v}_2)/8](1 - 2K_z + L_z) \tag{7.73}$$

$$F_{XZ}(\tilde{v}_1,\tilde{v}_2) = F_{ZX}(\tilde{v}_1,\tilde{v}_2) = [C(\tilde{v}_1,\tilde{v}_2)/2](K_z - L_z) \tag{7.74}$$

where $C(\tilde{v}_1,\tilde{v}_2)$ is the proportionality factor introduced in (7.44). If the absorbing moment is perpendicular to the long axis z and the emitting transition moment lies parallel to it, the observed intensities become

$$F_{ZZ}(\tilde{v}_1,\tilde{v}_2) = [C(\tilde{v}_1,\tilde{v}_2)/2](K_z - L_z) \tag{7.75}$$

$$F_{XX}(\tilde{v}_1,\tilde{v}_2) = [C(\tilde{v}_1,\tilde{v}_2)/16](1 + 2K_z - 3L_z) \tag{7.76}$$

$$F_{XZ}(\tilde{v}_1,\tilde{v}_2) = [C(\tilde{v}_1,\tilde{v}_2)/4](K_z + L_z) \tag{7.77}$$

$$F_{ZX}(\tilde{v}_1,\tilde{v}_2) = [C(\tilde{v}_1,\tilde{v}_2)/4](1 - 2K_z + L_z) \tag{7.78}$$

$$F_{XY}(\tilde{v}_1,\tilde{v}_2) = [C(\tilde{v}_1,\tilde{v}_2)/16](3 - 2K_z - L_z) \tag{7.79}$$

Disclike Orientation. In the case of disclike molecules in which the y and z axes are equivalent, the orientation factors can again be expressed in terms of only two quantities. Choosing these as K_x and L_x, the other nonvanishing orientation factors (Table 4.1) are given by

$$K_z = K_y = (1/2)(1 - K_x) \tag{7.80}$$

$$L_y = L_z = (3/8)(1 - 2K_x + L_x) \tag{7.81}$$

$$L_{yz} = (1/8)(1 - 2K_x + L_x) \tag{7.82}$$

$$L_{xy} = L_{xz} = (1/2)(K_x - L_x) \tag{7.83}$$

Very Highly Symmetrical Disclike Molecules. Considerable simplifications again result if it is assumed that transition moment directions are dictated by symmetry. We shalll use this opportunity to illustrate the case of transitions involving degenerate states. We shall consider two situations.

Degenerate Absorption, Nondegenerate Emission. First, let the ground state be nondegenerate, the initially reached state doubly degenerate (j,j′), and the emitting excited state nondegenerate (f) (Figure 7.10). An example of this situation is out-of-plane polarized phosphorescence from highly symmetrical aromatic molecules excited into a degenerate $\pi\pi^*$ state. In-plane polarized absorption from the ground state into the doubly degenerate excited state involves a pair of mutually perpendicular transition moments **M**(j) and **M**(j′), which can be chosen to lie along the y and z axes, respectively. One of them contributes $A_y(\tilde{\nu}_1)$ to the y-polarized absorption intensity, the other an equal amount $A_z(\tilde{\nu}_1)$ to z-polarized absorption intensity [(7.28)]. It is assumed that the interaction of the molecule with its environment, which usually is of low symmetry, is negligible, so that j and j′ are exactly degenerate and the absorption band shapes exactly superimposable. It is also assumed that any memory of the polarization direction of the absorbed photon is lost on a time-scale of electronic dephasing, i.e., essentially instantaneously relative to the emission lifetime. The emitting moment **M**(f) is polarized along the x axis and contributes $B_x(\tilde{\nu}_2)$ to the emission

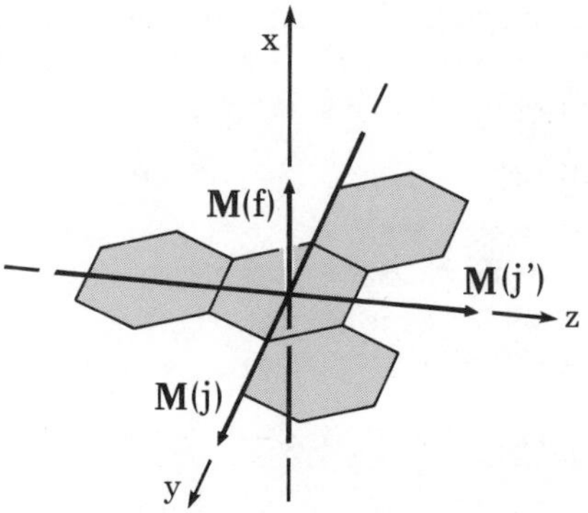

Figure 7.10 An example of transition moments responsible for in-plane polarized absorption into a degenerate excited state and out-of-plane polarized emission.

intensity. The results are obtained from equation (7.36) as

$$F_{UV}(\tilde{\nu}_1,\tilde{\nu}_2) = a(\tilde{\nu}_1)a'(\tilde{\nu}_2)\{[S_{UV}]_{zx}A_z(\tilde{\nu}_1) + [S_{UV}]_{yx}A_y(\tilde{\nu}_1)\}B_x(\tilde{\nu}_2)$$
$$= 2a(\tilde{\nu}_1)a'(\tilde{\nu}_2)[S_{UV}]_{zx}A_z(\tilde{\nu}_1)B_x(\tilde{\nu}_2) \quad (7.84)$$

Substituting from (7.31)–(7.35) and (7.80)–(7.83) and introducing the proportionality factor $C(\tilde{\nu}_1,\tilde{\nu}_2)$ as in (7.44), we obtain

$$F_{ZZ}(\tilde{\nu}_1,\tilde{\nu}_2) = C(\tilde{\nu}_1,\tilde{\nu}_2)(K_x - L_x) \quad (7.85)$$

$$F_{XX}(\tilde{\nu}_1,\tilde{\nu}_2) = [C(\tilde{\nu}_1,\tilde{\nu}_2)/8](1 + 2K_x - 3L_x) \quad (7.86)$$

$$F_{XZ}(\tilde{\nu}_1,\tilde{\nu}_2) = [C(\tilde{\nu}_1,\tilde{\nu}_2)/2](K_x + L_x) \quad (7.87)$$

$$F_{ZX}(\tilde{\nu}_1,\tilde{\nu}_2) = [C(\tilde{\nu}_1,\tilde{\nu}_2)/2](1 - 2K_x + L_x) \quad (7.88)$$

$$F_{XY}(\tilde{\nu}_1,\tilde{\nu}_2) = [C(\tilde{\nu}_1,\tilde{\nu}_2)/8](3 - 2K_x - L_x) \quad (7.89)$$

Degenerate Absorption, Degenerate Emission. Next we treat the case in which degenerate states are involved in both steps of the two-photon process. This obtains, for instance, in the observation of $\pi\pi^*$ fluorescence from a doubly degenerate state (f,f′) of a highly symmetrical aromatic molecule excited into a doubly degenerate $\pi\pi^*$ state (j,j′). This situation is similar to that depicted in Figure 7.10, except that the directions of **M**(f) and **M**(f′) coincide with those of **M**(j) and **M**(j′) and lie along y and z, respectively. Now there will be four terms in the sum in (7.36), since s and t can both acquire values y and z. By symmetry, $A_z(\tilde{\nu}_1) = A_y(\tilde{\nu}_1)$ and $B_z(\tilde{\nu}_2) = B_y(\tilde{\nu}_2)$, and we obtain

$$F_{UV}(\tilde{\nu}_1,\tilde{\nu}_2) = a(\tilde{\nu}_1)a'(\tilde{\nu}_2)\{[S_{UV}]_{yy}A_y(\tilde{\nu}_1)B_y(\tilde{\nu}_2) + [S_{UV}]_{zz}A_z(\tilde{\nu}_1)B_z(\tilde{\nu}_2)$$
$$+ [S_{UV}]_{yz}A_y(\tilde{\nu}_1)B_z(\tilde{\nu}_2) + [S_{UV}]_{zy}A_z(\tilde{\nu}_1)B_y(\tilde{\nu}_2)\}$$
$$= 2a(\tilde{\nu}_1)a'(\tilde{\nu}_2)\{[S_{UV}]_{zz}A_z(\tilde{\nu}_1)B_z(\tilde{\nu}_2) + [S_{UV}]_{yz}A_z(\tilde{\nu}_1)B_z(\tilde{\nu}_2)\} \quad (7.90)$$

where we have again ignored any possible symmetry-lowering interactions with the environment and assumed instantaneous electronic dephasing. The latter assumption makes the result valid even if the emitting state suffers a Jahn-Teller distortion, since such distortion will then not discriminate between the y and z directions. In practice, when deviations from the theory are observed, they will be most likely due to such discrimination by the low-symmetry environment. Evaluation of the expression (7.90), using (7.31)–(7.35) and introducing the proportionality constant $C(\tilde{\nu}_1,\tilde{\nu}_2)$ as in (7.44), yields

$$F_{ZZ}(\tilde{\nu}_1,\tilde{\nu}_2) = C(\tilde{\nu}_1,\tilde{\nu}_2)(1 - 2K_x + L_x) \quad (7.91)$$

$$F_{XX}(\tilde{\nu}_1,\tilde{\nu}_2) = [C(\tilde{\nu}_1,\tilde{\nu}_2)/8](3 + 2K_x + 3L_x) \quad (7.92)$$

$$F_{XZ}(\tilde{\nu}_1,\tilde{\nu}_2) = F_{ZX}(\tilde{\nu}_1,\tilde{\nu}_2) = [C(\tilde{\nu}_1,\tilde{\nu}_2)/2](1 - L_x) \quad (7.93)$$

$$F_{XY}(\tilde{\nu}_1,\tilde{\nu}_2) = [C(\tilde{\nu}_1,\tilde{\nu}_2)/8](1 + 6K_x + L_x) \quad (7.94)$$

7.3.5 Random Orientation Distribution (Ordinary Photoselection)

In practice, the most important limiting case of orientation distribution in photoluminescence and photodichroism has been that of isotropic orientation, which corresponds to an "ordinary" photoselection experiment. We have already encountered photoselection in Sections 4.6.3 and 4.6.4 but shall emphasize different aspects now.

In the present section, we assume negligible depletion of the ground state and an optically thin sample. First we consider the problem in its general form and outline two ways in which expressions for the polarized intensity can be derived. Then we consider special cases in which the results take a particularly simple form.

General. In the limit of an isotropic initial orientation distribution, only two independent polarized intensity measurements remain. This loss is partly compensated by the fact that the photoselected orientation factors of a solute in an isotropic solution are known (Sections 4.6.3 and 4.6.4). The two measurements are usually taken with linear polarizers parallel to each other, $F_{\parallel}^{iso}$, and perpendicular to each other, $F_{\perp}^{iso}$ (the case of Z photoselection discussed in Section 4.6.3):

$$F_{\parallel}^{iso} = F_{ZZ} = F_{YY} = F_{XX} \qquad F_{\perp}^{iso} = F_{XZ} = F_{ZX} = F_{XY} = F_{YZ} = F_{ZY} \qquad (7.95)$$

If the experiment is carried out with the angle θ between the exciting and the observed beams different from 0° and 90°, the polarization ratio $F_{\parallel}^{iso}/F_{\perp}^{iso}$ which corresponds to 90° can be derived for nondegenerate transitions in a way which still corrects for experimental polarization biases discussed in Section 3.2.6. It is obtained by solving the equation

$$(F_{\parallel}^{iso}/F_{\perp}^{iso})^2 \cos^2\theta + (F_{\parallel}^{iso}/F_{\perp}^{iso}) \sin^2\theta - (F'_{ZZ}/F'_{ZH})(F'_{YH}/F'_{YZ}) = 0 \qquad (7.96)$$

for the positive root of the unknown $F_{\parallel}^{iso}/F_{\perp}^{iso}$. Here the exciting beam is directed along X. It is polarized along Y in the measurement of F'_{YH} and F'_{YZ} and along Z in the measurement of F'_{ZZ} and F'_{ZH}. The emitted beam is observed along a direction in the XY plane forming an angle θ with X. The second polarizer is set along Z in the measurement of F'_{ZZ} and F'_{YZ} and perpendicular to Z in the measurement of F'_{ZH} and F'_{YH}.

Polarized Intensities—The General Case. ***The Simple Derivation.*** In order to derive the expressions for $F_{\parallel}^{iso}(\tilde{\nu}_1,\tilde{\nu}_2)$ and $F_{\perp}^{iso}(\tilde{\nu}_1,\tilde{\nu}_2)$, we can use (7.7), (7.10), (7.24), and (7.25) and substitute into $[^{(4)}P_{UV}]_{stuv}$ the values of K and L appropriate for an isotropic sample: $K_{uv}^{iso} = \delta_{uv}/3$, $L_{uuvv}^{iso} = L_{uvuv}^{iso} = L_{uvvu}^{iso} = (1 + 2\delta_{uv})/15$ and $L_{stuv}^{iso} = 0$ if x, y, or z occurs as subscript an odd number of times. This leads directly to (7.103) and (7.104).

The Complicated Derivation. Here we shall describe another possibility, which takes advantage of the explicit availability of formulas for the orientation factors of the photoselected molecular assembly derived in Section 4.6.3 for the limiting case of negligible depletion of the initial randomly oriented ground-state molecules. This is more complicated, but it prepares the way for the treatment of a fourth-rank process on a photoselected sample, i.e., a process of overall rank 6 (Section 7.3.6).

The size of the population photoselected in the first step with a beam linearly polarized along Z is proportional to the number of photons absorbed. Consider first the case of a single transition moment **M**(j) contributing to the initial excitation. Then the number of photons absorbed by the isotropic sample will be proportional to $|\mathbf{M}(j)|^2/3$. Using (5.12), and calling the proportionality constant k, the intensities observed in the second step will therefore be given by

$$\begin{pmatrix} F_{\perp}^{iso}(j,f) \\ F_{\parallel}^{iso}(j,f) \end{pmatrix} = \frac{k}{3}|\mathbf{M}(j)|^2|\mathbf{M}(f)|^2 \begin{pmatrix} [1 - K_f(j)]/2 \\ K_f(j) \end{pmatrix} \tag{7.97}$$

where $K_f(j) = \langle \cos^2 fZ \rangle$ is the orientation factor of the direction **M**(f) in the assembly photoselected by excitation of the transition moment **M**(j).

With the aid of (5.47), this can be written as

$$\begin{pmatrix} F_{\perp}^{iso}(j,f) \\ F_{\parallel}^{iso}(j,f) \end{pmatrix} = \frac{k}{3}|\mathbf{M}(j)|^2|\mathbf{M}(f)|^2 \begin{pmatrix} [1 - \sum_{u,v} \cos \phi_u^f K_{uv}(j) \cos \phi_v^f]/2 \\ \sum_{u,v} \cos \phi_u^f K_{uv}(j) \cos \phi_v^f \end{pmatrix} \tag{7.98}$$

The orientation factors $K_{uv}(j,Z)$ of the photoselected assembly are given by (4.70), in which the summations over j now reduce to a single term since only one transition moment **M**(j) is involved in the absorption step, and in which the argument Z indicates that the photoselecting light is linearly polarized along Z:

$$K_{uv}(j,Z) = (1/5)\left[\delta_{uv} + (1 + \delta_{uv})\frac{M_u(j)M_v(j)}{|\mathbf{M}(j)|^2}\right] \tag{7.99}$$

so that we obtain

$$K_{uv}(j,Z) = (1/5)[\delta_{uv} + (1 + \delta_{uv}) \cos \phi_u^j \cos \phi_v^j] \tag{7.100}$$

The resulting expressions for $F_{\perp}^{iso}(j,f)$ and $F_{\parallel}^{iso}(j,f)$ are

$$\begin{aligned} F_{\perp}^{iso}(j,f) = (k/30)&|\mathbf{M}(f)|^2|\mathbf{M}(j)|^2 \\ &\times \sum_{u,v} [2 \cos^2 \phi_u^f(2 - \delta_{uv}) \cos^2 \phi_v^j \\ &- \cos \phi_u^f \cos \phi_v^f(1 - \delta_{uv}) \cos \phi_u^j \cos \phi_v^j] \end{aligned} \tag{7.101}$$

$$\begin{aligned} F_{\parallel}^{iso}(j,f) = (k/15)&|\mathbf{M}(f)|^2|\mathbf{M}(j)|^2 \\ &\times \sum_{u,v} [\cos^2 \phi_u^f(1 + 2\delta_{uv}) \cos^2 \phi_v^j \\ &+ \cos \phi_u^f \cos \phi_v^f(1 - \delta_{uv}) \cos \phi_u^j \cos \phi_v^j] \end{aligned} \tag{7.102}$$

When the direction of the transition moment **M**(j) in the molecule is known, it can be taken for the z axis in the molecular system. The orientation tensor

K is then diagonal, and the second term in the brackets in (7.101) and (7.102) disappears.

In the case of spectral overlap, i.e., when several transition moments **M**(j) and/or several transition moments **M**(f) contribute, the orientation factors $K(\tilde{\nu}_1,Z)$ of the photoselected assembly must be obtained from the full form of (4.70), including the sum over all absorbing transitions. Even then, the second term in (7.101) and (7.102) will disappear when the direction of all the absorbing moments **M**(j) is alike or when the direction of all the monitored moments **M**(f) is alike, and can be taken for the z axis. Only when both the absorption and the monitoring occur at wavenumbers of mixed polarization will it be necessary to actually diagonalize the orientation tensor K(j) explicitly in order to remove the second term.

The Result. In the system of axes which diagonalizes $K(\tilde{\nu}_1,Z)$, we obtain for the observed intensities the same result as is obtained by direct use of ${}^{(4)}P_{UV}$:

$$F_{\perp}^{iso}(\tilde{\nu}_1,\tilde{\nu}_2) = a(\tilde{\nu}_1)a'(\tilde{\nu}_2)\sum_{jf} g_j'(\tilde{\nu}_1)g_f'(\tilde{\nu}_2)|\mathbf{M}(f)|^2|\mathbf{M}(j)|^2\left(2-\sum_u \cos^2\phi_u^f\cos^2\phi_u^j\right)\Big/15 \tag{7.103}$$

$$F_{||}^{iso}(\tilde{\nu}_1,\tilde{\nu}_2) = a(\tilde{\nu}_1)a'(\tilde{\nu}_2)\sum_{jf} g_j'(\tilde{\nu}_1)g_f'(\tilde{\nu}_2)|\mathbf{M}(f)|^2|\mathbf{M}(j)|^2\left(1+2\sum_u \cos^2\phi_u^f\cos^2\phi_u^j\right)\Big/15 \tag{7.104}$$

where (5.50), (7.25), (7.101), and (7.102) have been used.

The ratio of the two intensities observed at $\tilde{\nu}_1,\tilde{\nu}_2$ is then given by

$$\frac{F_{||}^{iso}(\tilde{\nu}_1,\tilde{\nu}_2)}{F_{\perp}^{iso}(\tilde{\nu}_1,\tilde{\nu}_2)} = \frac{\sum_{jf} r_j(\tilde{\nu}_1)q_f(\tilde{\nu}_2)\left(1+2\sum_u \cos^2\phi_u^f\cos^2\phi_u^j\right)}{\sum_{jf} r_j(\tilde{\nu}_1)q_f(\tilde{\nu}_2)\left(2-\sum_u \cos^2\phi_u^f\cos^2\phi_u^j\right)} \tag{7.105}$$

where $r_j(\tilde{\nu}_1)$ is the fraction of absorption at $\tilde{\nu}_1$ due to the j-th transition and $q_f(\tilde{\nu}_2)$ is the fraction of absorption or emission at $\tilde{\nu}_2$ due to the f-th transition.

The result can also be expressed in terms of the degree of polarization P,

$$P = (F_{||}^{iso} - F_{\perp}^{iso})/(F_{||}^{iso} + F_{\perp}^{iso}) \tag{7.106}$$

$$P(\tilde{\nu}_1,\tilde{\nu}_2) = \frac{\sum_{jf} r_j(\tilde{\nu}_1)q_f(\tilde{\nu}_2)\left(3\sum_u \cos^2\phi_u^f\cos^2\phi_u^j - 1\right)}{\sum_{jf} r_f(\tilde{\nu}_1)q_f(\tilde{\nu}_2)\left(\sum_u \cos^2\phi_u^f\cos^2\phi_u^j + 3\right)} \tag{7.107}$$

or the degree of anisotropy R,

$$R = (F_{||}^{iso} - F_{\perp}^{iso})/(F_{||}^{iso} + 2F_{\perp}^{iso}) \tag{7.108}$$

$$R(\tilde{\nu}_1,\tilde{\nu}_2) = \sum_{jf} r_j(\tilde{\nu}_1)q_f(\tilde{\nu}_2)\left(3\sum_u \cos^2\phi_u^f\cos^2\phi_u^j - 1\right)\Big/5 \tag{7.109}$$

where

$$\sum_{jf} r_j(\tilde{\nu}_1)q_f(\tilde{\nu}_2) = 1 \tag{7.110}$$

has been used.

Two special cases are of particular interest: first, purely polarized absorption and emission in molecules of any symmetry, and second, the case of molecules of high symmetry.

Polarized Intensities: Pure Polarization. If, in a molecule of any symmetry, all the absorbing transition moments **M**(j) are oriented along the same direction, that direction defines the z axis of the photoselected molecules, so that $\cos^2 \phi_u^j = \delta_{uz}$. If, in addition, all the monitored transition moments **M**(f) are oriented along one direction, the direction cosines $\cos \phi_u^f$ are independent of f. We then obtain a result originally due to Perrin:

$$\frac{F_{\parallel}^{iso}(j,f)}{F_{\perp}^{iso}(j,f)} = \frac{1 + 2\cos^2 \phi}{2 - \cos^2 \phi} \tag{7.111}$$

$$P(j,f) = \frac{3\cos^2 \phi - 1}{\cos^2 \phi + 3} \tag{7.112}$$

$$R(j,f) = (3\cos^2 \phi - 1)/5 = (\tfrac{2}{5})P_z(\cos \phi) \tag{7.113}$$

where ϕ is the angle between the directions of **M**(j) and **M**(f). It is customary to write the result in terms of P_z, the second-order Legendre polynomial, as shown. This result permits the determination of the angle between the absorbing and emitting transition moments, provided that both are purely polarized. If the transitions j and f are polarized parallel to each other, the polarization ratio, the degree of polarization, and the degree of anisotropy reach their maximum values

$$F_{\parallel}^{iso}(j,f)/F_{\perp}^{iso}(j,f) = 3 \tag{7.114}$$

$$P(j,f) = 1/2 \tag{7.115}$$

$$R(j,f) = 2/5 \tag{7.116}$$

If they are polarized perpendicular to each other, the minimum values are reached:

$$F_{\parallel}^{iso}(j,f)/F_{\perp}^{iso}(j,f) = 1/2 \tag{7.117}$$

$$P(j,f) = -1/3 \tag{7.118}$$

$$R(j,f) = -1/5 \tag{7.119}$$

Given (7.95), it is readily seen that equations (7.46)–(7.67) go to the proper limits if the values $K_u^{iso} = 1/3$ and $L_{uv}^{iso} = (1 + 2\delta_{uv})/15$ are used, with all other K's and L's vanishing, as is appropriate for the isotropic limit.

In practice, the individual jf contributions frequently overlap, and it is then not possible to use the above simple formulas in a quantitative fashion. Nevertheless, qualitative and even semiquantitative conclusions are frequently possible. In particular, one can often distinguish transitions polarized parallel from those polarized perpendicular to the emitting transition in molecules of high symmetry.

Polarized Intensities: High-Symmetry Molecules. The second case of particular interest is that of high-symmetry molecules, in which transition moment directions are limited to the symmetry axes x, y, and z (Figure 7.11). For nonoverlapping transitions, (7.114)–(7.119) represent the only possibilites, and the analysis is trivial. Now, however, it is also simple to analyze cases in which differently polarized absorption and/or emission bands overlap. Upon insertion of the proper values of K_u and L_{uv} for a random orientation distribution into (7.31)–(7.35) the expression (7.36) for the observed intensity yields the equations

$$F_{\perp}^{iso}(\tilde{\nu}_1,\tilde{\nu}_2) = (1/15) \sum_{u,v} (2 - \delta_{uv}) A_u(\tilde{\nu}_1) B_v(\tilde{\nu}_2) \tag{7.120}$$

$$F_{\parallel}^{iso}(\tilde{\nu}_1,\tilde{\nu}_2) = (1/15) \sum_{u,v} (1 + 2\delta_{uv}) A_u(\tilde{\nu}_1) B_v(\tilde{\nu}_2) \tag{7.121}$$

first exploited by Albrecht.

Alternatively, these expressions can be obtained from (7.103) and (7.104) by the use of (5.50), (7.28), and (7.29). They can also be written in terms of fractions of polarized absorption, $r_u(\tilde{\nu}_1)$, defined in (4.72), and of fractions of polarized emission (or photoinduced absorption), $q_v(\tilde{\nu}_2)$, defined in an analogous fashion:

$$q_v(\tilde{\nu}_2) = B_v(\tilde{\nu}_2)/[B_x(\tilde{\nu}_2) + B_y(\tilde{\nu}_2) + B_z(\tilde{\nu}_2)] = B_v(\tilde{\nu}_2)/3E_2^{iso}(\tilde{\nu}_2) \tag{7.122}$$

The subscript on E^{iso} indicates that it refers to the second photon. The quantities $r_u(\tilde{\nu}_1)$ and $q_v(\tilde{\nu}_2)$ are not to be confused with the fractions $r_j(\tilde{\nu}_1)$ and $q_f(\tilde{\nu}_2)$ used in (7.105) for low-symmetry molecules. Now (7.120) and (7.121) acquire the form

$$F_{\perp}^{iso}(\tilde{\nu}_1,\tilde{\nu}_2) = (3/5) E_1^{iso}(\tilde{\nu}_1) E_2^{iso}(\tilde{\nu}_2) \left[2 - \sum_u r_u(\tilde{\nu}_1) q_u(\tilde{\nu}_2) \right] \tag{7.123}$$

$$F_{\parallel}^{iso}(\tilde{\nu}_1,\tilde{\nu}_2) = (3/5) E_1^{iso}(\tilde{\nu}_1) E_2^{iso}(\tilde{\nu}_2) \left[1 + 2 \sum_u r_u(\tilde{\nu}_1) q_u(\tilde{\nu}_2) \right] \tag{7.124}$$

and the degree of anisotropy is

$$R(\tilde{\nu}_1,\tilde{\nu}_2) = (3/5) \left[\sum_u r_u(\tilde{\nu}_1) q_u(\tilde{\nu}_2) - \tfrac{1}{3} \right] \tag{7.125}$$

Relations between Degree-of-Anisotropy Curves. Up to two linearly independent curves for the degree of anisotropy $R(\tilde{\nu}_1,\tilde{\nu}_2)$ measured as a function of the wavenumber $\tilde{\nu}_1$ of the first photon can be obtained. The first, $R(\tilde{\nu}_1,\tilde{\nu}_2{}^{(1)})$, is measured by observing the second photon at a fixed wavenumber $\tilde{\nu}_2 = \tilde{\nu}_2^{(1)}$, where the fractions of polarized emission (or photoinduced absorption) are known to be $q_z(\tilde{\nu}_2^{(1)})$, $q_y(\tilde{\nu}_2^{(1)})$, and $q_x(\tilde{\nu}_2^{(1)}) = 1 - q_z(\tilde{\nu}_2^{(1)}) - q_y(\tilde{\nu}_2^{(1)})$. The second, $R(\tilde{\nu}_1,\tilde{\nu}_2^{(2)})$, is measured at another fixed wavenumber $\tilde{\nu}_2 = \tilde{\nu}_2^{(2)}$, characterized

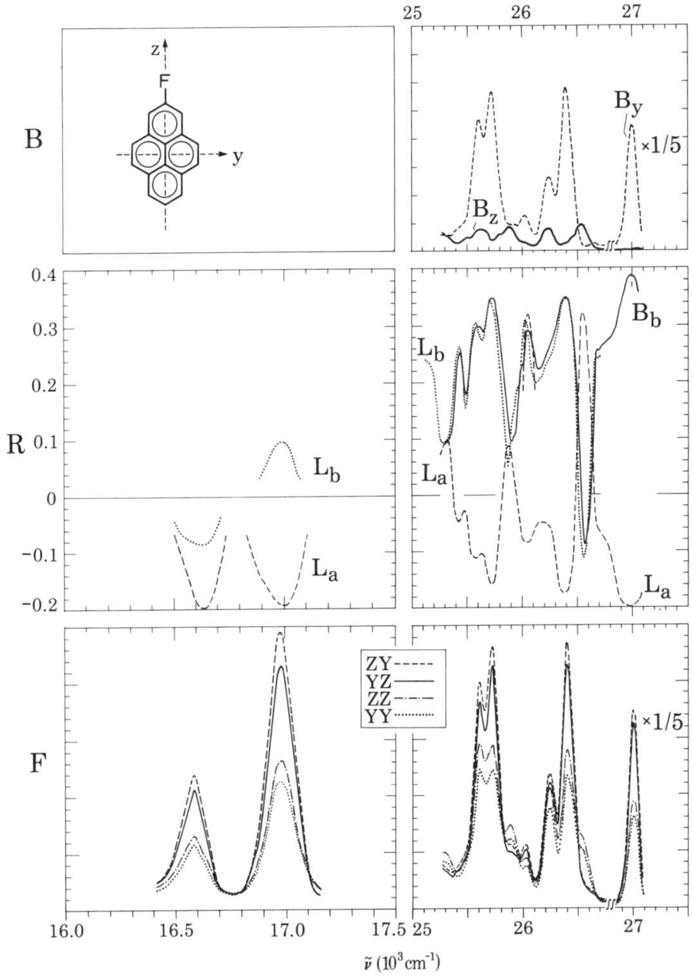

Figure 7.11 Polarized fluorescence and phosphorescence spectra of 2-fluoropyrene in glassy 3-methylpentane (77 K). Right: fluorescence; left: phosphorescence. Bottom, the four polarized emission curves excited at the L_a origin: full line, YZ; dashed line, ZY; dotted line, YY; dash-dot line, ZZ. The curves are not corrected for instrumental polarization biases nor for the 14.5° angle between the exciting and observed beam directions. Center, emission anisotropy: full line, excitation at the B_b origin; dashed line, excitation at the L_a origin; dotted line, excitation at the L_b origin. Top, reduced fluorescence curves excited at the L_a origin: full line, B_z; dashed line, B_y. Adapted by permission from F. W. Langkilde, E. W. Thulstrup, and J. Michl, *J. Chem. Phys.* **78**, 3372 (1983).

by known fractions $q_z(\tilde{\nu}_2^{(2)})$, $q_y(\tilde{\nu}_2^{(2)})$, and $q_x(\tilde{\nu}_2^{(2)}) = 1 - q_z(\tilde{\nu}_2^{(2)}) - q_y(\tilde{\nu}_2^{(2)})$. The measurement of $R(\tilde{\nu}_1,\tilde{\nu}_2)$ at any additional wavenumber $\tilde{\nu}_2 = \tilde{\nu}_2^{(3)}$ gives a result which can be related to the previous two through

$$C_3R(\tilde{\nu}_1,\tilde{\nu}_2^{(3)}) = C_0 + C_1R(\tilde{\nu}_1,\tilde{\nu}_2^{(1)}) + C_2R(\tilde{\nu}_1,\tilde{\nu}_2^{(2)}) \qquad (7.126)$$

provided that the emission fractions at $\tilde{\nu}_2^{(3)}$ are also known. Here,

$$5C_0 = \begin{vmatrix} q_x(\tilde{\nu}_2^{(1)}) & q_y(\tilde{\nu}_2^{(1)}) & q_z(\tilde{\nu}_2^{(1)}) \\ q_x(\tilde{\nu}_2^{(2)}) & q_y(\tilde{\nu}_2^{(2)}) & q_z(\tilde{\nu}_2^{(2)}) \\ q_x(\tilde{\nu}_2^{(3)}) & q_y(\tilde{\nu}_2^{(3)}) & q_z(\tilde{\nu}_2^{(3)}) \end{vmatrix} + C_1 - C_3 \qquad (7.127)$$

$$C_1 = \begin{vmatrix} 1 & 1 & 1 \\ q_x(\tilde{\nu}_2^{(2)}) & q_y(\tilde{\nu}_2^{(2)}) & q_z(\tilde{\nu}_2^{(2)}) \\ q_y(\tilde{\nu}_2^{(3)}) & q_y(\tilde{\nu}_2^{(3)}) & q_z(\tilde{\nu}_2^{(3)}) \end{vmatrix} \qquad (7.128)$$

$$C_2 = \begin{vmatrix} q_x(\tilde{\nu}_2^{(1)}) & q_y(\tilde{\nu}_2^{(1)}) & q_z(\tilde{\nu}_2^{(1)}) \\ 1 & 1 & 1 \\ q_x(\tilde{\nu}_2^{(3)}) & q_y(\tilde{\nu}_2^{(3)}) & q_z(\tilde{\nu}_2^{(3)}) \end{vmatrix} \qquad (7.129)$$

$$C_3 = \begin{vmatrix} q_x(\tilde{\nu}_2^{(1)}) & q_y(\tilde{\nu}_2^{(1)}) & q_z(\tilde{\nu}_2^{(1)}) \\ q_x(\tilde{\nu}_2^{(2)}) & q_y(\tilde{\nu}_2^{(2)}) & q_z(\tilde{\nu}_2^{(2)}) \\ 1 & 1 & 1 \end{vmatrix} \qquad (7.130)$$

This result, first formulated by Dörr, can be obtained from (7.125) using the relation

$$r_x(\tilde{\nu}_1) + r_y(\tilde{\nu}_1) + r_z(\tilde{\nu}_1) = 1 \qquad (7.131)$$

A completely analogous relation exists between the degrees of anisotropy $R(\tilde{\nu}_1^{(1)},\tilde{\nu}_2)$, $R(\tilde{\nu}_1^{(2)},\tilde{\nu}_2)$, and $R(\tilde{\nu}_1^{(3)},\tilde{\nu}_2)$ measured as a function of the wavenumber of the second photon $\tilde{\nu}_2$ for three different choices of the wavenumber of the first photon, $\tilde{\nu}_1^{(1)}$, $\tilde{\nu}_1^{(2)}$, and $\tilde{\nu}_1^{(3)}$. In either case, the three $R(\tilde{\nu}_1,\tilde{\nu}_2)$ curves obtained at the three fixed wavenumbers, either $\tilde{\nu}_2^{(1)}$, $\tilde{\nu}_2^{(2)}$, and $\tilde{\nu}_2^{(3)}$, or $\tilde{\nu}_1^{(1)}$, $\tilde{\nu}_1^{(2)}$, and $\tilde{\nu}_1^{(3)}$, have to be such that $C_1 \neq 0$, $C_2 \neq 0$, $C_3 \neq 0$.

If one of the fractions $r_u(\tilde{\nu}_1)$ or $q_v(\tilde{\nu}_2)$ vanishes, e.g., $r_x(\tilde{\nu}_1) = 0$ or $q_x(\tilde{\nu}_2) = 0$, the situation simplifies considerably. This case is frequently encountered in planar aromatic molecules which often have negligible out-of-plane polarized absorption intensity. Then the measurement of the degree of anisotropy $R(\tilde{\nu}_1,\tilde{\nu}_2^{(1)})$ at the fixed monitoring wavenumber $\tilde{\nu}_2^{(1)}$ and the knowledge of $q_z(\tilde{\nu}_2^{(1)})$, with $q_y(\tilde{\nu}_2^{(1)}) = 1 - q_z(\tilde{\nu}_2^{(1)})$, are sufficient to predict the curve $R(\tilde{\nu}_1,\tilde{\nu}_2^{(2)})$ monitored at any other wavenumber $\tilde{\nu}_2^{(2)}$ characterized by known values of $q_z(\tilde{\nu}_2^{(2)})$ and $q_y(\tilde{\nu}_2^{(2)}) = 1 - q_z(\tilde{\nu}_2^{(2)})$:

$$R(\tilde{\nu}_1,\tilde{\nu}_2^{(2)}) = C_0' + C_1' R(\tilde{\nu}_1,\tilde{\nu}_2^{(1)}) \qquad (7.132)$$

where

$$5C_0' = C_1'[1 - 3\, q_y(\tilde{\nu}_2^{(2)})] - [1 - 3\, q_y(\tilde{\nu}_2^{(1)})] \qquad (7.133)$$

$$C_1' = [q_z(\tilde{\nu}_2^{(1)}) - q_y(\tilde{\nu}_2^{(1)})]/[q_z(\tilde{\nu}_2^{(2)}) - q_y(\tilde{\nu}_2^{(2)})] \qquad (7.134)$$

and where it is assumed that $q_z(\tilde{\nu}_2^{(2)}) \neq q_y(\tilde{\nu}_2^{(2)})$. A similar relation holds between $R(\tilde{\nu}_1^{(1)},\tilde{\nu}_2)$ and $R(\tilde{\nu}_1^{(2)},\tilde{\nu}_2)$ obtained by scanning the monitoring wavelength. For instance, the degree of anisotropy measured by scanning the exciting wavenumber $\tilde{\nu}_1$, once monitored at a purely z-polarized spectral feature, $R(\tilde{\nu}_1,z)$, and once monitored at a purely y-polarized spectral feature, $R(\tilde{\nu}_1,y)$, will be related by

$$R(\tilde{\nu}_1,z) + R(\tilde{\nu}_1,y) = 1/5 \qquad (7.135)$$

Similar expressions can be obtained for the degree of polarization P and the polarization ratio $F_{\parallel}^{iso}/F_{\perp}^{iso}$ but are usually more complicated. For instance, (7.135) can be rewritten as

$$\left(3\frac{F_{\parallel}^{iso}(\tilde{\nu}_1,y)}{F_{\perp}^{iso}(\tilde{\nu}_1,y)} + 1\right)\left(3\frac{F_{\parallel}^{iso}(\tilde{\nu}_1,z)}{F_{\perp}^{iso}(\tilde{\nu}_1,z)} + 1\right) = 25 \qquad (7.136)$$

Reduced Spectra. The ultimate unknowns in measurements of polarized photoluminescence and photoinduced dichroism are the reduced absorption or emission spectra $A_u(\tilde{\nu}_1)$ and $B_v(\tilde{\nu}_2)$, related to the experimental observables by equations (7.120) and (7.121). The problem of deriving the reduced spectra from the measured quantities is simplified relative to the general problem discussed in Section 7.3.2 by the fact that the orientation factors of the photoselected molecular assembly are now known. However, this knowledge is insufficient in the absence of additional information. If either the three $A_u(\tilde{\nu}_1)$ or the three $B_v(\tilde{\nu}_2)$ spectra are known, the other set of three can be determined from (7.120) and (7.121). Alternatively, it is possible to work with the fractions $r_u(\tilde{\nu}_1)$ and $q_u(\tilde{\nu}_2)$ of (7.123) and (7.124). If this information is not available, the stepwise reduction method of Section 7.3.2 can be used. Partial solutions are also possible if it is known that one or more of the $A_u(\tilde{\nu}_1)$'s or $B_v(\tilde{\nu}_2)$'s vanish in the spectral region of interest, much like the situation in Section 5.1.2.

An example of this kind is provided in Figure 7.11.

Very High-Symmetry Limits. In very high-symmetry molecules with degenerate states, two cases are of importance as discussed above for more general types of orientation. First, if one of the transistions, say j, is polarized in the plane perpendicular to the high-order symmetry axis, and the other, f, is polarized parallel to that axis, then $F_{\parallel}^{iso}/F_{\perp}^{iso}$, P, and R reach their usual limits of 1/2, $-1/3$, and $-1/5$, respectively, for a mutually perpendicular pair of transitions. Second, if both transitions are polarized in the plane perpendicular to the symmetry axis, then $F_{\parallel}^{iso}/F_{\perp}^{iso} = 4/3$, $P = 1/7$, and $R = 1/10$. These limits can be derived from equations (7.85)–(7.89) and (7.91)–(7.95) when the isotropic values of the K's and L's are substituted. It needs to be emphasized that symmetry-lowering effects of the environment, ignored in our derivations, may cause deviations from this behavior.

7.3.6 Photoselected Orientation Distributions

General. Imagine that a polarized absorption or emission measurement is carried out on a small partially oriented fraction of molecules whose partial orientation is due to prior photoselection on a random assembly. If this prior photoselection were due to a single photon-molecule interaction using light linearly polarized along Z, the orientation factors K of the photoselected assembly are given by (4.70). We have already seen in Section 7.3.5 how the knowledge of the K's permits an evaluation of the observed intensities in the overall process of rank 4.

However, the higher orientation factors of the originally photoselected assembly can also be written explicitly. This can be done easily as long as only a small fraction of the molecules have been photoselected, and with some effort if the photoconversion was substantial. The results for the K's and L's for the special case of small depletion limit for highly symmetrical molecules have been given as an example in (4.74) and (4.78). This then permits the use of (7.6) to express the intensities in any process of rank 4 observed on the molecular assembly photoselected in the first step. This could be photoluminescence, photodichroism, Raman scattering, or simultaneous two-photon absorption due to the photoselected assembly; all of these are processes of rank 6 with respect to the original random assembly. Those processes of rank 6 on the original random assembly which involve a simultaneous interaction with three photons, such as three-photon absorption, cannot be described in this manner.

In the following, we shall illustrate the procedure for the case of a photoluminescence or photodichroism experiment on an assembly of high-symmetry molecules initially photoselected from an isotropic solution by light linearly polarized along Z. A similar procedure, based on (4.116) and (4.120), could be used to describe the results of an experiment on an assembly of molecules photoselected by unpolarized or circularly polarized light.

Moreover, the methods used to obtain the orientation factors of a molecular assembly upon selection from an isotropic solution by an electric dipole absorption process, i.e., a process of rank 2, can also be used to obtain the orientation factors of an assembly similarly photoselected by a process of rank 4. These orientation factors can then be used with (5.12) to obtain an alternative formulation of the results for processes which are of rank 6 on the original isotropic sample, or with (7.6) to obtain results for processes of rank 8 on the original sample. Examples of the latter are fluorescence-detected two-photon absorption by photoselected triplet molecules in a rigid glass or Raman scattering by an excited state initially populated by a simultaneous two-photon absorption event.

Finally, the orientation factors K, L, etc., produced by photoselection on a random assembly due to any combination of the above steps, i.e., by any combination of one-photon excitations and simultaneous successive two-photon excitations, can again be calculated in an explicit form by the method outlined in Section 4.6.3, so that even the results for many of the possible processes of ranks 10, 12, etc., can be obtained simply in this fashion. However, it should be emphasized that the procedure is applicable only to initially isotropic samples and is strictly valid only in the limit of negligible ground-state depletion.

Example: Successive Interactions with Three Photons. Of the multitude of possibilities, we shall consider in detail only the measurement of photoluminescence or photodichroism on a sample of molecules of high symmetry photoselected by a one-photon absorption event. This process is of rank 6 with respect to the original isotropic sample and involves successive interactions with three photons. The first is linearly polarized with electric vector along Z, and its wavenumber is $\tilde{\nu}_0$. It is followed by a photon at $\tilde{\nu}_1$ polarized along U and, finally, the monitored third photon has wavenumber $\tilde{\nu}_2$ and polarization V. The observed intensity is $G^{iso}_{ZUV}(\tilde{\nu}_0,\tilde{\nu}_1,\tilde{\nu}_2)$.

The molecular axes are chosen to be symmetry-adapted so that the orientation tensor K of the photoselected assembly is diagonal, and the orientation factors are given by (4.74) and (4.78). The relative size of the polarized fractional absorptions of the original molecules at $\tilde{\nu}_0$, $r_u(\tilde{\nu}_0)$, determines the labeling of the axes: $r_z(\tilde{\nu}_0) \geqslant r_y(\tilde{\nu}_0) \geqslant r_x(\tilde{\nu}_0)$.

In order to derive the expressions for $G^{iso}_{ZUV}(\tilde{\nu}_0,\tilde{\nu}_1,\tilde{\nu}_2)$, we recognize that the number of photoselected molecules is proportional to $E_0^{iso}(\tilde{\nu}_0)$:

$$G^{iso}_{ZUV}(\tilde{\nu}_0,\tilde{\nu}_1,\tilde{\nu}_2) = a''(\tilde{\nu}_0)E_0^{iso}(\tilde{\nu}_0)F_{UV}(\tilde{\nu}_1,\tilde{\nu}_2) \quad (7.137)$$

where $a''(\tilde{\nu}_0)$ reflects the intensity of the photoselecting beam and the quantum yield of photoselection, and $F_{UV}(\tilde{\nu}_1,\tilde{\nu}_2)$ is given by (7.25). We now use (7.36) to obtain

$$F_{UV}(\tilde{\nu}_1,\tilde{\nu}_2) = 9E_1^{iso}(\tilde{\nu}_1)E_2^{iso}(\tilde{\nu}_2)\sum_{uv}[\mathsf{S}_{UV}(\tilde{\nu}_0)]_{uv}q_u(\tilde{\nu}_1)p_v(\tilde{\nu}_2) \quad (7.138)$$

where

$$q_u(\tilde{\nu}_1) = A_u(\tilde{\nu}_1)/3E_1^{iso}(\tilde{\nu}_1) \quad (7.139)$$

$$p_v(\tilde{\nu}_2) = B_v(\tilde{\nu}_2)/3E_2^{iso}(\tilde{\nu}_2) \quad (7.140)$$

$$E_1^{iso}(\tilde{\nu}_1) = (1/3)\sum_{w} A_w(\tilde{\nu}_1) \quad (7.141)$$

$$E_2^{iso}(\tilde{\nu}_2) = (1/3)\sum_{w} B_w(\tilde{\nu}_2) \quad (7.142)$$

and where explicit recognition has been made of the fact that the tensor elements $[\mathsf{S}_{UV}(\tilde{\nu}_0)]_{uv}$ depend on $\tilde{\nu}_0$ through the orientation factors $K(\tilde{\nu}_0)$ and $L(\tilde{\nu}_0)$. The form of this dependence can be obtained by combining (4.74) and (4.78) with (7.31)–(7.35):

$$[\mathsf{S}_{ZZ}(\tilde{\nu}_0)]_{uv} = (1/35)(1 + 2\delta_{uv})[1 + 2r_u(\tilde{\nu}_0) + 2r_v(\tilde{\nu}_0)] \quad (7.143)$$

$$[\mathsf{S}_{XX}(\tilde{\nu}_0)]_{uv} = [\mathsf{S}_{YY}(\tilde{\nu}_0)]_{uv} = (1/35)(1 + 2\delta_{uv})[3 - r_u(\tilde{\nu}_0) - r_v(\tilde{\nu}_0)] \quad (7.144)$$

$$\begin{aligned}[\mathsf{S}_{XZ}(\tilde{\nu}_0)]_{uv} &= [\mathsf{S}_{YZ}(\tilde{\nu}_0)]_{uv}\\ &= (1/35)\{(3 - \delta_{uv})[1 + 2r_v(\tilde{\nu}_0)] - (1 + 2\delta_{uv})r_u(\tilde{\nu}_0)\}\end{aligned} \quad (7.145)$$

$$\begin{aligned}[\mathsf{S}_{ZX}(\tilde{\nu}_0)]_{uv} &= [\mathsf{S}_{ZY}(\tilde{\nu}_0)]_{uv}\\ &= (1/35)\{(3 - \delta_{uv})[1 + 2r_u(\tilde{\nu}_0)] - (1 + 2\delta_{uv})r_v(\tilde{\nu}_0)\}\end{aligned} \quad (7.146)$$

$$\begin{aligned}[\mathsf{S}_{XY}(\tilde{\nu}_0)]_{uv} &= [\mathsf{S}_{YX}(\tilde{\nu}_0)]_{uv}\\ &= (1/35)\{(4\delta_{uv} - 5)[r_u(\tilde{\nu}_0) + r_v(\tilde{\nu}_0)] + 8 - 5\delta_{uv}\}\end{aligned} \quad (7.147)$$

The values given in (7.143)–(7.147) can now be used in the relation

$$G^{iso}_{ZUV}(\tilde{\nu}_0,\tilde{\nu}_1,\tilde{\nu}_2) = 9a''(\tilde{\nu}_0)E^{iso}_0(\tilde{\nu}_0)E^{iso}_1(\tilde{\nu}_1)E^{iso}_2(\tilde{\nu}_2)\sum_{uv}[S_{UV}(\tilde{\nu}_0)]_{uv}q_u(\tilde{\nu}_1)p_v(\tilde{\nu}_2) \quad (7.148)$$

to obtain the final results for the intensities observed in this process of rank 6.

Frequently, only intensity ratios are measured; for these, we have

$$\frac{G^{iso}_{ZUV}(\tilde{\nu}_0,\tilde{\nu}_1,\tilde{\nu}_2)}{G^{iso}_{ZU'V'}(\tilde{\nu}_0,\tilde{\nu}_1,\tilde{\nu}_2)} = \frac{\sum_{uv}[S_{UV}(\tilde{\nu}_0)]_{uv}q_u(\tilde{\nu}_1)p_v(\tilde{\nu}_2)}{\sum_{uv}[S_{U'V'}(\tilde{\nu}_0)]_{uv}q_u(\tilde{\nu}_1)p_v(\tilde{\nu}_2)} \quad (7.149)$$

If the transition involved in the photoselection is purely polarized, i.e., $r_z(\tilde{\nu}_0) = 1$, $r_y(\tilde{\nu}_0) = r_x(\tilde{\nu}_0) = 0$, the K's and L's acquire values given in (4.75) and (4.79), yielding particularly simple values for the elements of $S_{UV}(\tilde{\nu}_0)$, which are now labeled $S_{UV}(z)$ to indicate that they are evaluated at a $\tilde{\nu}_0$ at which the photoselecting absorption is purely z polarized:

$35[S_{ZZ}(z)]_{uv} =$ (7.150)

u \ v	x	y	z
x	3	1	3
y	1	3	3
z	3	3	15

$35[S_{XX}(z)]_{uv} =$ (7.151)

u \ v	x	y	z
x	9	3	2
y	3	9	2
z	2	2	3

$35[S_{XZ}(z)]_{uv} =$ (7.152)

u \ v	x	y	z
x	2	3	9
y	3	2	9
z	2	2	3

$35[S_{ZX}(z)]_{uv} =$ (7.153)

u \ v	x	y	z
x	2	3	2
y	3	2	2
z	9	9	3

$35[S_{XY}(z)]_{uv} =$ (7.154)

u \ v	x	y	z
x	3	8	3
y	8	3	3
z	3	3	1

If the processes involving the second and third photons also occur in spectral regions with pure polarization, simple expressions for the four independent ratios $G^{iso}_{ZUV}/G^{iso}_{ZU'V'}$ in the limiting cases can be written. Particularly extreme limits are observed for the two ratios

$$\frac{G^{iso}_{ZZZ}(z,z,z)}{G^{iso}_{ZYX}(z,z,z)} = 15 \tag{7.155}$$

and

$$\frac{G^{iso}_{ZZZ}(z,y,x)}{G^{iso}_{ZYX}(z,y,x)} = \frac{1}{8} \tag{7.156}$$

which can be compared to the limiting ratios of 3 and 1/2 observed in ordinary photoselection [(7.114) and (7.117)].

7.4 Successive Events: Photoluminescence and Photodichroism. The Dynamic Case

A study of systems in which the excited molecule observed in the second step lies at an orientation different from that originally imparted in the first, photoselection step, naturally falls into three parts. It can usually be assumed that the fundamental spectroscopic properties of the molecule such as transition polarizations, reduced absorption and emission spectra, etc., have already been determined under static conditions. The remaining two steps are first, the evaluation of the rotational correlation functions for all accessible times and second, their interpretation in terms of a suitable model for diffusion. Here, we concentrate on the evaluation of the correlation functions from the raw data (Section 3.2.7), which is very similar to the analysis already described for the static case in Section 7.3. The dynamics of the processes involved in the changes of orientation of an excited molecule in the period of time τ between the photoselection and observation events represents a field of considerable current interest. However, the details of the nature of molecular rotational motion and of energy-transfer mechanisms definitely lie outside the scope of this book. In view of the important role which polarized optical spectroscopy plays in the study of these phenomena, we have nevertheless considered it advisable to mention some of the basics at least briefly. More thorough treatments can be found in the list of references provided in Section 7.6.

As already indicated, there are two mechanisms for the change of orientation of the excited molecule in the time interval τ which elapses before the interaction with the second photon takes place. The excited molecule can rotate to a new orientation, or the excitation can be transferred to another molecule with a different orientation. In the limit of large τ, this leads to a loss of memory of the orientational photoselection induced in the first step, and the two processes are known as rotational depolarization and concentration depolarization, respectively. In the following, we shall refer to rotational depolarization explicitly, but it should be remembered that the two processes are formally equivalent in that their effects on the observed polarized intensities are described by the same set of equations.

The lifetime of the species produced in the first step is generally finite; introducing $F(\tau)$ as its decay function, often exponential, one usually writes

$$F_{UV}(\tilde{\nu}_1,\tilde{\nu}_2,\tau) = F(\tau)J_{UV}(\tilde{\nu}_1,\tilde{\nu}_2,\tau) \qquad (7.157)$$

where all of the orientational behavior has been collected in the quantity $J_{UV}(\tilde{\nu}_1,\tilde{\nu}_2,\tau)$.

When E_{UV} is replaced by J_{UV} in the general formula (7.6), with the definitions (7.7)–(7.14), the formula remains valid for UV = ZZ, XZ, ZX, YZ, or ZY and for the average $(J_{XX} + J_{XY})/2$, provided that the first and third subscripts on the elements of ${}^{(4)}P_{UV}$ refer to the time of excitation, while the second and fourth subscripts refer to the time of observation τ. The orientation factors K and L acquire a time dependence accordingly. These dynamic orientation factors, or correlation functions, then relate the value of the direction cosine of an axis at time zero, say $\cos s(0)$, to that of an axis at time τ, say $\cos t(\tau)$ (Figure 7.12), and are defined by

$$K_{s(0)t(\tau)} = \langle \cos s(0) \cos t(\tau) \rangle \qquad (7.158)$$

$$L_{s(0)t(\tau)u(0)v(\tau)} = \langle \cos s(0) \cos t(\tau) \cos u(0) \cos v(\tau) \rangle \qquad (7.159)$$

Expressions for the time dependence of $E_{XX}(\tilde{\nu}_1,\tilde{\nu}_2)$, $E_{XY}(\tilde{\nu}_1,\tilde{\nu}_2)$, and $E_{YY}(\tilde{\nu}_1,\tilde{\nu}_2)$ require the introduction of additional orientation factors containing the averages of cosines of molecular axes with respect to the X or Y axis of the sample (cf. Section 4.7). They reflect those dynamical aspects of the orientation distribution which have to do with rotation around the Z axis of the sample and which do not change the magnitudes of the projections of unit vectors along molecular axes into the Z direction, and thus leave the usual K's and L's constant. This problem will not be considered here.

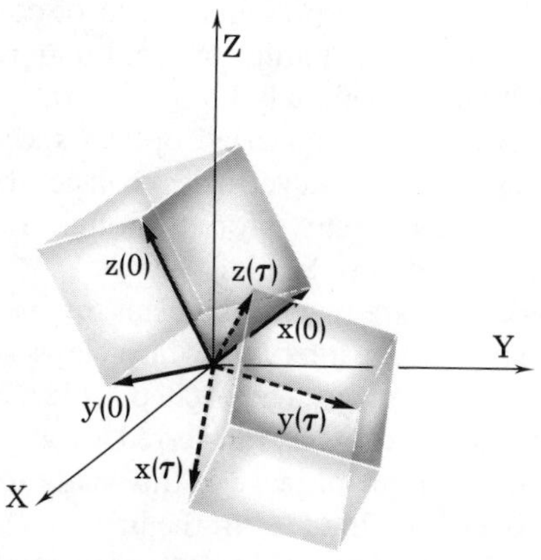

Figure 7.12 Rotation of the molecule-fixed system of axes between time $t = 0$ and time $t = \tau$.

If τ can be made infinitely long relative to the time scale of the rotational motion, the memory of the photoselection due to the photon will be lost, so that

$$\langle \cos u(0) \cos v(\infty) \rangle = \langle \cos u(0) \rangle \langle \cos v(\infty) \rangle = 0 \tag{7.160}$$

and similarly for other averages of quantities taken at different times. It can be verified that the general expressions for ${}^{(4)}\mathrm{P}_{\mathrm{UV}}$, equations (7.7)–(7.14), go to the expected time-independent limits:

$$[{}^{(4)}\mathrm{P}_{\mathrm{ZZ}}]_{\mathrm{s}(0)\mathrm{t}(\infty)\mathrm{u}(0)\mathrm{v}(\infty)} = \mathrm{K}_{\mathrm{s}(0)\mathrm{u}(0)}\mathrm{K}_{\mathrm{t}(\infty)\mathrm{v}(\infty)} \tag{7.161}$$

$$\begin{aligned}[{}^{(4)}\mathrm{P}_{\mathrm{XZ}}]_{\mathrm{s}(0)\mathrm{t}(\infty)\mathrm{u}(0)\mathrm{v}(\infty)} &= [{}^{(4)}\mathrm{P}_{\mathrm{YZ}}]_{\mathrm{s}(0)\mathrm{t}(\infty)\mathrm{u}(0)\mathrm{v}(\infty)} \\ &= (1/2)[\delta_{\mathrm{su}} - \mathrm{K}_{\mathrm{s}(0)\mathrm{u}(0)}]\mathrm{K}_{\mathrm{t}(\infty)\mathrm{v}(\infty)}\end{aligned} \tag{7.162}$$

$$\begin{aligned}[{}^{(4)}\mathrm{P}_{\mathrm{ZX}}]_{\mathrm{s}(0)\mathrm{t}(\infty)\mathrm{u}(0)\mathrm{v}(\infty)} &= [{}^{(4)}\mathrm{P}_{\mathrm{ZY}}]_{\mathrm{s}(0)\mathrm{t}(\infty)\mathrm{u}(0)\mathrm{v}(\infty)} \\ &= (1/2)[\delta_{\mathrm{tv}} - \mathrm{K}_{\mathrm{t}(\infty)\mathrm{v}(\infty)}]\mathrm{K}_{\mathrm{s}(0)\mathrm{u}(0)}\end{aligned} \tag{7.163}$$

$$\begin{aligned}&(1/2)\{[{}^{(4)}\mathrm{P}_{\mathrm{XX}}]_{\mathrm{s}(0)\mathrm{t}(\infty)\mathrm{u}(0)\mathrm{v}(\infty)} + [{}^{(4)}\mathrm{P}_{\mathrm{XY}}]_{\mathrm{s}(0)\mathrm{t}(\infty)\mathrm{u}(0)\mathrm{v}(\infty)}\} \\ &\quad = (1/2)\{[{}^{(4)}\mathrm{P}_{\mathrm{YY}}]_{\mathrm{s}(0)\mathrm{t}(\infty)\mathrm{u}(0)\mathrm{v}(\infty)} + [{}^{(4)}\mathrm{P}_{\mathrm{YX}}]_{\mathrm{s}(0)\mathrm{t}(\infty)\mathrm{u}(0)\mathrm{v}(\infty)}\} \\ &\quad = (1/2)\{[{}^{(4)}\mathrm{P}_{\mathrm{XX}}]_{\mathrm{s}(0)\mathrm{t}(\infty)\mathrm{u}(0)\mathrm{v}(\infty)} + [{}^{(4)}\mathrm{P}_{\mathrm{YX}}]_{\mathrm{s}(0)\mathrm{t}(\infty)\mathrm{u}(0)\mathrm{v}(\infty)}\} \\ &\quad = (1/2)\{[{}^{(4)}\mathrm{P}_{\mathrm{YY}}]_{\mathrm{s}(0)\mathrm{t}(\infty)\mathrm{u}(0)\mathrm{v}(\infty)} + [{}^{(4)}\mathrm{P}_{\mathrm{XY}}]_{\mathrm{s}(0)\mathrm{t}(\infty)\mathrm{u}(0)\mathrm{v}(\infty)}\} \\ &\quad = (1/2)\{[\delta_{\mathrm{su}} - \mathrm{K}_{\mathrm{s}(0)\mathrm{u}(0)}](1/2)[\delta_{\mathrm{tv}} - \mathrm{K}_{\mathrm{t}(\infty)\mathrm{v}(\infty)}]\}\end{aligned} \tag{7.164}$$

Here $\mathrm{K}_{\mathrm{s}(0)\mathrm{u}(0)}$ reflects the initial molecular orientation distribution and $\mathrm{K}_{\mathrm{t}(\infty)\mathrm{v}(\infty)}$ reflects the orientation distribution function after return to orientational equilibrium. If the excitation caused no permanent photochemical change, the two sets of K's will be the same:

$$\mathrm{K}_{\mathrm{s}(0)\mathrm{u}(0)} = \mathrm{K}_{\mathrm{s}(\infty)\mathrm{u}(\infty)} \tag{7.165}$$

The same formalism applies if the excitation changes the orientation distribution permanently, say by changing the molecular shape.

In the following, we shall first introduce some of the basic concepts encountered in the description of rotational diffusion on the simple but common case of photoselection on an isotropic sample. Then we shall proceed to the treatment of rotational diffusion in a uniaxial potential, such as that of solutes in liquid crystals, stretched polymers, or lipid membranes. Finally, we shall mention the case of photoselection in a macroscopically isotropic but locally uniaxial sample, such as a dispersion of very slowly moving uniaxial membranes suspended randomly in a solution.

7.4.1 Isotropic Samples

The measurement of polarized photoluminescence or photodichroism of an initially isotropic sample as a function of time after a δ-pulse excitation produces the time-dependent quantities $\mathrm{F}_{\|}^{\mathrm{iso}}(\tilde{v}_1,\tilde{v}_2,\tau)$ and $\mathrm{F}_{\perp}^{\mathrm{iso}}(\tilde{v}_1,\tilde{v}_2,\tau)$ analogous to the time-independent $F_{\parallel}^{iso}(\tilde{\nu}_1,\tilde{\nu}_2)$ and $F_{\perp}^{iso}(\tilde{\nu}_1,\tilde{\nu}_2)$ of the static case discussed in Section 7.3.5. These are to be determined over as long a time interval as possible. The limitation is given by the decay of the excited molecules $F(\tau)$; the method can provide information only on those rotations which occur on the time scale of the lifetime of the excited molecules or shorter.

The quantities usually determined in time-resolved or phase-modulation experiments are the dichroic ratio $F_{\parallel}^{iso}(\tilde{\nu}_1,\tilde{\nu}_2,\tau)/F_{\perp}^{iso}(\tilde{\nu}_1,\tilde{\nu}_2,\tau)$, the degree of polarization $P(\tilde{\nu}_1,\tilde{\nu}_2,\tau)$, or the degree of anisotropy $R(\tilde{\nu}_1,\tilde{\nu}_2,\tau)$, defined as in (7.106) and (7.108), respectively. The decay function $F(\tau)$ has been eliminated in these quantities. The values observed under the steady-state conditions of stationary illumination are referred to as $\bar{P}(\tilde{\nu}_1,\tilde{\nu}_2)$ and $\bar{R}(\tilde{\nu}_1,\tilde{\nu}_2)$ and are related to the above through (3.36).

Rotational Correlation Functions. We shall now address the problem of deriving the rotational correlation functions from the raw data. The most complicated case in which the solute molecule is of low symmetry and several overlapping transitions of different polarizations contribute at $\tilde{\nu}_1$ and $\tilde{\nu}_2$ has not been solved in its full generality. It is unlikely that a solution will be useful, since too many unknown quantities will appear in the expression.

Pure Polarization or High Symmetry. We now consider the case of a single transition j contributing to the absorption at $\tilde{\nu}_1$, with transition moment components $\mathbf{M}(j)\cos\phi_u^j$ along the three molecular axes u = x, y, and z, and of a single transition f contributing at $\tilde{\nu}_2$, with transition moment components $\mathbf{M}(f)\cos\phi_u^f$. The molecular axes will be chosen so as to diagonalize the diffusion tensor $\mathfrak{D}$ (see below) and not necessarily the optical tensors. The results are also applicable to the case of a high-symmetry molecule with several differently polarized transitions contribution at $\tilde{\nu}_1$ and at $\tilde{\nu}_2$. In this case the diffusion tensor as well as the optical tensors follow the symmetry of the molecule and are diagonalized simultaneously when the symmetry axes are chosen as the molecular system. Then the polarized fractional absorptions $r_u(\tilde{\nu}_1)$ and $q_v(\tilde{\nu}_2)$ will replace the quantities $\cos^2\phi_u^j$ and $\cos^2\phi_v^f$, respectively, and the cross-terms $\cos\phi_u^j\cos\phi_v^f$ will disappear.

If the depletion of the initial ground state is negligible, the polarized intensities are given by (7.24) and (7.25), where the elements of ${}^{(4)}\mathrm{P}_{UV}$ are now time-dependent. When (j,f) is the only contribution, we use (7.7) and (7.10) to obtain for the observed polarization ratio

$$\frac{F_{\parallel}^{iso}(\tilde{\nu}_1,\tilde{\nu}_2,\tau)}{F_{\perp}^{iso}(\tilde{\nu}_1,\tilde{\nu}_2,\tau)} = 2\frac{\sum\limits_{stuv} L_{s(0)t(\tau)u(0)v(\tau)}\cos\phi_s^j\cos\phi_t^f\cos\phi_u^j\cos\phi_v^f}{\sum\limits_{stuv}(\delta_{tv}\delta_{su}/3 - L_{s(0)t(\tau)u(0)v(\tau)})\cos\phi_s^j\cos\phi_t^f\cos\phi_u^j\cos\phi_v^f} \qquad (7.166)$$

where $\delta_{su}/3$ has been substituted for $K_{s(0)u(0)}$ since the initial solution is isotropic. The denominator can be further simplified using (5.50). In terms of the degree of anisotropy, the result is

$$R(\tilde{\nu}_1,\tilde{\nu}_2,\tau) = \left(9\sum_{stuv} L_{s(0)t(\tau)u(0)v(\tau)}\cos\phi_s^j\cos\phi_t^f\cos\phi_u^j\cos\phi_v^f - 1\right)\Big/2 \qquad (7.167)$$

This equation is a generalized version of (7.125). With the substitutions $L_{s(0)t(0)u(0)v(0)} = \delta_{|s \times t|,|u \times v|}(1 + 2\delta_{uv})/15$, $\cos^2 \phi_u^j = r_u(\tilde{\nu}_1)$, $\cos^2 \phi_u^f = q_u(\tilde{\nu}_1)$, and the deletion of the terms containing cross-products of direction cosines in (7.167), the two become equivalent.

A relation between the two observed intensities follows:

$$F_{\perp}^{iso}(\tilde{\nu}_1,\tilde{\nu}_2,\tau) = (1/6)F(\tau) - (1/2)F_{\|}^{iso}(\tilde{\nu}_1,\tilde{\nu}_2,\tau) \tag{7.168}$$

Symmetry of the isotropic orientation distribution demands that all $L_{s(0)t(\tau)u(0)v(\tau)}$'s in whose subscript an index x, y, or z appears an odd number of times vanish. Of the remaining nine correlation functions, $L_{u(0)v(\tau)u(0)v(\tau)}$ and $L_{u(0)u(\tau)v(0)v(\tau)}$, u,v = x,y,z, only three are independent (we work in the principal axes system of the diffusion tensor). In favorable cases of molecules with known optical properties, the methods of Section 7.3 can be used to determine them from the degree-of-anisotropy functions $R(\tilde{\nu}_1,\tilde{\nu}_2,\tau)$ observed for suitable choices of $\tilde{\nu}_1$ and $\tilde{\nu}_2$.

The limiting values can be obtained from the properties of an isotropic solution and from (7.161):

$$L_{u(0)v(0)u(0)v(0)} = L_{u(0)u(0)v(0)v(0)} = L_{uv} = (1 + 2\delta_{uv})/15 \tag{7.169}$$

$$L_{u(0)v(\infty)u(0)v(\infty)} = K_{u(0)}K_{v(\infty)} = 1/9 \tag{7.170}$$

$$L_{u(0)u(\infty)v(0)v(\infty)} = K^2_{u(0)v(\infty)} = 0, \qquad u \neq v \tag{7.171}$$

The three rotational correlation functions L represent the primary experimental information on the molecular reorientation or energy-transfer processes.

An alternative formulation of the problem can be derived in terms of the orientation factors $K_{uv}(\tilde{\nu}_1,\tau)$ of the photoselected assembly introduced in Section 4.6.3, along the lines discussed in Section 7.3.5. Now, it is not necessary to assume a negligible depletion of the ground state. If this assumption can be made, the expressions for the initial values $K_{uv}(\tilde{\nu}_1,0)$ simplify and acquire the form (4.70). They simplify further to (4.73) and (4.74) when the molecular axes system is chosen so as to diagonalize the photoselecting absorption tensor. In the general case, this choice will not diagonalize the diffusion tensor. It is preferable to work in the principal axes system of the diffusion tensor, but then the expressions for the initial value of $K_{uv}(\tilde{\nu}_1,0)$ remain more complicated. For molecules of high symmetry, this dilemma does not occur, since both principal systems coincide.

The values of $K_{uv}(\tilde{\nu}_1,\tau)$ can be determined from $F_{\|}^{iso}(\tilde{\nu}_1,\tilde{\nu}_2,\tau)$ and $F_{\perp}^{iso}(\tilde{\nu}_1,\tilde{\nu}_2,\tau)$ at any time τ by the methods of Section 5.1.1 for high-symmetry molecules and by the methods of Section 5.1.3 in favorable cases of low-symmetry molecules. The same methods are used to obtain the information necessary to calculate $K_{uv}(\tilde{\nu}_1,0)$ from (4.70). A comparison of (7.98) and (7.166) shows that the $K(\tilde{\nu},\tau)$'s are related to the $L_{s(0)t(\tau)u(0)v(\tau)}$'s by

$$K_{tv}(\tau) = \sum_{su} L_{s(0)t(\tau)u(0)v(\tau)} \cos \phi_s^j \cos \phi_u^j \tag{7.172}$$

if the depletion of the initial ground state is negligible.

Diffusion Tensor. The next step is the interpretation of the results in terms of a model of molecular rotational motion. The liquids and solids of interest as environments for solute molecules are all relatively viscous, and the solute molecules are relatively large. Then it is reasonable to ignore molecular free rotation and consider the reorientation to result from a sequence of collisions, each of which changes the molecular orientation only to a small degree (the diffusion model). For very highly viscous media, the diffusion model is no longer applicable and needs to be replaced by the rotational jump model, in which reorientation occurs by large-angle molecular jumps. However, the rotational jump model leads to similar results as the diffusion model, with the diffusion constants replaced by average jump angles.

The diffusion equation for the photoselected assembly is given by

$$\partial f(\Omega,\tau)/\partial\tau = -\hat{H}f(\Omega,\tau) \tag{7.173}$$

where $f(\Omega,\tau)$ is the time-dependent orientation distribution function, uniaxial with respect to the Z axis at all times, and the Hamiltonian is given by

$$\hat{H} = \sum_{u,v} \hat{\mathscr{L}}_u \mathscr{D}_{uv} \hat{\mathscr{L}}_v \tag{7.174}$$

where $\hat{\mathscr{L}}$ is formally equivalent to the quantum-mechanical angular momentum operator and $\mathfrak{D}$ is the diffusion tensor of the solute molecule in the environment of interest. Its principal axes may be difficult to find if molecular symmetry is low, since intermolecular interactions may displace them away from the principal axes of inertia. It is normally taken to be time-independent although this condition need not be satisfied if the nature of the solute–solvent interaction is substantially different in the ground and in the excited state of the solute, due to gradual solvent relaxation.

In the principal axes system, the Hamiltonian depends only on the diagonal elements of the diffusion tensor, $\mathscr{D}_x$, $\mathscr{D}_y$, and $\mathscr{D}_z$, and is represented by a single sum, $\sum_u \mathscr{D}_u \mathscr{L}_u^2$. Solution of the differential equation (7.173) produces an explicit result for $f(\Omega,\tau)$, which can then be used to express the observed polarized intensities. This yields

$$\begin{aligned} L_{u(0)u(\tau)u(0)u(\tau)} = {} & (1/9) + (1/15)\{[2/3 + (\mathscr{D} - \mathscr{D}_u)/\Delta] \exp[-2(3\mathscr{D} + \Delta)\tau] \\ & + [2/3 - (\mathscr{D} - \mathscr{D}_u)/\Delta] \exp[-2(3\mathscr{D} - \Delta)\tau]\} \end{aligned} \tag{7.175}$$

$$\begin{aligned} L_{u(0)v(\tau)u(0)v(\tau)} = {} & (1/9) - (1/15)\{[1/3 + (2\mathscr{D} - \mathscr{D}_u - \mathscr{D}_v)/\Delta] \exp[-2(3\mathscr{D} + \Delta)\tau] \\ & + [1/3 - (2\mathscr{D} - \mathscr{D}_u - \mathscr{D}_v)/\Delta] \exp[-2(3\mathscr{D} - \Delta)\tau]\}, \qquad u \neq v \end{aligned} \tag{7.176}$$

$$L_{u(0)u(\tau)v(0)(\tau)} = (1/15) \exp[-3(4\mathscr{D} - \mathscr{D}_u - \mathscr{D}_v)\tau], \qquad u \neq v \tag{7.177}$$

where $\mathscr{D}$ and Δ are defined in terms of $\mathscr{D}_x$, $\mathscr{D}_y$, and $\mathscr{D}_z$ as

$$\mathscr{D} = (1/3) \sum_u \mathscr{D}_u \tag{7.178}$$

$$\Delta = [(1/2) \sum_{u,v} (3\delta_{uv} - 1)\mathscr{D}_u\mathscr{D}_v]^{1/2} \tag{7.179}$$

Thus, it is seen that five different exponential decays occur simultaneously in the general case in addition to the trivial time dependence $F(\tau)$.

If a sufficient number of the correlation functions L have been obtained experimentally, equations (7.175)–(7.179) can be used to derive the values of the three diffusion constants $\mathscr{D}_u$. The system is overdetermined, and it may be possible to work backwards and obtain information about the direction cosines $\cos\phi_u^i$ and $\cos\phi_u^f$ or the fractional absorption probabilities $r_u(\tilde{\nu}_1)$ and $q_u(\tilde{\nu}_2)$. However, no new optical information will be obtained in this way which would not be available from the static measurement.

Isotropic Diffusion Tensor. Several special cases are of interest. If the diffusion tensor is isotropic, $\mathscr{D}_x = \mathscr{D}_y = \mathscr{D}_z = \mathscr{D}$, and

$$L_{u(0)u(\tau)u(0)u(\tau)} = 1/9 + (4/45)\exp[-6\mathscr{D}\tau] \qquad (7.180)$$

$$L_{u(0)v(\tau)u(0)v(\tau)} = 1/9 - (2/45)\exp[-6\mathscr{D}\tau], \qquad u \neq v \qquad (7.181)$$

$$L_{u(0)u(\tau)v(0)v(\tau)} = (1/15)\exp[-6\mathscr{D}\tau], \qquad u \neq v \qquad (7.182)$$

and from (7.167),

$$R(\tilde{\nu}_1,\tilde{\nu}_2,\tau) = R(\tilde{\nu}_1,\tilde{\nu}_2,0)\exp(-6\mathscr{D}\tau) \qquad (7.183)$$

where only one exponential remains. The rotational relaxation time τ_R is then given by

$$\tau_R = 1/6\mathscr{D} \qquad (7.184)$$

This is related to the viscosity η, molecular hydrodynamic volume V, the Boltzmann constant k, and absolute temperature T by the Debye-Stokes-Einstein relation

$$\tau_r = \eta V/kT$$

$$\tau_r(\text{ns}) = \eta(\text{cP})V(\text{Å}^3)/3.3T(\text{K}) \qquad (7.185)$$

In this limit, a simple expression also results for the degree of polarization $\bar{P}(\tilde{\nu}_1,\tilde{\nu}_2)$ observed under conditions of stationary illumination, as shown in a relation first obtained by Perrin:

$$1/\bar{P}(\tilde{\nu}_1,\tilde{\nu}_2) - 1/3 = [1/P(\tilde{\nu}_1,\tilde{\nu}_2) - 1/3](1 + 6\mathscr{D}\tau) \qquad (7.186)$$

where $P(\tilde{\nu}_1,\tilde{\nu}_2)$ is the degree of polarization observed in the absence of molecular rotation. For purely polarized absorbing and emitting transitions, this is related by (7.112) to the angle between the two transition moments.

Pure Polarization in a High-Symmetry Molecule. Another important limiting case is the rotation of highly symmetrical molecules with purely polarized transitions at $\tilde{\nu}_1$ and $\tilde{\nu}_2$. If their polarizations are u and v, respectively, we obtain from (7.167), (7.175), and (7.176)

$$R(u,u,\tau) = (3/10)\{[2/3 + (\mathscr{D} - \mathscr{D}_u)/\Delta]\exp[-2(3\mathscr{D} + \Delta)\tau] + [2/3 - (\mathscr{D} - \mathscr{D}_u)/\Delta]\exp[-2(3\mathscr{D} - \Delta)\tau]\} \qquad (7.187)$$

$$R(u,v,\tau) = (3/10)\{[1/3 + (2\mathscr{D} - \mathscr{D}_u - \mathscr{D}_v)/\Delta] \exp[-2(3\mathscr{D} + \Delta)\tau] + [1/3 - (2\mathscr{D} - \mathscr{D}_u - \mathscr{D}_v)/\Delta] \exp[-2(3\mathscr{D} - \Delta)\tau]\}, \quad u \neq v \tag{7.188}$$

Now only two exponential decays occur in addition to $F(\tau)$, with weights which are a function of $\mathscr{D}_x$, $\mathscr{D}_y$, and $\mathscr{D}_z$, permitting the determination of the latter quantities in favorable cases.

7.4.2 Uniaxial Samples

Time-resolved or phase-modulation measurements of polarized photoluminescence or photodichroism produce the five time-dependent quantities $F_{UV}(\tilde{\nu}_1,\tilde{\nu}_2,\tau)$ analogous to the time-independent $F_{UV}(\tilde{\nu}_1,\tilde{\nu}_2)$ of the static case discussed in Section 7.3. The methods described there can be used to evaluate the tensor elements ${}^{(4)}P_{UV}$ as a function of time. In the following, we consider only the case of highly symmetrical molecules whose symmetry axes represent the principal axes system for tensor diagonalization. Several differently polarized transitions are assumed to contribute both at $\tilde{\nu}_1$ and at $\tilde{\nu}_2$, with the usual fractional polarized intensities $r_u(\tilde{\nu}_1)$ and $q_v(\tilde{\nu}_2)$.

If the depletion of the initial ground state is negligible, the polarized intensities are again given by (7.23) and (7.25). The four measurements which are most readily interpreted are

$$F_{ZZ}(\tilde{\nu}_1,\tilde{\nu}_2,\tau) = C(\tilde{\nu}_1,\tilde{\nu}_2) \sum_{uv} L_{u(0)v(\tau)u(0)v(\tau)} r_u(\tilde{\nu}_1) q_v(\tilde{\nu}_2) \tag{7.189}$$

$$F_{XZ}(\tilde{\nu}_1,\tilde{\nu}_2,\tau) = C(\tilde{\nu}_1,\tilde{\nu}_2) \sum_{uv} (1/2)(K_{u(0)} - L_{u(0)v(\tau)u(0)v(\tau)}) r_u(\tilde{\nu}_1) q_v(\tilde{\nu}_2) \tag{7.190}$$

$$F_{ZX}(\tilde{\nu}_1,\tilde{\nu}_2,\tau) = C(\tilde{\nu}_1,\tilde{\nu}_2) \sum_{uv} (1/2)(K_{v(\tau)} - L_{u(0)v(\tau)u(0)v(\tau)}) r_u(\tilde{\nu}_1) q_v(\tilde{\nu}_2) \tag{7.191}$$

$$F_{XX}(\tilde{\nu}_1,\tilde{\nu}_2,\tau) + F_{XY}(\tilde{\nu}_1,\tilde{\nu}_2,\tau) = C(\tilde{\nu}_1,\tilde{\nu}_2) \sum_{uv} (1/2)[(1 - K_{u(0)} - K_{v(\tau)} + L_{u(0)v(\tau)u(0)v(\tau)}] r_u(\tilde{\nu}_1) q_v(\tilde{\nu}_2) \tag{7.192}$$

where

$$C(\tilde{\nu}_1,\tilde{\nu}_2) = 9a(\tilde{\nu}_1)a'(\tilde{\nu}_2)E_1^{iso}(\tilde{\nu}_1)E_2^{iso}(\tilde{\nu}_2) \tag{7.193}$$

Assuming that $r_u(\tilde{\nu}_1)$ and $q_v(\tilde{\nu}_2)$ are known from the analysis of static measurements discussed in Section 7.3.2, the ratios of the $F_{UV}(\tilde{\nu}_1,\tilde{\nu}_2,\tau)$'s can in principle be used to evaluate the rotational correlation functions. Their values at $\tau = 0$ are known from measurements under static conditions. Also, their values at $\tau = \infty$ [cf. (7.161)–(7.164)] are all accessible from the measurement of ordinary dichroism on the uniaxial sample under static conditions.

The next step is the solution of the diffusion equation in a uniaxial orienting potential, which would permit the determination of the constants characterizing molecular rotational motion and the orienting potential. This task is far more complicated than was the case for free diffussion, and a solution has apparently not yet been attempted in its full generality. Methods similar to those described briefly in Section 7.4.1 but now including the orienting potential in the Hamiltonian have been used to obtain results for the special case of symmetric top molecules. These were not obtained in a closed form except for short-time and long-time limits. References can be found in Section 7.6.

7.4.3 Macroscopically Isotropic Locally Uniaxial Samples

A system of considerable interest in biochemistry is suspension of membranes containing fluorescent molecules. The ensemble of approximately planar membrane segments is isotropic in a macroscopic sense, since all membrane orientations occur with equal probability. The rotation of the individual membrane segments on the time scale of the experiment is negligible; in that sense, the measurement of polarized fluorescence would correspond to the "static" case discussed in Section 7.3. However, the rotational motion of the fluorescent molecules within the membrane segments occurs on a time scale which can be comparable with the fluorescent lifetime. It is anisotropic because of the uniaxial orienting potential within each membrane. Thus, the effects of photoselection linger for times long on the time scale of motion within the membranes, but they eventually disappear on the time scale of the rotation of the membrane segments. In the following, it is assumed that the membrane segments do not rotate at all.

The two independent observable quantities are $F_{\parallel}^{mem}(\tilde{\nu}_1,\tilde{\nu}_2,\tau)$ and $F_{\perp}^{mem}(\tilde{\nu}_1,\tilde{\nu}_2,\tau)$, analogous to the $F_{\parallel}^{iso}(\tilde{\nu}_1,\tilde{\nu}_2,\tau)$ and $F_{\perp}^{iso}(\tilde{\nu}_1,\tilde{\nu}_2,\tau)$ of the ordinary dynamic case of photoselection on an isotropic solution (Section 7.4.1). In these experiments, reduced absorption and emission spectra are usually not sought and the two observed quantities are combined into the degree of anisotropy $R^{mem}(\tilde{\nu}_1,\tilde{\nu}_2,\tau)$ in the usual way [(7.108)].

At time $\tau = 0$, the degree of anisotropy is given by the static result (7.109). If a single transition j is responsible for absorption at $\tilde{\nu}_1$ and a single transition f is responsible for emission at $\tilde{\nu}_2$, (7.113) yields

$$R(\tilde{\nu}_1,\tilde{\nu}_2,0) = (3\cos^2\phi - 1)/5 = (2/5)P_2(\cos\phi) \qquad (7.194)$$

The results for the limit of long times have been obtained for the most general case of orientation within the membrane segments. If this orientation is uniaxial, the result acquires the form

$$R(\tilde{\nu}_1,\tilde{\nu}_2,\infty) = (9/10)\left[\sum_{u,v}\cos\phi_u^j\cos\phi_v^j K_{uv} - 1/3\right]\left[\sum_{u,v}\cos\phi_u^f\cos\phi_v^f K_{uv} - 1/3\right] \qquad (7.195)$$

where the orientation factors K_{UV} are those of the fluorescent probe relative to the uniaxial axis of the membrane (normal to its surface). If the orientation factors of the excited molecules are different from those of ground state molecules, the latter have to be used in the first bracket and the former in the second bracket in (7.195)–(7.197).

In the case of high-symmetry molecules with polarized absorption fractions $r_u(\tilde{\nu}_1)$ and polarized emission fractions $q_u(\tilde{\nu}_2)$ along their three symmetry axes $u = x,y,z$ [cf. (4.72)], the result becomes

$$R(\tilde{\nu}_1,\tilde{\nu}_2,\infty) = (9/10)\left[\sum_u r_u(\tilde{\nu}_1)K_u - 1/3\right]\left[\sum_u q_u(\tilde{\nu}_2)K_u - 1/3\right] \qquad (7.196)$$

and bears a certain resemblance to the analogous expression for the degree of anisotropy of an ordinary isotropic sample, (7.125).

If the absorbing transition j is purely polarized along u and the emitting transition is purely polarized along v, the result for the long-time limit simplifies to

$$R(\tilde{\nu}_1,\tilde{\nu}_2,\infty) = (1/10)(3K_u - 1)(3K_v - 1) \qquad (7.197)$$

At intermediate times τ, the observed anisotropy degree $R(\tilde{\nu}_1,\tilde{\nu}_2,\tau)$ is related to rotational correlation functions as in Section 7.4.2. The solution of the equation for diffusion of a symmetric-top molecule in a uniaxial potential alluded to there can again be applied to describe the observed time dependence. The situation is fairly complicated; references are given in Section 7.6.

7.5 Simultaneous Two-Photon Events: Two-Photon Absorption and Raman Scattering

We now turn to the discussion of processes involving a simultaneous (coherent) interaction of a solute molecule with two photons from the radiation field. If both photons are annihilated and their energy taken up by the molecule, the process is known as two-photon absorption and the molecular transition is characterized by the second-rank tensor $T(\tilde{\nu}_1,\tilde{\nu}_2)$ (Section 1.3.1). The measured quantity is the absorption intensity or cross section $I_{UV}(\tilde{\nu}_1,\tilde{\nu}_2)$. If one photon is annihilated and the other one created at a different energy, with the difference taken up by or taken away from the molecule, the process is known as Raman scattering, and the molecular transition is characterized by the second-rank tensor $\alpha'(\tilde{\nu}_1,\tilde{\nu}_2)$ (Section 1.3.1). The measured quantity is the Raman scattering intensity, $I_{UV}(\tilde{\nu}_1,\tilde{\nu}_2)$.

The coherence (simultaneity) requirement means that molecular rotational motion cannot now enter into the results for the observed intensities $I_{UV}(\tilde{\nu}_1,\tilde{\nu}_2)$ in the time-dependent way which we had to consider in the case of incoherent (successive) two-photon events; thus we do not distinguish the static and dynamic cases. However, the rotational motion of the molecules still affects the observed spectra through line shapes in a similar way as it does in one-photon spectroscopy. Fourier transformation of the frequency-domain line shapes into the time domain provides information comparable to that resulting from the picosecond time-domain measurements of rotational depolarization of photoinduced dichroism (Section 7.4). Vibrational Raman spectra are particularly suitable for this purpose because of their narrow line widths and relatively rare line overlap. A more detailed discussion lies outside the scope of this book.

In the following, we first provide general outlines of the two kinds of spectroscopic measurement, as they are most commonly performed on isotropic samples. Subsequently, we shall describe polarization measurements jointly for both, since the same formulas apply. Finally, at the end of the section, we shall discuss the reduction of the general formulas for the isotropic case. This will again be done separately for the two kinds of spectroscopic measurement, although the same formulas apply, since they are usually written differently in the two cases.

Two-Photon Absorption. It appears fair to say that full two-dimensional two-photon absorption spectra $I_{UV}(\tilde{\nu}_1,\tilde{\nu}_2)$ have not yet been obtained for any large molecule. The greatest number of available measurements have been performed on isotropic samples, in one of two modes. In the simpler arrangement, both photons are taken from the same beam and $\tilde{\nu}_1$ is scanned to yield a one-dimensional spectrum $I_{UU}(\tilde{\nu}_1,\tilde{\nu}_1)$, which shows absorptions due to various excited states f at energies $2h\nu_1$. Then only I_{ZZ}, I_{YY}, and I_{CON} are accessible. The more complicated arrangement uses two laser beams. The wavenumber of one of these ($\tilde{\nu}_1$) is kept constant at $\tilde{\nu}_0$; that of the other ($\tilde{\nu}_2$) is scanned to yield a one-dimensional spectrum $I_{UV}(\tilde{\nu}_0,\tilde{\nu}_2)$. This could be repeated at various settings of $\tilde{\nu}_0$ until ultimately a full two-dimensional spectrum $I_{UV}(\tilde{\nu}_1,\tilde{\nu}_2)$ would result.

Because of the equivalent way in which the two photons enter into the experiment, the function $I_{UV}(\tilde{\nu}_1,\tilde{\nu}_2)$ is symmetric with respect to the interchange of $\tilde{\nu}_1$ and $\tilde{\nu}_2$. Both $\tilde{\nu}_1$ and $\tilde{\nu}_2$ are limited in practice to the region of transparency in ordinary one-photon absorption. Resonant two-photon absorption, in which this condition is not fulfilled, is also possible but will be difficult to separate in practice from the simultaneously occurring one-photon absorption. Such measurements can be performed indirectly in experiments simultaneously involving more than two photons, which shall not be discussed here.

Raman Scattering. In the ordinary Raman experiment, a photon of wavenumber $\tilde{\nu}_1$ and polarization U is annihilated by interaction with the molecule, and a photon of a different wavenumber $\tilde{\nu}_2$ and polarization V is created and detected. The difference in the energies of the two photons, $h\,\Delta\nu = h\nu_1 - h\nu_2$, is conveyed to the molecule (or removed from it in the case of anti-Stokes scattering) in the form of vibrational, rotational, or, much less commonly, electronic excitation. The Raman spectrum $I_{UV}(\tilde{\nu}_1,\tilde{\nu}_2)$ then consists of a plot of the intensity of scattered radiation against $\tilde{\nu}_2$ or, more commonly, $\Delta\nu$.

The quantity which characterizes the elementary contribution of a given vibration f to the total intensity of the Raman effect is the Raman polarizability tensor $\alpha'(\tilde{\nu}_1,f)$ The intensity of the Raman effect is rather small, and it is now customary to use lasers as exciting sources. This makes it rather inconvenient to scan $\tilde{\nu}_1$ as is commonly done in fluorescence or to obtain a full two-dimensional spectrum. A plot of the Raman intensity of a given vibrational mode f against the wavenumber $\tilde{\nu}_1$ (its "Raman excitation profile") is therefore usually obtained only in a point-by-point mode. In obtaining it, both $\tilde{\nu}_1$ and $\tilde{\nu}_2$ have to be changed in order to keep $\Delta h\nu$ equal to the energy of vibration f.

Like the two-photon absorption tensor T [cf. (2.20)], the Raman polarizability tensor depends on $\tilde{\nu}_1$ in a complex fashion in that all excited electronic states of the molecule contribute [cf. (2.24)]. Now, however, it is quite convenient to make measurements in which $\tilde{\nu}_1$ coincides with the position of one of the electronic absorption bands of the molecule (resonance Raman). Then its contribution usually dominates, and all the others may be neglected. The tremendous increase of Raman intensity which occurs when $\tilde{\nu}_1$ gets into resonance with an electronic transition of the molecule makes resonance Raman a much more sensitive method than ordinary Raman spectroscopy.

From now nearly to the end of Section 7.5, we shall consider two-photon absorption and Raman scattering together, since they show the same dependence on molecular orientation.

7.5.1 Uniaxial Samples: Two-Photon Absorption and Raman Scattering

Notation. We shall deal with seven distinct arrangements of the polarizations of the two photons, given by the combinations

$$I_{ZZ}, \quad I_{XX} = I_{YY}, \quad I_{XZ} = I_{YZ}, \quad I_{ZX} = I_{ZY}, \quad I_{XY} = I_{YX},$$

$$I_{CON}, \quad I_{DIS} \quad (\text{or } I_{SYN} \text{ and } I_{ANTI}, \text{ see below}) \tag{7.198}$$

In two-photon absorption experiments, all of these possibilities are available only if the two photons are taken from two different beams. If both photons are taken from the same polarized beam, only the measurement of $I_{ZZ}(\tilde{\nu}_1,\tilde{\nu}_1)$, $I_{XX}(\tilde{\nu}_1,\tilde{\nu}_1) = I_{YY}(\tilde{\nu}_1,\tilde{\nu}_1)$, and $I_{CON}(\tilde{\nu}_1,\tilde{\nu}_1)$ is possible, and much less information is obtained. In Raman scattering experiments, all seven possibilities are available. However, in ordinary (nonresonant) Raman measurements, $I_{UZ} = I_{ZU}$.

If the two light beams do not propagate in the same direction, special care is required in the use of the symbols describing the arrangement of circular polarization. We consider only circularly polarized light propagating parallel to the Z axis of the sample, so that the two possible propagation directions are $+Z$ and $-Z$. The observed intensity I_{CON} is obtained in a measurement in which the tip of the electric vector rotates in the same sense for both photons when viewed along the $+Z$ direction, regardless of whether the photons propagate parallel or antiparallel to each other. The observed intensity I_{DIS} is obtained when the electric vector rotates in opposite sense for the two photons when viewed along the $+Z$ direction, again regardless of whether the photons propagate parallel or antiparallel.

For instance, in a two-photon absorption experiment, the use of a left-handed circularly polarized beam at $\tilde{\nu}_1$ propagating along $+Z$ and a left-handed circularly polarized beam at $\tilde{\nu}_2$ propagating along $+Z$ yields $I_{CON}(\tilde{\nu}_1,\tilde{\nu}_2)$, as does the use of a left-handed beam at $\tilde{\nu}_1$ propagating along $+Z$ and a right-handed beam propagating along $-Z$. In a Raman scattering experiment, the use of a left-handed exciting beam at $\tilde{\nu}_1$ propagating along $+Z$ and observation of the left-handed circularly polarized component of the forward scattered light at $\tilde{\nu}_2$ propagating along $+Z$, or, alternatively, the observation of the right-handed circularly polarized component of the backward-scattered light at $\tilde{\nu}_2$ propagating along $-Z$, both correspond to $I_{CON}(\tilde{\nu}_1,\tilde{\nu}_2)$. In the following, we shall also need to refer to intensities observed when both photons are of like handedness, regardless of their propagation direction, I_{SYN}, and to those observed when the two photons are of opposed handedness, I_{ANTI}. For two photons propagating in the same direction, SYN = CON, ANTI = DIS. For two photons propagating in antiparallel directions, the reverse holds, SYN = DIS, ANTI = CON. In the literature, the symbol CO is occasionally used for either SYN or CON and the symbol CONTRA for either ANTI or DIS. We shall avoid their use.

Polarized Intensity. The general expression for the observed intensity is obtained using (7.3) and (7.6) and summing over all contributing final states f:

$$I_{UV}(\tilde{\nu}_1,\tilde{\nu}_2) = a(\tilde{\nu}_1,\tilde{\nu}_2) \sum_f g_f'(\tilde{\nu}_2) I_{UV}(\tilde{\nu}_1,f) \qquad (7.199)$$

where $I_{UV}(\tilde{\nu}_1,f)$ is the elementary contribution due to the f-th transition, $g_f'(\tilde{\nu}_2)$ is the line shape, and the quantity $a(\tilde{\nu}_1,\tilde{\nu}_2)$ depends on the intensity of the exciting beam or beams, the number of solute molecules in the observed region, and the detection efficiency.

For linear polarizations U and V, the elementary contribution $I_{UV}(\tilde{\nu}_1,f)$ is given by

$$I_{UV}(\tilde{\nu}_1,f) = \sum_{stuv} [^{(4)}\mathsf{P}_{UV}]_{stuv} T_{st}(\tilde{\nu}_1,f) T^*_{uv}(\tilde{\nu}_1,f) \qquad \text{(two-photon absorption)} \quad (7.200)$$

$$I_{UV}(\tilde{\nu}_1,f) = \sum_{stuv} [^{(4)}\mathsf{P}_{UV}]_{stuv} \alpha'_{st}(\tilde{\nu}_1,f) \alpha'^*_{uv}(\tilde{\nu}_1,f) \qquad \text{(Raman)} \quad (7.201)$$

where the tensor elements $[^{(4)}\mathsf{P}_{UV}]_{stuv}$ are given by (7.7)–(7.11) [(7.7a)–(7.11a)] and (7.14).

Circular Polarization. The elements of P_{UV} have not yet been defined for UV = CON, DIS, SYN, and ANTI. This must be done before (7.200) and (7.201) can be used for experiments with circularly polarized light as well. The definitions of P_{CON} and P_{DIS} differ in the two kinds of spectroscopy:

$$\mathsf{P}_{CON} = \mathsf{P}_{++}, \qquad \mathsf{P}_{DIS} = \mathsf{P}_{+-} \qquad \text{(two-photon absorption)} \quad (7.202)$$

$$\mathsf{P}_{CON} = \mathsf{P}_{+-}, \qquad \mathsf{P}_{DIS} = \mathsf{P}_{++} \qquad \text{(Raman)} \quad (7.203)$$

where the elements of P_{++} and P_{+-} are as defined in (7.12) and (7.13) [(7.12a), (7.13a)].

The reason for the difference is interesting: In two-photon absorption, both photons are annihilated and the expression which has to be averaged over all molecular orientations is $|\varepsilon_U^* T \varepsilon_V^*|^2$. In Raman scattering, one photon is annihilated and the other created. The creation and annihilation operators are complex conjugates of each other, as are the electric vectors of right-handed and left-handed circularly polarized light. Thus, the quantity to be averaged in the case of Raman scattering is $|\varepsilon_U^* \alpha' \varepsilon_V|^2$ (cf. Table 7.1). The need for the difference can be seen qualitatively on the example of totally symmetric molecular tensors T and α', with all diagonal elements equal. Transitions with such tensors have no capacity to affect the angular momentum of the molecule. If two circularly polarized photons propagating along Z have the same sense of rotation and thus the same component of angular momentum along the Z direction, they cannot be absorbed, since the total angular momentum must be conserved, so that $I_{CON} = 0$. The problem does not occur if the two photons have opposed angular momenta which then cancel, and absorption is possible, $I_{DIS} \neq 0$. The opposite is true in allowed Raman scattering: The created circularly polarized photon carries the angular momentum of the destroyed circularly polarized photon, so that $I_{CON} \neq 0$, $I_{DIS} = 0$.

In the following, we shall specialize our discussion of circularly polarized intensities to the cases of two-photon absorption with the two photons propagating parallel to each other along Z ($I_{SYN} = I_{CON}$, $I_{ANTI} = I_{DIS}$) and of Raman scattering in the backward direction, with the scattered photon propagating antiparallel to the incident photon, i.e., in the $-Z$ direction ($I_{SYN} = I_{DIS}$, $I_{ANTI} = I_{CON}$). Results for other arrangements are easily derived from those given below. The advantage of this specialization is that now the same set of equations will describe the results of two-photon absorption spectroscopy and those of Raman spectroscopy.

Low Symmetry. The observables in the polarized two-photon absorption or Raman scattering experiments are the seven intensities $I_{UV}(\tilde{\nu}_1,\tilde{\nu}_2)$ as a function of $\tilde{\nu}_1$ and $\tilde{\nu}_2$; the unknowns are the five orientation factors K, nine independent orientation factors L, the nine $\tilde{\nu}_1$-dependent elements of the two-photon absorption tensor $\mathsf{T}(\tilde{\nu}_1,f)$ or the Raman scattering tensor $\alpha'(\tilde{\nu}_1,f)$ for each transition f which contributes, and the associated line shapes $g'_f(\tilde{\nu}_2)$. The most easily measured case of two-photon absorption, $\tilde{\nu}_1 = \tilde{\nu}_2$, and the case of nonresonant Raman are somewhat simpler in that the tensors $\mathsf{T}(\tilde{\nu}_1,f)$ and $\alpha'(\tilde{\nu}_1,f)$ then are symmetric and have only six independent elements. However, even if it is assumed that the orientation factors K are known from other measurements, the general problem remains formidable. In the following, we shall consider the simplifications which high molecular symmetry brings about in (i) the orientation distribution and (ii) the tensor $\mathsf{T}(\tilde{\nu}_1,f)$ or $\alpha'(\tilde{\nu}_1,f)$.

High Symmetry: Effect of Simplifications in Orientation Tensors. The arguments of Section 7.2 show that the presence of high symmetry will define the principal axes system (up to a permutation of labels), reduce the number of independent orientation factors in this system of axes to two K's and three L's, and thus reduce the number of nonzero elements of ${}^{(4)}\mathsf{P}_{UV}$ from 81 to 21: those of the types $[{}^{(4)}\mathsf{P}_{UV}]_{sstt}$, $[{}^{(4)}\mathsf{P}_{UV}]_{stst}$, and $[{}^{(4)}\mathsf{P}_{UV}]_{stts}$. The assignment of labels to the symmetry axes so as to satisfy the condition $K_z \geqslant K_y \geqslant K_x$ is usually easily accomplished by consideration of molecular shape or from other spectral measurements (Section 4.6).

The F, G, and H parts of polarized intensity. The quadruple sum over s, t, u, and v encountered in the expressions for an elementary contribution to absorption intensity (7.200) and (7.201) now reduces to three double sums over s and t. For two-photon absorption, these have the form

$$\text{Type F:} \quad \sum_{st} [{}^{(4)}\mathsf{P}_{UV}]_{sstt} T_{ss}(\tilde{\nu}_1,f) T^*_{tt}(\tilde{\nu}_1,f) \tag{7.204}$$

$$\text{Type G:} \quad \sum_{st} [{}^{(4)}\mathsf{P}_{UV}]_{stst} T_{st}(\tilde{\nu}_1,f) T^*_{st}(\tilde{\nu}_1,f) \tag{7.205}$$

$$\text{Type H:} \quad \sum_{st} [^{(4)}\mathsf{P}_{UV}]_{stts} T_{st}(\tilde{\nu}_1,f) T^*_{ts}(\tilde{\nu}_1,f) \tag{7.206}$$

where the designation F, G, and H corresponds to the standard usage in the case of isotropic solutions. For Raman scattering, their form is entirely analogous, with the elements of $\alpha'(\tilde{\nu}_1,f)$ replacing those of $\mathsf{T}(\tilde{\nu}_1,f)$.

When s and t are both allowed to acquire each of the three values x, y, z freely, the diagonal terms s = t will appear once in each sum. In order to make the sum of the three types of contribution, (7.204)–(7.206), equal to the elementary contribution to intensity $I_{UV}(\tilde{\nu}_1,f)$ according to (7.200) and (7.201), we multiply the diagonal terms in each sum by 1/3. It is useful to introduce the shorthand notation

$$[\mathsf{R}_{F,UV}]_{st} = [1 - (2\delta_{st}/3)][^{(4)}\mathsf{P}_{UV}]_{sstt} \tag{7.207}$$

$$[\mathsf{R}_{G,UV}]_{st} = [1 - (2\delta_{st}/3)][^{(4)}\mathsf{P}_{UV}]_{stst} = [1 - (2\delta_{st}/3)][\mathsf{S}_{UV}]_{st} \tag{7.208}$$

$$[\mathsf{R}_{H,UV}]_{st} = [1 - (2\delta_{st}/3)][^{(4)}\mathsf{P}_{UV}]_{stts} \tag{7.209}$$

where the three tensors R_{UV} are analogous to the tensor S_{UV} which we introduced in the discussion of successive two-photon events in high-symmetry molecules [(7.31)–(7.35)].

Using (7.16)–(7.22) for the elements of $[^{(4)}\mathsf{P}_{UV}]_{stuv}$, we can write some general relations for the second-rank tensors R_{UV}:

$$\mathsf{R}_{F,UU} = \mathsf{R}_{G,UU} = \mathsf{R}_{H,UU} = \mathsf{R}_{UU} \tag{7.210}$$

$$\mathsf{R}_{F,UV} = \mathsf{R}_{F,VU} = \mathsf{R}_{H,UV} = \mathsf{R}_{H,VU} \qquad \text{if } V \neq U \tag{7.211}$$

$$\mathsf{R}_{F,SYN} = \mathsf{R}_{H,ANTI} = 2\mathsf{R}_{F,XY} = 2\mathsf{R}_{H,XY} \tag{7.212}$$

$$\mathsf{R}_{F,ANTI} = \mathsf{R}_{G,ANTI} = \mathsf{R}_{G,SYN} = \mathsf{R}_{H,SYN} \tag{7.213}$$

and also write down explicitly their dependence on the orientation factors K and L:

$$[\mathsf{R}_{ZZ}]_{st} = [\mathsf{R}_{ZZ}]_{ts} = (1/3)(3 - 2\delta_{st})L_{st} \tag{7.214}$$

$$[\mathsf{R}_{XX}]_{st} = [\mathsf{R}_{XX}]_{ts} = (1/8)[1 - K_s - K_t + (3 - 2\delta_{st})L_{st}] \tag{7.215}$$

$$[\mathsf{R}_{F,XZ}]_{st} = [\mathsf{R}_{F,XZ}]_{ts} = (1/6)[\delta_{st}K_s - (3 - 2\delta_{st})L_{st}] \tag{7.216}$$

$$[\mathsf{R}_{G,XZ}]_{st} = (1/6)(3 - 2\delta_{st})(K_t - L_{st}) \tag{7.217}$$

$$[\mathsf{R}_{G,ZX}]_{st} = (1/6)(3 - 2\delta_{st})(K_s - L_{st}) \tag{7.218}$$

$$\begin{aligned}[\mathsf{R}_{F,XY}]_{st} &= [\mathsf{R}_{F,XY}]_{ts} \\ &= (1/24)[(4\delta_{st} - 3)(1 - K_s - K_t) + (3 - 2\delta_{st})L_{st}]\end{aligned} \tag{7.219}$$

$$[\mathsf{R}_{G,XY}]_{st} = [\mathsf{R}_{G,XY}]_{ts}$$
$$= (1/24)[(9 - 8\delta_{st})(1 - K_s - K_t) + (3 - 2\delta_{st})L_{st}] \qquad (7.220)$$

$$[\mathsf{R}_{F,ANTI}]_{st} = [\mathsf{R}_{F,ANTI}]_{ts} = (1/12)(3 - 2\delta_{st})[1 - K_s - K_t + L_{st}] \qquad (7.221)$$

Now the expression for polarized two-photon absorption intensity (7.200) becomes

$$I_{UV}(\tilde{\nu}_1,f) = \sum_{st} \{[\mathsf{R}_{F,UV}]_{st} T_{ss}(\tilde{\nu}_1,f)T^*_{tt}(\tilde{\nu}_1,f) + [\mathsf{R}_{G,UV}]_{st} T_{st}(\tilde{\nu}_1,f)T^*_{st}(\tilde{\nu}_1,f)$$
$$+ [\mathsf{R}_{H,UV}]_{st} T_{st}(\tilde{\nu}_1,f)T^*_{ts}(\tilde{\nu}_1,f)\} \qquad (7.222)$$

and the expression for polarized Raman scattering intensity (7.201) becomes

$$I_{UV}(\tilde{\nu}_1,f) = \sum_{st} \{[\mathsf{R}_{F,UV}]_{st} \alpha'_{ss}(\tilde{\nu}_1,f)\alpha'^*_{tt}(\tilde{\nu}_1,f) + [\mathsf{R}_{G,UV}]_{st} \alpha'_{st}(\tilde{\nu}_1,f)\alpha'^*_{st}(\tilde{\nu}_1,f)$$
$$+ [\mathsf{R}_{H,UV}]_{st} \alpha'_{st}(\tilde{\nu}_1,f)\alpha'^*_{ts}(\tilde{\nu}_1,f)\} \qquad (7.223)$$

Expressions (7.222) and (7.223) show that the contribution of the f-th transition to the experimental intensity is composed of three parts, whose relative weights depend on the choice of U and V, in a way dictated by the form of the tensors R_{UV}. Thus, a measurement for a sufficient number of choices of U and V may permit the determination of the magnitude of each part separately.

Relation of the F, G, and H parts of intensity to molecular tensor elements. Each of the parts samples the matrix elements of the molecular transition tensor $\mathsf{T}(\tilde{\nu},f)$ or $\alpha'(\tilde{\nu},f)$ in a different way, for two reasons.

(i) The first reason is the different arrangement of the subscripts on the tensor elements in the three cases. The F part is summed over the terms with index pairs (ss),(tt) and is thus sensitive only to the magnitudes of the diagonal elements of the tensor. The G part provides a measure of the squares of the absolute magnitudes of all the tensor elements by summing over the pairs (st),(st) and thus provides the best indication of the total strength of the transition. The difference between the G part and the H part, which sums over the pairs (st),(ts), provides information about the degree of asymmetry of the tensor. If the tensor is symmetric, there will be no difference between the G part and the H part; and if the tensor is antisymmetric, the difference of the two will tend to a maximum.

This first difference between the three parts is intrinsic to simultaneous two-photon events and reflects the effects of the relative polarization of the two photons. As such, it persists even in an isotropic solution (for instance, for an antisymmetric tensor, the G and H parts are then equal in magnitude and opposite in sign). By the same token, however, it is incapable of discriminating between the molecular x, y, and z directions. Figure 7.13 provides a schematic overview of the weights with which the various products of tensor elements $T_{st}T^*_{uv}$ or $\alpha'_{st}\alpha'^*_{uv}$ enter the total intensity through the F, G, and H parts in the case of an isotropic sample (isotropic K and L factors). On such a sample, $I_{\|}$, $I_{\perp}$, I_{SYN}, and I_{ANTI} represent the only possible distinct measurements [cf. (7.95)].

(ii) In partially aligned samples there is also a second reason why the elements of the transition tensor $\mathsf{T}(\tilde{\nu},f)$ or $\alpha'(\tilde{\nu},f)$ are sampled in a different way in

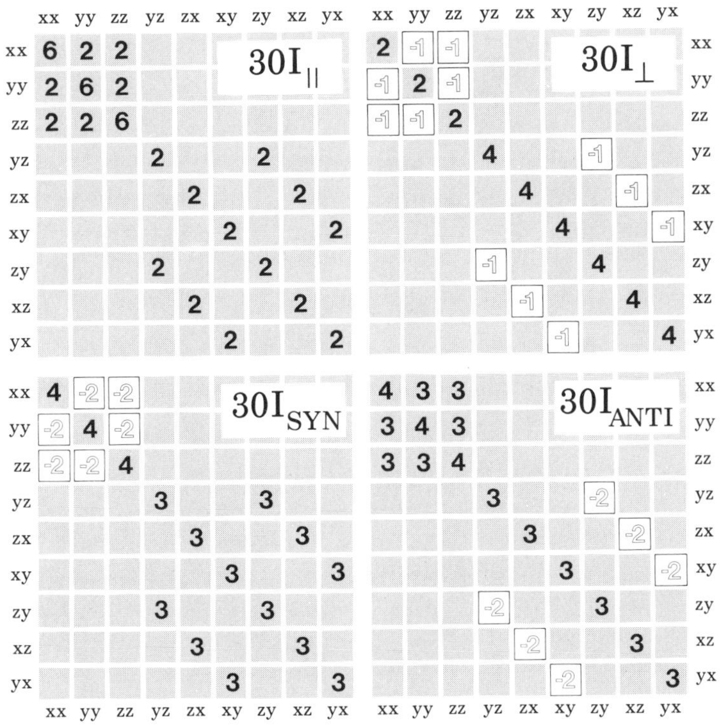

Figure 7.13 Two-photon absorption or Raman scattering by an isotropic solution: The numbers shown give 30 times the weights with which tensor element products $T_{st}T^*_{uv}$ or $\alpha'_{st}\alpha'^*_{uv}$ enter the expression for the total polarized intensity $I_{UV}(\tilde{\nu}_1,f)$, (7.222) or (7.223). The black contributions are positive, the white ones negative. See (7.210)–(7.221). Terms of type $T_{uu}T^*_{uu}$ contribute equally to F, G, and H parts. Terms of type $T_{uu}T^*_{vv}$ provide the F part, those of type $T_{uv}T^*_{uv}$ the G part, and those of type $T_{uv}T^*_{vu}$ the H part of the total intensity.

the three separable parts of the expression for the total intensity. This resides in the elements $[R_{UV}]_{st}$, which are in general different for the three parts. Through their dependence on the orientation factors K and L [(7.210)–(7.221)], these discriminate between the three molecular axes and thus, say, between the three diagonal elements of $T(\tilde{\nu}_1,f)$ or $\alpha'(\tilde{\nu}_1,f)$. Figure 7.14 provides a schematic overview of the weights with which the various products of tensor elements $T_{st}T^*_{uv}$ or $\alpha'_{st}\alpha'^*_{uv}$ enter the total intensity I_{UV} through the F, G, and H parts in the case of a partially aligned sample. The particular example chosen is a molecular assembly oriented by photoselection with Z-polarized light, assuming pure z-polarized absorption and negligible depletion of the originally isotropic ground state [K and L factors given by (4.75) and (4.79)]. Such an orientation distribution does not discriminate between the molecular x and y directions. Comparison with Figure 7.13 is particularly informative when the weights shown are normalized to I_{UV} through division by 70 in Figure 7.14 and division by 30 in Figure 7.13. It is then seen that the sum of the weights within each part,

$70I_{ZZ}$

	xx	yy	zz	yz	zx	xy	zy	xz	yx
xx	6	2	6						
yy	2	6	6						
zz	6	6	30						
yz				6			6		
zx					6			6	
xy						2			2
zy				6			6		
xz					6			6	
yx						2			2

$70I_{XX}$

	xx	yy	zz	yz	zx	xy	zy	xz	yx
xx	18	6	4						
yy	6	18	4						
zz	4	4	6						
yz				4			4		
zx					4			4	
xy						6			6
zy				4			4		
xz					4			4	
yx						6			6

$70I_{ZX}$

	xx	yy	zz	yz	zx	xy	zy	xz	yx
xx	4	-1	-3						
yy	-1	4	-3						
zz	-3	-3	6						
yz				4			-3		
zx					18			-3	
xy						6			-1
zy				-3			18		
xz					-3			4	
yx						-1			6

$70I_{XZ}$

	xx	yy	zz	yz	zx	xy	zy	xz	yx
xx	4	-1	-3						
yy	-1	4	-3						
zz	-3	-3	6						
yz				18			-3		
zx					4			-3	
xy						6			-1
zy				-3			4		
xz					-3			18	
yx						-1			6

$70I_{XY}$

	xx	yy	zz	yz	zx	xy	zy	xz	yx
xx	6	-5	-1						
yy	-5	6	-1						
zz	-1	-1	2						
yz				6			-1		
zx					6			-1	
xy						16			-5
zy				-1			6		
xz					-1			6	
yx						-5			16

$70I_{SYN}$

	xx	yy	zz	yz	zx	xy	zy	xz	yx
xx	12	-10	-2						
yy	-10	12	-2						
zz	-2	-2	4						
yz				5			5		
zx					5			5	
xy						11			11
zy				5			5		
xz					5			5	
yx						11			11

$70I_{ANTI}$

	xx	yy	zz	yz	zx	xy	zy	xz	yx
xx	12	11	5						
yy	11	12	5						
zz	5	5	4						
yz				5			-2		
zx					5			-2	
xy						11			-10
zy				-2			5		
xz					-2			5	
yx						-10			11

Figure 7.14 Two-photon absorption or Raman scattering by a sample photoselected from a random assembly with Z-polarized light absorbed by a purely z-polarized transition (negligible ground-state depletion): The numbers shown give 70 times the weights with which tensor element products $T_{st}T^*_{uv}$ or $\alpha'_{st}\alpha'^*_{uv}$ enter the expression for the total polarized intensity $I_{UV}(\tilde{\nu}_1,f)$, (7.222) or (7.223). The black contributions are positive, the white ones negative. See caption to Figure 7.13 for division into F, G, and H parts.

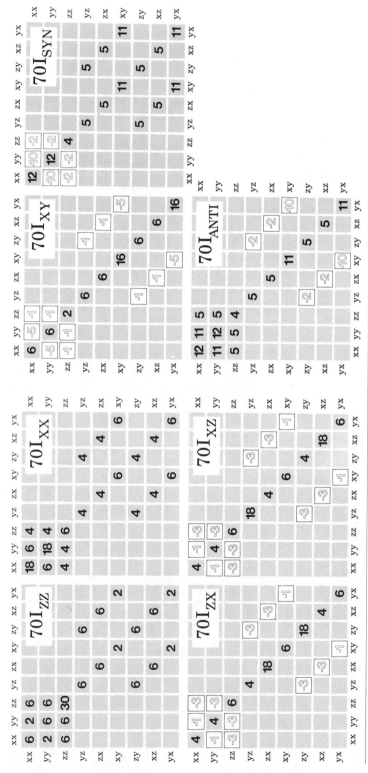

F, G, and H, is the same in the isotropic case of Figure 7.13 and in the partially aligned photoselected case of Figure 7.14. In the former, the subscripts x, y, and z play an equivalent role; in the latter, x and y are still equivalent (rodlike distribution), but the weights are shifted toward or away from the tensor elements with one or both subscripts equal to z, depending on U and V.

Up to now, we have considered only the simplifications which result from the symmetry of the molecular orientation distribution, and the results actually apply to all molecules in their principal orientation axes system as long as the orientation factors such as L_{xxxz}, in which two of the axes occur an odd number of times, vanish, as happens automatically in uniaxial assemblies of molecules of high symmetry and in isotropic solutions of molecules of any symmetry.

High Symmetry: Effect of Simplifications in Molecular Tensors α' and T. We shall now proceed to take advantage of the simplified form of the molecular transition tensor in high-symmetry molecules. For a given symmetry of the ground state and of the final state f, only certain tensor elements can be different from zero, and among these some may of necessity be equal. A list of possible forms which the tensors can take in the most commonly occurring molecular symmetry groups is given in Appendix I. For those irreducible representations not shown, all tensors elements are zero and the transitions are forbidden. This happens in particular for all u $\leftrightarrow$ g transitions in centrosymmetric molecules (the LaPorte rule of one-photon spectroscopy, which allows only u $\leftrightarrow$ g transitions, is thus reversed in two-photon absorption spectroscopy and in Raman scattering spectroscopy).

Isotropic, Symmetric, and Antisymmetric Parts of the Molecular Tensors. A customary way of organizing the nine matrix elements of the transition tensor is to write it as a sum of a spherically symmetric isotropic part $T^{(o)}$, or $\alpha'^{(o)}$, a symmetric traceless ("electric quadrupole-like") part $T^{(s)}$ or $\alpha'^{(s)}$, and an antisymmetric ("magnetic dipole-like") part $T^{(as)}$ or $\alpha'^{(as)}$, which have the following forms:

$$T^{(o)} \text{ or } \alpha'^{(o)}: \quad \begin{pmatrix} a & 0 & 0 \\ 0 & a & 0 \\ 0 & 0 & a \end{pmatrix} \tag{7.224}$$

$$T^{(s)} \text{ or } \alpha'^{(s)}: \quad \begin{pmatrix} b & c & d \\ c & e & f \\ d & f & -b-e \end{pmatrix} \tag{7.225}$$

$$T^{(as)} \text{ or } \alpha'^{(as)}: \quad \begin{pmatrix} 0 & g & h \\ -g & 0 & i \\ -h & -i & 0 \end{pmatrix} \tag{7.226}$$

$$T = T^{(0)} + T^{(s)} + T^{(as)} \tag{7.227}$$

$$\alpha' = \alpha'^{(o)} + \alpha'^{(s)} + \alpha'^{(as)} \tag{7.228}$$

In the high-symmetry molecules of interest to us presently, the transition tensor always has either (i) only two symmetrically disposed but possibly unequal off-diagonal elements with indices uv and vu, u $\neq$ v, in which case it can be written as $T^{(s)} + T^{(as)}$ or $\alpha'^{(s)} + \alpha'^{(as)}$, with b = e = 0, and either c,g $\neq 0$ or d,h $\neq 0$ or f,i $\neq 0$, with the other elements vanishing, or (ii) up to three possibly different diagonal elements and no off-diagonal elements, in which case it can be written as $T^{(o)} + T^{(s)}$ or $\alpha'^{(o)} + \alpha'^{(s)}$, with c = d = f = g = h = i = 0.

In the absence of spectral overlap in a high-symmetry molecule, the elementary contributions $T_{UV}(\tilde{\nu}_1,f)$ can be observed directly, and expressions (7.210)–(7.223) can then be used to derive the values of orientation factors and transition tensor elements. We shall consider the above cases (i) and (ii).

Molecular Tensor with Two Off-Diagonal Elements. In case (i), the uv and vu elements will be be given by c + g and c − g, respectively [cf. (7.225), (7.226)]. Then the F part does not contribute, and we find

$$I_{ZZ}(\tilde{\nu}_1,f) = 4L_{uv}c^2 \tag{7.229}$$

$$I_{XX}(\tilde{\nu}_1,f) = (1/2)(K_{|u\times v|} + 3L_{uv})c^2 \tag{7.230}$$

$$I_{XZ}(\tilde{\nu}_1,f) = (1/2)[K_v(c + g)^2 + K_u(c - g)^2 - 4L_{uv}c^2] \tag{7.231}$$

$$I_{ZX}(\tilde{\nu}_1,f) = (1/2)[K_u(c + g)^2 + K_v(c - g)^2 - 4L_{uv}c^2] \tag{7.232}$$

$$I_{XY}(\tilde{\nu}_1,f) = (1/2)[K_{|u\times v|}(c^2 + 2g^2) + L_{uv}c^2] \tag{7.233}$$

$$I_{SYN}(\tilde{\nu}_1,f) = (K_{|u\times v|} + L_{uv})c^2 \tag{7.234}$$

$$I_{ANTI}(\tilde{\nu}_1,f) = K_{|u\times v|}g^2 + L_{uv}c^2 \tag{7.235}$$

These equations overdetermine the two independent possibly unknown orientations factor K_u and the factor L_{uv}, as well as the two matrix elements c and g (up to an arbitrary phase factor). The two matrix elements will typically already be known from a measurement on an isotropic sample, and the K factors may already be known from one-photon measurements on the oriented sample. The use of circular polarization is helpful but not mandatory.

In the single-beam measurement of two-photon absorption with $\tilde{\nu}_1 = \tilde{\nu}_2$, T is symmetric and g = 0. Now, only I_{ZZ}, I_{XX}, and I_{SYN} are available, and only two of them are independent, since

$$2I_{SYN} = 4I_{XX} - I_{ZZ} \tag{7.236}$$

If the orientation factors K are known, L_{UV} can still be determined from the ratio $I_{ZZ}:I_{XX}$ or $I_{ZZ}:I_{SYN}$, and c^2 from an absolute measurement of one of the intensities. In nonresonant Raman measurements, all seven I_{UV}'s can be measured but are strongly interrelated since α' is symmetric and g = 0. Now,

$$I_{XZ} = I_{ZX}, \qquad I_{ZZ} = 4I_{ANTI}, \qquad I_{SYN} = 2I_{XY}, \qquad I_{XX} = I_{XY} + (1/4)I_{ZZ} \tag{7.237}$$

and the three independent intensities still determine $K_{|u\times v|}$, L_{uv}, and c^2.

If the transition tensor is antisymmetric, c = 0, and $I_{ZZ} = I_{XX} = I_{SYN} = 0$. Further,

$$I_{XZ} = I_{ZX} \quad \text{and} \quad I_{XY} = I_{ANTI} \tag{7.238}$$

and the intensities are independent of the orientation factors L. From their ratio, $K_{|u \times v|}$ can be determined, while g^2 can be obtained from a measurement of any of the absolute intensities.

The potential value of measurements of this kind is twofold. They could be used to determine L_{uv} but are perhaps more likely to be used in the assignment of spectroscopic transitions. In such cases, the real unknowns will be u and v. Since the K's will be generally known and the L's can usually at least be estimated, expressions (7.229)–(7.235) will allow us to identify uv with one of the pairs xy, xz, or yz, and thus will identify the symmetry of the excited state f.

Diagonal Molecular Tensor. In case (ii), a transition with a diagonal tensor, the presence of up to three different nonvanishing matrix elements makes the result more complicated. It can be written in a compact form when the three elements are collected into a vector:

$$\mathbf{q}(\tilde{\nu}_1,f) \equiv [T_{xx}(\tilde{\nu}_1,f), T_{yy}(\tilde{\nu}_1,f), T_{zz}(\tilde{\nu}_1,f)] \tag{7.239}$$

or

$$\mathbf{q}(\tilde{\nu}_1,f) \equiv [\alpha'_{xx}(\tilde{\nu}_1,f), \alpha'_{yy}(\tilde{\nu}_1,f), \alpha'_{zz}(\tilde{\nu}_1,f)] \tag{7.240}$$

since then the polarized intensity can be written as

$$I_{UV}(\tilde{\nu}_1,f) = \mathbf{q}(\tilde{\nu}_1,f)R^0_{UV}\mathbf{q}^+(\tilde{\nu}_1,f) \tag{7.241}$$

where $\mathbf{q}^+$ is the transpose of $\mathbf{q}$ and the elements of R^0_{UV} are

$$[R^0_{ZZ}]_{st} = L_{st} \tag{7.242}$$

$$[R^0_{XX}]_{st} = (1/8)[(2\delta_{st} + 1)K_{|s \times t|} + 3L_{st}] \tag{7.243}$$

$$[R^0_{XZ}]_{st} = [R^0_{ZX}]_{st} = (1/2)(\delta_{st}K_s - L_{st}) \tag{7.244}$$

$$[R^0_{XY}]_{st} = (1/8)[(2\delta_{st} - 1)K_{|s \times t|} + L_{st}] \tag{7.245}$$

$$[R^0_{SYN}]_{st} = [R^0_{ANTI}]_{st} = (1/4)(K_{|s \times t|} + L_{st}) \tag{7.246}$$

The solution of the equations implied in (7.239)–(7.246) for the unknown vectors $\mathbf{q}(\tilde{\nu}_1,f)$ and the orientation factors may be difficult but is possible in principle.

If several transitions overlap in a molecule of high symmetry, the overall intensity can be expressed through (7.199) as a sum of several elementary contributions, each of which will show either the behavior described by (7.229)–(7.238) or the behavior described by (7.239)—(7.246) as a function of U and V.

The Isotropic Limit. In practice, a measurement on an isotropic solution will virtually always be possible and will provide a good starting point for assigning transition symmetries by comparison with the results on the oriented sample. While a measurement on an isotropic sample cannot distinguish the x, y, and z axes of the molecule, it does permit the separation of the F, G, and H parts of the intensity and thus permits us to decide whether the transition tensor has off-diagonal or diagonal elements (as in the two examples just outlined). This may frequently be sufficient for an assignment.

As already noted in connection with Figure 7.13, in the isotropic limit, four distinct measurements remain of those discussed here: $I_{||} = I_{ZZ} = I_{XX}$, $I_{\perp} = I_{ZX} = I_{XZ} = I_{XY}$, I_{SYN}, and I_{ANTI}. Now, $K^{iso}_{uv} = \delta_{uv}/3$, $L^{iso}_{stuv} = \delta_{|s \times t|,|u \times v|}[(1 + 2\delta_{st})/15]$, so that (7.222) and (7.223) hold for molecules of any symmetry and simplify considerably. For historical reasons, the resulting expressions for the case of isotropic solutions are usually written in a different form for the two kinds of spectroscopy, and we shall discuss them separately in Sections 7.5.2 and 7.5.3.

7.5.2 Isotropic Samples: Two-Photon Absorption

Inspection of (7.214) and (7.215) shows that in the isotropic limit the elements $[\mathsf{R}_{UU}]_{st}$ in the expression for the intensity (7.222) all become equal to 1/15 and can be taken outside of the summation. It then becomes useful to define the three rotational invariants of the tensor $\mathsf{T}(\tilde{v}_1,f)$, namely, δ_F, δ_G, and δ_H, as follows (these should not be confused with the Kronecker δ symbol such as δ_{uv}):

$$\delta_F = \sum_{st} T_{ss}(\tilde{v}_1,f)T^*_{tt}(\tilde{v}_1,f) \qquad (7.247)$$

$$\delta_G = \sum_{st} T_{st}(\tilde{v}_1,f)T^*_{st}(\tilde{v}_1,f) \qquad (7.248)$$

$$\delta_H = \sum_{st} T_{st}(\tilde{v}_1,f)T^*_{st}(\tilde{v}_1,f) \qquad (7.249)$$

Also, the other three polarized intensities can be expressed in terms of these invariants, as substitution of K^{iso} and L^{iso} into (7.211)–(7.213) and (7.216)–(7.222) will show:

$$I_{||}(\tilde{v}_1,f) = (1/30)(2\delta_F + 2\delta_G + 2\delta_H) \qquad (7.250)$$

$$I_{\perp}(\tilde{v}_1,f) = (1/30)(-\delta_F + 4\delta_G - \delta_H) \qquad (7.251)$$

$$I_{SYN}(\tilde{v}_1,f) = (1/30)(-2\delta_F + 3\delta_G + 3\delta_H) \qquad (7.252)$$

$$I_{ANTI}(\tilde{v}_1,f) = (1/30)(3\delta_F + 3\delta_G - 2\delta_H) \qquad (7.253)$$

The results show clearly that only three of the four observed polarized intensities are linearly independent.

A determination of all three invariants of the tensor, δ_F, δ_G, and δ_H, is possible when not only linearly but also circularly polarized light is used ("complete polarization study"). The required three measurements could in principle also be made with other polarizations (elliptical), but this would offer no particular advantages. When both photons are taken from a single beam, only two independent measurements remain, and it is not possible to determine all three invariants. Even this may suffice for symmetry assignments in favorable cases.

The three tensor types defined in (7.224)–(7.227), $T^{(o)}$, $T^{(s)}$, and $T^{(as)}$, have characteristic signatures in the δ values, and therefore also in the two-photon absorption intensities. For polarized intensity ratios, the factor 1/30 cancels in (7.250)–(7.253), and we obtain

$$T^{(o)}: \quad \delta_F = 3\delta_G, \qquad \delta_H = \delta_G, \qquad I_{||}:I_{\perp}:I_{SYN}:I_{ANTI} = 10:0:0:10 \tag{7.254}$$

$$T^{(s)}: \quad \delta_F = 0, \qquad \delta_H = \delta_G, \qquad I_{||}:I_{\perp}:I_{SYN}:I_{ANTI} = 4:3:6:1 \tag{7.255}$$

$$T^{(as)}: \quad \delta_F = 0 \qquad \delta_H = -\delta_G, \qquad I_{||}:I_{\perp}:I_{SYN}:I_{ANTI} = 0:5:0:5 \tag{7.256}$$

Frequently, only the ratio $\Omega = I_{SYN}/I_{||}$ is measured. It varies from a minimum of 0 found for isotropic transition tensors $T^{(o)}$ to a maximum of 3/2 found for the traceless tensor $T^{(s)}$. It contains no information on the possible presence of antisymmetric contributions to the tensor. This is immaterial in measurements with a single beam for which $T(\tilde{\nu}_1,\tilde{\nu}_2)$ is symmetric.

If one diagonal element is far larger than any other element of the transition tensor $T(\tilde{\nu}_1,f)$, $\delta_F = \delta_G = \delta_H$ and Ω reaches the value 2/3. This value has been observed fairly frequently. In order to determine which one of the three diagonal elements it is, measurements on an oriented sample are needed. This is true in general, whenever it is necessary to distinguish between the individual elements of a tensor of a given type. Nevertheless, measurements on isotropic samples will frequently permit at least a decision as to the class to which a transition tensor T belongs. At times, this may permit an unequivocal transition symmetry assignment. For example, all $\pi\pi^*$ transitions in D_{2h} or C_{2v} aromatics with yz as the molecular plane, T must transform like $x^2 + y^2 + z^2$ (A_1) or yz (B_{3g} in D_{2h} group, B_2 in C_{2v}), and one has $T_{xz}(\tilde{\nu}_1,\tilde{\nu}_1) = T_{yz}(\tilde{\nu}_1,\tilde{\nu}_1) = 0$. Then the lowest possible value of Ω is 1/4 and is reached when $T_{zz}(\tilde{\nu}_1,\tilde{\nu}_1) = T_{yy}(\tilde{\nu}_1,\tilde{\nu}_1)$ and $T_{yz}(\tilde{\nu}_1,\tilde{\nu}_1) = T_{xx}(\tilde{\nu}_1,\tilde{\nu}_1) = 0$ (excited state of A_1 symmetry), while the maximum value 3/2 is reached for an excited state of B_{3g} symmetry (in C_{2v}, of B_2 symmetry). The opportunity to obtain symmetry assignments for excited states from measurements on isotropic samples alone represents an advantage for two-photon absorption spectroscopy over one-photon spectroscopy on such samples.

7.5.3 Isotropic Samples: Raman Scattering

Raman spectroscopists prefer to define the three rotational invariants in a way which parallels the decomposition of the general tensor α' in (7.228). They are the isotropic part of the Raman tensor $\bar{\alpha}^2$,

$$\bar{\alpha}^2 = (1/9) \sum_{st} \alpha'_{ss}\alpha'^*_{tt} \tag{7.257}$$

the symmetric anisotropy γ_s^2,

$$\gamma_s^2 = (1/2) \sum_{st} |\alpha'_{ss} - \alpha'_{tt}|^2 + (3/4) \sum_{\substack{sr \\ (s \neq t)}} |\alpha'_{st} + \alpha'_{ts}|^2 \tag{7.258}$$

and the antisymmetric anisotropy γ_{as}^2,

$$\gamma_{as}^2 = (3/4) \sum_{st} |\alpha'_{st} - \alpha'_{ts}|^2 \qquad (7.259)$$

and they are related to the invariants δ_F, δ_G, and δ_H through

$$\bar{\alpha}^2 = (1/9)\delta_F \qquad (7.260)$$

$$\gamma_s^2 = (3/4)(\delta_G + \delta_H) - (1/2)\delta_F \qquad (7.261)$$

$$\gamma_{as}^2 = (3/4)(\delta_G - \delta_H) \qquad (7.262)$$

Their relation to the three tensor types defined by (7.224)–(7.226) is apparent from the values they acquire when the Raman scattering tensor $\alpha'(\tilde{\nu}_1,f)$ belongs to one of these three basic types:

$$\alpha'^{(o)}:\quad \bar{\alpha}^2 = (1/3)\delta_G, \qquad \gamma_s^2 = 0, \qquad \gamma_{as}^2 = 0 \qquad (7.263)$$

$$\alpha'^{(s)}:\quad \bar{\alpha}^2 = 0, \qquad \gamma_s^2 = (3/2)\delta_G, \qquad \gamma_{as}^2 = 0 \qquad (7.264)$$

$$\alpha'^{(as)}:\quad \bar{\alpha}^2 = 0, \qquad \gamma_s^2 = 0, \qquad \gamma_{as}^2 = (3/2)\delta_G \qquad (7.265)$$

In terms of these invariants, the polarized intensities observed on an isotropic sample become

$$I_{||}(\tilde{\nu}_1,f) = \bar{\alpha}^2 + (4/45)\gamma_s^2 \qquad (7.266)$$

$$I_{\perp}(\tilde{\nu}_1,f) = (1/15)\gamma_s^2 + (1/9)\gamma_{as}^2 \qquad (7.267)$$

$$I_{SYN}(\tilde{\nu}_1,f) = (2/15)\gamma_s^2 \qquad (7.268)$$

$$I_{ANTI}(\tilde{\nu}_1,f) = \bar{\alpha}^2 + (1/45)\gamma_s^2 + (1/9)\gamma_{as}^2 \qquad (7.269)$$

The characteristic pattern of intensity ratios $I_{||}:I_{\perp}:I_{SYN}:I_{ANTI}$ for the three tensor types listed in (7.254)–(7.256) is immediately apparent from (7.266)–(7.269). Here again, the four measurable intensities overdetermine the three invariants. Only when it is known in advance that $\alpha'(\tilde{\nu}_1,f)$ is symmetric and $\gamma_{as}^2 = 0$ (as in nonresonant Raman scattering) is it possible to give up the use of circularly polarized light. Then $\bar{\alpha}^2$ and γ_s^2 can be determined from $I_{||}$ and $I_{\perp}$ alone. This has been commonly used for discriminating between totally symmetric and other vibrations. For the limiting case $\alpha' = \alpha'^{(o)}$, the depolarization ratio is $I_{\perp}/I_{||} = 0$; for the limiting case $\alpha' = \alpha'^{(s)}$, it is $I_{\perp}/I_{||} = 3/4$. Thus, if α' is symmetric and has only off-diagonal elements, the depolarization ratio must be equal to 3/4. If it has diagonal elements, the depolarization ratio lies between 0 and 3/4. If $\gamma_{as}^2 \neq 0$, $I_{\perp}/I_{||}$ can exceed the value 3/4, and such Raman lines are said to have anomalous polarization.

Once again, the measurements on isotropic samples alone provide useful information on symmetry assignments which goes well beyond that offered by one-photon measurements on such samples, but Raman measurements on aligned samples are required if the molecular x, y, and z directions are to be distinguished.

7.6 Comments and References

The general expressions for linearly polarized intensities in two-photon processes used in this chapter were obtained by J. Michl and E. W. Thulstrup, *J. Chem. Phys.* **72**, 3999 (1980). There is a misprint in this paper in the equation analogous to the present (7.20); also, the parentheses on the left-hand side of the equations for the matrix elements ${}^{(4)}P^{UV}_{stuv}$ should be deleted in the appendix to the paper [analogous to (7.7a)–(7.11a)].

The principles of photoselection go back to F. Weigert, *Ann. Phys.* **63**, 681 (1920); *Verhandl. Deut. Phys. Ges.* **1**, 100 (1920); S. I. Vavilov and V. L. Levshin, *Z. Physik* **16**, 135 (1923); F. Perrin, *J. Phys. Radium* **7**, 390 (1926), *Ann. Phys.* **12**, 169 (1929); A. Jabłoński, *Z. Physik* **96**, 236 (1935); and G. N. Lewis and D. Lipkin, *J. Am. Chem. Soc.* **64**, 2801 (1942). More recent leading references are P. P. Feofilov, *The Physical Basis of Polarized Emission*, Consultants Bureau, New York, 1961; A. C. Albrecht, *J. Mol. Spectrosc.* **6**, 84 (1961); *Progr. React. Kinet.* **5**, 301 (1970); A. H. Kalantar and A. C. Albrecht, *Ber. Bunsenges. Physik. Chem.* **68**, 361 (1964); J. Czekalla, W. Liptay, and E. Döllefeld, *Ber. Bunsenges. Physik. Chem.* **68**, 80 (1964); F. Dörr, in *Creation and Detection of the Excited State*, Vol. 1A, A. A. Lamola, Ed., Marcel Dekker, New York, 1971, Chap. 2 (pages 108 and 109 provide an introduction to the studies of energy transfer by concentration depolarization), and J. R. Lakowicz, *Principles of Fluorescence Spectroscopy*, Plenum, New York, 1983. The formula for a general angle of observation [equation (7.96)] was derived by J. B. Gallivan, J. S. Brinen, and J. G. Koren, *J. Mol. Spectrosc.* **26**, 24 (1968); see also A. H. Kalantar, *J. Chem. Phys.* **48**, 4992 (1968). Formulas for three-step photoselection were first worked out by P. M. Johnson and A. C. Albrecht, *J. Chem. Phys.* **48**, 851 (1968). An alternative view of the origin of the deviations from the theoretically expected behavior in the case of degenerate absorption and degenerate emission (Sections 7.3.4 and 7.3.5) observed upon excitation in the red edge of the absorption band is offered in P. S. Braterman, G. A. Heath, and L. J. Yellowlees, *J. Chem. Soc. Dalton Trans.* **1985**, 1081, who assign no role to the solvent environment, while we believe that it is dominant.

The early work on polarized photoluminescence measurements on partially aligned rigid samples is surveyed, and a concise survey of the theory is given by D. I. Bower, in *Structure and Properties of Oriented Polymers*, I. M. Ward, Ed., Applied Science Publishers, London, 1975, Chapter 5; see also L. Margulies and A. Yogev, *Chem. Phys.* **27**, 89 (1978). More recent reviews of applications in polymer science can be found in S. W. Beaven, J. S. Hargreaves, and D. Phillips, *Adv. Photochem.* **11**, 207 (1979), and I. Soutar, in *Developments in Polymer Photochemistry*, N. S. Allen, Ed., Applied Science Publishers, London, 1982, Vol. 3, p. 125.

For a brief review of the use of fluorescence for studying membrane dynamics, see J. Schlessinger and E. L. Elson, in *Biophysics*, G. Ehrenstein and H. Lecar, Eds., Academic Press, New York, 1982, Chap. 5. A general discussion of rotational dynamics in partially aligned media is given by P. L. Nordio and U. Segre in *The Molecular Physics of Liquid Crystals*, G. R. Luckhurst and G. W. Gray, Eds., Academic Press, New York, 1979. Further references to the theory of the effect of rotational diffusion on fluorescence polarization and transient dichroism can be found in the following:

Isotropic media: F. Perrin, *J. Phys. Radium* **7**, 390 (1926); *Ann. Phys.* **12**, 169 (1929); *J. Physique* **5**, 497 (1934); **7**, 1 (1936); T. J. Chuang and K. B. Eisenthal *J. Chem. Phys.* **57**, 5094 (1972); A. J. Cross, D. H. Waldeck, and G. R. Fleming, *J. Chem. Phys.* **78**, 6455 (1983); J. R. Lakowicz, *Biophys. Chem.* **19**, 13 (1984). For a treatment of depolarization by energy transfer, see J. Kaminski, *Acta Phys. Polon.* **A67**, 679 (1985).

Uniaxial media: A. Szabo, *J. Chem. Phys.* **72**, 4620 (1980); C. Zannoni, *Mol. Phys.* **38**, 1813 (1979); J. J. Fisz, *Chem. Phys.* **99**, 177 (1985).

Macroscopically isotropic media with locally anisotropic rotational diffusion: C. Zannoni, *Mol. Phys.* **42**, 1303 (1981); W. van der Meer, H. Pottel, W. Herreman, M. Ameloot, H. Hendrickx, and H. Schröder, *Biophys. J.* **46**, 515 (1984); A. Szabo, *J. Chem. Phys.* **81**, 150 (1984).

For a discussion of the rotational jump model, see P. E. Zinsli, *Chem. Phys.* **20**, 299 (1977).

For reviews of experimental work, see C. Zannoni, A. Arcioni, and P. Cavatorta, *Chem. Phys. Lipids* **32**, 179 (1983); R. B. Cundall and R. E. Dale, Eds., *Time-Resolved Fluorescence Spectroscopy in Biochemistry and Biology*, Plenum, New York, 1980; see also R. P. H. Kooyman and Y. K. Levine, *J. Mol. Struct.* **115**, 289 (1984) and R. P. H. Kooyman, M. H. Vos, and Y. K. Levine, *Chem. Phys.* **81**, 461 (1983); J. L. Viovy, C. W. Frank, and L. Monnerie, *Macromol.* **18**, 2606 (1985); E. V. Anufrieva, *Pure Appl. Chem.* **54**, 533 (1982).

The orientation factors of a molecule in its ground and excited states in the same anisotropic medium have traditionally been assumed to be equal. This has now been questioned by L. B.-Å. Johansson, *Chem. Phys. Lett.* **118**, 516 (1985), who compared linear dichroism of perylene, 9,10-dimethylanthracene, and 1,6-diphenyl-1,3,5-hexatriene with their fluorescence anisotropy in the long-time limit, using a lyotropic liquid crystal for alignment. We suspect that the discrepancies, found for the latter two molecules, are due to mixed polarization of the emission of 9,10-dimethylanthracene and to the presence of two differently oriented subpopulations of the hexatriene, suggested by the biexponential decay of its fluorescence. We expect large differences between the orientation factors of the ground and the excited state of a molecule only if they have different conformations or in the presence of special interactions, such as hydrogen bonds, which may be strongly state-dependent.

In some studies, no attempt is made to separate the effects of photoselection and of the originally present orientation on the observed fluorescence polarization; rather, the composite orientation is described by methods discussed in Chapter 5: J. J. Dekkers, W. P. Cofino, G. Ph. Hoornweg, C. Maclean, and N. H. Velthorst, *Chem. Phys.* **47**, 369 (1980) and references therein.

The theory of the time-dependent fluorescence polarization of solutes in uniaxial media oriented by saturating laser pulses is described by K. R. Naqvi, *J. Chem. Phys.* **74**, 2658 (1981).

A discussion of polarization effects in two-photon absorption experiments is found in W. M. McClain and R. A. Harris, in *Excited States*, Vol. 3, E. C.

Lim, Ed., Academic Press, New York, 1977, p. 2; for an analogous discussion of Raman polarizations see D. A. Long, *Raman Spectroscopy*, McGraw-Hill, New York, 1977.

Few Raman studies of solutes in uniaxial anisotropic media have been published, and the media themselves have been studied more extensively. For an overview of the early work on stretched polymers, see D. I. Bower, in *Structure and Properties of Oriented Polymers*, I. M. Ward, Ed., Applied Science Publishers, London, 1975, Chapter 5. See also J. Maxfield, R. S. Stein, and M. C. Chen, *J. Polym. Sci., Polym. Phys.* **16**, 37 (1978); S. K. Satija and C. H. Wang, *J. Chem. Phys.* **69**, 2739 (1978). For an example of Raman work on a doped liquid crystal, see S. Jen, N. A. Clark, P. S. Pershan, and E. B. Priestley, *J. Chem. Phys.* **66**, 4635 (1977). This paper presents the general expressions for polarized nonresonant Raman measurements on uniaxial samples in terms of order parameters but the formulas contain several misprints.

8

Examples of Application

This chapter illustrates the use of optical spectroscopy of partially oriented solutes to address problems in chemistry. We have not attempted any exhaustive review of the work which has been done, nor do we give references to all papers which deal with a particular molecule or subject. Rather, we present selected illustrative examples and mostly quote only our immediate sources.

We begin by taking a closer look at pyrene partially aligned in stretched polyethylene. This system is already familiar from Chapter 2, where it was used to demonstrate qualitatively the results of measurements to be discussed in subsequent chapters. We now return to the results and apply the quantitative methods of analysis developed in those chapters to obtain spectroscopic information about the pyrene molecule, as well as some information about its orientation distribution function in stretched polyethylene. We find that pyrene is not only a useful model for illustrating the procedures which can be used, but is also of considerable interest in its own right: The transition moment of the origin of its L_b band is so sensitive to perturbation by the environment that its direction deviates detectably from the symmetry-determined molecular axes.

The following three sections deal with the use of linear dichroism for the assignment of vibrational and electronic transitions in molecules. Most of the molecules are organic, some are of interest in biochemistry, and a few contain transition metal atoms. This imbalance reflects the tastes and preferences of the investigators who have been using the method rather than any intrinsic difference between organic and inorganic molecules. The next section presents examples of measurements of optical activity of oriented molecules. Progress in this experimentally rather difficult area has been slow so far.

Then we move to nonspectroscopic applications. We first present several examples in which problems of molecular structure and conformation are addressed. All of our examples are from organic chemistry, but once again inorganic applications are easy to envisage. This area is likely to witness a boom, particularly as measurements of infrared dichroism become more widespread. We then proceed to a discussion of selected investigations of molecular complexation and aggregation, exemplified on species of interest in organic chemistry and biochemistry.

In the two subsequent sections, focus is shifted from the solute molecule to the orienting medium, and we describe studies in which the solute was used as a probe embedded in a uniaxial orienting medium of interest, or in which it naturally occurs as such a probe. The first of these sections deals with the problem of the static structure of the medium; the second, with molecular motion of the solute within the medium. Work of this kind is of considerable current interest, both in the field of man-made polymers, liquid crystals, and membranes, and in the field of naturally occurring systems, primarily biological membranes.

Finally, we mention briefly two practical applications of the concepts outlined in this book. One of these, dichroic sheet polarizers, has been in use for decades. The other, optical information storage in birefringent sheets, is recent, and its commercial usefulness remains to be proved.

8.1 Section 2.3 Revisited: Pyrene

In Section 2.3 we provided several illustrative examples of optical spectra of partially oriented samples of pyrene (**1**) and discussed their prominent features in a qualitative fashion. We shall now summarize the information which has been obtained from these measurements by the methods outlined in this book.

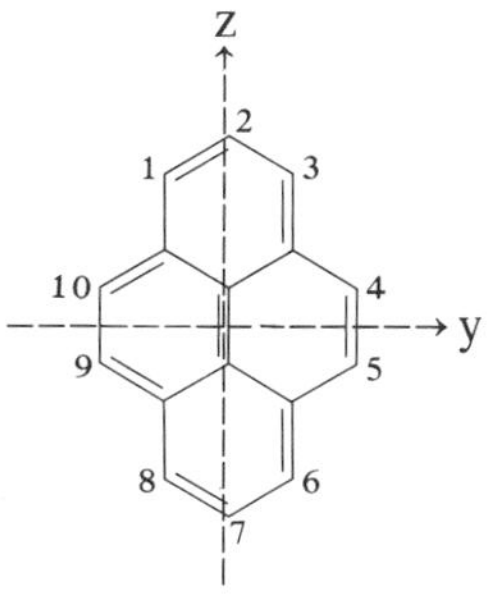

1

8.1.1 Infrared Dichroism.[1] *Orientation Factors K*

The three groups of vibrational transitions in pyrene contained in stretched polyethylene identified in Section 2.3 were characterized by room-temperature dichroic ratios of 2.17 ± 0.17, 1.03 ± 0.05, and 0.27 ± 0.03 in Figures 2.4 and 2.5 and assigned to z-, y-, and x-polarized transitions, respectively. Comparison with earlier crystal data and, independently, the information summarized in Section 4.6.1, allow the assignment of z as the long in-plane axis, y as the short in-plane axis, and x as the out-of-plane axis of the pyrene molecule, as shown in formula **1**. The total number of vibrational transitions assigned in this manner and by resolution of a few overlapping bands using the reduction procedure of

Section 5.1.2 is several dozen. The assignments are compared with previously available data in Table 8.1. Since these results can be obtained for several molecules simultaneously in a day or two, the potential of the method in mass-producing IR assignments is apparent.

The orientation factors derived from the dichroic ratios using equation (5.29) are $K_z = 0.52 \pm 0.02$, $K_y = 0.34 \pm 0.01$, and $K_x = 0.12 \pm 0.01$. Their sum is 0.98,

Table 8.1
Observed Vibrations of Pyrene (**1**) and Their Symmetries[a]

	Wavenumber				
Species	Lit.	CH_2Cl_2	Polyethylene	K	
B_{1u}	496	498	498	0.53	Fundamental[c]
	674	667[d]	677[e]	0.48	Fundamental[c]
	819	821	821	0.50	Fundamental[c]
		997	(989)	0.53	$542(B_{2u}) + 450(B_{3g}) = 992$ or $773(B_{2g}) + 219(B_{3u}) = 992$
		1002	1002	0.55	Fundamental
	1062	1063	1061		$842(B_{3u}) + 221(B_{2g}) = 1063$[f]
		1085	1085	0.55	$594(A_g) + 498(B_{1u}) = 1092$ or $970(B_{2g}) + 124(B_{3u}) = 1094$
	1095	1095	1095	0.55	Fundamental[c]
		(1143)	(1143)	0.48	
	1241	1243	1242	0.52	Fundamental[c]
		1289	1284	0.53	
		1409	1409	0.54	$821(B_{1u}) + 594(A_g) = 1415$
	1446	1449	1448	0.50	Fundamental[c]
	1462	1467	g		Fundamental[c]
		1524	1514	0.54	$773(B_{2g}) + 744(B_{3u}) = 1517$
		1572	(1562)	0.55	$1066(A_g) + 498(B_{1u}) = 1564$
	1584	1587	1578	0.50	f
		1597	1593	0.49	
		1600	1600	0.49	Fundamental
		1673	1663	0.49	
		1703	1690	0.50	$1095(B_{1u}) + 594(A_g) = 1689$ or $1182(B_{2u}) + 505(B_{3g}) = 1687$

[a] Reproduced by permission from ref. 1. Values in parentheses refer to weak bands overlapping with stronger bands and are less accurate. The Raman frequencies were taken from ref. b (B_{2g}) and ref. c (A_g,B_{3g}). The literature IR frequencies were taken from ref. c unless indicated otherwise. All data were obtained at room temperature unless indicated otherwise.

[b] A. Bree, R. A. Kydd, T. N. Misra, and V. V. B. Vilkos, *Spectrochim. Acta Part A* **27**, 2315 (1971).

[c] S. J. Cyvin, B. N. Cyvin, J. Brunvoll, J. C. Whitmer, P. Klaeboe, and J. E. Gustavsen, *Z. Naturforsch. A* **34**, 876 (1979).

[d] $CHCl_3$ solution.

[e] Observed at low temperature only (20 K).

[f] Assigned as fundamental in ref. c.

[g] Could not be observed because of strong polyethylene absorption.

[h] No peak was observed at this position.

[i] This vibration is presently assigned as B_{3u}.

[j] This vibration is presently assigned as B_{1u}.

[k] Lies outside the presently investigated frequency region.

[l] S. Califano and G. Abbondanza, *J. Chem. Phys.* **39**, 1016 (1963).

Table 8.1. (Continued)

Species	Wavenumber Lit.	CH_2Cl_2	Polyethylene	K	
		1722	1709	0.51	$1359(B_{3g}) + 352(B_{2u}) = 1711$ or $1174(B_{3g}) + 542(B_{2u}) = 1716$ or $970(B_{2g}) + 744(B_{3u}) = 1714$
		1769	1756	0.52	$1311(B_{2u}) + 450(B_{3g}) = 1761$
		1804	1789	0.53	$1276(B_{2u}) + 505(B_{3g}) = 1781$
		1873	1861	0.47	
		1927	1915	0.50	$1371(B_{3g}) + 542(B_{2u}) = 1913$
		1941	1928	0.51	$1433(B_{2u}) + 505(B_{3g}) = 1938$
	3039	3043	3040	0.47	Fundamental[c]
		3046	3046	0.48	
	3080[b]	3082[d]	3080	0.50	Fundamental[b]
	3098[b]	3103[d]	3102	0.48	Fundamental[b]
B_{2u}	351	353	352	0.33	Fundamental[c]
	540	542	542	0.33	Fundamental[c]
		812	(800)		
	963				[h]
	891				[i]
		955	955	0.36	Fundamental or $542(B_{2u}) + 408(A_g) = 950$ or $498(B_{1u}) + 450(B_{3g}) = 948$
		(1136)	(1135)	0.33	$594(A_g) + 542(B_{2u}) = 1136$
	1185	1184	1182	0.33	Fundamental[c]
		1192	1191	0.34	Fundamental
	1206				[h]
	1272	1276[d]	1272	0.34	Fundamental[c] or $821(B_{1u}) + 450(B_{3g}) = 1271$ or $775(B_{3g}) + 498(B_{1u}) = 1273$
	1312	1312	1311	0.34	Fundamental[c]
		1419[d]	1418	0.33	$1066(A_g) + 352(B_{2u}) = 1418$
	1433	1434	1433	0.34	Fundamental[c]
	1484	1487	1485	0.33	Fundamental[c]
		1499	1490	0.33	$1142(A_g) + 352(B_{2u}) = 1494$
		1551	1542	0.32	$1095(B_{1u}) + 450(B_{3g}) = 1545$
	1593				[j]
		1610	1611	0.36	Fundamental
		1652	1642	0.33	
		1671	1659	0.35	
		1755	1742	0.32	$1242(B_{1u}) + 505(B_{3g}) = 1746$
		1856	1844	0.33	$1174(B_{3g}) + 677(B_{1u}) = 1851$ or $1107(B_{3g}) + 821(B_{1u}) = 1928$
B_{3u}	124[b]	[k]	[k]		Fundamental[b]
	219[b]	220	219	0.13	Fundamental[b]
	484[b]	489	487	0.12	Fundamental[b]
		706[d]	(703)	0.11	
	708[l]	713[d]	711	0.12	Fundamental[l]
	745[l]	744[d]	744	0.13	Fundamental[l]
	840[l]	847	842	0.13	Fundamental[l]
	963[l]	969	962	0.12	Fundamental[l]

within experimental error of unity. They correspond to average angles of deviation from the stretching direction of 44° for the z axis, 54° for the y axis, and 70° for the x axis, but they provide us with no certainty concerning the spread of these angles from their average values and about the most probable—as opposed to the average—location and orientation of the molecule in the stretched polymer.

At liquid nitrogen temperature, the orientation factors become $K_z = 0.56 \pm 0.02$, $K_y = 0.33 \pm 0.01$, and $K_x = 0.10 \pm 0.01$, their sum equals unity within the experimental error, and the average angles of deviation from the stretching direction are 42°, 55°, and 72° for the z, y, and x axis, respectively.

8.1.2 Ultraviolet Dichroism.[2–4] *A Fly in the Ointment?*

The polarized UV absorption spectra of pyrene in stretched polyethylene at the temperature of liquid nitrogen[3] and their stepwise reduction were shown in Figure 2.6. The reduced spectra were shown in Figure 2.7. More complete results are now shown in Figure 8.1. The low-temperature reduction factors given earlier, $d_z = 2.78$ and $d_y = 1.0$, lead to orientation factors $K_z = 0.58$ and $K_y = 0.33$. By difference to unity, $K_x = 0.09$. Within experimental error limits,

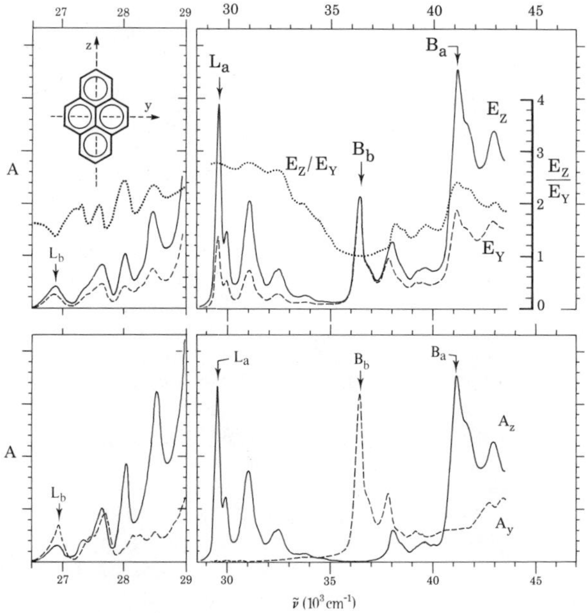

Figure 8.1 Polarized UV absorption of pyrene in stretched polyethylene at 77 K. Bottom: reduced absorption spectra $A_z = E_Z - 1.0E_Y$ and $A_y = 0.625(2.8E_Y - E_Z)$. Top: baseline-corrected polarized absorption spectra and the dichroic ratio E_Z/E_Y. The absorbance scale is in arbitrary units, different for the different sections of the spectrum. Note the change of horizontal scale in the left section. Adapted by permission from ref. 3.

these values are equal to the orientation factors obtained from the IR measurement at liquid nitrogen temperature, although the pyrene concentration in the stretched polymer was quite different in the two experiments. Similar concentration independence of the orientation factors has been observed for well over a dozen other aromatic solutes in stretched polyethylene.

The identity of the K values obtained from IR and UV measurements confirms that the basic assumption of the reduction procedure is fulfilled in this case: The origins of intense electronic transitions are purely polarized.

The reduced spectra permit symmetry assignment for the singlet excited electronic states of pyrene which were discussed in Section 2.3. The assignments are summarized in Table 8.2. They are well known today and require no further comment. Additional states are believed to be present in the low-energy region, but they are too weak in one-photon absorption to be observable.

The L_b band, included in Table 8.2 as no. 1, requires additional comment. It is too weak to be observable in Figures 2.6 and 2.7. Its polarized spectra can, however, be obtained when the concentration of pyrene in the stretched polyethylene sheet is increased to a level still well below that used in IR spectroscopy. The spectra shown in Figure 8.1 were taken at 77 K in order to enhance the resolution of the interesting vibrational structure of the L_b band.[3] This is seen to consist of a series of peaks of both in-plane polarizations (see Section 2.3.1). The weak y-polarized vibronic components originate in the Franck-Condon envelope of the transition and/or vibronic intensity borrowing from stronger y-polarized electronic transitions such as B_b. The vibrations involved are of a_g symmetry. The z-polarized components originate in vibronic mixing with z-polarized electronic transitions such as L_a and B_a and involve b_{3g} symmetry vibrations.

A striking observation can be made concerning the origin of the L_b band in Figure 8.1. Instead of exhibiting the same dichroic ratio as the presumably equally y-polarized origin of the B_b band, namely $d_y = 1.0$, its dichroic ratio is $d_\alpha = 1.45$, so that it is present in both reduced spectra, $A_z(\tilde{\nu})$ and $A_y(\tilde{\nu})$. The same discrepancy is found at room temperature[4] and at various pyrene concentrations.

There is no doubt about the identity of the observed peak as the L_b origin, since it coincides exactly with the origin of the fluorescence. It is too narrow

Table 8.2
Electronic States of Pyrene (**1**)[2]

No.	Energy (cm^{-1})	Symmetry	Polarization	Platt's Notation	Clar's Notation	Dominant Configurations
1	~27000	B_{2u}	y	L_b	α	$^1D_{b_{1g}\to b_{3u}} - {}^1D_{b_{2g}\to a_u}$
2	~29500	B_{1u}	z	L_a	p	$^1D_{b_{1g}\to a_u}$
3	~36500	B_{2u}	y	B_b	β	$^1D_{b_{1g}\to b_{3u}} + {}^1D_{b_{2g}\to a_u}$
4	~41000	B_{1u}	z	B_a	β'	$^1D_{b_{2g}\to b_{3u}}$

to contain any vibronic peaks based on intramolecular vibrations, the frequencies of all of which lie above 120 cm^{-1}. Thus, its strange dichroic ratio is not due to overlap with vibronic bands of z polarization.

The inevitable conclusion is that the origin of the L_b band is not purely polarized along any one of the symmetry axes of the molecule, x, y, or z. This certainly puts a fly in the ointment, since the assumption that the choice of possible polarization directions is strictly limited by molecular symmetry lies at the heart of the methods developed for the treatment of linear dichroism. Yet here is a transition in a D_{2h} molecule whose measured dichroism leads to the average value of $37 \pm 10°$ for the deviation of the polarization direction of its origin from the y axis!

This result does not violate any physical laws. Even in the nonpolar polyethylene, the molecule is far from isolated. The immediate environment of any one solute molecule is almost certain to have no symmetry at all, so that taken together with its solvation shell the total chromophoric unit belongs to the C_1 group and there are no rigorous restrictions on the polarization direction of any transitions. The immediate environment of different pyrene molecules can be quite different, so that the polarization directions of a given transition relative to the molecular framework can in principle also be quite different. The "observed" polarization directions are then statistical averages. The fact that solvent-solute interactions are quite strong even in nonpolar solvents is hardly surprising when one recalls the quite substantial energy shifts on going from the gas phase to nonpolar solution, particularly for electronic transitions, but to a lesser degree also for vibrational transitions. We can further recall that solvents are known to modify the intensity of symmetry-allowed bands[5] and to cause the appearance of symmetry-forbidden bands (e.g., the symmetry-forbidden origin of the lowest B_{2u} band of benzene appears weakly in its solution absorption spectrum although it is absent in the spectrum of gaseous benzene). The intensity dependence of the vibronically borrowed components of the weak L_b band of benzenoid hydrocarbons, including pyrene, as a function of solvent polarity and polarizability is well known as the Ham effect.[6]

The picture just drawn is certainly different from that carefully nurtured in Chapters 1 through 7. All the developments in this book so far and all the interpretations in hundreds of papers on polarized spectroscopy in dense media assumed that the solvent effects on polarization directions in molecules, although present in principle, are negligible.

Indeed, the picture we have just painted portrays only a straw man, in that the observations made on the origin of the L_b band of pyrene represent the rare exception rather than the rule (even for this molecule, the L_a, B_b, and B_a bands, the z-polarized vibronic peaks of the L_b band, and all the vibrational bands in the IR spectrum behave as expected). Yet it is useful to be aware of the possible complication the existence of such exceptions represents.

Before we summarize the evidence in favor of this happy conclusion and provide a rationale for the exceptional behavior of the pyrene L_b band as well as a prediction of other exceptions, we shall, however, assemble further evidence that the anomalous behavior of the origin of this transition is real.

The first pieces of relevant data were obtained from the measurement of the linear dichroism of 2-fluoropyrene in stretched polyethylene[3] (see formula **1** for pyrene numbering). This molecule is of C_{2v} symmetry, and x, y, and z are again the only polarization directions permitted for the isolated solute. Its orientation properties are very similar to those of pyrene (at 77 K, from UV dichroism of the L_a, B_b, and B_a bands: $K_z = 0.60 \pm 0.02$, $K_y = 0.30 \pm 0.02$, $K_x = 0.10 \pm 0.04$; from IR dichroism: $K_z = 0.63 \pm 0.015$, $K_y = 0.29 \pm 0.01$, $K_x = 0.08 \pm 0.005$) and its UV spectrum is almost identical to that of pyrene, except that the L_b band is now considerably stronger, particularly the origin. In 2-fluoropyrene, the dichroic ratio for the origin of the L_b band, 0.82, is identical within experimental error to that of the B_b band, $d_y = 0.86$. This argues strongly that the discrepancy found with pyrene is not due to some experimental artifact such as a strange concentration effect. It also provides a clue to the origin of the exceptional behavior shown by pyrene.

Further evidence can be obtained from polarized fluorescence measurements, which we shall discuss next.

8.1.3 Luminescence: Ordinary Photoselection[7]

The ordinary fluorescence polarization spectrum of pyrene in frozen 3-methylpentane at 77 K was shown in Figure 2.8 for one choice of excitation frequency. In Figure 8.2 we show a more complete set of spectra obtained under these conditions. They are the fluorescence and phosphorescence spectra, with their anisotropies R as obtained by exciting into the origin of the L_b, L_a, and B_b bands. Excitation into the strong z-polarized vibronic component of the L_b band (the fourth distinct vibronic peak in the L_b absorption band in Figure 8.1) yields essentially the same anisotropy curve as excitation into the origin of the L_a band. The measurements were performed under nearly ideal depolarization-free experimental conditions, yielding essentially exactly the theoretical limits for fluorescence anisotropy, $R = 0.4$ and $R = -0.2$, for a related molecule, 2-fluoropyrene (Figure 7.11, $R = 0.39$ for excitation at B_b and $R = -0.20$ for excitation at L_a, both monitored at the L_b origin).[7] This result also shows that molecules of this size do not undergo rotational depolarization on the nanosecond time scale at 77 K in 3-methylpentane glass.

Inspection of Figure 8.2 and comparison with expectations based on Section 7.3.4 are in perfect agreement with the results for pyrene discussed in Section 8.1.2. Judging from the anisotropy observed at the fourth emission peak in the fluorescence, the origin of the L_a transition is polarized parallel to this emission peak ($R = 0.3$) and parallel to the fourth vibronic peak in the L_b absorption, while the origin of the B_b transition is polarized perpendicular to the above ($R = -0.07$). From the ordinary photoselection experiment alone, one cannot tell which of these directions is y and which z, but this is already known (Table 8.2). This result establishes the polarization of the fourth fluorescence peak as z, and the deviation from the theoretical value of $R = 0.4$ suggests that it is not entirely purely polarized but rather overlaps with some y-polarized emission intensity. All z-polarized peaks in the fluorescence spectrum can be

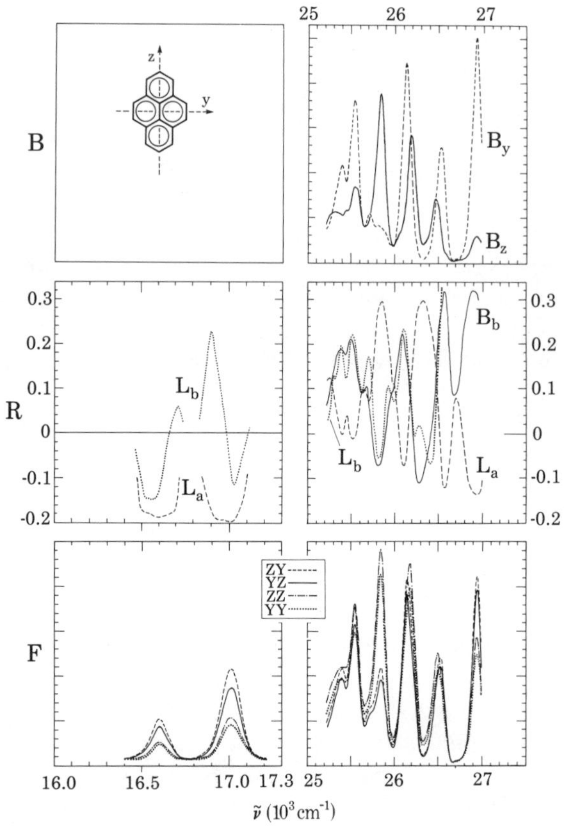

Figure 8.2 Polarized emission of pyrene in glassy 3-methylpentane at 77 K. Right: fluorescence; left: phosphorescence. Bottom, the four polarized emission curves excited at the L_a origin: full line, YZ; dashed line, ZY; dotted line, YY; dash-dot line, ZZ. The curves are not corrected for instrumental polarization biases nor for the 14.5° angle between the exciting and observed beam directions. Center, emission anisotropy: full line, excitation at the B_b origin; dashed line, excitation at the L_a origin; dotted line, excitation at the L_b origin. Top, reduced fluorescence curves excited at the L_a origin; full line, B_z; dashed line, B_y. Adapted by permission from ref. 7.

recognized similarly. They are associated with b_{3g} symmetry vibrational levels of the ground electronic state (Section 2.3.1). Oppositely polarized peaks are assigned as y-polarized and are associated with a_g vibrational levels of the ground electronic state. The polarized fluorescence curves obtained upon excitation into the purely and oppositely polarized L_a and B_b origins can be combined using (7.120) and (7.121) to obtain the y- and z-polarized components of the fluorescence, i.e., the reduced fluorescence spectra. These are shown in Figure 8.2.

The fluorescence polarization curve recorded while exciting at the origin of the L_b band, which is not overlapped by other vibronic peaks, is inverted relative to that obtained while exciting at the origin of the L_a band, as expected. However, in several regions it is also distinctly different from that obtained exciting at the origin of the y-polarized B_b band as if the transition moments at the L_b and the B_b origins were indeed not exactly parallel. The best evidence for this is obtained from the fluorescence anisotropy observed at the L_b origin upon excitation into the L_a and B_b origins, already known to be purely z- and y-polarized, respectively (Sections 8.1.1 and 8.1.2).

The anisotropy of the fluorescence spectrum monitored at the origin of the fluorescence never reaches the limits expected for two parallel ($R = 0.4$) or perpendicular ($R = -0.2$) transition moments. Since the origin is not overlapped by any other peaks, it follows that the emitting L_b moment is not exactly parallel or exactly perpendicular to any of the absorbing moments (a measurement in which the origin of the L_b band is both excited and monitored was not possible due to scattered light). The anisotropy observed when exciting at the z-polarized origin of the L_a band or at the fourth absorption peak in the L_b band, $R = -0.14$, formally corresponds to a value of 19° for the absolute value of the angle between the emitting L_b transition moment and the y axis when (7.113) is used. The anisotropy observed when exciting at the y-polarized origin of the B_b band, $R = 0.32$, formally corresponds to a value of 22° for the absolute value of this angle. The correspondence is only formal, since the equation assumes that the angle between the absorbing and emitting moments is the same in all solute molecules and this is unlikely if the magnitude of the angle is dictated by random differences in the solvent environment from one site to another. Still, the results leave no doubt that in 3-methylpentane glass the transition moment of the L_b origin deviates significantly from the y axis, a rough estimate of the angle of deviation being about 20°. This value is somewhat smaller than that deduced from dichroic absorption in stretched linear low-density polyethylene in Section 8.1.2. This may simply be due to the difference in the solvent environment in the two media.

The polarization of pyrene phosphorescence is also shown in Figure 8.2. Upon excitation into the z-polarized L_a origin, the observed anisotropy value equals the theoretical limit $R = -0.2$ for perpendicular absorbing and emitting moments. This shows that the polarization of the phosphorescence is constrained to the xy plane, and this is compatible with the ^3L_a assignment for the emitting triplet state (Section 2.3.1). The total symmetries of its three components are A_u, B_{3u}, and B_{2u}. Emission into the totally symmetric vibrational levels of the ground state, such as the lowest level (phosphorescence origin, observed at $\sim 17\,000\ cm^{-1}$), cannot take place by the electric dipole mechanism from the A_u component, while it can take place with x polarization from the B_{3u} component and with y polarization from the B_{2u} component. All three components of the triplet state will be populated at 77 K, and more than one can emit independently, with rates dominated by the spin-orbit coupling matrix elements discussed in Section 2.2.2. The relative rates are likely to be different for each vibronic peak in the phosphorescence. As a result, the anisotropy R to be expected upon excitation into a y-polarized absorption band is also likely to be different for each peak in the phosphorescence and can range from 0.4 if the

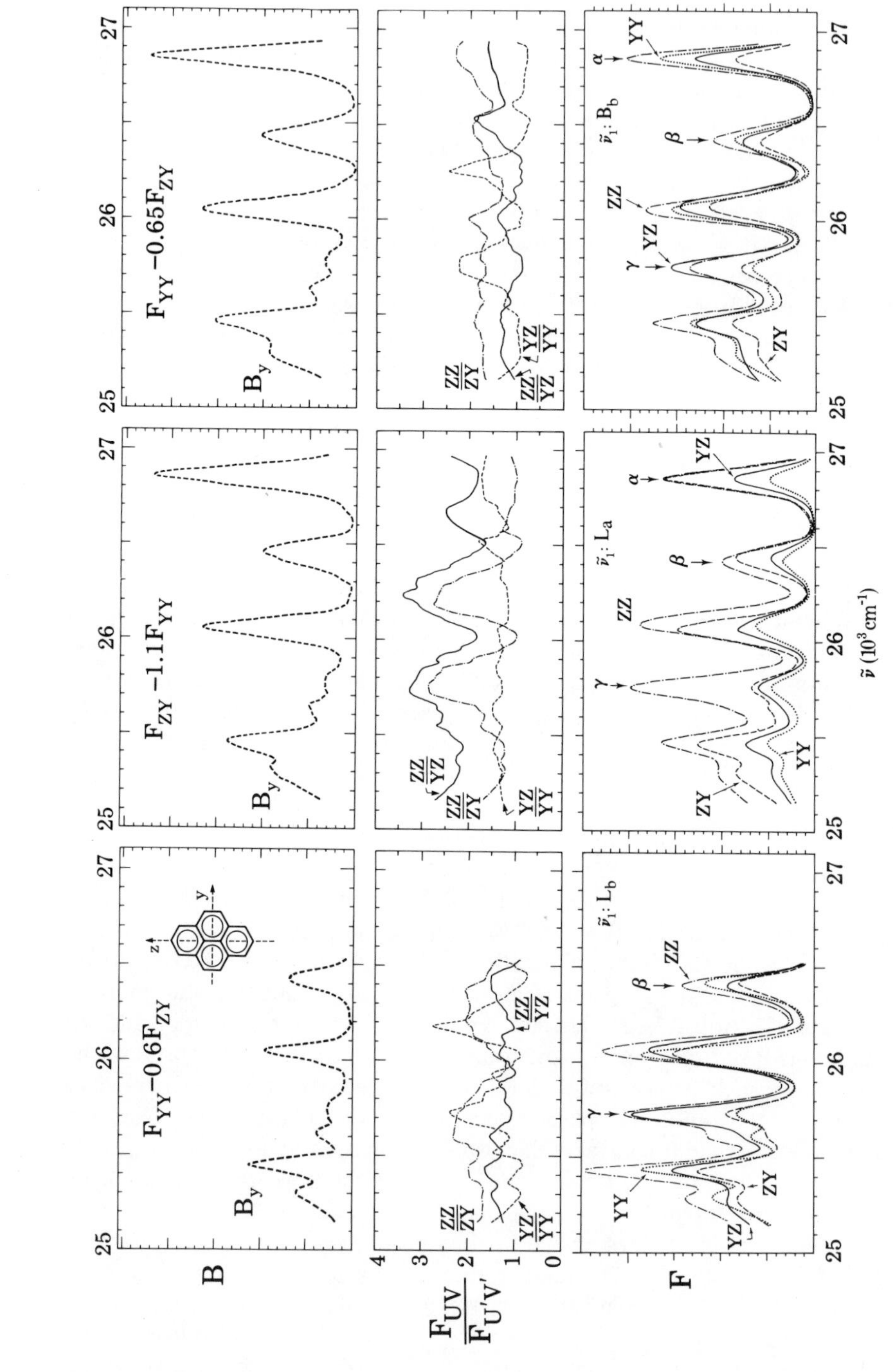
B
F_UV / F_U'V'
F
F_YY −0.6F_ZY
F_ZY −1.1F_YY
F_YY −0.65F_ZY
B_y
ν̃₁: L_b
ν̃₁: L_a
ν̃₁: B_b
ZZ
YZ
YY
ZY
α
β
γ
ν̃ (10³ cm⁻¹)
25
26
27
0
1
2
3
4

Figure 8.3 Polarized fluorescence of pyrene in stretched polyethylene at 77 K. Left: excited at the L_b origin; center: excited at the L_a origin; right: excited at the B_b origin. Bottom, the four polarized emission curves: full line, YZ; dashed line, ZY; dotted line, YY; dash-dot line, ZZ. The curves are corrected for instrumental polarization biases but not for the 14.5° angle between the exciting and observed beam directions. Reproduced by permission from ref. 3. Center, the polarization ratios: full line, F_{ZZ}/F_{YZ}; dashed line, F_{YZ}/F_{YY}; dash-dot line, F_{ZZ}/F_{ZY}. Top, the reduced fluorescence spectrum B_y as obtained at the three different choices of excitation (on three arbitrary scales).

B_{2u} component of the triplet emits much faster than the B_{3u} component to -0.2 if the opposite is true. This is the situation observed in 2-fluoropyrene (Figure 7.11), where the origin (17 000 cm^{-1}) is mostly y-polarized ($r = 0.09$), while the next peak (16 600 cm^{-1}) is mostly x-polarized ($R = -0.07$). However, in pyrene itself, matters are more complicated. As Figure 8.2 shows, phosphorescence anisotropy varies significantly across each vibronic peak, indicating that the relative rates of emission from the B_{2u} and the B_{3u} components of the excited triplet state are a function of the detailed arrangement of the solvent environment (site effects). For instance, those molecules whose environment causes them to have a relatively high energy for the phosphorescence origin ($>17\,000$ cm^{-1}) emit mostly from the x-polarized B_{3u} component ($R < 0$), while the presumably more strongly perturbed ones with a lower-energy phosphorescence origin ($<17\,000$ cm^{-1}) emit primarily from the y-polarized B_{2u} component ($R > 0$). Clearly, because of the presence of more than one emitting excited state component, the polarization of phosphorescence spectra is more difficult to interpret than was the case for fluorescence spectra.

8.1.4 Luminescence[3]: Photoselection on an Oriented Assembly and the Orientation Factors L

Four of the five distinct polarized fluorescence spectra of pyrene in stretched polyethylene at 77 K were shown in Figure 2.9 for excitation into the L_a origin. Figure 8.3 shows a more complete set of polarized fluorescence results obtained on this sample upon excitation into the L_b, L_a, and B_b origins. These include the polarized intensity curves, the three polarization ratios F_{ZZ}/F_{ZY}, F_{ZZ}/F_{YZ}, and F_{ZZ}/F_{YY}, analogous to the single polarization ratio $F_{\|}/F_{\perp}$ obtained in experiments on isotropic samples, and finally the y-polarized reduced fluorescence spectrum as obtained by the stepwise reduction procedure for each of the three excitation wavelengths. This agrees well with the y-polarized reduced fluorescence spectrum obtained in 3-methylpentane glass (Figure 8.2).

The polarization curves differ clearly from those obtained in the ordinary photoselection experiment in a way which is qualitatively compatible with the notion that the pyrene z axis is preferentially aligned with the sample Z axis. The polarization directions for all the peaks in the spectra are apparent upon closer inspection, including the intermediate position of the transition moment of the L_b origin in absorption as well as emission, which is again clearly not parallel to that of the B_b origin. Thus, the three fluorescence polarization curves depend on whether the L_b or the B_b origin is excited, and the intensities F_{YZ} and F_{ZY} obtained with excitation along y (B_b) and observation along the origin of L_b differ [if the latter were directed along y, they would have to be the same, and indeed they are when both the excitation (L_a origin) and the observation (fourth peak in the emission) involve a z-oriented transition moment]. Once again, the same measurements were performed on 2-fluoropyrene and all polarization ratios behaved as expected for fluorescence transition moments oriented exactly along y and z.

In order to evaluate the polarized spectra of pyrene quantitatively, it is best to perform stepwise reductions by producing the combinations $F_{UV} - \{[S_{UV}]_{st}/[S_{ST}]_{st}\}F_{ST}$, where s = z for L_a excitation and s = y for B_b excitation, and t = z if z-polarized emission features have just disappeared from the linear combination. The procedure is outlined in detail in Section 7.3.1 [cf. expression (7.39)]. The reductions are easy to do even in the absence of "two-dimensional" spectra, since regions of pure polarization can be found in the excitation (fourth peak in L_b and the origin of L_a along z, the origin of B_b along y) and emission (fourth peak along z) spectra. Examples of such reductions have already been shown in Figures 7.6 and 7.7. The resulting reduced excitation spectra are in agreement with the reduced absorption spectra obtained from the measurement of linear dichroism.

The resulting reduction factors for pyrene can be substituted into equations (7.47) to produce an overdetermined set of equations for L_{zz} and L_{yz} (note that the K's are known). The resulting orientation factors for pyrene in stretched polyethylene are listed in Table 8.3. The value of L_{xz} can be obtained using (4.25) and (4.31). In order to obtain the remaining L factors, it would be necessary to have reduction factors for x- or y-polarized emission features. Because of the complicated mixed nature of the phosphorescence (and also its weakness), and because of the mixed polarization of the fluorescence origin in pyrene (**1**), these are not available.

The situation is more favorable in the case of 2-fluoropyrene. Now the origin of the fluorescence is purely y-polarized, so that an equation for L_{yy} is also available (excitation in the origin of B_b, observation at origin of fluorescence), and all L factors can be obtained (Table 8.3).

The most striking property of the L_u factors for pyrene and 2-fluoropyrene is their closeness to the K_u^2 values. The limit $L_u = K_u^2$ is reached when the angle *u* which the molecular axis u makes with the stretching direction Z is identical for all solute molecules in the sample. This situation occurs for the out-of-plane axis x within the experimental error, and the limit is nearly reached for the other two axes as well, indicating that the fraction of molecules whose orientation deviates substantially from the average must be quite small. Of the infinite

Table 8.3
Orientation Factors[a] in Stretched Polyethylene from Linear Dichroism in the IR and UV and from Polarized Luminescence[3]

	K_x	K_y	K_z	L_x	L_y
Pyrene (**1**)	0.10 ± 0.01	0.33 ± 0.01	0.56 ± 0.02	—	—
2-Fluoropyrene	0.08 ± 0.005	0.29 ± 0.01	0.63 ± 0.015	0.00 ± 0.04	0.12 ± 0.02
	L_z	L_{xy}	L_{yz}	L_{zx}	
Pyrene (**1**)	0.37 ± 0.04	0.14 ± 0.02	0.05 ± 0.04	—	
2-Fluoropyrene	0.42 ± 0.04	0.15 ± 0.02	0.06 ± 0.04	0.02 ± 0.02	

[a] 77 K

number of orientation distributions compatible with the experimental K and L factors, the simplest representative distribution would have all the molecules of 2-fluoropyrene oriented alike, with the angles 37.5°, 57.5°, and 73.5°, respectively, between the z, y, and x molecular axis and the stretching direction. That this cannot be strictly correct is clear from the small but real difference between K_u^2 and L_u for u = y and z. The less complete data for pyrene itself are compatible with a similar orientation distribution with slightly different orientation angles, 41.5°, 55°, and 71.5° for the z, y, and x axis, respectively. The detailed nature of the molecular environment in stretched polyethylene which is responsible for this behavior is still unknown; some of the proposed possibilities are discussed in ref. 3.

An evaluation of all the above evidence from linear dichroism, polarized fluorescence in 3-methylpentane glass, and polarized fluorescence in stretched polyethylene pertinent to the transition moment direction of the origin of the L_b transition leaves no reasonable doubt: On the average, the polarization direction of this transition deviates significantly from the y axis, both in 3-methylpentane (~20°) and in stretched polyethylene (~40°) at 77 K. Since neither the other pyrene transitions nor any of the transitions of 2-fluoropyrene show any anomaly, this cannot be due to a trivial cause such as rotational diffusion[8] or experimental artifacts. Another reason why the L_b origin anomaly in **1** cannot be attributed to rotational diffusion is the very high anisotropy of the long-lived phosphorescence emission (R = −0.2) observed upon excitation of **1** in 3-methylpentane into the L_a origin.

The question then is, why is the 0-0 component of the L_b transition of pyrene anomalous? We believe that two factors play a role: First, its unusual weakness and, second, the energetic proximity of the intense L_a transition. The perturbation of the molecule by interaction with the unsymmetrical solvent environment can be described by perturbation theory as causing a mixing of the pure states of the isolated molecule among themselves, their mixing with locally excited states of the unsymmetrical solvent cage, and their mixing with charge-transfer states involving the solute and the cage. If a pure state ψ_1, such as L_b, is connected with the ground state ψ_0 by only a very small transition moment, it will be particularly sensitive to the admixture of a small amount of a state ψ_2, such as L_a, which is connected with the ground state by a very large transition moment:

$$\langle\psi_0|\hat{\mathbf{M}}|c_1\psi_1 + c_2\psi_2\rangle = c_1\langle\psi_0|\hat{\mathbf{M}}|\psi_1\rangle + c_2\langle\psi_0|\hat{\mathbf{M}}|\psi_2\rangle \qquad (8.1)$$

If $c_1 \gg c_2$ and $\langle\psi_0|\hat{\mathbf{M}}|\psi_1\rangle \ll \langle\psi_0|\hat{\mathbf{M}}|\psi_2\rangle$, $c_1\langle\psi_0|\hat{\mathbf{M}}|\psi_1\rangle$ and $c_2\langle\psi_0|\hat{\mathbf{M}}|\psi_2\rangle$ may well be comparable. Perturbation theory suggests that the mixing will be facilitated when the state to be mixed in is close in energy.

The admixture of ψ_2 into ψ_1 caused by solvent-induced perturbations will cause an intensification of the transition from the ground state to ψ_1. This is well known as the Ham effect.[6] At the same time, if $\langle\psi_0|\hat{\mathbf{M}}|\psi_1\rangle$ and $\langle\psi_0|\hat{\mathbf{M}}|\psi_2\rangle$ are not collinear, as is the case of L_b and L_a of pyrene, the admixture will also cause a twisting of the transition moment from the orientation it would have in an isolated molecule.

It is likely that the solvent-solute interactions are not too different for pyrene and 2-fluoropyrene, at least for a nonpolar and relatively nonpolarizable solvent. The larger intrinsic absorption intensity of the L_b origin in 2-fluoropyrene, understandable in terms of an increased deviation from alternancy and Pariser's plus-minus symmetry,[9] makes the twisting caused by the interaction unobservably small. If this explanation is correct, one can expect that the L_b bands of other alternant aromatic hydrocarbons, such as napthalene, will show similarly anomalous polarization behavior. We believe that also other anomalies in photoselection results, such as polarization ratios in excess of 1/7 upon irradiation into the long-wavelength tails of degenerate absorption bands followed by observation of degenerate emission,[9a] are due to similar symmetry-lowering effects of the environment.

The twisting of weak transition moments by solvent-solute interactions may have contributed to the failure of the polarization ratios recorded in many measurements of polarized fluorescence (ordinary photoselection) to come close to the theoretical limits. This failure had been previously ascribed to factors such as in-plane rotation of the aromatic ring in the rigid solvent at 77 K on a time scale comparable to that of fluorescence,[8] which appears rather implausible considering that pyrene and even much smaller aromatics are known from studies of photoinduced dichroism[10–12] to remain oriented for minutes or hours in organic glasses. The existence of the phenomenon may have other observable consequences. For instance, the rotational depolarization of pyrene fluorescence in solvents should appear anomalous, since the fluorescing transition moment may rotate faster than the molecule itself as the asymmetric environment changes. In terms ordinarily used by those interested in rotational diffusion studies (Section 8.9), the "electronic delocalization" of the emitting transition moment of pyrene is not decoupled from the rotational motion of the molecule (usually it is assumed to be).

8.1.5 Raman Spectra: Present Experimental Difficulties

In principle, independent information on the orientation factors L of pyrene in polyethylene is available from polarized Raman intensities (Chapter 7). In practice, it is difficult to obtain. Although the signals due to the polyethylene itself are very easy to observe, those due to the relatively small amount of pyrene present are too weak to measure the polarized intensity ratios with the required accuracy. Traces of colored impurities, easily formed from pyrene by oxidation in air, are sufficient to cause problems due to fluorescence and to excessive sample heating when high laser intensities are used. Increasing the thickness of the polyethylene sheet permits much stronger pyrene Raman lines to be observed, but only at the expense of partial depolarization of the visible exciting laser light and loss of accuracy in the polarized intensity ratios. Since the signals are so weak, it is necessary to collect the scattered radiation from a fairly wide angle using a low f-number lens, and this requires corrections which introduce further uncertainties. After a fair amount of effort, we have concluded that we cannot obtain reliable L factors for pyrene in stretched polyethylene from Raman measurements using a currently standard commercial spectrometer.

The situation is more favorable with other samples. For instance, it has been possible to obtain useful results for 2,3-dimethylnapthalene in stretched polyethylene (Section 2.3.2) and for the C≡N stretching vibration of a cyano group in a complicated molecule dissolved in a liquid crystal.[13]

8.1.6 Ordinary Photodichroism: Photodetachment from the Radical Anion[12]

When aromatic hydrocarbons dissolved in dry degassed ether solvents are allowed to come in contact with an alkali metal mirror, they are reduced to their anions carrying a single or even a multiple negative charge. The extra electrons can be detached again upon irradiation with sufficiently energetic photons so that the parent hydrocarbon is regenerated from the monoanion, the monoanion is produced from the dianion, etc. When the irradiated solution is rigid, the ejected electrons are trapped by the solvent for very long periods of time. If the irradiation wavelength is chosen to coincide with a purely polarized absorption band, or at least a fairly purely polarized one, photoselection occurs, and both the starting anion and the final neutral hydrocarbon molecular assemblies will be partially oriented. In this fashion, photodichroism (Sections 6.2, 6.3) is induced in the originally isotropic solid solution.

This phenomenon has been utilized to determine the relative polarization directions of electronic transitions in pyrene radical anion and to relate them to those in the neutral parent hydrocarbon. Figure 8.4 shows the absorption curves for pyrene and its radical anion. In the top row are the differential curves $E_Y(\tilde{\nu}) - E_Z(\tilde{\nu})$ obtained by measurement on a rigid solution of the radical anion which had been irradiated with Z-polarized light of energy indicated in the spectrum until about half the radical anion was converted into neutral pyrene. The sign of the photoinduced dichroism is positive for the L_a band of pyrene near 30 000 cm^{-1} and negative for its B_b band at 36 300 cm^{-1}. In 1960, when these experiments were performed, it was not known with certainty what the relative polarizations of these bands were, but the data showed clearly that they are mutually perpendicular. The L_b band, whose origin was then already known with certainty to be polarized along the short in-plane axis of the molecule (y) from single-crystal studies, unfortunately is too weak to be observed in the spectra. By analogy to other aromatics such as naphthalene, the authors concluded that L_b and L_a were mutually perpendicular and thus arrived at an assignment of absolute polarizations in pyrene, L_b: y, L_a: z, B_b: y. Today this is well known to be correct (Table 8.2).

Now, if the z-polarized peaks of the product appear more strongly in $E_Y(\tilde{\nu})$ than in $E_Z(\tilde{\nu})$ and the y-polarized peaks of pyrene appear more strongly in $E_Z(\tilde{\nu})$ than in $E_Y(\tilde{\nu})$, the absorbing transition of the radical anion must have been polarized along y (the exciting light was polarized along Z). The pyrene product is then preferentially aligned with its y axis along Z. On the other hand, this orientation will be preferentially depleted in the remaining radical anion. Therefore, those peaks of the radical anion which show $E_Y(\tilde{\nu}) > E_Z(\tilde{\nu})$, i.e., are positive in Figure 8.4, must correspond to y-polarized transitions, and those which show

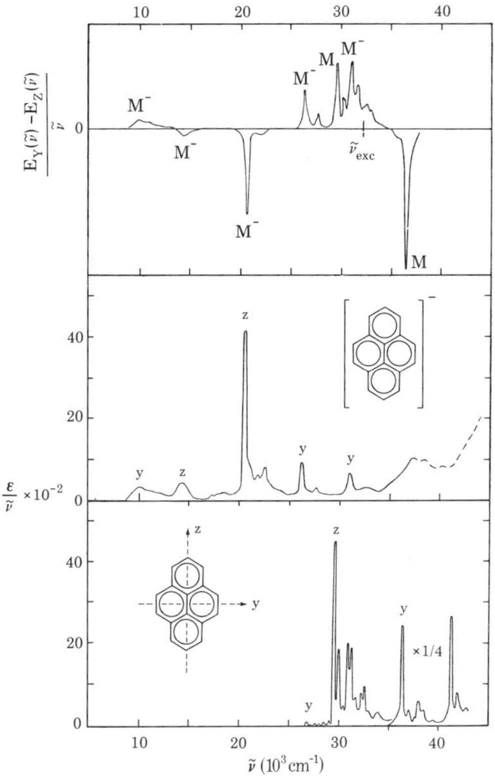

Figure 8.4 The UV-visible absorption spectra of pyrene (bottom) and its radical anion (center) and the differential dichroic spectrum obtained after partial photobleaching of the radical anion with UV light polarized along Z (31 900 cm^{-1}, as indicated in the top row by $\tilde{\nu}_{exc}$). Glassy 2-methyltetrahydrofuran at the temperature of liquid air. Adapted by permission from ref. 12.

$E_Y(\tilde{\nu}) < E_Z(\tilde{\nu})$ and are negative must belong to z-polarized transitions. This consideration permitted the peak assignments shown in Figure 8.4.

8.1.7 UV-Visible Dichroism: Radical Cation and Radical Anion[14]

γ-Irradiation of rigid solutions in solvents capable of trapping either electrons or positive charges is well known to produce solute radical ions.[15] Solvents such as halogenated hydrocarbons trap electrons and permit the observation of solute radical cations; solvents such as ethers trap positive charges and permit the observation of solute radical anions.

It is possible to produce radical ions in stretched polymers in a similar fashion, provided that a suitable solvent is also present. Poly(vinyl chloride) sheets containing *sec*-butyl chloride or carbon tetrachloride have been used for the preparation of oriented radical cations; poly(vinyl alcohol) containing *sec*-butyl amine has been used for the preparation of oriented radical anions.

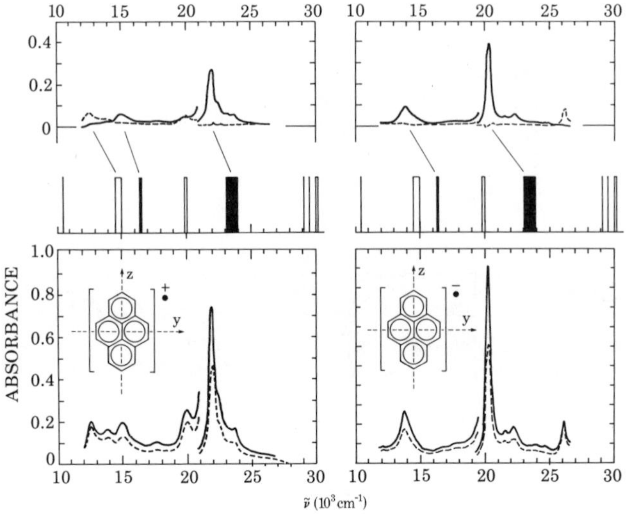

Figure 8.5 The polarized (E_Z: full, E_Y: dashed) UV-visible absorption spectra of pyrene radical cation [left, in γ-irradiated stretched poly(vinyl chloride) film] and radical anion [right, in γ-irradiated stretched poly(vinyl alcohol) film)] shown at the bottom, adapted by permission from ref. 14. Center, results of PPP calculations[14] of transition energies, polarizations (black, z; white, y), and oscillator strengths (thick, f = 0.3; medium, f = 0.06; thin, $0.02 > f \geqslant 0$; very thin, symmetry-forbidden). Top, reduced spectra (full, A_z; dashed, A_y) obtained by stepwise reduction of the experimental curves of ref. 14.

The polarized absorption spectra of pyrene radical anion and pyrene radical cation obtained in this manner are shown at the bottom in Figure 8.5. Absolute assignments are possible, since in pyrene the long axis is known to be the orientation axis z (this can be easily checked by performing a measurement on the neutral molecule prior to irradiation). The purely polarized reduced spectra obtained by the stepwise reduction procedure of Chapter 5 are shown on top and can be compared with the results of π-electron calculations in the center. The assigned state symmetries also agree with the results obtained by the method of ordinary photoinduced dichroism, discussed in Section 8.1.6. The stretched sheet method is simpler and more generally capable of providing absolute assignments.

8.2 IR Transition Moment Directions and Assignment of Vibrational Transitions

The use of partially aligned samples for the assignment of symmetries of vibrational states has received relatively little attention. This may well be due to the fact that until the advent of Fourier transform instrumentation it was quite difficult to obtain reliable results for any but the most intense bands, and for many of those there was never any doubt about the assignment in the first place (nitrile or carbonyl stretches in organic molecules; out-of-plane C—H bends in aromatics). The assignment of several dozen strong and weak fundamentals and overtones in pyrene, discussed in Section 8.1.1, shows that modern instrumentation permits the use of polymers and liquid crystals as a simple and effective tool for IR assignments.

8.2.1 An Introduction: Linear Triatomic Molecules[16]

It has been known for some time from studies of UV-visible linear dichroism that even molecules as small as triatomics can be oriented in stretched polyethylene to a sufficient degree to permit polarization assignments.[17]

Here we illustrate the alignment on three linear triatomic molecules, CO_2, OCS, and N_2O. Figure 8.6 shows two segments of the polarized IR spectra of these molecules in stretched polyethylene, with two of the three expected fundamental modes for each of the molecules. The assignments are, of course, well known, but for illustrative purposes we shall for the moment pretend that this is not so. We shall further pretend that the third vibration has not been observed.

The observed K values are collected in Table 8.4. We note first of all that the degree of orientation is quite remarkable and second, that for each molecule the sum of the higher K value plus twice the lower K value equals unity. It

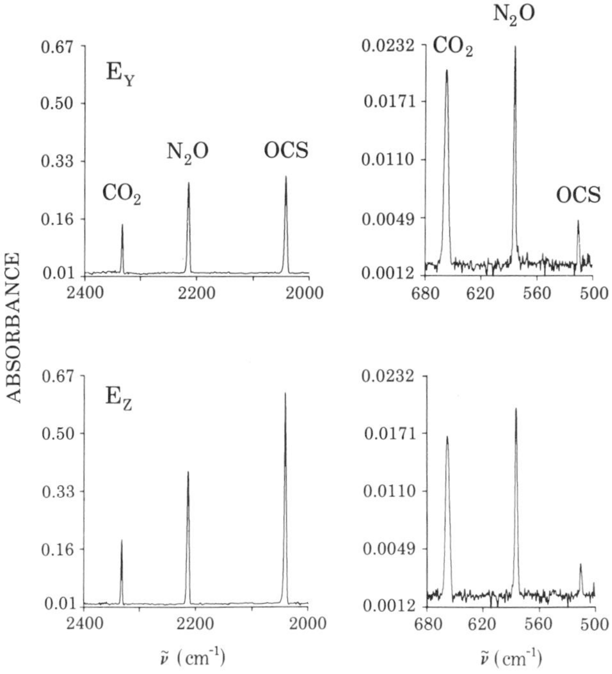

Figure 8.6 Segments of polarized IR spectra[16] of CO_2, OCS, and N_2O in stretched polyethylene at 14 K. Top: E_Y, bottom: E_Z.

Table 8.4
Orientation Factors[a] of Some Triatomic Molecules in Stretched Polyethylene from Integrated Polarized IR Intensities[16]

CO_2		OCS		N_2O	
$\tilde{\nu}$ (cm^{-1})	K	$\tilde{\nu}$ (cm^{-1})	K	$\tilde{\nu}$ (cm^{-1})	K
2333	0.40	2042	0.52	2214	0.42
658	0.29	516	0.24	585	0.30

[a] Measured in a low-density polyethylene sheet in contact with a CsI window held at 14 K.

follows that the former is K_z and that $K_x = K_y$ is equal to the latter. Thus, the distribution is of the rodlike type, and this is plausible only if the molecules are linear.

In a linear triatomic molecule, the bending mode will be degenerate and polarized perpendicular to the molecular axis, in the xy plane. The stretching modes, if allowed, will be polarized along the molecular axis z, and both will show the same dichroism. Since we observe two different K values for the two bands for each molecule, we must be observing one bending and one stretching mode. We conclude that the lower frequency band ($K_x = K_y$) is the bending mode and the higher frequency band (K_z) is a stretching mode.

8.2.2 A Simple Case: The Haloforms[16]

Our next example was chosen to help dispel the notion that only molecules of highly anisotropic shape can be sufficiently oriented in a stretched polymer such as polyethylene.

In Figure 8.7 we show portions of the polarized absorption spectra of chloroform and bromoform in a stretched polyethylene sheet. The K values calculated from the observed dichroism are given in Table 8.5. There is a slight but reproducible difference in the orientation factors of $CHCl_3$ and $CDCl_3$.

The situation is similar to that of the previous case. If the symmetry assignments to the allowed a_1 and e species of the C_{3v} group to which the haloforms belong were not already known, they could be obtained from these spectra. Vibrational transitions of the former symmetry species are polarized along the threefold axis, those of the latter perpendicular to it. As expected from the C_{3v} molecular symmetry, only two distinct K values are observed. Since K_z must be larger than 1/3 and K_x smaller than 1/3, by definition, z-polarized transitions show a positive dichroism and x-polarized transitions a negative dichroism in the absence of overlap. The question is only, does the orientation axis z lie in the threefold symmetry axis ($K_z > 1/3 > K_y = K_x$) or perpendicular to it ($K_z = K_y > 1/3 > K_x$)?

The resolution of the problem is the same as before. If the former were true, $\sum_u K_u = K_z + 2K_y = 0.924$ ($CHCl_3$) and 0.939 ($CHBr_3$), which differs from unity by more than the experimental error; if the latter were true, $\sum_u K_u = 2K_z + K_x = 0.999$ ($CHCl_3$) and 0.998 ($CHBr_3$), and this is within experimental

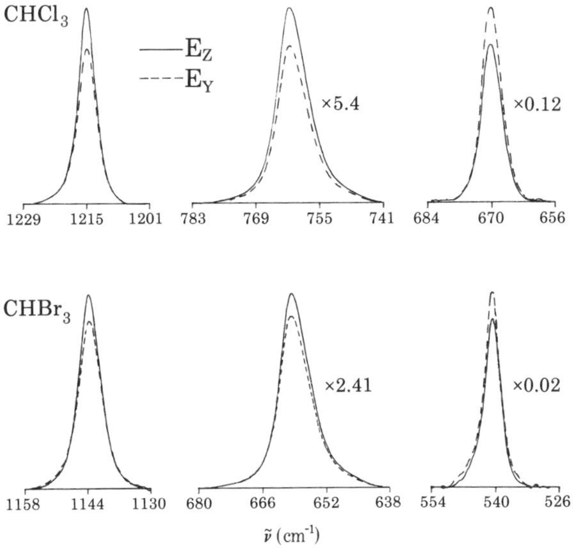

Figure 8.7 Segments of polarized IR spectra[16] of $CHCl_3$ and $CHBr_3$ in stretched polyethylene at room temperature. Full line, E_Z; dashed line, E_Y.

Table 8.5
Orientation Factors[a] of the Haloforms in Stretched Polyethylene from Integrated Polarized IR Intensities[16]

$CHCl_3$		$CHBr_3$		CHI_3	
$\tilde{\nu}$ (cm^{-1})	K	$\tilde{\nu}$ (cm^{-1})	K	$\tilde{\nu}$ (cm^{-1})	K
1215	0.357	1144	0.353	1066	0.349
762	0.358	659	0.352	580	0.351
670	0.283	541	0.293		

[a] Room temperature, average of 10 measurements.

error of unity. Thus, the threefold symmetry axis is seen to lie perpendicular to the orientation axis in all three molecules, e vibrations have the larger K value and exhibit positive dichroism, and a_1 vibrations have the smaller K value and exhibit negative dichroism. This is the orientation expected from molecular shape.

The averaged values are $K_z = K_y = 0.358 \pm 0.002$ and $K_x = 0.283 \pm 0.004$ for $CHCl_3$, $K_z = K_y = 0.353 \pm 0.005$ and $K_x = 0.293 \pm 0.007$ for $CHBr_3$, and $K_z = K_y = 0.350 \pm 0.005$ for CHI_3. In this type of measurement the three K values generally add up to unity with an accuracy better than ± 0.005 when results of several measurements are averaged and when the dichroic ratios are calculated from integrated intensities.

Although somewhat trivial, these two examples illustrate the promise which the method holds for the study of small molecules. As more becomes known about the relation between the orientation factors of such molecules and their shape or other properties, the reliability of absolute symmetry assignments will increase. Of course, even relative assignments can be very useful in complicated spectra.

8.2.3 Anthracene, Acridine, Phenazine: Ground-State Vibrations[18]

This example illustrates the use of the stretched sheet method for the assignment of symmetries of vibrational transitions in organic molecules of high symmetry and moderate complexity.

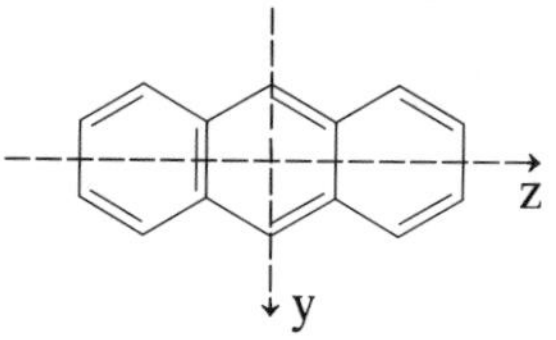

2

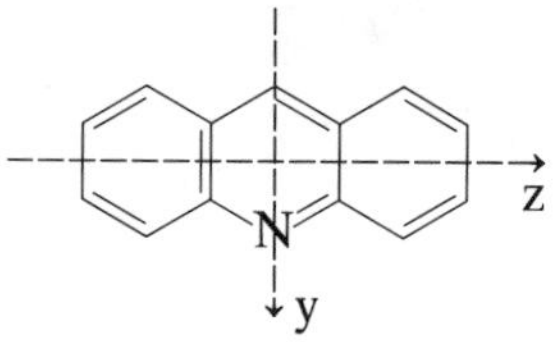

3

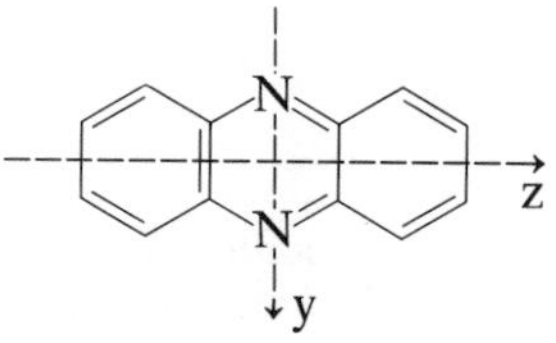

4

The dichroic IR spectra of anthracene (**2**), acridine (**3**), and phenazine (**4**) in linear low-density polyethylene and in perdeuterated high-density polyethylene were similar in that each showed three and only three distinct dichroic ratios E_Z/E_Y for the multitude of peaks present. These are clearly associated with the orientation factors of the three symmetry-adapted molecular axes, x, y, and z, shown in formulas **2**–**4** (Table 8.6).

In each case, the three K values add up to unity, as expected for three mutually orthogonal directions. Their differences are sufficiently large that an unambiguous assignment of each peak in the spectrum to one of the three directions is possible. This would not be the case if the orientation were rodlike, $K_x = K_y$, as might have been suspected on the basis of molecular shape. In general, it is desirable to avoid rodlike distributions if complete symmetry assignments are to be reached. A certain amount of control over the situation can be exercised by the choice of polyethylene used. The linear low-density type used in ref. 18 seems to be particularly good at avoiding rodlike distributions.

The absolute assignment of the x, y, and z axes to actual directions in the molecule stated in formulas **2**–**4** was based on the general shape-orientation relations (Section 4.6.1), on linear dichroism in the UV region ($\pi\pi^*$ transitions are in-plane polarized), and on comparison with prior crystal work. The simi-

Table 8.6
Orientation Factors[a] of Anthracene (**2**), Acridine (**3**), and Phenazine (**4**) in Stretched Polyethylene from Integrated Polarized IR Intensities[18]

Polyethylene	**2**	**3**	**4**
Linear low-density			
K_x	0.11	0.14	0.21
K_y	0.26	0.26	0.26
K_z	0.63	0.58	0.52
High-density perdeuterated			
K_x	0.12	0.14	0.22
K_y	0.26	0.28	0.28
K_z	0.60	0.54	0.48

[a] Room temperature.

larity of the orientation factors among the three molecules is in accord with the similarity of their shapes. The differences are larger than experimental error, however, and leave no doubt that molecular shape, though dominant, is not the only factor which determines the orientation.

In order to assign vibrational symmetries throughout the fundamental region, measurements in ordinary polyethylene were combined with measurements in perdeuterated polyethylene. Only high-density perdeuterated polyethylene appears to be commercially available at present, so that the orientation factors cannot be expected to be exactly the same even if there are no isotope effects on the alignment forces. Still, as Table 8.6 shows, they are not very different.

Figure 8.8 shows the dichroic spectra of acridine (**3**) from 80 to 3200 cm^{-1}, demonstrating the complete coverage obtainable by the combined use of the two polymers. Sixty-eight bands appear with sufficient intensity for a safe symmetry assignment, given in the figure relative to the axis labeling of formula **3**. Only a fraction of these bands are due to fundamental vibrations.

Similarly complete spectral coverage based on the use of ordinary and perdeuterated polyethylene has been obtained for nitrobenzene.[19]

8.2.4 Acridine: Triplet State Vibrations[20]

The direct observation of the IR spectra of metastable excited species is difficult but possible. The assignment of such spectra will be greatly facilitated if polarization directions can be determined. Recently, the first successful measurement of this type has been reported, using a substrate partially oriented in a stretched polyethylene sheet.

A sample of acridine (**3**) in stretched polyethylene mounted between two CsI plates on the cold tip of a closed-cycle helium refrigerator was irradiated by the chopped unpolarized full output of a 1 kW Hg-Xe high-pressure lamp. At the same time, a continuous IR beam was passed through the sample. The

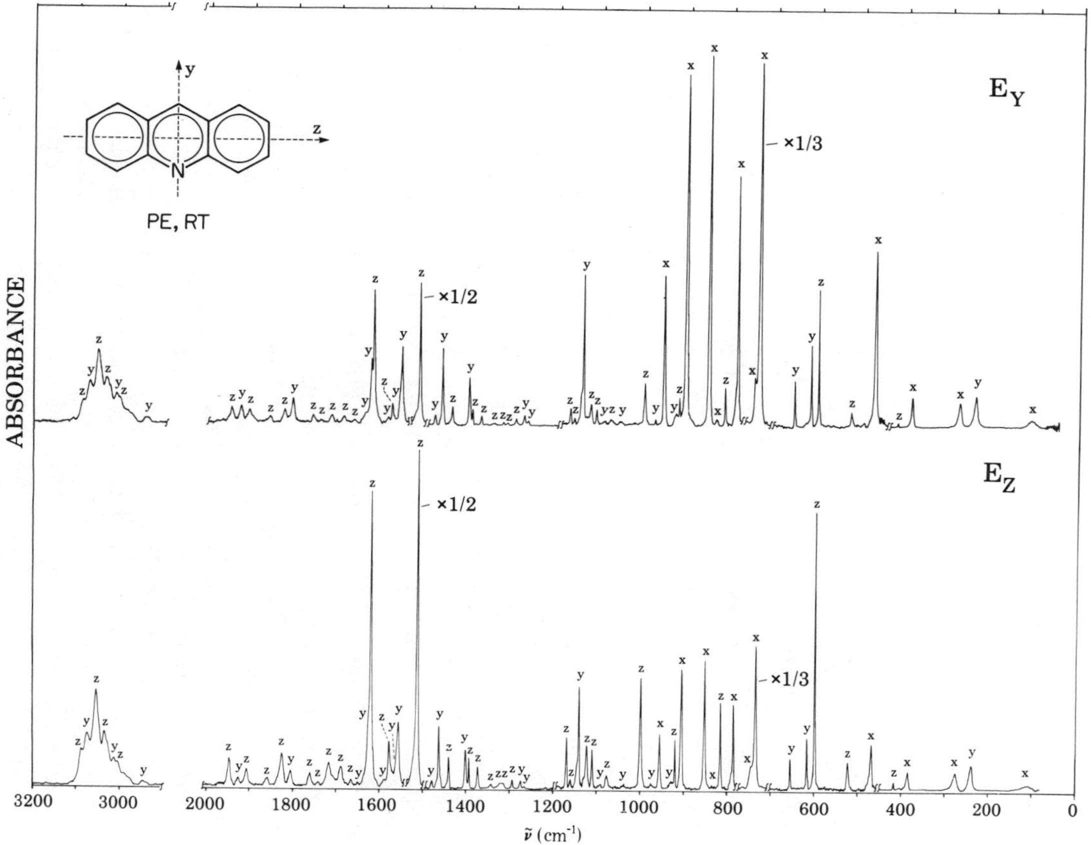

Figure 8.8 Polarized IR spectra of acridine in stretched polyethylene at room temperature. Top, E_Y; bottom; E_Z. Composed from segments measured in ordinary linear low-density polyethylene and segments measured in perdeuterated high-density polyethylene. The 2000–2900 cm^{-1} region was deleted since it showed no peaks of intensity observable on this scale. Reproduced by permission from ref. 18.

transmitted IR intensity was modulated due to absorption by the periodically varying triplet and ground-state concentrations. The ac part of the transmitted intensity was detected using a lock-in amplifier. A detailed mathematical model of the experiment, including a Fourier expansion of the input signal to the lock-in amplifier, led to equations (3.43)–(3.45) and permitted the evaluation of the dichroic ratio for the IR peaks of triplet acridine from the intensities in the induced spectra I_{T_1} and the intensities of the IR monitoring beam at the detector in the absence of the sample $I^0(\tilde{\nu})$

$$\frac{[E_{T_1}(i)]_Z}{[E_{T_1}(i)]_Y} = \frac{\int_i [I_{T_1}(\tilde{\nu})]_Z/[I^0(\tilde{\nu})]_Z \, d\tilde{\nu}}{\int_i [I_{T_1}(\tilde{\nu})]_Y/[I^0(\tilde{\nu})]_Y \, d\tilde{\nu}} \qquad (8.2)$$

where Y and Z refer to the polarization of the monitoring beam.

Acridine has both y- and z-polarized absorption in the region of excitation. Moreover, its concentration was high, and triplet energy transfer probably contributed to the loss of all photoselection by the exciting beam. For isotropic excitation, the same IR dichroic ratios would be expected for the nonoverlapping peaks of the triplet state as are found in the ordinary polarized IR spectra of the ground state.

This was indeed found. At 77 K, the dichroic ratios were 0.2, 1.0, and 3.5 for x-, y-, and z-polarized IR transitions in both the triplet and ground states of **3** (i.e., $K_x = 0.1$, $K_y = 0.3$, $K_z = 0.6$).

An example of the observed and processed spectra is found in Figure 8.9. This displays the most intense triplet-state IR absorption band of **3** at 751 cm^{-1}. This transition is out-of-plane polarized.

8.2.5 Acetic Acid Dimer[20a]

In stretched polymers, simple carboxylic acids exist almost exclusively in the dimeric form (Section 8.7.1). The IR transition moment directions in the dimer of acetic acid (**4A**) and its various deuterated analogs have been determined from IR linear dichroism in stretched polyethylene and perdeuterated polyethylene.

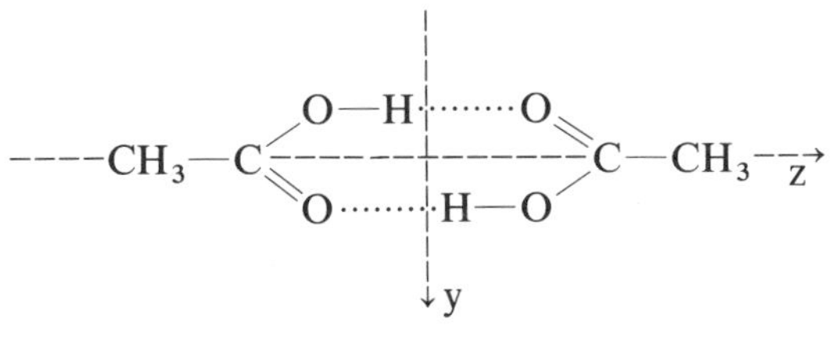

4A

The C_s symmetry of **4A** permitted an immediate determination of the orientation factor K_x for the out-of-plane axis from the dichroism of the out-of-plane bending of the hydroxyl group, and of the orientation factor K_z for the long axis from the dichroism of the O–H or O–D stretching band. For instance, in ordinary polyethylene, $K_x = 0.15$ and $K_z = 0.69$, so that $K_y = 0.16$ and the orientation is almost rod-like. The expression (5.64) was used to convert the

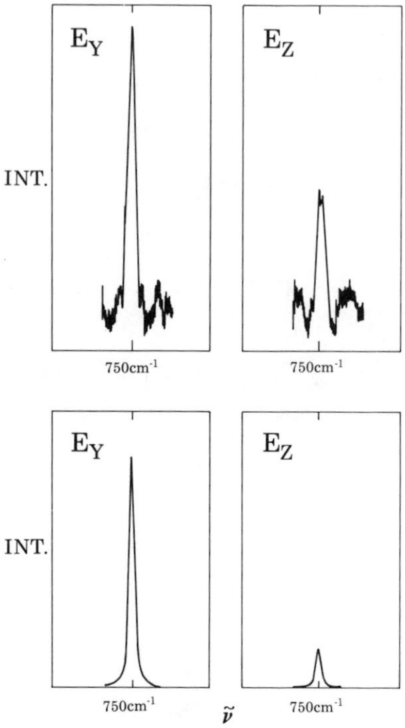

Figure 8.9 The 751 cm^{-1} band of triplet acridine in stretched polyethylene Top, experimental signal intensity (average of four scans); bottom, intensity after deconvolution of the IR beam intensity and of the slit function. Adapted by permission from ref. 20.

dichroic ratios for all other observed IR bands into absolute values for the angles between their transition moments and the z axis. Out-of-plane (x) and nearly short-in-plane-axis (y) polarized IR transitions had similar K values but were distinguished on the basis of their frequencies and the known vibrational assignments. Transition moment directions for in-plane polarized vibrations of the carboxylic ring are collected in Table 8.6a and compared with calculations by the fixed partial charge method using the formic acid dimer as a model. The presence of angles intermediate between 0° and 90° is compatible with the generally accepted inequality of the –O–H and =O· · · ·H bond lengths in the ring.

8.2.6 Dirhenium Decacarbonyl[21]

Liquid crystals tend to have numerous absorption bands throughout the usual IR region and are generally less suited as orienting solvents for comprehensive IR spectral studies of solutes than polymers of simple structure. They do have some transparent windows in the IR, however, and at times can be used as well, as the present example will show.

One region of transparency generally occurs between 1800 and 2800 cm^{-1}. Metal carbonyl stretching vibrations occur in the lower part of this region and

Table 8.6a
Transition Moment Directions of Carboxylic Acid Dimer Ring In-Plane Vibrations[a]

	Transition Moment Direction, deg				
Vibration	$(CH_3COOH)_2$	$(CH_3COOD)_2$	$(CD_3COOH)_2$	$(CD_3COOD)_2$	$(HCOOH)_2$calcd.
ν(O—H)(D)	0	0	0	0	−8
ν(C=O)	−75	−60	−60	−60	−75
ν(C—O)	+60	+60	+55	+45	+56
δ(O—H)(D)	65[b]	70[b]	35[b]	85[b]	±90
δ(OCO)	+30	+30	+30	+30	+36
δ(CCO)	0	10[b]	5[b]	10[b]	(−16)[c]

[a] The signs refer to a counterclockwise direction in formula **4A**.
[b] Sign undetermined.
[c] Value for δ(HCO).

have been studied in the nematic phase of *p*-(*p*′-ethoxybenzoxy)phenyl butyl carbonate, aligned by rubbing the cell plates (BaF_2) in one direction before introducing the sample.

Figure 8.10 shows the polarized IR absorption spectra of $Re_2(CO)_{10}$ in the C≡O stretch region, 1900–2100 cm^{-1}. This molecule has D_{4d} symmetry (**5**), permitting only two rather than three different K values (either $K_z = K_y > 1/3 > K_x$ or $K_z > 1/3 > K_y = K_x$).

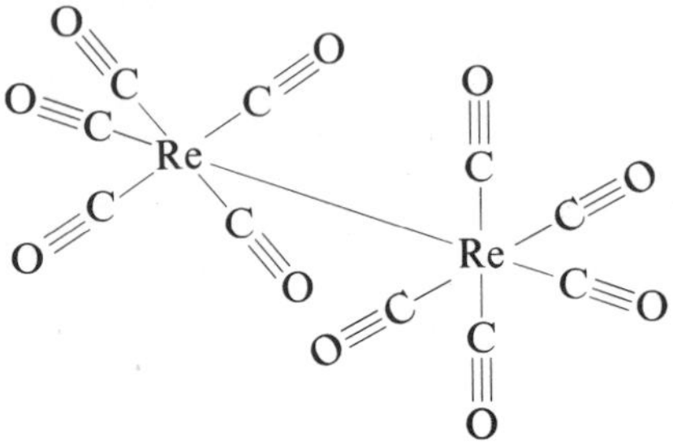

5

Three bands are observed. Those at 1980 and 2045 cm^{-1} exhibit positive dichroism, $E_Z(\tilde{\nu}) > E_Y(\tilde{\nu})$, and are z-polarized by definition; the one at 2010 cm^{-1} shows negative dichroism, $E_Z(\tilde{\nu}) < E_Y(\tilde{\nu})$, and is x-polarized by definition. As in Sections 8.2.1 and 8.2.2, the question is, do we have $K_y = K_z > 1/3$ or $K_y = K_x < 1/3$, i.e., does the orientation axis z lie perpendicular to or parallel to the symmetry axis of the molecule? The measurements were performed on an older dispersive instrument; the estimated experimental error is large enough and the degree of orientation small enough that it would be difficult to base the decision between the two possible directions of the orientation axis z on the condition $\sum K_u = 1$ as we did earlier.

However, this is not necessary; there are three IR-allowed normal modes for C≡O stretches in $Re_2(CO)_{10}$, two of b_2 symmetry and one of e_1 symmetry.

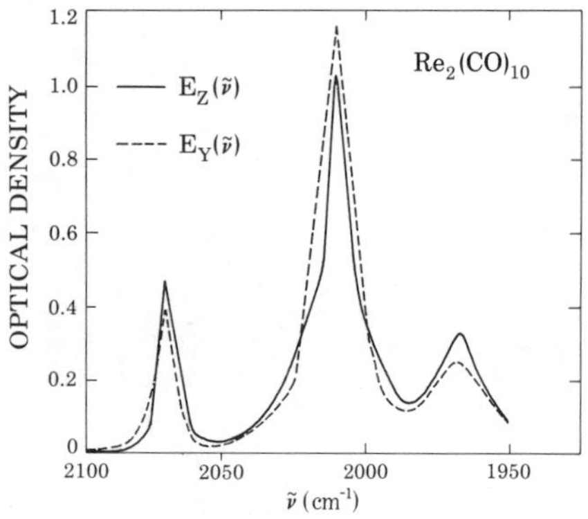

Figure 8.10 Polarized IR spectra of a 2.7×10^{-2} *M* nematic solution of $Re_2(CO)_{10}$, corrected for baseline. Adapted by permission from ref. 21.

The b_2 modes are polarized parallel and the e_1 mode perpendicular to the symmetry axis. It follows that the two peaks with positive dichroism, 1980 and 2045 cm^{-1}, belong to the b_2 species, and the one with negative dichroism, 2010 cm^{-1}, belongs to the e_1 species. Therefore, the Re—Re axis is the orientation axis z of the molecule. This is not surprising, given that the ratio of the longer to the shorter molecular dimension is about 1.5.

8.2.7 Photoselection in a Rare Gas Matrix: Manganese Pentacarbonyl[22]

Photoselection techniques have been applied with considerable success to the elucidation of the IR and UV-visible spectra of unstable metal carbonyls isolated in low-temperature matrices. Here we describe the results from only one of a series of papers dealing with these compounds.

The pentacarbonyl manganese(0) radical, $Mn(CO)_5$ (**6**), can be prepared by irradiation of $HMn(CO)_5$ isolated in a matrix of solid CO. The by-product is the HCO radical. The highly reactive $Mn(CO)_5$ shows characteristic IR bands at 1978.4, 1987.6, and 1992.7 cm^{-1} in the carbonyl stretch region (Figure 8.11), a visible band at 798 nm, and end absorption in the UV region below 340 nm. On the basis of annealing experiments, the band at 1992.7 cm^{-1} was assigned as a matrix side band of the strong band at 1987.6 cm^{-1}. The analysis of the IR spectra, using also ^{13}CO enrichment, suggested strongly that the molecule has a square pyramidal geometry (C_{4v}), as shown in formula **6**, with an axial-equatorial bond angle of 96 ± 3°. Two carbonyl stretch bands of a_1 symmetry and one of e symmetry are then expected. One of the a_1 bands is presumably not observed. It should be weak and is expected to occur in a region obscured

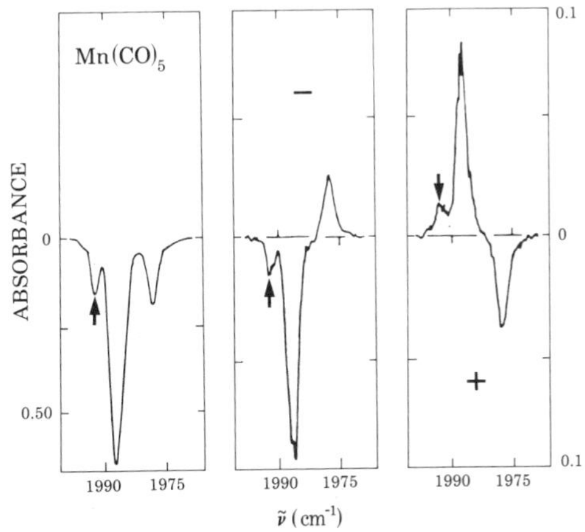

Figure 8.11 Infrared spectra of $Mn(CO)_5$ in a CO matrix at 20 K: left, ordinary absorbance of unoriented sample; center, linear dichroism $E_Y - E_Z$ after irradiation with Z-polarized light at 330 ± 10 nm (Z horizontal); right, linear dichroism $E_Z - E_Y$ after irradiation with Z-polarized light at 330 ± 10 nm (Z vertical). Adapted by permission from ref. 22.

by the CO matrix. The 1978.4 cm^{-1} band is assigned to a_1 symmetry, and the 1987.6 cm^{-1} band to e symmetry.

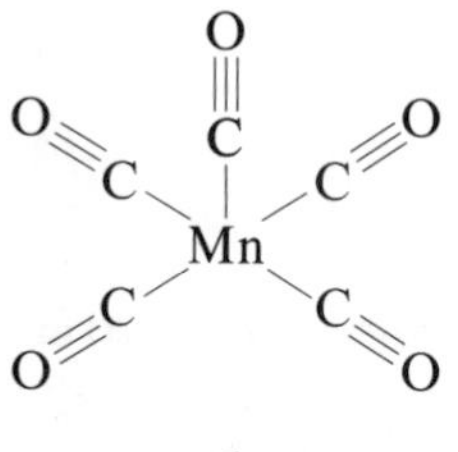

6

When unpolarized light is used in the photochemical production of matrix-isolated $Mn(CO)_5$, it shows no linear dichroism in IR or UV. (The band in the visible was too weak to measure reliably.) When such an unoriented sample is irradiated in the region of UV absorption with Z-polarized light (313 or 330 nm), linear dichroism is induced. Not surprisingly, the UV region now shows negative dichroism, $E_Z < E_Y$. The linear dichroism in the IR region is shown in Figure 8.11. It is positive for the 1978.4 cm^{-1} band ($E_Z > E_Y$) and negative for the 1987.6 and 1992.7 cm^{-1} bands ($E_Z < E_Y$). This is in full agreement with the above assignments and demonstrates that the UV absorption is polarized in the same way as the 1987.6 cm^{-1} IR band, perpendicular to the fourfold axis.

This photoorientation effect is believed to be due to the repeated loss of a CO ligand and recapture of the same or another CO molecule from the matrix.

8.2.8 Photoselection in a Rare Gas Matrix: 1,3-Perinaphthadiyl[22a]

Bonding in organic 1,3-biradicals has been a subject of considerable interest. A recent IR study of argon-matrix isolated triplet ground state methylene-bridged 1,8-naphthoquinodimethanes **6A–6D** (Figure 8.11a) depended heavily on the use of linear dichroism.

The isotopic labels present pinpointed the locations of the four IR bands associated with the C—H stretches in the trimethylene moiety. If the carbon skeleton is planar as shown in Figure 8.11a, one expects the α C—H bonds to provide a z-polarized symmetric and a y-polarized antisymmetric combination, and the β C—H bonds to provide a z-polarized symmetric and an x-polarized

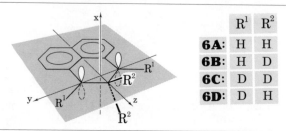

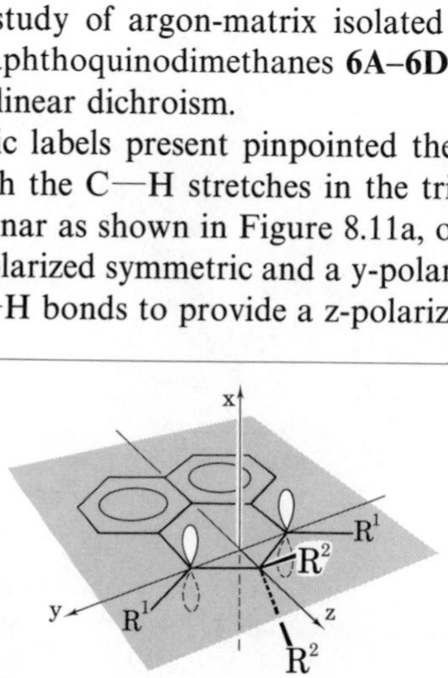

	R^1	R^2
6A:	H	H
6B:	H	D
6C:	D	D
6D:	D	H

Figure 8.11a The geometry of triplet 1,3-perinaphthadiyl biradicals.

antisymmetric combination. In order to assign the four observed bands, oriented samples of matrix-isolated **6A** and **6C** were prepared by photoselection, taking advantage of the ready photoisomerization of the biradical by visible light, and of the availability of two mutually perpendicularly polarized visible transitions.

The absolute polarizations of the visible transitions were known from previous work on related naphthoquinodimethanes and were confirmed by a measurement of the UV-visible and IR dichroic absorption of **6A** in stretched polyethylene after it was produced at 10 K by unpolarized irradiation of a cyclopropane precursor at a wavelength where the absorption is of strongly mixed polarization. The orientation of the precursor was known from the measurement of the UV dichroism of its naphthalene chromophore.

Irradiation with linearly polarized laser light at 496.5 nm was absorbed by a y-polarized peak of **6A** or **6C**, after partial photoconversion yielding a sample with orientation factors K(y) for the remaining triplet biradical. Similar irradiation at 488.0 nm was absorbed by a z-polarized peak, yielding a biradical sample with orientation factors K(z) (cf. Section 4.6.3).

Measurement of the dichroic ratios of the approximately twenty IR bands on the four samples of the biradical (two different orientations for each of **6A** and **6C**) permitted a determination of the three orientation factors, $K_u(y)$ or $K_u(z)$ (u = x,y,z), for each (Table 8.6b). For biradical molecules remaining after

Table 8.6b
IR Vibrations of Triplet 1,3-Perinaphthadiyl and Their Polarizations

	Experimental				Calculated[a]	
Cmpd	$\tilde{\nu}$ (cm^{-1})	K(y)	K(z)	pol	$\tilde{\nu}$ (cm^{-1})	pol
6A	2742	0.36	0.44	x	2824	x
	2761	(0.33)[b]	0.22	z	2885	z
	3056	0.28	0.39	y	3011	y
	3066	0.35	0.19	z	3016	z
	$[K_y(y) = 0.26, K_x(y) = K_z(y) = 0.37]$[c]					
	$[K_z(z) = 0.23, K_x(z) = K_y(z) = 0.39]$[c]					
6C	2046	0.38	0.41	x	2081	x
	2061	0.35	0.26	z	2088	z
	2237	0.22	0.36	y	2208	y
	2253	0.39	0.23	z	2212	z
	$[K_y(y) = 0.24, K_x(y) = K_z(y) = 0.41]$[c]					
	$[K_z(z) = 0.22, K_x(z) = K_y(z) = 0.38]$[c]					

[a] MNDO, corrected by a multiplicative factor of 0.885, chosen so as to bring the calculated and experimental frequency of the CH_2 as stretch in propane into coincidence.

[b] This value is subject to a large experimental uncertainty.

[c] Derived from the dichroic ratios of about twenty IR bands. Some of the small difference between $K_x(y)$ and $K_z(y)$ and between $K_x(z)$ and $K_y(z)$ is probably due to slightly mixed polarization of the absorption at 496.5 and 488 nm. This has no effect on the proposed assignments nor on the conclusion that the molecule is of C_{2v} symmetry.

photoselection on an absorbing visible $\pi\pi^*$ transition moment oriented in the naphthalene plane along z, $K_z(z) < K_y(z) = K_x(z)$. For either C_{2v} or C_s symmetry $K_y(z)$ is determined unequivocally as 0.39 in **6A** and 0.36 in **6C** for the particular samples used, and this calls for $K_x(z) = 0.39$, $K_z(z) = 0.22$ in **6A** and for $K_x(z) = 0.36$, $K_z(z) = 0.28$ in **6C**. The dichroism of the symmetric and antisymmetric CH_2 or CD_2 vibrations reflects the orientation of the methylene group relative to z and to the out-of-plane axis x; for a transition polarized in the xz plane at angle ϕ to z, equation (5.64) applies. Within the experimental error of 0.02–0.03, the data in Table 8.6b demand a z polarization of the CH_2 and CD_2 symmetric stretches ($\phi = 0°$) and an x polarization of the antisymmetric stretches ($\phi = 90°$).

These results are only compatible with C_{2v} symmetry for the triplet molecule, and lead to unequivocal absolute assignment of the four C—H or C—D stretching vibrations of the —$\dot{C}$H—CH_2—$\dot{C}$H— and —$\dot{C}$D—CD_2—$\dot{C}$D— moieties (Table 8.6b).

8.2.9 Photoselection in a Rare Gas Matrix: 1-Methylsilene[22b]

In the preceding example, the same matrix-isolated species offered both independent UV-visible transition moments which need to be used for the photoselecting absorption if absolute assignments in three dimensions are to result. Now, we consider a case of two photochemically interconverting argon matrix-isolated isomers, **6E** and **6F**, each of which carries one transition moment used for photoselection. In such a case, not only the orientation of the remaining molecules but also that of the photoproduct plays an essential role. It is therefore important that the overall orientation of the molecular framework be modified in the phototransformation process in a predictable fashion or not at all. This is most likely for photoreactions which involve a minimal degree of skeletal rearrangement and which release a minimal amount of heat into the matrix, thereby minimizing rotation of the product molecule in its momentarily warm environment.

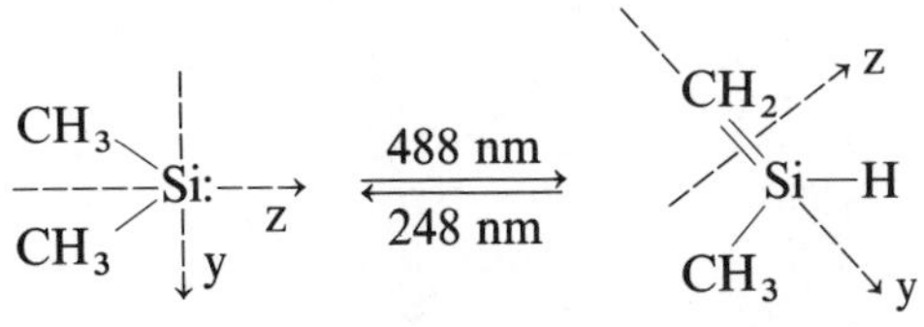

6E **6F**

The present case is favorable in one respect, since **6E** and **6F** differ by a mere hydrogen shift. However, while the 488 nm irradiation of matrix-isolated **6E** yields oriented **6F**, the 248 nm irradiation of matrix-isolated **6F** yields unoriented **6E**, presumably since the photon energy is nearly doubled whereas the heats of formation of **6E** and **6F** are comparable. Of course, the remaining substrate molecules are aligned in either case.

Based on calculations and analogy to similar molecules, the relevant transition moment in **6E** is due to an n → p excitation and is therefore directed along x, perpendicular to the yz (CSiC) plane while that in **6F** is due to a $\pi\pi^*$ excitation and lies along the C—Si bond. According to Section 4.6.3, after irradiation of **6E** with light linearly polarized along Z, one expects $K_x < 1/3 < K_y = K_z = (1 - K_x)/2$ for the remaining sample. Indeed only two distinct dichroic ratios were observed and were related as expected. This permits an immediate assignment of the out-of-plane (x) polarized IR bands of **6E**. In-plane polarized IR transitions can also be discerned, but no distinction is possible between y and z polarizations. Since irradiation of **6F** produced an unoriented sample of **6E** as noted above, this part of the assignment remained incomplete.

The assignments in **6F**, on the other hand, can be made complete, and we shall consider them in detail. Since the sample of **6F** formed by partial photoconversion of matrix-isolated **6E** is oriented, Section 4.6.3 leads one to expect $K_x > 1/3 > K_y = K_z = (1 - K_x)/2$. As before, only two distinct dichroic ratios were observed. Once again, out-of-plane (x) polarized IR transitions are identified immediately (Table 8.6c), as are in-plane polarized ones, but for the latter a distinction between the various possible directions within the yz plane is not yet possible.

Such differentiation can be obtained from a photoselection experiment on **6F**, since the relevant $\pi\pi^*$ transition moment is not parallel to the one used so far (x) but rather, lies along the C=Si bond direction, which we shall take for the y axis in **6F**. After partial photodestruction with linearly polarized 248 nm light, the orientation factors of the remaining **6F** will be $K_y < 1/3 < K_x = K_z = (1 - K_y)/2$. As expected for a molecule of C_s symmetry, a whole series of different K values was observed for the IR transitions. The IR LD spectrum is shown in Figure 8.11b.

The orientation factors of the molecular axes, K_x, K_y, and K_z were then obtained in two ways: (i) The dichroism of a 614 cm^{-1} band which was already known to be out-of-plane (x) polarized from the photoselection work on **6E** yields $K_x = 0.38$ and thus $K_y = 0.24$, $K_z = 0.38$. (ii) While the dichroism of the $\pi\pi^*$ transition could not be measured with sufficient accuracy to yield K_y, this value could be obtained from the dichroic ratio of the Si=C stretch at 989 cm^{-1} as $K_y = 0.26$; then, $K_x = K_z = 0.37$.

Both sets of K values were used in (5.64) to evaluate the absolute values of the angles ϕ between the IR transition moments of the in-plane polarized vibrations of **6F** and the C=Si bond directions. The signs of the angles were obtained by comparison with MNDO calculations. The results for both limiting assumptions are collected in Table 8.6c. It is remarkable how simply most of the IR transition moment directions are related to molecular structure, and Figure 8.11c shows how they map onto the molecular structural formula.

8.3 Detection and Assignment of Electronic Transitions in High-Symmetry Molecules

There have been well over 100 studies in which symmetry assignments of transitions in symmetrical molecules were deduced from their linear dichroism in a uniaxial medium. In many of these, previously undetected "hidden" transitions were uncovered by constructing the purely polarized reduced spectra.

Table 8.6c
1-Methylsilene Vibrations

Calculated (MNDO, planarity enforced)						Observed				
			Assignment	$\tilde{\nu}$ (cm^{-1})	Polarization	$\tilde{\nu}$ (cm^{-1})	Polarization	K_i	ϕ(i)[a]	ϕ(ii)[a]
1	ν_1	a′	s CH_2 stretch	3427	+1°	3018 (1–2)	[b]			
2	ν_2	a′	a CH_2 stretch	3396	−89°					
3	ν_3	a′	s CH_3 stretch	3354	+59°	2976 (3–5)				
4	ν_4	a′	a CH_3 stretch	3289	+34°		[b]			
5	ν_{15}	a″	a CH_3 stretch	3283	a″					
6	ν_5	a′	Si—H stretch	2294	−62°	2182, 2187	In-plane	0.38	90°	90°
7	ν_{16}	a″	a CH_3 def	1430	a″	1412	Out-of-plane			
8	ν_6	a′	a CH_3 def	1427	+89°	1394	In-plane			
9	ν_7	a′	CH_2 scissor	1422	+7°	1296	In-plane	0.26	(−)22°	0°
10	ν_8	a′	CH_3 def	1383	+50°	1255	In-plane	0.36	(+)68°	(+)72°
11	ν_9	a′	Si═C stretch	1078	+6°	(996[c]), 989	In-plane	0.28	(−)32°	(−)25°
12	ν_{17}	a″	CH_2 out-of-plane wag	994	a″	831	Out-of-plane			
13	ν_{10}	a′	CH_2 + SiH in-plane bend	882	+3°	(889)[c], 880	In-plane	0.30	(+)41°	(+)37°
14	ν_{11}	a′	in-plane CH_3 rock	833	−56°	(801)[c], 811	In-plane	0.39	90°	90°
15	ν_{18}	a″	out-of-plane CH_3 rock	772	a″	711	Out-of-plane			
16	ν_{12}	a′	Si—C stretch	754	−62°	729	[b]			
17	ν_{13}	a′	SiH + CH_2 in-plane bend	689	+26°	678	In-plane	0.29	(+)37°	(+)31°
18	ν_{19}	a″	CH_2 twist + SiH out-of-plane bend	609	a″	614	Out-of-plane	0.38	90°	90°
19	ν_{20}	a″	SiH out-of-plane bend + CH_2 twist	(241i)						
20	ν_{14}	a′	C-Si-C bend	253	+66°					
21	ν_{21}	a″	CH_3 torsion	25	a″					

[a] The angle between the IR transition and the $\pi\pi^*$ transition moment direction, measured counterclockwise in Figure 8.11c. The signs in parentheses are not experimental but are chosen to agree with the MNDO calculations. The two choices of K_x, K_y and K_z values are labeled (i) and (ii), see text.
[b] Too weak to obtain reliable dichroic data.
[c] Weak satellites.

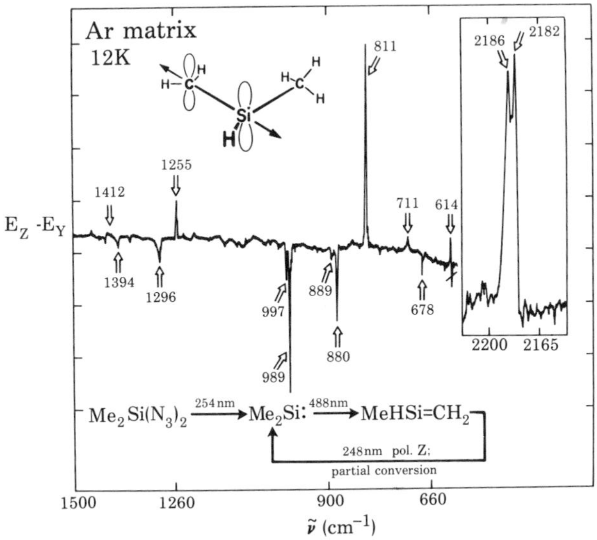

Figure 8.11b The IR linear dichroism $E_Z(\tilde{\nu}) - E_Y(\tilde{\nu})$ of an argon matrix isolated sample of 1-methylsilene after partial photoconversion into dimethylsilylene with Z polarized 248 nm light (arbitrary units). Reproduced by permission from ref. 22b.

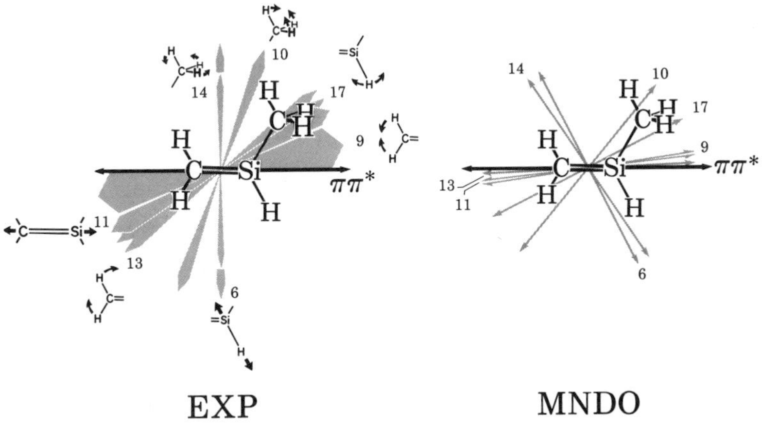

Figure 8.11c Transition moment directions of in-plane polarized transitions of 1-methylsilene in the UV (black arrows) and IR (grey arrows) regions. Left, measured; right, calculated (MNDO for IR and INDO/S for UV transitions). See Table 8.6c for the numbering and assignment of IR transitions. Reproduced by permission from ref. 22b.

Only a few examples will be given here; many additional references can be found in ref. 23.

8.3.1 The Azines[24,25]

Our first example demonstrates the use of compensated nematic liquid crystals and the equivalence of the measurement of linear dichroism and of liquid-crystal-induced circular dichroism (ICD). It also illustrates the use of the Saupe order matrix elements S_{uu} common in liquid crystal work. The nπ^* transitions of the diazines pyrazine (**7**), pyrimidine (**8**), and pyridazine (**9**) are well separated from their first $\pi\pi^*$ transition (L_b) and are known from vapor-phase studies to be out-of-plane polarized.

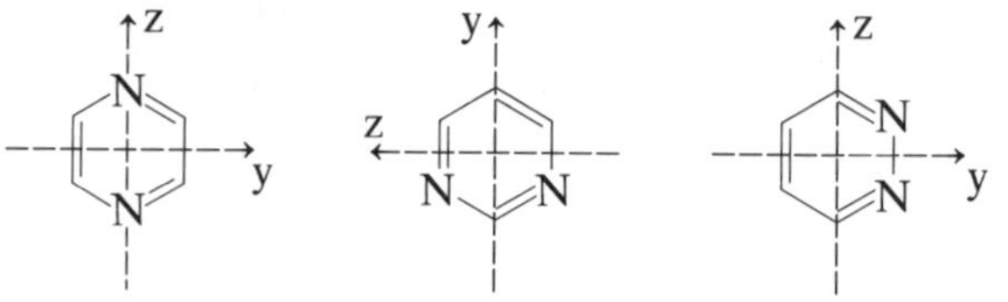

7 **8** **9**

The compensated mixture was cholesteryl chloride – cholesteryl laurate (1.85:1.0 w/w, T_{nem} = 30°C, in an alternating electric field, 1200 Hz, up to 16 kV across a distance of 11.5 mm), the cholesteric mixture was cholesteryl chloride – cholesteryl nonanoate (1:1 w/w, pitch band at 6900 cm^{-1} reflecting left-handed circularly polarized light; mean refractive index n = 1.50, linear birefringence Δn = 0.05, pitch length P = 0.97 μm).

The orientation matrix elements S_{uu} of the three diazines had been determined by NMR in a similar nematic solvent and are given in Table 8.7. They show that the orientation is very nearly disclike. In each case, the x axis is perpendicular to the molecular plane and the y axis is the twofold rotation axis which interchanges the positions of the two nitrogen atoms. The difference between the two in-plane axes, y and z, is so small that it can hardly be used for reliable polarization assignments.

Given these S_{uu} values, one expects the linear dichroism (LD) spectra to show the nπ^* transitions of the diazines with negative dichroism and all of their $\pi\pi^*$ transitions with a similar degree of positive dichroism. This is borne out by the observations as shown on the left in Figure 8.12. Quantitative evaluation, assuming pure polarizations for the transitions, yields the S_{uu} elements also listed in Table 8.7. Again, we note $S_{yy} \doteq S_{zz}$ ($K_y \doteq K_z$), i.e., a very nearly disc-like type of orientation. The Fraser-Beer model could be used in this instance. The differences in the S_{uu} values derived from the optical experiments and those derived from NMR give the reader an indication of concentration effects and of the degree of transferability between methods and, perhaps, laboratories.

Induced circular dichroism (ICD) was measured in solutions of the diazines in a left-handed cholesteric liquid crystal where individual nematic layers should be quite similar to those used in the above LD measurements. The ICD spectra

Table 8.7
S_{uu} Values[a] of the Diazines in a Liquid Crystal[24,25]

Nitrogen Positions	S_{xx}	S_{yy}	S_{zz}
NMR:			
1,2 (**9**)	−0.099	0.035	0.064
1,3 (**8**)	−0.092	0.028	0.064
1,4 (**7**)	−0.097	0.041	0.056
LD:			
1,4(1,2; 1,3)[b]	−0.144 (nπ*)	0.063 (L_b)	0.081[c]
ICD:			
1,2 (**9**)	−0.15 (nπ*)	0.07 (L_b)	0.08[c]
1,3 (**8**)	−0.14 (nπ*)	0.07 (L_b?)[d]	0.07[c] (L_b?)[d]
1,4 (**7**)	−0.145 (nπ*)	0.065 (L_b)	0.085[c]

[a] See the text for the composition of the liquid crystal.
[b] Stated to be "similar to 1,4" in ref. 25.
[c] Determined from $S_{zz} = -(S_{xx} + S_{yy})$.
[d] The L_b transition is polarized along z, but this cannot be determined from the ICD spectra.

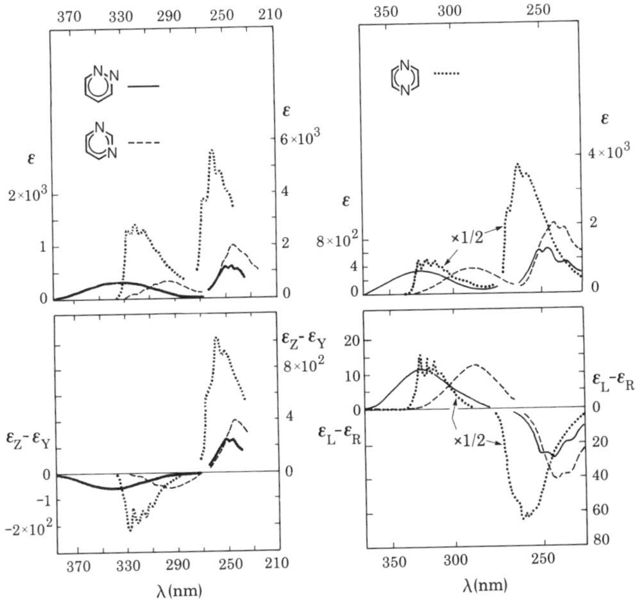

Figure 8.12 The absorption (top), linear dichroism (bottom left) and liquid-crystal-induced circular dichroism (bottom right) of the three diazines. The linear dichroism was measured in a compensated nematic mixture of cholesteryl chloride and cholesteryl laurate (1.85:1 by weight, T_{nem} = 30°C). The induced dichroism was measured in the cholesteric plane texture formed by a mixture of cholesteryl chloride and cholesteryl nonanoate (1:1 by weight, 25°C). On the right-hand side, the intensities for pyrazine (**7**) have been divided by 2. Adapted by permission from refs. 25 (left-hand side) and 24 (right-hand side).

are shown on the right-hand side of Figure 8.12. The matrix elements S_{uu} derived from the circular dichroism of the $n\pi^*$ and L_b bands using equation (3.29) and neglecting $\tilde{\nu}_0^2$ next to $\tilde{\nu}^2$ compare very well with those obtained from the linear dichroism (Table 8.7). Since the crystal is left-handed (P is negative), transitions which show positive (negative) linear dichroism $[E_Z(\tilde{\nu}) - E_Y(\tilde{\nu})]$ in the nematic crystal exhibit negative (positive) circular dichroism $[E_L(\tilde{\nu}) - E_R(\tilde{\nu})]$ in the cholesteric crystal.

The authors of this study did not consider it necessary to obtain the reduced spectra $A_x(\tilde{\nu})$ and $A_y(\tilde{\nu}) + A_z(\tilde{\nu})$, which could have been done using equations (5.36) and (5.37) or their equivalents expressed in terms of the S_{uu} matrix elements. The reduction would probably have revealed minor components of in-plane polarization in the $n\pi^*$ band.

8.3.2 Perylene[26]

Polycyclic benzenoid hydrocarbons have provided a fertile field of investigation for the spectroscopist, and many applications of the stretched sheet method to them have been reported. Our first example from this area, perylene (**10**) in stretched ($R_s = 4.0$) poly(vinyl alcohol), illustrates the fine experimental work which has been coming from the laboratories of Tanizaki and his collaborators.

As we shall see, the orientation of perylene observed in these measurements happens to satisfy equation (5.84) nearly exactly. Also, the molecule has only in-plane polarized absorption of any intensity in the spectral region of interest, so that the use of Tanizaki's equations (Section 5.2.2) should produce essentially correct reduced spectra.

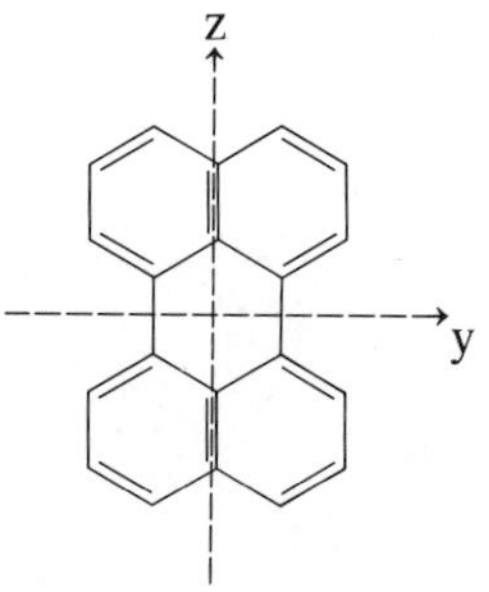

10

The dichroic spectra and the measured dichroic ratio R_d are shown at the top in Figure 8.13 and make it clear that the visible band is long-axis polarized—as had been known from prior single-crystal work—and that the strong transition near 40 000 cm^{-1} is short-axis polarized. Assuming that the latter is polarized purely, the authors used its R_d value and $R_s = 4$ to derive the value 57.3° for the angle between the direction of the short axis and the direction of Tanizaki's orientation axis ($|\theta_y| = 57.3°$ in the notation of Chapter 5). A similar consideration of the R_d value of the long-axis-polarized transition near 23 000 cm^{-1} provided the value 33.3° for the angle between the long axis

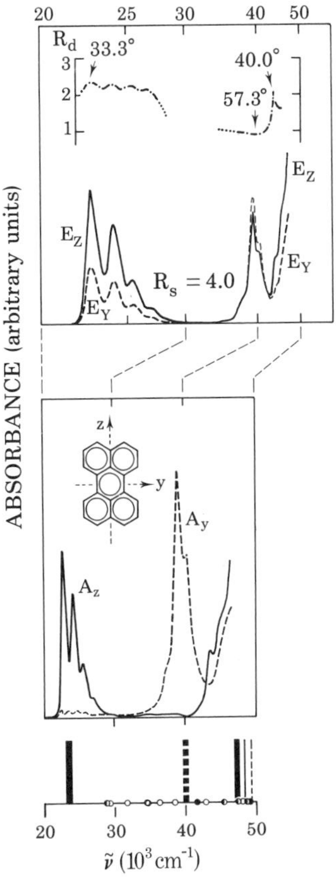

Figure 8.13 Top, dichroic spectra of perylene in stretched poly(vinyl alcohol) at stretch ratio $R_s = 4.0$: solid line, $E_Z(\tilde{\nu})$; dashed line, $E_Y(\tilde{\nu})$; above, the dichroic ratio R_d. Center, reduced spectra of perylene: solid line $A_z(\tilde{\nu})$; dashed line, $A_y(\tilde{\nu})$. Bottom, calculated (PPP) excitation frequencies, polarizations (full line, z; broken line, y), and intensities (thick lines, oscillator strength $f \geqslant 0.2$; thin lines, $f < 0.2$; white dots, symmetry-forbidden transitions; black dots, z-polarized alternant pairing symmetry-forbidden transitions; half-filled dots, y-polarized alternant pairing symmetry-forbidden transitions). Adapted by permission from ref. 26.

and the orientation axis if the transition is assumed to be purely polarized. The two angles add up to 90.6° rather than 90° as hoped for, and the authors conclude that the transition in the visible is not quite purely polarized. Since the deviation from 90° is small, the short-axis-polarized intensity which needs to be assumed in the visible is quite weak. The reduced spectra derived by the Tanizaki procedure described in Section 5.2.2 are shown at the bottom in Figure 8.13.

The spectra which one would have derived using the TEM model as described in Section 5.1.2 are essentially identical, except that it would have been natural to assume that the visible band, too, is purely polarized, and the stepwise reduction procedure would have reduced the $A_y(\tilde{\nu})$ curve to zero (except for some remaining wiggles to which no simple physical significance would be attributed). The assumption of pure polarization for both transitions would be quite reasonable considering that they are very strong ($\varepsilon \doteq 40\,000$). Also, the weak intensity shown for $A_y(\tilde{\nu})$ in Figure 8.13, as obtained by the Tanizaki reduction procedure, does not have the shape expected for a vibronically borrowed component, in that it extends to the 0-0 component of this strong transition. As the authors point out, there is no theoretical reason to expect the presence of a separate y-polarized electronic transition in the region, in agreement with the results of the π-electron calculations which the authors performed to complete their assignments (Figure 8.13).

On the whole, however, both methods of procedure agree quite nicely in this instance, as expected when $|\theta_y| + |\theta_z| \cong 90°$. In terms of the TEM model, the observed dichroic ratios lead to $K_y = 0.31$ and $K_z = 0.54$. These values add up to 0.85. According to (5.84), Tanizaki's equations will hold if K_y and K_z add up to $(1 + T)/2$. However, for $R_s = 4.0$, we have $T = 0.70$ according to (5.71) and $(1 + T)/2$ equals 0.85. Thus, both the Tanizaki and the TEM models will give the same results in this case, and the corresponding point in the orientation triangle shown in Figure 5.8 will lie almost exactly on the dashed line appropriate for $T = 0.70$. A larger deviation from this situation would have been revealed by a larger deviation between the baseline and the $A_y(\tilde{\nu})$ curve obtained for the visible region by the Tanizaki procedure. In this region, the $A_y(\tilde{\nu})$ curve would then have been of the same shape as the $A_z(\tilde{\nu})$ curve, only weaker in intensity (incompletely reduced spectrum in the terminology of the TEM model).

Note that $K_x = 1 - K_y - K_z = 0.15$. An assumption of rodlike ($K_x = K_y$) or disclike ($K_y = K_z$) behavior (Fraser-Beer model) would clearly be quite wrong in this instance.

The assignment of both intense transitions (23 000 and 40 000 cm^{-1}) to the calculated transitions shown at the bottom in Figure 8.13 is straightforward. The reduced spectra show the presence of both y- and z-polarized intensity above 42 000 cm^{-1}, and both of these transitions can be confidently associated with their counterparts in the calculated spectra located above 45 000 cm^{-1}. The spectra further show weak y-polarized intensity near 30 000 cm^{-1} and weak z-polarized intensity starting at about 33 000 cm^{-1} and extending to higher wavenumbers. The former would stand out much more clearly if the $A_y(\tilde{\nu})$ spectrum had been obtained using the TEM reduction, which would have removed the y-polarized intensity at lower energies, where R_d is higher, but not near 30 000 cm^{-1}. This transition had also been detected previously by polarized fluorescence[27] and can be assigned to the second calculated transition, whose calculated intensity vanishes in the usual PPP approximation but which is not actually symmetry-forbidden (it is y-polarized). Finally, the weak z-polarized intensity at 33 000 cm^{-1} and above is probably due to vibronic activity by one or more of the symmetry-forbidden transitions calculated for this region.

Detailed analyses of this kind are now available for numerous benzenoid hydrocarbons, and the agreement with π-electron theory has been found to be excellent. As is the case for perylene, the theory generally predicts more transitions than one-photon experiments reveal for hydrocarbons with a center of symmetry. This is understandable considering the g ↔ u symmetry selection rule of one-photon transitions and the g symmetry of the ground state. It is only through recent progress in the recording of two-photon absorption spectra, which reveal excited states of g symmetry, that the number of observed transitions has been brought into agreement with the number of calculated transitions (cf. Section 8.3.7).

8.3.3 Terphenyl and its Aza Analogs[28]

Our next example of a benzenoid hydrocarbon illustrates a case in which the orientation is rodlike and the Fraser-Beer model is applicable. The three compounds considered, p-terphenyl (**11**), 3,6-diphenylpyridazine (**12**), and 3,6-diphenyltetrazine (**13**), differ only by aza nitrogen replacement. Their dichroic

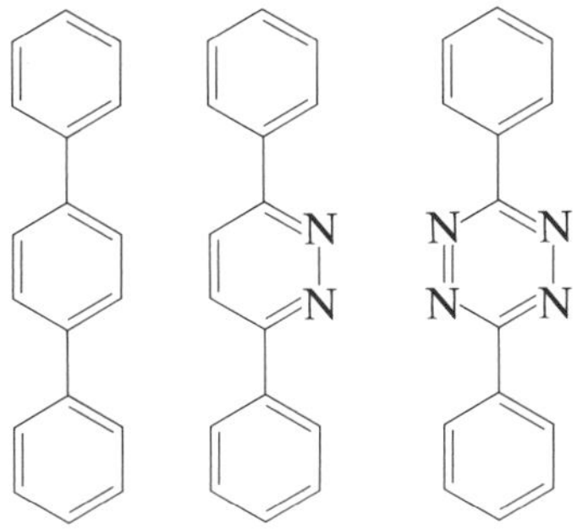

11 **12** **13**

spectra were measured in stretched polyethylene at 77 K and reduced by the trial-and-error method (Figure 8.14). The reduction was based on the assumption that the strong long-axis-polarized absorption peak near 32 000 cm^{-1} and the strong short-axis-polarized absorption peak near 45 000 cm^{-1} represent purely polarized spectral features (in the latter case, superimposed on structureless absorption intensity of the opposite polarization). The resulting orientation factors, $K_z = 0.85$ and $K_y = K_x = 0.08$, are the same within experimental error for all three molecules. In the case of **13**, an independent determination of K_x was possible from the dichroic ratio of the very weak $n\pi^*$ absorption band at 17 000 cm^{-1}. It yielded the value $K_x = 0.07$–0.08, thus confirming the validity of the assumption made in the reduction procedure based on the intense $\pi\pi^*$ transitions.

Since the orientation is rodlike, the reduction procedure yields $A_z(\tilde{\nu})$ and $[A_y(\tilde{\nu}) + A_x(\tilde{\nu})]/2$; these are the curves shown in Figure 8.14. The molecules are not necessarily strictly planar, but the authors found that calculations based on the assumption of planarity gave very good results, and it is indeed reasonable to consider the intense absorption peaks as $\pi\pi^*$ transitions polarized along z or y, and the extremely weak $n\pi^*$ transition at 17 000 cm^{-1} as polarized along

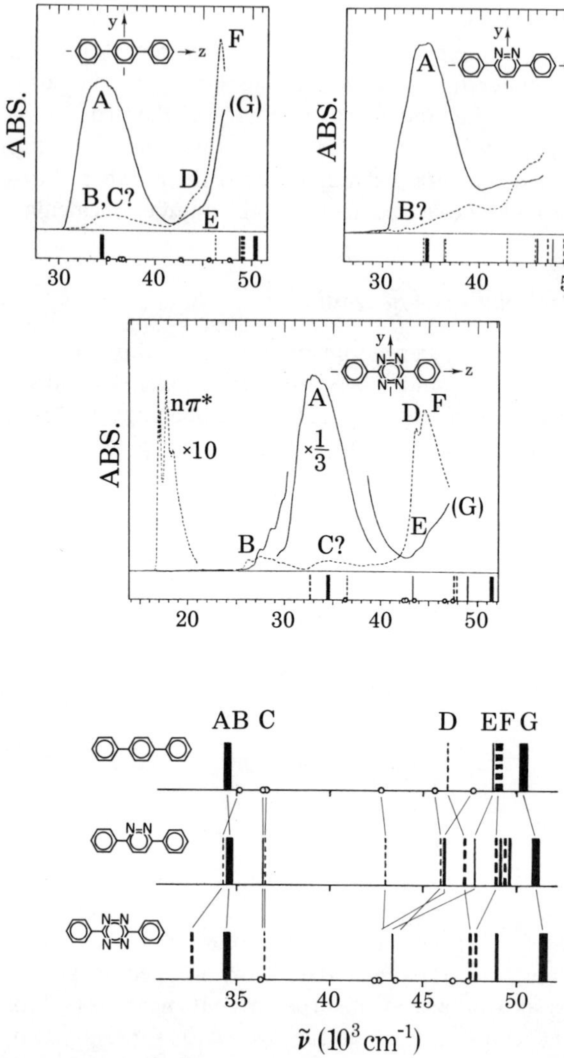

Figure 8.14 Top, the reduced spectra A_z (full) and $A_x + A_y$ (dashed) of **11–13** obtained from linear dichroism in stretched polyethylene (77 K). Below each spectrum, results of a PPP calculation of transition energies, polarizations (full, z; dashed, y) and oscillator strengths (thick, $f \geqslant 0.4$; medium, $0.4 \geqslant f \geqslant 0.1$; thin, $0.1 > f > 0$; circles, $f = 0$). Bottom, a correlation diagram for $\pi\pi^*$ transitions in the three molecules. Adapted by permission from ref. 28.

x, although strictly speaking $A_y(\tilde{\nu})$ and $A_x(\tilde{\nu})$ cannot be separated. If one assumes that other x-polarized transitions ($n\pi^*$, $\sigma\pi^*$, etc.) are of similarly weak intensity, all observed transitions in the reduced spectra can be assigned to calculated transitions as shown at the bottom in Figure 8.14. A larger number of $\pi\pi^*$ transitions are computed than are actually observed, particularly at higher energies, and this is not surprising considering that many of them are calculated to have zero intensities. The assignment of the second transition in **13** (labeled B in Figure 8.14) to a y-polarized $\pi\pi^*$ transition has recently been questioned, and reassignment as an x-axis polarized second $n\pi^*$ transition has been proposed.[28a]

The reduced spectra of all three compounds show great similarity, except that the $n\pi^*$ transition, which is clearly separated in **13**, is present only as an indistinct shoulder at 29 000 cm^{-1} in **12** and is absent in **11**. The similarity of the $\pi\pi^*$ spectra of **11**–**13** is reflected in the calculated spectra, which nicely reproduce the experimentally observed trends (bottom of Figure 8.14). In particular, the red shift of the state labeled B along the series **11**, **12**, **13** is striking. The authors found that it is readily rationalized in terms of the shapes of the MOs involved in the excitation, and the interested reader is referred to ref. 28.

The progression of long-axis polarized absorption peaks on the high-energy side of the origin of the short-axis-polarized B band in the spectrum of **13** was assigned by the authors to vibronic intensity borrowing by the B band from the much more intense A band. We point out this case as an example of a situation which is probably fairly common: It is important to resist the temptation to assign every separate clearly polarized series of peaks to a separate electronic transition. An extreme case of the resulting uncertainty is provided by the L_a–L_b region of anthracene (Section 8.3.7).

8.3.4 Acenaphthylene,[29] Fluoranthene,[30] and Benzo[k]fluoranthene[31]

We have selected this series of nonalternant hydrocarbons as our next example for two reasons. First, it demonstrates the ease with which "hidden" transitions are detected by this type of measurement; second, it shows how the orientation factors in stretched polyethylene vary smoothly in a manner quite compatible with molecular shape (Table 8.8): for acenaphthylene (**14**), $K_z > K_y > 1/3$ (this is a nice counterexample to the claims[32] that the Fraser-Beer model is generally applicable, since the assumption $K_x = K_y$ leads to the absurd result that $\sum K_u > 1$). The longer fluoranthene molecule (**15**) behaves somewhat more nearly like a rod but still clearly has $K_x \neq K_y$ in spite of some early assertions[32] to the contrary. Benzo[*k*]fluoranthene (**16**) already fulfills the condition for rod-like molecules within experimental error.

Table 8.8
Orientation Factors of Non-alternant Hydrocarbons **14**–**16** in Stretched Polyethylene from UV Linear Dichroism

	Temp.[a]	K_z	K_y	K_x[b]
Acenaphthylene[29] (**14**)	L.T.	0.43	0.38	0.19
Fluoranthene[30] (**15**)	L.T.	0.60	0.27	0.13
	R.T.	0.56	0.29	0.15
Benzo[*k*]fluoranthene[31] (**16**)	R.T.	0.71	0.15	0.14

[a] L.T., 77 K or below, R.T., room temperature.
[b] Determined from $K_x = 1 - K_y - K_z$.

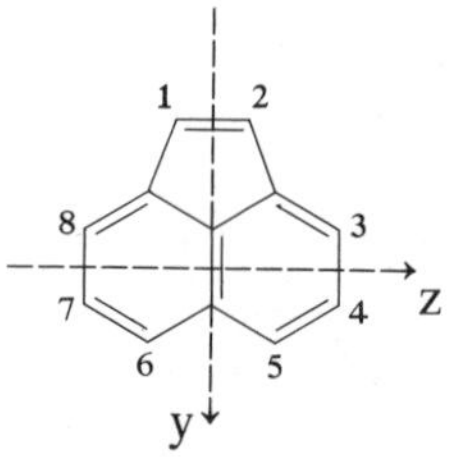

14

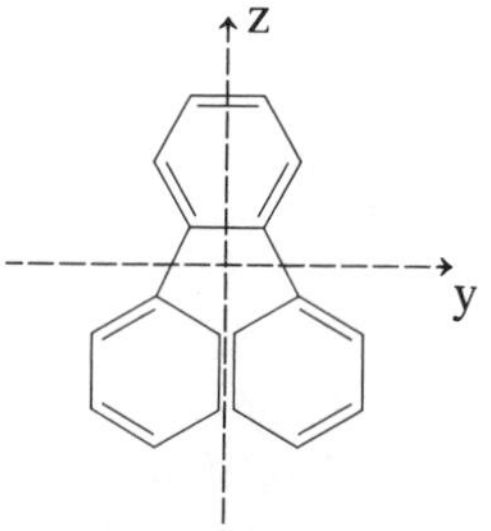

15

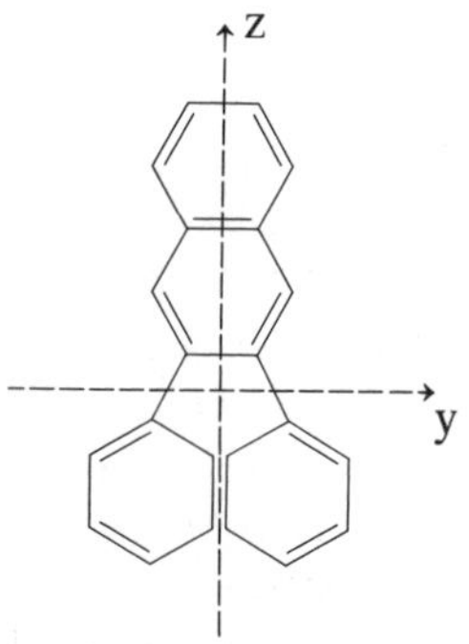

16

The dichroic UV-visible spectra of **14**–**16** obtained in stretched polyethylene were reduced by the standard TEM procedure with the results shown in Figure 8.15. The reduction factors for **14** were determined from simultaneous disappearance of the peaks due to transitions labeled 1, 3, and 5 in the linear combination $E_Z(\tilde{\nu}) - 1.5E_Y(\tilde{\nu})$ and of the peaks due to transitions labeled 2 and 4 in the linear combination $E_Z(\tilde{\nu}) - 1.25E_Y(\tilde{\nu})$. The reduction factors for **15** were obtained from the simultaneous disappearance of the peaks due to transitions labeled 2, 4, and 9 in one linear combination of $E_Z(\tilde{\nu})$ and $E_Y(\tilde{\nu})$ and of the peaks labeled 3 and 7 in another. The reduction factors for **16** were obtained similarly using peaks of transitions 1 and 3 for one and the peak of the transition 4 for the other polarization direction (in the last-named case, an independent check was obtained from a measurement on an unstretched sheet as described in ref. 31).

The assignment of directions in the molecular framework to the z and y axes was straightforward for the elongated molecules **15** and **16**. Comparison with single-crystal data permitted the assignment of the better aligned z axis to the long axis of **14** and of the less well aligned y axis to the short (twofold symmetry) axis of **14**. Clearly, even the nearly circular shape **14** deviates sufficiently from true disklike symmetry that the usual rule, "the longer molecular axis aligns better," is applicable. This conclusion was reached independently by measuring the linear dichroism of 1,2-dibromoacenaphthylene.[29] In this molecule, whose absorption spectrum is very similar to that of **14**, the positions of the longer and shorter in-plane axes z and y are interchanged relative to **14**, and indeed all transitions which were z-polarized in **14** are now y-polarized, and all those which were y-polarized in **14** are now z-polarized.

The assignment of the transitions observed in the reduced spectra of **14**–**16** to those calculated by standard π-electron methods is straightforward for **14** and **15**, in which all transitions predicted up to quite high frequencies are also observed, with correct polarizations and relative intensities. The situation is less favorable for **16**, in which the positions and polarizations of the calculated strong transitions agree quite well with those observed, but several weak transitions are also predicted and those are not apparent in the observed spectra. In

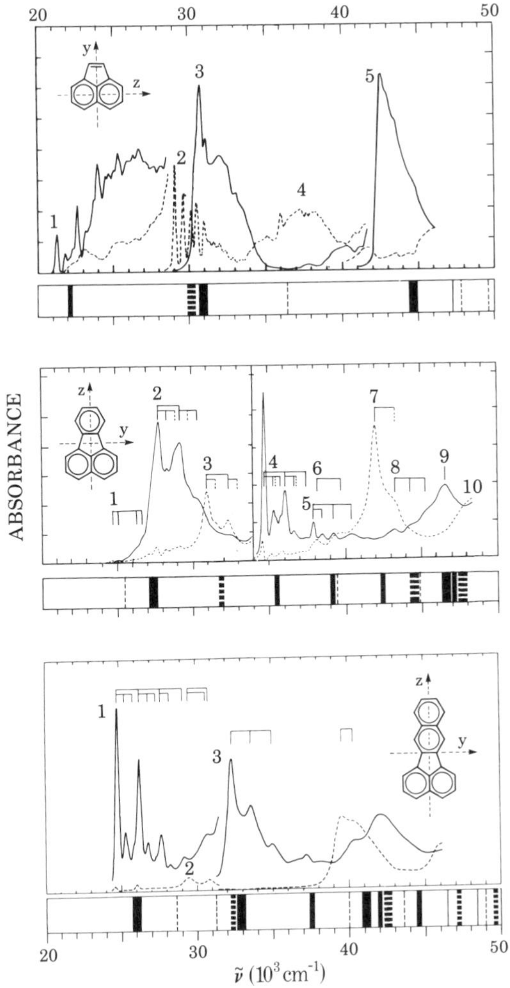

Figure 8.15 Reduced polarized UV-visible spectra of acenaphthylene (**14**), fluoranthene (**15**), and benzo[*k*]fluoranthene (**16**): $A_z(\tilde{\nu})$, full line; $A_y(\tilde{\nu})$, dashed line. The measurements were performed at room temperature for **16** and at low temperature (77 K or less) for **14** and **15**. Results of π-electron PPP calculations are shown at the bottom as full bars for z-polarized and dashed bars for y-polarized transitions. Increasing thickness of the bars corresponds to increasing calculated oscillator strengths. Adapted by permission from refs. 29–31.

all three cases, the measurements of linear dichroism were helpful in revealing hidden transitions and establishing their identity, at times settling previous controversies. Further details concerning the systematics of electronic transitions in this class of hydrocarbons, the use of the perimeter model, and the use of magnetic circular dichroism for additional assignments can be found in ref. 33 and 34.

In general, the agreement of the number, energies, relative intensities, and polarizations of the observed and calculated transitions in molecules of this type is excellent. Their C_{2v} symmetry and, in particular, the absence of a center of symmetry remove the difficulties with the detection of half of the excited states mentioned earlier, even without resorting to two-photon spectroscopy. Also, the absence of alternant pairing symmetry in nonalternants reduces greatly the number of symmetry-allowed transitions with nearly vanishing intensity which plagues the investigations of alternant hydrocarbons.

8.3.5 Azulene[35] *and Azulenophenalenes*[36,37]

Among cyclic π-electron hydrocarbons, azulene (**17**) and its analogs have been among the most popular subjects for spectroscopic investigation. The three molecules chosen here, **17–19**, provide us with further examples of orientation distributions in stretched polyethylene which are of a general type and cannot be properly described by the Fraser-Beer model (Table 8.9; only **19** begins to approach the required limit $K_x = K_y$).

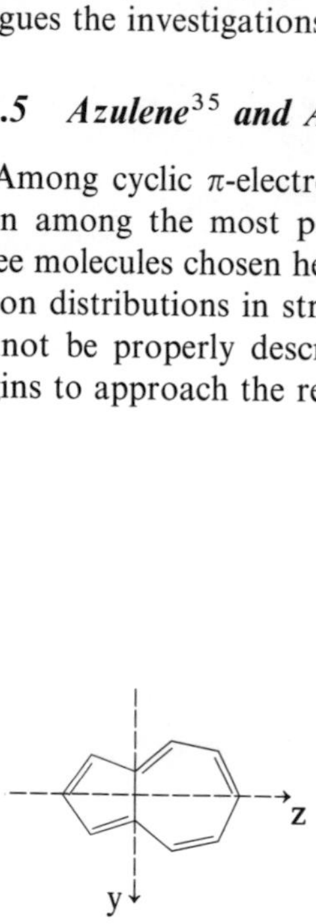

17

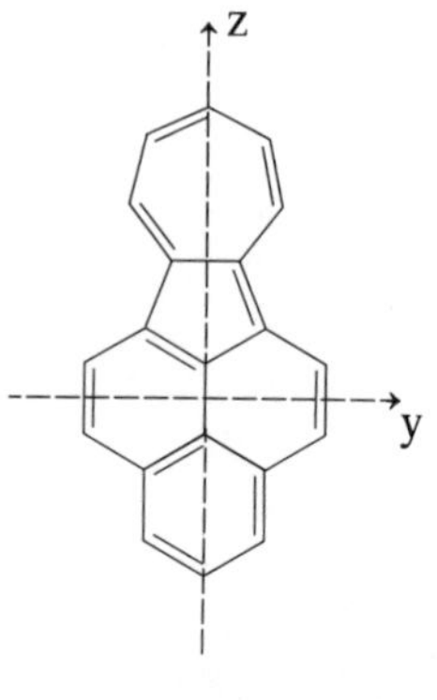

18

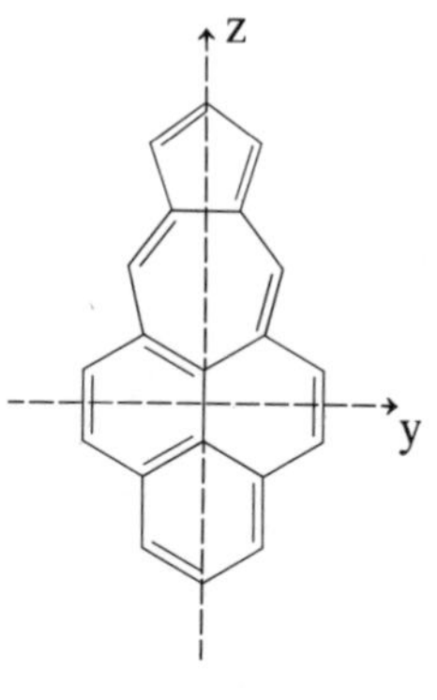

19

Table 8.9
Orientation Factors[a] of Non-alternant Hydrocarbons **17–19** in Stretched Polyethylene from UV Linear Dichroism

	K_z	K_y	K_x[b]
Azulene[35] (**17**)[c]	0.48	0.32	0.20
Azuleno[1,2,3-*cd*]phenalene[36] (**18**)	0.74	0.20	0.06
Azuleno[5,6,7-*cd*]phenalene[37] (**19**)	0.64	0.20	0.16

[a] At 77 K.
[b] Determined from $K_x = 1 - K_y - K_z$.
[c] The same orientation factors were obtained in a ethylene-vinyl acetate copolymer (for larger molecules, the copolymer tends to yield a somewhat lower degree of alignment for the y and z axes than polyethylene; for **19**, $K_z = 0.63$, $K_y = 0.24$, $K_x = 0.13$).

The dichroic UV-visible spectra of **17**–**19** obtained in a stretched polymer were reduced by the stepwise method, yielding the results shown in Figure 8.16. Polyethylene and a copolymer of ethylene with vinyl acetate produced the same reduced spectra. The reduction procedure was based on the assumption that the 0-0 bands of the first two transitions are purely polarized. The assignment of absolute directions of the y and z axes was based on molecular shape,

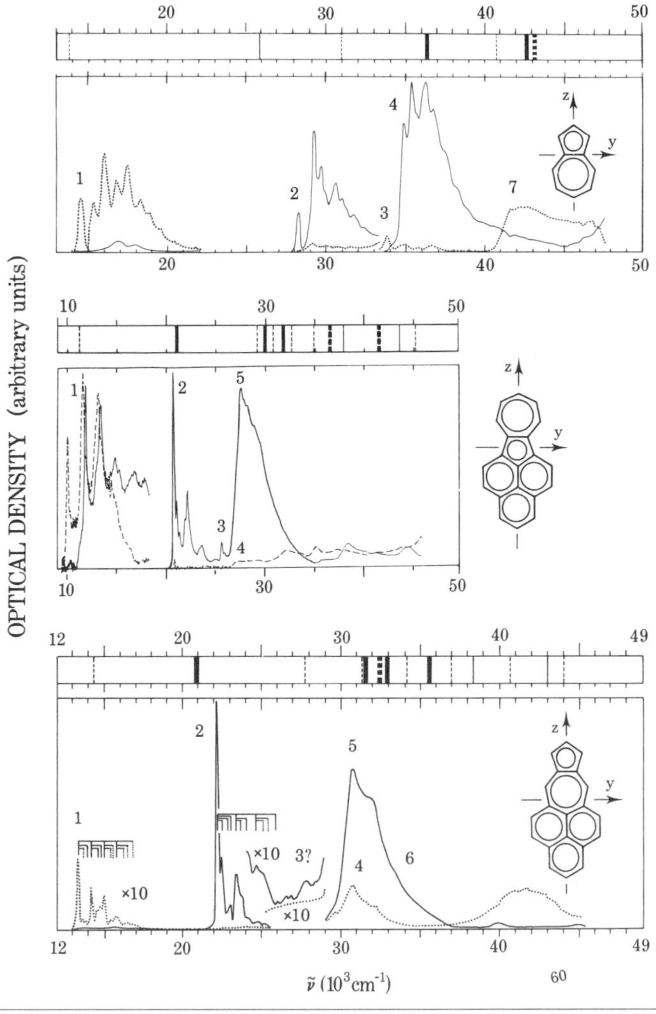

Figure 8.16 Reduced polarized UV-visible spectra of azulene (**16**) and the azulenophenalenes **17** and **18**. $A_z(\tilde{\nu})$, full line; $A_y(\tilde{\nu})$, dashed line. The measurements were performed in stretched polyethylene at 77 K. Results of PPP π-electron calculations are shown on the top of each spectrum as full bars for z-polarized and dashed bars for y-polarized transitions. Increasing thickness of the bars corresponds to increasing calculated oscillator strengths. Adapted by permission from refs. 35–37.

with z lying along the longer molecular dimension. In the case of azulene, independent support for this assignment was available from single-crystal work.[38]

The assignment of observed to calculated transitions was facilitated by the use of magnetic circular dichroism in the case of **17** and **19**. A more complete and more definitive analysis of the high-energy region (above 35 000 cm^{-1}) in the spectrum of azulene (**17**) is now available from a study of its alkyl-substituted derivatives,[39] and the results agree well with π-electron calculations. A reader interested in further detail is directed to ref. 39. An interesting aspect of these results is the fact that the combination of LD and MCD spectroscopy revealed the presence of previously unsuspected excited states at the higher energies, at least one of which is best described as a "doubly excited" state in the usual MO formalism.

The spectra of **18** and **19** are superficially very similar to each other but differ when inspected more closely. For instance, the intensity of the weak first transition in **19** is nearly all short-axis polarized, while in **18** this transition has a short-axis-polarized origin and several vibrational components but also contains many vibronic peaks of long-axis polarization. The presence of two oppositely polarized intense transitions near 31 000 cm^{-1} in the spectra would not be suspected from ordinary absorption curves but is made very clear by the strongly mixed polarization in this region: here, the dichroic ratio is distinctly different from its value at the origins of either of the first two transitions.

Once again, the reduced spectra appear to be in very good agreement with the results of π-electron calculations (Figure 8.16). Among low-energy transitions, only the third transition in **18** is unaccounted for. The authors suggested[36] that it may be due to an excited state with a large weight of doubly excited configurations, which were not included in the calculation. (Subsequently, this proposal was shown to be correct.[40]) A more detailed examination of the region of weak absorption in **19** (18 000–22 000 cm^{-1}) at low temperatures indicates the presence of yet another electronic transition, not apparent in the spectra shown in Figure 8.16, in a region where even calculations with extensive configuration interaction fail to predict any additional transitions.[40a] If the interpretation of the experimental data proposed in ref. 40a is correct, this represents a unique failure of simple semiempirical models and is worthy of further investigation.

8.3.6 Fluorene: S-S and T-T Transitions

Fluorene (**20**) has been selected for discussion here as one of the very few molecules for which stretched sheet spectra have been reported for both singlet-singlet[41] and triplet-triplet[42] transitions. We shall also use it to illustrate the difficulties encountered when Tanizaki's model is applied to a compound whose orientation in stretched poly(vinyl alcohol) does not happen to satisfy relation (5.84).

The stretched sheet UV spectrum of **20** has been measured repeatedly, partly because of interest [43,44] in the "double fluorene chromophore," 9,9′-spirobi[9*H*-fluorene] **21**, whose degenerate excited states are Jahn-Teller distorted to produce enantiomeric species discriminated in energy in a chiral environment.

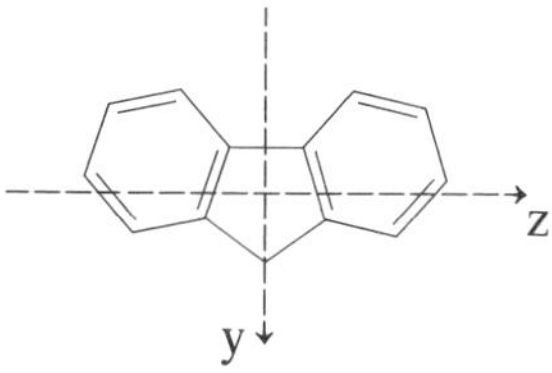

20

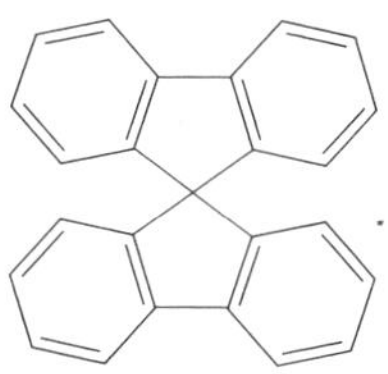

21

Absolute polarization directions of electronic transitions are generally needed in attempts to understand natural optical activity (cf. ref. 45).

The singlet-singlet absorption region in the linear dichroism of **20** (Figure 8.17) is rather uninformative up to about 43 000 cm^{-1}, since it appears uniformly polarized along the long axis z; the first short-axis-polarized transition

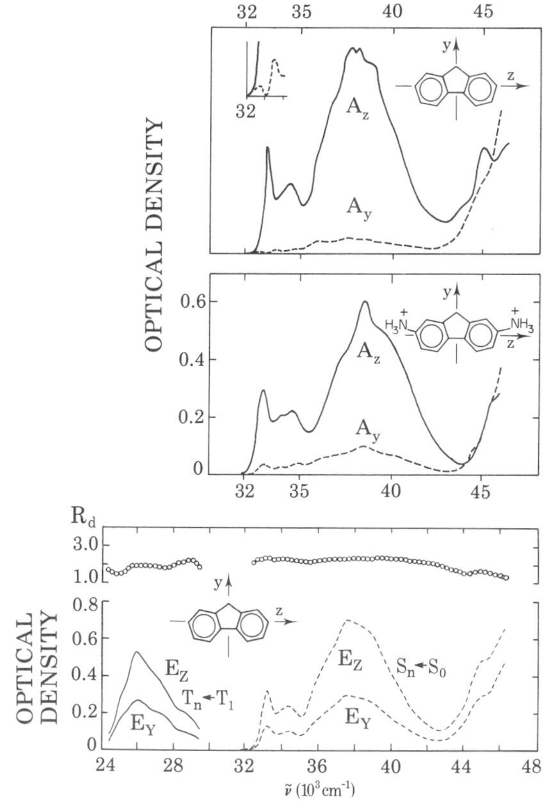

Figure 8.17 Spectra in stretched poly(vinyl alcohol). Above, reduced polarized spectra: top, fluorene, reduced correctly[41]; center, diprotonated 2,7-diaminofluorene, reduced incorrectly.[48] Note the wiggle in the A_y curve for fluorene shown in the insert. Below: observed[42] polarized spectra of fluorene in the $S_0 - S_n$ and $T_1 - T_n$ transition regions and the dichroic ratio $R_d = E_Z/E_Y$. Adapted by permission from refs. 41, 42, and 48.

is apparent at 45 000 cm^{-1}. Other types of spectroscopy, such as magnetic circular dichroism, indicate the presence of y-polarized transitions in the lower-energy region.[46] However, these transitions are so weak that their presence is hard to detect in linear dichroic spectra. Some y-polarized absorption is seen weakly at about 36 000 cm^{-1} and above in the reduced spectrum at the top of Figure 8.17. The insert clearly shows that the dichroic ratio is not constant across the 0-0 component of the first absorption band of **20** so that "wiggles" result in the reduced spectrum A_y. The authors of ref. 41 assign this to the presence of a weak y-polarized transition on the low-energy side of the first z-polarized transition. We believe that this conclusion is not warranted. Rather, we ascribe the lack of constancy of the dichroic ratio over the 0-0 band of the first z-polarized transition to site effects: Molecules in those environments which yield the lowest excitation energies are aligned somewhat worse, on the average, than the others (see Section 5.1.1).

The essentially pure long-axis polarization in the low-energy region of fluorene below 35 000 cm^{-1} is well documented from single-crystal work,[47] and this molecule, or rather its diprotonated 2,7-diamino derivative, which has almost the same spectrum,[48] can serve as a convincing demonstration of the lack of generality of the Tanizaki model. The orientation factors of doubly protonated 2,7-diaminofluorene in poly(vinyl alcohol) do not satisfy equation (5.84). An attempt to derive the reduced spectra A_z and A_y using (5.75) and (5.76) yields two curves almost exactly proportional to each other, i.e., a uniformly mixed degree of polarization (Figure 8.17). When taken seriously, this would lead one to conclude[48] incorrectly that for each z-polarized transition a y-polarized one is present as well, at equal energy and possessing an identical spectral shape.

Figure 8.17 shows also the experimental curves $E_Z(\tilde{\nu})$ and $E_Y(\tilde{\nu})$ for singlet-singlet and triplet-triplet absorption of **20** measured in stretched poly(vinyl alcohol) sheets in ref. 42. The T-T absorption curves were obtained using a flash spectrophotometer. The exciting light was polychromatic, but the bulk of the absorption of the flash light probably was due to z-polarized S-S transitions since they dominate the absorption spectrum of **20**. However, as the exciting flash light was not polarized, and as neither a substantial ground-state depletion nor concentration depolarization in the sample can be ruled out, it would be difficult to obtain any information on the orientation factors L from the published data. Still, the near equality of the dichroic ratio in the T-T and S-S regions suggests that the bulk of the T-T absorption near 26 000 cm^{-1} is long-axis polarized. The variation of the observed dichroic ratio across this band, clearly indicated in Figure 8.17, can be due to the presence of two electronic transitions in this region or to vibronic borrowing of intensity by the z-polarized transition observed. A decision between the two possibilities is difficult at this level of resolution. The authors of ref. 42 argue in favor of the former on the basis of their MO calculations (which are of questionable reliability, since the CH_2 group is assumed to have very strong electron-withdrawing inductive effect).

We believe that the work presented in ref. 42 is significant in that it demonstrates the ease with which absolute polarization assignments for T-T transitions can be obtained by the stretched sheet method.

8.3.7 Anthracene

Our third and last example of an alternant hydrocarbon, anthracene (**2**), has been selected to demonstrate the complications caused in the interpretation of electronic spectra by vibronic interaction between purely electronic states. It also provides an example of the use of polarized two-photon spectroscopy to detect excited states of the gerade type and to assign their symmetry.

Anthracene has been the favorite test case in investigations of various kinds of orienting media and of various techniques of dichroic measurement. A few[49–54] of the numerous published dichroic UV spectra E_Z and E_Y along with a left-handed (negative) cholesteric liquid-crystal-induced circular dichroism (ICD) spectrum[54] and a directly measured LD[55] spectrum $E_Z - E_Y$ are reproduced in Figure 8.18, which shows the great similarity of the results obtained under a variety of conditions. Given the $\pi\pi^*$ nature of the observed UV transitions of **2**, the spectra obtained in media transparent to at least 250 nm, such as polyethylene, poly(vinyl alcohol), and saturated nematic liquid crystals, permit an unambiguous recognition of the intense B_b transition near 255 nm as long-axis polarized (z), based on its high positive dichroism. All the spectra permit the assignment of the 0-0 component of the L_a band as in-plane short-axis polarized (y). Both of these assignments are supported by a wealth of other evidence and are not in doubt.

For years, however, there has been argument about the assignment of a third symmetry-allowed $\pi\pi^*$ transition in the low-energy region. This is the L_b transition, predicted by theory to be nearly degenerate with the L_a transition but far weaker in intensity. At the simple PPP level, it is actually expected to have zero intensity as a result of the validity of the pairing theorem.[9]

One of the proposed assignments of the L_b origin can be dismissed out of hand: The weak positive peak at 390 nm in the absorption and LD spectra of ref. 55 is due to crystalline anthracene present on the surface of the polyethylene sheets used in the experiment. We have included the spectra in parts I and J of Figure 8.18 partly as a warning against this kind of complication, which is likely to be encountered with many samples unless the surface of the polyethylene sheet is carefully rinsed before measurement.

All the spectra make it quite clear that the polarization in the bulk of the L_a band is mixed, in that the dichroic ratio steadily increases as one proceeds from the band origin to higher energies. This is equally clearly reflected in the ICD and LD curves, which change sign near 350 nm. It is tempting to identify the peaks in these spectra or the points where they change sign with peaks or origins of individual spectral transitions (e.g., refs. 54 and 55; note that in ref. 55 the B_b and L_b bands are erroneously assigned different symmetries). This procedure can be quite treacherous, suggesting, for instance, that the peak of the L_b

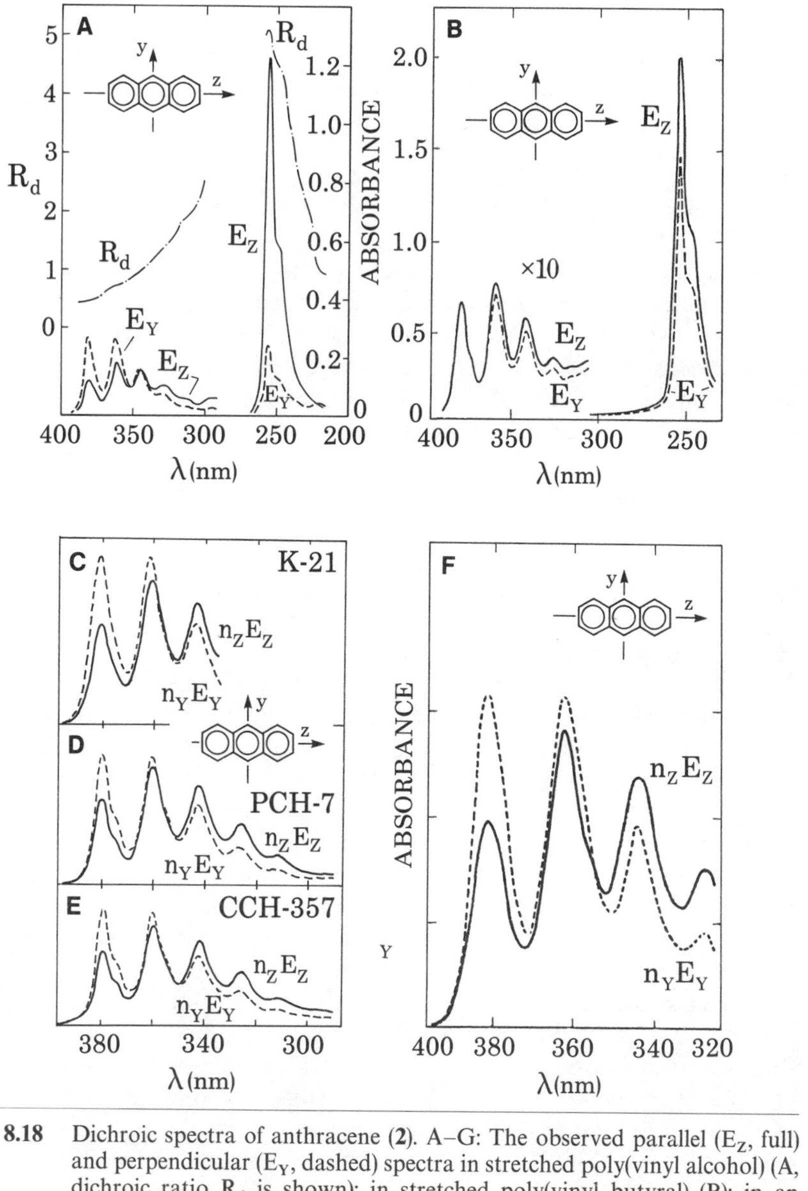

Figure 8.18 Dichroic spectra of anthracene (**2**). A–G: The observed parallel (E_Z, full) and perpendicular (E_Y, dashed) spectra in stretched poly(vinyl alcohol) (A, dichroic ratio R_d is shown); in stretched poly(vinyl butyral) (B); in an aligned 4-cyano-4′-*n*-heptylbiphenyl nematic liquid crystal (C); 4-*p*-cyano-phenyl-1-heptylcyclohexane nematic liquid crystal (D); the CCH-357 nematic liquid crystal (E, mixture of substituted bicyclohexyls); the N9 nematic liquid crystal (F, mixture of nematic esters); and the 1.85:1 by weight nematic mixture of cholesteryl chloride and cholesteryl laurate (G). The observed absorption and linear dichroism ($E_Z - E_Y$) of anthracene in stretched polyethylene are shown in parts I and J, respectively. The surface

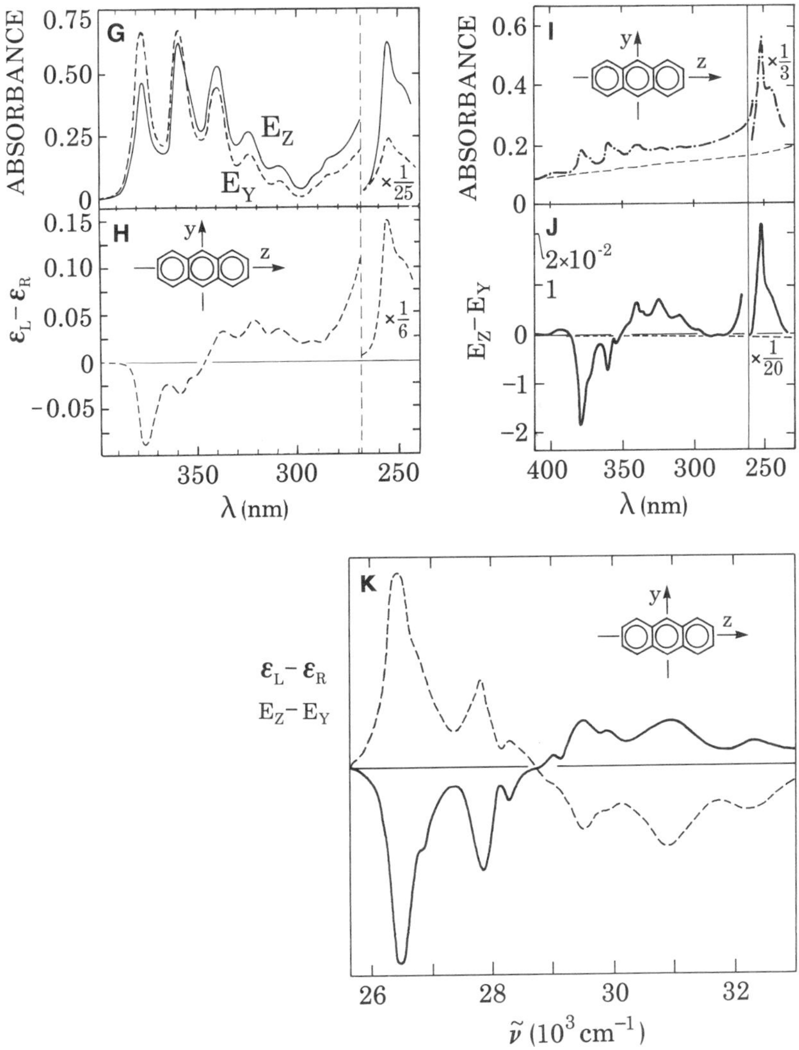

of the polymer was contaminated with anthracene crystals, and these are responsible for the weak first peak in absorption and LD. The full curve in part K also shows LD of anthracene in stretched polyethylene. The dashed curves in parts H and K show liquid-crystal-induced circular dichroism $\varepsilon_L - \varepsilon_R$ of anthracene; in H, a right-handed cholesteric liquid crystal, 55.5 mole percent cholesteryl nonanoate and 44.5 mole percent cholesteryl chloride (pitch 800 nm); and K, a left-handed cholesteric liquid crystal, 1:1 by weight cholesteryl nonanoate and cholesteryl chloride (pitch $\sim 1\ \mu m$). Adapted by permission from refs. 49 (A), 51 (B), 52 (C,D,E), 53 (F), 54 (G,H), 55 (I,J), and 50 (K).

band is located near 330 nm or its origin near 350 nm, in disagreement with most recent work, and it is far preferable to base band assignments on reduced spectra. The reduced spectra obtained in poly(vinyl alcohol)[49] and in polyethylene[56] are shown in Figure 8.19.

Unfortunately, even a clean decomposition of the absorption intensity into its purely polarized components does not answer the question as to whether the z-polarized component of absorption in the 300–380 nm region is due to the L_b band or whether it is vibronically borrowed by the L_a band from the B_b band. This is a general type of limitation, typical of all forms of linearly polarized spectroscopy discussed in this book. Indeed, at times this will not even be a meaningful question to ask. This happens when the simplest zero-order approach to electronic and vibronic states is not applicable: Zero-order vibrational levels built on the L_a origin will mix with zero-order vibrational levels built on the L_b origin if their symmetries are the same, and borrow intensity from the B_b band.

It is not our intention to reproduce here the many arguments which have been made concerning the location of the zero-order L_b level of anthracene,

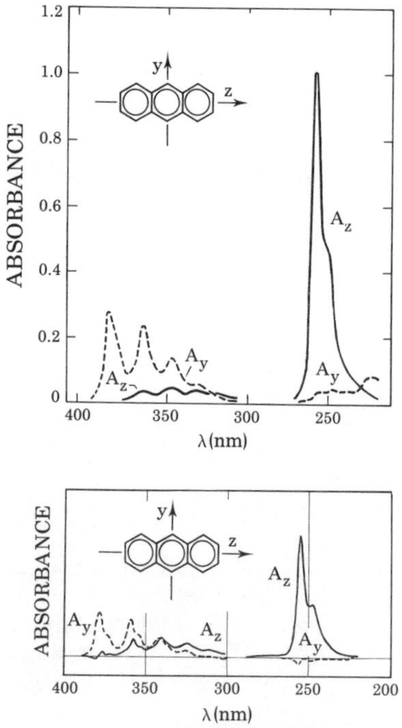

Figure 8.19 Reduced polarized spectra of anthracene A_z (full lines) and A_y (dashed lines) obtained from linear dichroism in stretched poly(vinyl alcohol) (top) and polyethylene (bottom) sheets. Adapted by permission from refs. 49 (top) and 56 (bottom).

and we shall merely refer the reader to refs. 56 and 57 for a survey of the earlier work and to ref. 58 for evidence from magnetic circular dichroism which favors strongly the assignment of a z-polarized peak near 360 nm as the origin of the L_b band, probably mixed with z-polarized vibrational levels of the L_a band and obtaining electronic intensity from the B_b band. Additional evidence for such an assignment resulted from CD investigations of a molecule containing two anthracene chromophores.[58a]

Another problem in the anthracene spectrum which polarized one-photon spectroscopy does not answer is the assignment of electronic states of gerade symmetry. This is a general problem for molecules with a center of symmetry and has been already alluded to in Section 8.3.2.

A general answer lies in the use of two-photon spectroscopy. To our knowledge, this has not yet been performed on partially aligned solutes, and such an application of this tool would be of considerable interest both for spectroscopic assignments and for the determination of the orientation factors L.

However, since two-photon absorption is a process of rank 4, measurements on isotropic samples provide polarization information, as in ordinary photoselection. The simultaneous nature of the absorption process for both photons makes the situation similar to ordinary Raman spectroscopy in that rapid molecular rotation does not cause a loss of the polarization information. Thus, it is not necessary to work with highly viscous or rigid solutions as is the case in ordinary one-photon photoselection or photoinduced dichroism.

The fluorescence-detected two-photon absorption spectrum of anthracene in ethanol at room temperature[59] is displayed in Figure 8.20. The polarization degree $\Omega = I_{SYN}/I_{||}$ is plotted on top. The one-photon absorption spectrum is shown for comparison.

Among $\pi\pi^*$ excited states, A_g and B_g symmetry states are accessible by two-photon absorption from the A_g ground state. The value of Ω for transitions into an A_g state is less than 1.5 and can be as low as 0.25. For transitions into a B_g state, $\Omega = 1.5$. Inspection of the two-photon absorption spectrum and the polarization curve led to the assignments shown in Figure 8.20. The question mark labels a very low-intensity transition which appears to be two-photon forbidden and apparently acquires A_g vibronic intensity. The authors wonder whether it could represent a high-energy tail of the L_b band.[59] Table 8.10 summarizes the best presently available experimental information on the excited singlet states of anthracene and compares it with results of a π-electron calculation using singly and doubly excited configurations. The use of doubly excited configurations is important for proper description of the positions of the gerade states. It is seen that the agreement is excellent up to quite high energies.

8.3.8 Acridine Dyes: S-S and T-T Transitions

Several protonated derivatives of acridine (**3**), in particular aminoacridines, have been of interest in the studies of the binding of small molecules to DNA. For optical measurements using aligned DNA, it is important to understand

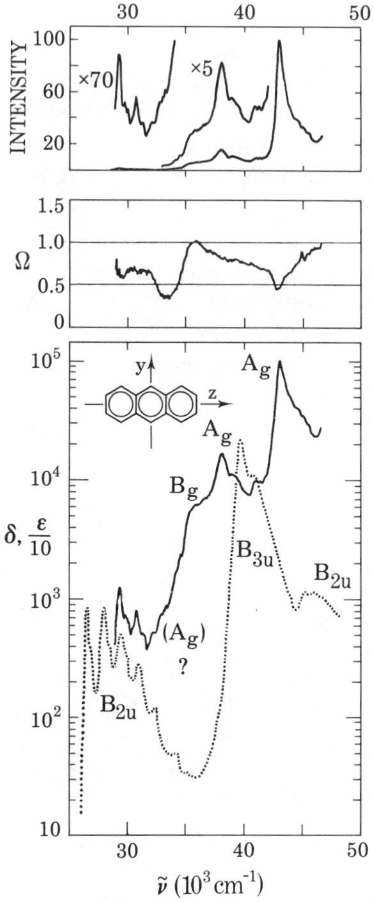

Figure 8.20 One-photon (dotted line) and fluorescence-detected two-photon (full line) absorption spectra of anthracene in ethanol. Top, linear scale; bottom, logarithmic scale. The polarization degree $\Omega = I_{SYN}/I_{||}$ is shown in the center. Adapted by permission from ref. 59.

the optical properties of the dyes themselves, in particular their transition moment directions. These have been investigated by numerous authors, as cited in ref. 60, and we have selected a few of the results obtained on oriented aminoacridines using linear dichroism in stretched poly(vinyl alcohol).

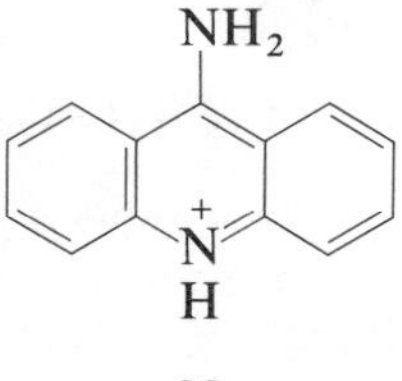

22

Table 8.10
Excited Singlet States of Anthracene (**2**)[59]

		Experimental				Calculated[c]				
Label[a]	Symmetry[b]	E (cm^{-1})	f^d	δ^e_{rel}	Ω^f	E (cm^{-1})	f^d	δ^e	Ω^f	%D^g
L_a	B_{2u}	26 600	0.1	1.3	0.75	29 700	0.1	—	—	8
L_b	B_{3u}	28 000	—		(0.35)	30 500	0.001	—		6
	B_{1g}	35 800	—	6.3	1.00	38 600	—	1.2	1.5	13
	A_g	38 000	—	16.6	0.80	40 000	—	2.2	0.9	43
	B_{1g}	39 000	—	11.1	0.80	41 700	—	1.7	1.5	15
	B_{2u}					42 600	0.005	—	—	13
B_b	B_{3u}	42 300	2.3	—	—	45 300	2.25	—	—	5
	A_g	43 000		100	0.45	45 400	—	26	0.39	21
B_a	B_{2u}	45 300	0.23	—		46 500	0.1	—	—	9
	B_{3u}					48 800	0.001	—	—	37
	A_g	50 000	—	—		50 200	—	30	1.03	61
	B_{2u}					51 200	0.08	—	—	12
	B_{2u}	53 700	0.65			59 300	0.8	—	—	21

[a] Platt's nomenclature.
[b] With z as the long axis and y as the in-plane short axis.
[c] CNDO/SDCI.
[d] Oscillator strength.
[e] Two-photon absorption cross section. The experimental values are only relative.
[f] $\Omega = I_{SYN}/I_{||}$.
[g] Percent contribution from doubly excited configurations to the state wavefunction.

An illustration of the various kinds of spectra encountered in polarized absorption studies is given in Figure 8.21, using protonated 9-aminoacridine (**22**) as an example.[60] At the top (A), the measured polarized spectra $E_Z(\tilde{\nu})$ and $E_Y(\tilde{\nu})$ are shown. The stepwise reduction procedure (Section 5.1.2) leads directly to the reduced spectra $A_z(\tilde{\nu})$ and $A_y(\tilde{\nu})$ shown at the bottom (E) and obtained from the assumption that the structured band between 340 and 460 nm is purely y-polarized and that the intense band at 260 nm as well as the weak double band at 320 nm are purely z-polarized (note that both of the latter disappear simultaneously in the reduction procedure). The orientation factors obtained from the reduction procedure are $K_y = 0.31$, $K_z = 0.55$. Since both strong bands reduce essentially exactly to zero, the stepwise reduction is actually not necessary. From the dichroic ratios in the region of the two bands, $d_y = 0.89$ and $d_z = 2.40$, one obtains immediately both the K's and the reduced spectra $A_z(\tilde{\nu})$ and $A_y(\tilde{\nu})$.

The authors have also displayed the linear dichroism curve $LD(\tilde{\nu}) = E_Z(\tilde{\nu}) - E_Y(\tilde{\nu})$, the isotropic absorption curve $E^{iso}(\tilde{\nu}) = E_Z(\tilde{\nu}) + 2E_Y(\tilde{\nu})$, and the value of

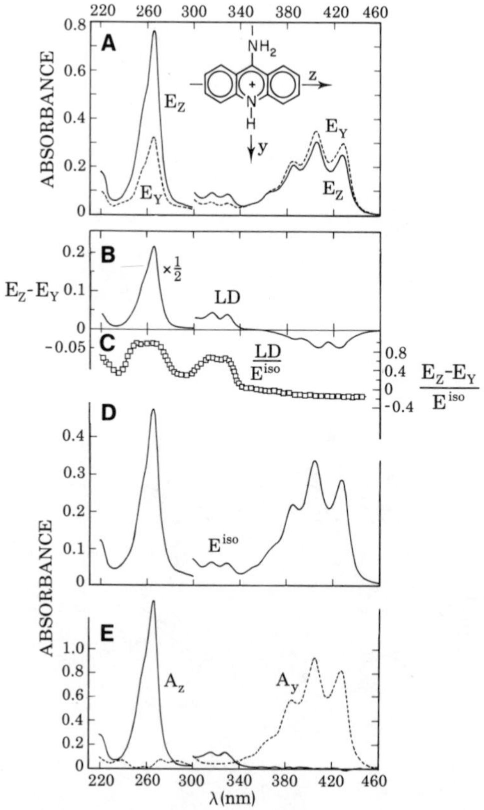

Figure 8.21 Protonated 9-aminoacridine in stretched poly(vinyl alcohol). Polarized absorption E_Z and E_Y (A), linear dichroism $E_Z - E_Y$ (B), reduced dichroism $(E_Z - E_Y)/E^{iso}$ (C), isotropic absorption E^{iso} (D), and the reduced polarized spectra (E) A_z (full) and A_y (dashed). Adapted by permission from ref. 60.

reduced dichroism $LD(\tilde{\nu})/E^{iso}(\tilde{\nu})$. This provides the reader with a useful comparison of the various forms in which the same information can be cast.

The three prominent bands in the spectrum correlate in the order of increasing energy with the L_a, L_b, and B_b bands of the parent chromophore, anthracene. Quite analogous bands have been found in the spectra of various other protonated aminoacridine dyes.[60,61] The orientation factors of a series of such dyes in stretched poly(vinyl alcohol) stretched to the ratio $R_s = 4.3$ are displayed in the orientation triangle shown in Figure 8.22. As in Figures 4.15–4.17, there is a general relation between solute shape and orientation, but the details of the point positions are hard to rationalize.

Finally, we note that one of these dyes, the proflavin cation, has also been studied by the stretched sheet method in its triplet state.[63] Its solution in stretched poly(vinyl alcohol) film was irradiated with an intense unpolarized continuous light source at low temperature, so that a significant population in the triplet state resulted, as revealed by a decrease in the S-S absorption bands and the appearance of new absorption in the near-IR. The use of unpolarized light in the excitation precludes any attempt at the extraction of the orientation factors K and L, particularly since the absorption polarization of the molecule is mixed in the region where the intensity of the lamp is concentrated (365 nm).

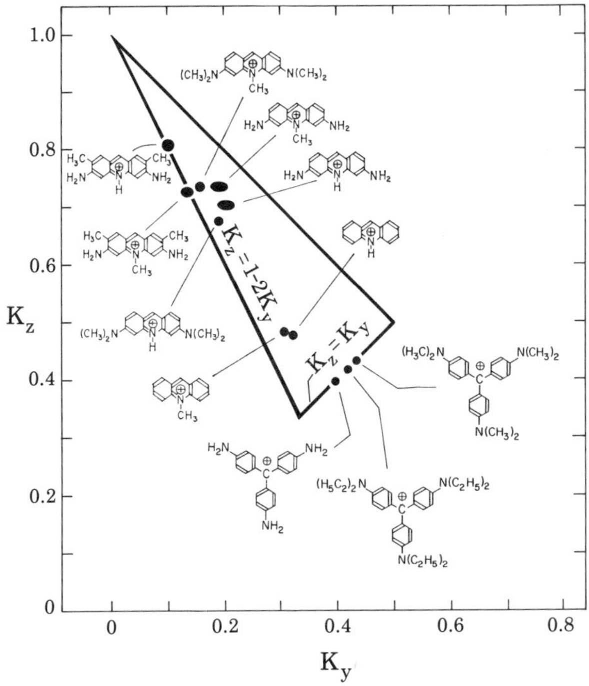

Figure 8.22 Orientation factors of some acridine and triphenylmethane dyes in stretched poly(vinyl alcohol). The degree of stretching was $R_s = 4.3$ for the acridine dyes and $R_s = 3.7$ for the triphenylmethane dyes. Adapted by permission from refs. 61 and 62.

However, the authors' goal, assigning the polarization of the near-IR T-T transition, can readily be achieved: While the S-S transition at 480 nm shows a strong positive dichroism, the T-T transition at 1100 nm shows a strong negative dichroism. The former is polarized along z, the latter along y (or possibly x if it is not a $\pi\pi^*$ transition, but this is highly unlikely).

8.3.9 Triphenylmethane Dyes[62]

This is another common class of dyes which is of interest in connection with binding to biopolymers. The orientation of symmetrical dyes of this type in stretched poly(vinyl alcohol) can be expected to be disclike, and these molecules will be one of the few examples of this type of behavior selected for this book.

Because of the propeller-like shape of the triphenylmethyl cation, its nominal $\pi\pi^*$ transitions may be both in-plane (E) and out-of-plane (A_1) polarized. Simple π-electron theory describes the intense absorption in the visible region as due to a superposition of three $\pi\pi^*$ transitions of approximately the same energy: two in-plane polarized degenerate transitions into two excited states of E symmetry and one out-of-plane polarized transition to an excited state of B symmetry in the D_3 group. The presence of the two degenerate transitions was detected in magnetic circular dichroism[64]; it is interesting to inquire whether any evidence for out-of-plane polarized absorption can be obtained from measurements of linear dichroism in a stretched polymer sheet. Such measurements were reported for several dyes of this type. Here we use the data obtained in a recent study[62] which provides references to earlier work.

The polarized spectra of tris(*p*-*N*,*N*-diethylaminophenyl)carbenium ion (**23**, ethyl violet) in stretched poly(vinyl alcohol) are shown in the top part of Figure 8.23. The wavelength dependence of the dichroic ratio shows that out-of-plane polarized intensity indeed contributes measurably to the spectra. A reduction of the spectra by the stepwise procedure of Section 5.1 yielded the reduced spectra shown at the bottom in Figure 8.23 and permitted an unequivocal detection of the weak out-of-plane intensity expected for a nonplanar triphenylmethyl cation derivative.

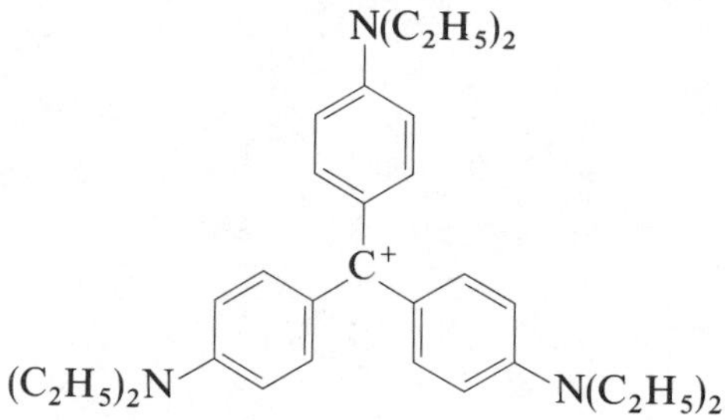

23

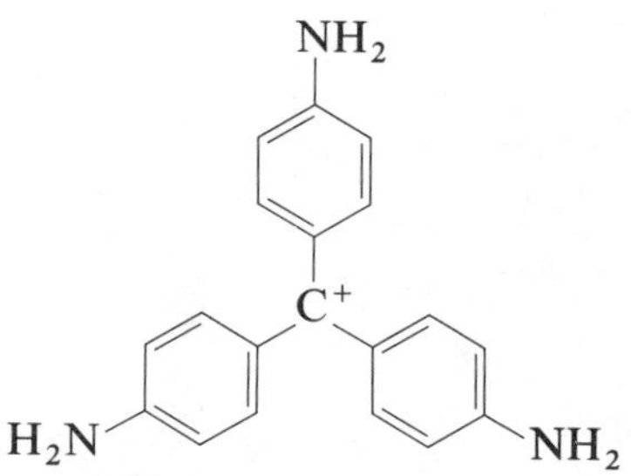

24

Similar spectra were obtained[62] for other dyes of this class. Their reduction factors obtained at the degree of stretching $R_s = 3.7$ have been plotted in the orientation triangle in Figure 8.22 along with those of the acridine dyes.

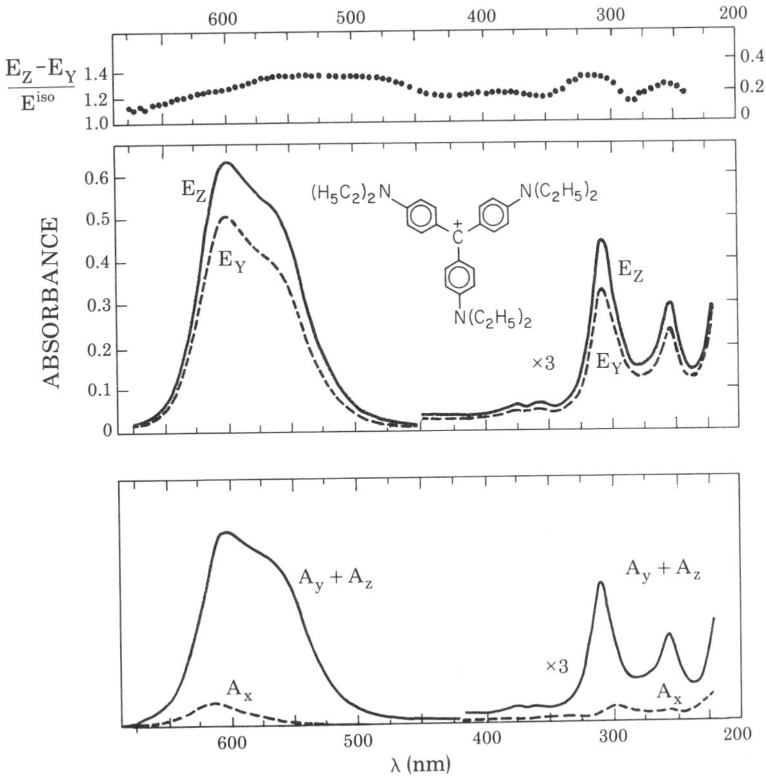

Figure 8.23 Ethyl violet in stretched poly(vinyl alcohol) ($R_s = 3.7$). Top, reduced dichroism $(E_Z - E_Y)/E^{iso}$; center, polarized absorbance E_Z (full) and E_Y (dashed); bottom, reduced polarized spectra $A_y + A_z$ (full) and A_x (dashed). Adapted by permission from ref. 62.

8.3.10 Trithiocarbonate Esters and Related Heterocycles[65]

Most of the applications of the stretched sheet method have been to aromatics. However, the method has also been successfully used for other structural classes of compounds, and the trithiocarbonate esters have been selected here to provide an example of this. They also illustrate the helpful role which spectroscopy with polarized light can play in attempts to interrelate the spectra of similar compounds.

Figure 8.24 shows the reduced spectra of a simple trithiocarbonate **25** and of an analog **26** with cyclic conjugation. The original polarized absorption spectra, which are not shown, were obtained in stretched polyethylene and reduced by the stepwise method of Section 5.1. The reduction rests on the assumption that out-of-plane polarized intensity is negligible.

The spectra of both **25** and **26** are dominated by two strong bands of opposite polarization. Several weak absorption bands are also present, and at least some of them are likely to be of the $n\pi^*$ type. The very weak first $n\pi^*$ transition, which is responsible for the yellow color of these compounds, is not seen

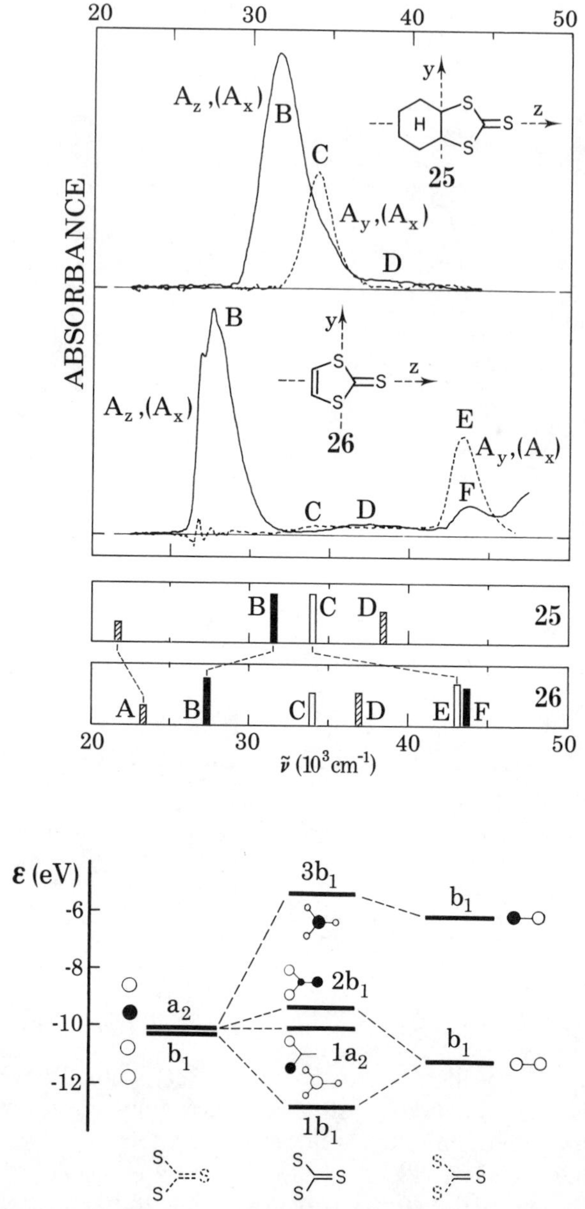

Figure 8.24 Cyclic trithiocarbonates in stretched polyethylene. The spectra shown on top represent an unknown linear combination of A_z and A_x (full line) and an unknown linear combination A_y and A_x (dashed line) as obtained from the assumption that **B** is purely z-polarized and C and E purely y-polarized. Since x-polarized absorption is believed to have negligible intensity, these curves should be good approximations to A_z and A_y, respectively. Center: a correlation diagram for observed transition energies in **25** and **26**: $\pi\pi^*$ z-polarized (full bars), $\pi\pi^*$ y-polarized (empty bars), and $n\pi^*$ (shaded bars). The height of the bars is proportional to log ε_{max}. Bottom: molecular orbital correlation diagram for the trithiocarbonate chromophore. Adapted by permission from ref. 65.

at this concentration but is included in the correlation diagram in the center of Figure 8.24. The intensity labeled D is believed to be in-plane polarized and to be due to vibronic intensity borrowing by an $n\pi^*$ transition located at this energy.

The two nearly degenerate transitions in the simple trithiocarbonate ester **25** can be understood in terms of the orbital energy diagram also shown in Figure 8.24. The diagram indicates how the orbital energies of **25** relate to those of the π and π^* orbitals of a thiocarbonyl group, C=S, and those of the two π-symmetry lone pairs on the other two sulfur atoms. This simple scheme is supported by photoelectron spectra. π-Electron calculations assign the two transitions as A_1 ($2b_1 \rightarrow 3b_1$, polarized along the C=S axis and labeled B in the figure) and B_2 ($1a_2 \rightarrow 3b_1$, in-plane polarized, perpendicular to the C=S axis, labeled C in the figure). The presence of two transitions would not be clear from ordinary absorption spectra alone.

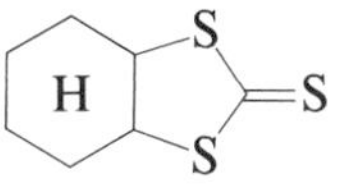

25

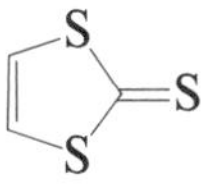

26

Upon formation of a cyclic π system by ethylene bridging to yield **26**, the two transitions move far apart. The polarized spectra show that the A_1 transition moves to lower energies and the B_2 transition to higher energies. This behavior is understood simply upon inspection of the MO diagram from the cyclic interaction of the orbitals of **25** with the π and π^* orbitals of ethylene to produce **26** (Figure 8.24). In addition to shifting the bands already present, the interaction also doubles their number. The results of π-electron calculations mimic these observations very nicely.

8.3.11 The Bis(benzene-1,2-dithiolato)cobaltate Anion[66]

There are a large number of interesting organometallic complexes whose electronic spectra are understood only poorly if at all. The use of the simple and rapid experimental techniques discussed in this book would appear indicated, yet only relatively few measurements on transition metal complexes have been reported. Limited solubility of ionic species in suitable uniaxial media may be a part of the reason.

We have selected two of the published studies, hoping to encourage an expanded use of dichroic techniques for this class of compounds. The first of these, involving the use of a liquid crystal as the orienting medium, is discussed in this section; the other is treated in Section 8.3.12.

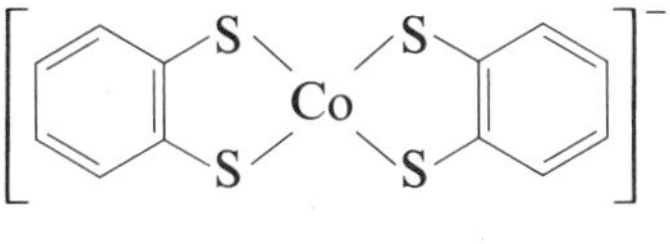

27

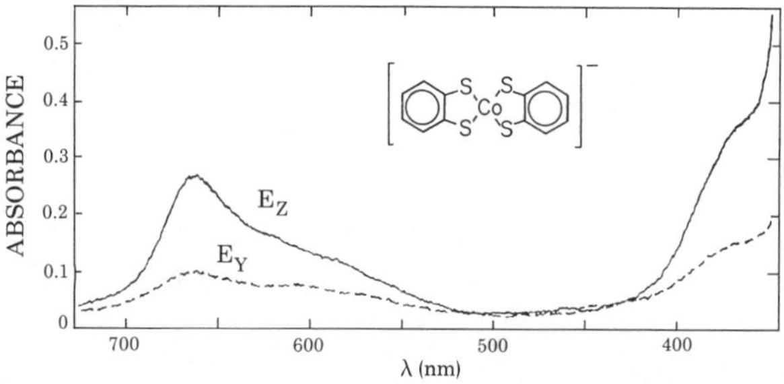

Figure 8.25 Polarized absorption spectra E_Z (full line) and E_Y (dashed line) of the bis(benzene-1,2-dithiolato)cobaltate anion in *p*-(*p*′-ethoxybenzoxy)phenyl butyl carbonate nematic liquid crystal. Adapted by permission from ref. 66.

Figure 8.25 shows the polarized absorption spectra of the tetra-*n*-butylammonium salt of bis(benzene-1,2-dithiolato)cobaltate (**27**) in a nematic liquid crystal, *p*-(*p*′-ethoxybenzoxy)phenyl butyl carbonate. Two transitions are present (at 660 and 380 nm), and both are long-axis polarized. The authors point out that this agrees with a previous analysis of the electronic structure of the complex **27** in which the bands were assigned to ligand-to-metal charge-transfer transitions.

8.3.12 Porphyrins and Metalloporphyrins

These high-symmetry complexes have been investigated repeatedly in uniaxially orienting media in the hope that out-of-plane polarized transitions will be discernible against the intense $\pi\pi^*$ background and assignable to $n\pi^*$ excitations. Some encouraging initial observations were indeed used to make such an assignment,[67] but more recent work indicates that the situation is more complicated than originally thought.[68,69] As a result, the location of $n\pi^*$ transitions is still uncertain. We use this example to illustrate one class of ambiguities which can be encountered in LD work.

It now appears (Figure 8.26, top) that free-base tetraphenylporphyrin (**28**) and its metal salts yield a constant positive dichroic ratio throughout the UV and visible absorption range when observed in solution in nematic liquid crystals.[68,69] The degree of orientation is quite high, with $K_y \cong K_z = 0.4$. The uniform degree of dichroism implies that within experimental error all absorption intensity is in-plane polarized, and the orientation factors are those expected for a disc. If out-of-plane-polarized transitions such as $n\pi^*$ are present in the accessible wavelength region, they are so weak relative to the in-plane-polarized absorption that they do not affect the observed dichroic ratio.

However, newer observations in stretched polyethylene and polypropylene[69] (Figure 8.26, bottom) confirm the original claim[67] that in a stretched polymer the shoulder at the high-energy side of the Soret band has a much lower dichroic ratio than the rest of the spectrum (slightly less than 1). The discrepancy with

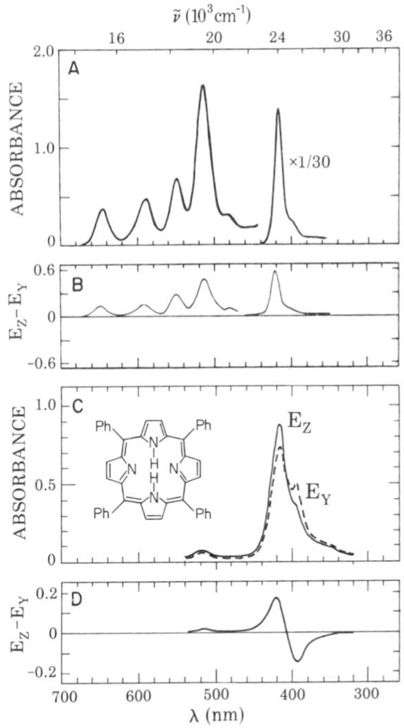

Figure 8.26 Tetraphenylporphyrin free base. (A) Isotropic absorption E^{iso} in $CHCl_3$; (B) linear dichroism $E_Z - E_Y$ in a compensated nematic mixture of cholesteryl chloride and cholesteryl myristate (1.787:1 by weight); (C) polarized absorption spectra E_Z (full line) and E_Y (dashed line) in stretched polypropylene; (D) linear dichroism $E_Z - E_Y$ in stretched polypropylene. Adapted by permission from ref. 69.

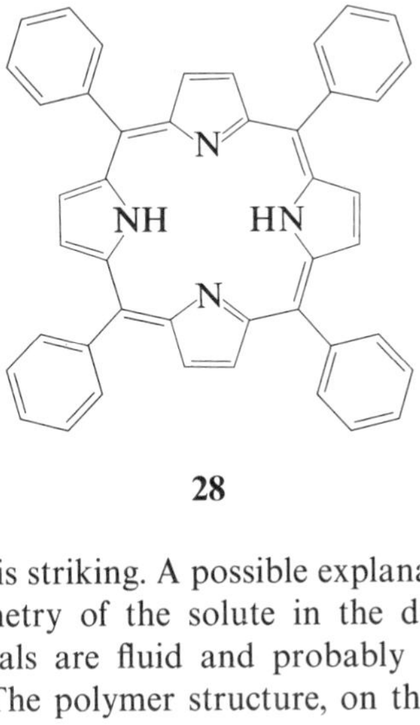

28

the liquid-crystal results is striking. A possible explanation is a difference in the average molecular geometry of the solute in the different media. Although viscous, the liquid crystals are fluid and probably adapt well to any steric demands of the solute. The polymer structure, on the other hand, is likely to be much more rigid and unyielding. The solubility of these tetraphenylporphyrins in polymer sheets is very low, and it is possible that the available sites are such that a higher degree of out-of-plane twist is forced on the four phenyl substituents. Their participation in a nominal $\pi\pi^*$ transition could be responsible for the out-of-plane-polarized intensity component.

8.3.13 Dibenzo[a,j]chrysene[69a]

We conclude Section 8.3 with a slightly amusing story which also has some pedagogical value. The UV-visible linear dichroism of dibenzo[*a,j*]chrysene (**28A**) shown in the top part of Figure 8.26a was among the first measured in a stretched polyethylene sheet. It was reported by J. H. Eggers and one of us at the 8th European Congress on Molecular Spectroscopy in Copenhagen in 1965 along with the reduced spectra shown in the bottom part of Figure 8.26a, some of the first ever obtained from such measurements.

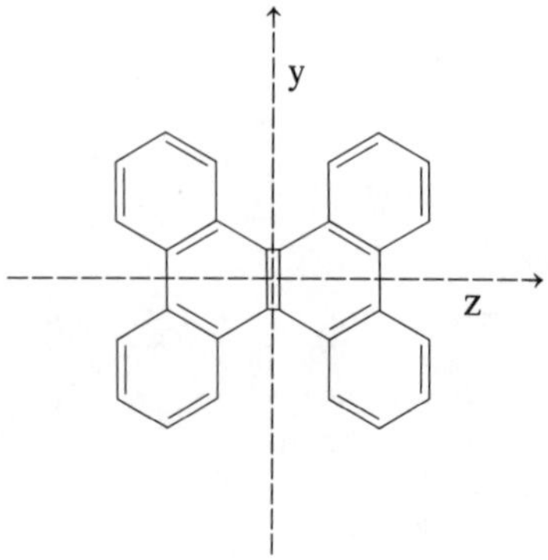

28A

At the time it was assumed that the weak reduced spectrum represented absorption polarized along the short in-plane axis (A_y) and that the strong reduced spectrum represented long-axis polarized absorption (A_z), based on a consideration of the molecular shape and on the assumption that only in-plane polarized $\pi\pi^*$ transitions had observable intensity. It was considered peculiar at the time that the degree of alignment of the two in-plane axes should be so different for a molecule of nearly circular shape, and also the huge difference in the intensities of the two reduced spectra was puzzling. Because of these doubts, the results were not published.

Twenty years later, the spectra of **28A** in stretched polyethylene were remeasured and re-reduced. The results agreed perfectly with the old ones, but no longer appeared puzzling. With the present knowledge that the stretched polymer sample is uniaxial and with the expressions of the TEM model (Section 5.1), the two different observed K values, 0.42 and 0.16, clearly had to correspond to $K_y = K_z = 0.42$ and $K_x = 0.16$, since $K_x + K_y + K_z = 1$. Thus, the orientation is disclike with two equivalent in-plane axes. A_y cannot be separated

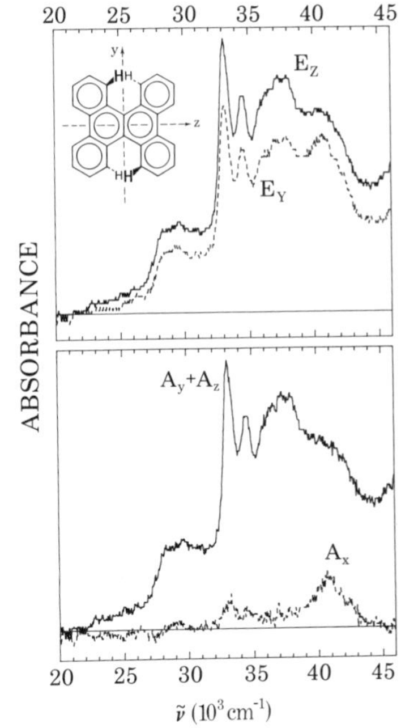

Figure 8.26a Top, dichroic spectra of dibenzo[a, j]chrysene in stretched polyethylene: solid line, $E_Z(\tilde{\nu})$; dashed line, $E_Y(\tilde{\nu})$. Bottom, reduced spectra: solid line, $A_y(\tilde{\nu}) + A_z(\tilde{\nu})$; dashed line, $A_x(\tilde{\nu})$. Adapted by permission from ref. 69a.

from A_z, and the weakness of A_x is no longer surprising since it is the out-of-plane polarized absorption. Actually, for an out-of-plane polarized absorption, A_x is unusually intense.

It is likely that the molecule is distorted from planarity (D_{2h}) to D_2 symmetry as indicated in the formula shown in Figure 8.26a, due to the crowding of the hydrogens in the bay areas. This will provide out-of-plane polarized intensity to nominally $\pi\pi^*$ transitions that are forbidden in the D_{2h} symmetry group, accounting for the appearance of relatively high A_x intensity.

8.4 Detection and Assignment of Electronic Transitions in Low-Symmetry Molecules

The determination of polarization directions in molecules whose symmetry is so low that two or all three of the symmetry-adapted axes x, y, z transform according to the same one-dimensional irreducible representation is quite difficult by any means. When attempted by the methods described in this book, it requires far more information than was the case for high-symmetry molecules. So far, quantitative attempts have been mostly limited to molecules with at least some symmetry, such as planar ones. We shall illustrate four cases. In the first two, the orientation factors and the location of the principal orientation axes were obtained from comparison with high-symmetry molecules of similar shape. In the third, they were obtained from a combination of linear dichroism induced by mechanical orientation, by electric field orientation, and by photoselection or, alternatively, from IR linear dichroism. In the fourth instance, the orientation factors were obtained from linear dichroism in the IR region complemented by a similar argument based on analogy to related compounds of high symmetry. To our knowledge, an assumption-free derivation of the orientation factors and the location of the principal orientation axes for a low-symmetry molecule has not yet been performed, although the work on flavin (Section 8.4.3) comes quite close to this ideal.

8.4.1 Dibenz[a,h]anthracene[70]

The first molecule selected as an example of the low symmetry case (C_{2h}) is dibenz[*a,h*]anthracene **29**. Its large size and limited solubility make it relatively difficult to attain large concentrations in stretched sheets, so polarized IR spectra are not easy to obtain. Moreover, while K_x could be obtained from the linear dichroism of out-of-plane-polarized vibrations such as C—H bends, the linear dichroism of the in-plane-polarized IR transitions would not be very useful for the purpose of establishing the other orientation factors and the principal orientation axes since their absolute polarization directions are not known.

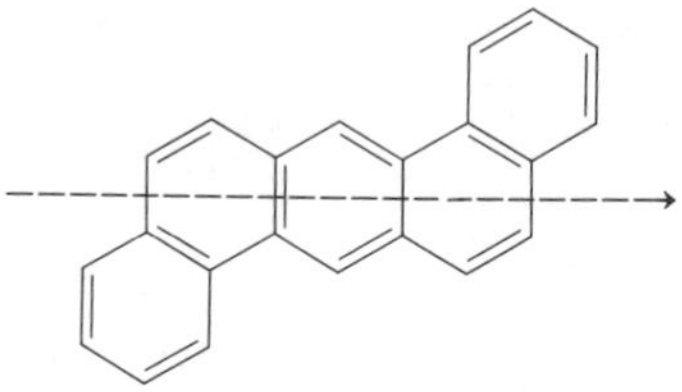

29

Fortunately, polynuclear aromatics are probably the most favorable group for the application of the empirical relations among molecular shape, orientation factors, and location of the principal orientation axes. A glance at Figure 4.15 suggests that the orientation factor K_z in stretched polyethylene is at least 0.7 but less than 0.8, leaving little freedom of choice for K_y and K_x.

Therefore, **29** represents a suitable example of the use of the empirical solute shape–solute orientation relationship. We shall see that for a molecule which orients as well as **29**, the uncertainty in the K values translates into only very small uncertainty in the deduced angles between transition moment directions and the principal orientation axis.

Figure 8.27 shows the family of curves $aE_Z(\tilde{\nu}) + bE_Y(\tilde{\nu})$ for various choices of a and b ($|a| + |b| = 1$) obtained from the linear dichroic spectra $E_Z(\tilde{\nu})$ and $E_Y(\tilde{\nu})$ of **29**. The dots indicate the locations in which a particular spectral feature just disappears. The reduction factors b/a obtained from this stepwise reduction range from a minimum of $d_{min} = 1.0$ to a maximum of $d_{max} = 5.7$. Assuming that the observed absorption intensity in the near-UV region is of $\pi\pi^*$ nature and

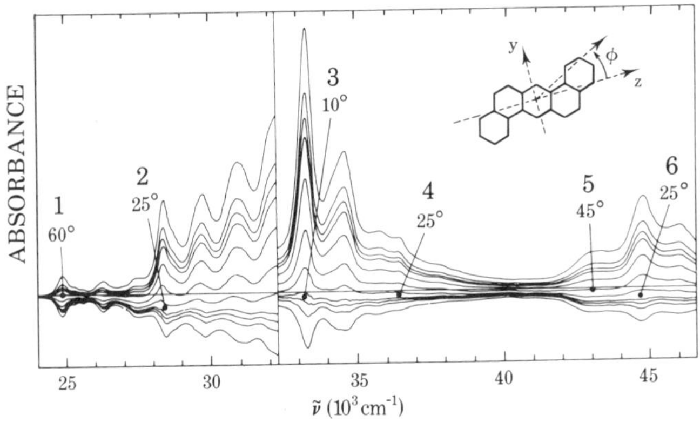

Figure 8.27 Dibenz[*a,h*]anthracene (**29**) in stretched polyethylene. From top to bottom: linear combinations $aE_Z + bE_Y$ for (a,b) = (1.00,0.00), (0.80, −0.20), (0.72, −0.28), (0.66, −0.34), (0.53, −0.47), (0.37, −0.63), (0.25, −0.75), (0.17, −0.83), (0.14, −0.86), (0.1, −0.90), (0.00, −1.00). Assuming (K_z,K_y) = (0.75,0.16), the curves with (a,b) values between (0.72, −0.28) and (0.14, −0.86) correspond to the following values of the angle ϕ between the molecular z axis and the transition moment: 90°, 75°, 60°, 45°, 30°, 15°, and 0°. For instance, spectral features for which $\phi = 75°$ disappear in the linear combination $0.66E_Z - 0.34E_Y$. Adapted by permission from ref. 70.

in-plane polarized, we have $K_y \leqslant d_{min}/(d_{min} + 2) \leqslant d_{max}/(d_{max} + 2) \leqslant K_z$ and thus $K_y \leqslant 0.33$ and $K_z \geqslant 0.74$. Comparison with molecules of similar shape in Figure 4.15 suggests that K_y is about twice the size of K_x (in the rod-shaped limit, $K_y = K_x$). In conclusion, the set of values $K_z = 0.75$, $K_y = 0.16$, $K_x = 0.09$ would appear to represent a reasonable first guess. In order to obtain a feeling of the sensitivity of the results to the choice of these numbers, we shall also use the set $K_z = 0.80$, $K_y = 0.10$, $K_x = 0.10$ for comparison.

The angles of deviation of the transition moments of the individual absorption peaks from the orientation axis z, obtained using formula (5.64) and the two sets of orientation factors, are listed in Table 8.11. Also listed are the results of PPP calculations for **29**. The comparison is based on the assumption that the orientation axis z is located at 15° from the long axis of the anthracene part of the chromophore, as shown in the formula, and is perpendicular to the smallest cross section of the molecule. The signs of the angles which the transition moments form with the axis z were derived by comparison with polarized fluorescence results[71] obtained in an isotropic rigid glass. The differences between the results obtained with the two assumed sets of values for K_z and K_y are miniscule, suggesting that for fairly well oriented molecules, rough estimates of orientation factors from molecular shape may well be useful. The agreement with the calculated polarization directions is also quite nice, at least at the lower excitation energies. No calculated value is available for the first transition which has zero intensity in the simple PPP approximation (it is forbidden by alternant pairing symmetry).

Table 8.11
Polarization of Electronic Transitions in Dibenz[*a,h*]anthracene (**29**)

Transition	E[a]	Intensity/osc. strength	MCD sign	ϕ_i[b]	ϕ_i[c]
1 Observed	25.0	weak	−	+60°	+57°
Calculated	27.5	0	0	—	
2 Observed	28.2	medium	−	+25°	+28°
Calculated	28.9	0.8	−	+21°	
3 Observed	33.5	v. strong	−	±10°	±18°
Calculated	34.9	2.0	−	−6°	
4 Observed	36.5	strong	+	+25°	+28°
Calculated	38.4	0.9	+	+65°	
5 Observed	43.0	medium	−	−45°	−45°
Calculated	44.7	0.6	+	+80°	
6 Observed	44.7	strong	−	±25°	±28°
Calculated	48.1	0.7	−	+2°	

[a] Transition energy in 10^3 cm^{-1}.
[b] Angle between the orientation axis (z) and the transition moment direction measured counterclockwise from z. The z axis is assumed to lie as shown in Figure 8.27, i.e., to deviate 15° counterclockwise from the long axis of the anthracene moiety shown in formula **29**, and the orientation distribution is assumed to correspond to $(K_z,K_y) = (0.75,0.16)$. The observed signs of ϕ_i have been determined by comparison with fluoresence polarization data of ref. 71.
[c] Angles which would result from the assumption $(K_z,K_y) = (0.80,0.10)$.
Source: Reproduced by permission from ref. 70.

The use of polarized fluorescence results to determine the signs of the angle of deviation of the transition moments from the orientation axis is of general validity and is illustrated in Figure 8.28. Suppose that two transition moments deviate from the z axis by equal but opposite acute angles ϕ and $-\phi$, producing identical results in measurements of linear dichroism. If the emitting transition moment deviates from the z axis by angle β, it will form two different angles $\alpha_1 = \beta - \phi$ and $\alpha_2 = \beta + \phi$ with the two transition moments in question. There-

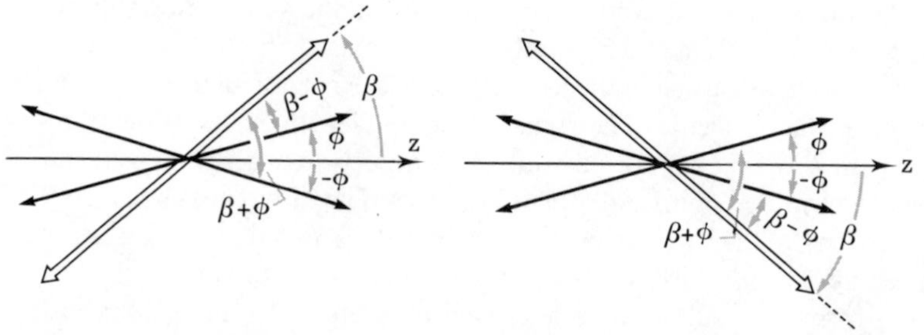

Figure 8.28 The relation of the emitting transition moment (double arrows) to two absorbing transition moments (single arrows) which deviate by the angles ϕ and $-\phi$ from the orientation axis z. All moments are assumed to lie in the same plane.

fore, the values of $\cos^2 \alpha_1$ and $\cos^2 \alpha_2$ which enter into the formula for the polarization degree of luminescence [(7.112)] will be in general also different. The measured value of fluorescence polarization thus permits one to distinguish between transitions with deviation angles ϕ and $-\phi$. The method is not applicable if the emitting transition moment lies in the orientation axis z.

Unless the absolute orientation of the emitting transition moment, i.e., the sign of β, is known, a degree of ambiguity will still be left in the assignments of signs to the angles of deviation from the z axis for all other transitions: All the signs will be relative to the sign of β, so that two choices are possible. Often, MO calculations can be used to select one or the other choice as vastly more probable. This is the case in the instances discussed here.

Other possibilities for the determination of the sign of the ϕ's exist, and these are particularly important for nonfluorescent molecules. If the molecule has a large permanent dipole moment along a direction different from z, electric dichroism can yield the necessary information (see Section 8.4.3 for an example). If it is possible to twist the location of the orientation axis z by a substituent which does not affect transition moment directions but changes the orientation tensor, the same purpose is served. All these procedures work on the basis of the principle illustrated in Figure 8.28.

Using similar arguments, it is at times possible to narrow down the possible choices for the location of the orientation axis z from a combination of linear dichroism with polarized fluorescence measurements.[72]

The values of a and b used for the curves shown in Figure 8.27 were chosen so that, for $(K_z,K_y) = (0.75,0.16)$, in the third curve from the bottom all features polarized along the orientation axis z will just disappear, in the fourth curve those polarized at $\phi = \pm 15°$ from z will disappear, in the fifth those at $\phi = \pm 30°$, etc., until in the ninth curve those polarized at $\phi = 90°$ to z will disappear. This requires the choice

$$a = \frac{1 - K_z + (K_z - K_y)\sin^2 \phi}{1 + K_z - (K_z - K_y)\sin^2 \phi} \qquad b = \frac{-2[K_z - (K_z - K_y)\sin^2 \phi]}{1 + K_z - (K_z - K_y)\sin^2 \phi} \tag{8.3}$$

In this presentation, the polarization direction of a transition can be read off directly by inspection of the set of stepwise reduction curves.

8.4.2 The Aminoanthracenes[56]

The isomeric aminoanthracenes **30–32** provide our second example of the use of equation (5.64) for low-symmetry solutes. In this case, the characteristic vibrations of the NH_2 group, taken along with the out-of-plane bends of the

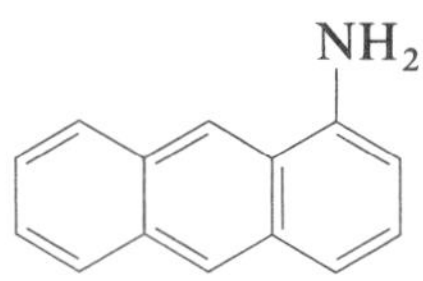

30

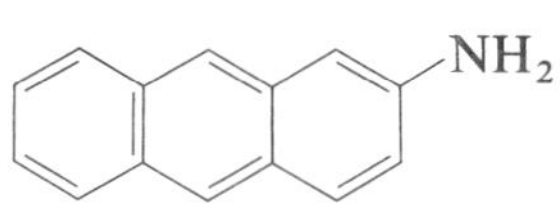

31

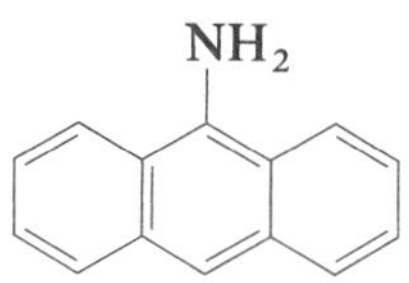

32

C—H bonds, are in principle capable of providing the information needed to establish the orientation tensor from a measurement of IR dichroism. Once again, however, the solubility of these materials is limited and an IR measurement not particularly easy.

Since the orientation tensor elements are known for many singly substituted anthracenes of high symmetry, it is possible to proceed as in Section 8.4.1 and to estimate the position of the orientation axis as well as the orientation factors for **30** and **31**, albeit with some sacrifice in accuracy. The effective symmetry of 9-aminoanthracene (**32**) is C_{2v}, which is high enough for a straightforward reduction procedure, with results shown in Figure 8.29. Note that the reduction is facilitated considerably by the diffuse nature of one but not the other band in the 300–450 nm region. This helpful feature is characteristic of all the aminoanthracenes and removes any difficulty which might otherwise be connected with the strong overlap of the first two transitions. Note that in an instance such as this, the dichroic ratio $E_Z(\tilde{\nu})/E_Y(\tilde{\nu})$ is so severely distorted by the overlap of the two differently polarized transitions as to be almost worthless, and yet the stepwise reduction procedure works well.

Figure 4.15 indicates strongly that the orientation factors of **32**, quite similar to those of 9-methylanthracene, represent a lower limit for the degree of orientation for 1-aminoanthracene (**30**) and 2-aminoanthracene (**31**): $K_z = 0.55$, $K_y = 0.29$. The estimated ranges for the orientation factors **30** and **31** based on Figure 4.15 are further supported by comparison with the results for 1- and 2-substituted methyl and fluoro derivatives of anthracene, in which one can reasonably assume that the perturbation of the substituent is too weak to affect the short-axis orientation of the L_a band and the long-axis orientation of the B_b band of the parent, which are both quite intense. This notion is confirmed by the very small effect which methyl substitution in anthracene has on its polarized fluorescence excitation spectra.

The most likely possible values of orientation factors were estimated as $0.59 \leqslant K_z \leqslant 0.61$, $0.25 \leqslant K_y \leqslant 0.27$ for **30** and $0.71 \leqslant K_z \leqslant 0.77$, $0.11 \leqslant K_y \leqslant 0.15$ for **31**. Given these assumptions, it is possible to use the reduction factor K_i of each observed transition in each molecule, obtained by the stepwise reduction procedure as shown in Figure 8.29, to determine the possible range for the magnitude $|\phi|$ of its angle of deviation from the orientation axis z, using formula (5.64). Less optimistic estimates of the likely accuracy with which the orientation factors can be guessed would produce correspondingly larger error limits on the resulting angles.

The next step is the assignment of signs to the angles ϕ, and two possible choices of sign sequences were obtained using the polarized fluorescence results of ref. 73 in a way described in Section 8.4.1. Finally the position of the principal orientation axis in **30** and **31** must be estimated. From a plot of the size of the cross section of the molecule against the angle of inclination from the anthracene long axis, it was concluded that the effective orientation axis is inclined from the long axis of anthracene by about 8° for **30** and by about 5° for **31**. Thus, two sets of polarization directions with respect to the framework of each molecule were obtained. A decision between them was reached on the

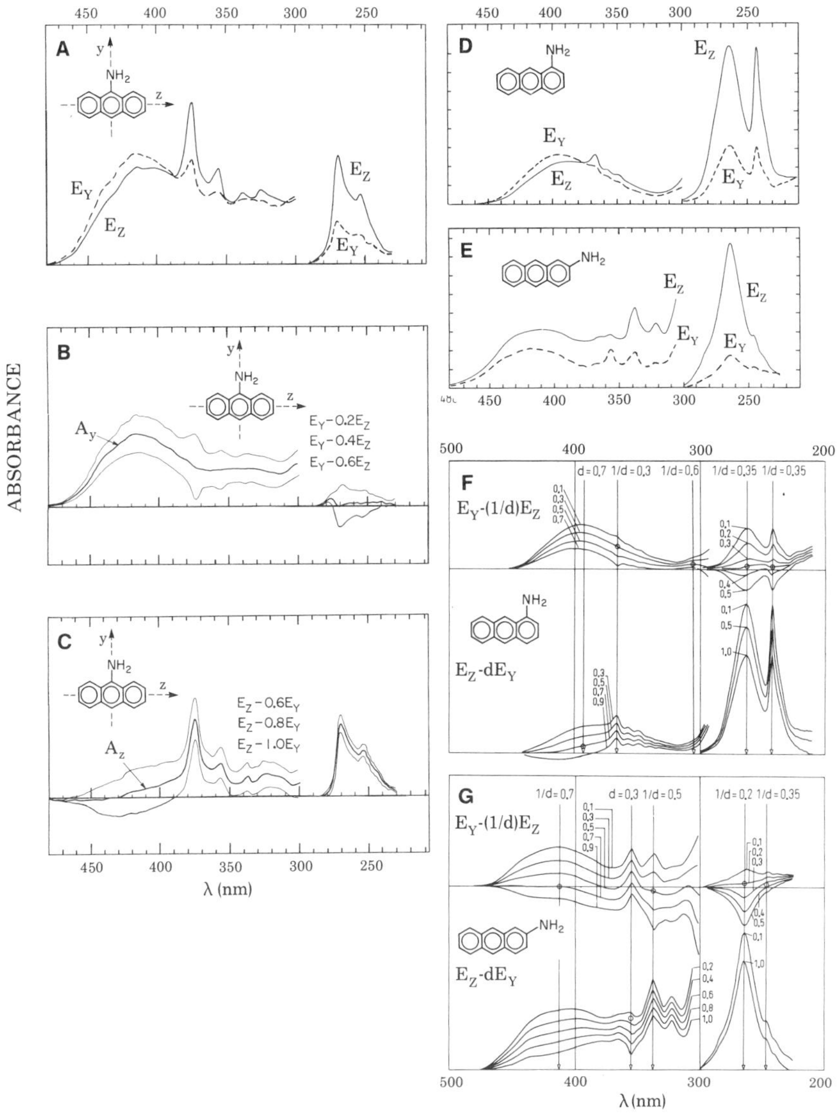

Figure 8.29 Aminoanthracenes in stretched polyethylene: polarized absorption E_Z (full lines) and E_Y (dashed lines), and stepwise reduction. (A,B,C) 9-Aminoanthracene; (D,F) 1-aminoanthracene; (E,G) 2-aminoanthracene. Adapted by permission from ref. 56.

basis of MO calculations which indicated that the plus sign should be used for the sign of the deviation of the transition moment of the lowest energy transition in **30** and that the minus sign should be used in **31**. This produced a set of final results for the angles ϕ' between the transition moment directions and the long axis of the anthracene chromophore which are reproduced in Table 8.12

Table 8.12
Polarization Directions in the Aminoanthracenes

	λ (nm), exp.	Intensity, exp.	f, calc.	E(eV)		d_j exp.	ϕ (°), exp.	ϕ'(°),[a] exp.	ϕ'(°),[a] calc.
				exp.	calc.				
9-Aminoanthracene	415	m (diffuse)	0.5	3.0	3.1	0.8	90	90	90
32	374	m	0.2	3.3	3.5	2.5	0	0	0
	(356)[b]								
	(340)[b]								
	[c]		0.1		3.9				0
	277	m	0.01	4.5	4.8	0.9	90	90	90
	270	s	1.4	4.6	4.8	2.5	0	0	0
	254	s	0.4	4.9	5.0	2.5	0	0	0
1-Aminoanthracene	392	m (diffuse)	0.3	3.2	3.0	0.7	+84 ± 10	+88 ± 10	+80
30	366	w-m	0.1	3.4	3.6	3.3	8 ± 10	0 ± 10	
	(357)[b]							or	
	(349)[b]							+16 ± 10	+20
	[c]		0.1		4.1				−56
	307	w	0.4	4.0	4.5	~2	34 ± 10	+42 ± 10	+31
								or	
								−26 ± 10	

	263	s	0.7	4.7	4.5	3.0	(−)8 ± 10	0 ± 10	−10
	243	s (sharp)	1.3	5.1	5.1	3.0	(+)8 ± 10	+16 ± 10	0
	220	m	0.1	5.6	5.3	~1	64 ± 10	+72 ± 10 or −56 ± 10	−47
2-Aminoanthracene **31**	410	m (diffuse)	0.3	3.0	3.1	0.7	−62 ± 5	−57 ± 5	−58
	354	w-m	0.1	3.5	3.6	0.3	85 ± 10	+(75–90)	+86
	(337)[b]								
	(321)[b]								
	284	m	0.6	4.4	4.3	3	29 ± 10	+34 ± 10 or −24 ± 10	+17
	[c]		0.1		4.5				+35
	265	s	0.9	4.7	4.8	5.5	(−)8 ± 10	−3 ± 10	+14
	246	s	1.1	5.0	4.8	3.3	(−)26 ± 5	−21 ± 5	+1

[a] Angle between the transition moment direction and the long axis of the anthracene chromophore, measured counterclockwise in formulas **30–32**.
[b] Vibrational structure.
[c] Unobserved.
Source: Reproduced by permission from ref. 56.

and can be compared with the results of a PPP calculation (it should be understood that the opposite sign assignment is also compatible with the experimental data).

8.4.3 *Flavins*[74,75]

Our next example illustrates, first, the simultaneous use of three techniques for the assignment of polarization directions of electronic transitions in a molecule of low symmetry[74]: linear dichroism induced in an oriented membrane, linear dichroism induced by electric field (electric dichroism), and polarized fluorescence measurement (ordinary photoselection). Second, it illustrates how the same problem can be tackled using linear dichroism in the IR region to establish the orientation tensor.[75]

The chromophore flavin is of considerable interest in the study of the structure of biological membranes. It is the prosthetic group of many flavoproteins which occur in biological membranes and form a part of the respiratory chain. It is felt that many of the functions of the membranes are intimately tied to the spatial arrangement of membrane constituents and that information about molecular orientation within a membrane would be useful for the understanding of the functional mechanism of the membrane. Such information can be obtained from optical studies of the membranes provided that the directional optical properties of the individual chromophores contained therein are well understood (cf. Section 8.8).

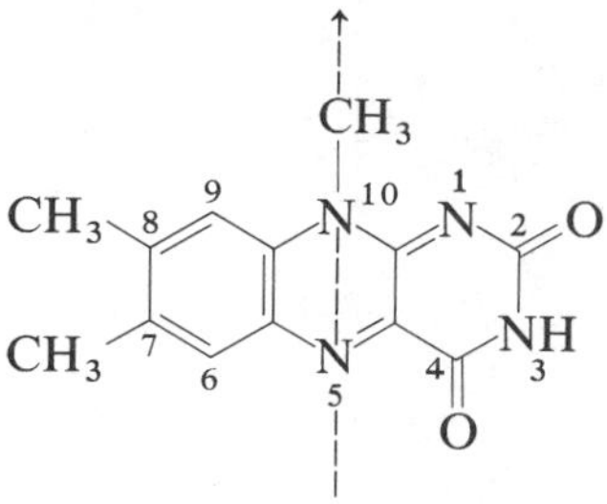

33

The authors of the first study which we shall describe[74] note that lumiflavin (**33**), the oxidized form of 7,8,10-trimethylisoalloxazine, has three $\pi\pi^*$ absorption bands located at 450 nm (I), 375–330 nm (II), and 270 nm (III). Their polarization directions relative to the N—N axis shown in formula **33** will be referred to as ϕ^{I}, ϕ^{II}, and ϕ^{III} (measured counterclockwise). These transitions are clearly apparent in Figure 8.30, which shows the linear dichroic spectra of **33** aligned mechanically or by strong electric field.

From the polarization of fluorescence, the authors conclude that the angle $|\phi^{\mathrm{I}} - \phi^{\mathrm{III}}|$ between transitions III and I and the angle $|\phi^{\mathrm{I}} - \phi^{\mathrm{II}}|$ between transitions II and I fulfill the relation $|\phi^{\mathrm{I}} - \phi^{\mathrm{III}}| > |\phi^{\mathrm{I}} - \phi^{\mathrm{II}}| > 30°$. If the transitions did not overlap, one would obtain $|\phi^{\mathrm{I}} - \phi^{\mathrm{II}}| = 29 \pm 1°$ and $|\phi^{\mathrm{I}} - \phi^{\mathrm{III}}| = 48 \pm 3°$,

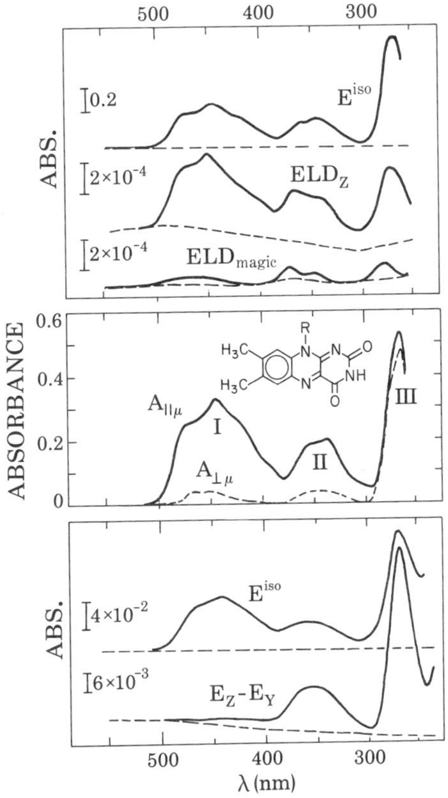

Figure 8.30 Top, lumiflavin: isotropic absorption E^{iso} in $CHCl_3$—CCl_4, electric dichroism ELD_Z (ELD_{magic}) with light electric vector parallel (at magic angle, 54.7°) to the electric field direction. Center: lumiflavin absorption polarized parallel ($A_{\|\mu}$) and perpendicular ($A_{\perp\mu}$) to the direction of the molecular dipole moment, as deduced from the above. Bottom: isotropic absorption E^{iso} of flavin mononucleotide solubilized in a lamellar liquid crystal and the corresponding linear dichroism $E_Z - E_Y$ for the aligned mesophase. Adapted by permission from ref. 74.

but the actual values are probably higher. Since lumiflavin, riboflavin (in which the *N*-methyl group is replaced by a ribose residue), and the even more complicated flavin mononucleotide all gave the same polarized fluorescence and excitation spectra, they are believed to have the same transition moment directions relative to the aromatic framework.

The key measurement in this study[74] was the determination of linear dichroism of lumiflavin induced by an electric field. The authors assumed a pure electric dipole orientation and assumed further that the direction of the permanent electric dipole moment makes an angle of 75° counterclockwise from the N—N axis shown in formula **33**, as calculated previously by a semiempirical MO method. The overall consistency of the results of the three types of measurement was taken as evidence for the correctness of this calculated result.

The measurement consisted of determining the linear dichroism curves

$$\mathrm{ELD_Z}(\lambda) = \mathrm{E_Z}(\lambda) - \mathrm{E^{iso}}(\lambda) \qquad \mathrm{ELD_{magic}}(\lambda) = \mathrm{E_{magic}}(\lambda) - \mathrm{E^{iso}}(\lambda) \tag{8.4}$$

where $E_Z(\lambda)$ is the absorbance measured with the electric field oriented parallel to the electric vector of the light and $E_{magic}(\lambda)$ is the absorbance measured with the electric field forming the angle of 54.7° with the electric vector of the light.

A simplified version of the theory of electric dichroism was used to obtain absorption parallel $[A_{||\mu}(\lambda)]$ and perpendicular $[A_{\perp\mu}(\lambda)]$ to the direction of the dipole moment μ:

$$A_{||\mu}(\lambda) = E^{iso}(\lambda) + (2/3)\Delta(\lambda) \qquad A_{\perp\mu}(\lambda) = E^{iso}(\lambda) - (1/3)\Delta(\lambda) \tag{8.5}$$

where

$$\Delta(\lambda) = (45k^2T^2/2\mu^2E^2)[\mathrm{ELD_Z}(\lambda) - (9/5)\mathrm{ELD_{magic}}(\lambda)] \tag{8.6}$$

Here, T is the absolute temperature, k is the Boltzmann constant, and E is the electric field at the molecule. If μ^2E^2 were known, this set of expressions would permit the calculation of the angle Ω between the direction of each transition moment and the direction of the permanent dipole moment μ:

$$\Omega = \pm\tan^{-1}[A_{\perp\mu}(\lambda)/A_{||\mu}(\lambda)]^{1/2} \tag{8.7}$$

In actual fact, μ^2E^2 is not known. The authors treated it as an adjustable parameter whose value is to be optimized until agreement with the fluorescence polarization results is obtained. For $|\phi^I - \phi^{II}| < 39 \pm 4°$, the criterion $A_{||\mu}(\lambda) > 0$, $A_{\perp\mu}(\lambda) > 0$ could not be met, and the value of μ^2E^2 which led to $|\phi^I - \phi^{II}| = 39°$ was accepted as the best compromise between the fluorescence polarization and electric dichroism results. Its use produced the angles $\Omega^I = \pm 17°$, $\Omega^{II} = \pm 22°$, and $\Omega^{III} = \pm 44°$. Given the assumption about the direction of the dipole moment, this corresponds to $\phi^I = 92 \pm 4°$ or $58 \pm 4°$, $\phi^{II} = 97 \pm 3°$ or $53 \pm 3°$, and $\phi^{III} = 119 \pm 2°$ or $31 \pm 2°$. Six of the eight possible choices of the three signs are incompatible with the previous conclusion $|\phi^I - \phi^{III}| > |\phi^I - \phi^{II}| > 30°$. The two remaining possibilities are $\phi^I = 92°$, $\phi^{II} = 53°$, $\phi^{III} = 31°$ and $\phi^I = 58°$, $\phi^{II} = 97°$, $\phi^{III} = 119°$.

The linear dichroism measured on a sample of flavin mononucleotide dissolved in a lamellar liquid crystal (monooctanoin/water) allowed the authors of ref. 74 to distinguish between these alternatives. The measurements were performed on a sample contained between two quartz plates with light incidence angle of 45° and were evaluated as described in Section 5.3.1. Each transition can be characterized by its K_i value; when the three ϕ values are known, it is possible to calculate the three independent nonzero elements of the orientation tensor K_y, K_z, and K_{yz} using equation (5.47) (the authors actually worked with the equivalent S_{yy}, S_{zz}, S_{yz} values). Of the two possible choices of the angles ϕ, only the set $\phi^I = 58°$, $\phi^{II} = 97°$, and $\phi^{III} = 119°$ gave physically meaningful values for the orientation parameters. The authors then compared this result with previously published work on single crystals and reconciled all of the available information which appeared reliable.[74]

More recently, riboflavin and the 3-methyl derivative of lumiflavin (**33**, with NH replaced by NCH_3) were investigated by a combination of IR and UV-visible linear dichroism using stretched poly(vinyl alcohol) as the orienting medium.[75] The aim of the study was to determine the UV-visible transition moment directions without having to rely on assumptions concerning the detailed nature of orientation by an electric field, on which the previous work[74] relied heavily.

The authors used the orientation tensor formalism described in Section 5.1.3 but chose to work with the Saupe matrix elements $S_{uv} = (3K_{uv} - \delta_{uv})/2$. We shall provide a summary of their arguments in terms of the equivalent orientation factors K_{uv}.

The first step is the determination of the orientation factors and the positions of the orientation axes in the molecular framework using IR dichroism of transitions with known transition moments. Both for 3-methyllumiflavin and for riboflavin, the authors took one of the axes (x) to lie perpendicular to the plane of the aromatic rings. This follows from symmetry in the case of 3-methyllumiflavin but represents a questionable assumption in the case of riboflavin, whose ribityl side chain need not have the plane of the aromatic rings as a symmetry element. In the following, we shall restrict our attention to 3-methyllumiflavin.

Comparison with the orientation properties of related molecules of high symmetry in stretched poly(vinyl alcohol) suggests that the out-of-plane axis lines up worst with the stretching direction and therefore should be labeled x in the standard nomenclature. By symmetry, $K_{xy} = K_{xz} = 0$, and this leaves three independent unknowns, say K_y, K_z, and K_{yz}, in a system of axes in which y and z are both located in the aromatic plane and are mutually orthogonal but otherwise arbitrary.

The IR band due to the $C_4{=}O$ stretching vibration lies at 1700 cm^{-1} and its dichroic ratio is 1.07 ± 0.03. The $C_2{=}O$ carbonyl stretch lies at 1645 cm^{-1}, and its dichroic ratio is virtually identical, 1.11 ± 0.02. It is reasonable to assume that the transition moments of these IR bands lie within a few degrees of the respective C=O bond directions. Relative to the N—N axis of formula **33**, they then lie at the counterclockwise angles $\phi(1700) = 0°$ and $\phi(1645) = 120°$, respectively. The fortuitous identity of their dichroic ratios permits an effortless determination of the possible positions of the effective orientation axis z. Within quite narrow error limits, this in-plane axis must have the same relation to both C=O directions, i.e., it must bisect the angle formed by the two C=O directions (N_3—N_{10}) or else be perpendicular (C_2—C_4) to the bisectrix. In view of the molecular shape, we accept the former (N_3—N_{10}) as correct for the z axis and assign the latter (C_2—C_4) to the y axis.

Under less fortunate circumstances, a third IR band with an accurately known transition moment direction would be needed for the determination of the axes which diagonalize the orientation tensor. This was not available to the authors of ref. 75. This piece of information is also missing for the completion of the task, namely, the determination of the orientation factors. From (5.64), K_y and K_z are related by the observed IR dichroism of the carbonyl stretches. Since

$d_i = 1.09$ implies $K_i = 0.35$, and since $\tan^2 60° = 3$, this relation is $3(0.35 - K_y) = K_z - 0.35$, i.e., $K_y = (1.40 - K_z)/3$ and $K_x = 0.53 - 2K_z/3$. However, this information is insufficient for the actual determination of K_y and K_z, and an assumption is necessary.

The authors took advantage of the relation $0 \leqslant K_x \leqslant d/(d + 2) \leqslant K_z \leqslant 1$, which must hold for the dichroic ratios d of all transitions. Since the dichroic ratio in the visible region is $d = 2.97$, they obtained $K_z \geqslant 0.60$. This leaves little freedom for the choice of K_y and K_z. Obviously, $K_z < 0.80$, since K_x must be positive. However, it is likely that K_x is substantially larger than zero. For instance, for acridine orange in similarly stretched poly(vinyl alcohol), $K_x = 0.13$ (cf. Section 8.3.8). Such a value would require $K_z = 0.60$, $K_y = 0.27$. Thus, it is very likely that K_z is not much larger than 0.60, and the authors proposed the values $K_z = 0.609$, $K_y = 0.268$, and $K_x = 0.123$ as the best guess.

They also considered the dichroism observed for the amide II IR band at 1555 cm^{-1}, whose transition moment direction in the molecular framework is not known with certainty but is believed to be in the range $\phi(1555) = 100 \pm 20°$, and thus at $40 \pm 20°$ from the effective orientation axis z (N_3—N_{10}). The dichroic ratio observed for this band, $d(1555) = 1.70$, i.e., $K(1555) = 0.46$, is substantially larger than the value of 1.09 observed for the C=O stretching bands whose polarization deviates from the orientation axis by 60°. The exact value which will result for this angle from (5.64) depends on the choice of the orientation factors. For $K_z = 0.60$, and therefore $K_y = 0.27$, $K_x = 0.13$, the value 40.6° results; for $K_z = 0.80$, and therefore $K_y = 0.20$, $K_x = 0$, the value 48.8° results. One can conclude that the polarization of the amide II band is at $45 \pm 5°$ from the N_3—N_{10} axis, but this does not help much in the determination of the orientation factors. For the values recommended by the authors of ref. 75, the angle from the N_3—N_{10} axis is 41°.

With the "best" set of orientation factors proposed by the authors of ref. 75, the linear dichroism observed in the UV-visible region immediately produces the absolute values of the angles between electronic transition moments and the orientation axis z, using (5.64). The sign of the angles was determined by comparison with polarized fluorescence data and single-crystal data. For the three main transitions, the authors obtained $\phi^{I} = 50°$, $\phi^{II} = 74°$, $\phi^{III} = 90°$. A careful consideration of the regions of weaker absorption permitted them to assign additional weaker transitions at 313 nm ($\phi = 103°$) and 249 nm ($\phi = 95°$), and a short-wavelength transition at 225 nm ($\phi = 40°$). They pointed out that the presence of the additional transitions is also supported by earlier CD and MCD measurements.

The polarization directions ϕ^{I}–ϕ^{III} proposed in this study[75] are in approximate agreement with those proposed in prior work.[74] Some of the discrepancy may well be due to the presence of an additional methyl group on the flavin chromophore in the more recent study. The remainder must be blamed on inadequacies in the assumptions on which the two studies were based: the uncertainties in the exact nature of the orientation by an electric field, particularly in the static dipole moment direction,[74] and those in the estimate of the orientation factors necessitated by the lack of a third IR transition with a known polarization direction.[75] Finally, experimental error undoubtedly also contributes, but probably no more than ± 3 to ± 4 degrees.[74,75]

8.4.4 *DNA Bases*

Polarization directions of electronic transitions of nucleic acid bases have been of considerable interest in biophysics, and their determination by the stretched sheet method represents our next example. Our account is adapted from ref. 76, which deals with thymine, uracil, cytosine, cytidine, guanine, guanosine, and adenine; we shall limit our description to thymine (**34**). Previous stretched sheet studies were less extensive[77] or used questionable assumptions[78] [such as estimation of the principal orientation axes for this circular molecule from molecular shape, or use of Tanizaki's model for molecules which probably do not satisfy equation (5.84); note that the statement in ref. 76 that the Tanizaki model is valid only for rodlike but not for planar molecules is incorrect, since the orientation function of molecules of either kind may accidentally fulfill 5.84)].

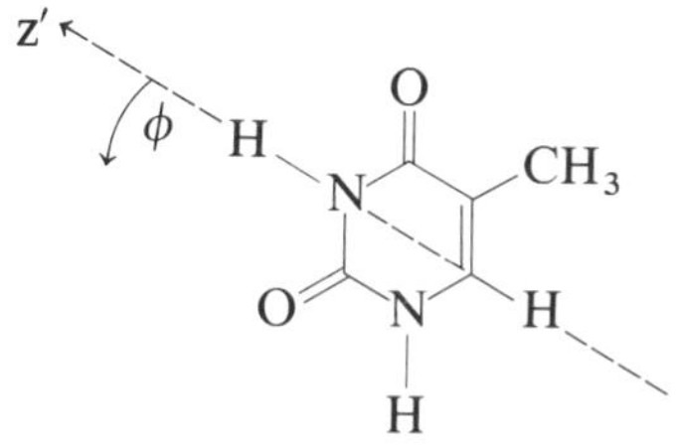

34

The low symmetry of the bases (C_s at most) and their rather circular shape suggest that in this case it will be difficult to obtain a reliable estimate of the location of their principal orientation axes and their orientation factors from molecular shape by comparison with related high-symmetry molecules. Such an attempt would be further complicated by the very poor solubility of these polar molecules in polyethylene, in which specific solute-solvent interactions such as hydrogen bonding are minimized. Indeed, all stretched sheet studies of these molecules used poly(vinyl alcohol) instead.

The other obvious choice is to use the polarization directions of vibrational transitions to determine the elements of the orientation tensor. Since the molecules have effective C_s symmetry, only three vibrational transitions of known polarization directions are required. Such data are available from previous IR work on single crystals. Unfortunately, poly(vinyl alcohol) absorbs strongly in much of the infrared, and only the 1400–1900 cm^{-1} region could be used to observe the IR linear dichroism of the solutes. As this contained only two useful vibrational transitions, the authors decided to estimate the average alignment of the out-of-plane axis by comparison with related symmetrical molecules. As a result, some uncertainty still remains, but a comparison with MO calculations and with polarized UV spectroscopy on single crystals indicated that the remaining margin of error is relatively small.

Figure 8.31 shows the ordinary UV absorption spectrum of thymine in unstretched poly(vinyl alcohol) and the value of the reduced dichroism determined in a sheet stretched to a ratio $R_s = 3.9$. Like the simple dichroic ratio, reduced dichroism has no simple physical significance in regions of overlapping transitions of differing polarization, but its constancy across the absorption band with a maximum at 266 nm, assigned to a $\pi\pi^*$ transition, suggests the presence of little if any overlap and a constant transition moment direction across the band. This permits the immediate determination of the orientation factor for this transition: For $R_s = 4.3$, $K(266) = 0.42$.

Figure 8.31 also shows the dichroic IR spectra with peaks at 1670 cm^{-1} (C_4=O stretch) and 1705 cm^{-1} (C_2=O stretch), taken in a sheet stretched to $R_s = 4.3$. The values of their dichroic ratios yield $K(1670) = 0.40$ and $K(1705) = 0.43$. (The authors of ref. 76 actually performed their measurements at several values of R_s, extrapolated to $R_s = \infty$, and worked with the orientation factors which would describe this hypothetical orientation distribution; this procedure

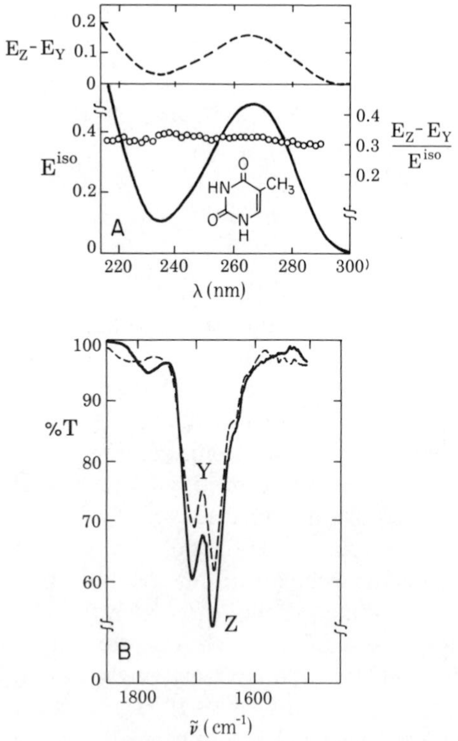

Figure 8.31 Thymine in stretched poly(vinyl alcohol). The linear dichroism $E_Z - E_Y$ (dashed), the isotropic absorption spectrum E^{iso} (full) and the reduced dichroism $(E_Z - E_Y)/E^{iso}$ are shown in part A ($R_s = 3.9$). Polarized IR transmittance in the carbonyl stretch region is shown in part B ($R_s = 4.3$). Adapted by permission from ref. 76.

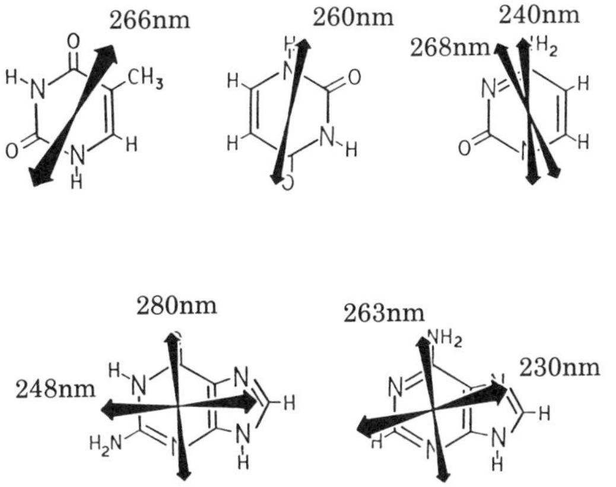

Figure 8.32 Proposed polarization moment directions in the DNA bases. Adapted by permission from ref. 76.

offers the advantage of more severely limiting the possible choices of K_x, but it introduces a likely error in the extrapolation.)

The transition moment directions of the 1670 and 1705 cm^{-1} transitions are $\phi(1670) = -63°$ and $\phi(1705) = 72°$ measured from the z′ axis of formula **34**. This gives one equation too few for solving the equation set (5.47) for K_y, K_{yz}, and K_z, which would permit a diagonalization of the orientation tensor (because of the effective C_s symmetry, $K_{xy} = K_{xz} = 0$). Various guesses can now be made for K_x; this will remove one unknown, since $\sum_u K_u = 1$, and produce a solution for the principal values of the orientation tensor.

The polarization directions obtained will depend on the actual value of K_x. The value preferred by the authors led to the polarization direction assignments shown in Figure 8.32. The selection of signs for the angle of deviation from the orientation axis was based on comparison with MO calculations and with results of measurements of polarized fluorescence. The accuracy of the results shown is difficult to assess in the absence of additional information on the K_x value.

8.5 Optical Activity of Oriented Molecules

Measurement of optical activity of molecules in uniaxial orienting media is much more difficult than the measurement of dichroism discussed so far. Indeed, the stage of development of this field is now such that a successful measurement on any molecule is in itself a publishable result. Advancement to a stage where the results obtained for natural circular dichroism are actually interpretable in terms of molecular structure is hampered by the fact that only low-symmetry molecules have been investigated so far. The interpretations should be easier for magnetic circular dichroism, where high-symmetry molecules can be used easily, but studies of this kind have just barely begun to appear.

We shall illustrate the natural CD spectra obtained so far on two examples, without attempting any detailed analysis.

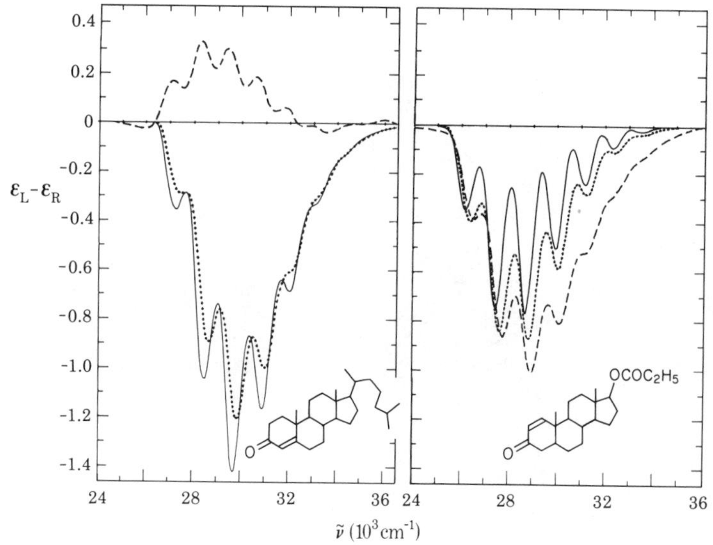

Figure 8.33 CD spectra of cholest-4-en-3-one (**35**, left) and 17β-propionyloxy-5α-androst-1-en-3-one (**36**, right). Full line, in n-heptane and dioxane, respectively; dotted line, in a mixture of cholesteryl chloride and cholesteryl laurate (1.8:1 by weight) in its isotropic state at 80°C; dashed line, in its oriented compensated nematic liquid crystal state at 35.3°C. Adapted by permission from ref. 80.

8.5.1 α,β-Unsaturated Ketosteroids[79]

Steroids represent a large class of molecules of rather similar shape, all of which one could expect to orient similarly in a given uniaxial solvent. This circumstance and the availability of naturally optically active samples are favorable for studies of oriented CD.

Pioneering experimental work in this area has been done by Kuball and collaborators. Figure 8.33 shows two examples of the kind of spectra they have been able to obtain for α,β-unsaturated steroidal ketones dissolved in a compensated nematic liquid crystal. The molecules chosen for this purpose are cholest-4-en-3-one (**35**) and 17β-propionyloxy-5α-androst-1-en-3-one (**36**). Diffi-

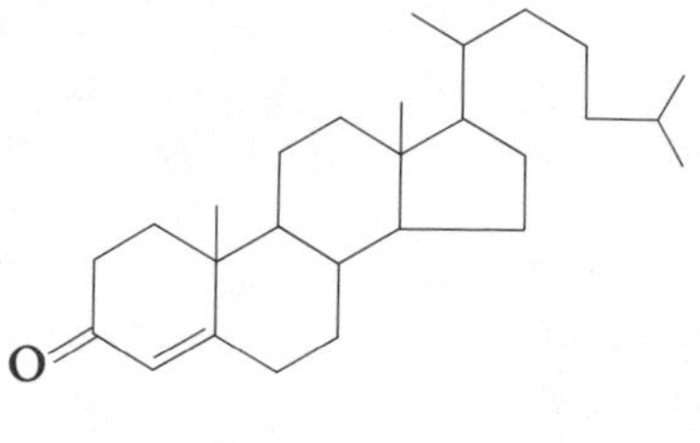

35

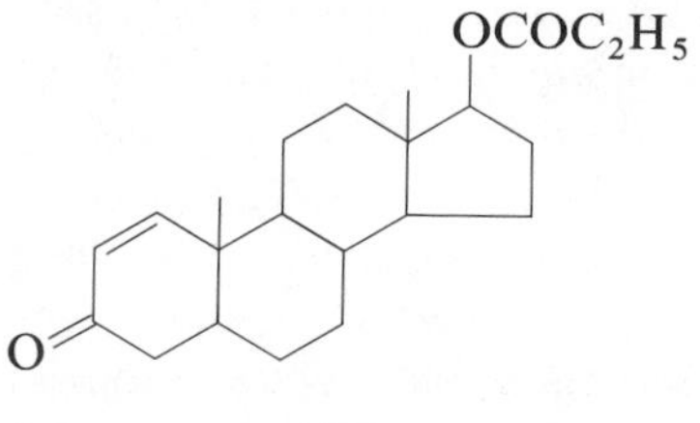

36

culties with an attempted quantitative evaluation in terms of the structure of the chromophore are discussed in some detail in the original article.[79] They are due primarily to an excessive number of unknowns, which include the elements of the orientation tensor. The latter are hard to estimate reliably for molecules with no symmetry, although some information is available from independent measurements of UV linear dichroism. These studies are likely to benefit significantly from measurements of linear dichroism in the IR region, which could establish the orientation tensor unambiguously.

It has been shown convincingly that the experimental results are not subject to significant systematic errors due to the linear birefringence of the sample, since the measured circular dichroism and the measured optical rotatory dispersion are related by the Kramers-Kronig transform as they should be.[80] This paper also gives a list of references to previous work.

8.5.2 Rhodopsin[81] and Bacteriorhodopsin[82]

The optical activity of natural photosensitive pigments is of considerable interest, since it can provide clues to the nature of the protein-chromophore interaction as well as to the conformational changes which occur subsequent to light absorption. The molecular basis of this optical activity is still poorly understood, and in an attempt to rectify this situation both the CD of rhodopsin[81] and that of bacteriorhodopsin[82] have been investigated recently in an oriented state. Uniaxial orientation was achieved by depositing uniform multilamellar films of the biological membrane on a substrate, and the measurement was performed in a direction normal to the surface.

Figure 8.34 shows the absorption and the CD spectra of rhodopsin in ordinary isotropic solution (1% digitonin) and in an oriented multilayer film. A small drop in the 280/500 nm absorption ratio can be attributed to a preferential orientation of the retinylidene chromophore (500 nm) in the film plane compared with the aromatic amino acid chromophores (280 nm). The CD spectra show much larger differences, and their appearance will place considerable constraints on future efforts to explain the optical activity of rhodopsin at a molecular level.

The absorption and CD spectra of oriented purple membrane films containing bacteriorhodopsin were obtained similarly and were amenable to an interpretation in structural terms.[82] Using a trimer exciton interaction description of the origin of the CD of bacteriorhodopsin in the visible region and an exciton interaction of the far-UV region, the authors concluded that (i) a relatively strong in-plane monomeric interaction occurs between the retinyl chromophore and apoprotein; (ii) the helical axes of the native and regenerated membrane protein are oriented primarily normal to the membrane plane; (iii) the helical axes of the bleached membrane proteins are tilted more in-plane than the axes of the native or regenerated membrane. The arguments leading to these and several additional conclusions from the observed spectra rely on the exciton theory of optical activity and are too involved for detailed description here.

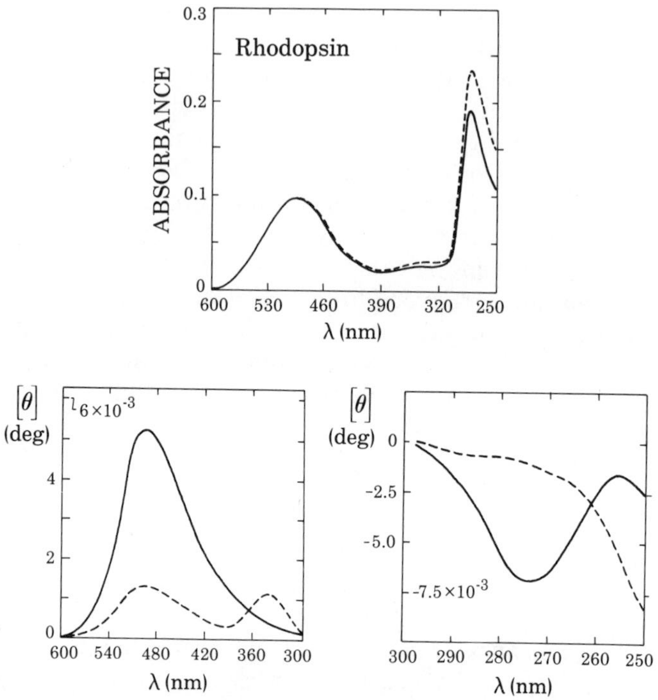

Figure 8.34 Photoreceptor membrane in digitonin solution (dashed lines) and in the form of an oriented multilayer film (full lines). Top, absorption; bottom, circular dichroism (ellipticity) at the same concentration. Adapted by permission from ref. 81.

8.6 Molecular Structure and Conformation

Once the absolute polarization directions of transitions within individual electronic or vibrational chromophores are known, measurements of linear dichroism of partially aligned molecules containing two or more such chromophores will provide information about their relative orientation. If the orientation tensor can be estimated, information on the absolute orientation of such a chromophore or chromophores relative to the principal orientation axes of the molecule can also be obtained. Examples of some of the chromophores which can be used for this purpose are linear polyenes or aromatic rings in UV-visible spectra and carbonyl, nitrile, or nitro groups in IR spectra. Sometimes, a mere measurement of the dichroic ratios for a series of transitions provides structural information: if more than three distinct ratios are observed, the observed species cannot be of high symmetry, such as C_{2v} or D_{2h}.

Up to now, most of the work on molecular structure and conformation by the methods of polarized optical spectroscopy discussed in this text has dealt with the constituent parts of anisotropic media such as stretched polymers or liquid crystals and not with small-molecule solutes contained within such media.

As such, it lies outside the scope of this book; it is covered in specialized monographs.[83]

Application of these methods to small molecules dissolved in uniaxial anisotropic media is of considerable interest in organic chemistry and biochemistry, particularly in conformational studies. Up to now, relatively few studies of this kind have appeared, primarily from the laboratory of Mazur, Yogev, and collaborators. This is most likely due to the limited number of UV chromophores which occur in the systems of greatest interest for this type of work. It seems to us that recent instrumental advances in measurements of IR linear dichroism of solutes in stretched polymers are likely to spur increased activity in this area.

The desirability of knowledge of the orientation tensor of the solute places a premium on one's ability to select a class of molecules all of which orient very much the same way. Steroids represent a natural choice, and most of our examples indeed deal with this important class of molecules.

8.6.1 Steroids: Orientation of Conjugated Enone[84,85] *and Diene*[85,86] *Chromophores*

Measurements of UV linear dichroism on a large number of steroids containing an α,β-unsaturated ketone or a 1,3-diene chromophore, using stretched polyethylene as the uniaxial aligning medium, are in excellent agreement with the notions (i) that the effective orientation axis z coincides with the longitudinal axis of inertia of the steroidal skeleton as shown in Figure 8.35, (ii) that the orientation factors are rodlike ($K_x = K_y$) and nearly identical for all steroids with a C_8 side chain such as cholestenone **35** ($f = S_{zz} \simeq 0.5$ or $K_z \simeq 0.67$) and that the alignment is poorer but all orientation factors again rodlike and nearly identical for all steroids without this side chain such as the androstenone ester **36** ($f = S_{zz} \simeq 0.3$, $K_z \simeq 0.53$), and (iii) that the transition moment direction of the first $\pi\pi^*$ transition is quite accurately given by the line which joins the two ends of the conjugated chromophores used in these studies. This last point was derived from the observation that the measured dichroic ratios, which immediately yield the angles of deviation $|\phi|$ of the transition moment from

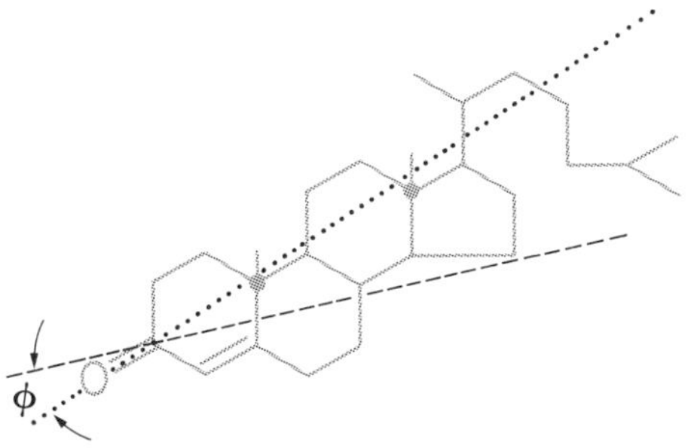

Figure 8.35 The principal orientation axis z of a cholestenone (dotted) and the direction of the $\pi\pi^*$ transition moment of the conjugated enone chromophore (dashed). Adapted by permission from ref. 84.

the orientation axis by use of equation (5.63) (or equivalent expressions based on f or S_{zz} and the rodlike orientation behavior), produce the same angles as a simple measurement on a molecular model. As a result, a measurement of linear dichroism can now be used for the assignment of possible locations of this type of chromophore in any new steroid molecule of a similar shape. This is illustrated in Figure 8.36, which shows the plot of the dichroic ratio of a transition with a moment directed at $|\phi|$ from the orientation axis expected from equation (5.63) for $K_z = 0.68$ and $K_x = K_y$ as a function of $|\phi|$. Superimposed are points showing the observed dichroic ratios plotted against the angles $|\phi|$ measured on a molecular model.

These results might have been as well discussed in Section 8.4 as an example of determination of transition moment directions, since actual applications to structure determination, although now possible in principle, have not been reported. Such an attitude was taken by the authors of ref. 85, who also determined the polarization directions of the second $\pi\pi^*$ transition in many of these molecules and compared their results with those of MO calculations. These authors found the value of $K_i = 0.75$ for the $\pi\pi^*$ transition in a steroidal dienone with a C_8 side chain (and therefore $K_z \geqslant 0.75$) and concluded that the degree of alignment for these steroids is even somewhat higher than that given by the value $K_z = 0.67$ adopted earlier.[84] Such a change would not change the angles

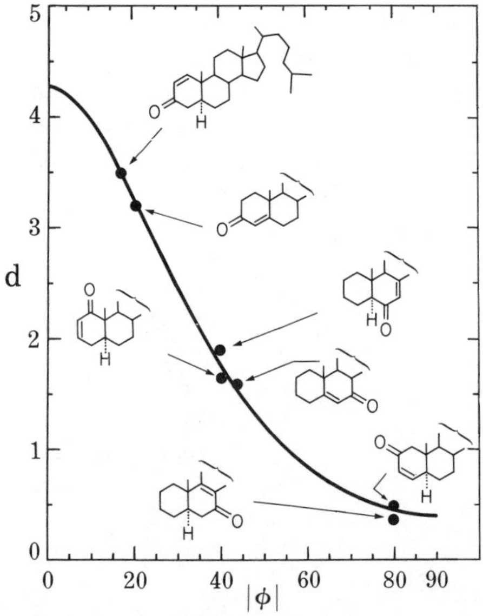

Figure 8.36 The dichroic ratio expected from (5.63) plotted against the angle $|\phi|$ between the transition moment direction and the orientation axis z. The values $K_z = 0.68$, $K_x = K_y = 0.16$ are assumed. The points indicate the observed dichroic ratios of $\pi\pi^*$ enone transitions in stretched polyethylene as a function of the angle $|\phi|$ measured on a molecular model, assuming the transition moment to lie along the line connecting the end atoms of the chromophore. Adapted by permission from ref. 84.

$|\phi|$ calculated from the observed dichroic ratios (e.g., Figure 8.36) substantially; indeed, even the assumption $K_z = 1.0$ would cause a change of only a few degrees except for transition moments oriented nearly perfectly along the z axis (here, the error could be 10–20°). This lack of sensitivity to the exact values assumed for the orientation factors has already been noted in Section 8.4. It is particularly pronounced if the degree of molecular alignment is very high. In such a case, even the assumption $K_x = K_y$ cannot be too wrong (it is likely that it does not hold exactly for the steroids in question, but this will have little effect on the transition moment directions derived from the measurements). The situation is quite different when the degree of molecular alignment is relatively low, and much more caution is necessary then.

8.6.2 Steroids: Vitamin D Analogs[86]

The understanding of the orientation of steroids in stretched polyethylene obtained in the studies described in the preceding section has been used to address the problem of the conformation of vitamin D and its analogs. In the flexible molecules **37** and **38**, the dichroic ratio of the $\pi\pi^*$ transition at 220–

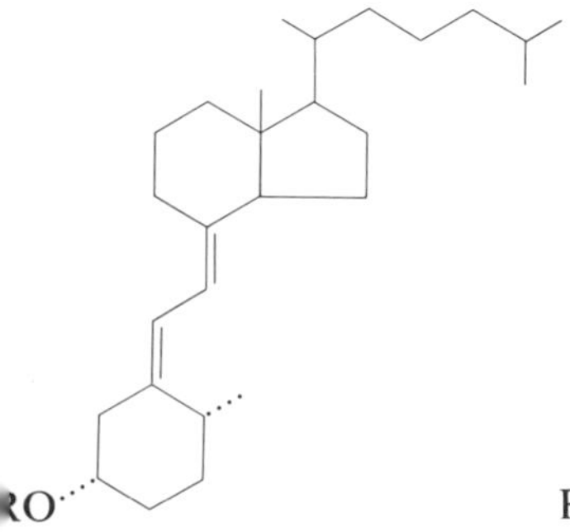

37: R = H
39: R = $OCOC_6H_4N(CH_3)_2$

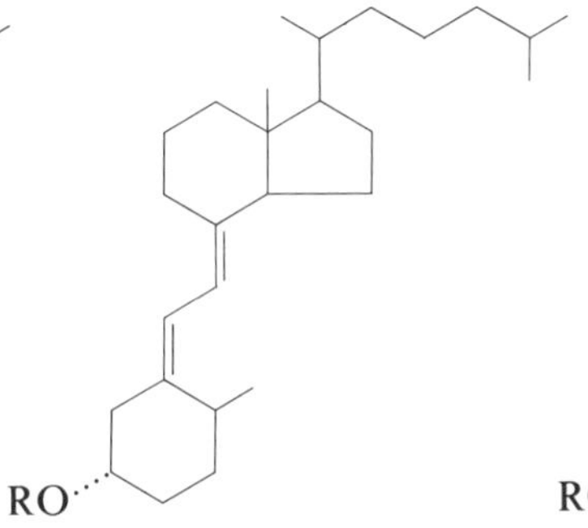

38: R = H

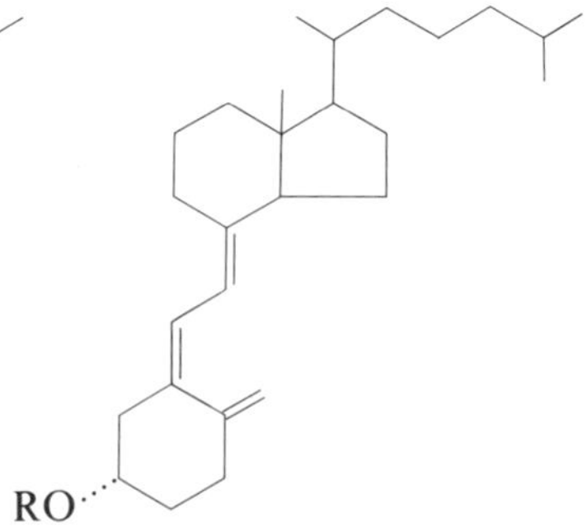

40: R = H

270 nm is far lower than expected, d = 1.0–1.2, corresponding to an angle $\phi \simeq 50°$ between the transition moment and the z axis if the orientation distribution is rodlike and the degree of alignment as high as expected from the shape. Yet molecular models suggest that the angle ϕ is near 30° (Figure 8.37). The authors suggest that this discrepancy would be accounted for if the s-trans diene moiety were twisted out of planarity. They found support for this proposal in the observation that the *p*-*N*,*N*-dimethylaminobenzoyloxy substituent, whose L_a transition near 310 nm is polarized along its long axis, shows a far higher dichroic ratio (d = 2.8) when located in the equatorial position of the bottom ring as in **39** (Figure 8.38). In the planar geometry of **39**, the directions of the L_a and diene $\pi\pi^*$ transition moments form angles of 20° and 30°, respectively, with the z axis and should show similar dichroic ratios. However, on twisting the central single bond, the orientation axis as well as the transition moment

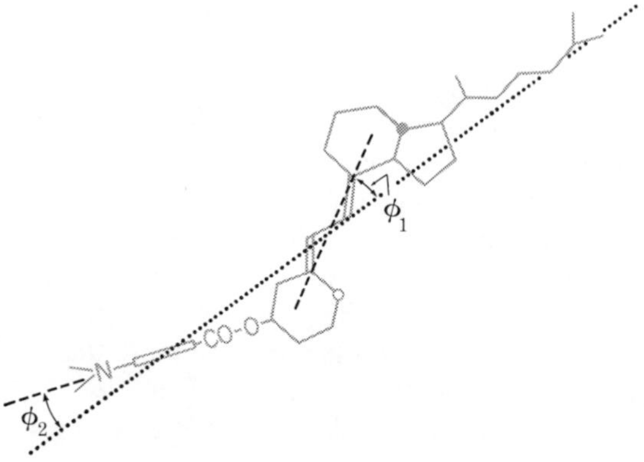

Figure 8.37 Dihydrovitamin D *p-N,N*-dimethylaminobenzoate. Transition moment directions of the *p*-dimethylaminobenzoate and diene chromophores (dashed) and the assumed direction of the principal orientation axis z (dotted). Adapted by permission from ref. 86.

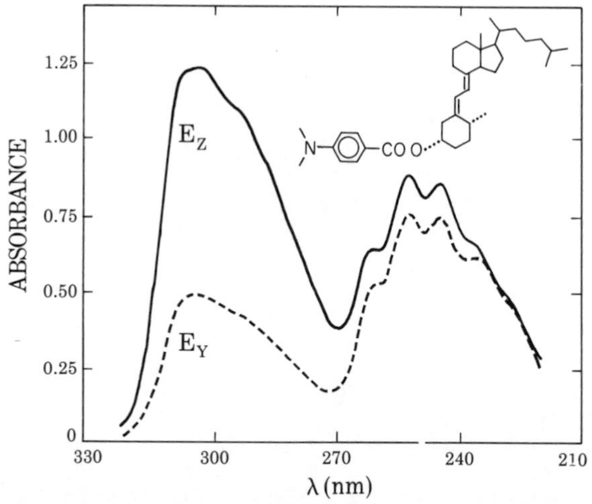

Figure 8.38 Polarized absorption spectra E_Z (full) and E_Y (dashed) of the *p-N,N*-dimethylaminobenzoate ester of dihydrovitamin D (**39**) in stretched polyethylene. Adapted by permission from ref. 86.

directions of both chromophores change in a way compatible with their observed dichroic ratios.

The polarized spectra of vitamin D_3(**40**) and its *p-N,N*-dimethylaminobenzoate ester are analogous to those of the dihydro derivatives, and the authors propose that these molecules are also twisted around the central single bond.

The authors measured the linear dichroism of several additional s-trans dienes in stretched polyethylene in an attempt to secure additional evidence for their interpretation. Although their proposal appears to be plausible, the uncertainties associated with the orientational behavior of these flexible low-symmetry molecules are sufficient to call for further evidence before the conclusions are accepted as firm. It is possible that IR dichroism would provide such data.

8.6.3 Steroids: Conformational Preferences of Acyloxy Substituents[87]

The information on the alignment of steroids in stretched polyethylene described in Section 8.6.1 can be exploited for the determination of rotameric conformations of substituents on the rigid polycyclic skeleton.

Our first example deals with a series of *p*-substituted benzoyl derivatives of cholesterol. Measurements on the cyclohexanecarboxy enones **41** and **42** permitted the determination of the orientation factor $K_z = 0.77$ ($f = S_{zz} = 0.65$) from the linear dichroism of the enone $\pi\pi^*$ transition ($d = 1.8$ and 0.35, respectively), recognizing that the transition moment directions in the two molecules differ by 120°, and using equation (5.63) as in Section 8.6.1, assuming again $K_x = K_y$. The alignment is thus somewhat better than in steroids without a bulky equatorial group in position 2, as would be expected from the differences in shapes. It was then assumed that the removal of the keto group and the replacement of the cyclohexanecarboxy group by the benzoyloxy group have negligible effects on the orientation, so that the esters **43** also have $K_z = 0.77$.

The benzoate chromophore has three transitions in the accessible region. If their mutual overlap can be neglected and their transition moment directions are known, the measured dichroic ratios of **43** can be used to determine the orientation of the benzoate chromophore relative to the effective orientation axis in **43**. Measurements on *para*-substituted benzoic acids, which are present in polyethylene in their dimeric form **44**, indicated that the 255 nm transition in the *p*-methoxybenzoate ester is polarized nearly exactly along the long axis of the molecule (the orientation factors of the dimeric benzoic acids were obtained by combining their dichroic ratios with those of the corresponding esters **43**).

Now, the linear dichroism of the 255 nm transition in **43**, X = CH_3O—, was measured. The observed dichroic ratio, $d = 6.6$, corresponds to $K_i = 0.77$, so

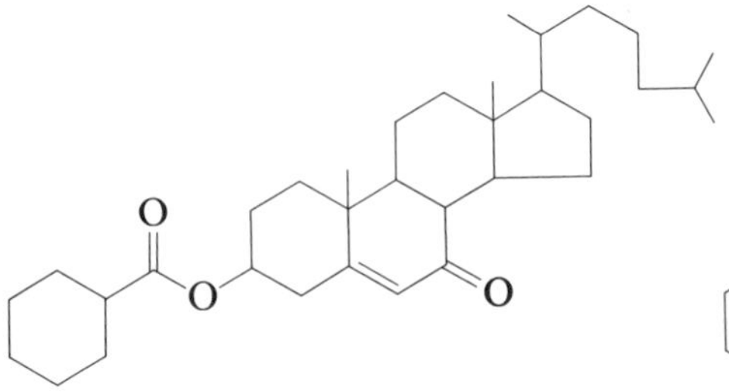

41

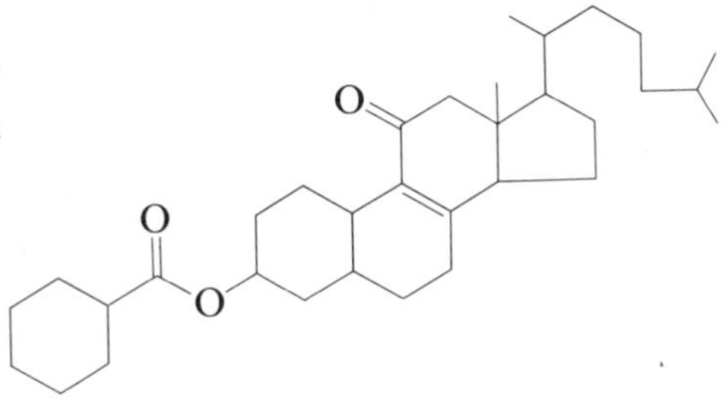

42

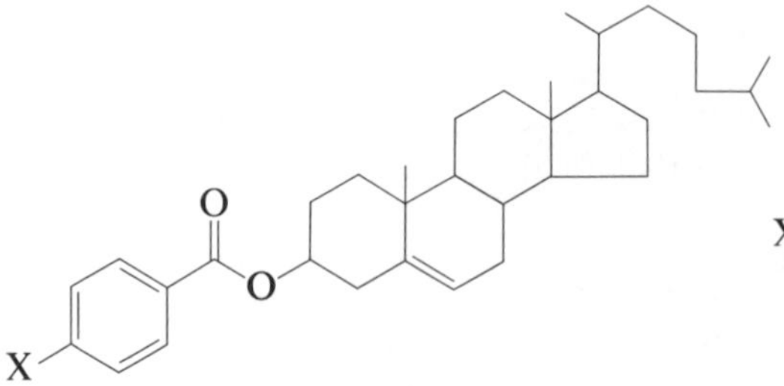

43

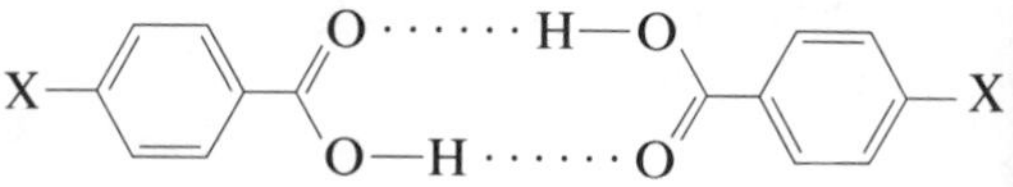

44

that it is exactly equal to K_z. This means that in **43** the long axis of the *p*-methyoxybenzoate chromophore is aligned within experimental error with the effective orientation axis z of the molecule. Thus the conformation of **43** is as shown in Figure 8.39.

Similar results were obtained for 17β-benzoyloxyandrost-5-ene (**45**). Its degree of alignment was established as $K_z = 0.58$ ($f = S_{zz} = 0.37$, assuming $K_x = K_y$) by measuring the dichroic ratios of the $\pi\pi^*$ transition of the enone chromophore in **46** and **47** ($d = 1.64$ and $d = 2.14$, respectively), recognizing that the transition moments in the two molecules lie 60° apart, and using equation (5.63).

It was assumed that K_z and the direction of the z axis do not change upon going to **45**, and the measured dichroic ratio in **45** at 230 nm was combined with the known orientation of the transition moment of the benzoate chromophore to deduce that the conformation of **45** is as shown in Figure 8.39.

While some of the assumptions made in this study may not be totally valid, this will have no significant effect on the conclusions, which do not require

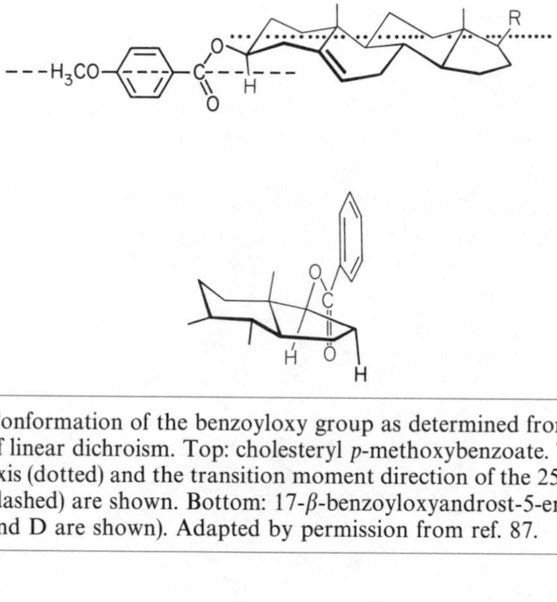

Figure 8.39 Conformation of the benzoyloxy group as determined from measurement of linear dichroism. Top: cholesteryl *p*-methoxybenzoate. The orientation axis (dotted) and the transition moment direction of the 255 nm transition (dashed) are shown. Bottom: 17-β-benzoyloxyandrost-5-ene (only rings C and D are shown). Adapted by permission from ref. 87.

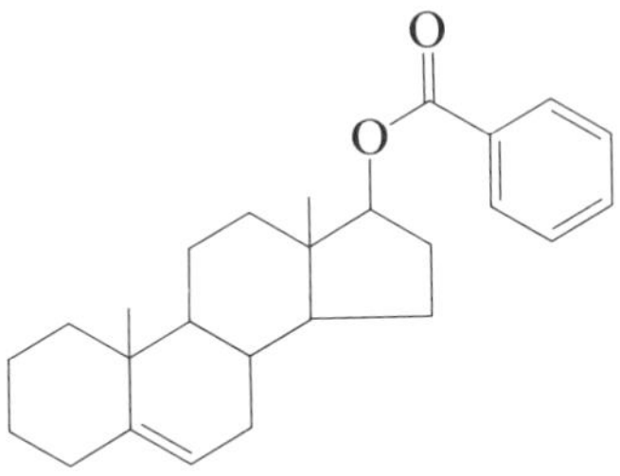

45

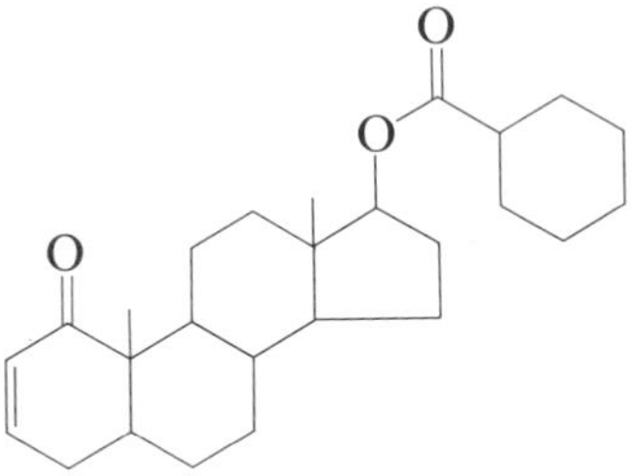

46

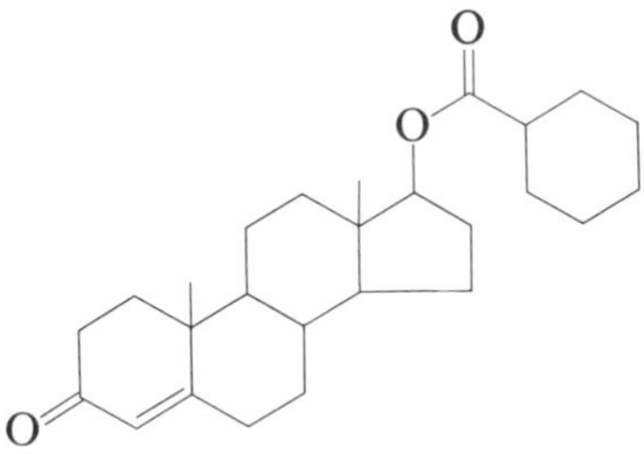

47

highly accurate knowledge of the transition moment directions. This is generally true in conformational studies.

8.6.4 Steroids: Conformational Preferences of the Acetyl Substituent[88]

This second example of the use of linear dichroism in stretched polyethylene for establishing the preferred rotameric conformation of a substituent on the steroidal skeleton deals with two compounds (**48**, **49**) from a series of steroidal methyl ketones. This study relied on the finding by the same authors that the short-wavelength transition moment of the carbonyl chromophore ($\lambda_{max} \cong$ 180 nm) is directed along the C=O bond axis.

The degree of alignment for **48** and **49** was taken to be $K_z = 0.69$ ($f = S_{zz} = 0.54$), assuming $K_x = K_y$. This was based on previous results for the linear

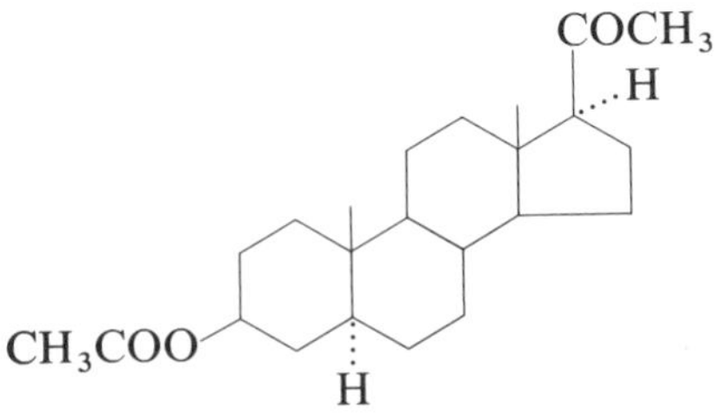

48

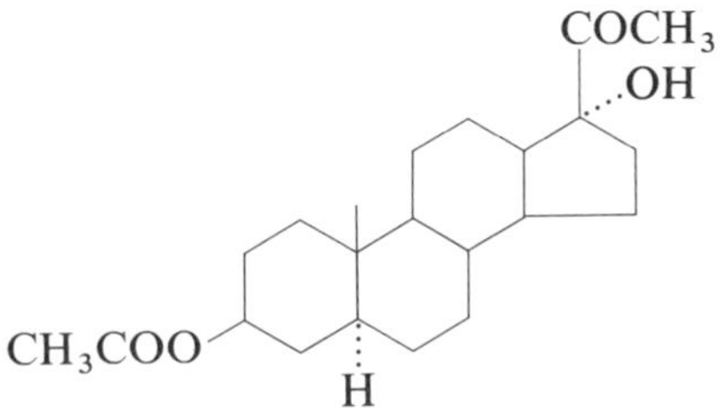

49

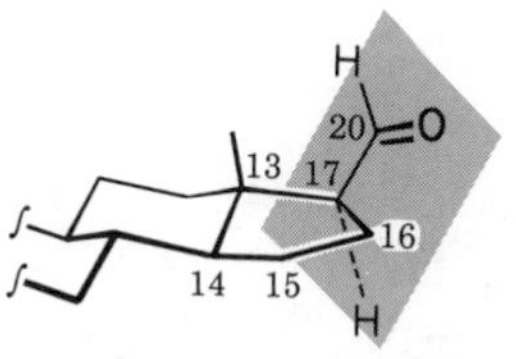

48

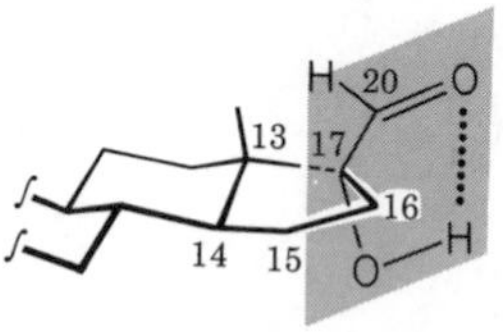

49

Figure 8.40 Conformations of the 17-acetyl group in two steroids in the pregnane series proposed on the basis of measurement of linear dichroism. Adapted by permission from ref. 88.

dichroism of the $\pi\pi^*$ transition of enones with similar molecular shapes in a manner analogous to that discussed in the preceding sections.

The dichroic ratios of the short-wavelength transitions of the carbonyl group in **48** and **49** were d = 2.1–2.8 and d = 1.1–1.6, respectively. The variation of the dichroic ratio across the absorption band, ascribed to overlap with a shorter-wavelength transition, reduces the accuracy with which the angle between the C═O bond axis and the orientation axis z can be determined using equation (5.63). The results are 24–33° for **48** and 42–53° for **49** and suggest that the conformations of the acetyl substituent are different in the two compounds. For **48**, the angle is compatible with a conformation in which the C═O bond axis is parallel to the plane passing through the C_{16}, C_{17}, C_{20} carbon atoms (Figure 8.40). For **49**, the angle is compatible with a conformation in which the C═O bond is eclipsed with the C_{17}—α-OH bond, and this can be ascribed to internal hydrogen bonding (Figure 8.40).

With present-day Fourier transform IR instrumentation, this study could be repeated in the IR region with considerably higher accuracy. In general, the field of stereochemistry should benefit considerably from the introduction of IR linear dichroism techniques. We shall illustrate this potential in the following three examples.

8.6.5 *Aniline: Pyramidalization at Nitrogen*[16]

It is instructive to constrast the IR linear dichroism of aniline (**49A**) in stretched polyethylene (Figure 8.40a) with that of nitrobenzene (**49B**). Only three different dichroic ratios were observed for the thirty observed IR bands of the latter. They yield $K_z = 0.445$, $K_y = 0.315$ and $K_x = 0.230$ with an error of

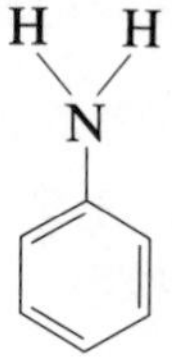

49A

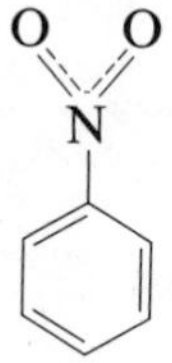

49B

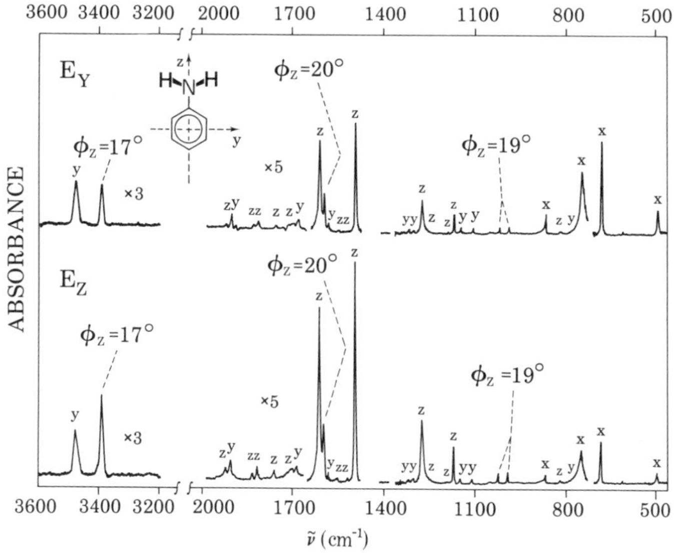

Figure 8.40a Polarized IR spectra of aniline in stretched polyethylene (12K). Polarization directions are shown; ϕ_z is the angle between the transition moment and the molecular z axis. Reproduced by permission from ref. 16.

±0.005. This is as expected for a molecule of C_{2v} symmetry, in which all transition moments lie along x, y, or z, as long as the interaction with the environment is negligible.

In the case of aniline (**49A**), four distinct dichroic ratios were observed although the degree of orientation was similar to that observed for nitrobenzene. A detailed investigation at 77 K gave the results shown in Table 8.12a.

The fact that vibrations of aniline exhibit four different K values proves that on the IR time scale its symmetry cannot be as high as C_{2v}. The immediate suspicion is that the amino group is pyramidal in the solid solution similarly as it is known to be in the gas phase, so that the symmetry is C_s. Closer

Table 8.12a
Orientation Factors for IR Transitions in Aniline (77 K)

	K[a]	No. of bands	Assignment
x	0.180	4	a′
y	0.325	15	a″
z	0.495	14	a′
996 cm^{-1}	0.460	1	a′
1028 cm^{-1}	0.460	1	a′
1602 cm^{-1}	0.460	1	a′ NH_2 scissoring
3393 cm^{-1}	0.470	1	a′ NH_2 sym. stretch

[a] Uncertainty: $\pm$ 0.005

inspection of Table 8.12a confirms this conclusion. The orientation factor of the short-in-plane axis y is determined by the dichroism of the antisymmetric NH_2 stretching vibration and other vibrations of this polarization, for all of which $K_y = 0.325$. Several out-of-plane polarized vibrations have the orientation factor $K_x = 0.180$. The orientation factor of the long axis should therefore be $K_z = 0.495$ and this value is indeed observed for the C–N stretching vibration and quite a few others.

However, the orientation factors of four other vibrations differ from the above three values by more than the experimental error of ± 0.005. In particular, the orientation factors of the symmetric NH_2 stretch, $K_\phi = 0.47$, and of the NH_2 scissoring vibration, 0.46, deviate from the value 0.495 which would be expected if the NH_2 group were coplanar with the ring. This result does not only permit a determination of molecular symmetry as C_s rather than C_{2v}, but also allows a rough estimate of the degree to which the molecular geometry deviates from planarity.

If it is assumed that the pyramidal inversion of the NH_2 group is relatively rapid on the molecular interaction scale (cf. the umbrella motion in NH_3), the principal orientation axes are dictated by the locations of the heavy atoms, with the z axis along the C—N bond and the x axis perpendicular to the aromatic ring. Such shape averaging would however be slow on the IR time scale, so that IR transition moment directions would reflect the instantaneous symmetry C_s.

The use of (5.64) for the angle ϕ between the z axis and the moment direction of a transition f characterized by the orientation factor K_f and polarized in the xz plane yields $\phi = 17° \pm 5°$ for the NH_2 symmetric stretch at 3393 cm^{-1} and $\phi = 20° \pm 5°$ for the NH_2 scissoring vibration at 1602 cm^{-1}. If one assumes in the first approximation that these transitions are polarized along the bisectrix of the HNH angle, these results lead to $19° \pm 7°$ for the angle between the plane of the amino group and the plane of the ring.

This value is lower than the gas-phase microwave value of $37.5° \pm 2°$. Some of the discrepancy may be due to the effects of the polyethylene environment, but most of it is probably due to a deviation of the IR polarization directions from the bisectrix of the HNH angle.

8.6.6 The Symmetry of a Reactive Intermediate: Tricyclo[3.3.2.0^{3,7}]dec-3(7)-ene.[88a]

The structure of highly reactive molecules is difficult to determine by standard structural tools such as X-ray diffraction. Many have only been observed under conditions of matrix isolation, where microwave spectroscopy and electron diffraction are also inapplicable. At times, the methods of polarized optical spectroscopy can provide information on molecular symmetry and possibly even finer structural details for such matrix-isolated species.

A recent example is provided by the highly strained and extremely reactive olefin, tricyclo[3.3.2.0^{3,7}]dec-3(7)-ene (**49C**). Molecular models suggest that the tendency of the strained double bond to become planar can be expected to pull

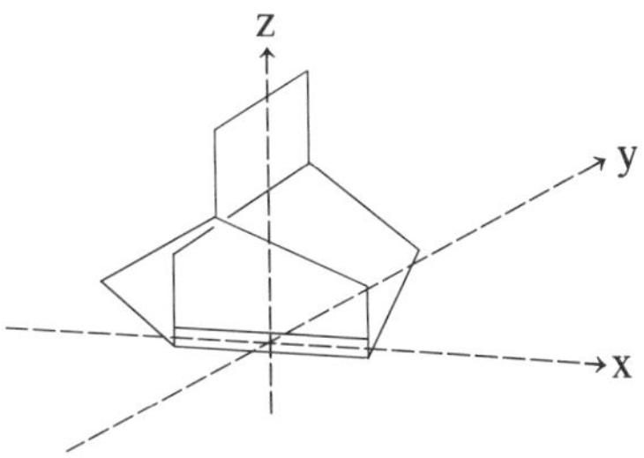

49C

the ethano bridge toward planarity, eclipsing its two CH_2 groups and endowing the molecule with C_{2v} symmetry. Since such eclipsing is normally resisted, the molecule also might be chiral, of C_2 symmetry. The issue can be resolved by means of polarized optical spectroscopy.

The UV spectrum of **49C** contains a broad band centered at 245 nm, compatible with the expected lowering of the $\pi\pi^*$ excitation energy by the strong pyramidalization. Irradiation with 248 nm light causes a gradual conversion into an isomer, and when linearly polarized light is used, a uniaxially aligned sample of remaining **49C** is obtained after partial photodestruction. Assuming that the UV transition is polarized along x, we shall have $K_x < 1/3 < K_y = K_z = (1 - K_x)/2$ (cf. Section 4.6.3).

The IR spectrum of **49C** contains a series of peaks. A relatively weak one at 1557 cm^{-1} is a good candidate for a C=C stretch since it is very strong in the Raman spectrum. For such an assignment to be correct, this IR band would have to be polarized along the twofold symmetry axis z (in a planar symmetrical tetrasubstituted olefin this IR transition moment would vanish, but a deviation from planarity will induce a nonvanishing z-polarized component).

The observed IR dichroic ratios fall in three categories, separated by margins in excess of the experimental error (about ± 0.03): (i) $K = 0.17$–0.19, (ii) $K = 0.36$–0.39, (iii) $0.24 < K < 0.31$. Thus, (i) and (ii) must clearly correspond to K_x and $K_y = K_z$, respectively. The orientation factors therefore lie between the limits $K_x = 0.18$, $K_y = K_z = 0.41$ and $K_x = 0.20$, $K_y = K_z = 0.40$. The K value observed for the 1557 cm^{-1} vibration is 0.37, compatible with z polarization. This band is thus indeed assignable to the C=C stretching mode.

If only mutually perpendicular polarization directions x, y, z, were possible, as in a C_{2v} molecule, only two experimentally distinct K values could be observed. Since one observes five fairly narrow lines in category (iii), with K values very different from both K_x and $K_y = K_z$, and since it is highly unlikely that all of them are due to accidentally overlapping lines of differing polarization, the question posed in the beginning can be answered: The symmetry is lower than C_{2v}. The most likely candidate that still preserves orthogonality among many of the moments in the C_2 symmetry group. This also is the symmetry predicted by quantum chemical and molecular mechanics calculations.

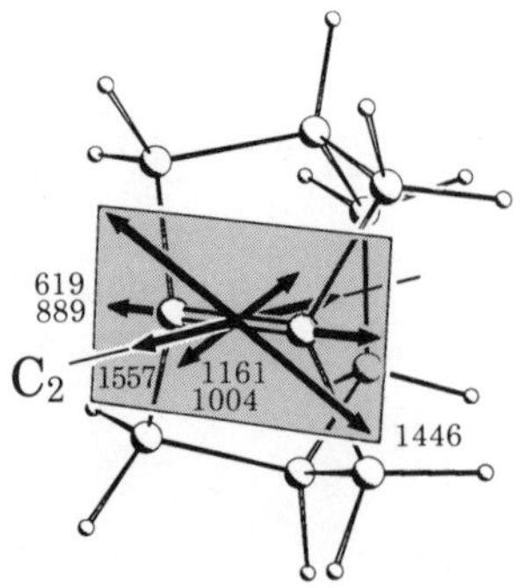

Figure 8.40b Transition moment directions of several IR transitions in tricyclo[3.3.2.0^{3,7}]dec-3(7)-ene (vibration frequencies and the location of the two-fold axis of symmetry are shown).

The angles between the C—C axis and the IR transition moment directions obtained by means of (5.64) are shown in Figure 8.40b. The signs of the angles cannot be determined from the data and the choices shown are arbitrary.

8.6.7 *Peptides*[89]

A fair number of oligopeptides and polypeptides were investigated in stretched polyoxyethylene, which permits the observation of IR linear dichroism in the region of the amide N—H stretching bands between 3000 and 3500 cm^{-1}, of amide I and II bands between 1500 and 1700 cm^{-1}, and of far-IR bands below 800 cm^{-1}. Polyethylene, which has even more favorable IR transmission properties, is too nonpolar to accept most peptides.

The authors of this study used an ingenious hydrogen-to-deuterium exchange technique to obtain additional information about the various spectrally distinct N—H groups. They observed the dependence of the dichroism of the N—H stretch band as a function of the extent of the N—H to N—D conversion as the latter was occurring in the stretched polymer film. The rate of exchange, which reflects accessibility to the solvent medium, and the orientation of each spectrally distinct N—H group were determined simultaneously.

Quantitative analysis of the observed IR linear dichroism is difficult since the molecules are of low symmetry and contain many amide groups. On the positive side, the transition moment directions for IR bands of the amide groups are known: The N—H stretch and the amide I band are polarized roughly along the C=O bond, and the amide II band nearly perpendicular to it in a typical s-trans amide conformation. The authors assumed a rodlike orientation behavior ($K_x = K_y$) for all their peptides (the Fraser-Beer model). Since the observed degree of alignment is relatively small and at least some of the molecules apparently aggregated into β-sheet structures which are not at all rod-shaped but rather plate-shaped, this assumption is most likely wrong and the results are of questionable quantitative significance. Even at a qualitative level, however, interesting conclusions are possible with due caution (it is to be remembered

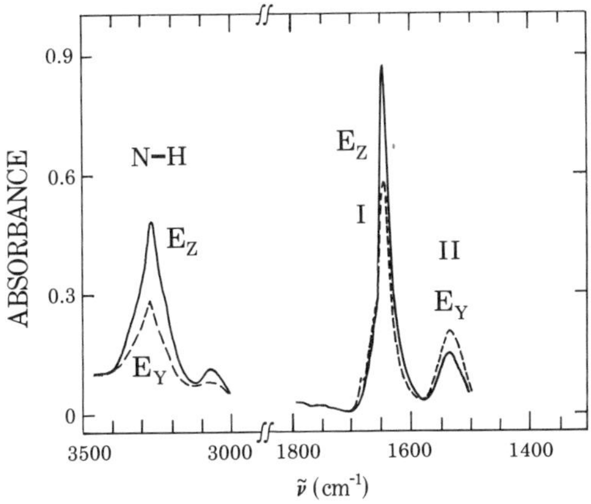

Figure 8.41 Polarized IR absorption E_Z (full lines) and E_Y (dashed lines) in the N—H stretch, amide I, and amide II regions of gramicidin S in stretched polyoxyethylene. Adapted by permission from ref. 89.

that even transitions polarized perpendicular to the orientation axis may have positive dichroism if $K_x \neq K_y$). Thus, gramicidin S, a cyclic decapeptide with the sequence cyclo(L-Pro-L-Val-L-Orn-L-Leu-D-Phe)$_2$, yields an IR spectrum with a single uniformly polarized peak in the N—H stretch region, which exchanges slowly into N—D without changing its dichroic ratio (in isotropic solution, there are at least two kinetically distinct classes of N—H bands). The dichroic ratio of the amide II band is also unaffected by the isotopic exchange. These results suggest that the amide groups of gramicidin S comprise a single class, all protected against isotope exchange by involvement in strong hydrogen bonding and all oriented roughly alike. The positive dichroism of the N—H stretch (d = 2.0) and the negative dichroism of the amide II band (d = 0.7) shown in Figure 8.41 led the authors to suggest that gramicidin S is present in the stretched polyoxyethylene in the form of an antiparallel β-sheet structure (Figure 8.42), with the long dimension of the sheet (vertical in the figure) aligned with the stretching direction, and the long dimension of each individual gramicidin S molecule, as well as its C_2 axis of symmetry, aligned roughly perpendicular to the stretching direction. It is necessary to assume that at least five gramicidin S molecules are contained in the average aggregate in order for the vertical axis of the aggregate shown in Figure 8.42 to be its long axis.

The dichroism of the amide I band is compatible with this interpretation in that the band seems to contain two components (Figure 8.41): a shoulder at 1680 cm^{-1} with negative dichroism assigned to the two tertiary Phe-Pro amide groups which lack a N—H stretch, and a peak at 1640 cm^{-1} with positive dichroism, assigned to the other amide groups, all of which are oriented alike.

Other oligopeptides and polypeptides also orient in stretched polyoxyethylene, both in their α-helical and β conformations. The latter can be oriented with the peptide chains either parallel or perpendicular to the stretching direction of the polymer film.

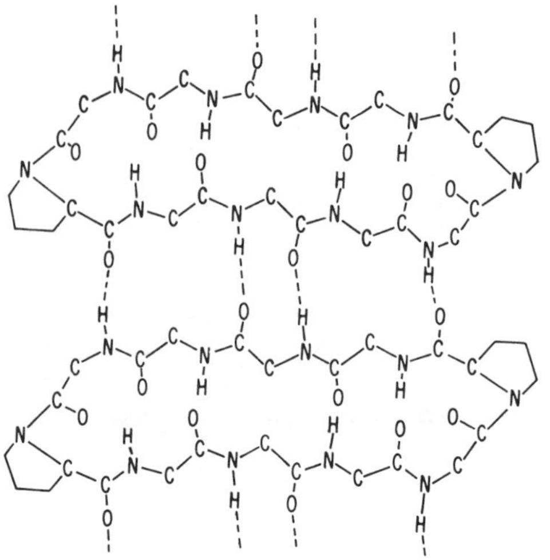

Figure 8.42 Cross-β-type aggregate structure proposed for gramicidin S in stretched polyoxyethylene on the basis of IR linear dichroism. Reproduced by permission from ref. 89.

8.6.8 Z-DNA in Solution[90]

Sequence effects on conformation and dynamics in DNA are important in view of their potential relation to gene regulation. Although the structure of Z-DNA has been established in oligonucleotide crystals, the solution structure and the mechanism of its formation from B-DNA are poorly understood. The present example addresses the solution structure of Z-DNA and the salt-induced B to Z conversion, using flow linear dichroism.

A convenient tool for obtaining fast flow needed for adequate alignment of long molecules (>50 nm) such as DNA is the Couette cell. It consists of an outer cylinder with two opposing quartz windows and an inner cylinder of quartz. One of the cylinders rotates, the other is static. Rotating the outer cylinder is preferable in principle, in that laminar flow then persists up to higher rotation speeds, but in the present example, the inner cylinder was rotated. The light beam passes radially through the cylinders. The solution to be studied is contained in the space between the two cylinders. The gap between these is quite narrow (~0.5 mm), and the flow gradient that results when the inner cylinder rotates is the aligning force. Velocities up to several thousand cycles per second are used and produce an alignment of long solute molecules. The alignment of low-molecular weight organic molecules such as anthracene is too weak to be observed. The measurements are performed with linearly polarized light with the electric vector parallel and perpendicular to the cylinder axis, producing E_Y and E_Z respectively. Baselines are recorded similarly but with both cylinders kept static.

The reduced linear dichroism $(E_Z - E_Y)/E^{iso}$ of the B and Z forms of DNA was measured in the region of $\pi\pi^*$ transitions in the base pairs. The discussion was based on equation (4.43). The reduced LD of the Z form was found to be about twice as large as that of the B form, indicating a better alignment of the helix axis or a better average alignment of the base planes, or both. The results were compatible with the notion that the solution conformation of Z-DNA is similar to that known from the crystal. Moreover, the time dependence of the reduced LD during the slow salt-induced process of B to Z interconversion indicated that a fraction of the base-pairs roll around a pivot axis parallel to the hydrogen bonds.

8.7 Molecular Association

Three types of phenomena which fall into this category have been studied by the techniques of polarized optical spectroscopy. First, the formation of complexes or aggregates of small molecules can be detected if it has a significant effect on the shape of the orientable entities. Second, the polarization directions of charge-transfer or other transitions involving both components of a complex can be studied. Third, the directional aspects of the incorporation of small chromophores into large orientable molecules such as DNA can be investigated if the transition moment directions within the isolated chromophore are known. In the following, we provide examples of all three kinds.

8.7.1 Dimerization of Benzoic Acid and Dimedone[90a]

Benzoic acid and its substituted derivatives are well known to associate in nonhydroxylic solvents in the form of cyclic dimers such as **44**. The dimeric form is present in stretched polyethylene, as is easily demonstrated by comparing the linear dichroic spectra of the acids and their methyl esters. For instance, for *p*-methoxybenzoic acid, the dichroic ratio of the long-axis-polarized transition near 260 nm is $d = 8.8$, while for its methyl ester it is $d = 3.3$ (Figure 8.43). The linear dichroism of the acids in stretched polyethylene is concentration-independent, indicating that no other association forms are present. On the other hand, the dichroic ratios observed for *o*-methoxybenzoic acid and its methyl ester in stretched polyethylene are almost identical, showing that this acid prefers intramolecular hydrogen bonding to dimer formation in this medium. Benzoic acids exhibit much smaller dichroic ratios in stretched poly(vinyl alcohol) than in stretched polyethylene, suggesting that their hydrogen bonding with the polymer competes with dimer formation.

More recently, several carboxylic acids have been shown by IR linear dichroism to be present as dimers in poly(chlorotrifluoroethylene)[90b]; see also Section 8.2.5.

The behavior of dimedone (**50**) has also been studied[90a] and found to be concentration-dependent. Its methyl and ethyl esters show no detectable UV linear dichroism in stretched polyethylene, and it is reasonable to assume the same of parent **50** in its monomeric form.

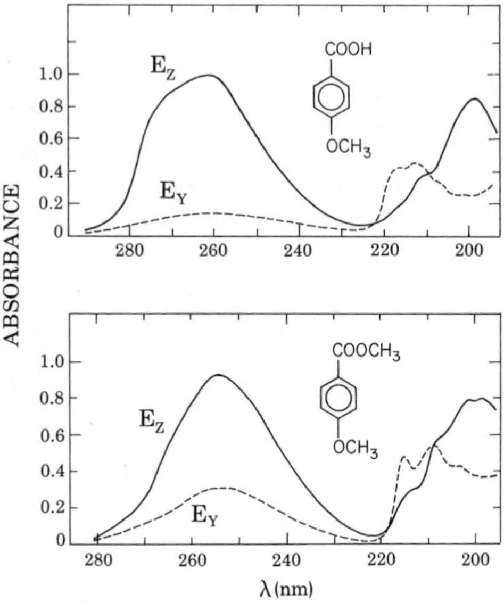

Figure 8.43 Polarized UV absorption E_Z (full lines) and E_Y (dashed lines) of *p*-methoxybenzoic acid (top) and its methyl ester (bottom) in stretched polyethylene. Adapted by permission from ref. 90a.

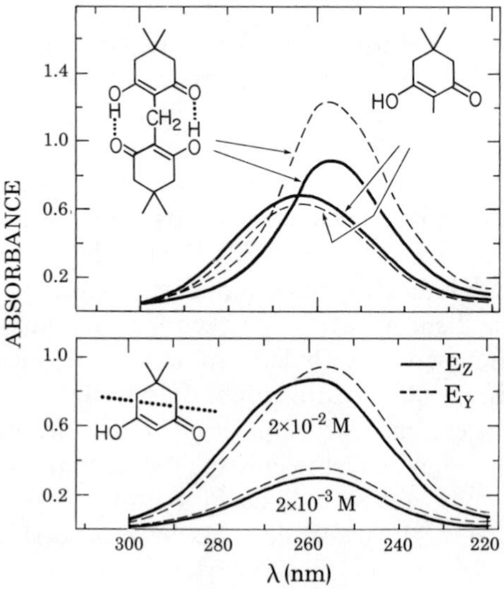

Figure 8.44 Polarized absorption E_Z (full lines) and E_Y (dashed lines) in stretched polyethylene. Top, methylene-bis-dimedone **52** and 2-methyldimedone **54**. Bottom, dimedone at two concentrations, determined from the measured absorbance by assuming $\varepsilon = 15\,000$ for the enol form (sheet thickness, 0.09 mm). Adapted by permission from ref. 90a.

However, below 2×10^{-3} *M*, measurements on **50** in stretched polyethylene reveal negative linear dichroism for the band at 253 nm, d = 0.87, independent of concentration (Figure 8.44). It is reasonable to assign this to the dimeric form of dimedone, **51**. The transition moment of the 253 nm band makes an angle of 55° or more with the orientation axis z, presumably coincident with the long axis of the dimer. A molecule of related shape, methylene-bis-dimedone **52**, shows d = 0.72 at $\lambda_{max} = 258$ nm, in agreement with these conclusions (Figure 8.44).

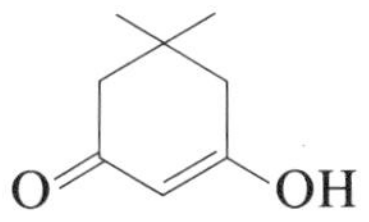

50

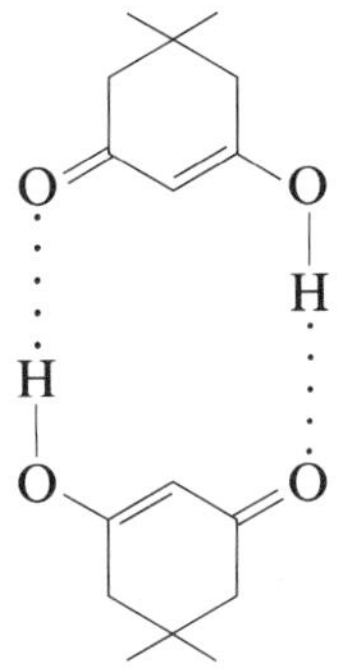

51

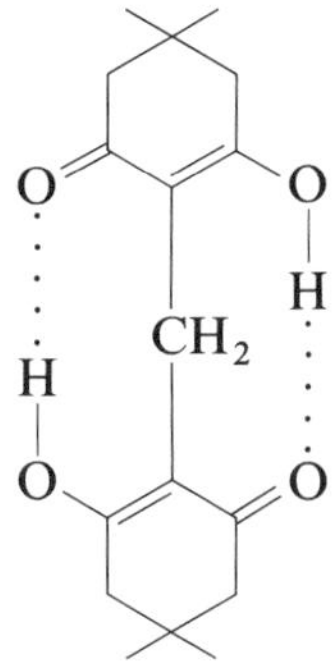

52

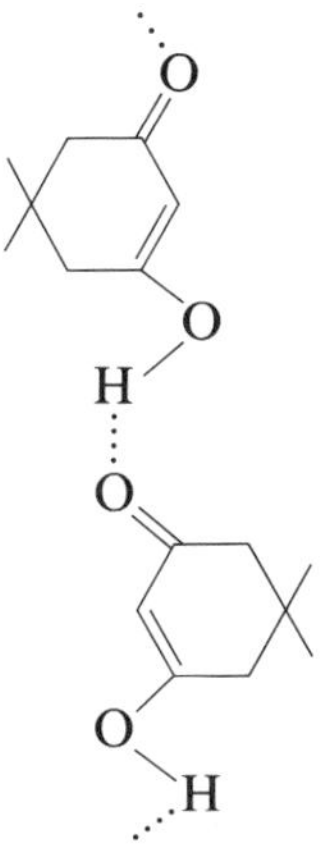

53

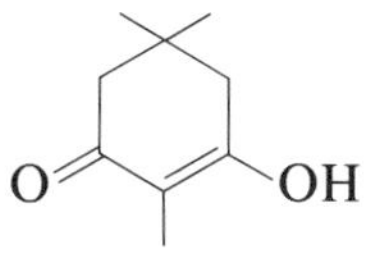

54

At higher concentrations (2×10^{-3} to 2×10^{-2} *M*), the UV spectrum of **50** remains almost unchanged but its linear dichroism changes. At concentrations of 2×10^{-2} *M* and higher, the longer-wavelength part of the absorption band exhibits positive dichroism (Figure 8.44). The authors suggest that at these concentrations a noncyclic form of aggregation, with a slightly red-shifted λ_{max}, prevails. Its probable form is shown in **53**, based on the observed positive dichroism.

Independent support for these interpretations was obtained from measurements on 2-methyldimedone (**54**), which is incapable of forming a cyclic dimer due to steric hindrance by the methyl group in position 2. At all concentrations measured, this molecule exhibits a dichroism (d = 1.07 at λ_{max} = 265 nm) similar to that observed for **50** in the high concentration limit. Incapable of forming a dimer, **54** apparently exists in stretched polyethylene in the polymeric association form analogous to **53**.

In stretched poly(vinyl alcohol), dimedone behaves in the same way as in hydroxylic solvents. It absorbs at λ_{max} = 283 nm and shows concentration-independent linear dichroism (d = 1.1). It is presumably present in its monomeric form **50**, hydrogen bonded to the polymer and thus weakly oriented.

8.7.2 DNA Base–Metal Ion Complexation[90c]

The structures of nucleic acids change upon interaction with silver ions in ways that are not clearly understood. In an attempt to unravel some of the interactions involved, the complexation of guanosine (**54A**) and 7-methylguanine (**54B**) with Ag^+ ions was investigated.

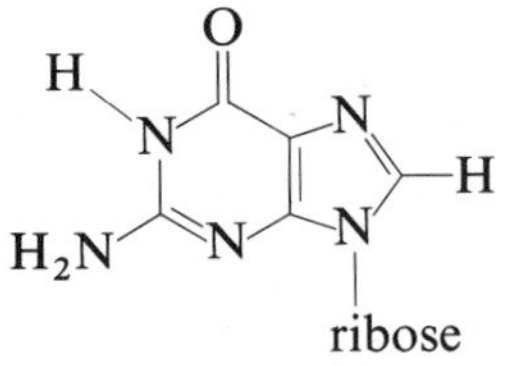

54A

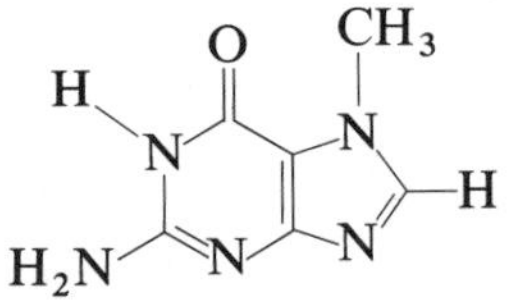

54B

One of the tools used was a comparison of the linear dichroism of the uncomplexed bases and of their Ag^+ complexes in stretched poly(vinyl alcohol) sheets, both in the UV and IR regions. A very large increase in the degree of alignment of **54A** in the presence of Ag^+ ions suggests that the complex very likely is of dimeric form.

In the case of 7-methylguanine (**54B**), the measured dichroism of the C=O stretching vibration in the IR spectrum, combined with the condition that the

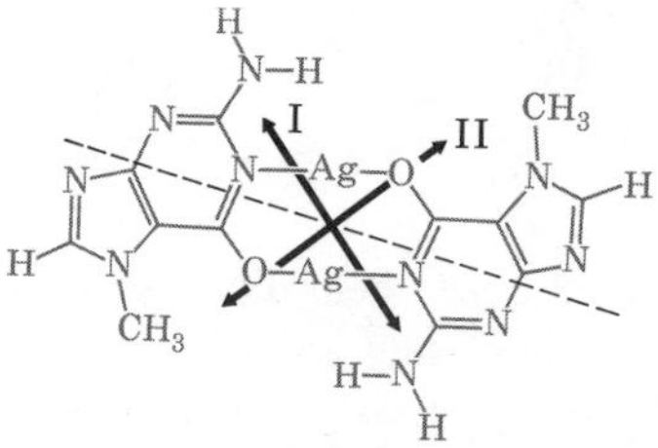

Figure 8.44a Proposed UV transition moment directions in a complex of 7-methylguanine with Ag^+ ions. Reproduced by permission from ref. 90c.

orientation factors of no in-plane polarized transitions can exceed K_z nor be smaller than K_y, restricts the possible range of orientation factors K_x, K_y, and K_z. This permitted the authors to propose a dimer structure for the complex and to obtain a rough estimate of the UV transition moment directions (Figure 8.44a).

A similar investigation of the complexes of Ag^+ with adenine and 1-methyladenine by dichroism in stretched poly(vinyl alcohol) and by flow dichroism led to the discovery of linear Ag^+-adenine polymers in dilute solutions.[90d]

8.7.3 Charge-Transfer Complexes and Exciplexes

Molecular complexes characterized by a low-energy transition which is absent in the separate components are referred to as charge-transfer complexes, although the transfer of charge is usually not particularly important for their ground-state bonding. Intermolecular charge transfer is, however, largely responsible for the low-energy transition which gave these complexes their name. The polarization direction of this transition can be determined on a partially aligned solution in the same way as the polarization directions of any intramolecular transition, and in that sense the determination does not particularly warrant separate treatment. An example of such measurement is the determination of the transition moment direction of the charge-transfer band in the complex of pyrene with pyromellitic dianhydride **55**, which was performed in a supercooled compensated nematic liquid crystal.[54] A measurement of polarized fluorescence showed that the transition is polarized perpendicular to the orientation axis. The direction of the polarization axis was determined by an ESR measurement of photoexcited triplet of the complex and was found to lie parallel to the planes of the two constituent molecules of the complex.[91] The conclusion was that the charge-transfer transition is polarized perpendicular to the ring planes of the two components, in agreement with measurements on single crystals, with those using photoselection, and with theory.

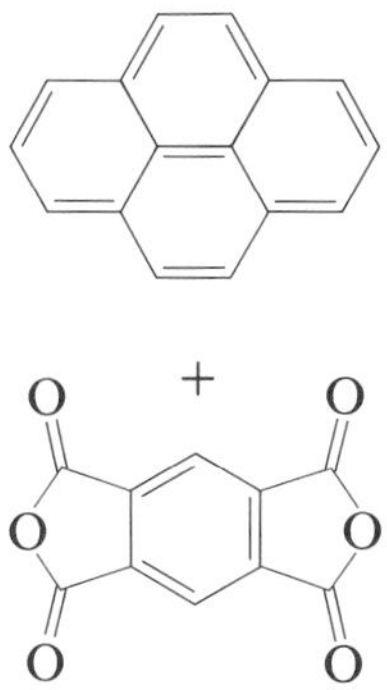

55

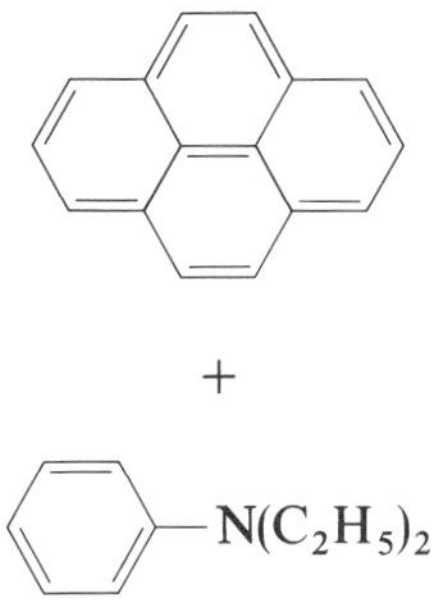

56

Molecular complexes which are not significantly bound in the ground state but which are bound in the excited state are referred to as exciplexes (excimers if the two components are identical). They, too, are characterized by a low-energy excited state. Its wavefunction is believed to contain both intermolecular charge-transfer and interacting local excitation contributions. In the general case, it is difficult to measure the polarization of the emissive transition from this bound state of the exciplex to the dissociative ground state by methods such as photoselection or single-crystal measurements, since the complex can only be prepared in its short-lived excited state. (Some of these complexes can be prepared by photodissociation of sandwich dimers in rigid media and kept together in their repulsive ground state indefinitely as long as the solvent remains rigid.)

The investigation of the polarization of fluorescence of exciplexes contained in fluid uniaxially oriented liquid crystals provides a simple answer to the problem. Such experiments are described for a few exciplexes in ref. 92. They were performed in a 1.85:1 (w/w) mixture of cholesteryl chloride and cholesteryl laurate which is nematic at 35°C. The liquid-crystal solution was aligned with electric fields of 3–4 × 10^4 V/cm. Molecular motion removed any residual photoselection effects due to the directional nature of the (unpolarized) exciting light. The polarization of the emitted light was analyzed; this experiment arrangement corresponds to the case of "spontaneous luminescence" of Section 5.1, for which the same equations are applicable as for polarized absorption.

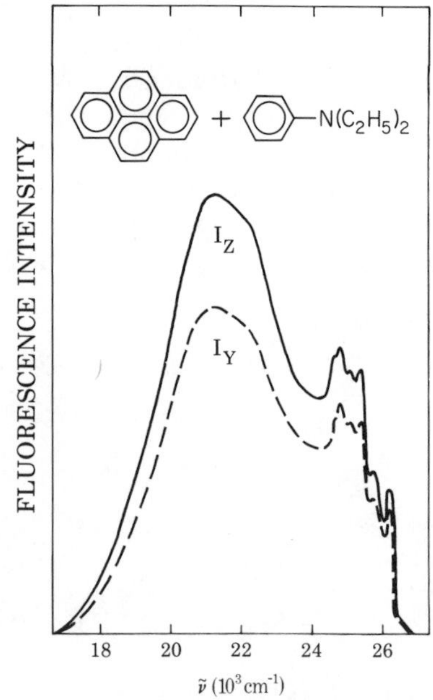

Figure 8.45 Polarized emission I_Z (full curve) and I_Y (dashed curve) from 8 × 10^{-4} *M* pyrene and 6 × 10^{-2} *M* diethylaniline in a 1.85:1 by weight nematic mixture of cholesteryl chloride and cholesteryl laurate aligned by electric field at 35°C. Adapted by permission from ref. 92.

Information about the orientation of molecular complexes in the liquid crystal solvent was obtained from ESR spectra of excited triplet states of charge-transfer complexes dissolved in undercooled liquid crystals.[91] These studies produced the following generalization: Complexes of molecules such as naphthalene, anthracene, or pyrene (with one long and one short in-plane axis) orient in the liquid crystal in such a way that the long molecular axis is preferably parallel to the unique axis of the nematic liquid crystal.

Given this information, the observed dichroism of the exciplex emissions immediately produces results for the polarization direction of the intermolecular transition. The example given in Figure 8.45 shows the polarized emission of the pyrene-diethylaniline exciplex **56**. The structured band at higher frequencies is due to fluorescence of uncomplexed pyrene, which is mostly polarized parallel to the long axis of the pyrene molecule as discussed in Section 8.1 and exhibits a positive dichroic ratio in the liquid crystal. The broad band at $\tilde{\nu} < 24\,000\ cm^{-1}$ is due to the exciplex emission. Its emission dichroism is also positive, $d = 1.35$. The conclusion is that the transition moment of the exciplex emission is also directed parallel to the long molecular axis of pyrene. Similar results were obtained with other donor molecules, *N*,*N*-dimethylaniline, *p*-methoxydimethylaniline, and *p*-*N*,*N*-dimethyltoluidine. This result agrees with theoretical expectations and supports the current ideas about the nature of the exciplex state.

8.7.4 *Binding of Dyes to DNA*

The nature of the binding of small molecules such as dyes to DNA has been the subject of intense interest for a long time. Polarized spectroscopy on partially aligned samples of DNA is one of the primary tools which have been used to elucidate the geometrical nature of the DNA-dye interaction. The alignment has been achieved by flow, by electric field, or in stretched sheets (leading references can be found on p. 201 in ref. 93). Polarized measurements on stretched fibers under a microscope have been reported as well.[94]

Some of the dyes have been found to bind to DNA by intercalation, i.e., insertion between base pairs in such a way that the plane of the dye is parallel to those of the DNA bases; others bind onto the surface of the helix. In several instances, it has been found that the binding is by intercalation at low concentrations and by nonintercalated association at higher concentrations. In keeping with the spirit of this chapter, we shall not provide a review of the work that has been done but, rather, will illustrate it on two examples.

For the first of these we have chosen the work on the binding of 9-aminoacridine hydrochloride (**57**) studied on stretched films of dyed DNA deposited on DNA-coated poly(vinyl alcohol) support.[95] Unlike orientation by flow or electric field, this method produces high dichroic ratios. The dye to DNA phosphate ratio was 10^{-2} to 10^{-3}. Under these conditions the only dye species present in significant quantity is the intercalated form. The stretched films ($R_s = 2.8$) were equilibrated at 93% relative humidity in order to ensure that DNA would be in its ordinary equilibrium form.

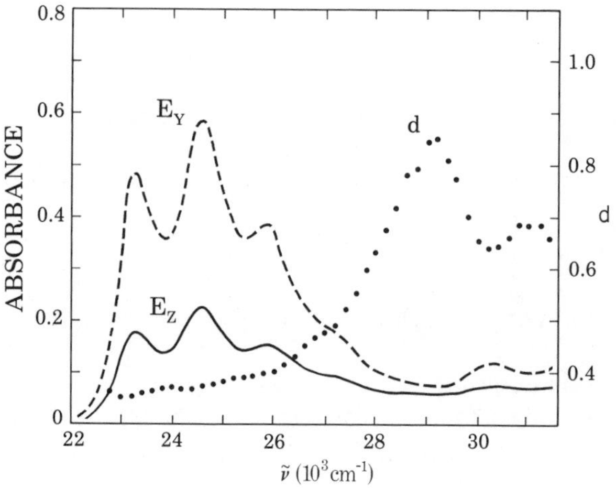

Figure 8.46 Polarized absorption E_Z (full line) and E_Y (dashed line) of a DNA–9-aminoacridine complex in a stretched film, and the dichroic ratio d (dotted line). Adapted by permission from ref. 95.

The dichroic spectrum of the DNA-9-aminoacridine hydrochloride complex is shown in Figure 8.46. The intense absorption band of 9-aminoacridine hydrochloride at 23 000–26 000 cm^{-1} is known to be due to a short-axis-polarized absorption, analogous to the L_a band of anthracene. A weaker long-axis-polarized band is found at higher energies and corresponds to the L_b band of anthracene. The authors assigned its origin at 30 300 cm^{-1}, at the first observable vibronic band. The dichroic ratio for the first band is $d = 0.35 \pm 0.03$. Due to partial overlap of the two bands, the dichroic ratio measured at 30 300 cm^{-1} does not have a simple physical significance. The authors corrected for the overlap using a sensible but unproven set of assumptions and concluded that in the absence of overlap the dichroic ratio would be $d = 0.54 \pm 0.07$. A more reliable value could have been obtained if the reduced spectra of protonated 9-aminoacridine had been available from the standard reduction procedure, and this example provides a good case for the need for such measurements on simple dyes. In fact, the reduced spectra of **57** have now been obtained, and they suggest strongly that the high values of d in the region 28 000–30 000 cm^{-1} are due to experimental inaccuracies (since this absorption is still of $\pi\pi^*$ nature,

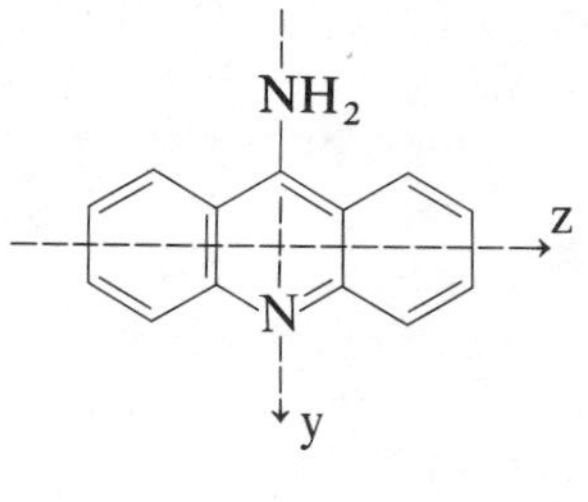

57

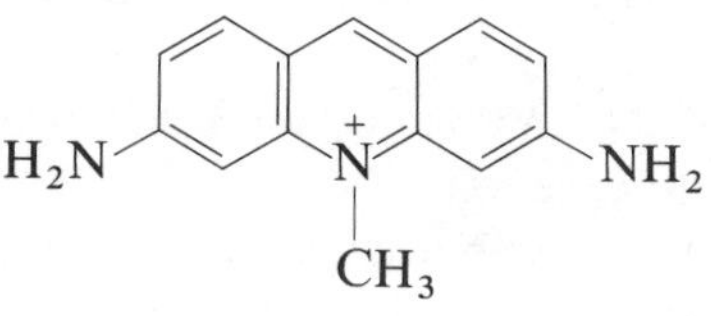

58

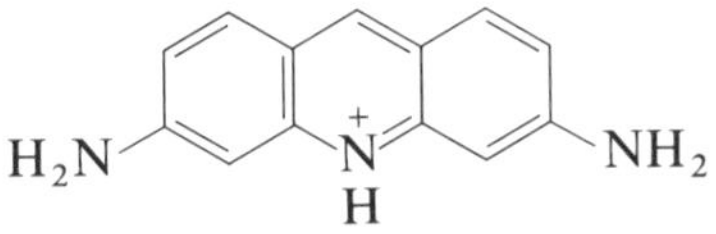

59

the correct d value must lie between those for purely polarized short- and long-axis-polarized absorption, i.e., presumably in the range 0.35–0.54).

In order to deduce the relative orientation of the dye and the helix axis of the DNA, the average alignment of the DNA axis relative to the stretching direction of the polymer must be known. Due to the shape of the DNA, one can safely assume the Fraser-Beer model for it ($K_x = K_y$). In this model, the dichroic ratio in the limit of perfect alignment of the DNA axes d_0 is related to the measured dichroic ratio d obtained with incompletely aligned DNA axes by

$$(d_0 - 1)/(d_0 + 2) = (1/f)(d - 1)/(d + 2) \tag{8.8}$$

where $f = S_{zz} = (3K_z - 1)/2$ characterizes the alignment of the helix axes with the stretching direction. In this expression, $(d - 1)/(d + 2)$ will be recognized as equal to the reduced dichroism $(E_Z - E_Y)/E^{iso}$.

Since $d_0 \geqslant 0$, the lower limit for f is $f = 0.67$. An upper limit was obtained from an examination of the dichroic ratios reported for anisotropic pure DNA samples. The lowest dichroic ratio found by at least two independent groups was $d = 0.21$, yielding an upper limit of $f = 0.72$. Allowing both for the experimental error in d and for the uncertainty in f, the authors obtained the result $d_0 = 0.08$–0.20 for the short-axis-polarized transition and $d_0 = 0.27$–0.48 for the long-axis-polarized transition, i.e., $K_{y'}^0 = 0.04$–0.09 and $K_{z'}^0 = 0.12$–0.19, where $K_{y'}^0$ and $K_{z'}^0$ are the orientation factors of the y′ (short) and z′ (long) axes of the dye molecule **57** relative to the helix axis of DNA. If all dye molecules are bound in like manner, these are equal to $\cos^2 y'$ and $\cos^2 z'$, respectively, where y' and z' stand for the angles formed by the y' and z' axes of the dye molecule with the helix axis. The angle between the long axis of the dye and the helix axis ("angle of tilt") then is $z' = 67 \pm 3°$, and the angle between the short axis of the dye and the helix axis is $y' = 75 \pm 3°$. The inaccuracy in the former value may be larger than shown if the assumption which was made about the overlap of the second with the first transition was incorrect. The authors also define the angle δ between the normal to the plane of the dye and the projection of this normal into the plane perpendicular to the helix axis ("angle of twist"). If all dye molecules are oriented alike, it can be determined from $\cos^2 y' = \sin^2 z' \cos^2 \delta$ as $\delta = 74 \pm 3°$ or $\delta = 106 \pm 3°$. Similar results were obtained for acriflavine **58** and proflavine **59**, and it was found that each dye oriented somewhat differently although they were all intercalated.[95]

Our second example concerns work on the binding of two metabolites of benz[*a*]pyrene on DNA, using a Couette cell (Section 8.6.8) for alignment.

The structure of DNA-carcinogen complexes is of great importance for the understanding of the molecular mechanisms responsible for the induction of

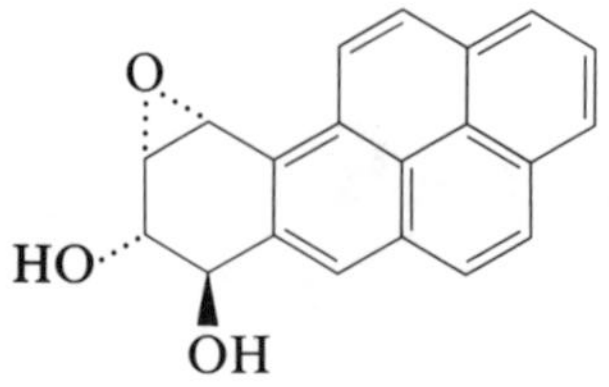

60

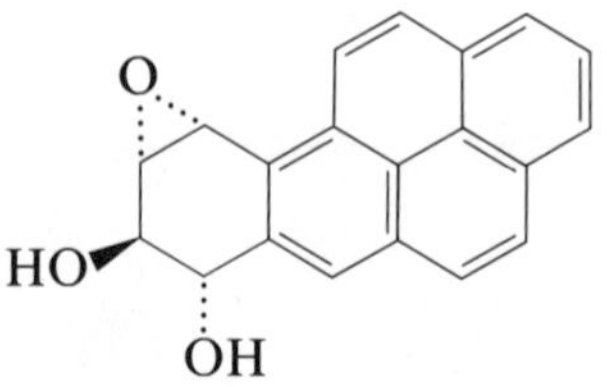

61

cancer. Recently, two derivatives of the well-known carcinogen benzo[*a*]pyrene have been studied[96]: 7β,8α-dihydroxy-9α,10α-epoxy-7,8,9,10-tetrahydrobenz[*a*]-pyrene **60** (*anti*-BPDE) and 7α,8β-dihydroxy-9α,10α-epoxy-7,8,9,10-tetrahydrobenz[*a*]pyrene **61** (*syn*-BPDE). Of the two derivatives, *anti*-BPDE, which is a main product when enzymes metabolize benzo[*a*]pyrene in vivo, is a strong carcinogen, whereas *syn*-BPDE is a weak one. In the investigation, both isomers were allowed to react with a calf thymus DNA to form covalent complexes. The amount of covalently bound BPDE was one molecule per 250 DNA bases.

The absorption and flow linear dichroism spectra are shown in Figure 8.47. While the LD for the long-wavelength band is positive for *anti*-BPDE, it is clearly negative for *syn*-BPDE. Note that the negative LD around 250 nm is due

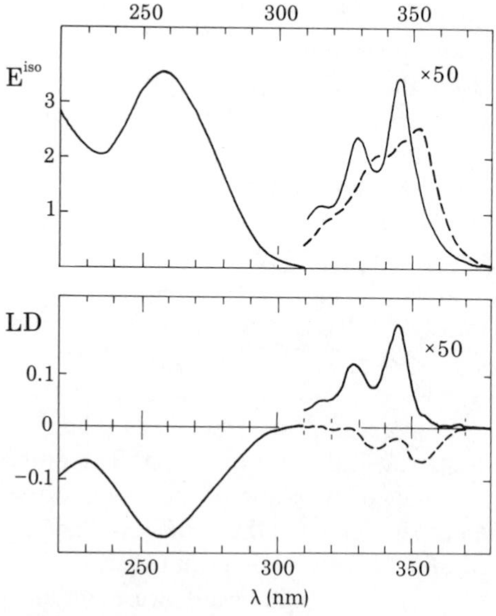

Figure 8.47 Absorbance E^{iso} (top) and Couette cell linear dichroism LD (bottom, G = 2050 s^{-1}) of 7β,8α-dihydroxy-9α,10α-epoxy-7,8,9,10-tetrahydrobenzo[*a*]-pyrene (*anti*-BPDE, **60**, full line) and 7α,8β-dihydroxy-9α,10α-epoxy-7,8,9,10-tetrahydrobenzo[*a*]pyrene (*syn*-BPDE, **61**, dashed line) in 10^{-3} *M* sodium cacodylate at pH = 7. The spectra are normalized to 1 cm optical path length. Reproduced by permission from ref. 96.

almost exclusively to $\pi\pi^*$ transitions in the DNA bases since only a very small amount of BPDE was incorporated. These transitions are polarized perpendicular to the long axis Z′ of the DNA helix. The measurement of their dichroism permits the evaluation of the Saupe matrix element $S_{Z'Z'}$ from (4.41) and

$$S_{X'X'} = LD(250)/E^{iso} \tag{8.9}$$

The difference in the LD for the two compounds is explained by their different orientation relative to the Z′ axis of the DNA helix. Since the transitions that contribute significantly to the long-wavelength absorption are all of $\pi\pi^*$ type and are polarized in the plane of the BPDE molecule, the observed LD signs indicate that *anti*-BPDE is oriented with its molecular plane forming a smaller angle with the helix than *syn*-BPDE. The observed linear dichroism of the i-th transition of the BPDE chromophore provides the Saupe matrix element S_{ii} describing the average orientation of its transition moment relative to the sample axis Z. From (4.43), the alignment of the transition moment relative to the DNA axis Z′ is given by the Saupe matrix element $S'_{ii} = (3\langle\cos^2 iZ'\rangle - 1)/2$:

$$S'_{ii} = S_{ii}/S_{Z'Z'} \tag{8.10}$$

Since S_{ii} can be measured and $S_{Z'Z'}$ is already known, $\langle\cos^2 iZ'\rangle$ can be evaluated. It is concluded that the plane of the strong carcinogen *anti*-BPDE is bound to DNA at an angle to the helix axis which is below 35°, whereas that of the weak carcinogen *syn*-BPDE is bound at an angle larger than 65°.

8.8 Structure of Anisotropic Media

As already mentioned, polarized spectroscopy work on anisotropic media such as stretched polymers, natural (e.g., DNA or polypeptides) or man-made [e.g., polyethylene, poly(ethylene terephthalate), poly(vinyl chloride)], liquid crystals, membranes, etc., has been excluded from this book so as to keep its size within limits, and the reader is referred elsewhere for this type of information.[83,93] However, even measurements on solutes contained within anisotropic media provide information about their environment, and they have been used as probes in many investigations. Their disadvantage is the uncertainty concerning the geometric relation of the probe molecules and the host molecules, which is usually considerable. Their advantage is the tremendous variety available in their choice: For instance, many polymers are nonfluorescent but accept fluorescent probe molecules.

In many natural systems, such as biological membranes, the probe molecules have been built in by nature. The investigation of the organization of such media involves inextricably both the small-molecule chromophores and their host environment. The subdivision of the biological system into "host" and "chromophores" (UV, IR, CD, etc.) is frequently artificial. We have elected to include an example from the study of biological membranes, although it might be argued that strictly speaking here one does not deal with solutes in an anisotropic medium but with an anisotropic medium as such.

8.8.1 Fluorescent Probes in Polymers

A useful method for the investigation of amorphous regions of linear polymers consists of doping them with a fluorescent probe and then measuring the second and fourth moments of the probe orientation distribution. In investigations performed so far it was assumed that the probe molecule is rodlike so that its two short axes x and y are equivalent, and that the moments of the absorbing and emitting transitions are both parallel to the long axis z. Measurements of fluorescence intensity are performed with two settings of the polarizer for the exciting beam and two settings of the polarizer in the emitted beam. This permits the evaluation of K_z and L_z, and with the assumptions made, these values then determine those of the other orientation factors. A careful discussion of the method is given in ref. 97, in which a series of resulting K_z values at various degrees of stretching of a poly(ethylene terephthalate) doped with 4,4′-(dibenzoxazolyl)stilbene (**62**) are compared with those obtained from linear dichroism. Excellent agreement is found. The values of L_z for the same samples are then obtained. It is noteworthy that on the average the probe orients better than polymer chains themselves.

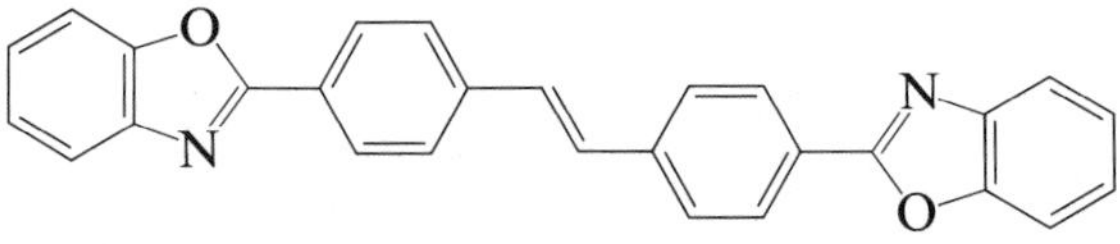

62

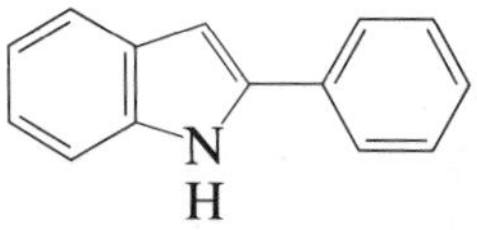

63

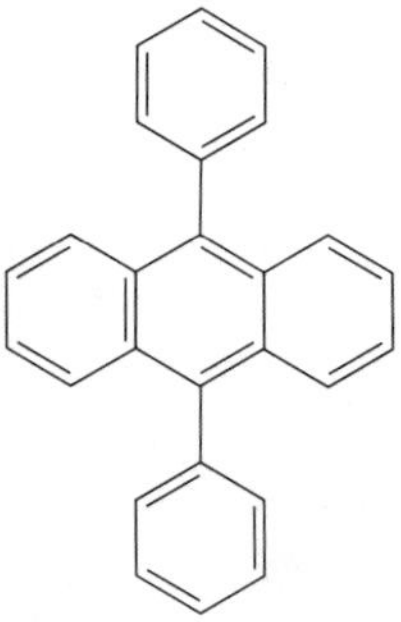

64

The example which we have selected to illustrate the method deals with stretched polyethylene sheets doped with a variety of fluorescent probes.[98,98a] The property of interest was the orientational distribution in the noncrystalline regions and its changes during a uniaxial deformation of the polymer. The procedure used for the derivation of the values of K_z and L_z was more complicated than that given in Chapter 6, and the reader is referred to the original article for details. The results for the various probe molecules used are collected in Table 8.13 as a function of the degree of stretching $\Delta l/l_0$. A plot of L_z against

Table 8.13
Orientational Mean Values and Total Fluorescence Intensity[a] of Different Probes in Stretched PE 1810 Films

Probe Molecule			$\Delta l/l_0$[b]				
			0%	25%	50%	100%	150%
2-Phenylindole	**63**	I_{F1}	100	95	88,5	72	58
		$\langle\cos^2\theta\rangle$	0,348	0,386	0,406	0,432	0,475
		$\langle\cos^4\theta\rangle$	0,21	0,22	0,22	0,24	0,30
9,10-Diphenyl-anthracene	**64**	I_{F1}	100	95	92	92	89
		$\langle\cos^2\theta\rangle$	0,331	0,325	0,33	0,36	0,395
		$\langle\cos^4\theta\rangle$	0,18	0,20	0,21	0,22	0,27
BIBUQ[c]	**65**	I_{Fl}	100	107	126	127	121
		$\langle\cos^2\theta\rangle$	0,34	0,612	0,742	0,83	0,909
		$\langle\cos^4\theta\rangle$	0,21	0,45	0,58	0,68	0,83
trans-Stilbene	**66**	I_{Fl}	100	107	110	129	145
		$\langle\cos^2\theta\rangle$	0,35	0,41	0,502	0,613	0,69
		$\langle\cos^4\theta\rangle$	0,22	0,27	0,34	0,46	0,60
1,4-Diphenyl-1,3-butadiene	**67**	I_{Fl}	100	113	116	115	116
		$\langle\cos^2\theta\rangle$	0,33	0,44	0,545	0,695	0,75
		$\langle\cos^4\theta\rangle$	0,18	0,31	0,41	0,55	0,64
1,6-Diphenyl-1,3,5-hexatriene	**68**	I_{Fl}	100	124	136	142	140
		$\langle\cos^2\theta\rangle$	0,32	0,49	0,58	0,71	0,78
		$\langle\cos^4\theta\rangle$	0,20	0,36	0,43	0,58	0,67
1,4-Bis-(*o*-methylstyryl) benzene	**69**	I_{Fl}	100	129	158	217	270
		$\langle\cos^2\theta\rangle$	0,338	0,52	0,66	0,84	0,88
		$\langle\cos^4\theta\rangle$	0,19	0,37	0,46	0,68	0,75
1,1,4,4-Tetra-phenyl-1,3-butadiene	**70**	I_{Fl}	100	89	79	63	54
		$\langle\cos^2\theta\rangle$	0,337	0,36	0,368	0,365	0,366
		$\langle\cos^4\theta\rangle$	0,19	0,20	0,22	0,22	0,22
POPOP[d]	**71**	I_{Fl}	100	102	104	101	96
		$\langle\cos^2\theta\rangle$	0,34	0,47	0,55	0,62	0,66
		$\langle\cos^4\theta\rangle$	0,21	0,34	0,40	0,47	0,52

[a] Referred to $I_{Fl} = 100$ at $\Delta l/l_0 = 0\%$.
[b] The degree of stretching; the values listed for $\Delta l/l_0 = 0\%$ were obtained by extrapolation.
[c] BIBUQ = 4,4''''-Bis(2-butyloctyloxy)-*p*-quaterphenyl.
[d] POPOP = 1,4-Bis(4-methyl-5-phenyl-2-oxazolyl)benzene.
Source: Reproduced by permission from ref. 98.

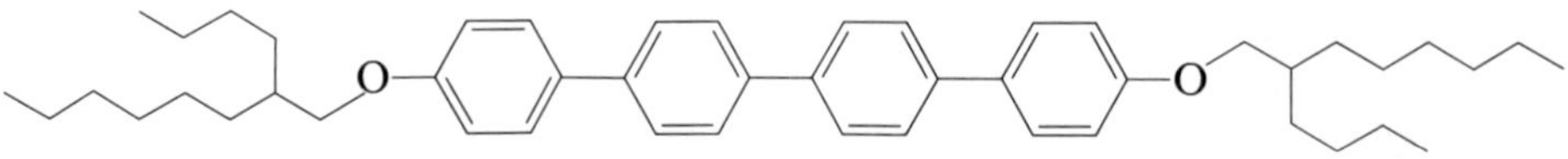

65

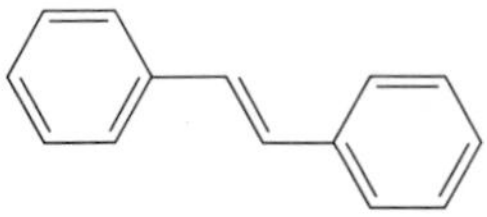

66

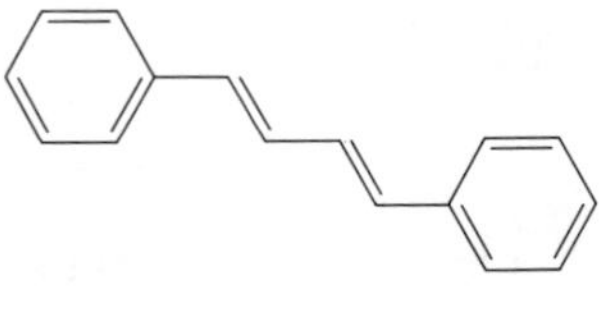

67

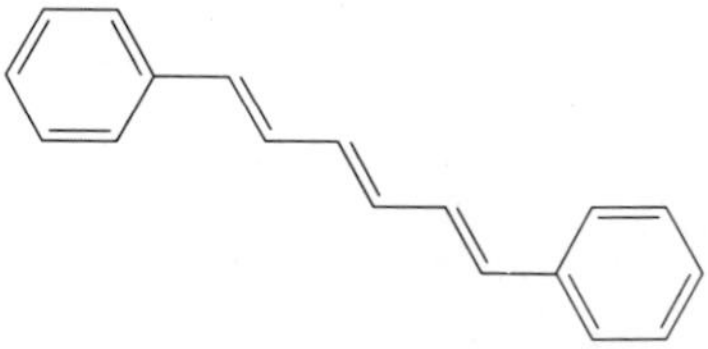

68

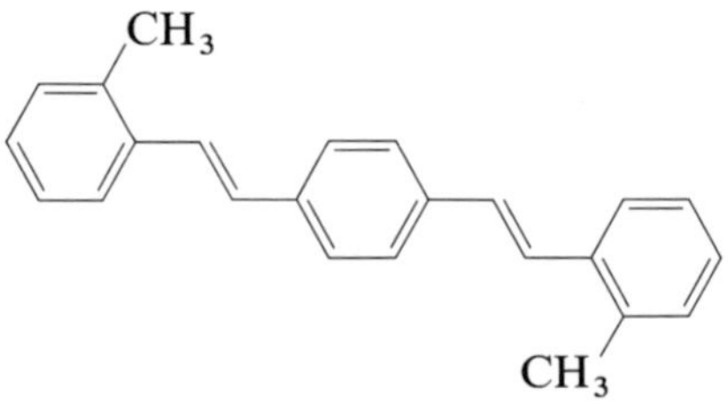

69

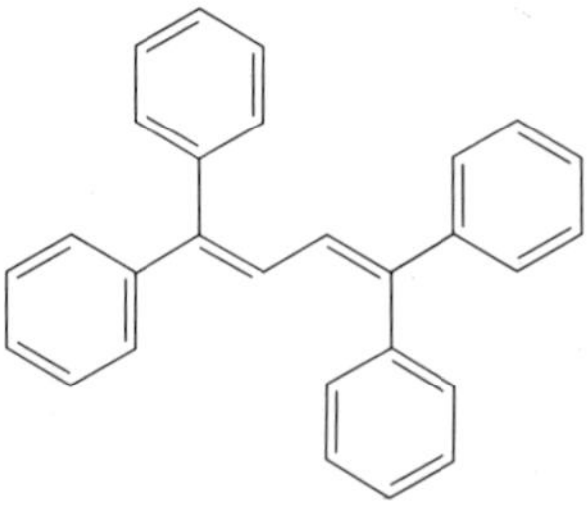

70

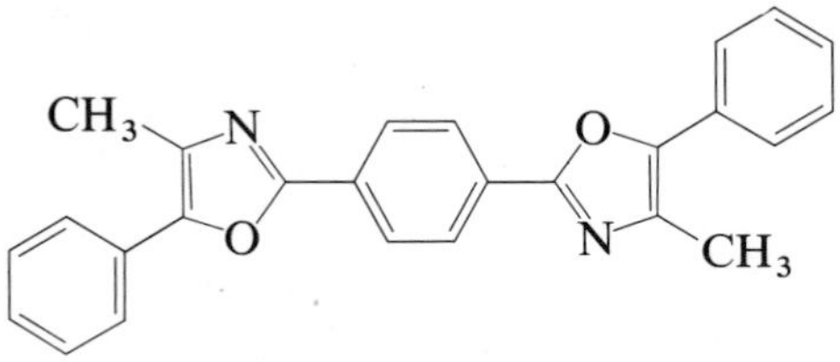

71

K_z for three of the probe molecules and for different degrees of stretching is shown in Figure 8.48. The plot also shows L_z as a function of K_z for the floating-rod model in an affine deformed matrix[99] as a dashed line. The full lines represent the limits of possible L_z values for given K_z values. Although the observed points do not lie very far from the dashed line, the authors[98] conclude that in polyethylene stretched more than about 100%, both the floating-rod approximation and a two-phase model composed of a randomly oriented phase and a uniaxially perfectly aligned phase are incorrect. They believe that rodlike probes reflect the alignment of the polymer in their immediate vicinity.

The authors note that the fluorescence quantum yield of several of the probes changes as a function of the degree of stretching and assign this behavior to conformational changes or restrictions of internal motion of the probes. They

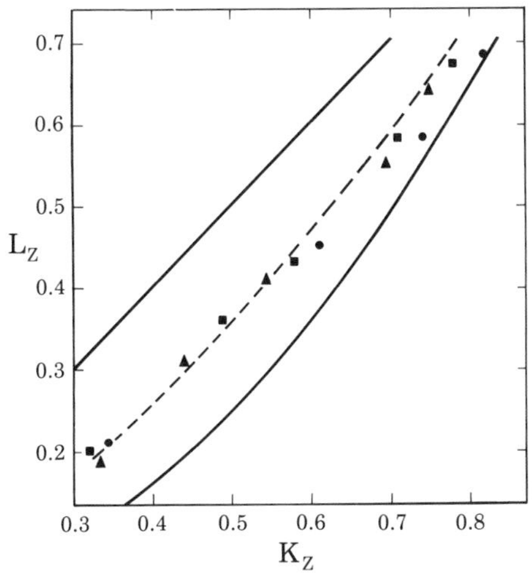

Figure 8.48 L_z plotted against K_z for 1,4-diphenyl-1,3-butadiene **67** (triangles), 1,6-diphenyl-1,3,5-hexatriene **68** (squares), and 4,4‴-bis(2-butyloctyloxy)-*p*-quater-phenyl **65** (circles) in polyethylene stretched to various degrees. Dashed line: the floating rod model. Full lines: $K_z = L_z$ and $K_z^2 = L_z$. Adapted by permission from ref. 98.

believe, as is usual, that the probes are located outside the crystalline regions of the polymer. A correlation of the enhancement of the quantum yield with the orientation factor K_z suggests that the increase of the fluoresence quantum yield upon stretching is due to those molecules which are oriented the best and restricted in their motion by being clamped among the polymer chains. In such an instance, the orientation factors derived from the fluorescence measurement do not represent uniformly weighted averages over the whole polymer, but weight preferentially those regions of the polymer which are oriented best. From a comparison of several such probes, the authors estimate that only about 25% of the amorphous part of the polymer is oriented.

The detailed nature of the orientation mechanism of solutes in stretched polyethylene still remains unknown; in view of the general usefulness of the stretched sheet method this is a rather lamentable situation.

8.8.2 Adsorbed Monolayers[100]

There is considerable current interest in the construction and characterization of organized molecular assemblies on the surface of various materials. Such organization can be detected and investigated by the techniques of polarized optical spectroscopy. The present example deals with monolayers adsorbed on surfaces of uniaxially stretched polymers which are anisotropic with respect to their normal direction. The author was interested in characterizing this anisotropy and found that it is transferred onto the adsorbed monolayer, causing a measurable degree of alignment of dye chromophores incorporated into the monolayer.

The first experiments were done with stretched poly(vinyl alcohol), on whose surface mixed monolayers of *n*-octadecyltrichlorosilane and long-chain substituted cyanine dyes such as **72** were deposited. The latter exhibited considerable dichroism, with light polarized with its electric vector parallel to the film stretching direction absorbed much more strongly than light polarized with its electric vector perpendicular to this direction. The degree of dichroism depended on the particular dye used but was not a simple function of its structure.

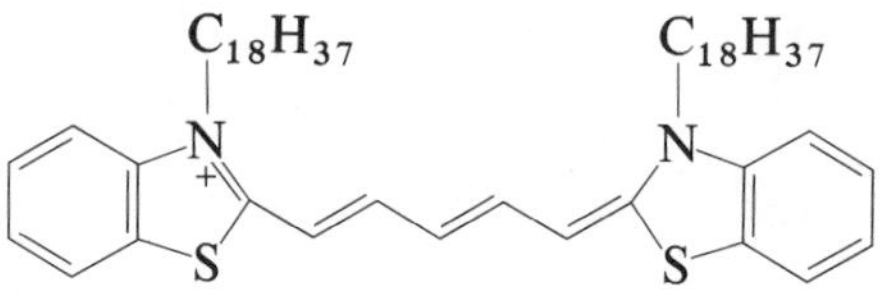

72

In a search for a water-resistant polar surface with a simple uniaxial molecular organization, the author tested surface-oxidized polyethylene. He found that the surface of stretched polyethylene sheets oxidized with fuming sulfuric acid to a light gray tint was wettable with water (films which were oxidized first and then stretched had a water-repellent surface, apparently because bulk material is brought to the surface during the stretching process). The sheets were highly birefringent, but the surface polar groups were apparently totally random, since no dichroism was induced in dyes incorporated in a monolayer deposited on the surface.

A substrate which possessed the desired characteristics was finally obtained by depositing an ultrathin layer of poly(vinyl alcohol) on the surface of the oxidized polyethylene sheet and subsequently stretching. Dyed monolayers of *n*-octadecyltrichlorosilane were deposited and were found to be oriented in the same fashion as on sheets of pure poly(vinyl alcohol). The surface of this material is stable to water at room temperature, but exposure to water reduces the degree of surface alignment as shown by the reduction in the observed dichroism (Figure 8.49).

Experiments with the transfer of preformed Langmuir monolayers containing these dyes from a water-air interface showed much lower degrees of dye orientation, particularly when cadmium arachidate was added to the pure dye (in these monolayers the dye was aggregated). These results provided information on molecular mobility of the dye molecules in the various monolayers: Langmuir monolayers of pure dye have a much looser molecular packing and lower viscosity than those of dye mixed with the arachidate where the dye molecules are embedded within a stiff matrix of aggregated cadmium arachidate. On the other hand, in the adsorbed monolayers formed with *n*-octadecyltrichlorosilane and the dye, the dye is apparently almost completely free to rearrange during the adsorption process, so that the most favorable orientation on the surface of the substrate is reached.

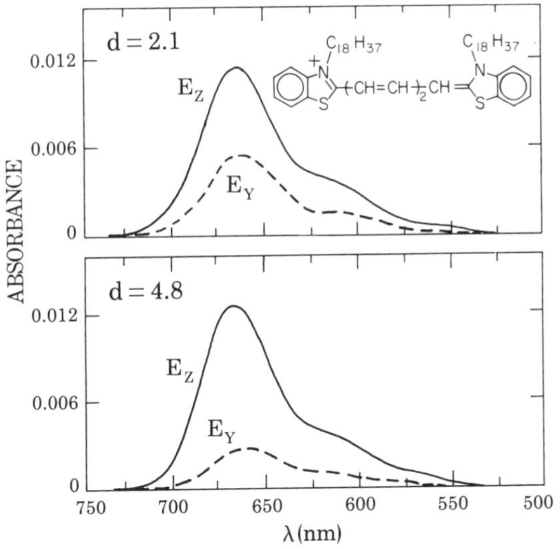

Figure 8.49 Polarized absorption spectra E_Z (full curves) and E_Y (dashed curves) of the cyanine dye **72** in mixed monolayers with *n*-octadecyltrichlorosilane adsorbed on a stretched poly(vinyl alcohol)-coated polyethylene film before (below) and after (above) a 3-minute treatment with water. Adapted by permission from ref. 100.

8.8.3 Chlorophyll in Lipid Bilayers[101]

In addition to their intrinsic interest, artificial lipid bilayers are also often used as models for true biological membranes. Dichroic measurements on artificial membranes will be illustrated on the case of chlorophyll *a* incorporated into several phospholipids (dioleyllecithin, phosphatidylserine, phosphatidylethanolamine).

Measurements of dichroic absorption were performed at a series of oblique angles of incidence using a sensitive polarization-modulation method. Chlorophyll *a* possesses two mutually perpendicular $\pi\pi^*$ transition moments lying in the plane of the porphyrin ring, one at 440 nm and the other at 670 nm. The measurement of their dichroism, corrected for polarized reflection, gave almost identical results for the three lipids used. The dichroic ratio was independent of chlorophyll *a* concentration in the membrane up to molar ratio [chl]/[lec] = 0.7. The dichroism at 670 nm is much higher than that at 440 nm. The average angle of alignment with the membrane plane was found to be 34–36° for the 670 nm transition moment, depending on the lipid, and 23–29° for the 440 nm transition moment, again depending on the lipid, with uncertainties of the order of 1–2°. From these values, the angle between the porphyrin plane and the membrane can be calculated to be 44–49°. Similar but less conclusive results were obtained for the orientation of chlorophyll *b*, for which there is less certainty that the two visible transition moments are mutually perpendicular.

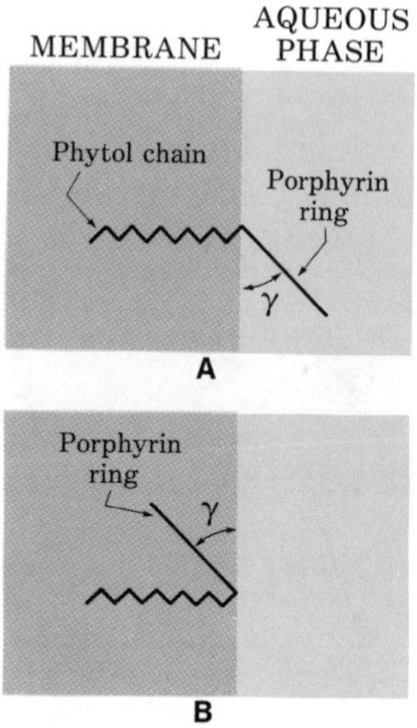

Figure 8.50 Two proposed possible conformations of the chlorophyll molecule with respect to the membrane surface. Reproduced by permission from ref. 101.

The authors suggest that there is very little spread in the actual orientation angles and all the chlorophyll molecules in the sample are oriented alike. If this were true, the above angles would not only represent average values but also describe this orientation for each molecule. Of the two possibilities shown in Figure 8.50, they consider B to be more probable, since the larger part of the porphyrin ring is hydrophobic.

The authors note that the very weak dichroism shown by chlorophyll in oriented chloroplasts or isolated chloroplast lamellae need not be due to its being randomly oriented (except for chlorophyll 695, which has a special environment and shows noticeable dichroism). It had been pointed out previously that dichroism would also be absent if both transition moments lie at angles of 35° from the plane of the lamella (i.e., the porphyrin ring forms an angle of 55° with the lamella). The values obtained in this study with three different lipids are not far from 35°, indicating that this may well be the right interpretation and that chlorophyll is not randomly oriented in the lamellae after all.

8.8.4 Photoreceptor Membrane

An understanding of the structure of biological membranes is a prerequisite for a detailed understanding of their function. In this section, we shall provide an example of the use of dichroic measurements to obtain information about the structure of rhodopsin in intact membranes.

In these experiments, uniform multilayer arrays of membranes (i.e., smectic liquid crystals) were produced by slowly evaporating the solvent from a suspension of membrane fragments while simultaneously ultracentrifuging the fragments onto an isopotential surface; this was found more satisfactory than the many other previously published methods of orienting membrane fragments.[102] In particular, the layers produced by the isopotential dry-spin method are much thicker (over 2000 stacked layers).

The alignment of the stacked membranes with the surface of the substrate is nearly perfect, as evidenced by the measurement of the linear dichroism at the 498 nm absorption maximum of rhodopsin. Its retinylidene chromophore is known to lie at 17° from the membrane plane from previous measurements[103] of linear dichroism on retinal cells ($S_{zz} = -0.372$); measurements of linear dichroism on the layer of stacked membranes at an oblique angle of incidence yield essentially the same degree of orientation ($S_{zz} = -0.355$).[102] Irradiation experiments indicated that the rhodopsin in these aligned membranes is structurally intact and capable of photobleaching and regeneration.

The thickness of the membrane stack permitted polarized Fourier-transform IR measurements to be performed and combined with far-UV CD measurements.[104] It was already known from previous Raman, IR, and CD work that rhodopsin contains extensive α-helices and little β structure. The question addressed in this study was, what is the orientation of the α-helix IR and CD "chromophores" relative to the membrane plane?

Polarized IR spectra were obtained at a series of oblique angles of incidence, CD spectra at normal incidence. In the IR region, the dichroism of the amide I (1657 cm^{-1}), amide II (1545 cm^{-1}), and amide A (3300 cm^{-1}, N—H stretch) bands was measured, and it was concluded that the helix axis deviated less than 40° from the normal to the membrane plane. In the CD region, the disappearance of an extremum at 210 nm, which is observed in isotropic solutions and is active only for light incident perpendicular to the helix axis, confirmed the conclusion that the α helices are aligned approximately perpendicular to the membrane surface.

Similar UV-visible and CD measurements were performed on the purple membrane whose main constituent is bacteriorhodopsin.[82]

8.8.5 *Sickle-Cell Hemoglobin in Individual Erythrocytes*[104a]

Red blood cells from patients suffering from sickle cell anemia contain the mutant hemoglobin S. Upon deoxygenation, the hemoglobin condenses inside the cell and forms polymers which then aggregate in specific orientations. This distorts the shape of the cell into various forms, particularly the sickle form and the holly-leaf form.

A detailed characterization of the state of alignment of the hemoglobin S polymers within individual erythrocytes has been accomplished recently using a microscope that produces an image proportional to linear dichroism. Monochromatic light at 415 nm, corresponding to the extremely intense Soret band in the absorption spectrum of the hemoglobin, was modulated between two mutually perpendicular polarizations, and passed through a microscope and the observed individual cell. Average as well as modulated light intensity was detected and used to obtain an ordinary image of the cell as well as an image reflecting only its linear dichroism.

It was found that the sickle-shaped cells exhibit the highest degree of polymer orientation, with the main axis parallel to the morphological axis of the cell. In some irreversibly sickled cells there are regions of oriented polymeric hemoglobin next to the cell membrane. The authors speculate that this could be related to the mechanism by which the cell membrane is deformed and damaged. It is possible that images of this kind will be useful as fingerprints that are specific for the age, metabolic state, and type of the cell.

8.9 Anisotropic Rotational Motion

The decay of fluorescence polarization after short-pulse excitation has often been used for the measurement of rotational diffusion rates. Here we have selected an example of work in which the anisotropy of the rotational diffusion of perylene (**10**) in glycerol between 10 and 40°C was determined in this manner, using a nanosecond flash lamp and single-photon-counting techniques.[105]

Perylene was excited either into the long-axis-polarized first absorption band at 430 nm or into the short-axis-polarized band at 256 nm (see Section 8.3.2). The fluorescence, which is long-axis polarized, was observed at 448 nm. The initial emission anisotropies R(430,0) = 0.35 and R(256,0) = 0.15 agreed with values measured by the static method in propylene glycol at −50°C, where rotational depolarization is negligible, and were close to the theoretical values $R_{\parallel}(0) = 0.4$ and $R_{\perp}(0) = 0.2$ expected for purely polarized transitions.

The time dependence of both anisotropies, R(430,t) and R(256,t), could be fitted satisfactorily at each temperature by a biexponential function. The curve R(256,t) is shown in Figure 8.51 as an example. The negative emission anisotropy at early times is due to the orthogonality of the absorption and emission dipoles. The rapid increase to positive values demonstrates that rotation around the out-of-plane axis is fast relative to rotation around in-plane axes. Those molecules whose short in-plane axes were approximately parallel to the electric vector of the exciting light, i.e., vertical, were photoselected by the initial pulse. The direction of their emitting moments is therefore approximately horizontal at first and the fluorescence initially shows negative anisotropy. Fast rotation in the aromatic plane tends to randomize the distribution of the short and long axes in the plane of the molecule, thus reducing the number of emitting horizontally oriented transition moments and increasing the number of emitting vertically oriented transition moments. If this randomization were the only type of rotational motion occurring, the anisotropy would reach the value +0.1. This does not quite happen, since the slower rotations which change the orientation of the aromatic plane intervene. Eventually, the combined action of all the rotations reduces the anisotropy to zero.

The authors assumed that the rotational motions of the perylene molecule in glycerol can be described as a rotational diffusion of an oblate ellipsoid of rotation, with diffusion coefficients $D_{\parallel}$ for rotation around the unique out-of-plane axis and $D_{\perp}$ for rotation around either of the two in-plane axes. The diffusion axes and the transition moment directions coincide with the symmetry axes in the molecule; then the anisotropy decay can be at most biexponential.

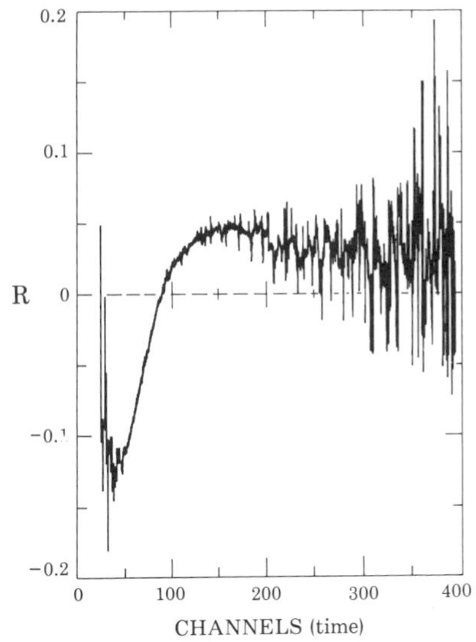

Figure 8.51 Decay of the observed fluorescence anisotropy R of perylene in glycerol (30°C, $\lambda_{exc} = 256$ nm, $\lambda_{obs} = 448$ nm). See equation (7.108) for the definition of R; the observed intensities were corrected for instrumental bias. Adapted by permission from ref. 105.

Computer fitting of the experimental data showed that they are compatible with the assumed diffusional model. The analysis yielded the values of $D_{\parallel}$ and $D_{\perp}$ at each temperature (e.g., at 25°C, $D_{\parallel} = 5.8 \times 10^7\ s^{-1}$ and $D_{\perp} = 6.0 \times 10^6\ s^{-1}$. Their ratio is very high, $D_{\parallel}/D_{\perp} = 10 \pm 1$, and essentially temperature-independent. More recent values obtained by multifrequency phase-modulation fluorometry[105a] are in substantial agreement.

The authors[105] noted that the resulting $D_{\parallel}$ is about six times as large as expected under sticking boundary conditions but is in rough agreement with expectations for slipping boundary conditions. However, $D_{\parallel}$ is much lower than the free rotor value of about $10^{11}\ s^{-1}$. The authors pointed out that in reality perylene is not a perfect disc, so that even the in-plane rotation causes some displacement of the surrounding solvent molecules. They found no evidence for any "local temperature" anomalies due to the conversion of excess energy deposited in the molecule upon 256 nm excitation into heat in the immediate environment.

It is noteworthy that measurements on tetracene in ethylene glycol and in 1-dodecanol suggest that the relative values of the rotational diffusion constants of a molecule may depend strongly on the nature of the solvent, so that they should not be thought of as molecular characteristics.[105b]

Very recently, the three diffusion coefficients of 9,10-dimethylanthracene in glyceryl tripropanoate have been determined between 235 and 338K by accurate fitting of the fluorescence anisotropy decay excited at a single wavelength using a synchrotron source.[105c] We suspect that much of the temperature-dependent "electronic delocalization" of the emitting transition moment postulated in this study was due to the presence of two overlapping differently polarized bands at the emission wavelength and to the temperature dependence of their profiles.

8.10 Practical Applications

Although the emphasis in this book is clearly on the fundamental aspects of optical spectroscopy with polarized light, we shall conclude Chapter 8 with a brief mention of two practical aspects of the use of polarized light.

8.10.1 Dichroic Polarizers

Polarized light is used for numerous purposes in daily life. It may be generated directly in a number of ways, e.g., in a laser or in a synchrotron, but for normal uses polarized light is produced from unpolarized light by dividing it into two orthogonally polarized components and eliminating one of them. The most popular polarizers of this type are dichroic polarizers of the sheet type. The story of sheet polarizers is in many ways quite fascinating.

In the middle of the 19th century, Mr. Phelps, a pupil of Dr. William Bird Herapath, a Bristol physician, happened to drop iodine into the urine of a dog that had been fed quinine. Mr. Phelps noticed little green crystals forming in the liquid, and when Herapath later studied these under a microscope he noticed that where the crystals overlapped they were sometimes dark and sometimes light. Dr. Herapath realized that he had found a new polarizing material. Unfortunately, when he tried to grow the crystals large enough for use in a microscope, he did not succeed; the crystals became too thin and fragile.

Sir David Brewster wrote a book on kaleidoscopes in 1858. In this book he mentioned the herapathite crystals, and this was noted by an undergraduate student at Harvard College, D. H. Land, in 1926. Land was interested in polarized light and had studied the work of Ambronn, which also involved iodine. In the late 19th century, Ambronn had stained natural tissues with iodine and noticed that some became darker and that these were dichroic. He probably realized the possibility of making a polarizer using this technique but apparently did not find it worthwhile to continue work along these lines (cf. Figure 1.3).

Land was impressed by the work of Herapath and considered the possibility of using a large number of small crystals instead of one large crystal. Within some months, Land was ready for his first successful experiment. Herapathite, which is a crystalline form of quinine sulfate periodide, was ground in a ball mill in a solution of nitrocellulose lacquer. After sufficiently small crystals had been prepared in this way, the suspension was placed in a cell, which in turn was put in the gap of a 10-kilogauss electromagnet. The magnet was turned on, and slowly the color of the liquid changed from opaque reddish black to a lighter and transparent color. The transmitted light was investigated with a Nicol prism and was found to be polarized.

The next problem was to make a solid polarizer from the liquid. This was done by dipping a plastic sheet into the test cell, placing it all in the magnetic field, and removing the cell but leaving a coating of the polarizing crystals on the sheet. The sheet was then allowed to dry in the magnetic field; a solid polarizer had been produced.

Land continued his experiments by looking for other methods of producing orientation. Electric field clearly was a possibility, but a simpler method was designed when he placed a suspension of herapathite on a rubber sheet stretched to several times its original length. The sheet was then allowed to contract and was stretched in the perpendicular direction. The result was satisfactory, but still the method was not suited for mass production of sheet polarizers.

Considerable progress was made in 1928 when Land produced a colloidal dispersion of submicroscopic needle-shaped crystals of herapathite, which was extruded through long, narrow slits. This resulted in a satisfactory orientation of the needles, and the method led to the production of the world's first commercial sheet polarizer, the J sheet. The J sheets became very popular but had some drawbacks, primarily their haziness, caused by light scattering from the millions of small crystals. Therefore Land continued to work in the field, and the most important result appeared in 1938: the H sheet.

When producing the H sheet, Land took advantage of a material that was new at that time, poly(vinyl alcohol) (PVA). As we have seen in Chapter 3, PVA is easy to stretch when heated and is completely transparent in the visible region. H sheets are produced simply by absorbing iodine in a stretched PVA sheet. This causes the iodine molecules to align in long chains like the host molecules. Before the PVA sheet is dipped into the iodine-rich solution, relaxation is prevented by fixing the sheet to another sheet of a material such as cellulose acetate.

The H polarizers became immensely popular; in the visible region, they transmit around 90% of light polarized perpendicular to the stretching direction and less than 0.1% of the light polarized along the stretching direction.

Other types of polarizers, similar to H sheets, soon appeared, for example the K polarizer, which is chemically extremely simple: Stretched PVA is dehydrated, producing oriented polyvinylene, $(-CH{=}CH-)_x$, which is a polarizer with optical properties close to those of the H sheet. It has the advantage that it can withstand high temperatures, but on the other hand it is more expensive. It was designed for use in a glareless headlight system. The idea was that both headlights and windshields on cars could be covered with K sheet polarizers tilted parallel to each other, 45° to vertical. This would lead to a situation of crossed polarizers for opposing cars, allowing the drivers to use light comparable to the ordinary upper headlight beam without disturbing drivers in approaching cars.

In spite of the fact that Land considered this project very important, it never became a success. There were several reasons: The system did not help people in the traffic other than the driver, it would have to be installed on all cars, and the first drivers using it would gain nothing from it before this had happened. There were also several technical problems: The polarizer at the windshield absorbed half the light, some windshields were birefringent and altered the polarization of the light, and some windshields were tilted so much that this also led to changes in polarization and glare leaks.

However, K sheets are used for several other purposes today, such as the production of polarized light from the strong (and hot) lamps used for stereoscopic pictures.

Another type of sheet similar to the H sheet is the L polarizer, which works as a polarizer in only part of the spectrum, transmitting light at other wavelengths unpolarized. Any stretched sheet such as those described in Chapter 3 containing an oriented solute with strong absorption in the visible region will act as an L polarizer.

The most common sheet polarizers are of the H type. They are used in sunglasses, where they are placed so that horizontally polarized light is not transmitted. This prevents light reflected from horizontal surfaces, which is usually horizontally polarized and often of unpleasant brightness, from reaching the eye. For use in sunglasses, the polarizer is usually combined with an isotropic absorber, which works like ordinary sunglasses, removing part of the light intensity, especially in the UV region.

Recently, there has been considerable interest in the development of sheet polarizers for liquid-crystal displays, for instance for the dashboards of automobiles. Such polarizers have to withstand high temperature and humidity, and the standard H sheets are not very suitable.

8.10.2 Permanent Information Storage[107]

Our second example of a practical utilization of the concepts discussed in this book is quite different from the first. While polarizing sheets have been an article of commerce for a long time, the notion that dyed stretched polymers can be used as optical storage media has barely passed the stage of feasibility demonstration. Although it is not yet obvious whether it will ever find practical use in competition with the many other extant and future possible media, its illustrative value makes it worth mentioning here.

Optical information storage is based on the use of laser light to write in and read from a suitable medium. Particularly attractive among optical storage media are those whose optical properties are modified by the thermal effects of absorbed focused laser light, such as ablation of spots in the surface and production of holes or depressions. The recording media are thin layers of materials of low melting point and low thermal conductivity and diffusivity, such as tellurium and organic polymers loaded with light-absorbing material. The readout is based on detecting the intensity of light reflected or transmitted by the recording layer.

Some of the desirable properties are high sensitivity, high resolution, high signal-to-noise ratio, real-time recording, instant playback, high immunity to defects, longevity of stored information, and low cost.

It has now been proposed to use a thin sheet of stretched polymer, doped by a suitable dye molecule, as the recording medium. The dye and its concentration are selected so as to absorb a substantial fraction of the laser radiation. The polymer is selected so as to exhibit large birefringence, to accept the dye in an appropriate concentration, and to have suitable mechanical properties. It must be possible to produce it in the form of very thin sheets, it must remain in the stretched state for a long time at room temperature, and it must relax rapidly to a more or less completely isotropic state when heated.

The writing beam causes a local partial or complete relaxation of the birefringence at power levels well below those needed for melting. The polarization state of the reading beam is changed if it passes through those areas of the recording sheet which have not been recorded upon and is changed differently or not at all when it passes through a spot whose birefringence had been modified by writing. Many possibilities exist for the detection of polarization change in the emerging beam.

Two read-out arrangements have been analyzed in detail. In the transmission mode, the stretched sheet is located between two polarizers (the first polarizer can be omitted if the laser light is already polarized). In the reflection mode, the sheet is rigidly sandwiched between a mirror and a linear polarizer.

Qualitatively, it is easy to see how the two arrangements work, and we shall show this for the transmission mode. Suppose that the birefringent sheet acts as a half-wave plate at the wavelength of the laser. Let the two polarizers be crossed, and let the stretching direction of the sheet form a 45° angle with their polarization directions. Then the polarization direction of the light which passed through the first polarizer will be rotated by 90° by the sheet and will pass freely through the second polarizer. However, the laser beam passes through a spot on the sheet whose birefringence has been totally relaxed by the recording beam; its polarization will not be affected by the sheet at all, and it will not be able to pass the second polarizer. Thus, the recorded areas will appear as dark spots on a light background.

More quantitatively, the optical properties of a uniaxially stretched sheet can be characterized by its mean absorbance $\mathscr{A}_e$, its linear dichroism $\mathscr{LD}$, and its linear birefringence $\mathscr{LB}$. The other elements of the Mueller matrix (Appendix III) vanish: $\mathscr{LD}' = \mathscr{LB}' = \mathscr{CD} = \mathscr{CB} = 0$. We take Z to be the stretching direction and view the sheet in transmission between two crossed polarizing elements. We let the polarization direction of the polarizer form an angle of 45° with Z and that of the analyzer an angle θ with Z.

$$\mathscr{A}_e = (\ln 10)(E_{dye} + E_{polym})/2 \tag{8.11}$$

$$\mathscr{LD} = (\ln 10)(\Delta E_{dye} + \Delta E_{polym})/2 \tag{8.12}$$

$$\mathscr{LB} = 2\pi(\Delta n_{dye} + \Delta n_{polym})l/\lambda_{vac} \tag{8.13}$$

where $E = E_Z + E_Y$, $\Delta E = E_Z - E_Y$, $\Delta n = n_Z - n_Y$, and the subscripts indicate the contributions from the dye and the polymer.

The use of the Mueller calculus (Appendix III) leads to the result for the fraction of intensity transmitted

$$I/I_0 = (1/2) \exp(-2\mathscr{A}_e)(\cosh \mathscr{LD} + \cos \mathscr{LB} \sin 2\theta - \sinh \mathscr{LD} \cos 2\theta) \tag{8.14}$$

The decrease in the degree of orientation of the polymer and dye molecules caused by the act of recording is capable of changing $\mathscr{LB}$, $\mathscr{LD}$, and $\mathscr{A}_e$ and therefore also the measured intensity I.

The term in $\mathscr{A}_e$ is dissipative. A change in $\mathscr{A}_e$ upon recording is a reflection of chemical changes in the dye and of changes in its orientation. To appreciate the existence of the latter contribution, consider a dye of D_{2h} symmetry and write its contribution to the absorbance polarized along X and Y in terms of the reduced spectra A_x, A_y, and A_z:

$$\text{Oriented state:}\quad A_Y + A_Z = (1/2)\sum_u (1 + K_u)A_u \tag{8.15}$$

$$\text{Random state:}\quad A_Y + A_Z = (2/3)\sum_u A_u \tag{8.16}$$

The terms in $\mathscr{LD}$ reflect the rotation of the plane of the polarized light by the linear dichroism of the sample (Figure 1.14).

The term in $\mathscr{LB}$ is due to the birefringence of the dyed stretched polymer which makes it act as a retardation plate. It appears at all wavelengths, even those at which the dye and the polymer are transparent, $\mathscr{A}_e = 0$, $\mathscr{LD} = 0$. At these wavelengths, the transmitted intensity fraction is simply

$$I/I_0 = (1/2)(1 + \sin 2\theta \cos \mathscr{LB}) \tag{8.17}$$

In the case of crossed polarizers, $\theta = -45°$, one obtains $I = 0$ if $\mathscr{LB} = 0$, as appropriate for an isotropic medium, and $I = I_0$ if $\mathscr{LB} = \pi$, as expected for a half-wave plate. A recording medium with $\mathscr{LB}$ equal to an odd multiple of π whose birefringence relaxes after recording in such a way that $\mathscr{LB}$ equals an even multiple of π (e.g., effectively, $\mathscr{LB} = 0$) should therefore produce a series of dark dots on a light background.

In the case of parallel polarizers, $\theta = 45°$, one obtains $I = I_0$ if $\mathscr{LB} = 0$ and $I = 0$ if $\mathscr{LB} = \pi$. The same recording medium will then produce a series of light dots on a dark background.

Figure 8.52 provides an example of a stretched polymer sheet on which small dots (~ 3 μm across) have been recorded using pulses from a 5 mW near-IR diode laser. In a real application, the laser focus would have to be improved, a thinner sheet would have to be used, and collection of dust particles during its preparation and use would have to be avoided. However, a qualitative demonstration of the principle has clearly been accomplished.[108]

8.10.3 Erasable Information Storage

Proceeding further within the realm of practical applications from the already commercial to the hypothetical, we note that materials which permit nondestructive photoorientation of solutes appear promising as media for erasable optical recording.

As outlined at the end of Section 4.6.3, partial solute alignment upon absorption of linearly polarized light imprints memory of the direction of light polarization. If the solute molecules cannot rotate spontaneously, and if the

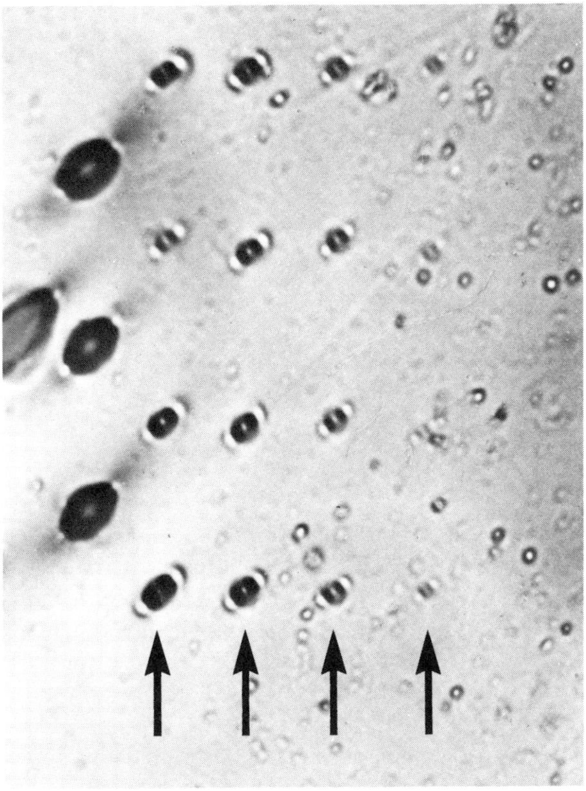

Figure 8.52 Optical recording on a birefringent polymer. A dyed stretched sheet of poly(vinyl chloride)-poly(vinylidene chloride) copolymer (2.5 μm thick) viewed between crossed polarizers. Four series of dark dots 25 μm apart (indicated by arrows) have been recorded by pulses of focused 823 nm GaAlAs diode laser light (5 mW). From left to right, the pulse lengths were 1000, 750, 500, and 250 μs. The larger black dots are holes in the sheets produced by longer pulses and used as markers. The irregularly placed gray dots are due to dust particles.

photochemically induced rotation or pseudorotation which is responsible for the alignment is not thermally reversible, this memory will be kept until the next period of irradiation (recording). Conditions which favor long-term stability thus are high viscosity, low temperature, and a rotation or pseudorotation process that is facile photochemically but forbidden thermally.

Memory readout is based on detecting the linear dichroism or linear birefringence in the small aligned area ($\sim 1\mu \times 1\mu$) with light whose wavelength (and/or intensity) are such that no significant alteration of the stored alignment occurs.

Although erasable storage systems based on nondestructive alignment of molecules dispersed in rigid matrices such as polymers can readily be envisaged in principle, we are not aware of a system that provides a stable but erasable memory at room temperature using this concept. At low temperatures, reversible nondestructive photodichroism has been demonstrated for media such as octaethylporphin (**73**) in various plastics and matrices.[109] Here, the pseudorotation process is the translocation of the internal protons between the two pairs of diagonally opposed nitrogen atoms:

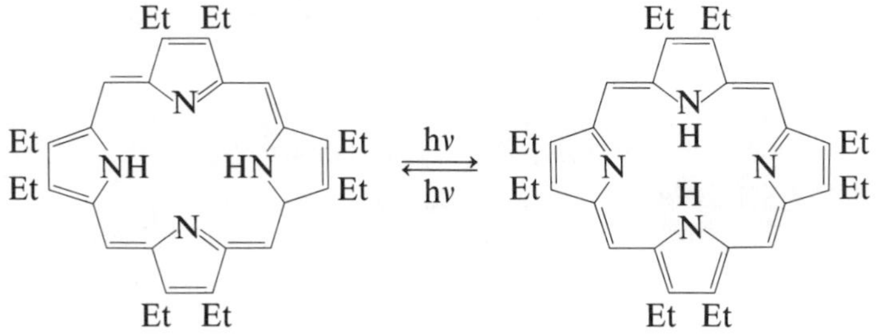

73

This induces a 90° rotation of the transition moment, and the theory of Section 4.6.3 predicts a dichroic ratio of 0.4 in the photostationary state for the transition that is being pumped. The value 0.46 has been observed in a neon matrix at 3 K and remained stable indefinitely. Although the system seems perfectly reversible, it is of limited practical interest since the central protons redistribute rapidly above 100K and the memory is therefore not stable at ordinary temperatures.

An example of a system based on nondestructive photoorientation that comes close to meeting the practical requirements, and which has indeed been used successfully to store holograms,[110] is one that lies slightly outside the scope of this monograph in that it is not based on a rigid solution doped with molecules but rather on a single crystal doped with defects. Since this difference is immaterial when it comes to demonstrating the principle, we shall describe it briefly.

The recording medium is a KCl single crystal doped with Na and the orientable absorbing defect is an F_A center. As shown in Figure 8.53, this consists

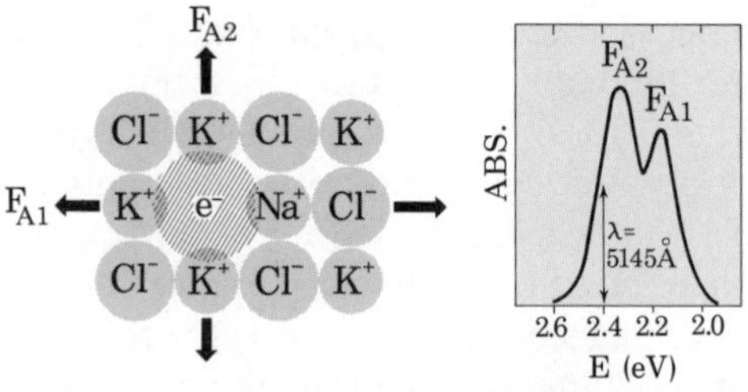

Figure 8.53 Left: An F_A center in a Na-doped KCl single crystal. Polarization directions for the F_{A1} and F_{A2} transitions are shown. Right: the absorption spectrum of the F_A center and the wavelength used for reversible information storage. Adapted by permission from ref. 110.

of a Na^+ impurity ion next to an anionic vacancy occupied by an electron. It is associated with a degenerate absorption band (F_{A2}) polarized perpendicular to the Na–e axis and a nondegenerate absorption band at lower energy (F_{A1}) polarized along this axis. At temperatures slightly below ambient ($\sim -20°$ C) the orientation of the axis is stable essentally indefinitely in the dark, but it can be readily randomized upon absorption of a 574.5 nm photon from an argon ion laser except at very low temperatures, providing a means for erasable information storage by nondestructive photoorientation. The authors point out that the readout of their holograms is practically nondestructive when performed at temperatures below 55 K.

8.11 References

1. J. G. Radziszewski and J. Michl, *J. Phys. Chem.* **85**, 2934 (1981).
2. E. W. Thulstrup, J. Michl, and J. H. Eggers, *J. Phys. Chem.* **74**, 3868 (1970).
3. F. W. Langkilde, M. Gisin, E. W. Thulstrup, and J. Michl, *J. Phys. Chem.* **87**, 2901 (1983).
4. T. Yoshinaga, H. Hiratsuka, and Y. Tanizaki, *Bull. Chem. Soc. Japan* **50**, 3096 (1977).
5. A. B. Myers and R. R. Birge, *J. Chem. Phys.* **73**, 5314 (1980).
6. J. S. Ham, *J. Chem. Phys.* **21**, 756 (1953); J. R. Platt, *J. Mol. Spectrosc.* **9**, 288 (1962); A. Nakajima, *Spectrochim. Acta* **A30**, 860 (1974).
7. F. W. Langkilde, E. W. Thulstrup, and J. Michl, *J. Chem. Phys.* **78**, 3372 (1983).
8. J. J. Dekkers, W. P. Cofino, G. Ph. Hoornweg, C. Maclean, and N. H. Velthorst, *Chem. Phys.* **47**, 369 (1980).
9. R. Pariser, *J. Chem. Phys.* **24**, 250 (1956).

9a. C. M. Carlin and M. K. DeArmond, *Chem, Phys. Lett.* **89**, 297 (1982).

10. A. C. Albrecht, *Progr. React. Kinet.* **5**, 301 (1970); D. M. Friedrich and A. C. Albrecht, *J. Chem. Phys.* **58**, 4766 (1973).
11. J. Kolc, J. W. Downing, A. P. Manzara, and J. Michl, *J. Am. Chem. Soc.* **98**, 930 (1976).
12. G. J. Hoijtink and P. J. Zandstra, *Mol. Phys.* **3**, 371 (1960).
13. S. Jen, N. A. Clark, P. S. Pershan, and E. B. Priestley, *J. Chem. Phys.* **66**, 4635 (1977).
14. H. Hiratsuka and Y. Tanizaki, *J. Phys. Chem.* **83**, 2501 (1979), cf. H. Hiratsuka, Y. Hatano, Y. Tanizaki, and Y. Mori, *J. Chem. Soc.*, Faraday Trans. 2, **81**, 1653 (1985).
15. T. Shida and S. Iwata, *J. Am. Chem. Soc.* **95**, 3473 (1973), and references therein.
16. J. G. Radziszewski and J. Michl, *J. Am. Chem. Soc.*, **108**, 3289 (1986).
17. B. Nordén, *Chem. Scripta* **1**, 145 (1971).
18. J. G. Radziszewski and J. Michl, *J. Chem. Phys.* **82**, 3527 (1985).
19. M. Ovaska and A. Kivinen, *J. Mol. Struct.* **101**, 255 (1983).

20. M. B. Mitchell, W. A. Guillory, J. Michl, and J. Radziszewski, *Chem. Phys. Lett.* **96**, 413 (1983).
20a. M. Ovaska, *J. Phys. Chem.* **88**, 5981 (1984).
21. G. P. Ceasar, R. A. Levenson, and H. B. Gray, *J. Am. Chem. Soc.* **91**, 772 (1969).
22. S. P. Church, M. Poliakoff, J. A. Timney, and J. J. Turner, *J. Am. Chem. Soc.* **103**, 7515 (1981); for a related study of $Mn_2(CO)_9$ see I. R. Dunkin, P. Härter, and C. J. Shields, *J. Am. Chem. Soc.* **106**, 7248 (1984).
22a. J. J. Fisher, J. H. Penn, D. Döhnert, and J. Michl, *J. Am. Chem. Soc.*, **108**, 1715 (1986).
22b. G. Raabe, H. Vančik, R. West, and J. Michl, *J. Am. Chem. Soc.*, **108**, 671 (1986).
23. E. W. Thulstrup and J. Michl, *J. Phys. Chem.* **84**, 82 (1980).
24. W. G. Hill, S. F. Mason, and R. D. Peacock, *J. Chem. Soc.* Perkin *II* **1977**, 1262.
25. G. Gottarelli, B. Samorì, and R. D. Peacock, *J. Chem. Soc. Perkin II* **1977**, 1208.
26. Y. Tanizaki, T. Yoshinaga, and H. Hiratsuka, *Spectrochim. Acta* **34A**, 205 (1978).
27. H. Zimmerman and N. Joop, *Z. Elektrochem.* **65**, 138 (1961).
28. E. W. Thulstrup, J. Spanget-Larsen, and R. Gleiter, *Mol. Phys.* **37**, 1381 (1979).
28a. S. Ghosh and M. Chowdhury, *Chem. Phys. Lett.* **85**, 233 (1982).
29. E. W. Thulstrup and J. Michl, *J. Am. Chem. Soc.* **98**, 4533 (1976).
30. J. Kolc, E. W. Thulstrup, and J. Michl, *J. Am. Chem. Soc.* **96**, 7188 (1974).
31. J. Michl and J. H. Eggers, *Tetrahedron* **30**, 813 (1974).
32. A. Yogev, L. Margulies, and Y. Mazur, *Chem. Phys. Lett.* **8**, 157 (1971).
33. J. Michl and J. F. Muller, *J. Am. Chem. Soc.* **98**, 4550 (1976).
34. J. W. Kenney III, D. A. Herold, J. Michl, and J. Michl, *J. Am. Chem. Soc.* **100**, 6884 (1978); G. P. Dalgaard and J. Michl, *Ibid.*, p. 6887; M. A. Souto, D. Otteson, and J. Michl, *Ibid.*, p. 6892.
35. E. W. Thulstrup, P. L. Case, and J. Michl, *Chem. Phys.* **6**, 410 (1974).
36. R. Gleiter, J. Spanget-Larsen, E. W. Thulstrup, I. Murata, K. Nakasuji, and C. Jutz, *Helv. Chim. Acta* **59**, 1459 (1976).
37. E. W. Thulstrup, J. Michl, and C. Jutz, *J. Chem. Soc. Faraday II* **71**, 1618 (1975).
38. J. W. Sidman and D. S. McClure, *J. Chem. Phys.* **24**, 757 (1956).
39. W. Gerhartz and J. Michl, *J. Am. Chem. Soc.* **100**, 6877 (1978).
40. J. Spanget-Larsen and R. Gleiter, *Helv. Chim. Acta* **61**, 2999 (1978).
40a. U. P. Wild, in *Excited States in Organic Chemistry and Biochemistry*, B. Pullman and N. Goldblum, Eds., D. Reidel Publishing Co., Dordrecht, Holland, 1977, p. 387.
41. T. Yoshinaga, H. Hiratsuka, and Y. Tanizaki, *Bull. Chem. Soc. Japan* **51**, 996 (1978).
42. K. Ota, K. Murofushi, T. Hoshi, E. Shibuya, and J. Yoshino, *Z. Phys. Chem. N. F.* **104**, 181 (1977).

43. J. Sagiv, A. Yogev, and Y. Mazur, *J. Am. Chem. Soc.* **99**, 6861 (1977); J. Spanget-Larsen, R. Gleiter, and R. Haider, *Helv. Chim. Acta* **66**, 1441 (1983).
44. P. Palmieri and B. Samorì *J. Am. Chem. Soc.* **103**, 6818 (1981).
45. N. Harada and K. Nakanishi, *Accts. Chem. Res.* **5**, 257 (1972).
46. D. A. Schooley, E. Bunnenberg, and C. Djerassi, *Proc. Nat. Acad. Sci. U.S.A.* **56**, 1377 (1966); A. Tajiri, H. Uchimura, and M. Hatano, *Chem. Phys. Lett.* **34**, 1021 (1975).
47. A. Bree and R. Zwarich, *J. Chem. Phys.* **51**, 903 (1969).
48. T. Hoshi, H. Inoue, J. Shiraishi, and Y. Tanizaki, *Bull. Chem. Soc. Japan* **44**, 1743 (1971).
49. H. Inoue, T. Hoshi, T. Masamoto, J. Shiraishi, and Y. Tanizaki, *Ber. Bunsenges. Physik. Chem.* **75**, 441 (1971).
50. S. F. Mason and R. D. Peacock, *Chem. Phys. Lett.* **21**, 406 (1973).
51. K. R. Popov, *Opt. Spektrosk.* **3**, 579 (1957).
52. H. Wedel and W. Haase, *Chem. Phys. Lett.* **55**, 96 (1978).
53. W. Haase and H. Wedel, *Mol. Cryst. Liq. Cryst.* **38**, 61 (1977).
54. E. Sackmann and H. Möhwald, *J. Chem. Phys.* **58**, 5407 (1973).
55. Å. Davidsson and B. Nordén, *Tet. Lett.* **1972**, 3093.
56. J. Michl, E. W. Thulstrup, and J. H. Eggers, *Ber. Bunsenges. Physik. Chem.* **78**, 575 (1974).
57. D. M. Friedrich, R. Mathies, and A. C. Albrecht, *J. Mol. Spectrosc.* **51**, 166 (1974).
58. R. P. Steiner and J. Michl, *J. Am. Chem. Soc.* **100**, 6861 (1978).
58a. M. Pawlikowski and M. Z. Zgierski, *J. Chem. Phys.* **76**, 4789 (1982).
59. B. Dick, H. Gonska, and G. Hohlneicher, *Ber. Bunsenges. Phys. Chem.* **85**, 746 (1981); B. Dick and G. Hohlneicher, *Chem. Phys. Lett.* **83**, 615 (1981).
60. Y. Matsuoka and B. Nordén, *Chem. Phys. Lett.* **85**, 302 (1982).
61. Y. Matsuoka and K. Yamaoka, *Bull. Chem. Soc. Japan* **52**, 3163 (1979).
62. Y. Matsuoka and Y. Yamaoka, *Bull. Chem. Soc. Japan* **52**, 2244 (1979).
63. V. Zanker and J. Thies, *Z. Physik. Chem.* N. F. **33**, 46 (1962).
64. H. P. J. M. Dekkers and E. C. M. Kielman-Van Luyt, *Mol. Phys.* **31**, 1001 (1976).
65. J. Spanget-Larsen, R. Gleiter, M. Kobayashi, E. M. Engler, P. Shu, and D. O. Cowan, *J. Am. Chem. Soc.* **99**, 2855 (1977).
66. G. P. Ceasar and H. B. Gray, *J. Am. Chem. Soc.* **91**, 191 (1969).
67. Å. Davidsson, M. Gouterman, L. Y. Johansson, R. Larsson, B. Nordén, and M. Sundbom, *Acta Chem. Scand.* **26**, 840 (1972).
68. R. Gale, R. D. Peacock, and B. Samorì, *Chem. Phys. Lett.* **37**, 430 (1976).
69. N. Fischer, E. V. Goldammer, and J. Pelzl, *J. Mol. Struct.* **56**, 95 (1979).
69a. J. Waluk and E. W. Thulstrup, *Chem. Phys. Lett.* **123**, 102 (1986).
70. P. B. Pedersen, E. W. Thulstrup, and J. Michl, *Chem. Phys.* **60**, 187 (1981).
71. R. Kiessling, G. Hohlneicher, and F. Dörr, *Z. Naturforsch.* **22a**, 1097 (1967).
72. J. Kolc, E. W. Thulstrup, and J. Michl, *J. Am. Chem. Soc.* **96**, 7188 (1974).
73. K. Rotkiewicz and Z. R. Grabowski, *Trans. Faraday Soc.* **65**, 3263 (1969).

74. L. B.-Å. Johansson, Å. Davidsson, G. Lindblom, and K. R. Naqvi, *Biochemistry* **18**, 4249 (1979).
75. Y. Matsuoka and B. Nordén, *J. Phys. Chem.* **87**, 220 (1983).
76. Y. Matsuoka and B. Nordén, *J. Phys. Chem.* **86**, 1378 (1982).
77. C. C. Bott and T. Kurucsev, in *Molecular Optical Dichroism and Chemical Applications of Polarized Spectroscopy*, B. Nordén, Ed., Lund University Press, Lund, Sweden, 1977, p. 81; *Spectrosc. Lett.* **10**, 495 (1977).
78. A. F. Fucaloro and L. S. Forster, *J. Am. Chem. Soc.* **93**, 6443 (1971).
79. H.-G. Kuball, M. Acimis, and J. Altschuh, *J. Am. Chem. Soc.* **101**, 20 (1979).
80. H.-G. Kuball and J. Altschuh, *Chem. Phys. Lett.* **87**, 599 (1982).
81. T.-L. Hsiao and K. J. Rothschild, *Biochem. Biophys. Res. Commun.* **94**, 618 (1980).
82. D. D. Muccio and J. Y. Cassim, *Biophys. J.* **26**, 427 (1979).
83. I. M. Ward, Ed., *Structure and Properties of Oriented Polymers*, Applied Science Publishers Ltd., London, 1975; H. Kelker and R. Hatz, *Handbook of Liquid Crystals*, Verlag Chemie, Weinheim, 1980; K. Mittal, Ed., *Solution Chemistry of Surfactants*, Plenum, New York, 1979.
84. A. Yogev, J. Riboid, J. Marero, and Y. Mazur, *J. Am. Chem. Soc.* **91**, 4559 (1969).
85. J. Gawronski, T. Liljefors, and B. Nordén, *J. Am. Chem. Soc.* **101**, 5515 (1979).
86. M. Sheves, N. Friedman, D. Levendis, L. Margulies, and Y. Mazur, *Israel J. Chem.* **18**, 359 (1979).
87. A. Yogev, L. Margulies, and Y. Mazur, *J. Am. Chem. Soc.* **93**, 249 (1971).
88. A. Yogev, J. Sagiv, and Y. Mazur, *J. Chem. Soc., Chem. Commun.* **1973**, 943.
88a. J. G. Radziszewski, T.-K. Yin, F. Miyake, G. E. Renzoni, W. T. Borden, and J. Michl, *J. Am. Chem. Soc.* **108**, 3544 (1986).
89. R. T. Ingwall, C. Gilon, and M. Goodman, *J. Am. Chem. Soc.* **97**, 4356 (1975).
90. M. Eriksson, B. Nordén, P.-O. Lycksell, A. Gräslund, and B. Jernström, *J. Chem. Soc., Chem. Commun.* **1985**, 1300.
90a. A. Yogev, L. Margulies, and Y. Mazur, *J. Am. Chem. Soc.* **92**, 6059 (1970).
90b. A. Kivinen and M. Ovaska, *J. Phys. Chem.* **87**, 3809 (1983).
90c. Y. Matsuoka, B. Nordén, and T. Kurucsev, *J. Phys. Chem.* **88**, 971 (1984).
90d. Y. Matsuoka, B. Nordén, and T. Kurucsev, *J. Cryst. Spectrosc. Res.* **15**, 545 (1985).
91. P. Krebs, E. Sackmann, and J. Schwarz, *Chem. Phys. Lett.* **8**, 417 (1971).
92. H. Beens, H. Möhwald, D. Rehm, E. Sackmann, and A. Weller, *Chem. Phys. Lett.* **8**, 341 (1971).
93. B. Nordén, *Appl. Spectrosc. Rev.* **14**, 157 (1978).
94. E. Ohmes, J. Pauluhn, J.-U. Weidner, and H. W. Zimmermann, *Ber. Bunsenges. Physik. Chem.* **84**, 23 (1980).
95. G. R. Kelly and T. Kurucsev, *Biopolymers* **15**, 1481 (1976).

96. F. Tjerneld, thesis, Chalmers University of Technology, Lund, Sweden, 1982; O. Undeman, P.-O. Lycksell, A. Gräslund, T. Astlind, A. Ehrenberg, B. Jernström, F. Tjerneld, and B. Nordén, *Cancer Res.* **43**, 1851 (1983).
97. J. H. Nobbs, D. I. Bower, I. M. Ward, and D. Patterson, *Polymer* **15**, 287 (1974).
98. M. Hennecke and J. Fuhrmann, *Colloid Polymer Sci.* **258**, 219 (1980).
98a. J. Fuhrmann and M. Hennecke, *Makromol. Chem.* **181**, 1685 (1980).
99. O. Kratky, *Kolloid-Z.* **64**, 213 (1933).
100. J. Sagiv, *Israel J. Chem.* **18**, 339 (1979).
101. A. Steinemann, G. Stark, and P. Läuger, *J. Membrane Biol.* **9**, 177 (1972).
102. K. J. Rothschild, K. M. Rosen, and N. A. Clark, *Biophys. J.* **31**, 45 (1980).
103. P. A. Liebman, *Ann. N.Y. Acad. Sci.* **157**, 250 (1969).
104. K. J. Rothschild, R. Sanches, T. L. Hsiao, and N. A. Clark, *Biophys. J.* **31**, 53 (1980).
104a. W. C. Mickols, C. Bustamante, M. F. Maestre, I. Tinoco, Jr., and S. H. Embury, *Biotechnology* **3**, 711 (1985).
105. M. D. Barkley, A. A. Kowalczyk, and L. Brand, *J. Chem. Phys.* **75**, 3581 (1981).
105a. J. R. Lakowicz, H. Cherek, B. P. Maliwal, and E. Gratton, *Biochemistry* **24**, 376 (1985).
105b. M. J. Sanders and M. J. Wirth, *Chem. Phys. Lett.* **101**, 361 (1983).
105c. J. L. Viovy, *J. Phys. Chem.* **89**, 5465 (1985).
106. E. H. Land, *J. Opt. Soc. Am.* **30**, 230 (1940); **41**, 957 (1951); W. A. Shurcliff and S. S. Ballard, *Polarized Light*, Van Nostrand, Princeton, NJ, 1964.
107. C. Puebla and J. Michl, *Appl. Phys. Lett.*, **42**, 570 (1983); J. Michl and C. Puebla, U.S. Patent No. 4,551,829, Nov. 5, 1985.
108. P. S. Murthy, K. A. Klingensmith, and J. Michl, *J. Appl. Polymer. Sci.*, **31**, 2331 (1986).
109. J. G. Radziszewski, F. A. Burkhalter, and J. Michl, submitted for publication.
110. H. Blume, T. Bader, and F. Luty, *Opt. Commun.* **12**, 147 (1974).

APPENDIX I

Molecular Symmetry and Elementary Group Theory

This appendix provides a brief and very simple review of group theory and its application to molecular optical spectroscopy. A summary of the properties of groups is presented first, followed by a set of character tables for the most commonly encountered molecular point symmetry groups. Then the forms which the transition moments **M**(0f), $\mathscr{M}$(0f), α′(0f), and T(0f) take in the important symmetry groups are tabulated. Finally, multiplication of irreducible representations is discussed, and the products are listed for the most important point symmetry groups.

General

Fundamental Properties of Groups

A mathematical *group* is defined as a set of elements (A,B,C, . . .) with the following properties:

1 The product AB of any two elements, A and B, is defined and is also contained in the group.
2 The group contains the identity element E: EA = AE = A.
3 Every element A has a reciprocal A^{-1} in the group: $AA^{-1} = E$.
4 The elements obey the associative law: (AB)C = A(BC).

The number of elements in a group is called the *order of the group.*

Two group elements A and B are said to belong to the same *class* if there exists a third element X of the group such that $X^{-1}AX = B$.

A very important concept is the *multiplication table* which lists all products of two elements from the group. The multiplication table is symmetrical across its diagonal when all group elements commute: AB = BA. Such groups are called Abelian.

For a molecule, a set of symmetry operations may be defined: identity (E), reflection through a plane (σ), inversion through a point (i), rotation around an axis (C_n), and rotation around an axis followed by reflection through a plane perpendicular to the axis (S_n). The subscript n shows the number of times a rotation must be applied before the identity element results: $(C_n)^n = E$. The set of symmetry operations for a molecule form a *symmetry point group*. A multiplication of two symmetry operations is defined as their successive application. In keeping with the usage in this text, the operators such as E or σ should be written with carets—$\hat{E}$, $\hat{\sigma}$, etc.—but this is not customary and we shall not do it except in a few instances where confusion might otherwise result. Some authors prefer to let the symmetry operations act on the coordinates rather than the physical objects or functions, but we shall not do so.

Representation of Groups. If a number (or a matrix), $\Gamma(A)$, $\Gamma(B)$, . . . , is assigned to each group element, A,B, . . . , in such a way that the multiplication table of the numbers (or matrices) corresponds to that of the group elements:

$$AB = C \rightarrow \Gamma(A)\Gamma(B) = \Gamma(C) \tag{I.1}$$

then the set of numbers (or matrices) $\Gamma(A)$, $\Gamma(B)$, . . . , is said to be a *representation* of the group.

For example, when the twofold symmetry axis of the H_2O molecule at its equilibrium geometry is labeled z, the in-plane axis perpendicular to it is labeled y, and the out-of-plane axis is labeled x, the water molecule contains the following symmetry operations (Figure I.1, part A): E, $C_2(z)$, σ_{xz}, σ'_{yz}. The multiplication table for this symmetry point group, which is called C_{2v}, is given in Table I.1.

One obvious representation (A_1) is the set of numbers 1,1,1,1, but three other representations (A_2,B_1,B_2) may also easily be constructed from the numbers 1 and -1 (Table I.2). Such representations are said to be one-dimensional. In general, the *dimensionality* of the representation is given by the dimension of its member matrices.

An example of a two-dimensional representation of the C_{2v} group is the set $\Gamma = \{\Gamma(E), \Gamma(C_2), \Gamma(\sigma), \Gamma(\sigma')\}$:

$$\begin{pmatrix} 1 & 0 \\ 0 & 1 \end{pmatrix}, \quad \begin{pmatrix} 1 & 0 \\ 0 & -1 \end{pmatrix}, \quad \begin{pmatrix} 1 & 0 \\ 0 & -1 \end{pmatrix}, \quad \begin{pmatrix} 1 & 0 \\ 0 & 1 \end{pmatrix} \tag{I.2}$$

Other such multidimensional representations can be obtained by combining two or more representations of lower dimension in this fashion, i.e., adding blocks along the diagonal. This is known as forming a *direct sum* of representations,

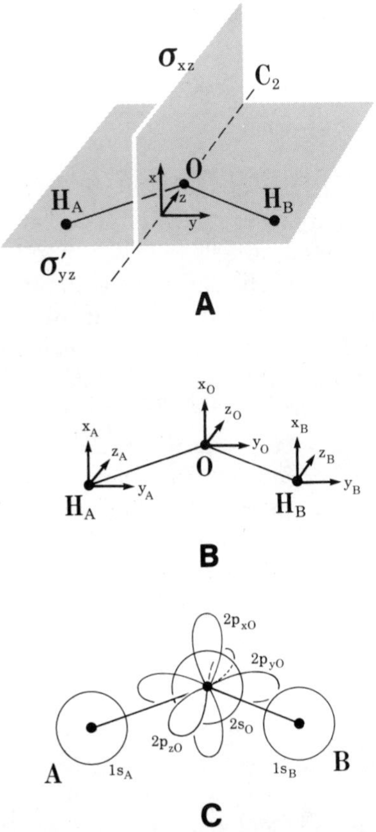

Figure I.1 The water molecule. (A) The symmetry operations; (B) the atomic displacement vectors; (C) the valence AO basis set.

Table I.1
Multiplication Table for the C_{2v} Group

C_{2v}	E	C_2	σ	σ'
E	E	C_2	σ	σ'
C_2	C_2	E	σ'	σ
σ	σ	σ'	E	C_2
σ'	σ'	σ	C_2	E

Table I.2
The One-Dimensional Representations of the C_{2v} Group

C_{2v}	E	C_2	σ_{xz}	σ'_{yz}
A_1	1	1	1	1
A_2	1	1	−1	−1
B_1	1	−1	1	−1
B_2	1	−1	−1	1

and one writes, e.g., $A_1 + B_2 = \Gamma$ for (I.2). Conversely, representations of higher dimension in which every matrix has the same block structure can be decomposed into two or more representations of lower dimensions, $\Gamma = A_1 + B_2$.

Multidimensional representations which can be brought into a common block structure by procedures to be discussed below are called *reducible*; others, including all one-dimensional ones, are called *irreducible*.

The *character* $\chi(R)$ of a representation for a given symmetry operation R is given by the trace (sum of diagonal elements) in the matrix for that operation. One-dimensional representations are their own characters; thus the table of C_{2v} given above is also a *character table*. Such tables for the most important point groups are listed later in this appendix.

A *direct product* of representations Γ_1 and Γ_2 is a representation $\Gamma = \Gamma_1 \times \Gamma_2$ whose matrix for each group element is obtained as the direct product of matrices of Γ_1 and Γ_2 for that element: $\Gamma(R) = \Gamma_1(R) \times \Gamma_2(R)$. The characters of a direct product representation Γ are equal to the product of characters of Γ_1 and Γ_2: $\chi(R) = \chi_1(R)\chi_2(R)$.

A representation may be *generated from a set of basis functions*. For example, the representation Γ in (I.2) results from application of the symmetry operations to the set of functions $\binom{z}{y}$:

$$E\begin{pmatrix} z \\ y \end{pmatrix} = \begin{pmatrix} z \\ y \end{pmatrix} = \begin{pmatrix} 1 & 0 \\ 0 & 1 \end{pmatrix}\begin{pmatrix} z \\ y \end{pmatrix}, \qquad C_2\begin{pmatrix} z \\ y \end{pmatrix} = \begin{pmatrix} z \\ -y \end{pmatrix} = \begin{pmatrix} 1 & 0 \\ 0 & -1 \end{pmatrix}\begin{pmatrix} z \\ y \end{pmatrix}$$

$$\sigma_{xz}\begin{pmatrix} z \\ y \end{pmatrix} = \begin{pmatrix} z \\ -y \end{pmatrix} = \begin{pmatrix} 1 & 0 \\ 0 & -1 \end{pmatrix}\begin{pmatrix} z \\ y \end{pmatrix}, \qquad \sigma'_{yz}\begin{pmatrix} z \\ y \end{pmatrix} = \begin{pmatrix} z \\ y \end{pmatrix} = \begin{pmatrix} 1 & 0 \\ 0 & 1 \end{pmatrix}\begin{pmatrix} z \\ y \end{pmatrix} \qquad (I.3)$$

Here, $\binom{z}{y}$ is said to form a *basis* for Γ, or to "belong to" Γ, or to "transform like" Γ.

If a function f_1 forms a basis for the representation Γ_1 and a function f_2 forms a basis for the representation Γ_2, then the product f_1f_2 forms a basis for the direct product representation $\Gamma_1 \times \Gamma_2$.

The following properties of group representations are given without proof. The reader interested in more detailed information will find it in the references given at the end of this appendix.

(i) The number of irreducible representations of a group is equal to the number of classes of symmetry elements.

(ii) A reducible representation Γ can be decomposed into irreducible representations, $\Gamma = \sum_i n_i\Gamma_i$, either by inspection as above or by determining the number n_i of times each irreducible representation Γ_i is contained in Γ, using

$$n_i = \frac{1}{h}\sum_R \chi(R)\chi_i(R) \qquad (I.4)$$

where h is the order of the group, and $\chi(R)$ and $\chi_i(R)$ are characters of Γ and Γ_i, respectively. The summation runs over all group elements. Expression (I.4) has the form of a *scalar product* of vectors, $[\chi(R_1),\chi(R_2),\ldots]\cdot[\chi_i(R_1),\chi_i(R_2),\ldots]$.

The irreducible representation character vectors are orthogonal:

$$\sum_R \chi_i(R)\chi_{i'}(R) = h\delta_{ii'} \tag{I.5}$$

(iii) A *projection operator* may be defined for the irreducible representation Γ_i with dimension l_i:

$$\hat{P}_i = (l_i/h) \sum_R \chi_i(R)R \tag{I.6}$$

If $\hat{P}_i$ is applied to a function f_j which forms a basis for a one-dimensional irreducible representation Γ_j, we obtain

$$\hat{P}_i f_j = \frac{1}{h}\sum_R \chi_i(R)Rf_j = \frac{1}{h}\sum_R \chi_i(R)\chi_j(R)f_j = \delta_{ij}f_j = \delta_{ij}f_i \tag{I.7}$$

Then if $\hat{P}_i$ is applied to an arbitrary function f which may be written as a linear combination of basis functions for irreducible representations,

$$f = \sum_j c_j f_j \tag{I.8}$$

the result is

$$\hat{P}_i f = \sum_j c_j \hat{P}_i f_j = c_i f_i \tag{I.9}$$

$\hat{P}_i$ therefore projects out the component of f which forms a basis for the irreducible representation Γ_i.

Properties (ii) and (iii) may be illustrated by two very simple examples: First, the decomposition of Γ given in (I.2) may be calculated from (I.4):

$$n_{A_1} = \tfrac{1}{4}(2,0,0,2)(1,1,1,1) = 1 \tag{I.10}$$

$$n_{A_2} = \tfrac{1}{4}(2,0,0,2)(1,1,-1,-1) = 0 \tag{I.11}$$

$$n_{B_1} = \tfrac{1}{4}(2,0,0,2)(1,-1,1,-1) = 0 \tag{I.12}$$

$$n_{B_2} = \tfrac{1}{4}(2,0,0,2)(1,-1,-1,1) = 1 \tag{I.13}$$

and

$$\Gamma = A_1 + B_2 \tag{I.14}$$

The projection operators $\hat{P}_i$ may be applied to any function. Our second example is the application to the 1s atomic orbital on hydrogen atom A, $1s_A$, in the water molecule as shown in part C of Figure I.1 (the 1s atomic orbital on the other hydrogen atom, B, is labeled $1s_B$).

$$A_1{:}\quad \hat{P}_{A_1}1s_A = \frac{1}{4}\sum_R \chi_{A_1}R1s_A$$

$$= \tfrac{1}{4}(1s_A + 1s_B + 1s_B + 1s_A) = \tfrac{1}{2}(1s_A + 1s_B) \tag{I.15}$$

This combination of 1s orbitals forms a basis for the representation A_1.

$$A_2\colon\quad \hat{P}_{A_2}1s_A = \frac{1}{4}\sum_R \chi_{A_2}R1s_A = \tfrac{1}{4}(1s_A + 1s_B - 1s_B - 1s_A) = 0 \qquad (I.16)$$

$$B_1\colon\quad \hat{P}_{B_1}1s_A = \frac{1}{4}\sum_R \chi_{B_1}R1s_A = \tfrac{1}{4}(1s_A - 1s_B + 1s_B - 1s_A) = 0 \qquad (I.17)$$

These results show that $1s_A$ does not contain a component which forms a basis for either A_2 or B_1.

$$B_2\colon\quad \hat{P}_{B_2}\,1s_A = \frac{1}{4}\sum_R \chi_{B_3}R1s_A$$

$$= \tfrac{1}{4}(1s_A - 1s_B - 1s_B + 1s_A) = \tfrac{1}{2}(1s_A - 1s_B) \qquad (I.18)$$

This combination of the 1s orbitals forms a basis for the representation B_2.

Symmetry Properties of the Hamiltonian. The importance of symmetry operations is that they commute with molecular Hamiltonians. This is immediately clear if a complete set $\{\psi\}$ of nondegenerate eigenfunctions of a Hamiltonian is considered:

$$\hat{H}\psi = E\psi \qquad (I.19)$$

For a symmetry operation R,

$$|R\psi|^2 = |\psi|^2 \quad \text{and therefore} \quad R\psi = \pm\psi \qquad (I.20)$$

Then, for all ψ,

$$\hat{H}R\psi = \pm E\psi = R\hat{H}\psi \qquad (I.21)$$

and $\hat{H}R = R\hat{H}$. This means that any exact eigenfunction of $\hat{H}$ forms a basis for an irreducible representation of the molecular point group. This is true generally: for the usual molecular orbitals, for many-electron state wavefunctions, and for vibrational wavefunctions.

Calculation of Transition Moments. When quantum-mechanical integrals are calculated, group theoretical considerations are very useful. If the integrand f(x,y,z) in the integral

$$\int f(x,y,z)\,dx\,dy\,dz \qquad (I.22)$$

which describes a molecular property does not form a basis for a representation whose decomposition into irreducible representations contains the totally symmetric one, the integral must vanish. This is due to cancelling positive and negative contributions from different regions of space defined by one or more symmetry operations.

The electric dipole transition moment is defined by

$$\mathbf{M}(12) = \langle\psi_2|\hat{\mathbf{M}}|\psi_1\rangle \qquad (I.23)$$

This means that unless the *direct product* of the representations of ψ_1, ψ_2, and the electric dipole operator $\hat{\mathbf{M}}$ forms a basis for the totally symmetric irreducible representation or a reducible representation which contains the totally symmetric irreducible representation, **M**(12) must vanish. The three components of the electric dipole operator $\hat{\mathbf{M}}$, defined in (2.1), correspond to vectors directed along the coordinates x, y, and z, which are usually listed in character tables next to the irreducible representations for which they form a basis.

Therefore, **M**(12) will vanish unless the direct product of the irreducible representations for which ψ_1 and ψ_2 form bases contains one or more of the irreducible representations for which x, y, and z form bases. For molecules of symmetry such as C_{2v}, only one of the coordinates will contribute, and this coordinate determines the direction of **M**(12). In point groups with a center of symmetry, x,y,z are of ungerade parity, i.e., they change sign upon inversion. Therefore, ψ_1 and ψ_2 must differ in their parity if **M**(12) is to be nonzero (one changes, the other does not change sign upon inversion).

The symmetry of the magnetic dipole operator $\hat{\mathscr{M}}$, defined in (2.2), corresponds to the infinitesimal rotations $\mathscr{R}_x$, $\mathscr{R}_y$, and $\mathscr{R}_z$ around the molecular axes x, y, and z. An electronic transition between states of the same spin multiplicity is magnetic dipole allowed if the product of the representations of the two states contains one or more of the representations of $\mathscr{R}_x$, $\mathscr{R}_y$, and $\mathscr{R}_z$. These are usually also listed in character tables. In point groups with a center of symmetry, $\mathscr{R}_x$, $\mathscr{R}_y$, and $\mathscr{R}_z$ do not change sign upon inversion and are therefore of gerade parity. Thus, ψ_1 and ψ_2 must be of like parity if $\mathscr{M}(12)$ is to be nonzero.

Finally, the symmetry of the elements of the two-photon absorption tensor T(12) [defined in (2.20)] and the Raman scattering tensor $\alpha'(12)$ [defined in (2.24)] is given by the products of Cartesian coordinates, xx, xy, etc. A transition from the ground state ψ_1 to a final state ψ_2 is allowed if the direct product of the representations to which ψ_1 and ψ_2 belong contains the representation to which at least one of the products xx, xy, . . . belongs. If the molecule has a center of symmetry, ψ_1 and ψ_2 must have the same parity in order for the two-photon or Raman transition to be allowed. This means that in molecules with a center of symmetry, different electronic states are reached by two-photon spectroscopy than by one-photon (electric dipole) spectroscopy and that different vibrational states are reached in Raman than in IR absorption spectroscopy.

The transitions moments **M**(12) and $\mathscr{M}(12)$ are three-dimensional vectors, and the two-photon absorption and Raman tensors T(12) and $\alpha'(12)$ are of rank 2 and are represented by 3×3 matrices. Only in the case of groups for which all coordinates x, y, and z belong to the same one-dimensional irreducible representation are all three elements of the vectors and all nine elements of the tensors independent. In all other cases the experimental determination of these vectors and tensors is simplified by symmetry rules.

Example: The Vibrations of H_2O. A set of simple basis functions suitable for the description of nuclear motions in the H_2O molecule is represented by the nine possible unit displacement vectors of one of the three nuclei at a time along one of the Cartesian coordinates at a time. Using the previously introduced system of x, y, and z axes, these will be labeled x_O, y_O, and z_O for displacements of the oxygen atom; x_A, y_A, and z_A for those of the hydrogen atom

A; and x_B, y_B, and z_B for those of the hydrogen atom B, as shown in part B of Figure I.1.

This set of basis functions can be written as a nine-dimensional column vector **f**. It generates a nine-dimensional representation Γ_f. Application of the symmetry operations yields

$$\mathbf{Ef} = E\begin{pmatrix} x_O \\ y_O \\ z_O \\ x_A \\ y_A \\ z_A \\ x_B \\ y_B \\ z_B \end{pmatrix} = \begin{pmatrix} x_O \\ y_O \\ z_O \\ x_A \\ y_A \\ z_A \\ x_B \\ y_B \\ z_B \end{pmatrix} = \begin{pmatrix} 1 & 0 & 0 & 0 & 0 & 0 & 0 & 0 & 0 \\ 0 & 1 & 0 & 0 & 0 & 0 & 0 & 0 & 0 \\ 0 & 0 & 1 & 0 & 0 & 0 & 0 & 0 & 0 \\ 0 & 0 & 0 & 1 & 0 & 0 & 0 & 0 & 0 \\ 0 & 0 & 0 & 0 & 1 & 0 & 0 & 0 & 0 \\ 0 & 0 & 0 & 0 & 0 & 1 & 0 & 0 & 0 \\ 0 & 0 & 0 & 0 & 0 & 0 & 1 & 0 & 0 \\ 0 & 0 & 0 & 0 & 0 & 0 & 0 & 1 & 0 \\ 0 & 0 & 0 & 0 & 0 & 0 & 0 & 0 & 1 \end{pmatrix} \begin{pmatrix} x_O \\ y_O \\ z_O \\ x_A \\ y_A \\ z_A \\ x_B \\ y_B \\ z_B \end{pmatrix} = \Gamma_f(E)\mathbf{f} \tag{I.24}$$

$$C_2\mathbf{f} = \begin{pmatrix} -x_O \\ -y_O \\ z_O \\ -x_B \\ -y_B \\ z_B \\ -x_A \\ -y_A \\ z_A \end{pmatrix} = \begin{pmatrix} -1 & 0 & 0 & 0 & 0 & 0 & 0 & 0 & 0 \\ 0 & -1 & 0 & 0 & 0 & 0 & 0 & 0 & 0 \\ 0 & 0 & 1 & 0 & 0 & 0 & 0 & 0 & 0 \\ 0 & 0 & 0 & 0 & 0 & 0 & -1 & 0 & 0 \\ 0 & 0 & 0 & 0 & 0 & 0 & 0 & -1 & 0 \\ 0 & 0 & 0 & 0 & 0 & 0 & 0 & 0 & 1 \\ 0 & 0 & 0 & -1 & 0 & 0 & 0 & 0 & 0 \\ 0 & 0 & 0 & 0 & -1 & 0 & 0 & 0 & 0 \\ 0 & 0 & 0 & 0 & 0 & 1 & 0 & 0 & 0 \end{pmatrix} \begin{pmatrix} x_O \\ y_O \\ z_O \\ x_A \\ y_A \\ z_A \\ x_B \\ y_B \\ z_B \end{pmatrix}$$
$$= \Gamma_f(C_2)\mathbf{f} \tag{I.25}$$

$$\sigma\mathbf{f} = \begin{pmatrix} x_O \\ -y_O \\ z_O \\ x_B \\ -y_B \\ z_B \\ x_A \\ y_A \\ z_A \end{pmatrix} = \begin{pmatrix} 1 & 0 & 0 & 0 & 0 & 0 & 0 & 0 & 0 \\ 0 & -1 & 0 & 0 & 0 & 0 & 0 & 0 & 0 \\ 0 & 0 & 1 & 0 & 0 & 0 & 0 & 0 & 0 \\ 0 & 0 & 0 & 0 & 0 & 0 & 1 & 0 & 0 \\ 0 & 0 & 0 & 0 & 0 & 0 & 0 & -1 & 0 \\ 0 & 0 & 0 & 0 & 0 & 0 & 0 & 0 & 1 \\ 0 & 0 & 0 & 1 & 0 & 0 & 0 & 0 & 0 \\ 0 & 0 & 0 & 0 & -1 & 0 & 0 & 0 & 0 \\ 0 & 0 & 0 & 0 & 0 & 1 & 0 & 0 & 0 \end{pmatrix} \begin{pmatrix} x_O \\ y_O \\ z_O \\ x_A \\ y_A \\ z_A \\ x_B \\ y_B \\ z_B \end{pmatrix}$$
$$= \Gamma_f(\sigma)\mathbf{f} \tag{I.26}$$

$$\sigma'\mathbf{f} = \begin{pmatrix} -x_O \\ y_O \\ z_O \\ -x_A \\ y_A \\ z_A \\ -x_B \\ y_B \\ z_B \end{pmatrix} = \begin{pmatrix} -1 & 0 & 0 & 0 & 0 & 0 & 0 & 0 & 0 \\ 0 & 1 & 0 & 0 & 0 & 0 & 0 & 0 & 0 \\ 0 & 0 & 1 & 0 & 0 & 0 & 0 & 0 & 0 \\ 0 & 0 & 0 & -1 & 0 & 0 & 0 & 0 & 0 \\ 0 & 0 & 0 & 0 & 1 & 0 & 0 & 0 & 0 \\ 0 & 0 & 0 & 0 & 0 & 1 & 0 & 0 & 0 \\ 0 & 0 & 0 & 0 & 0 & 0 & -1 & 0 & 0 \\ 0 & 0 & 0 & 0 & 0 & 0 & 0 & 1 & 0 \\ 0 & 0 & 0 & 0 & 0 & 0 & 0 & 0 & 1 \end{pmatrix} \begin{pmatrix} x_O \\ y_O \\ z_O \\ x_A \\ y_A \\ z_A \\ x_B \\ y_B \\ z_B \end{pmatrix}$$

$$= \Gamma_f(\sigma')\mathbf{f} \qquad (I.27)$$

Since all irreducible representations of the C_{2v} group are one-dimensional, Γ_f must be a reducible representation. In order to find out how many times each of the irreducible representations, a_1, a_2, b_1, and b_2, is contained in Γ_f, we make use of (I.4).

We start by obtaining the characters by summing the diagonal elements of each matrix: $\chi(E) = 9$, $\chi(C_2) = -1$, $\chi(\sigma) = 1$, $\chi(\sigma') = 3$. The order of the group, h, is equal to 4. Then, using the characters of the irreducible representations from Table I.2, we have

$$\begin{aligned} n_{a_1} &= \tfrac{1}{4}[9 \times 1 + (-1) \times 1 + 1 \times 1 + 3 \times 1] = 3 \\ n_{a_2} &= \tfrac{1}{4}[9 \times 1 + (-1) \times 1 + 1 \times (-1) + 3 \times (-1)] = 1 \\ n_{b_1} &= \tfrac{1}{4}[9 \times 1 + (-1) \times (-1) + 1 \times 1 + 3 \times (-1)] = 2 \\ n_{b_2} &= \tfrac{1}{4}[9 \times 1 + (-1) \times (-1) + 1 \times (-1) + 3 \times 1] = 3 \end{aligned} \qquad (I.28)$$

The decomposition is

$$\Gamma_f = 3a_1 + a_2 + 2b_1 + 3b_2 \qquad (I.29)$$

Obviously, it was not necessary to construct the large matrices $\Gamma(R)$ in (I.24)–(I.27), since only the diagonal terms were needed for the computation of the characters $\chi(R)$. The nonzero diagonal terms are easily recognized by inspection. They occur for atoms which do not change place as a result of the symmetry operation.

The basis of displacement vectors permit the description of $3N = 9$ degrees of freedom in nuclear motion. These include the three degrees of translational freedom, of symmetries $a_1(z)$, $b_1(x)$, and $b_2(y)$, and the three degrees of rotational freedom $a_2(\mathscr{R}_z)$, $b_1(\mathscr{R}_y)$, and $b_2(\mathscr{R}_x)$. This leaves $2a_1 + b_2$ for the vibrational degrees of freedom. We conclude that the H_2O molecule has two totally symmetric vibrations and one vibration of symmetry b_2. The transitions from the ground vibrational state to states with one quantum of an a_1 vibration will be polarized along the z axis; those involving a quantum of a b_2 vibration will be polarized along the y axis.

Example: The Electronic States of H_2O. A minimum basis set of one-electron functions suitable for the description of electronic motion in the valence shell of the H_2O molecule is represented by a 1s atomic orbital on each of the hydrogen atoms, $1s_A$ and $1s_B$, and by atomic orbitals on the oxygen atom, $2s_O$, $2p_{xO}$, $2p_{yO}$, and $2p_{zO}$ (Figure I.1, part C). When these are combined, six valence shell molecular orbitals will result.

The six-dimensional reducible representation Γ_f generated by this basis set might be constructed as before, but once again only the characters are needed. Since $\Gamma(E)$ is a unit matrix, clearly $\chi(E) = 6$. To obtain $\chi(C_2)$, we note that only oxygen orbitals can contribute to the diagonal in $\Gamma(C_2)$. Among these, $2s_O$ and $2p_{zO}$ are unchanged by the C_2 operation and contribute $+2$. The orbitals $2p_{xO}$ and $2p_{yO}$ both change sign when C_2 is applied, contributing -2. Mutual cancellation thus leads to the result $\chi(C_2) = 0$. Similar considerations yield $\chi(\sigma) = 2$ and $\chi(\sigma') = 4$.

Now, the use of (I.4) yields

$$\Gamma_f = 3a_1 + b_1 + 2b_2 \qquad (I.30)$$

Thus, in the minimum basis set the valence shell MOs of H_2O consist of three orbitals of a_1 symmetry, one of b_1 symmetry, and two of b_2 symmetry. These accommodate the eight valence electrons. In the ground configuration, two electrons are in a $2s_O$ lone pair orbital on oxygen (a_1), two in a $2p_{xO}$ lone pair orbital on oxygen (b_1), two in the in-phase (a_1) and two in the out-of-phase (b_2) combination of the O—H bond σ orbitals. The remaining two orbitals, a_1 and b_2, are the in-phase and out-of-phase combinations of the O—H bond σ^* orbitals and are vacant. Thus, the ground configuration of the valence shell is $(1a_1)^2(2a_1)^2(1b_2)^2(b_1)^2$. Since $a_1 \times a_1 = b_2 \times b_2 = b_1 \times b_1 = a_1$, the total symmetry of this configuration is A_1.

Symmetries of excited configurations are obtained by forming the direct product of the irreducible representations for the singly occupied orbitals. For example, $(1a_1)^2(2a_1)^2(1b_2)^2(b_1)^1(2b_2)^1$ is of symmetry A_2, since $b_1 \times b_2 = a_2$.

Example: Two-Photon Absorption. We shall further illustrate the consequences of symmetry on the example of the two-photon absorption tensor operator $\hat{T}$ in the C_{2v} point group. We need to consider transitions from the ground state A_1 to excited states of symmetries A_1, A_2, B_1, and B_2.

The four irreducible representations of C_{2v} may be defined by the symmetry properties with respect to reflection in the two symmetry planes σ_{xz} and σ'_{yz}. Using "s" for symmetric and "a" for antisymmetric with respect to σ_{xz} and σ'_{yz}, respectively, we have

$$A_1\text{: ss} \qquad A_2\text{: aa} \qquad B_1\text{: sa} \qquad B_2\text{: as} \qquad (I.31)$$

In addition, we know that one-photon transitions to excited states of A_1, B_1, and B_2 symmetry are allowed and polarized along z, x, and y, respectively. Expression (2.20) for the elements of T(12) shows that a transition from the ground state ψ_1 to a final state ψ_2 can be imagined to proceed through any one of the infinitely many electronic states of the molecule ψ_j ("intermediate virtual state") as a sequence of two electric dipole allowed steps. In order for a final state ψ_2 of A_1 symmetry to be reached by a two-photon process from a ground state ψ_1 of A_1 symmetry, the transition moments for the two steps must be both symmetric or both antisymmetric with respect to each of the symmetry planes. This is fulfilled only if the two steps are polarized along the same molecular axis, x or y or z. Thus the two-photon tensor for A_1 states can be written:

$$A_1: \quad \begin{array}{c|ccc} & x & y & z \\ \hline x & s_1 & 0 & 0 \\ y & 0 & s_2 & 0 \\ z & 0 & 0 & s_3 \end{array} \quad \text{or} \begin{pmatrix} s_1 & 0 & 0 \\ 0 & s_2 & 0 \\ 0 & 0 & s_3 \end{pmatrix} \tag{I.32}$$

In order to reach an A_2 state (aa), which cannot be reached by one-photon processes at all, one and only one of the two steps must have a transition moment which is antisymmetric with respect to symmetry plane yz. The same holds for symmetry plane xz. This is fulfilled only if the transition moment for one step is polarized along x and the other along y. We have

$$A_2: \begin{pmatrix} 0 & s_4 & 0 \\ s_5 & 0 & 0 \\ 0 & 0 & 0 \end{pmatrix} \tag{I.33}$$

For B_1 states (sa), neither or both steps must be antisymmetric with respect to σ and one and only one antisymmetric with respect to σ'. Thus one step must be polarized along x and the other along z. We have

$$B_1: \begin{pmatrix} 0 & 0 & s_6 \\ 0 & 0 & 0 \\ s_7 & 0 & 0 \end{pmatrix} \tag{I.34}$$

For B_2 states (as), similar arguments lead to the condition that one step should be polarized along y and the other along z:

$$B_2: \begin{pmatrix} 0 & 0 & 0 \\ 0 & 0 & s_8 \\ 0 & s_9 & 0 \end{pmatrix} \tag{I.35}$$

Character Tables

In the following, character tables are given for selected point groups. To the right of the characters for each irreducible representation are listed the Cartesian coordinates (x,y,z), the infinitesimal rotations ($\mathcal{R}_x, \mathcal{R}_y, \mathcal{R}_z$), and some bilinear combinations of x, y, and z, whose transformation properties correspond to that irreducible representation.

The Groups C_1, C_s, C_i

C_1	E	
A	1	$x,y,z,\mathscr{R}_x\mathscr{R}_y\mathscr{R}_z$ x^2,y^2,z^2,xy,yz,xz

$C_s = C_h$	E	σ_h		
A′	1	1	$x,y,\mathscr{R}_z$	x^2,y^2
A″	1	−1	$z,\mathscr{R}_x,\mathscr{R}_y$	yz,xz

$C_i = S_2$	E	i		
A_g	1	1	$\mathscr{R}_x,\mathscr{R}_y,\mathscr{R}_z$	x^2,y^2,z^2 xy,xz,yz
A_u	1	−1	x,y,z	

The Groups C_n $(n = 2,3, \ldots ,8)$

C_2	E	C_2		
A	1	1	$z,\mathscr{R}_z$	x^2,y^2,z^2,xy
B	1	−1	$x,y,\mathscr{R}_x,\mathscr{R}_y$	yz,xz

C_3	E	C_3	C_3^2		$\varepsilon = \exp(2\pi i/3)$
A	1	1	1	$z,\mathscr{R}_z$	$x^2 + y^2,z^2$
E	$\begin{Bmatrix}1\\1\end{Bmatrix}$	$\begin{matrix}\varepsilon\\\varepsilon^*\end{matrix}$	$\begin{matrix}\varepsilon^*\\\varepsilon\end{matrix}$	$(x,y)(\mathscr{R}_x,\mathscr{R}_y)$	$(x^2 - y^2,xy)(yz,xz)$

C_4	E	C_4	C_2	C_4^3		
A	1	1	1	1	$z,\mathscr{R}_z$	$x^2 + y^2,z^2$
B	1	−1	1	−1		$x^2 - y^2,xy$
E	$\begin{Bmatrix}1\\1\end{Bmatrix}$	$\begin{matrix}i\\-i\end{matrix}$	$\begin{matrix}-1\\-1\end{matrix}$	$\begin{matrix}-i\\i\end{matrix}$	$(x,y)(\mathscr{R}_x,\mathscr{R}_y)$	(yz,xz)

C_5	E	C_5	C_5^2	C_5^3	C_5^4		$\varepsilon = \exp(2\pi i/5)$
A	1	1	1	1	1	$z,\mathscr{R}_z$	$x^2 + y^2,z^2$
E_1	$\begin{Bmatrix}1\\1\end{Bmatrix}$	$\begin{matrix}\varepsilon\\\varepsilon^*\end{matrix}$	$\begin{matrix}\varepsilon^2\\\varepsilon^{*2}\end{matrix}$	$\begin{matrix}\varepsilon^{*2}\\\varepsilon^2\end{matrix}$	$\begin{matrix}\varepsilon^*\\\varepsilon\end{matrix}$	$(x,y)(\mathscr{R}_x,\mathscr{R}_y)$	(yz,xz)
E_2	$\begin{Bmatrix}1\\1\end{Bmatrix}$	$\begin{matrix}\varepsilon^2\\\varepsilon^{*2}\end{matrix}$	$\begin{matrix}\varepsilon^*\\\varepsilon\end{matrix}$	$\begin{matrix}\varepsilon\\\varepsilon^*\end{matrix}$	$\begin{matrix}\varepsilon^{*2}\\\varepsilon^2\end{matrix}$		$(x^2 - y^2,xy)$

C_6	E	C_6	C_3	C_2	C_3^2	C_6^5		$\varepsilon = \exp(2\pi i/6)$
A	1	1	1	1	1	1	$z,\mathscr{R}_z$	$x^2 + y^2,z^2$
B	1	−1	1	−1	1	−1		
E_1	$\begin{Bmatrix}1\\1\end{Bmatrix}$	$\begin{matrix}\varepsilon\\\varepsilon^*\end{matrix}$	$\begin{matrix}-\varepsilon^*\\-\varepsilon\end{matrix}$	$\begin{matrix}-1\\-1\end{matrix}$	$\begin{matrix}-\varepsilon\\-\varepsilon^*\end{matrix}$	$\begin{matrix}\varepsilon^*\\\varepsilon\end{matrix}$	(x,y) $(\mathscr{R}_x,\mathscr{R}_y)$	(xz,yz)
E_2	$\begin{Bmatrix}1\\1\end{Bmatrix}$	$\begin{matrix}-\varepsilon^*\\-\varepsilon\end{matrix}$	$\begin{matrix}-\varepsilon\\-\varepsilon^*\end{matrix}$	$\begin{matrix}1\\1\end{matrix}$	$\begin{matrix}-\varepsilon^*\\-\varepsilon\end{matrix}$	$\begin{matrix}-\varepsilon\\-\varepsilon^*\end{matrix}$		$(x^2 - y^2,xy)$

The Groups C_n (n = 2,3, . . . ,8) (Continued)

C_7	E	C_7	C_7^2	C_7^3	C_7^4	C_7^5	C_7^6		$\varepsilon = \exp(2\pi i/7)$
A	1	1	1	1	1	1	1	$z, \mathscr{R}_z$	$x^2 + y^2, z^2$
E_1	1	ε	ε^2	ε^3	ε^{*3}	ε^{*2}	ε^*	(x,y)	(xz,yz)
	1	ε^*	ε^{*2}	ε^{*3}	ε^3	ε^2	ε	$(\mathscr{R}_x, \mathscr{R}_y)$	
E_2	1	ε^2	ε^{*3}	ε^*	ε	ε^3	ε^{*2}		$(x^2 - y^2, xy)$
	1	ε^{*2}	ε^3	ε	ε^*	ε^{*3}	ε^2		
E_3	1	ε^3	ε^*	ε^2	ε^{*2}	ε	ε^{*3}		
	1	ε^{*3}	ε	ε^{*2}	ε^2	ε^*	ε^3		

C_8	E	C_8	C_4	C_2	C_4^3	C_8^3	C_8^5	C_8^7		$\varepsilon = \exp(2\pi i/8)$
A	1	1	1	1	1	1	1	1	$z, \mathscr{R}_z$	$x^2 + y^2, z^2$
B	1	−1	1	1	1	−1	−1	−1		
E_1	1	ε	i	−1	$-i$	$-\varepsilon^*$	$-\varepsilon$	ε^*	(x,y)	(xz,yz)
	1	ε^*	$-i$	−1	i	$-\varepsilon$	$-\varepsilon^*$	ε	$(\mathscr{R}_x, \mathscr{R}_y)$	
E_2	1	i	−1	1	−1	$-i$	i	$-i$		$(x^2 - y^2, xy)$
	1	$-i$	−1	1	−1	i	$-i$	i		
E_3	1	$-\varepsilon$	i	−1	$-i$	ε^*	ε	$-\varepsilon^*$		
	1	$-\varepsilon^*$	$-i$	−1	i	ε	ε^*	$-\varepsilon$		

The Groups D_n (n = 2,3,4,5,6)

D_2	E	$C_2(z)$	$C_2(y)$	$C_2(x)$		
A	1	1	1	1		x^2, y^2, z^2
B_1	1	1	−1	−1	$z, \mathscr{R}_z$	xy
B_2	1	−1	1	−1	$y, \mathscr{R}_y$	xz
B_3	1	−1	−1	1	$x, \mathscr{R}_x$	yz

D_3	E	$2C_3$	$3C_2$		
A_1	1	1	1		$x^2 + y^2, z^2$
A_2	1	1	−1	$z, \mathscr{R}_z$	
E	2	−1	0	$(x,y)(\mathscr{R}_x, \mathscr{R}_y)$	$(x^2 - y^2, xy)(xz, yz)$

D_4	E	$2C_4$	$C_2(=C_4^2)$	$2C_2'$	$2C_2''$		
A_1	1	1	1	1	1		$x^2 + y^2, z^2$
A_2	1	1	1	−1	−1	$z, \mathscr{R}_z$	
B_1	1	−1	1	1	−1		$x^2 - y^2$
B_2	1	−1	1	−1	1		xy
E	2	0	−2	0	0	$(x,y)(\mathscr{R}_x, \mathscr{R}_y)$	(xz,yz)

D_5	E	$2C_5$	$2C_5^2$	$5C_2$		
A_1	1	1	1	1		$x^2 + y^2, z^2$
A_2	1	1	1	−1	$z, \mathscr{R}_z$	
E_1	2	2 cos 72°	2 cos 144°	0	$(x,y)(\mathscr{R}_x, \mathscr{R}_y)$	(xz,yz)
E_2	2	2 cos 144°	2 cos 72°	0		$(x^2 - y^2, xy)$

The Groups D_n (n = 2,3,4,5,6) (Continued)

D_6	E	$2C_6$	$2C_3$	C_2	$3C_2'$	$3C_2''$		
A_1	1	1	1	1	1	1		$x^2 + y^2, z^2$
A_2	1	1	1	1	−1	−1	$z, \mathscr{R}_z$	
B_1	1	−1	1	−1	1	−1		
B_2	1	−1	1	−1	−1	1		
E_1	2	1	−1	−2	0	0	$(x,y)(\mathscr{R}_x, \mathscr{R}_y)$	(xz,yz)
E_2	2	−1	−1	2	0	0		$(x^2 - y^2, xy)$

The Groups C_{nv} (n = 2,3,4,5,6)

C_{2v}	E	C_2	$\sigma_v(xz)$	$\sigma_v'(yz)$		
A_1	1	1	1	1	z	x^2, y^2, z^2
A_2	1	1	−1	−1	$\mathscr{R}_z$	xy
B_1	1	−1	1	−1	$x, \mathscr{R}_y$	xz
B_2	1	−1	−1	1	$y, \mathscr{R}_x$	yz

C_{3v}	E	$2C_3$	$3\sigma_v$		
A_1	1	1	1	z	$x^2 + y^2, z^2$
A_2	1	1	−1	$\mathscr{R}_z$	
E	2	−1	0	$(x,y)(\mathscr{R}_x, \mathscr{R}_y)$	$(x^2 - y^2, xy)(xz, yz)$

C_{4v}	E	$2C_4$	C_2	$2\sigma_v$	$2\sigma_d$		
A_1	1	1	1	1	1	z	$x^2 + y^2, z^2$
A_2	1	1	1	−1	−1	R_z	
B_1	1	−1	1	1	−1		$x^2 - y^2$
B_2	1	−1	1	−1	1		xy
E	2	0	−2	0	0	$(x,y)(\mathscr{R}_x, \mathscr{R}_y)$	(xz,yz)

C_{5v}	E	$2C_5$	$2C_5^2$	$5\sigma_v$		
A_1	1	1	1	1	z	$x^2 + y^2, z^2$
A_2	1	1	1	−1	R_z	
E_1	2	2 cos 72°	2 cos 144°	0	$(x,y)(\mathscr{R}_x, \mathscr{R}_y)$	(xz,yz)
E_2	2	2 cos 144°	2 cos 72°	0		$(x^2 - y^2, xy)$

C_{6v}	E	$2C_6$	$2C_3$	C_2	$3\sigma_v$	$3\sigma_d$		
A_1	1	1	1	1	1	1	z	$x^2 + y^2, z^2$
A_2	1	1	1	1	−1	−1	$\mathscr{R}_z$	
B_1	1	−1	1	−1	1	−1		
B_2	1	−1	1	−1	−1	1		
E_1	2	1	−1	−2	0	0	$(x,y)(\mathscr{R}_x, \mathscr{R}_y)$	(xz,yz)
E_2	2	−1	−1	2	0	0		$(x^2 - y^2, xy)$

The Groups C_{nh} (n = 2,3,4,5,6)

C_{2h}	E	C_2	i	σ_h		
A_g	1	1	1	1	$\mathscr{R}_z$	x^2,y^2,z^2,xy
B_g	1	−1	1	−1	$\mathscr{R}_x,\mathscr{R}_y$	xz,yz
A_u	1	1	−1	−1	z	
B_u	1	−1	−1	1	x,y	

C_{3h}	E	C_3	C_3^2	σ_h	S_3	S_3^5		$\varepsilon = \exp(2\pi i/3)$
A′	1	1	1	1	1	1	$\mathscr{R}_z$	$x^2 + y^2,z^2$
E′	1	ε	ε^*	1	ε	ε^*	(x,y)	$(x^2 - y^2,xy)$
	1	ε^*	ε	1	ε^*	ε		
A″	1	1	1	−1	−1	−1	z	
E″	1	ε	ε^*	−1	$-\varepsilon$	$-\varepsilon^*$	$(\mathscr{R}_x,\mathscr{R}_y)$	(xz,yz)
	1	ε^*	ε	−1	$-\varepsilon^*$	$-\varepsilon$		

C_{4h}	E	C_4	C_2	C_4^3	i	S_4^3	σ_h	S_4		
A_g	1	1	1	1	1	1	1	1	$\mathscr{R}_z$	$x^2 + y^2,z^2$
B_g	1	−1	1	−1	1	−1	1	−1		$x^2 - y^2,xy$
E_g	1	i	−1	$-i$	1	i	−1	$-i$	$(\mathscr{R}_x,\mathscr{R}_y)$	(xz,yz)
	1	$-i$	−1	i	1	$-i$	−1	i		
A_u	1	1	1	1	−1	−1	−1	−1	z	
B_u	1	−1	1	−1	−1	1	−1	1		
E_u	1	i	−1	$-i$	−1	$-i$	1	i	(x,y)	
	1	$-i$	−1	i	−1	i	1	$-i$		

C_{5h}	E	C_5	C_5^2	C_5^3	C_5^4	σ_h	S_5	S_5^7	S_5^3	S_5^9		$\varepsilon = \exp(2\pi i/5)$
A′	1	1	1	1	1	1	1	1	1	1	$\mathscr{R}_z$	$x^2 + y^2, z^2$
E_1'	1	ε	ε^2	ε^{*2}	ε^*	1	ε	ε^2	ε^{*2}	ε^*	(x,y)	
	1	ε^*	ε^{*2}	ε^2	ε	1	ε^*	ε^{*2}	ε^2	ε		
E_2'	1	ε^2	ε^*	ε	ε^{*2}	1	ε^2	ε^*	ε	ε^{*2}		$(x^2 - y^2, xy)$
	1	ε^{*2}	ε	ε^*	ε^2	1	ε^{*2}	ε	ε^*	ε^2		
A″	1	1	1	1	1	−1	−1	−1	−1	−1	z	
E_1''	1	ε	ε^2	ε^{*2}	ε^*	−1	$-\varepsilon$	$-\varepsilon^2$	$-\varepsilon^{*2}$	$-\varepsilon^*$	$(\mathscr{R}_x, \mathscr{R}_y)$	(xz,yz)
	1	ε^*	ε^{*2}	ε^2	ε	−1	$-\varepsilon^*$	$-\varepsilon^{*2}$	$-\varepsilon^2$	$-\varepsilon$		
E_2''	1	ε^2	ε^*	ε	ε^{*2}	−1	$-\varepsilon^2$	$-\varepsilon^*$	$-\varepsilon$	$-\varepsilon^{*2}$		
	1	ε^{*2}	ε	ε^*	ε^2	−1	$-\varepsilon^{*2}$	$-\varepsilon$	$-\varepsilon^*$	$-\varepsilon^2$		

C_{6h}	E	C_6	C_3	C_2	C_3^2	C_6^5	i	S_3^5	S_6^5	σ_h	S_6	S_3		$\varepsilon = \exp(2\pi i/6)$
A_g	1	1	1	1	1	1	1	1	1	1	1	1	$\mathscr{R}_z$	$x^2 + y^2, z^2$
B_g	1	−1	1	−1	1	−1	1	−1	1	−1	1	−1		
E_{1g}	1	ε	$-\varepsilon^*$	−1	$-\varepsilon$	ε^*	1	ε	$-\varepsilon^*$	−1	$-\varepsilon$	ε^*	$(\mathscr{R}_x, \mathscr{R}_y)$	(xz,yz)
	1	ε^*	$-\varepsilon$	−1	$-\varepsilon^*$	ε	1	ε^*	$-\varepsilon$	−1	$-\varepsilon^*$	ε		
E_{2g}	1	$-\varepsilon^*$	$-\varepsilon$	1	$-\varepsilon^*$	$-\varepsilon$	1	$-\varepsilon^*$	$-\varepsilon$	1	$-\varepsilon^*$	$-\varepsilon$		$(x^2 - y^2, xy)$
	1	$-\varepsilon$	$-\varepsilon^*$	1	$-\varepsilon$	$-\varepsilon^*$	1	$-\varepsilon$	$-\varepsilon^*$	1	$-\varepsilon$	$-\varepsilon^*$		
A_u	1	1	1	1	1	1	−1	−1	−1	−1	−1	−1	z	
B_u	1	−1	1	−1	1	−1	−1	1	−1	1	−1	1		
E_{1u}	1	ε	$-\varepsilon^*$	−1	$-\varepsilon$	ε^*	−1	$-\varepsilon$	ε^*	1	ε	$-\varepsilon^*$	(x,y)	
	1	ε^*	$-\varepsilon$	−1	$-\varepsilon^*$	ε	−1	$-\varepsilon^*$	ε	1	ε^*	$-\varepsilon$		
E_{2u}	1	$-\varepsilon^*$	$-\varepsilon$	1	$-\varepsilon^*$	$-\varepsilon$	−1	ε^*	ε	−1	ε^*	ε		
	1	$-\varepsilon$	$-\varepsilon^*$	1	$-\varepsilon$	$-\varepsilon^*$	−1	ε	ε^*	−1	ε	ε^*		

The Groups D_{nh} ($n = 2,3,4,5,6$)

D_{2h}	E	$C_2(z)$	$C_2(y)$	$C_2(x)$	i	$\sigma(xy)$	$\sigma(xz)$	$\sigma(yz)$		
A_g	1	1	1	1	1	1	1	1		x^2,y^2,z^2
B_{1g}	1	1	−1	−1	1	1	−1	−1	$\mathscr{R}_z$	xy
B_{2g}	1	−1	1	−1	1	−1	1	−1	$\mathscr{R}_y$	xz
B_{3g}	1	−1	−1	1	1	−1	−1	1	$\mathscr{R}_x$	yz
A_u	1	1	1	1	−1	−1	−1	−1		
B_{1u}	1	1	−1	−1	−1	−1	1	1	z	
B_{2u}	1	−1	1	−1	−1	1	−1	1	y	
B_{3u}	1	−1	−1	1	−1	1	1	−1	x	

D_{3h}	E	$2C_3$	$3C_2$	σ_h	$2S_3$	$3\sigma_v$		
A'_1	1	1	1	1	1	1		$x^2 + y^2,z^2$
A'_2	1	1	−1	1	1	−1	$\mathscr{R}_z$	
E'	2	−1	0	2	−1	0	(x,y)	$(x^2 - y^2,xy)$
A''_1	1	1	1	−1	−1	−1		
A''_2	1	1	−1	−1	−1	1	z	
E''	2	−1	0	−2	1	0	$(\mathscr{R}_x,\mathscr{R}_y)$	(xz,yz)

D_{4h}	E	$2C_4$	C_2	$2C'_2$	$2C''_2$	i	$2S_4$	σ_h	$2\sigma_v$	$2\sigma_d$		
A_{1g}	1	1	1	1	1	1	1	1	1	1		$x^2 + y^2,z^2$
A_{2g}	1	1	1	−1	−1	1	1	1	−1	−1	$\mathscr{R}_z$	
B_{1g}	1	−1	1	1	−1	1	−1	1	1	−1		$x^2 - y^2$
B_{2g}	1	−1	1	−1	1	1	−1	1	−1	1		xy
E_g	2	0	−2	0	0	2	0	−2	0	0	$(\mathscr{R}_x,\mathscr{R}_y)$	(xz,yz)
A_{1u}	1	1	1	1	1	−1	−1	−1	−1	−1		
A_{2u}	1	1	1	−1	−1	−1	−1	−1	1	1	z	
B_{1u}	1	−1	1	1	−1	−1	1	−1	−1	1		
B_{2u}	1	−1	1	−1	1	−1	1	−1	1	−1		
E_u	2	0	−2	0	0	−2	0	2	0	0	(x,y)	

D_{5h}	E	$2C_5$	$2C_5^2$	$5C_2$	σ_h	$2S_5$	$2S_5^3$	$5\sigma_v$		
A_1'	1	1	1	1	1	1	1	1		$x^2 + y^2, z^2$
A_2'	1	1	1	−1	1	1	1	−1	$\mathscr{R}_z$	
E_1'	2	2 cos 72°	2 cos 144°	0	2	2 cos 72°	2 cos 144°	0	(x,y)	
E_2'	2	2 cos 144°	2 cos 72°	0	2	2 cos 144°	2 cos 72°	0		$(x^2 - y^2, xy)$
A_1''	1	1	1	1	−1	−1	−1	−1		
A_2''	1	1	1	−1	−1	−1	−1	1	z	
E_1''	2	2 cos 72°	2 cos 144°	0	−2	−2 cos 72°	−2 cos 144°	0	$(\mathscr{R}_x, \mathscr{R}_y)$	(xz,yz)
E_2''	2	2 cos 144°	2 cos 72°	0	−2	−2 cos 144°	−2 cos 72°	0		

D_{6h}	E	$2C_6$	$2C_3$	C_2	$3C_2'$	$3C_2''$	i	$2S_3$	$2S_6$	σ_h	$3\sigma_d$	$3\sigma_v$		
A_{1g}	1	1	1	1	1	1	1	1	1	1	1	1		$x^2 + y^2, z^2$
A_{2g}	1	1	1	1	−1	−1	1	1	1	1	−1	−1	$\mathscr{R}_z$	
B_{1g}	1	−1	1	−1	1	−1	1	−1	1	−1	1	−1		
B_{2g}	1	−1	1	−1	−1	1	1	−1	1	−1	−1	1		
E_{1g}	2	1	−1	−2	0	0	2	1	−1	−2	0	0	$(\mathscr{R}_x, \mathscr{R}_y)$	(xz,yz)
E_{2g}	2	−1	−1	2	0	0	2	−1	−1	2	0	0		$(x^2 - y^2, xy)$
A_{1u}	1	1	1	1	1	1	−1	−1	−1	−1	−1	−1		
A_{2u}	1	1	1	1	−1	−1	−1	−1	−1	−1	1	1	z	
B_{1u}	1	−1	1	−1	1	−1	−1	1	−1	1	−1	1		
B_{2u}	1	−1	1	−1	−1	1	−1	1	−1	1	1	−1		
E_{1u}	2	1	−1	−2	0	0	−2	−1	1	2	0	0	(x,y)	
E_{2u}	2	−1	−1	2	0	0	−2	1	1	−2	0	0		

The Groups D_{nd} (n = 2,3,4,5,6)

$D_{2d} = V_d$	E	$2S_4$	C_2	$2C'_2$	$2\sigma_d$		
A_1	1	1	1	1	1		$x^2 + y^2, z^2$
A_2	1	1	1	−1	−1	$\mathcal{R}_z$	
B_1	1	−1	1	1	−1		$x^2 - y^2$
B_2	1	−1	1	−1	1	z	xy
E	2	0	−2	0	0	(x,y) $(\mathcal{R}_x, \mathcal{R}_y)$	(xz,yz)

D_{3d}	E	$2C_3$	$3C_2$	*i*	$2S_6$	$3\sigma_d$		
A_{1g}	1	1	1	1	1	1		$x^2 + y^2, z^2$
A_{2g}	1	1	−1	1	1	−1	$\mathcal{R}_z$	
E_g	2	−1	0	2	−1	0	$(\mathcal{R}_x, \mathcal{R}_y)$	$(x^2 - y^2, xy)$ (xz,yz)
A_{1u}	1	1	1	−1	−1	−1		
A_{2u}	1	1	−1	−1	−1	1	z	
E_u	2	−1	0	−2	1	0	(x,y)	

D_{4d}	E	$2S_8$	$2C_4$	$2S_8^3$	C_2	$4C'_2$	$4\sigma_d$		
A_1	1	1	1	1	1	1	1		$x^2 + y^2, z^2$
A_2	1	1	1	1	1	−1	−1	$\mathcal{R}_z$	
B_1	1	−1	1	−1	1	1	−1		
B_2	1	−1	1	−1	1	−1	1	z	
E_1	2	$\sqrt{2}$	0	$-\sqrt{2}$	−2	0	0	(x,y)	
E_2	2	0	−2	0	2	0	0		$(x^2 - y^2, xy)$
E_3	2	$-\sqrt{2}$	0	$\sqrt{2}$	−2	0	0	$(\mathcal{R}_x, \mathcal{R}_y)$	(xz,yz)

D_{5d}	E	$2C_5$	$2C_5^2$	$5C_2$	i	$2S_{10}^3$	$2S_{10}$	$5\sigma_d$		
A_{1g}	1	1	1	1	1	1	1	1		$x^2 + y^2, z^2$
A_{2g}	1	1	1	−1	1	1	1	−1	$\mathscr{R}_z$	
E_{1g}	2	2 cos 72°	2 cos 144°	0	2	2 cos 72°	2 cos 144°	0	$(\mathscr{R}_x, \mathscr{R}_y)$	(xz,yz)
E_{2g}	2	2 cos 144°	2 cos 72°	0	2	2 cos 144°	2 cos 72°	0		$(x^2 - y^2, xy)$
A_{1u}	1	1	1	1	−1	−1	−1	−1		
A_{2u}	1	1	1	−1	−1	−1	−1	1	z	
E_{1u}	2	2 cos 72°	2 cos 144°	0	−2	−2 cos 72°	−2 cos 144°	0	(x,y)	
E_{2u}	2	2 cos 144°	2 cos 72°	0	−2	−2 cos 144°	−2 cos 72°	0		

D_{6d}	E	$2S_{12}$	$2C_6$	$2S_4$	$2C_3$	$2S_{12}^5$	C_2	$6C_2'$	$6\sigma_d$		
A_1	1	1	1	1	1	1	1	1	1		$x^2 + y^2, z^2$
A_2	1	1	1	1	1	1	1	−1	−1	$\mathscr{R}_z$	
B_1	1	−1	1	−1	1	−1	1	1	−1		
B_2	1	−1	1	−1	1	−1	1	−1	1	z	
E_1	2	$\sqrt{3}$	1	0	−1	$-\sqrt{3}$	−2	0	0	(x,y)	
E_2	2	1	−1	−2	−1	1	2	0	0		$(x^2 - y^2, xy)$
E_3	2	0	−2	0	2	0	−2	0	0		
E_4	2	−1	−1	2	−1	−1	2	0	0		
E_5	2	$-\sqrt{3}$	1	0	−1	$\sqrt{3}$	−2	0	0	$(\mathscr{R}_x, \mathscr{R}_y)$	(xz,yz)

The Groups S_n (n = 4,6,8)

S_4	E	S_4	C_2	S_4^3		
A	1	1	1	1	$\mathscr{R}_z$	x^2+y^2, z^2
B	1	−1	1	−1	z	x^2-y^2, xy
E	$\begin{cases}1 \\ 1\end{cases}$	$\begin{matrix} i \\ -i\end{matrix}$	$\begin{matrix}-1 \\ -1\end{matrix}$	$\left.\begin{matrix}-i \\ i\end{matrix}\right\}$	$(x,y)(\mathscr{R}_x,\mathscr{R}_y)$	(xz,yz)

S_6	E	C_3	C_3^2	i	S_6^5	S_6		$\varepsilon = \exp(2\pi i/3)$
A_g	1	1	1	1	1	1	$\mathscr{R}_z$	x^2+y^2, z^2
E_g	$\begin{cases}1 \\ 1\end{cases}$	$\begin{matrix}\varepsilon \\ \varepsilon^*\end{matrix}$	$\begin{matrix}\varepsilon^* \\ \varepsilon\end{matrix}$	$\begin{matrix}1 \\ 1\end{matrix}$	$\begin{matrix}\varepsilon \\ \varepsilon^*\end{matrix}$	$\left.\begin{matrix}\varepsilon^* \\ \varepsilon\end{matrix}\right\}$	$(\mathscr{R}_x,\mathscr{R}_y)$	(x^2-y^2, xy) (xz,yz)
A_u	1	1	1	−1	−1	−1	z	
E_u	$\begin{cases}1 \\ 1\end{cases}$	$\begin{matrix}\varepsilon \\ \varepsilon^*\end{matrix}$	$\begin{matrix}\varepsilon^* \\ \varepsilon\end{matrix}$	$\begin{matrix}-1 \\ -1\end{matrix}$	$\begin{matrix}-\varepsilon \\ -\varepsilon^*\end{matrix}$	$\left.\begin{matrix}-\varepsilon^* \\ -\varepsilon\end{matrix}\right\}$	(x,y)	

S_8	E	S_8	C_4	S_8^3	C_2	S_8^5	C_4^3	S_8^7		$\varepsilon = \exp(2\pi i/8)$
A	1	1	1	1	1	1	1	1	$\mathscr{R}_z$	x^2+y^2, z^2
B	1	−1	1	−1	1	−1	1	−1	z	
E_1	$\begin{cases}1 \\ 1\end{cases}$	$\begin{matrix}\varepsilon \\ \varepsilon^*\end{matrix}$	$\begin{matrix}i \\ -i\end{matrix}$	$\begin{matrix}-\varepsilon^* \\ -\varepsilon\end{matrix}$	$\begin{matrix}-1 \\ -1\end{matrix}$	$\begin{matrix}-\varepsilon \\ -\varepsilon^*\end{matrix}$	$\begin{matrix}-i \\ i\end{matrix}$	$\left.\begin{matrix}\varepsilon^* \\ \varepsilon\end{matrix}\right\}$	(x,y) $(\mathscr{R}_x,\mathscr{R}_y)$	
E_2	$\begin{cases}1 \\ 1\end{cases}$	$\begin{matrix}i \\ -i\end{matrix}$	$\begin{matrix}-1 \\ -1\end{matrix}$	$\begin{matrix}-i \\ i\end{matrix}$	$\begin{matrix}1 \\ 1\end{matrix}$	$\begin{matrix}i \\ -i\end{matrix}$	$\begin{matrix}-1 \\ -1\end{matrix}$	$\left.\begin{matrix}-i \\ i\end{matrix}\right\}$		(x^2-y^2, xy)
E_3	$\begin{cases}1 \\ 1\end{cases}$	$\begin{matrix}-\varepsilon^* \\ -\varepsilon\end{matrix}$	$\begin{matrix}-i \\ i\end{matrix}$	$\begin{matrix}\varepsilon \\ \varepsilon^*\end{matrix}$	$\begin{matrix}-1 \\ -1\end{matrix}$	$\begin{matrix}\varepsilon^* \\ \varepsilon\end{matrix}$	$\begin{matrix}i \\ -i\end{matrix}$	$\left.\begin{matrix}-\varepsilon \\ -\varepsilon^*\end{matrix}\right\}$		(xz,yz)

The Cubic Groups

T	E	$4C_3$	$4C_3^2$	$3C_2$		$\varepsilon = \exp(2\pi i/3)$
A	1	1	1	1		$x^2 + y^2 + z^2$
E	$\begin{cases}1\\1\end{cases}$	ε ε^*	ε^* ε	$\begin{rcases}1\\1\end{rcases}$		$(x^2 - y^2, 2z^2 - x^2 - y^2)$
T	3	0	0	−1	(x,y,z) $(\mathscr{R}_x,\mathscr{R}_y,\mathscr{R}_z)$	(xy,xz,yz)

T_d	E	$8C_3$	$3C_2$	$6S_4$	$6\sigma_d$		
A_1	1	1	1	1	1		$x^2 + y^2 + z^2$
A_2	1	1	1	−1	−1		
E	2	−1	2	0	0		$(2z^2 - x^2 - y^2, x^2 - y^2)$
T_1	3	0	−1	1	−1	$(\mathscr{R}_x,\mathscr{R}_y,\mathscr{R}_z)$	
T_2	3	0	−1	−1	1	(x,y,z)	(xy,xz,yz)

T_h	E	$4C_3$	$4C_3^2$	$3C_2$	i	$4S_6$	$4S_6^2$	$3\sigma_d$		$\varepsilon = \exp(2\pi i/3)$
A_g	1	1	1	1	1	1	1	1		$x^2 + y^2 + z^2$
E_g	$\begin{cases}1\\1\end{cases}$	ε ε^*	ε^* ε	1 1	1 1	ε ε^*	ε^* ε	$\begin{rcases}1\\1\end{rcases}$		$(2z^2 - x^2 - y^2, x^2 - y^2)$
T_g	3	0	0	−1	3	0	0	−1	$(\mathscr{R}_x,\mathscr{R}_y,\mathscr{R}_z)$	(xy,yz,xz)
A_u	1	1	1	1	−1	−1	−1	−1		
E_u	$\begin{cases}1\\1\end{cases}$	ε ε^*	ε^* ε	1 1	−1 −1	$-\varepsilon$ $-\varepsilon^*$	$-\varepsilon^*$ $-\varepsilon$	$\begin{rcases}-1\\-1\end{rcases}$		
T_u	3	0	0	−1	−3	0	0	1	(x,y,z)	

(Continued)

The Cubic Groups (Continued)

O	E	$8C_3$	$3C_2$	$6C_4$	$6C_2'$		
A_1	1	1	1	1	1		$x^2 + y^2 + z^2$
A_2	1	1	1	−1	−1		
E	2	−1	2	0	0		$(2z^2 - x^2 - y^2, x^2 - y^2)$
T_1	3	0	−1	1	−1	(x,y,z) $(\mathcal{R}_x,\mathcal{R}_y,\mathcal{R}_z)$	
T_2	3	0	−1	−1	1		(xy,xz,yz)

O_h	E	$8C_3$	$6C_2$	$6C_4$	$3C_2$ $(= C_4^2)$	i	$6S_4$	$8S_6$	$3\sigma_h$	$6\sigma_d$		
A_{1g}	1	1	1	1	1	1	1	1	1	1		$x^2 + y^2 + z^2$
A_{2g}	1	1	−1	−1	1	1	−1	1	1	−1		
E_g	2	−1	0	0	2	2	0	−1	2	0		$(2z^2 - x^2 - y^2, x^2 - y^2)$
T_{1g}	3	0	−1	1	−1	3	1	0	−1	−1	$(\mathcal{R}_x,\mathcal{R}_y,\mathcal{R}_z)$	
T_{2g}	3	0	1	−1	−1	3	−1	0	−1	1		(xz,yz,xy)
A_{1u}	1	1	1	1	1	−1	−1	−1	−1	−1		
A_{2u}	1	1	−1	−1	1	−1	1	−1	−1	1		
E_u	2	−1	0	0	2	−2	0	1	−2	0		
T_{1u}	3	0	−1	1	−1	−3	−1	0	1	1	(x,y,z)	
T_{2u}	3	0	1	−1	−1	−3	1	0	1	−1		

The Groups I, I_h

I	E		$12C_5^2$	$20C_3$	$15C_2$		$\eta^{\pm} = \frac{1}{2}(1 \pm 5^{1/2})$
A	1	1	1	1	1		$x^2 + y^2 + z^2$
T_1	3	η^+	η^-	0	-1	(x,y,z) $(\mathscr{R}_x,\mathscr{R}_y,\mathscr{R}_z)$	
T_2	3	η^-	η^+	0	-1		
G	4	-1	-1	1	0		
H	5	0	0	-1	1		$(2z^2 - x^2 - y^2, x^2 - y^2, xy,yz,zx)$

I_h	E	$12C_5$	$12C_5^2$	$20C_3$	$15C_2$	i	$12S_{10}$	$12S_{10}^3$	$20S_6$	15σ		$\eta^{\pm} = \frac{1}{2}(1 \pm 5^{1/2})$
A_g	1	1	1	1	1	1	1	1	1	1		$x^2 + y^2 + z^2$
T_{1g}	3	η^+	η^-	0	-1	3	η^-	η^+	0	-1	$(\mathscr{R}_x,\mathscr{R}_y,\mathscr{R}_z)$	
T_{2g}	3	η^-	η^+	0	-1	3	η^+	η^-	0	-1		
G_g	4	-1	-1	1	0	4	-1	-1	1	0		
H_g	5	0	0	-1	1	5	0	0	-1	1		$(2z^2 - x^2 - y^2, x^2 - y^2, xy,yz,zx)$
A_u	1	1	1	1	1	-1	-1	-1	-1	-1		
T_{1u}	3	η^+	η^-	0	-1	-3	$-\eta^-$	$-\eta^+$	0	1	(x,y,z)	
T_{2u}	3	η^-	η^+	0	-1	-3	$-\eta^+$	$-\eta^-$	0	1		
G_u	4	-1	-1	1	0	-4	1	1	-1	0		
H_u	5	0	0	-1	1	-5	0	0	1	1		

The Groups $C_{\infty v}$ and $D_{\infty h}$

$C_{\infty v}$	E	$2C_{\infty}^{\phi}$	⋯	$\infty\sigma_v$		
$A_1 \equiv \Sigma^+$	1	1	⋯	1	z	$x^2 + y^2, z^2$
$A_2 \equiv \Sigma^-$	1	1	⋯	−1	$\mathscr{R}_z$	
$E_1 \equiv \Pi$	2	$2 \cos \phi$	⋯	0	$(x,y)(\mathscr{R}_x,\mathscr{R}_y)$	(xz,yz)
$E_2 \equiv \Delta$	2	$2 \cos 2\phi$	⋯	0		$(x^2 - y^2, xy)$
$E_3 \equiv \Phi$	2	$2 \cos 3\phi$	⋯	0		
⋯	⋯	⋯	⋯	⋯		

$D_{\infty h}$	E	$2C_{\infty}^{\phi}$	⋯	$\infty\sigma_i$	i	$2S_{\infty}^{\phi}$	⋯	∞C_2		
Σ_g^+	1	1	⋯	1	1	1	⋯	1		$x^2 + y^2, z^2$
Σ_g^-	1	1	⋯	−1	1	1	⋯	−1	$\mathscr{R}_z$	
Π_g	2	$2 \cos \phi$	⋯	0	2	$-2 \cos \phi$	⋯	0	$(\mathscr{R}_x,\mathscr{R}_y)$	(xz,yz)
Δ_g	2	$2 \cos 2\phi$	⋯	0	2	$2 \cos 2\phi$	⋯	0		$(x^2 - y^2, xy)$
⋯	⋯	⋯	⋯	⋯	⋯	⋯	⋯	⋯		
Σ_u^+	1	1	⋯	1	−1	−1	⋯	−1	z	
Σ_u^-	1	1	⋯	−1	−1	−1	⋯	1		
Π_u	2	$2 \cos \phi$	⋯	0	−2	$2 \cos \phi$	⋯	0	(x,y)	
Δ_u	2	$2 \cos 2\phi$	⋯	0	−2	$-2 \cos 2\phi$	⋯	0		
⋯	⋯	⋯	⋯	⋯	⋯	⋯	⋯	⋯		

Patterns in the Vector and Tensor Elements of the Operators $\hat{M}$, $\hat{\mathcal{M}}$, and $\hat{T}$ or $\hat{\alpha}'$

These are listed for transitions from a totally symmetric ground state to states of various symmetries.

The Groups C_1, C_s, C_i

C_1	$\hat{M}$	$\hat{\mathcal{M}}$	$\hat{T}$ or $\hat{\alpha}'$
A	$\begin{pmatrix} u_1 \\ u_2 \\ u_3 \end{pmatrix}$	$\begin{pmatrix} v_1 \\ v_2 \\ v_3 \end{pmatrix}$	$\begin{pmatrix} s_1 & s_2 & s_3 \\ s_4 & s_5 & s_6 \\ s_7 & s_8 & s_9 \end{pmatrix}$
C_s	$\hat{M}$	$\hat{\mathcal{M}}$	$\hat{T}$ or $\hat{\alpha}'$
A′	$\begin{pmatrix} u_1 \\ u_2 \\ 0 \end{pmatrix}$	$\begin{pmatrix} 0 \\ 0 \\ v_1 \end{pmatrix}$	$\begin{pmatrix} s_1 & s_4 & 0 \\ s_5 & s_2 & 0 \\ 0 & 0 & s_3 \end{pmatrix}$
A″	$\begin{pmatrix} 0 \\ 0 \\ u_3 \end{pmatrix}$	$\begin{pmatrix} v_2 \\ v_3 \\ 0 \end{pmatrix}$	$\begin{pmatrix} 0 & 0 & s_6 \\ 0 & 0 & s_7 \\ s_8 & s_9 & 0 \end{pmatrix}$
C_i	$\hat{M}$	$\hat{\mathcal{M}}$	$\hat{T}$ or $\hat{\alpha}'$
A_g		$\begin{pmatrix} v_1 \\ v_2 \\ v_3 \end{pmatrix}$	$\begin{pmatrix} s_1 & s_2 & s_3 \\ s_4 & s_5 & s_6 \\ s_7 & s_8 & s_9 \end{pmatrix}$
A_u	$\begin{pmatrix} u_1 \\ u_2 \\ u_3 \end{pmatrix}$		

The Groups C_n

C_2	$\hat{M}$	$\hat{\mathcal{M}}$	$\hat{T}$ or $\hat{\alpha}'$
A	$\begin{pmatrix} 0 \\ 0 \\ u_1 \end{pmatrix}$	$\begin{pmatrix} 0 \\ 0 \\ v_1 \end{pmatrix}$	$\begin{pmatrix} s_1 & s_4 & 0 \\ s_5 & s_2 & 0 \\ 0 & 0 & s_3 \end{pmatrix}$
B	$\begin{pmatrix} u_2 \\ u_3 \\ 0 \end{pmatrix}$	$\begin{pmatrix} v_2 \\ v_3 \\ 0 \end{pmatrix}$	$\begin{pmatrix} 0 & 0 & s_6 \\ 0 & 0 & s_7 \\ s_8 & s_9 & 0 \end{pmatrix}$

The Groups C_n (Continued)

C_3	$\hat{\mathbf{M}}$	$\hat{\mathcal{M}}$	$\hat{\mathrm{T}}$ or $\hat{\alpha}'$
A	$\begin{pmatrix} 0 \\ 0 \\ u_1 \end{pmatrix}$	$\begin{pmatrix} 0 \\ 0 \\ v_1 \end{pmatrix}$	$\begin{pmatrix} s_1 & s_3 & 0 \\ -s_3 & s_1 & 0 \\ 0 & 0 & s_2 \end{pmatrix}$
E	$\begin{pmatrix} u_2 \\ iu_2 \\ 0 \end{pmatrix}, \begin{pmatrix} u_2^* \\ -iu_2^* \\ 0 \end{pmatrix}$	$\begin{pmatrix} v_2 \\ iv_2 \\ 0 \end{pmatrix}, \begin{pmatrix} v_2^* \\ -iv_2^* \\ 0 \end{pmatrix}$	$\begin{pmatrix} s_4 & is_4 & s_5 \\ is_4 & -s_4 & -is_5 \\ s_6 & -is_6 & 0 \end{pmatrix}, \begin{pmatrix} s_4^* & -is_4^* & s_5^* \\ -is_4^* & -s_4^* & is_5^* \\ s_6^* & is_6^* & 0 \end{pmatrix}$
C_4	$\hat{\mathbf{M}}$	$\hat{\mathcal{M}}$	$\hat{\mathrm{T}}$ or α'
A	$\begin{pmatrix} 0 \\ 0 \\ u_1 \end{pmatrix}$	$\begin{pmatrix} 0 \\ 0 \\ v_1 \end{pmatrix}$	$\begin{pmatrix} s_1 & s_3 & 0 \\ -s_3 & s_1 & 0 \\ 0 & 0 & s_2 \end{pmatrix}$
B			$\begin{pmatrix} s_4 & s_5 & 0 \\ s_5 & -s_4 & 0 \\ 0 & 0 & 0 \end{pmatrix}$
E	$\begin{pmatrix} u_2 \\ iu_2 \\ 0 \end{pmatrix}, \begin{pmatrix} u_2^* \\ -iu_2^* \\ 0 \end{pmatrix}$	$\begin{pmatrix} v_2 \\ iv_2 \\ 0 \end{pmatrix}, \begin{pmatrix} v_2^* \\ -iv_2^* \\ 0 \end{pmatrix}$	$\begin{pmatrix} 0 & 0 & s_6 \\ 0 & 0 & -is_6 \\ s_7 & -is_7 & 0 \end{pmatrix}, \begin{pmatrix} 0 & 0 & s_6^* \\ 0 & 0 & is_6^* \\ s_7 & is_7^* & 0 \end{pmatrix}$
C_6	$\hat{\mathbf{M}}$	$\hat{\mathcal{M}}$	$\hat{\mathrm{T}}$ or α'
A_1	$\begin{pmatrix} 0 \\ 0 \\ u_1 \end{pmatrix}$	$\begin{pmatrix} 0 \\ 0 \\ v_1 \end{pmatrix}$	$\begin{pmatrix} s_1 & s_3 & 0 \\ -s_3 & s_1 & 0 \\ 0 & 0 & s_2 \end{pmatrix}$
E_1	$\begin{pmatrix} u_2 \\ iu_2 \\ 0 \end{pmatrix}, \begin{pmatrix} u_2^* \\ -iu_2^* \\ 0 \end{pmatrix}$	$\begin{pmatrix} v_2 \\ iv_2 \\ 0 \end{pmatrix}, \begin{pmatrix} v_2^* \\ -iv_2^* \\ 0 \end{pmatrix}$	$\begin{pmatrix} 0 & 0 & s_4 \\ 0 & 0 & -is_4 \\ s_5 & -is_5 & 0 \end{pmatrix}, \begin{pmatrix} 0 & 0 & s_4^* \\ 0 & 0 & is_4^* \\ s_5^* & is_5^* & 0 \end{pmatrix}$
E_2			$\begin{pmatrix} s_6 & -is_6 & 0 \\ -is_6 & -s_6 & 0 \\ 0 & 0 & 0 \end{pmatrix}, \begin{pmatrix} s_6^* & is_6^* & 0 \\ is_6^* & -s_6^* & 0 \\ 0 & 0 & 0 \end{pmatrix}$

The Groups D_n

D_2	$\hat{\mathbf{M}}$	$\hat{\mathcal{M}}$	$\hat{\mathsf{T}}$ or $\hat{\alpha}'$
A			$\begin{pmatrix} s_1 & 0 & 0 \\ 0 & s_2 & 0 \\ 0 & 0 & s_3 \end{pmatrix}$
B_1	$\begin{pmatrix} 0 \\ 0 \\ u_1 \end{pmatrix}$	$\begin{pmatrix} 0 \\ 0 \\ v_1 \end{pmatrix}$	$\begin{pmatrix} 0 & s_4 & 0 \\ s_5 & 0 & 0 \\ 0 & 0 & 0 \end{pmatrix}$
B_2	$\begin{pmatrix} 0 \\ u_2 \\ 0 \end{pmatrix}$	$\begin{pmatrix} 0 \\ v_2 \\ 0 \end{pmatrix}$	$\begin{pmatrix} 0 & 0 & s_6 \\ 0 & 0 & 0 \\ s_7 & 0 & 0 \end{pmatrix}$
B_3	$\begin{pmatrix} u_3 \\ 0 \\ 0 \end{pmatrix}$	$\begin{pmatrix} v_3 \\ 0 \\ 0 \end{pmatrix}$	$\begin{pmatrix} 0 & 0 & 0 \\ 0 & 0 & s_8 \\ 0 & s_9 & 0 \end{pmatrix}$

D_3	$\hat{\mathbf{M}}$	$\hat{\mathcal{M}}$	$\hat{\mathsf{T}}$ or $\hat{\alpha}'$
A_1			$\begin{pmatrix} s_1 & 0 & 0 \\ 0 & s_1 & 0 \\ 0 & 0 & s_2 \end{pmatrix}$
A_2	$\begin{pmatrix} 0 \\ 0 \\ u_1 \end{pmatrix}$	$\begin{pmatrix} 0 \\ 0 \\ v_1 \end{pmatrix}$	$\begin{pmatrix} 0 & s_3 & 0 \\ -s_3 & 0 & 0 \\ 0 & 0 & 0 \end{pmatrix}$
E	$\begin{pmatrix} u_2 \\ iu_2 \\ 0 \end{pmatrix}, \begin{pmatrix} u_2^* \\ -iu_2^* \\ 0 \end{pmatrix}$	$\begin{pmatrix} v_2 \\ iv_2 \\ 0 \end{pmatrix}, \begin{pmatrix} v_2^* \\ -iv_2^* \\ 0 \end{pmatrix}$	$\begin{pmatrix} s_4 & is_4 & s_5 \\ is_4 & -s_4 & -is_5 \\ s_6 & -is_6 & 0 \end{pmatrix}, \begin{pmatrix} s_4^* & -is_4^* & s_5^* \\ -is_4^* & -s_4^* & is_5^* \\ s_6^* & is_6^* & 0 \end{pmatrix}$

The Groups D_n (Continued)

D_4	$\hat{\mathbf{M}}$	$\hat{\mathcal{M}}$	$\hat{\mathrm{T}}$ or $\hat{\alpha}'$
A_1	$\begin{pmatrix} 0 \\ 0 \\ u_1 \end{pmatrix}$	$\begin{pmatrix} 0 \\ 0 \\ v_1 \end{pmatrix}$	$\begin{pmatrix} s_1 & 0 & 0 \\ 0 & s_1 & 0 \\ 0 & 0 & s_2 \end{pmatrix}$
A_2			$\begin{pmatrix} 0 & s_3 & 0 \\ -s_3 & 0 & 0 \\ 0 & 0 & 0 \end{pmatrix}$
B_1			$\begin{pmatrix} s_4 & 0 & 0 \\ 0 & -s_4 & 0 \\ 0 & 0 & 0 \end{pmatrix}$
B_2			$\begin{pmatrix} 0 & s_5 & 0 \\ s_5 & 0 & 0 \\ 0 & 0 & 0 \end{pmatrix}$
E	$\begin{pmatrix} u_2 \\ iu_2 \\ 0 \end{pmatrix}, \begin{pmatrix} u_2^* \\ -iu_2^* \\ 0 \end{pmatrix}$	$\begin{pmatrix} v_2 \\ iv_2 \\ 0 \end{pmatrix}, \begin{pmatrix} v_2^* \\ -iv_2^* \\ 0 \end{pmatrix}$	$\begin{pmatrix} 0 & 0 & s_6 \\ 0 & 0 & -is_6 \\ s_7 & -is_7 & 0 \end{pmatrix}, \begin{pmatrix} 0 & 0 & s_6^* \\ 0 & 0 & is_6^* \\ s_7^* & is_7^* & 0 \end{pmatrix}$

D_6	$\hat{\mathbf{M}}$	$\hat{\mathcal{M}}$	$\hat{\mathrm{T}}$ or $\hat{\alpha}'$
A_1			$\begin{pmatrix} s_1 & 0 & 0 \\ 0 & s_1 & 0 \\ 0 & 0 & s_2 \end{pmatrix}$
A_2	$\begin{pmatrix} 0 \\ 0 \\ u_1 \end{pmatrix}$	$\begin{pmatrix} 0 \\ 0 \\ v_1 \end{pmatrix}$	$\begin{pmatrix} 0 & s_3 & 0 \\ -s_3 & 0 & 0 \\ 0 & 0 & 0 \end{pmatrix}$
E_1	$\begin{pmatrix} u_2 \\ iu_2 \\ 0 \end{pmatrix}, \begin{pmatrix} u_2^* \\ -iu_2^* \\ 0 \end{pmatrix}$	$\begin{pmatrix} v_2 \\ iv_2 \\ 0 \end{pmatrix}, \begin{pmatrix} v_2^* \\ -iv_2^* \\ 0 \end{pmatrix}$	$\begin{pmatrix} 0 & 0 & s_4 \\ 0 & 0 & -is_4 \\ s_5 & -is_5 & 0 \end{pmatrix}, \begin{pmatrix} 0 & 0 & s_4^* \\ 0 & 0 & is_4^* \\ s_5^* & is_5^* & 0 \end{pmatrix}$
E_2			$\begin{pmatrix} s_6 & -is_6 & 0 \\ -is_6 & -s_6 & 0 \\ 0 & 0 & 0 \end{pmatrix}, \begin{pmatrix} s_6^* & is_6^* & 0 \\ is_6^* & -s_6^* & 0 \\ 0 & 0 & 0 \end{pmatrix}$

The Groups C_{nv}

C_{2v}	$\hat{\mathbf{M}}$	$\hat{\mathcal{M}}$	$\hat{T}$ or $\hat{\alpha}'$
A_1	$\begin{pmatrix} 0 \\ 0 \\ u_1 \end{pmatrix}$		$\begin{pmatrix} s_1 & 0 & 0 \\ 0 & s_2 & 0 \\ 0 & 0 & s_3 \end{pmatrix}$
A_2		$\begin{pmatrix} 0 \\ 0 \\ v_1 \end{pmatrix}$	$\begin{pmatrix} 0 & s_4 & 0 \\ s_5 & 0 & 0 \\ 0 & 0 & 0 \end{pmatrix}$
B_1	$\begin{pmatrix} u_2 \\ 0 \\ 0 \end{pmatrix}$	$\begin{pmatrix} 0 \\ v_2 \\ 0 \end{pmatrix}$	$\begin{pmatrix} 0 & 0 & s_6 \\ 0 & 0 & 0 \\ s_7 & 0 & 0 \end{pmatrix}$
B_2	$\begin{pmatrix} 0 \\ u_3 \\ 0 \end{pmatrix}$	$\begin{pmatrix} v_3 \\ 0 \\ 0 \end{pmatrix}$	$\begin{pmatrix} 0 & 0 & 0 \\ 0 & 0 & s_8 \\ 0 & s_9 & 0 \end{pmatrix}$

C_{3v}	$\hat{\mathbf{M}}$	$\hat{\mathcal{M}}$	$\hat{T}$ or $\hat{\alpha}'$
A_1	$\begin{pmatrix} 0 \\ 0 \\ u_1 \end{pmatrix}$		$\begin{pmatrix} s_1 & 0 & 0 \\ 0 & s_1 & 0 \\ 0 & 0 & s_2 \end{pmatrix}$
A_2		$\begin{pmatrix} 0 \\ 0 \\ v_1 \end{pmatrix}$	$\begin{pmatrix} 0 & s_3 & 0 \\ -s_3 & 0 & 0 \\ 0 & 0 & 0 \end{pmatrix}$
E	$\begin{pmatrix} u_2 \\ u_2 \\ 0 \end{pmatrix}$	$\begin{pmatrix} v_2 \\ v_2 \\ 0 \end{pmatrix}$	$\begin{pmatrix} s_4 & is_4 & s_5 \\ is_4 & -s_4 & -is_5 \\ s_6 & -is_6 & 0 \end{pmatrix}, \begin{pmatrix} s_4^* & -is_4^* & s_5^* \\ -is_4^* & -s_4^* & is_5^* \\ s_6^* & is_6^* & 0 \end{pmatrix}$

The Groups C_{nv} (Continued)

C_{4v}	$\hat{M}$	$\hat{\mathcal{M}}$	$\hat{T}$ or $\hat{\alpha}'$
A_1	$\begin{pmatrix} 0 \\ 0 \\ u_1 \end{pmatrix}$		$\begin{pmatrix} s_1 & 0 & 0 \\ 0 & s_1 & 0 \\ 0 & 0 & s_2 \end{pmatrix}$
A_2		$\begin{pmatrix} 0 \\ 0 \\ v_1 \end{pmatrix}$	$\begin{pmatrix} 0 & s_3 & 0 \\ -s_3 & 0 & 0 \\ 0 & 0 & 0 \end{pmatrix}$
B_1			$\begin{pmatrix} s_4 & 0 & 0 \\ 0 & -s_4 & 0 \\ 0 & 0 & 0 \end{pmatrix}$
B_2			$\begin{pmatrix} 0 & s_5 & 0 \\ s_5 & 0 & 0 \\ 0 & 0 & 0 \end{pmatrix}$
E	$\begin{pmatrix} u_2 \\ u_2 \\ 0 \end{pmatrix}$	$\begin{pmatrix} v_2 \\ v_2 \\ 0 \end{pmatrix}$	$\begin{pmatrix} 0 & 0 & s_6 \\ 0 & 0 & -is_6 \\ s_7 & -is_7 & 0 \end{pmatrix}, \begin{pmatrix} 0 & 0 & s_6^* \\ 0 & 0 & is_6^* \\ s_7^* & is_7^* & 0 \end{pmatrix}$

C_{6v}	$\hat{M}$	$\hat{\mathcal{M}}$	$\hat{T}$ or $\hat{\alpha}'$
A_1	$\begin{pmatrix} 0 \\ 0 \\ u_1 \end{pmatrix}$		$\begin{pmatrix} s_1 & 0 & 0 \\ 0 & s_1 & 0 \\ 0 & 0 & s_2 \end{pmatrix}$
A_2		$\begin{pmatrix} 0 \\ 0 \\ v_1 \end{pmatrix}$	$\begin{pmatrix} 0 & s_3 & 0 \\ -s_3 & 0 & 0 \\ 0 & 0 & 0 \end{pmatrix}$
E_1	$\begin{pmatrix} u_2 \\ u_2 \\ 0 \end{pmatrix}$	$\begin{pmatrix} v_2 \\ v_2 \\ 0 \end{pmatrix}$	$\begin{pmatrix} 0 & 0 & s_4 \\ 0 & 0 & -is_4 \\ s_5 & -is_5 & 0 \end{pmatrix}, \begin{pmatrix} 0 & 0 & s_4^* \\ 0 & 0 & is_4^* \\ s_5^* & is_5^* & 0 \end{pmatrix}$
E_2			$\begin{pmatrix} s_6 & -is_6 & 0 \\ -is_6 & -s_6 & 0 \\ 0 & 0 & 0 \end{pmatrix}, \begin{pmatrix} s_6^* & is_6^* & 0 \\ is_6^* & -s_6^* & 0 \\ 0 & 0 & 0 \end{pmatrix}$

The Groups C_{nh}

C_{2h}	$\hat{\mathbf{M}}$	$\hat{\mathcal{M}}$	$\hat{T}$ or $\hat{\alpha}'$
A_g		$\begin{pmatrix}0\\0\\v_1\end{pmatrix}$	$\begin{pmatrix}s_1 & s_4 & 0\\ s_5 & s_2 & 0\\ 0 & 0 & s_3\end{pmatrix}$
B_g		$\begin{pmatrix}v_2\\v_3\\0\end{pmatrix}$	$\begin{pmatrix}0 & 0 & s_6\\ 0 & 0 & s_7\\ s_8 & s_9 & 0\end{pmatrix}$
A_u	$\begin{pmatrix}0\\0\\u_1\end{pmatrix}$		
B_u	$\begin{pmatrix}u_2\\u_3\\0\end{pmatrix}$		

C_{3h}	$\hat{\mathbf{M}}$	$\hat{\mathcal{M}}$	$\hat{T}$ or $\hat{\alpha}'$
A'		$\begin{pmatrix}0\\0\\v_1\end{pmatrix}$	$\begin{pmatrix}s_1 & s_3 & 0\\ -s_3 & s_1 & 0\\ 0 & 0 & s_2\end{pmatrix}$
E'	$\begin{pmatrix}u_1\\iu_1\\0\end{pmatrix}, \begin{pmatrix}u_1^*\\-iu_1^*\\0\end{pmatrix}$		
A''	$\begin{pmatrix}0\\0\\u_2\end{pmatrix}$		
E''		$\begin{pmatrix}v_2\\iv_2\\0\end{pmatrix}, \begin{pmatrix}v_2^*\\-iv_2^*\\0\end{pmatrix}$	$\begin{pmatrix}s_4 & is_4 & s_5\\ is_4 & -s_4 & -is_5\\ s_6 & -is_6 & 0\end{pmatrix}, \begin{pmatrix}s_4^* & -is_4^* & s_5^*\\ -is_4^* & -s_4^* & is_5^*\\ s_6 & is_6^* & 0\end{pmatrix}$

The Groups C_{nh} (Continued)

C_{4h}	$\hat{M}$	$\hat{\mathcal{M}}$	$\hat{T}$ or $\hat{\alpha}'$
A_g		$\begin{pmatrix} 0 \\ 0 \\ v_1 \end{pmatrix}$	$\begin{pmatrix} s_1 & s_3 & 0 \\ -s_3 & s_1 & 0 \\ 0 & 0 & s_2 \end{pmatrix}$
B_g			$\begin{pmatrix} s_4 & s_5 & 0 \\ s_5 & -s_4 & 0 \\ 0 & 0 & 0 \end{pmatrix}$
E_g		$\begin{pmatrix} v_2 \\ iv_2 \\ 0 \end{pmatrix}, \begin{pmatrix} v_2^* \\ -iv_2^* \\ 0 \end{pmatrix}$	$\begin{pmatrix} 0 & 0 & s_6 \\ 0 & 0 & -is_6 \\ s_7 & -is_7 & 0 \end{pmatrix}, \begin{pmatrix} 0 & 0 & s_6^* \\ 0 & 0 & is_6^* \\ s_7^* & is_7^* & 0 \end{pmatrix}$
A_u	$\begin{pmatrix} 0 \\ 0 \\ u_1 \end{pmatrix}$		
E_u	$\begin{pmatrix} u_2 \\ iu_2 \\ 0 \end{pmatrix}, \begin{pmatrix} u_2^* \\ -iu_2^* \\ 0 \end{pmatrix}$		

C_{6h}	$\hat{M}$	$\hat{\mathcal{M}}$	$\hat{T}$ or $\hat{\alpha}'$
A_g		$\begin{pmatrix} 0 \\ 0 \\ v_1 \end{pmatrix}$	$\begin{pmatrix} s_1 & s_3 & 0 \\ -s_3 & s_1 & 0 \\ 0 & 0 & s_2 \end{pmatrix}$
E_{1g}		$\begin{pmatrix} v_2 \\ iv_2 \\ 0 \end{pmatrix}, \begin{pmatrix} v_2^* \\ -iv_2^* \\ 0 \end{pmatrix}$	$\begin{pmatrix} 0 & 0 & s_4 \\ 0 & 0 & -is_4 \\ s_5 & -is_5 & 0 \end{pmatrix}, \begin{pmatrix} 0 & 0 & s_4^* \\ 0 & 0 & is_4^* \\ s_5^* & is_5^* & 0 \end{pmatrix}$
E_{2g}			$\begin{pmatrix} s_6 & -is_6 & 0 \\ -is_6 & -s_6 & 0 \\ 0 & 0 & 0 \end{pmatrix}, \begin{pmatrix} s_6^* & is_6^* & 0 \\ is_6^* & -s_6^* & 0 \\ 0 & 0 & 0 \end{pmatrix}$
A_u	$\begin{pmatrix} 0 \\ 0 \\ u_1 \end{pmatrix}$		
E_{1u}	$\begin{pmatrix} u_2 \\ iu_2 \\ 0 \end{pmatrix}, \begin{pmatrix} u_2^* \\ -iu_2^* \\ 0 \end{pmatrix}$		

The Groups D_{nh}

D_{2h}	$\hat{M}$	$\hat{\mathscr{M}}$	$\hat{T}$ or $\hat{\alpha}'$
A_g			$\begin{pmatrix} s_1 & 0 & 0 \\ 0 & s_2 & 0 \\ 0 & 0 & s_3 \end{pmatrix}$
B_{1g}		$\begin{pmatrix} 0 \\ 0 \\ v_1 \end{pmatrix}$	$\begin{pmatrix} 0 & s_4 & 0 \\ s_5 & 0 & 0 \\ 0 & 0 & 0 \end{pmatrix}$
B_{2g}		$\begin{pmatrix} 0 \\ v_2 \\ 0 \end{pmatrix}$	$\begin{pmatrix} 0 & 0 & s_6 \\ 0 & 0 & 0 \\ s_7 & 0 & 0 \end{pmatrix}$
B_{3g}		$\begin{pmatrix} v_3 \\ 0 \\ 0 \end{pmatrix}$	$\begin{pmatrix} 0 & 0 & 0 \\ 0 & 0 & s_8 \\ 0 & s_9 & 0 \end{pmatrix}$
B_{1u}	$\begin{pmatrix} 0 \\ 0 \\ u_1 \end{pmatrix}$		
B_{2u}	$\begin{pmatrix} 0 \\ u_2 \\ 0 \end{pmatrix}$		
B_{3u}	$\begin{pmatrix} u_3 \\ 0 \\ 0 \end{pmatrix}$		

D_{3h}	$\hat{M}$	$\hat{\mathscr{M}}$	$\hat{T}$ or $\hat{\alpha}'$
A'_1			$\begin{pmatrix} s_1 & 0 & 0 \\ 0 & s_1 & 0 \\ 0 & 0 & s_2 \end{pmatrix}$
A'_2		$\begin{pmatrix} 0 \\ 0 \\ v_1 \end{pmatrix}$	$\begin{pmatrix} 0 & s_3 & 0 \\ -s_3 & 0 & 0 \\ 0 & 0 & 0 \end{pmatrix}$
E'	$\begin{pmatrix} u_1 \\ iu_1 \\ 0 \end{pmatrix}, \begin{pmatrix} u_1^* \\ -iu_1^* \\ 0 \end{pmatrix}$		$\begin{pmatrix} s_6 & -is_6 & 0 \\ -is_6 & -s_6 & 0 \\ 0 & 0 & 0 \end{pmatrix}, \begin{pmatrix} s_6^* & is_6^* & 0 \\ is_6^* & -s_6^* & 0 \\ 0 & 0 & 0 \end{pmatrix}$
A''_2	$\begin{pmatrix} 0 \\ 0 \\ u_2 \end{pmatrix}$		
E''		$\begin{pmatrix} v_2 \\ iv_2 \\ 0 \end{pmatrix}, \begin{pmatrix} v_2^* \\ -iv_2^* \\ 0 \end{pmatrix}$	$\begin{pmatrix} 0 & 0 & s_4 \\ 0 & 0 & -is_4 \\ s_5 & -is_5 & 0 \end{pmatrix}, \begin{pmatrix} 0 & 0 & s_4^* \\ 0 & 0 & is_4^* \\ s_5^* & is_5^* & 0 \end{pmatrix}$

The Groups D_{nh} (Continued)

D_{4h}	$\hat{M}$	$\hat{\mathscr{M}}$	$\hat{T}$ or $\hat{\alpha}'$
A_{1g}			$\begin{pmatrix} s_1 & 0 & 0 \\ 0 & s_1 & 0 \\ 0 & 0 & s_2 \end{pmatrix}$
A_{2g}		$\begin{pmatrix} 0 \\ 0 \\ v_1 \end{pmatrix}$	$\begin{pmatrix} 0 & s_3 & 0 \\ -s_3 & 0 & 0 \\ 0 & 0 & 0 \end{pmatrix}$
B_{1g}			$\begin{pmatrix} s_4 & 0 & 0 \\ 0 & -s_4 & 0 \\ 0 & 0 & 0 \end{pmatrix}$
B_{2g}			$\begin{pmatrix} 0 & s_5 & 0 \\ s_5 & 0 & 0 \\ 0 & 0 & 0 \end{pmatrix}$
E_g		$\begin{pmatrix} v_2 \\ iv_2 \\ 0 \end{pmatrix}, \begin{pmatrix} v_2^* \\ -iv_2^* \\ 0 \end{pmatrix}$	$\begin{pmatrix} 0 & 0 & s_6 \\ 0 & 0 & -is_6 \\ s_7 & -is_7 & 0 \end{pmatrix}, \begin{pmatrix} 0 & 0 & s_6^* \\ 0 & 0 & is_6^* \\ s_7^* & is_7^* & 0 \end{pmatrix}$
A_{2u}	$\begin{pmatrix} 0 \\ 0 \\ u_1 \end{pmatrix}$		
E_u	$\begin{pmatrix} u_2 \\ iu_2 \\ 0 \end{pmatrix}, \begin{pmatrix} u_2^* \\ -iu_2^* \\ 0 \end{pmatrix}$		

The Groups D_{nh} (Continued)

D_{6h}	$\hat{M}$	$\hat{\mathscr{M}}$	$\hat{T}$ or $\hat{\alpha}'$
A_{1g}			$\begin{pmatrix} s_1 & 0 & 0 \\ 0 & s_1 & 0 \\ 0 & 0 & s_2 \end{pmatrix}$
A_{2g}		$\begin{pmatrix} 0 \\ 0 \\ v_1 \end{pmatrix}$	$\begin{pmatrix} 0 & s_3 & 0 \\ -s_3 & 0 & 0 \\ 0 & 0 & 0 \end{pmatrix}$
E_{1g}		$\begin{pmatrix} v_2 \\ iv_2 \\ 0 \end{pmatrix}, \begin{pmatrix} v_2^* \\ -iv_2^* \\ 0 \end{pmatrix}$	$\begin{pmatrix} 0 & 0 & s_4 \\ 0 & 0 & -is_4 \\ s_5 & -is_5 & 0 \end{pmatrix}, \begin{pmatrix} 0 & 0 & s_4^* \\ 0 & 0 & is_4^* \\ s_5^* & is_5^* & 0 \end{pmatrix}$
E_{2g}			$\begin{pmatrix} s_6 & -is_6 & 0 \\ -is_6 & -s_6 & 0 \\ 0 & 0 & 0 \end{pmatrix}, \begin{pmatrix} s_6^* & is_6^* & 0 \\ is_6^* & -s_6^* & 0 \\ 0 & 0 & 0 \end{pmatrix}$
A_{2u}	$\begin{pmatrix} 0 \\ 0 \\ u_1 \end{pmatrix}$		
E_{1u}	$\begin{pmatrix} u_2 \\ iu_2 \\ 0 \end{pmatrix}, \begin{pmatrix} u_2^* \\ -iu_2^* \\ 0 \end{pmatrix}$		

The Groups D_{nd}

D_{2d}	$\hat{M}$	$\hat{\mathscr{M}}$	$\hat{T}$ or $\hat{\alpha}'$
A_1			$\begin{pmatrix} s_1 & 0 & 0 \\ 0 & s_1 & 0 \\ 0 & 0 & s_2 \end{pmatrix}$
A_2		$\begin{pmatrix} 0 \\ 0 \\ v_1 \end{pmatrix}$	$\begin{pmatrix} 0 & s_3 & 0 \\ -s_3 & 0 & 0 \\ 0 & 0 & 0 \end{pmatrix}$
B_1			$\begin{pmatrix} s_4 & 0 & 0 \\ 0 & -s_4 & 0 \\ 0 & 0 & 0 \end{pmatrix}$
B_2	$\begin{pmatrix} 0 \\ 0 \\ u_1 \end{pmatrix}$		$\begin{pmatrix} 0 & s_5 & 0 \\ s_5 & 0 & 0 \\ 0 & 0 & 0 \end{pmatrix}$
E	$\begin{pmatrix} u_2 \\ iu_2 \\ 0 \end{pmatrix}, \begin{pmatrix} u_2^* \\ -iu_2^* \\ 0 \end{pmatrix}$	$\begin{pmatrix} v_2 \\ iv_2 \\ 0 \end{pmatrix}, \begin{pmatrix} v_2^* \\ -iv_2^* \\ 0 \end{pmatrix}$	$\begin{pmatrix} 0 & 0 & s_6 \\ 0 & 0 & -is_6 \\ s_7 & -is_7 & 0 \end{pmatrix}, \begin{pmatrix} 0 & 0 & s_6^* \\ 0 & 0 & is_6^* \\ s_7^* & is_7^* & 0 \end{pmatrix}$

D_{3d}	$\hat{M}$	$\hat{\mathscr{M}}$	$\hat{T}$ or $\hat{\alpha}'$
A_{1g}			$\begin{pmatrix} s_1 & 0 & 0 \\ 0 & s_1 & 0 \\ 0 & 0 & s_2 \end{pmatrix}$
A_{2g}		$\begin{pmatrix} 0 \\ 0 \\ v_1 \end{pmatrix}$	$\begin{pmatrix} 0 & s_3 & 0 \\ -s_3 & 0 & 0 \\ 0 & 0 & 0 \end{pmatrix}$
E_g		$\begin{pmatrix} v_2 \\ iv_2 \\ 0 \end{pmatrix}, \begin{pmatrix} v_2^* \\ -iv_2^* \\ 0 \end{pmatrix}$	$\begin{pmatrix} s_4 & is_4 & s_5 \\ is_4 & -s_4 & -is_5 \\ s_6 & -is_6 & 0 \end{pmatrix}, \begin{pmatrix} s_4^* & -is_4^* & s_5^* \\ -is_4^* & -s_4^* & is_5^* \\ s_6^* & is_6^* & 0 \end{pmatrix}$
A_{2u}	$\begin{pmatrix} 0 \\ 0 \\ u_1 \end{pmatrix}$		
E_u	$\begin{pmatrix} u_2 \\ iu_2 \\ 0 \end{pmatrix}, \begin{pmatrix} u_2^* \\ -iu_2^* \\ 0 \end{pmatrix}$		

The Groups S_n

S_4	$\hat{\mathbf{M}}$	$\hat{\mathscr{M}}$	$\hat{T}$ or $\hat{\alpha}'$
A		$\begin{pmatrix} 0 \\ 0 \\ v_1 \end{pmatrix}$	$\begin{pmatrix} s_1 & s_3 & 0 \\ -s_3 & s_1 & 0 \\ 0 & 0 & s_2 \end{pmatrix}$
B	$\begin{pmatrix} 0 \\ 0 \\ u_1 \end{pmatrix}$		$\begin{pmatrix} s_4 & s_5 & 0 \\ s_5 & -s_4 & 0 \\ 0 & 0 & 0 \end{pmatrix}$
E	$\begin{pmatrix} u_2 \\ iu_2 \\ 0 \end{pmatrix}, \begin{pmatrix} u_2^* \\ -iu_2^* \\ 0 \end{pmatrix}$	$\begin{pmatrix} v_2 \\ iv_2 \\ 0 \end{pmatrix}, \begin{pmatrix} v_2^* \\ -iv_2^* \\ 0 \end{pmatrix}$	$\begin{pmatrix} 0 & 0 & s_6 \\ 0 & 0 & -is_6 \\ s_7 & -is_7 & 0 \end{pmatrix}, \begin{pmatrix} 0 & 0 & s_6^* \\ 0 & 0 & is_6^* \\ s_7^* & is_7^* & 0 \end{pmatrix}$

S_6	$\hat{\mathbf{M}}$	$\hat{\mathscr{M}}$	$\hat{T}$ or $\hat{\alpha}'$
A_g		$\begin{pmatrix} 0 \\ 0 \\ v_1 \end{pmatrix}$	$\begin{pmatrix} s_1 & s_3 & 0 \\ -s_3 & s_1 & 0 \\ 0 & 0 & s_2 \end{pmatrix}$
E_g		$\begin{pmatrix} v_2 \\ iv_2 \\ 0 \end{pmatrix}, \begin{pmatrix} v_2^* \\ -iv_2^* \\ 0 \end{pmatrix}$	$\begin{pmatrix} s_4 & is_4 & s_5 \\ is_4 & -s_4 & -is_5 \\ s_6 & -is_6 & 0 \end{pmatrix}, \begin{pmatrix} s_4^* & -is_4^* & s_5^* \\ -is_4^* & -s_4^* & is_5^* \\ s_6^* & is_6^* & 0 \end{pmatrix}$
A_u	$\begin{pmatrix} 0 \\ 0 \\ u_1 \end{pmatrix}$		
E_u	$\begin{pmatrix} u_2 \\ iu_2 \\ 0 \end{pmatrix}, \begin{pmatrix} u_2^* \\ -iu_2^* \\ 0 \end{pmatrix}$		

The Cubic Groups

T	$\hat{\mathbf{M}}$	$\hat{\mathscr{M}}$	$\hat{T}$ or $\hat{\alpha}'$
A			$\begin{pmatrix} s_1 & 0 & 0 \\ 0 & s_1 & 0 \\ 0 & 0 & s_1 \end{pmatrix}$
E			$\begin{pmatrix} s_2 & 0 & 0 \\ 0 & \omega s_2 & 0 \\ 0 & 0 & \omega^* s_2 \end{pmatrix}, \begin{pmatrix} s_2^* & 0 & 0 \\ 0 & \omega^* s_2^* & 0 \\ 0 & 0 & \omega s_2^* \end{pmatrix}$
T	$\begin{pmatrix} u_1 \\ 0 \\ 0 \end{pmatrix}, \begin{pmatrix} 0 \\ u_1 \\ 0 \end{pmatrix}, \begin{pmatrix} 0 \\ 0 \\ u_1 \end{pmatrix}$	$\begin{pmatrix} v_1 \\ 0 \\ 0 \end{pmatrix}, \begin{pmatrix} 0 \\ v_1 \\ 0 \end{pmatrix}, \begin{pmatrix} 0 \\ 0 \\ v_1 \end{pmatrix}$	$\begin{pmatrix} 0 & 0 & 0 \\ 0 & 0 & s_3 \\ 0 & s_4 & 0 \end{pmatrix}, \begin{pmatrix} 0 & 0 & s_4 \\ 0 & 0 & 0 \\ s_3 & 0 & 0 \end{pmatrix}, \begin{pmatrix} 0 & s_3 & 0 \\ s_4 & 0 & 0 \\ 0 & 0 & 0 \end{pmatrix}$

T_d	$\hat{\mathbf{M}}$	$\hat{\mathscr{M}}$	$\hat{T}$ or $\hat{\alpha}'$
A_1			$\begin{pmatrix} s_1 & 0 & 0 \\ 0 & s_1 & 0 \\ 0 & 0 & s_1 \end{pmatrix}$
E			$\begin{pmatrix} s_2 & 0 & 0 \\ 0 & \omega s_2 & 0 \\ 0 & 0 & \omega^* s_2 \end{pmatrix}, \begin{pmatrix} s_2^* & 0 & 0 \\ 0 & \omega^* s_2^* & 0 \\ 0 & 0 & \omega s_2^* \end{pmatrix}$
T_1		$\begin{pmatrix} v_1 \\ 0 \\ 0 \end{pmatrix}, \begin{pmatrix} 0 \\ v_1 \\ 0 \end{pmatrix}, \begin{pmatrix} 0 \\ 0 \\ v_1 \end{pmatrix}$	$\begin{pmatrix} 0 & s_3 & 0 \\ -s_3 & 0 & 0 \\ 0 & 0 & 0 \end{pmatrix}, \begin{pmatrix} 0 & 0 & -s_3 \\ 0 & 0 & 0 \\ s_3 & 0 & 0 \end{pmatrix}, \begin{pmatrix} 0 & 0 & 0 \\ 0 & 0 & s_3 \\ 0 & -s_3 & 0 \end{pmatrix}$
T_2	$\begin{pmatrix} u_1 \\ 0 \\ 0 \end{pmatrix}, \begin{pmatrix} 0 \\ u_1 \\ 0 \end{pmatrix}, \begin{pmatrix} 0 \\ 0 \\ u_1 \end{pmatrix}$		$\begin{pmatrix} 0 & s_4 & 0 \\ s_4 & 0 & 0 \\ 0 & 0 & 0 \end{pmatrix}, \begin{pmatrix} 0 & 0 & s_4 \\ 0 & 0 & 0 \\ s_4 & 0 & 0 \end{pmatrix}, \begin{pmatrix} 0 & 0 & 0 \\ 0 & 0 & s_4 \\ 0 & s_4 & 0 \end{pmatrix}$

T_h	$\hat{\mathbf{M}}$	$\hat{\mathscr{M}}$	$\hat{T}$ or $\hat{\alpha}'$
A_g			$\begin{pmatrix} s_1 & 0 & 0 \\ 0 & s_1 & 0 \\ 0 & 0 & s_1 \end{pmatrix}$
E_g			$\begin{pmatrix} s_2 & 0 & 0 \\ 0 & \omega s_2 & 0 \\ 0 & 0 & \omega^* s_2 \end{pmatrix}, \begin{pmatrix} s_2^* & 0 & 0 \\ 0 & \omega^* s_2^* & 0 \\ 0 & 0 & \omega s_2^* \end{pmatrix}$
T_g		$\begin{pmatrix} v_1 \\ 0 \\ 0 \end{pmatrix}, \begin{pmatrix} 0 \\ v_1 \\ 0 \end{pmatrix}, \begin{pmatrix} 0 \\ 0 \\ v_1 \end{pmatrix}$	$\begin{pmatrix} 0 & 0 & 0 \\ 0 & 0 & s_3 \\ 0 & s_4 & 0 \end{pmatrix}, \begin{pmatrix} 0 & 0 & s_4 \\ 0 & 0 & 0 \\ s_3 & 0 & 0 \end{pmatrix}, \begin{pmatrix} 0 & s_3 & 0 \\ s_4 & 0 & 0 \\ 0 & 0 & 0 \end{pmatrix}$
T_u	$\begin{pmatrix} u_1 \\ 0 \\ 0 \end{pmatrix}, \begin{pmatrix} 0 \\ u_1 \\ 0 \end{pmatrix}, \begin{pmatrix} 0 \\ 0 \\ u_1 \end{pmatrix}$		

The Cubic Groups (Continued)

O	$\hat{M}$	$\hat{\mathscr{M}}$	$\hat{T}$ or $\hat{\alpha}'$
A_1			$\begin{pmatrix} s_1 & 0 & 0 \\ 0 & s_1 & 0 \\ 0 & 0 & s_1 \end{pmatrix}$
E			$\begin{pmatrix} s_2 & 0 & 0 \\ 0 & \omega s_2 & 0 \\ 0 & 0 & \omega^* s_2 \end{pmatrix}, \begin{pmatrix} s_2^* & 0 & 0 \\ 0 & \omega^* s_2^* & 0 \\ 0 & 0 & \omega s_2^* \end{pmatrix}$
T_1	$\begin{pmatrix} u_1 \\ 0 \\ 0 \end{pmatrix}, \begin{pmatrix} 0 \\ u_1 \\ 0 \end{pmatrix}, \begin{pmatrix} 0 \\ 0 \\ u_1 \end{pmatrix}$	$\begin{pmatrix} v_1 \\ 0 \\ 0 \end{pmatrix}, \begin{pmatrix} 0 \\ v_1 \\ 0 \end{pmatrix}, \begin{pmatrix} 0 \\ 0 \\ v_1 \end{pmatrix}$	$\begin{pmatrix} 0 & s_3 & 0 \\ -s_3 & 0 & 0 \\ 0 & 0 & 0 \end{pmatrix}, \begin{pmatrix} 0 & 0 & -s_3 \\ 0 & 0 & 0 \\ s_3 & 0 & 0 \end{pmatrix}, \begin{pmatrix} 0 & 0 & 0 \\ 0 & 0 & s_3 \\ 0 & -s^3 & 0 \end{pmatrix}$
T_2			$\begin{pmatrix} 0 & s_4 & 0 \\ s_4 & 0 & 0 \\ 0 & 0 & 0 \end{pmatrix}, \begin{pmatrix} 0 & 0 & s_4 \\ 0 & 0 & 0 \\ s_4 & 0 & 0 \end{pmatrix}, \begin{pmatrix} 0 & 0 & 0 \\ 0 & 0 & s_4 \\ 0 & s_4 & 0 \end{pmatrix}$

O_h	$\hat{\mathbf{M}}$	$\hat{\mathscr{M}}$	$\hat{T}$ or $\hat{\alpha}'$
A_{1g}			$\begin{pmatrix} s_1 & 0 & 0 \\ 0 & s_1 & 0 \\ 0 & 0 & s_1 \end{pmatrix}$
E_g			$\begin{pmatrix} s_2 & 0 & 0 \\ 0 & \omega s_2 & 0 \\ 0 & 0 & \omega^* s_2 \end{pmatrix}, \begin{pmatrix} s_2^* & 0 & 0 \\ 0 & \omega^* s_2^* & 0 \\ 0 & 0 & \omega s_2^* \end{pmatrix}$
T_{1g}		$\begin{pmatrix} v_1 \\ 0 \\ 0 \end{pmatrix}, \begin{pmatrix} 0 \\ v_1 \\ 0 \end{pmatrix}, \begin{pmatrix} 0 \\ 0 \\ v_1 \end{pmatrix}$	$\begin{pmatrix} 0 & s_3 & 0 \\ -s_3 & 0 & 0 \\ 0 & 0 & 0 \end{pmatrix}, \begin{pmatrix} 0 & 0 & -s_3 \\ 0 & 0 & 0 \\ s_3 & 0 & 0 \end{pmatrix}, \begin{pmatrix} 0 & 0 & 0 \\ 0 & 0 & s_3 \\ 0 & -s_3 & 0 \end{pmatrix}$
T_{2g}			$\begin{pmatrix} 0 & s_4 & 0 \\ s_4 & 0 & 0 \\ 0 & 0 & 0 \end{pmatrix}, \begin{pmatrix} 0 & 0 & s_4 \\ 0 & 0 & 0 \\ s_4 & 0 & 0 \end{pmatrix}, \begin{pmatrix} 0 & 0 & 0 \\ 0 & 0 & s_4 \\ 0 & s_4 & 0 \end{pmatrix}$
T_{1u}	$\begin{pmatrix} u_1 \\ 0 \\ 0 \end{pmatrix}, \begin{pmatrix} 0 \\ u_1 \\ 0 \end{pmatrix}, \begin{pmatrix} 0 \\ 0 \\ u_1 \end{pmatrix}$		

The Groups $C_{\infty v}$ and $D_{\infty h}$

$C_{\infty v}$	$\hat{\mathbf{M}}$	$\hat{\mathcal{M}}$	$\hat{T}$ or $\hat{\alpha}'$
Σ^+	$\begin{pmatrix} 0 \\ 0 \\ u_1 \end{pmatrix}$		$\begin{pmatrix} s_1 & 0 & 0 \\ 0 & s_1 & 0 \\ 0 & 0 & s_2 \end{pmatrix}$
Σ^-		$\begin{pmatrix} 0 \\ 0 \\ v_1 \end{pmatrix}$	$\begin{pmatrix} 0 & s_3 & 0 \\ -s_3 & 0 & 0 \\ 0 & 0 & 0 \end{pmatrix}$
Π	$\begin{pmatrix} u_2 \\ iu_2 \\ 0 \end{pmatrix}, \begin{pmatrix} u_2^* \\ -iu_2^* \\ 0 \end{pmatrix}$	$\begin{pmatrix} v_2 \\ iv_2 \\ 0 \end{pmatrix}, \begin{pmatrix} v_2^* \\ -iv_2^* \\ 0 \end{pmatrix}$	$\begin{pmatrix} 0 & 0 & s_4 \\ 0 & 0 & -is_4 \\ s_5 & -is_5 & 0 \end{pmatrix}, \begin{pmatrix} 0 & 0 & s_4^* \\ 0 & 0 & is_4^* \\ s_5^* & is_5^* & 0 \end{pmatrix}$
Δ			$\begin{pmatrix} s_6 & -is_6 & 0 \\ -is_6 & -s_6 & 0 \\ 0 & 0 & 0 \end{pmatrix}, \begin{pmatrix} s_6^* & is_6^* & 0 \\ is_6^* & -s_6^* & 0 \\ 0 & 0 & 0 \end{pmatrix}$

$D_{\infty h}$	$\hat{\mathbf{M}}$	$\hat{\mathcal{M}}$	$\hat{T}$ or $\hat{\alpha}'$
Σ_g^+			$\begin{pmatrix} s_1 & 0 & 0 \\ 0 & s_1 & 0 \\ 0 & 0 & s_2 \end{pmatrix}$
Σ_g^-		$\begin{pmatrix} 0 \\ 0 \\ v_1 \end{pmatrix}$	$\begin{pmatrix} 0 & s_3 & 0 \\ -s_3 & 0 & 0 \\ 0 & 0 & 0 \end{pmatrix}$
Π_g		$\begin{pmatrix} v_2 \\ iv_2 \\ 0 \end{pmatrix}, \begin{pmatrix} v_2^* \\ -iv_2^* \\ 0 \end{pmatrix}$	$\begin{pmatrix} 0 & 0 & s_4 \\ 0 & 0 & -is_4 \\ s_5 & -is_5 & 0 \end{pmatrix}, \begin{pmatrix} 0 & 0 & s_4^* \\ 0 & 0 & is_4^* \\ s_5^* & is_5^* & 0 \end{pmatrix}$
Δ_g			$\begin{pmatrix} s_6 & -is_6 & 0 \\ -is_6 & -s_6 & 0 \\ 0 & 0 & 0 \end{pmatrix}, \begin{pmatrix} s_6^* & is_6^* & 0 \\ is_6^* & -s_6^* & 0 \\ 0 & 0 & 0 \end{pmatrix}$
Σ_u^+	$\begin{pmatrix} 0 \\ 0 \\ u_1 \end{pmatrix}$		
Π_u	$\begin{pmatrix} u_2 \\ iu_2 \\ 0 \end{pmatrix}, \begin{pmatrix} u_2^* \\ -iu_2^* \\ 0 \end{pmatrix}$		

For nondegenerate species the basis set used for vectors or tensors corresponding to $\hat{\mathbf{M}}$, $\hat{\mathsf{T}}$, and $\hat{\alpha}'$ is (x,y,z). In the case of $\hat{\mathcal{M}}$, $(\mathcal{R}_x,\mathcal{R}_y,\mathcal{R}_z)$ is used. For degenerate species the two or three entries corresponding to $\hat{\mathbf{M}}$, $\hat{\mathsf{T}}$, $\hat{\alpha}'$ or $\hat{\mathcal{M}}$ separated by commas correspond to different ways of reaching the various components of the same degenerate state, and their form depends on the basis sets used. For the species E, E_u, E_g, E_1, E_{1u}, E_{1g}, E_1', Π, Π_u, and Π_g we have chosen the basis set $(x + iy,\ x - iy)$ except for the cubic groups, where the set $[2z^2 - x^2 - y^2 + i\sqrt{3}(x^2 - y^2),\ 2z^2 - x^2 - y^2 - i\sqrt{3}(x^2 + y^2)]$ was used. For the species E_2, E_{2u}, E_{2g}, E_1'', Δ, Δ_u, Δ_g, we chose $[(x + iy)^2,\ (x - iy)^2]$. For T, T_u, T_g, T_{1u}, and T_{1g}, (x,y,z) was used, while (yz,zx,xy) was used for T_2, T_{2u}, and T_{2g}. It should be noted that for the cubic groups $\omega = \exp(2\pi i/3)$.

Direct Products of Irreducible Representations

It is often necessary to form direct products of irreducible representations. If a molecule is vibrating with one quantum in each of two modes with symmetry species s_1 and s_2, the resulting vibrational state will have symmetry species contained in the product $s_1 \times s_2$. If a molecule is in an electronic state with symmetry species s_1 and a vibrational state with symmetry species s_2, the total vibronic state will also have the species contained in $s_1 \times s_2$, and if an electronic configuration has two unpaired electrons in orbitals with symmetry species s_1 and s_2, the product $s_1 \times s_2$ will provide information on the symmetry species of the configuration.

The multiplication of two symmetry species s_1 and s_2 is simple as long as one of them is nondegenerate. In this case the result is another of the irreducible symmetry species with characters $\chi_{s_1 \times s_2}(R) = \chi_{s_1}(R)\chi_{s_2}(R)$. Thus the character tables immediately tell us what the resulting symmetry species is. For example, for C_{2v}, the product table is that of Table I.3.

For the multiplication a few general rules exist. If the symmetry species are of g or u types we have:

$$g \times g = g \qquad u \times u = g \qquad g \times u = u \tag{I.36}$$

If the symmetry species have single or double primes, we have

$$(') \times (') = (') \qquad ('') \times ('') = (') \qquad (') \times ('') = ('') \tag{I.37}$$

Table I.3

C_{2v}	A_1	A_2	B_1	B_2
A_1	A_1	A_2	B_1	B_2
A_2	A_2	A_1	B_2	B_1
B_1	B_1	B_2	A_1	A_2
B_2	B_2	B_1	A_2	A_1

In point groups where all subscripts to species A and B are smaller than 3 (or nonexistent), we have

$$A \times A = A \qquad B \times B = A \qquad A \times B = B \tag{I.38}$$

and the subscripts for such groups follow the rule

$$1 \times 1 = 1 \qquad 2 \times 2 = 1 \qquad 1 \times 2 = 2 \tag{I.39}$$

An illustration of the latter rules is the C_{2v} point group discussed above.

In the important point group D_{2h}, the following rules hold for subscripts 1, 2, and 3:

$$1 \times 2 = 3 \qquad 2 \times 3 = 1 \qquad 1 \times 3 = 2 \tag{I.40}$$

while subscript products of the form I × I lead to either A_u or A_g.

A generalization of multiplication to three or more species is simple:

$$s_1 \times s_2 \times s_3 = (s_1 \times s_2) \times s_3 \tag{I.41}$$

For example, for C_{2v},

$$A_2 \times B_1 \times B_2 = (A_2 \times B_1) \times B_2 = B_2 \times B_2 = A_1 \tag{I.42}$$

The difficult part of multiplication of symmetry species occurs in connection with the multiplication of two degenerate species d_1 and d_2. In such a case the characters $\chi_{d_1 \times d_2}(R) = \chi_{d_1}(R)\chi_{d_2}(R)$ represent a reducible representation, and the corresponding linear combination of irreducible representations must be found by the methods discussed in the beginning of this appendix. For the D_{6h} point group we have, for example, for the multiplications $E_{1g} \times E_{2u}$, $E_{1g} \times E_{1g}$, and a number of relevant irreducible representations, the character table of Table I.4. Thus we obtain

$$E_{1g} \times E_{2u} = B_{1u} + B_{2u} + E_{1u} \tag{I.43}$$

$$E_{1g} \times E_{1g} = A_{1g} + A_{2g} + E_{2g} \tag{I.44}$$

This method will be even more useful when rules are given for applications to specific problems. For *vibrational states* the rules are reasonably straight-

Table I.4

D_{6h}	E	C_6	C_3	C_2	C_2'	C_2''	i	S_3	S_6	σ_h	σ_d	σ_v
E_{1g}	2	1	−1	−2	0	0	2	1	−1	−2	0	0
E_{2u}	2	−1	−1	2	0	0	−2	1	1	−2	0	0
$E_{1g} \times E_{2u}$	4	−1	1	−4	0	0	−4	1	−1	4	0	0
$E_{1g} \times E_{1g}$	4	1	1	4	0	0	4	1	1	4	0	0
A_{1g}	1	1	1	1	1	1	1	1	1	1	1	1
A_{2g}	1	1	1	1	−1	−1	1	1	1	1	−1	−1
E_{2g}	2	−1	−1	2	0	0	2	−1	−1	2	0	0
B_{1u}	1	−1	1	−1	1	−1	−1	1	−1	1	−1	1
B_{2u}	1	−1	1	−1	−1	1	−1	1	−1	1	1	−1
E_{1u}	2	1	−1	−2	0	0	−2	−1	1	2	0	0

forward. If a molecule has two quanta of vibration in two different modes with symmetry species d_1 and d_2, the product $d_1 \times d_2$ represents the possible symmetry species of the resulting vibrational states. If the two quanta are in the same vibrational mode, the resulting states are restricted to what may be called the "symmetric" part of the product. For the product $E_{1g} \times E_{1g}$ this is $A_{1g} + E_{2g}$. The "antisymmetric" part, a vibrational state of A_{2g} species, cannot be reached by supplying two quanta of vibration in the same mode of E_{1g} symmetry.

There are no such restrictions in the possible symmetries of *vibronic states* determined from the symmetry species of the electronic and vibrational part, but the rules for symmetry species of *electron configurations* are quite complicated. This is due to electron spin and the Pauli principle. First of all, when an orbital (or a set of degenerate orbitals) is completely filled (a filled shell), it does not contribute to the net electron spin, and it represents a totally symmetric species. For example, in the benzene (point group D_{6h}) electronic ground-state configuration, two electrons are in an A_{2u} orbital and four in an E_{1g} orbital. Clearly, the product $A_{2u} \times A_{2u}$ is A_{1g}, totally symmetric. The product $E_{1g} \times E_{1g} \times E_{1g} \times E_{1g}$ is much more complicated. Fortunately, the rule tells us that of all the irreducible representations contained in the product only the species A_{1g} should be used for a completely filled orbital. Thus the benzene ground state becomes $^1A_{1g}$, like all closed-shell ground states.

Another very useful rule is that looking at the species defined by the "holes" in a shell is as good as looking at the electrons when the symmetry species of a configuration is determined. Thus, the configuration for $C_6H_6^{3+}$, which has one electron and three holes in E_{1g}, has the same symmetry (E_{1g}) and spin (doublet) as the configuration for $C_6H_6^{+}$, which has one hole and three electrons in E_{1g}.

The possible symmetry species of states of an electron configuration with two single electrons (or holes) in different degenerate orbitals are given by the product of the species of the two orbitals. The electron spin may be singlet or triplet. For example, the lowest excited states of benzene have the configuration $(A_{2u})^2(E_{1g})^3(E_{2u})^1$, that is, one hole in E_{1g}, one electron in E_{2u}. The states may have the symmetry species

$$E_{1g} \times E_{2u} = B_{1u} + B_{2u} + E_{1u} \tag{I.45}$$

and triplet or singlet electron spin.

If two electrons are in the same degenerate orbital, the contribution to symmetry and spin is determined by more complicated rules. The resulting singlet states (corresponding to antisymmetric spin functions) may have any of the symmetry species given by the "symmetric" part of the product, while the resulting triplets (corresponding to symmetric spin functions) are restricted to the "antisymmetric" part. A well-known example is the oxygen molecule, O_2 (point group $D_{\infty h}$). The ground-state configuration has one partly filled orbital, π_g, with two electrons. The product

$$\Pi_g \times \Pi_g = \Sigma_g^- + \Sigma_g^+ + \Delta_g \tag{I.46}$$

has the symmetric part $\Sigma_g^+ + \Delta_g$ and the antisymmetric part Σ_g^-. The resulting states are $^3\Sigma_g^-$ (the ground state according to Hund's rule), $^1\Delta_g$, and $^1\Sigma_g^+$.

In the following, products of two degenerate symmetry species are listed for a number of important point groups. The "antisymmetric" part (which in some cases may be the totally symmetric species) of the result for the multiplication of a symmetry species with itself is given in parentheses; it does not contribute to a vibrational state if the two vibrational quanta occur in the same mode, and it will contribute a triplet state to the states resulting from the presence of two electrons in the same degenerate orbital.

The Groups C_n

C_3: $E \times E = A + E + (A)$

C_4: $E \times E = A + 2B + (A)$

C_5: $E_1 \times E_1 = A + E_2 + (A)$
$E_2 \times E_2 = A + E_1 + (A)$
$E_1 \times E_2 = E_1 + E_2$

C_6: $E_1 \times E_1 = A + E_2 + (A)$
$E_2 \times E_2 = A + E_2 + (A)$
$E_1 \times E_2 = 2B + E_1$

C_7: $E_1 \times E_1 = A + E_2 + (A)$
$E_2 \times E_2 = A + E_3 + (A)$
$E_3 \times E_3 = A + E_1 + (A)$
$E_1 \times E_2 = E_1 + E_3$
$E_1 \times E_3 = E_2 + E_3$
$E_2 \times E_3 = E_1 + E_2$

C_8: $E_1 \times E_1 = A + E_2 + (A)$
$E_2 \times E_2 = A + 2B + (A)$
$E_3 \times E_3 = A + E_2 + (A)$
$E_1 \times E_2 = E_1 + E_3$
$E_1 \times E_3 = 2B + E_2$
$E_2 \times E_3 = E_1 + E_3$

The Groups D_n

D_3: $E \times E = A_1 + E + (A_2)$

D_4: $E \times E = A_1 + B_1 + B_2 + (A_2)$

D_5: $E_1 \times E_1 = A_1 + E_2 + (A_2)$
$E_2 \times E_2 = A_1 + E_1 + (A_2)$
$E_1 \times E_2 = E_1 + E_2$

D_6: $E_1 \times E_1 = A_1 + E_2 + (A_2)$
$E_2 \times E_2 = A_1 + E_2 + (A_2)$
$E_1 \times E_2 = B_1 + B_2 + E_1$

The Groups C_{nv}

C_{3v}: $E \times E = A_1 + E + (A_2)$

C_{4v}: $E \times E = A_1 + B_1 + B_2 + (A_2)$

C_{5v}: $E_1 \times E_1 = A_1 + E_2 + (A_2)$
$E_2 \times E_2 = A_1 + E_1 + (A_2)$
$E_1 \times E_2 = E_1 + E_2$

C_{6v}: $E_1 \times E_1 = A_1 + E_2 + (A_2)$
$E_2 \times E_2 = A_1 + E_2 + (A_2)$
$E_1 \times E_2 = B_1 + B_2 + E_1$

The Groups C_{nh}

C_{3h}: $E' \times E' = A' + E' + (A')$
$E'' \times E'' = A' + E' + (A')$
$E' \times E'' = 2A'' + E''$

C_{4h}: $E_g \times E_g = A_g + 2B_g + (A_g)$
$E_u \times E_u = A_g + 2B_g + (A_g)$
$E_g \times E_u = 2A_u + 2B_u$

C_{5h}: $E_1' \times E_1' = A' + E_2' + (A')$
$E_2' \times E_2' = A' + E_1' + (A')$
$E_1'' \times E_1'' = A' + E_2' + (A')$
$E_2'' \times E_2'' = A' + E_1' + (A')$
$E_1' \times E_1'' = 2A'' + E_2''$
$E_1' \times E_2' = E_1' + E_2'$
$E_1' \times E_2'' = E_1'' + E_2''$
$E_1'' \times E_2' = E_1'' + E_2''$
$E_1'' \times E_2'' = E_1' + E_2'$
$E_2' \times E_2'' = 2A'' + E_1''$

C_{6h}: $E_{1g} \times E_{1g} = A_g + E_{2g} + (A_g)$
$E_{2g} \times E_{2g} = A_g + E_{2g} + (A_g)$
$E_{1u} \times E_{1u} = A_g + E_{2g} + (A_g)$
$E_{2u} \times E_{2u} = A_g + E_{2g} + (A_g)$
$E_{1g} \times E_{1u} = 2A_u + E_{2u}$
$E_{1g} \times E_{2g} = 2B_g + E_{1g}$
$E_{1g} \times E_{2u} = 2B_u + E_{1u}$
$E_{1u} \times E_{2g} = 2B_u + E_{1u}$
$E_{1u} \times E_{2u} = 2B_g + E_{1g}$
$E_{2g} \times E_{2u} = 2A_u + E_{2u}$

The Groups D_{nh}

D_{3h}: $E' \times E' = A_1' + E' + (A_2')$
$E'' \times E'' = A_1' + E' + (A_2')$
$E' \times E'' = A_1'' + A_2'' + E''$

D_{4h}: $E_g \times E_g = A_{1g} + B_{1g} + B_{2g} + (A_{2g})$
$E_u \times E_u = A_{1g} + B_{1g} + B_{2g} + (A_{2g})$
$E_g \times E_u = A_{1u} + A_{2u} + B_{1u} + B_{2u}$

D_{5h}: $E_1' \times E_1' = A_1' + E_2' + (A_2')$
$E_1'' \times E_1'' = A_1' + E_2' + (A_2')$
$E_2' \times E_2' = A_1' + E_1' + (A_2')$
$E_2'' \times E_2'' = A_1' + E_1' + (A_2')$
$E_1' \times E_1'' = A_1'' + A_2'' + E_2''$
$E_1' \times E_2' = E_1' + E_2'$
$E_1' \times E_2'' = E_1'' + E_2''$
$E_1'' \times E_2' = E_1'' + E_2''$
$E_1'' \times E_2'' = E_1' + E_2'$
$E_2' \times E_2'' = A_1'' + A_2'' + E_1''$

D_{6h}: $E_{1g} \times E_{1g} = A_{1g} + E_{2g} + (A_{2g})$
$E_{2g} \times E_{2g} = A_{1g} + E_{2g} + (A_{2g})$
$E_{1u} \times E_{1u} = A_{1g} + E_{2g} + (A_{2g})$
$E_{2u} \times E_{2u} = A_{1g} + E_{2g} + (A_{2g})$
$E_{1g} \times E_{2g} = B_{1g} + B_{2g} + E_{1g}$
$E_{1g} \times E_{1u} = A_{1u} + A_{2u} + E_{2u}$
$E_{1g} \times E_{2u} = B_{1u} + B_{2u} + E_{1u}$
$E_{2g} \times E_{1u} = B_{1u} + B_{2u} + E_{1u}$
$E_{2g} \times E_{2u} = A_{1u} + A_{2u} + E_{2u}$
$E_{1u} \times E_{2u} = B_{1g} + B_{2g} + E_{1g}$

The Groups D_{nd}

D_{2d}: $E \times E = A_1 + B_1 + B_2 + (A_2)$

D_{3d}: $E_g \times E_g = A_{1g} + E_g + (A_{2g})$
$E_u \times E_u = A_{1g} + E_g + (A_{2g})$
$E_g \times E_u = A_{1u} + A_{2u} + E_u$

D_{4d}: $E_1 \times E_1 = A_1 + E_2 + (A_2)$
$E_2 \times E_2 = A_1 + B_1 + B_2 + (A_2)$
$E_3 \times E_3 = A_1 + E_2 + (A_2)$
$E_1 \times E_2 = E_1 + E_3$
$E_1 \times E_3 = B_1 + B_2 + E_2$
$E_2 \times E_3 = E_1 + E_3$

D_{5d}: $E_{1g} \times E_{1g} = A_{1g} + E_{2g} + (A_{2g})$
$E_{2g} \times E_{2g} = A_{1g} + E_{1g} + (A_{2g})$
$E_{1u} \times E_{1u} = A_{1g} + E_{2g} + (A_{2g})$
$E_{2u} \times E_{2u} = A_{1g} + E_{1g} + (A_{2g})$
$E_{1g} \times E_{2g} = E_{1g} + E_{2g}$
$E_{1g} \times E_{1u} = A_{1u} + A_{2u} + E_{2u}$
$E_{1g} \times E_{2u} = E_{1u} + E_{2u}$
$E_{2g} \times E_{1u} = E_{1u} + E_{2u}$

$E_{2g} \times E_{2u} = A_{1u} + A_{2u} + E_{1u}$
$E_{1u} \times E_{2u} = E_{1g} + E_{2g}$

D_{6d}: $E_1 \times E_1 = A_1 + E_2 + (A_2)$
$E_2 \times E_2 = A_1 + E_4 + (A_2)$
$E_3 \times E_3 = A_1 + B_1 + B_2 + (A_2)$
$E_4 \times E_4 = A_1 + E_4 + (A_2)$
$E_5 \times E_5 = A_1 + E_2 + (A_2)$
$E_1 \times E_2 = E_1 + E_3$
$E_1 \times E_3 = E_2 + E_4$
$E_1 \times E_4 = E_3 + E_5$
$E_1 \times E_5 = B_1 + B_2 + E_4$
$E_2 \times E_3 = E_1 + E_5$
$E_2 \times E_4 = B_1 + B_2 + E_2$
$E_2 \times E_5 = E_3 + E_5$
$E_3 \times E_4 = E_1 + E_5$
$E_3 \times E_5 = E_2 + E_4$
$E_4 \times E_5 = E_1 + E_3$

The Groups S_n

S_4: $E \times E = A + 2B + (A)$

S_6: $E_g \times E_g = A_g + E_g + (A_g)$
$E_u \times E_u = A_g + E_g + (A_g)$
$E_g \times E_u = A_u + E_u + A_u$

S_8: $E_1 \times E_1 = A + E_2 + (A)$
$E_2 \times E_2 = A + 2B + (A)$
$E_3 \times E_3 = A + E_2 + (A)$
$E_1 \times E_2 = E_1 + E_3$
$E_1 \times E_3 = 2B + E_2$
$E_2 \times E_3 = E_1 + E_3$

The Cubic Groups

T: $E \times E = A + E + (A)$
$T \times T = A + E + T + (T)$
$E \times T = 2T$

T_d: $E \times E = A_1 + E + (A_2)$
$T_1 \times T_1 = A_1 + E + T_2 + (T_1)$
$T_2 \times T_2 = A_1 + E + T_2 + (T_1)$
$E \times T_1 = T_1 + T_2$
$E \times T_2 = T_1 + T_2$
$T_1 \times T_2 = A_2 + E + T_1 + T_2$

T_h: $E_g \times E_g = A_g + E_g + (A_g)$
$T_g \times T_g = A_g + E_g + T_g + (T_g)$
$E_u \times E_u = A_g + E_g + (A_g)$

$T_u \times T_u = A_g + E_g + T_g + (T_g)$
$E_g \times T_g = 2T_g$
$E_g \times E_u = 2A_u + E_u$
$E_g \times T_u = 2T_u$
$T_g \times E_u = 2T_u$
$T_g \times T_u = A_u + E_u + 2T_u$
$E_u \times T_u = 2T_g$

O: $E \times E = A_1 + E + (A_2)$
$T_1 \times T_1 = A_1 + E + T_2 + (T_1)$
$T_2 \times T_2 = A_1 + E + T_2 + (T_1)$
$E \times T_1 = T_1 + T_2$
$E \times T_2 = T_1 + T_2$
$T_1 \times T_2 = A_2 + E + T_1 + T_2$

O_h: $E_g \times E_g = A_{1g} + E_g + (A_{2g})$
$E_u \times E_u = A_{1g} + E_g + (A_{2g})$
$T_{1g} \times T_{1g} = A_{1g} + E_g + T_{2g} + (T_{1g})$
$T_{2g} \times T_{2g} = A_{1g} + E_g + T_{2g} + (T_{1g})$
$T_{1u} \times T_{1u} = A_{1g} + E_g + T_{2g} + (T_{1g})$
$T_{2u} \times T_{2u} = A_{1g} + E_g + T_{2g} + (T_{1g})$
$E_g \times E_u = A_{1u} + A_{2u} + E_u$
$E_g \times T_{1g} = T_{1g} + T_{2g}$
$E_g \times T_{2g} = T_{1g} + T_{2g}$
$E_g \times T_{1u} = T_{1u} + T_{2u}$
$E_g \times T_{2u} = T_{1u} + T_{2u}$
$E_u \times T_{1g} = T_{1u} + T_{2u}$
$E_u \times T_{2g} = T_{1u} + T_{2u}$
$E_u \times T_{1u} = T_{1g} + T_{2g}$
$E_u \times T_{2u} = T_{1g} + T_{2g}$
$T_{1g} \times T_{2g} = A_{2g} + E_g + T_{1g} + T_{2g}$
$T_{1g} \times T_{1u} = A_{1u} + E_u + T_{1u} + T_{2u}$
$T_{1g} \times T_{2u} = A_{2u} + E_u + T_{1u} + T_{2u}$
$T_{2g} \times T_{1u} = A_{2u} + E_u + T_{1u} + T_{2u}$
$T_{2g} \times T_{2u} = A_{1u} + E_u + T_{1u} + T_{2u}$
$T_{1u} \times T_{2u} = A_{2g} + E_g + T_{1g} + T_{2g}$

The Groups $C_{\infty v}$ and $D_{\infty h}$

$C_{\infty v}$: $\Pi \times \Pi = \Sigma^+ + \Delta + (\Sigma^-)$
$\Delta \times \Delta = \Sigma^+ + \Gamma + (\Sigma^-)$
$\Pi \times \Delta = \Pi + \Phi$

$D_{\infty h}$: $\Pi_g \times \Pi_g = \Sigma_g^+ + \Delta_g + (\Sigma_g^-)$
$\Pi_u \times \Pi_u = \Sigma_g^+ + \Delta_g + (\Sigma_g^-)$
$\Delta_g \times \Delta_g = \Sigma_g^+ + \Gamma_g + (\Sigma_g^-)$
$\Delta_u \times \Delta_u = \Sigma_g^+ + \Gamma_g + (\Sigma_g^-)$
$\Pi_g \times \Pi_u = \Sigma_u^+ + \Sigma_u^- + \Delta_u$

$$\Pi_g \times \Delta_g = \Pi_g + \Phi_g$$
$$\Pi_g \times \Delta_u = \Pi_u + \Phi_u$$
$$\Pi_u \times \Delta_g = \Pi_u + \Phi_u$$
$$\Pi_u \times \Delta_u = \Pi_g + \Phi_g$$
$$\Delta_g \times \Delta_u = \Sigma_u^+ + \Sigma_u^- + \Gamma_u$$

References

Group theory is treated in many excellent books, e.g., M. Tinkham, *Group Theory and Quantum Mechanics*, McGraw-Hill, New York, 1964. Several detailed self-contained presentations of group theory for spectroscopists are available: S. B. Piepho and P. N. Schatz, *Group Theory in Spectroscopy*, Wiley-Interscience, New York, 1983, emphasizes magnetic circular dichroism; P. R. Bunker, *Molecular Symmetry and Spectroscopy*, Academic Press, New York, 1979, emphasizes the techniques useful for nonrigid molecules (not treated in the present text).

Less demanding introductory presentations useful for applications to molecular spectroscopy are M. Orchin and H. H. Jaffé, *Symmetry, Orbitals and Spectra*, Wiley-Interscience, New York, 1971; F. A. Cotton, *Chemical Applications of Group Theory*, Wiley-Interscience, New York, 1971; D. C. Harris and M. D. Bertolucci, *Symmetry and Spectroscopy*, Oxford University Press, Oxford, 1978; and S. F. A. Kettle, *Symmetry and Structure*, Wiley, New York, 1985. A simple introduction to molecular symmetry including a discussion of direct products of representations is found in J. M. Hollas, *Symmetry in Molecules*, Chapman and Hall, London, 1972. Group tables can be found in these and many other books, and in P. W. Atkins, M. S. Child, and C. S. G. Phillips, *Tables for Group Theory*, Oxford University Press, 1970.

The Cartesian tensor patterns for two-photon processes are discussed in detail by W. M. McClain and R. A. Harris in *Excited States*, Vol. 3, E. C. Lim, Ed., Academic Press, New York, 1977.

APPENDIX II

Cartesian Tensors and Polarized Intensities

In the theory of polarized spectroscopy, tensors are used extensively. This appendix contains a summary of important tensor properties followed by a detailed example of orientational averaging.

Tensors

A tensor is a multilinear function of direction which is independent of the basis chosen. The number of directions involved is called the rank of the tensor (some authors refer to it as the order of the tensor). The dimensionality of space in which the directions are defined is called the order of the tensor. Thus a scalar may be considered a tensor of rank 0, and a vector a tensor of rank 1. We shall need to work only with directions in ordinary space of dimensionality 3, i.e., with tensors of order 3.

A special example of an element of a tensor of rank 2 is the Kronecker symbol δ_{st} ($\delta_{st} = 0$ for $s \neq t$ and $\delta_{st} = 1$ for $s = t$; s,t = x,y,z) and a similar example of an element of a tensor of rank 3 is the symbol ε_{stu}($\varepsilon_{xyz} = \varepsilon_{yzx} = \varepsilon_{zxy} = 1$; $\varepsilon_{yxz} = \varepsilon_{zyx} = \varepsilon_{xzy} = -1$; and $\varepsilon_{stu} = 0$ if any two of s, t, and u are the same). Note that the number of subscripts agrees with the rank of the tensor and the possible number of values which they can acquire; x, y, and z, agrees with its order.

A tensor is represented in a given basis (in our case, based on the Cartesian coordinate system) by its coordinates or elements. These are usually arranged into a rectangular array, called a matrix. In a matrix the components are ordered in rows and columns; if the number of rows and columns are the same, the matrix is said to be square, if they are different the matrix is called rectangular.

The representation of a tensor by a matrix can be illustrated for the well-known case of a tensor of rank 1, a vector **V** in ordinary space. Its coordinates in the basis of unit vectors along x, y, and z, which we label $\boldsymbol{\varepsilon}_x$, $\boldsymbol{\varepsilon}_y$, $\boldsymbol{\varepsilon}_z$, are

V_x, V_y, V_z, where $V_s = \cos(\mathbf{V},\varepsilon_s)$, and s = x,y,z. Here, $(\mathbf{V},\varepsilon_s)$ is the angle between $\mathbf{V}$ and ε_s. This can also be written as $V_s = \mathbf{V} \cdot \varepsilon_s$, where the expression $\mathbf{V} \cdot \varepsilon_s$ is known as the scalar or inner product of the vectors $\mathbf{V}$ and ε_s. The three coordinates, V_x, V_y, and V_z, can be arranged into a 1×3 matrix. In a different basis, say, $\varepsilon_{x'},\varepsilon_{y'},\varepsilon_{z'}$, the coordinates of $\mathbf{V}$ would have different values, given by $\mathbf{V} \cdot \varepsilon_{x'}$, etc. Thus, the matrix representation will be different although the vector $\mathbf{V}$ is the same. Clearly, before the vector is specified by its coordinates, the basis must be defined.

The components of $\mathbf{V}$ are the vectors $V_x\varepsilon_x$, $V_y\varepsilon_y$, and $V_z\varepsilon_z$. They are directed along the coordinate axes, and their sum is equal to $\mathbf{V}$. The terms "coordinates of a tensor," "components of a tensor," and "elements of a tensor" are often used interchangeably when there is no danger of misunderstanding.

Various operations can be carried out on tensors. Before describing these in detail, we again provide an illustration on the familiar example of vectors.

The scalar product of two arbitrary vectors $\mathbf{U}$ and $\mathbf{V}$ with components U_s and V_s is given by

$$\mathbf{U} \cdot \mathbf{V} = \sum_s U_s V_s \tag{II.1}$$

Another kind of product of two general vectors is the vector product $\mathbf{U} \times \mathbf{V}$. The result of this multiplication is a vector $\mathbf{W}$ with components

$$\mathbf{W} = \mathbf{U} \times \mathbf{V} = (U_yV_z - U_zV_y, U_zV_x - U_xV_z, U_xV_y - U_yV_x) \tag{II.2}$$

Finally, the tensor product or outer product of the two vectors $\mathbf{U}$ and $\mathbf{V}$ produces a second-rank tensor, $\mathsf{T} = \mathbf{UV}$, whose elements are defined by

$$T_{st} = U_s V_t \tag{II.3}$$

We shall now proceed to the description of operations which can be performed on tensors of arbitrary rank. We shall use tensors of rank 2 and order 3 for illustration and shall represent them by 3×3 square matrices.

A tensor may be multiplied by a constant c

$$\mathsf{B} = c\mathsf{A} \quad \text{if} \quad B_{st} = cA_{st} \quad \text{for all s,t} \tag{II.4}$$

Two tensors may be added

$$\mathsf{C} = \mathsf{A} + \mathsf{B} \quad \text{if} \quad C_{st} = A_{st} + B_{st} \quad \text{for all s,t} \tag{II.5}$$

Several different ways of multiplying tensors (and matrices) are defined. The *double inner product* $\mathsf{A}:\mathsf{B}$ of two tensors A and B is equal to a scalar. For tensors of rank 2,

$$\mathsf{A}:\mathsf{B} = \sum_{st} A_{st} B_{ts} \tag{II.6}$$

The *inner product* $\mathsf{A} \cdot \mathsf{B}$ of two tensors of rank 2, A and B, is a new tensor of rank 2:

$$\mathsf{C} = \mathsf{A} \cdot \mathsf{B} \quad \text{if} \quad C_{st} = \sum_u A_{su} B_{ut} \quad \text{for all s,t} \tag{II.7}$$

Finally the *outer* (*or direct*) *product* AB of two tensors of rank 2, A and B, is a tensor of rank 4:

$$^{(4)}\mathsf{C} = \mathsf{AB} \quad \text{if} \quad C_{stuv} = A_{su}B_{tv} \quad \text{for all s,t} \tag{II.8}$$

The tensor $^{(4)}\mathsf{C}$ can be represented by a matrix whose dimension will be mn if the dimension of A is n and that of B is m. In our case $m = n = 3$ and $^{(4)}\mathsf{C}$ will be represented by a 9×9 matrix.

A number of special matrices deserve mentioning:

A *diagonal matrix* D has zero elements everywhere except on the diagonal. Its elements A_{st} fulfill the relation

$$D_{st} = D_s\delta_{st} \tag{II.9}$$

A special case of a diagonal matrix is a *constant matrix* C, for which

$$C_{st} = C\delta_{st} \tag{II.10}$$

A very important constant matrix is the *unit matrix* 1

$$1_{st} = \delta_{st} \tag{II.11}$$

A matrix A which has a nonvanishing determinant is called *nonsingular*. It has an *inverse* A^{-1}, which is defined by the relation

$$\mathsf{A}^{-1}\mathsf{A} = 1 \tag{II.12}$$

The inverse of a product of two matrices A and B is

$$[\mathsf{AB}]^{-1} = \mathsf{B}^{-1}\mathsf{A}^{-1} \tag{II.13}$$

The *transpose* A′ of a matrix A is defined by the interchange of rows and columns:

$$A'_{st} = A_{ts} \tag{II.14}$$

The transpose of a product of matrices A and B is

$$(\mathsf{AB})' = \mathsf{B}'\mathsf{A}' \tag{II.15}$$

The complex conjugate A* of a matrix A is defined by

$$[A^*]_{st} = [A_{st}]^* \tag{II.16}$$

Now a number of special properties of matrices can be defined. A matrix A is *symmetric* and represent a symmetric tensor if

$$\mathsf{A}' = \mathsf{A} \quad \text{or} \quad A'_{st} = A_{ts} = A_{st} \quad \text{for all s,t} \tag{II.17}$$

A matrix A is *antisymmetric* if

$$\mathsf{A}' = -\mathsf{A} \quad \text{or} \quad A'_{st} = -A_{st} \quad \text{for all s,t} \tag{II.18}$$

Diagonal elements of an antisymmetric matrix vanish.

A matrix A is *Hermitian* if

$$\mathsf{A}^* = \mathsf{A}' \quad \text{or} \quad A^*_{st} = A_{ts} \quad \text{for all s,t} \tag{II.19}$$

A matrix A is *orthogonal* if

$$\mathsf{A}^{-1} = \mathsf{A}' \quad \text{or} \quad [A^{-1}]_{st} = A_{ts} \quad \text{for all s,t} \tag{II.20}$$

Finally, a matrix A is *unitary* if

$$A^{-1} = A'^* \quad \text{or} \quad [A^{-1}]_{st} = A^*_{ts} \quad \text{for all s,t} \tag{II.21}$$

The matrix A'^* is called the *conjugate transpose* of A and is often written as $A^\dagger$.

Vectors may be written as row matrices [with just one row, as in (II.2)] or column matrices (with just one column). The transpose of a column matrix is a row matrix, and vice versa. Then if **U** and **V** are row matrices and A is a square matrix, the expression

$$\mathbf{U}A\mathbf{V}' = \sum_{st} U_s A_{st} V_t \tag{II.22}$$

is a scalar and is called a *quadratic form.* Its magnitude is equal to the double inner product of the tensors **UV** and A.

Linear Transformations

Linear transformations can be written in a very compact form using matrix notation.

Consider the transformation

$$\begin{aligned} X &= A_{11}x + A_{12}y + A_{13}z \\ Y &= A_{21}x + A_{22}y + A_{23}z \\ Z &= A_{31}x + A_{32}y + A_{33}z \end{aligned} \tag{II.23}$$

between the molecular axes x,y,z and the laboratory axes X,Y,Z. If we define

$$\mathbf{v} = \begin{pmatrix} x \\ y \\ z \end{pmatrix} \qquad \mathbf{V} = \begin{pmatrix} X \\ Y \\ Z \end{pmatrix}$$

and

$$A = \begin{pmatrix} A_{11} & A_{12} & A_{13} \\ A_{21} & A_{22} & A_{23} \\ A_{31} & A_{32} & A_{33} \end{pmatrix} \tag{II.24}$$

the transformation and the inverse transformation may be written

$$\mathbf{V} = A\mathbf{v} \qquad \mathbf{v} = A^{-1}\mathbf{V} \tag{II.25}$$

Diagonalization

The eigenvalues of a square matrix A are defined as the values of λ for which the determinant $|A - \lambda 1|$ vanishes:

$$\begin{vmatrix} A_{11} - \lambda & A_{12} & A_{13} \\ A_{21} & A_{22} - \lambda & A_{23} \\ A_{31} & A_{32} & A_{33} - \lambda \end{vmatrix} = 0 \tag{II.26}$$

Equation (II.26) is called the secular equation. It is of a degree in λ which is equal to the number of columns of the matrix. This determines the number of eigenvalues (some of which may be repeated). To each eigenvalue $\lambda^{(s)}$ corresponds an eigenvector. This is a column vector $\mathbf{V}^{(s)}$ with the same number of rows as A:

$$\mathbf{V}^{(s)} = \begin{pmatrix} V_x^{(s)} \\ V_y^{(s)} \\ V_z^{(s)} \end{pmatrix} \tag{II.27}$$

which satisfies the equation

$$\mathsf{A}\mathbf{V}^{(s)} = \lambda^{(s)}\mathbf{V}^{(s)} \tag{II.28}$$

If the eigenvectors $\mathbf{V}^{(1)}$, $\mathbf{V}^{(2)}$, $\mathbf{V}^{(3)}$ are written as columns next to each other they form a square matrix V with the same number of rows and columns as A.

A similarity transformation is defined as a multiplication to the right with a matrix and to the left with its inverse. Many matrices can be diagonalized, i.e., brought to a diagonal form by a similarity transformation. In particular, all symmetric matrices can be diagonalized. If we define the diagonal matrix Λ in such a way that

$$\Lambda_{st} = \lambda_s \delta_{st} \tag{II.29}$$

we have

$$\mathsf{AV} = \mathsf{V}\Lambda \tag{II.30}$$

or

$$\mathsf{V}^{-1}\mathsf{AV} = \Lambda \tag{II.31}$$

The matrix needed for this similarity transformation is the matrix V formed by the collection of the eigenvectors of A.

The directions defined by the eigenvectors of a matrix which represents a tensor are called the principal axes of the tensor. When they are used for the coordinate system, the tensor is said to be diagonal and the eigenvalues are called the principal values of the tensor.

A general tensor T need not be symmetric. It can, however, be written as a sum of a symmetric part and an antisymmetric part, $\mathsf{T} = \mathsf{T}^{(s)} + \mathsf{T}^{(a)}$, such that $\mathsf{T}^{(s)\prime} = \mathsf{T}^{(s)}$ and $\mathsf{T}^{(a)\prime} = -\mathsf{T}^{(a)}$. Its principal axes are then defined by the eigenvectors of the symmetric part $\mathsf{T}^{(s)}$.

Diagonalization of the Orientation Tensor

As an example let us consider the determination of the principal orientation axes for a molecule. Suppose that an experiment produced the following K values, $K'_{s't'} = \langle\cos(s'Z)\cos(t'Z)\rangle$, with the initial choice of molecular axes ($s',t' = x',y',z'$):

$$\mathsf{K}' = \begin{pmatrix} 0.30 & 0.20 & 0.10 \\ 0.20 & 0.50 & 0.00 \\ 0.10 & 0.00 & 0.20 \end{pmatrix} \tag{II.32}$$

This leads to the secular equation

$$\begin{vmatrix} 0.30-\lambda & 0.20 & 0.10 \\ 0.20 & 0.50-\lambda & 0.00 \\ 0.10 & 0.00 & 0.20-\lambda \end{vmatrix} = 0 \tag{II.33}$$

which gives

$$\begin{aligned} &-0.01(0.50-\lambda) + (0.20-\lambda)[(0.30-\lambda)(0.50-\lambda) - 0.04] = 0 \\ &-\lambda^3 + \lambda^2 - 0.26\lambda + 0.017 = 0 \end{aligned} \tag{II.34}$$

One of the three solutions is clearly $\lambda^{(3)} = 0.10$, and we have

$$(0.10-\lambda)(\lambda^2 - 0.90\lambda + 0.17) = 0 \tag{II.35}$$

The remaining two solutions are

$$\lambda^{(1)} = 0.45 + \sqrt{0.033} = 0.63 \qquad \lambda^{(2)} = 0.45 - \sqrt{0.033} = 0.27 \tag{II.36}$$

The principal values of the orientation tensor, i.e., the orientation factors K in the principal system of axes are

$$K_z = 0.63 \qquad K_y = 0.27 \qquad K_x = 0.10 \tag{II.37}$$

The eigenvectors can now be determined, e.g.,

$$\mathbf{V}^{(1)} = \begin{pmatrix} 0.54 \\ 0.83 \\ 0.13 \end{pmatrix} \tag{II.38}$$

We note that (II.28) is fulfilled:

$$\begin{pmatrix} 0.30 & 0.20 & 0.10 \\ 0.20 & 0.50 & 0.00 \\ 0.10 & 0.00 & 0.20 \end{pmatrix} \begin{pmatrix} 0.54 \\ 0.83 \\ 0.13 \end{pmatrix} = \begin{pmatrix} 0.34 \\ 0.53 \\ 0.08 \end{pmatrix} = 0.63 \begin{pmatrix} 0.54 \\ 0.83 \\ 0.13 \end{pmatrix} \tag{II.39}$$

Similarly,

$$\mathbf{V}^{(2)} = \begin{pmatrix} 0.52 \\ -0.44 \\ 0.74 \end{pmatrix} \tag{II.40}$$

and

$$\mathbf{V}^{(3)} = \begin{pmatrix} 0.67 \\ -0.33 \\ -0.67 \end{pmatrix} \tag{II.41}$$

The eigenvectors define the principal orientation axes system (x,y,z) in terms of the originally chosen coordinate system (x′,y′,z′). We have

$$\begin{aligned} \boldsymbol{\varepsilon}_x &= 0.67\boldsymbol{\varepsilon}_{x'} - 0.33\boldsymbol{\varepsilon}_{y'} - 0.67\boldsymbol{\varepsilon}_{z'} \\ \boldsymbol{\varepsilon}_y &= 0.52\boldsymbol{\varepsilon}_{x'} - 0.44\boldsymbol{\varepsilon}_{y'} + 0.74\boldsymbol{\varepsilon}_{z'} \\ \boldsymbol{\varepsilon}_z &= 0.54\boldsymbol{\varepsilon}_{x'} + 0.83\boldsymbol{\varepsilon}_{y'} + 0.13\boldsymbol{\varepsilon}_{z'} \end{aligned} \tag{II.42}$$

Tensors of Higher Rank

The properties of tensors of higher ranks, such as tensors of rank 4, are similar to those of tensors of rank 2. A rank 4 tensor of order 3 has $81 = 3^4$ elements which may be listed in a 9×9 matrix as shown in Chapter 7. The expressions for polarized intensities in processes of rank 4 involve double inner products of two symmetric rank 4 tensors. For this purpose (II.6) is easily generalized:

$$ {}^{(4)}\mathsf{A}:{}^{(4)}\mathsf{B} = \sum_{stuv} {}^{(4)}A_{stuv}\,{}^{(4)}B_{stuv} \qquad (II.43) $$

Other procedures may be generalized similarly for use on tensors of higher ranks.

Orientational Averaging—An Example

The evaluation of polarized intensities observed on partially oriented molecular ensembles requires properly weighted averaging over molecular orientations. The final result can be written in the form of a double inner product of two tensors:

$$ I_{UV\ldots} = {}^{(n)}\mathsf{P}^{UV\ldots}:{}^{(n)}\mathsf{A} = \sum_{stuv} [{}^{(n)}\mathsf{P}]^{UV\ldots}_{stuv\ldots}[{}^{(n)}\mathsf{A}]_{stuv\ldots} \qquad (II.44) $$

where the total number of subscripts stu ... equals the rank n of the process in question. $[\mathsf{P}]^{UV\ldots}_{stuv\ldots}$ are elements of a tensor ${}^{(n)}\mathsf{P}^{UV\ldots}$ of rank n which contains only information on the orientation distribution (i.e., depends on the orientation factors K,L,M, ...) and on the polarization of the photons involved in the experiment, U,V, ... The superscripts U,V, ... can acquire the values X, Y, or Z and refer to the laboratory set of axes. The indices s,t,u,v, ... can acquire the values x, y, or z and refer to the molecular set of axes. ${}^{(n)}\mathsf{A}$, with elements $[{}^{(n)}\mathsf{A}]_{stuv\ldots}$ is a tensor of rank n which contains only information on molecular properties and whose detailed form is a function of the process involved in the experiment.

The problem to be discussed is the evaluation of the tensor elements $[\mathsf{P}]^{UV\ldots}_{stuv\ldots}$. It will be illustrated on the case of a process of rank 4.

First, we recall the exact definition of the quantity to be averaged. Using the abbreviated notation introduced in (1.140), the observable intensity for photon polarizations U and V is given by

$$ I_{UV} = \langle|\mathsf{T}_{UV}:\mathsf{A}|^2\rangle \qquad (II.45) $$

where

$$ \mathsf{T}_{UV} = \varepsilon_U\varepsilon_V \qquad (II.46) $$

$$ \mathsf{A} = \sum_{uv} A_{uv}\varepsilon_u\varepsilon_v \qquad (II.47) $$

and where ε_U, ε_V, ε_u, and ε_v stand for unit vectors directed along the axes U,

V, u, and v, respectively. Then,

$$I_{UV} = \sum_{stuv} \langle(\varepsilon_U \cdot \varepsilon_s)(\varepsilon_V \cdot \varepsilon_t)(\varepsilon_U \cdot \varepsilon_u)(\varepsilon_V \cdot \varepsilon_v)\rangle A_{st}A_{uv} \qquad (II.48)$$

so that

$$[{}^{(4)}P_{UV}]_{stuv} = \langle(\varepsilon_U \cdot \varepsilon_s)(\varepsilon_V \cdot \varepsilon_t)(\varepsilon_U \cdot \varepsilon_u)(\varepsilon_V \cdot \varepsilon_v)\rangle \qquad (II.49)$$

Second, we proceed to the evaluation of this average for various choices of the subscripts. The expression is evaluated readily if $U = V = Z$:

$$\begin{aligned}[{}^{(4)}P_{ZZ}]_{stuv} &= \langle(\varepsilon_Z \cdot \varepsilon_s)(\varepsilon_Z \cdot \varepsilon_t)(\varepsilon_Z \cdot \varepsilon_u)(\varepsilon_Z \cdot \varepsilon_v)\rangle \\ &= \langle\cos s \cos t \cos u \cos v\rangle = L_{stuv} \qquad (II.50)\end{aligned}$$

If U or V or both are equal to X or Y, matters become slightly more complicated. For instance, for $U = Z$, $V = X$, and $t = v$,

$$\begin{aligned}[{}^{(4)}P_{ZX}]_{stuv} &= \langle(\varepsilon_Z \cdot \varepsilon_s)(\varepsilon_X \cdot \varepsilon_t)(\varepsilon_Z \cdot \varepsilon_u)(\varepsilon_X \cdot \varepsilon_v)\rangle \\ &= \langle\cos s \sin t \cos \theta \cos u \sin v \cos \theta\rangle \\ &= \tfrac{1}{2}\langle\cos s \cos u \sin^2 t\rangle \\ &= \tfrac{1}{2}\langle\cos s \cos u - \cos s \cos u \cos^2 t\rangle = \tfrac{1}{2}(K_{su} - L_{stuv}) \qquad (II.51)\end{aligned}$$

where θ is the angle between the X axis and the projection of the t axis into the XY plane. The averaging $\langle\cos^2 \theta\rangle$ yields a factor of 1/2, since the distribution around Z is uniform (uniaxial). If $t \neq v$,

$$\begin{aligned}[{}^{(4)}P_{ZX}]_{stuv} &= \langle(\varepsilon_Z \cdot \varepsilon_s)(\varepsilon_X \cdot \varepsilon_t)(\varepsilon_Z \cdot \varepsilon_u)(\varepsilon_X \cdot \varepsilon_v)\rangle \\ &= \langle\cos s \sin t \cos \theta \cos u \sin v \cos \theta'\rangle \qquad (II.52)\end{aligned}$$

where θ and θ' are the angles between the X axis and the projections of the t and v axes, respectively, into the XY plane (Figure II.1). The averaging of $\cos \theta \cos \theta'$ can still be performed explicitly, since the distribution around Z is uniform. We write $\theta' = \theta + \phi$, where ϕ is the dihedral angle from the plane tZ to the plane vZ, measured counterclockwise. From trigonometry,

$$\cos \phi = -\frac{\cos t \cos v}{\sin t \sin v} \qquad \sin \phi = \frac{\cos w}{\sin t \sin v} \qquad (II.53)$$

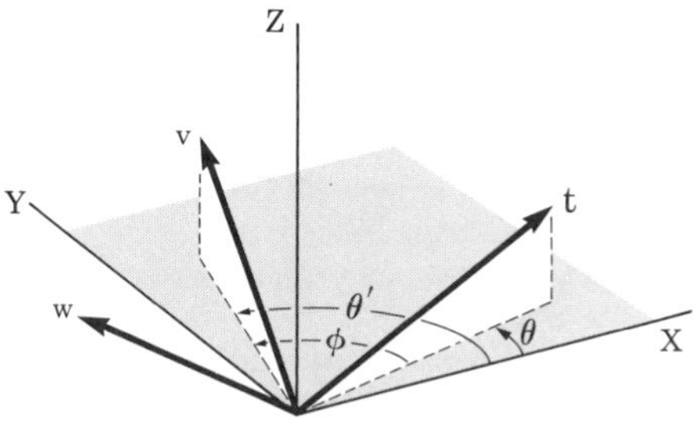

Figure II.1 Angles involved in averaging over the positions of the molecular axes.

where w is the third molecular axis (w $\neq$ t, w $\neq$ v), and

$$\cos\theta\cos(\theta+\phi) = \cos\theta\cos\theta\cos\phi - \cos\theta\sin\theta\sin\phi$$
$$= -(\cos^2\theta\cos s\cos u + \cos\theta\sin\theta\cos v)/(\sin s\sin u) \quad \textit{(II.54)}$$

Now, for a uniaxial distribution,

$$[{}^{(4)}\mathrm{P}_{ZX}]_{stuv} = -\langle\cos s\cos u\cos t\cos v\rangle\langle\cos^2\theta\rangle$$
$$-\langle\cos s\cos u\cos w\rangle\langle\cos\theta\sin\theta\rangle$$
$$= -\tfrac{1}{2}\mathrm{L}_{stuv} \quad \textit{(II.55)}$$

since $\langle\cos\theta\sin\theta\rangle = 0$.

Similar considerations lead to the results for the other nontrivial cases in which U and V are X or Y:

$$\langle(\mathbf{X}\cdot\mathbf{u})^2\rangle = \langle(\mathbf{Y}\cdot\mathbf{u})^2\rangle = (1/2)\langle 1-\cos^2 u\rangle \quad \textit{(II.56)}$$

$$\langle(\mathbf{X}\cdot\mathbf{u})(\mathbf{X}\cdot\mathbf{v})\rangle = \langle(\mathbf{Y}\cdot\mathbf{u})(\mathbf{Y}\cdot\mathbf{v})\rangle = -(1/2)\langle\cos u\cos v\rangle \quad \textit{(II.57)}$$

$$\langle(\mathbf{X}\cdot\mathbf{v})^2(\mathbf{Y}\cdot\mathbf{v})^2\rangle = (1/8)\langle(1-\cos^2 v)^2\rangle \quad \textit{(II.58)}$$

$$\langle(\mathbf{X}\cdot\mathbf{u})^2(\mathbf{Y}\cdot\mathbf{u})(\mathbf{Y}\cdot\mathbf{v})\rangle = \langle(\mathbf{Y}\cdot\mathbf{u})^2(\mathbf{X}\cdot\mathbf{u})(\mathbf{X}\cdot\mathbf{v})\rangle$$
$$= -(1/8)\langle\cos u\cos v(1-\cos^2 u)\rangle \quad \textit{(II.59)}$$

$$\langle(\mathbf{X}\cdot\mathbf{u})^2(\mathbf{Y}\cdot\mathbf{v})^2\rangle = (1/8)\langle\cos^2 u\cos^2 v + 3\cos^2 w\rangle \quad \textit{(II.60)}$$

$$\langle(\mathbf{X}\cdot\mathbf{u})(\mathbf{X}\cdot\mathbf{v})(\mathbf{Y}\cdot\mathbf{w})(\mathbf{Y}\cdot\mathbf{v})\rangle = (1/8)\langle\cos^2 u\cos^2 v - \cos^2 w\rangle \quad \textit{(II.61)}$$

$$\langle(\mathbf{X}\cdot\mathbf{u})^2(\mathbf{Y}\cdot\mathbf{v})(\mathbf{Y}\cdot\mathbf{w})\rangle = \langle(\mathbf{Y}\cdot\mathbf{u})^2(\mathbf{X}\cdot\mathbf{v})(\mathbf{X}\cdot\mathbf{w})\rangle$$
$$= (1/8)\langle\cos^2 u\cos v\cos w - 3\cos v\cos w\rangle \quad \textit{(II.62)}$$

$$\langle(\mathbf{X}\cdot\mathbf{u})(\mathbf{X}\cdot\mathbf{v})(\mathbf{Y}\cdot\mathbf{u})(\mathbf{Y}\cdot\mathbf{w})\rangle = (1/8)\langle\cos^2 u\cos v\cos w + \cos v\cos w\rangle \quad \textit{(II.63)}$$

$$\langle(\mathbf{X}\cdot\mathbf{v})^4\rangle = \langle(\mathbf{Y}\cdot\mathbf{v})^4\rangle = (3/8)\langle 1-2\cos^2 v + \cos^4 v\rangle \quad \textit{(II.64)}$$

$$\langle(\mathbf{X}\cdot\mathbf{u})^2(\mathbf{X}\cdot\mathbf{v})^2\rangle = \langle(\mathbf{Y}\cdot\mathbf{u})^2(\mathbf{Y}\cdot\mathbf{v})^2\rangle = (1/8)\langle\cos^2 w + 3\cos^2 u\cos^2 v)\rangle \quad \textit{(II.65)}$$

$$\langle(\mathbf{X}\cdot\mathbf{u})^3(\mathbf{X}\cdot\mathbf{v})\rangle = \langle(\mathbf{Y}\cdot\mathbf{u})^3(\mathbf{Y}\cdot\mathbf{v})\rangle = -(3/8)\langle\cos u\cos v(1-\cos^2 u)\rangle \quad \textit{(II.66)}$$

$$\langle(\mathbf{X}\cdot\mathbf{u})^2(\mathbf{X}\cdot\mathbf{v})(\mathbf{X}\cdot\mathbf{w})\rangle = \langle(\mathbf{Y}\cdot\mathbf{u})^2(\mathbf{Y}\cdot\mathbf{v})(\mathbf{Y}\cdot\mathbf{w})\rangle$$
$$= (1/8)\langle 3\cos^2 u\cos v\cos w - \cos v\cos w\rangle \quad \textit{(II.67)}$$

Entirely analogous procedures can be used to obtain the results for ${}^{(6)}\mathrm{P}_{UVW}$ and other tensors of even higher rank.

If the distribution is not uniaxial, none of the averaging can be carried out explicitly. The quantities

$$\langle(\varepsilon_S \cdot \varepsilon_s)(\varepsilon_T \cdot \varepsilon_t)(\varepsilon_U \cdot \varepsilon_u)(\varepsilon_V \cdot \varepsilon_v)\rangle = \langle \cos sS \cos tT \cos uU \cos vV \rangle$$
$$= L_{STUVstuv} \tag{II.68}$$

then can be treated as generalized orientation factors [cf. (4.120), (4.121)].

References

A useful summary of matrix notation can be found in E. B. Wilson, J. C. Decius, and P. C. Cross, *Molecular Vibrations*, McGraw-Hill, New York, 1955. A condensed introduction to Cartesian tensors is available, for instance, in G. Temple, *Cartesian Tensors*, Methuen, London, 1960; and H. Jeffreys, *Cartesian Tensors*, Cambridge University Press, Cambridge, England, 1931 (this author uses "order" where we use "rank").

APPENDIX III

Mueller-Stokes Formalism for Description of the Interaction between Polarized Light and Anisotropic Samples

Isotropic samples may exhibit absorbance, refraction, circular dichroism, and circular birefringence. Anisotropic samples may in addition exhibit linear dichroism, linear birefringence, anisotropic circular dichroism, and anisotropic circular birefringence. The passage of polarized light through anisotropic samples therefore becomes quite complicated. One efficient way of describing the interaction between polarized light and anisotropic matter is the Mueller-Stokes formalism.

Polarized light propagating along X is described by the Stokes vector:

$$\mathbf{S} = \begin{pmatrix} s_0 \\ s_1 \\ s_2 \\ s_3 \end{pmatrix} = \begin{pmatrix} a_Y^2 + a_Z^2 \\ 2a_Y a_Z \cos\delta \\ 2a_Y a_Z \sin\delta \\ a_Y^2 - a_Z^2 \end{pmatrix} \tag{III.1}$$

where a_Y and a_Z are electric field amplitudes along Y and Z and δ is the phase between them. The frequency dependence is not shown explicitly.

The effect of the sample on the light is expressed by the Mueller matrix M, which transforms the ingoing beam, described by $\mathbf{S}_0$, into the outgoing beam $\mathbf{S}_1$:

$$\mathbf{S}_1 = \mathsf{M}\mathbf{S}_0 \qquad (III.2)$$

The Mueller matrix M reflects the properties of the molecules in the sample and their orientation distribution. It is related to the matrix H:

$$\mathsf{M} = \exp(-\mathsf{H}) = -\mathsf{H} + \frac{1}{2!}\mathsf{H}^2 - \frac{1}{3!}\mathsf{H}^3 + \cdots \qquad (III.3)$$

$$\mathsf{H} = \begin{pmatrix} \mathscr{A}_e & \mathscr{L}\mathscr{D}' & -\mathscr{C}\mathscr{D} & \mathscr{L}\mathscr{D} \\ \mathscr{L}\mathscr{D}' & \mathscr{A}_e & \mathscr{L}\mathscr{B} & \mathscr{C}\mathscr{B} \\ -\mathscr{C}\mathscr{D} & -\mathscr{L}\mathscr{B} & \mathscr{A}_e & \mathscr{L}\mathscr{B}' \\ \mathscr{L}\mathscr{D} & -\mathscr{C}\mathscr{B} & -\mathscr{L}\mathscr{B}' & \mathscr{A}_e \end{pmatrix} \qquad (III.4)$$

The elements of H represent the optical properties of the sample.

Mean absorbance:

$$\mathscr{A}_e = (\ln 10)(E_Y + E_Z)/2 \qquad (III.5)$$

Linear dichroism (measured with polarization along Z and Y):

$$\mathscr{L}\mathscr{D} = (\ln 10)(E_Z - E_Y)/2 \qquad (III.6)$$

Linear dichroism measured at 45° and 135° to Z:

$$\mathscr{L}\mathscr{D}' = (\ln 10)(E_{45^\circ} - E_{135^\circ})/2 \qquad (III.7)$$

Linear birefringence (measured with polarization along Z and Y and for path length l and vacuum wavelength λ_{vac}):

$$\mathscr{L}\mathscr{B} = 2\pi(n_Z - n_Y)l/\lambda_{vac} \qquad (III.8)$$

Linear birefringence at 45° and 135° relative to Z (measured and for path length l and vacuum wavelength λ_{vac}):

$$\mathscr{L}\mathscr{B}' = 2\pi(n_{45^\circ} - n_{135^\circ})l/\lambda_{vac} \qquad (III.9)$$

Circular dichroism:

$$\mathscr{C}\mathscr{D} = (\ln 10)(E_L - E_R)/2 = (\ln 10) \cdot \Delta E/2 \qquad (III.10)$$

Circular birefringence (for pathlength l and vacuum wavelength λ_{vac}):

$$\mathscr{C}\mathscr{B} = 2\pi(n_L - n_R)l/\lambda_{vac} \qquad (III.11)$$

The exponential form of (III.3) is useful since H may be written as

$$\mathsf{H} = e^{-\mathscr{A}_e}(1 - \mathsf{F} + \tfrac{1}{2}\mathsf{F}^2 - \cdots) \qquad (III.12)$$

where

$$F = \begin{pmatrix} 0 & \mathscr{LD}' & -\mathscr{CD} & \mathscr{LD} \\ \mathscr{LD}' & 0 & \mathscr{LB} & \mathscr{CB} \\ -\mathscr{CD} & -\mathscr{LB} & 0 & \mathscr{LB}' \\ \mathscr{LD} & -\mathscr{CB} & -\mathscr{LB}' & 0 \end{pmatrix} \qquad \textit{(III.13)}$$

A simple example of an application follows.

The Stokes vector of a beam emerging from an acousto-optic polarization modulator may be written as

$$S_0 = \begin{pmatrix} 1 \\ 0 \\ \sin(\delta_0 \sin \omega t) \\ \cos(\delta_0 \sin \omega t) \end{pmatrix} \qquad \textit{(III.14)}$$

where ω is 2π times the frequency of the modulator oscillation and $\delta_0 \sin \omega t$ is the phase modulation.

The components may be expanded in a Fourier series:

$$\sin(\delta_0 \sin \omega t) = 2J_1(\delta_0) \sin \omega t + 2J_3(\delta_0) \sin 3\omega t + \cdots \qquad \textit{(III.15)}$$

$$\cos(\delta_0 \sin \omega t) = J_0(\delta_0) + 2J_2(\delta_0) \cos 2\omega t + \cdots \qquad \textit{(III.16)}$$

where $J_i(\delta_0)$ is a Bessel function of order i.

In an experiment the total transmitted light intensity is measured. This corresponds to the first component of the Stokes vector: $a_Y^2 + a_Z^2$. The electronic components connected to the photomultiplier separate the ac and dc components of the signal. The ac component is processed by a lock-in amplifier which measures the average value of the ac currents with frequencies ω and 2ω. The intensities of the dc and the two ac signals from the outgoing beam $\mathbf{S}_1$ may be expressed by inserting (III.13) into (III.12), (III.16) into (III.14), (III.14) into (III.1), and (III.12) and (III.1) into (III.2). We obtain

$$\begin{aligned} I_{dc} = e^{-\mathscr{A}_e}\{1 &+ \tfrac{1}{2}(\mathscr{LD}^2 + \mathscr{LD}'^2 + \mathscr{CD}^2) \\ &+ J_0(\delta_0)[-\mathscr{LD} + \tfrac{1}{2}(-\mathscr{LB}' \cdot \mathscr{CD} + \mathscr{CB} \cdot \mathscr{LD}')]\} \end{aligned} \qquad \textit{(III.17)}$$

$$\langle I_\omega \rangle = \frac{4}{\pi} J_1(\delta_0) e^{-\mathscr{A}_e}[\mathscr{CD} + \tfrac{1}{2}(\mathscr{LB} \cdot \mathscr{LD}' - \mathscr{LB}' \cdot \mathscr{LD})] \qquad \textit{(III.18)}$$

$$\langle I_{2\omega} \rangle = \frac{4}{\pi} J_2(\delta_0)[-\mathscr{LD} + \tfrac{1}{2}(-\mathscr{LB}' \cdot \mathscr{CD} + \mathscr{CB} \cdot \mathscr{LD}')] \qquad \textit{(III.19)}$$

The optical systems used will often contain small imperfections, such as linear birefringence preceding the sample. It is clear from (III.17)–(III.19) that these may have a considerable effect on the measured results. In order to take this into account, the Stokes vector of the incident light may be multiplied with a correction matrix.

References

Most of the optics texts listed in Section 1.4 describe the Mueller-Stokes formalism in general terms. A more complete description of its use in modulation spectroscopy may be found in H. P. Jensen, *Spectrosc. Lett.* **10**, 471 (1977); H. P. Jensen, J. A. Schellman, and T. Troxell, *Appl. Spectrosc.* **32**, 192 (1978); in H. P. Jensen, Thesis, Chalmers University of Technology, Gothenburg, Sweden, 1981, and particularly in J. A. Schellman and H. P. Jensen *Chem. Rev.*, in press (1987). For a brief review, see B. Nordén, *Appl. Spectrosc. Rev.* **14**, 157 (1978).

Subject Index

Compound Index

RETURN TO: CHEMISTRY LIBRARY
100 Hildebrand Hall • 510-642-3753

LOAN PERIOD 1	2 1 Month	3
4	5	6

ALL BOOKS MAY BE RECALLED AFTER 7 DAYS.

Renewals may be requested by phone ~~or, using GLADIS, type inv followed by your patron ID number.~~

DUE AS STAMPED BELOW.

~~JAN 20~~		

FORM NO. DD 10
3M 7-08

UNIVERSITY OF CALIFORNIA, BERKELEY
Berkeley, California 94720–6000